D1068403

Molecular Thermodynamics of Fluid-Phase Equilibria

Third Edition

PRENTICE HALL INTERNATIONAL SERIES
IN THE PHYSICAL AND CHEMICAL ENGINEERING SCIENCES

BALZHISER, SAMUELS, & ELLIASSEN *Chemical Engineering Thermodynamics*
BEQUETTE *Process Dynamics*
BIEGLER, GROSSMAN, & WESTERBERG *Systematic Methods of Chemical Process Design*
CROWL & LOUVAR *Chemical Process Safety*
CUTLIP & SHACHAM *Problem Solving in Chemical Engineering with Numerical Methods*
DENN *Process Fluid Mechanics*
ELLIOT & LIRA *Introductory Chemical Engineering Thermodynamics*
FOGLER *Elements of Chemical Reaction Engineering, 3rd Edition*
HANNA & SANDALL *Computational Methods in Chemical Engineering*
HIMMELBLAU *Basic Principles and Calculations in Chemical Engineering, 6th edition*
HINES & MADDOX *Mass Transfer*
KYLE *Chemical and Process Thermodynamics, 3rd edition*
NEWMAN *Electrochemical Systems, 2nd edition*
PRAUSNITZ, LICHTENTHALER, & DE AZEVEDO *Molecular Thermodynamics of Fluid-Phase Equilibria, 3rd edition*
PRENTICE *Electrochemical Engineering Principles*
SHULER & KARGI *Bioprocess Engineering*
STEPHANOPOULOS *Chemical Process Control*
TESTER & MODELL *Thermodynamics and Its Applications, 3rd edition*
TURTON, BAILIE, WHITING, & SHAEWITZ *Analysis, Synthesis and Design of Chemical Processes*
WILKES *Fluid Mechanics for Chemical Engineering*

Molecular Thermodynamics of Fluid-Phase Equilibria

Third Edition

John M. Prausnitz, *University of California, Berkeley*

Rüdiger N. Lichtenthaler, *University of Heidelberg*

Edmundo Gomes de Azevedo, *Instituto Superior Técnico, Lisbon*

Prentice Hall PTR
Upper Saddle River, New Jersey 07458
htpp://www.phptr.com

Acquisitions editor: *Bernard Goodwin*
Editorial/production supervision: *Joan L. McNamara*
Cover design director: *Jerry Votta*
Book Design and Layout: *Rüdiger N. Lichtenthaler and Edmundo Gomes de Azevedo*
Electronic Typesetting: *Rüdiger N. Lichtenthaler and Edmundo Gomes de Azevedo*
Figures and Artwork: *Rüdiger N. Lichtenthaler and Edmundo Gomes de Azevedo*
Manufacturing manager: *Alan Fischer*
Editorial Assistant: *Diane Spina*
Marketing Manager: *Kaylie Smith*

Prentice-Hall, Inc.
Upper Saddle River, New Jersey 07458

Prentice Hall books are widely used by corporations and
government agencies for training, marketing, and resale.
The publisher offers discounts on this book when ordered in bulk quantities.
For more information, contact Corporate Sales Department, Phone: 800-382-3419;
FAX: 201-236-7141; E-mail: corpsales@prenhall.com
Prentice Hall PTR, One Lake Street, Upper Saddle River, NJ 07458.

Printed in the United States of America
10 9 8 7 6 5 4 3 2

ISBN 0-13-977745-8

PRENTICE-HALL INTERNATIONAL (UK) LIMITED, *LONDON*
PRENTICE-HALL OF AUSTRALIA PTY. LIMITED, *SYDNEY*
PRENTICE-HALL CANADA INC., *TORONTO*
PRENTICE-HALL HISPANOAMERICANA, S.A., *MEXICO*
PRENTICE-HALL OF INDIA PRIVATE LIMITED, *NEW DELHI*
PRENTICE-HALL OF JAPAN, INC., *TOKYO*
PEARSON EDUCATION ASIA PTE. LTD., *SINGAPORE*
EDITORA PRENTICE-HALL DO BRASIL, LTDA., *RIO DE JANEIRO*

Contents

Preface

The first edition of this book appeared in 1969; the second edition in 1986. The purpose of this book remains unchanged: to present to senior or first-year graduate students in chemical engineering (and related sciences) a broad introduction to the thermodynamics of phase equilibria typically encountered in design of chemical products and processes, in particular, in separation operations. Thermodynamic tools are provided for efficient design and improvement of conventional and new separation processes including those that may be useful for environmental protection.

This book is suitable as a text for those students who have completed a first course in chemical engineering thermodynamics. While most of the material is based on classical thermodynamics, molecular properties are introduced to facilitate applications to real systems. Although no effort is made to teach statistical thermodynamics, useful results from statistical thermodynamics are included to connect thermodynamic and molecular properties.

The new edition presents an expanded discussion of theoretical concepts to describe and interpret solution properties, with emphasis on those concepts that bear promise for practical applications. Attention is given to a variety of models including the lattice-fluid theory and the statistical associated-fluid theory (SAFT).

A new chapter is devoted to polymer solutions including gas-polymer equilibria at ordinary and high pressures, polymer blends, polymeric membranes and gels. Other novel sections of the third edition include discussions of osmotic pressure and Donnan equilibria.

A serious omission in previous editions has now been corrected: the third edition contains an entirely new chapter on electrolyte solutions. This new chapter first gives the thermodynamic basis for describing activities of components in electrolyte solutions and then presents some semi-empirical models for solutions containing salts or volatile electrolytes. Also discussed are some applications of these models to phase-equilibrium calculations relevant to chemical, environmental and biochemical engineering.

All chapters have been updated primarily through presentation of some recent examples and some new homework problems.

It is a pleasure for the senior author to indicate here his thanks for the essential contributions of his two co-authors. Without their dedicated devotion and attention to numerous details, this third edition could not have been completed.

For helpful advice and comments, the authors are grateful to numerous colleagues, especially to Allan Harvey, Dan Kuehner, Huen Lee, Gerd Maurer, Van Nguyen, John O'Connell, and Jianzhong Wu.

Since 1986, the literature concerning fluid-phase thermodynamics has grown tremendously. To keep the book to a reasonable size, it has been necessary to omit many fine contributions. The authors apologize to their many colleagues whose important work could not be included lest the book become excessively long.

Chemical engineering thermodynamics is now in a state of transition. Classical thermodynamics is becoming increasingly replaced by new tools from applied statistical thermodynamics and molecular simulations. However, many – indeed most – of these new tools are not as yet sufficiently developed for practical applications. For the present and near future, it remains necessary to rely primarily on classical thermodynamics informed and extended through molecular physics and physical chemistry. Molecular thermodynamics, as presented here, is characterized by a combination of classical methods augmented by molecular science and supported by fundamental experimental data.

As in previous editions, this book is motivated by the authors' enthusiasm for explaining and extending the insights of thermodynamics towards useful applications in chemical engineering. If that enthusiasm can be communicated to students and to practicing engineers, the purpose of this book will be fulfilled. As in previous editions, the motto of the third remains, as before: *Felix qui potuit rerum cognescere causas*.

J. M. Prausnitz
Berkeley, California

Effective teamwork that did not leave anything to be desired was the corner-stone for completing the third edition of this book during the time we spent in Berkeley. We thank John for preparing the ground correspondingly, for his leadership as senior author and for his support.

Because we were responsible for electronic typesetting, layout, artwork and figures (many from the previous edition), it was a challenge to take care of all details, requiring from us skills that we learned by doing. We did our best to prepare this book to meet professional standards.

Coming back to Berkeley and to collaborate once again with John was for us an enriching experience. In doing so we asked from our families even more sacrifices than we usually request as academic scientists.

I (R.N.L.) express most sincere thanks to my wife Brigitte, and to my children, Ulrike, Heike, Felix, Philipp, and Martin who provided the support necessary to commit myself in Berkeley exclusively to the revision of this book. I am also grateful to my colleagues at the Institute of Physical Chemistry of the University of Heidelberg for stepping in to meet my teaching duties while I was in Berkeley. Further, I want to thank Siegfried Kraft, Chancellor of the University of Heidelberg, whose understanding and advice made part of my stay in Berkeley possible.

I (E.G.A.) am thankful to my family, Cristina, Miguel and Marta, for their encouragement, understanding and support during the long and difficult months we were apart. Also I am grateful to Instituto Superior Técnico for a leave of absence, and to the Fulbright Program, Programa de Bolsas de Estudos da OTAN, and Fundação Luso-Americana para o Desenvolvimento for financial support of my residence in Berkeley during the academic year 1992/93 (when the preparation of the present edition started) and during the first semester of 1998.

R. N. Lichtenthaler
Heidelberg, Germany

E. Gomes de Azevedo
Lisbon, Portugal

Preface to the Second Edition

Molecular thermodynamics is an engineering science in the sense that its goal is to provide quantitative estimates of equilibrium properties for mixtures as required for chemical process design. To provide these estimates, molecular thermodynamics uses not only classical thermodynamics but also concepts from statistical thermodynamics and chemical physics; the operational procedure can be summarized by these steps:

1. Use statistical thermodynamics whenever possible, at least as a point of departure.
2. Apply appropriate concepts from molecular science.
3. Construct physically grounded models for expressing (abstract) thermodynamic functions in terms of (real) measurable properties.
4. Obtain model parameters from a few, but representative, experimental measurements.
5. Reduce the model to practice through a computer program that efficiently interfaces with engineering-design calculations.

The second edition, like the first, attempts to provide some guidance toward establishing the principles of molecular thermodynamics. This guidance is intended primarily for seniors or first-year graduate students in chemical engineering, but practicing engineers also may find it useful.

In preparing the second edition, I have taken a position of compromise between on the one hand, a "scientific" book that stresses molecular theory and on the other, an "engineering" book that gives practical advice toward specific design procedures. As in the first edition, emphasis is placed on fundamental concepts and how they can be reduced to practice to yield useful results.

Like the earlier edition, the second edition contains ten chapters and several appendices. All chapters have been partially revised and updated. Major changes are in Chapters 4, 6, 7, and 8, and much of Chapter 10 is totally new. Some earlier appendices have been removed and others have been added: Appendix II gives a brief introduction to statistical thermodynamics, while Appendices VIII and IX present summaries of some special aspects of the theory of solutions as addenda to Chapter 7.

Many new problems have been added. Solving problems is essential for serious students. Numerical answers to numerous problems are given in the final Appendix.

Since work for the first edition ceased in 1968, there have been formidable developments in a variety of areas that bear on molecular thermodynamics. It is not possible, in a reasonable number of pages, to do justice to all or even a major part of these developments. I have had to omit much that might have been included, lest this book become even larger; I can only ask my colleagues to forgive me if some of their contributions are not here mentioned for reasons of economy.

Perhaps the most promising development in the last fifteen years is in the statistical thermodynamics of fluids and fluid mixtures, especially through perturbation theory and computer simulation. There is little doubt that these developments will continue toward eventual direct application in engineering design. However, it is also likely that such direct application is not in the immediate future and that therefore, the semi-empirical methods discussed in this book will be utilized for many more years. Nevertheless, chemical-engineering students should now receive at least some introduction to the statistical thermodynamics of fluids, not only because of utility in the future, but also because idealized results from contemporary statistical thermodynamics are already now of much use in guiding development of semi-theoretical models toward thermodynamic-property correlations. Therefore, some limited discussion of applied statistical thermodynamics is now included in Chapters 4, 7, and 10.

I am deeply grateful to many colleagues who have contributed to my understanding of molecular thermodynamics and its applications, and thereby to this book; perhaps the most helpful of these has been B. J. Alder. In addition to those mentioned in the Preface to the First Edition, I want to record here my thanks to R. A. Heidemann, E. U. Franck, K. E. Gubbins, R. C. Reid, the late T. K. Sherwood, H. Knapp, F. Kohler, C. Tsonopoulos, L. C. Claitor, H. C. van Ness, F. Selleck, and C. J. King. Further, I owe much to my numerous co-workers (graduate students and post-doctoral visitors) who have provided me with new information, stimulating questions and good fellowship. I am, however, especially grateful to my two co-authors, R. N. Lichtenthaler and E. G. Azevedo, who ably assisted me in making revisions and additions to the original manuscript. Their contributions to the second edition are considerable and they deserve much credit for whatever success the second edition may achieve. All three authors are particularly indebted to P. Rasmussen for his critical review, to S. F. Barreiros for preparing the index and to R. Spontak for assistance in proof-reading.

Almost all of the new and revised sections of the second edition were prepared in the period 1978-80. It is unfortunate that, for a variety of reasons, publication was so long delayed. The final manuscript was sent to the publisher in February 1983.

The second edition maintains the pragmatic (engineering-science) philosophy that characterized the first edition: it is useful and ultimately economic to utilize whatever theoretical concepts may be suitable, but it is also important consistently to bear in mind the ultimate applied objective. To attain that objective, theory is rarely sufficient and inevitably at least some experimental data are required. The goal must always be to attain a healthy balance between theory and experiment, to avoid extreme emphasis in either direction.

This need for balance was recognized many years ago by a pioneer in applied science, Sir Francis Bacon, who used an analogy between scientific enterprise and the world of insects. In "Novum Organum" (1620), Bacon wrote about ants, spiders, and bees:

> Those who have handled sciences, have been either men of experiment or men of dogmas. The men of experiment are like the ant; they only collect and use. The reasoners resemble spiders who make cobwebs out of their own substance. But the bee takes a middle course: it gathers its material from the flowers of the garden and of the field, and transforms and digests it by a power of its own. Therefore, from a closer and purer league between these two faculties, the experimental and the rational, much may be hoped.

Finally, as in the Preface to the First Edition, I want to stress once again that studying, practicing and extending molecular thermodynamics is not only a useful activity but also one that provides a sense of joy and satisfaction. I shall be glad if some of that sense is infectious so that the reader may attain from molecular thermodynamics the same generous rewards that it has given to me.

J. M. Prausnitz
Berkeley, California

About 14 years ago I met J. M. Prausnitz for the first time. He immediately stimulated my interest in the exciting science of phase-equilibrium thermodynamics and ever since he has strongly sustained my work in this field. Throughout the years, we usually agreed quickly on how to approach and to solve problems but when we did not, open, honest and sometimes tough discussions always brought us to mutual agreement. To be one of the co-authors of this book is the culminating point so far in our joint effort to establish molecular thermodynamics as a useful engineering science for practical application. Thank you, John!

A scientist demands a lot of sacrifice from those who share his life. Therefore I owe many, many thanks to my wife Brigitte, and to my children, Ulrike, Heike, Felix and Philipp who give me enduringly all the support I need to pursue my scientific work in the way I do it.

R. N. Lichtenthaler
Heidelberg, Federal Republic of Germany

Preface to the First Edition

Since the generality of thermodynamics makes it independent of molecular considerations, the expression "molecular thermodynamics" requires explanation.

Classical thermodynamics presents broad relationships between macroscopic properties, but it is not concerned with quantitative prediction of these properties. Statistical thermodynamics, on the other hand, seeks to establish relationships between macroscopic properties and intermolecular forces through partition functions; it is very much concerned with quantitative prediction of bulk properties. However, useful configurational partition functions have been constructed only for nearly ideal situations and, therefore, statistical thermodynamics is at present insufficient for many practical purposes.

Molecular thermodynamics seeks to overcome some of the limitations of both classical and statistical thermodynamics. Molecular phase-equilibrium thermodynamics is concerned with application of molecular physics and chemistry to the interpretation, correlation, and prediction of the thermodynamic properties used in phase-equilibrium calculations. It is an engineering science, based on classical thermodynamics but relying on molecular physics and statistical thermodynamics to supply insight into the behavior of matter. In application, therefore, molecular thermodynamics is rarely exact; it must necessarily have an empirical flavor.

In the present work I have given primary attention to gaseous and liquid mixtures. I have been concerned with the fundamental problem of how best to calculate fugacities of components in such mixtures; the analysis should therefore be useful to engineers engaged in design of equipment for separation operations. Chapters 1, 2, and 3 deal with basic thermodynamics and, to facilitate molecular interpretation of thermodynamic properties, Chapter 4 presents a brief discussion of intermolecular forces. Chapter 5 is devoted to calculation of fugacities in gaseous mixtures and Chapter 6 is concerned with excess functions of liquid mixtures. Chapter 7 serves as an introduction to the theory of liquid solutions with attention to both "physical" and "chemical" theories. Fugacities of gases dissolved in liquids are discussed in Chapter 8 and those of solids dissolved in liquids in Chapter 9. Finally, Chapter 10 considers fluid-phase equilibria at high pressures.

While it is intended mainly for chemical engineers, others interested in fluid-phase equilibria may also find the book useful. It should be of value to university seniors or first-year graduate students in chemistry or chemical engineering who have completed a standard one-year course in physical chemistry and who have had some previous experience with classical thermodynamics.

The subjects discussed follow quite naturally from my own professional activities. Phase-equilibrium thermodynamics is a vast subject, and no attempt has been made to be exhaustive. I have arbitrarily selected those topics with which I am familiar and have omitted others which I am not qualified to discuss; for example, I do not consider solutions of metals or electrolytes. In essence, I have written about those topics which interest me, which I have taught in the classroom, and which have comprised much of my research. As a result, emphasis is given to results from my own research publications, not because they are in any sense superior, but because they encompass material with which I am most closely acquainted.

In the preparation of this book I have been ably assisted by many friends and former students; I am deeply grateful to all. Helpful comments were given by J. C. Berg, R. F. Blanks, P. L. Chueh, C. A. Eckert, M. L. McGlashan, A. L. Myers, J. P. O'Connell, Otto Redlich, Henri Renon, F. B. Sprow, and H. C. Van Ness. Generous assistance towards improvement of the manuscript was given by R. W. Missen and by C. Tsonopoulos who also prepared the index. Many drafts of the manuscript were cheerfully typed by Mrs. Irene Blowers and Miss Mary Ann Williams and especially by my faithful assistant for over twelve years, Mrs. Edith Taylor, whose friendship and conscientious service deserve special thanks.

Much that is here presented is a reflection of what I have learned from my teachers of thermodynamics and phase equilibria: G. J. Su, R. K. Toner, R. L. Von Berg, and the late R. H. Wilhelm; and from my colleagues at Berkeley: B. J. Alder, Leo Brewer, K. S. Pitzer and especially J. H. Hildebrand, whose strong influence on my thought is evident on many pages.

I hope that I have been able to communicate to the reader some of the fascination I have experienced in working on and writing about phase-equilibrium thermodynamics. To think about and to describe natural phenomena, to work in science and engineering – all these are not only useful but they are enjoyable to do. In writing this book I have become aware that for me phase-equilibrium thermodynamics is a pleasure as well as a profession; I shall consider it a success if a similar awareness can be awakened in those students and colleagues for whom this book is intended. *Felix qui potuit rerum cognoscere causas.*

Finally, I must recognize what is all too often forgotten – that no man lives or works alone, but that he is molded by those who share his life, who make him what he truly is. Therefore I dedicate this book to Susie, who made it possible, and to Susi and Toni, who prepared the way.

J. M. Prausnitz
Berkeley, California

Nomenclature

a	Parameter in a cubic equation of state; activity
A	Helmholtz energy; constant in Margules equation; Debye-Hückel constant
b	Parameter in a cubic equation of state
B	Second virial coefficient
B^*	Osmotic second virial coefficient
c	Molar concentration
c_p	Constant-pressure molar heat capacity
c_v	Constant-volume molar heat capacity
C	Third virial coefficient
C^*	Osmotic third virial coefficient
D	Fourth virial coefficient; diffusion coefficient
e	Electron charge
E	Enhancement factor; electrical field strength
f	Fugacity
F	Force; number of degrees of freedom; Faraday constant
g_{ij}, g_{ji}	Binary parameter in NRTL
g	Molar Gibbs energy
g^E	Molar excess Gibbs energy
$\bar{g}_i^E$	Partial molar excess Gibbs energy of component i
G	Gibbs energy
$\Delta_r G$	Gibbs energy change of reaction
h	Molar enthalpy; Planck's constant
h^E	Molar excess enthalpy
$\bar{h}$	Partial molar enthalpy
H	Enthalpy
$H_{i,j}$	Henry's constant of solute i in solvent j
I	Ionization potential; ionic strength
J	Flux
k	Boltzmann's constant
k_{ij}	Pair i-j interaction parameter
K	Equilibrium constant; solubility product; K factor
l_{ij}	Pair i-j interaction parameter
m	Number of components; molecular mass; molality
m_M	Molality of molecular (non-dissociated) component
M	Molar mass
$\bar{M}$	Average molecular weight
n	Number of moles; index of refraction
n_T	Total number of moles
N_A	Avogadro's constant
N_{12}	Number of 1-2 contacts for a real (non-random) mixture
N_{12}^*	Number of 1-2 contacts for a non-random mixture

p_i	Partial pressure of component i
P	Pressure
P_i^s	Saturation pressure of pure i
q_i	Effective volume of molecule i
Q	Heat; quadrupole moment; partition function
r	Number of segments
r_{ij}	Distance between molecules i and j
R	Gas constant
$\Re$	Proportionality constant; residual quantity
s	Molar entropy
s^E	Molar excess entropy
$\bar{s}_i$	Partial molar entropy of component i
S	Entropy; solubility coefficient
T	Absolute temperature
u	Molar internal energy
u^E	Molar excess internal energy
$\Delta u_{ij}, \Delta u_{ji}$	Energy parameter in UNIQUAC
U	Internal energy
v	Molar volume; specific volume
v^E	Molar excess volume
$\bar{v}$	Partial molar volume
V	Total volume
x	Liquid-phase mole fraction
X_{12}	Interaction parameter
y	Vapor-phase mole fraction
w	Interchange energy
W	Work
z	Compressibility factor; coordination number; effective volume fraction; ionic valence
$\mathfrak{z}$	Overall ("true") mole fraction
Z_N	Configuration integral
Z_M	Total lattice-site coordination number

Greek Symbols

α	Polarizability; fraction of molecules; separation factor
α_p	Thermal expansion coefficient
χ	Flory-Huggins interaction parameter
δ	Solubility parameter
δ_M	Membrane thickness
$\bar{\delta}$	Volume-fraction average of solubility parameters
ε	Energy parameter; permittivity
ε_{ij}	Lennard-Jones interaction
ϕ	Osmotic coefficient; electric potential
γ	Activity coefficient; thermal pressure coefficient
γ^*	Unsymmetrically normalized activity coefficient
η	Reduced density
φ	Fugacity coefficient
Φ	Volume fraction
Φ^*	Segment fraction
κ	Inverse of Debye length
κ_T	Isothermal compressibility
$\lambda_{ij}, \lambda_{ji}$	Energy parameters in Wilson equation
Λ	de Broglie wavelength
$\Lambda_{ij}, \Lambda_{ji}$	Binary parameters in Wilson model
μ	Chemical potential; dipole moment
ν	Frequency; ionic charge
π	Number of independent variables; osmotic pressure
θ	Surface fraction; theta temperature
θ_i	Area fraction of component i
ρ	Molar density
σ	Distance parameter
τ_{ij}, τ_{ji}	Binary parameters in NRTL
Γ_{ij}	Potential energy for pair i-j

Γ	Intermolecular potential energy
ω	Acentric factor
ω_{ij}	Pair interchange energy

Subscripts

a	Anion
c	Critical property; cation
cp	Closest packing
i	Component i
F	Feed
hs	Hard chain
M	Molecular
M	Membrane
m	Molar property
mix	Mixing property
mixt	Property of a mixture
P	Permeate
r	Reaction
R	Reduced property
s	Solvent
sp	solubility product
sub	Sublimation
T	Total
vap	Vaporization
w	Water

Superscripts

0	Reference state; standard state
conf	Configurational property
E	Excess property
fv	Free volume
id	Ideal gas
L	Liquid phase
m	Number of components
R	Residual property
$\mathcal{S}$	Solid phase
s	Saturated property
V	Vapor phase
α	Phase α
β	Phase β
*	Equilibrium or normalized property; complete randomness; hard-core property
∞	Infinite dilution
~	Reduced property

CHAPTER 1

The Phase-Equilibrium Problem

We live in a world of mixtures – the air we breathe, the food we eat, the gasoline in our automobiles. Wherever we turn, we find that our lives are linked with materials that consist of a variety of chemical substances. Many of the things we do are concerned with the transfer of substances from one mixture to another; for example, in our lungs, we take oxygen from the air and dissolve it in our blood, while carbon dioxide leaves the blood and enters the air; in our coffee maker, water-soluble ingredients are leached from coffee grains into water; and when someone stains his tie with gravy, he relies on cleaning fluid to dissolve and thereby remove the greasy spot. In each of these common daily experiences, as well as in many others in physiology, home life, industry, and so on, there is a transfer of a substance from one phase to another. This occurs because when two phases are brought into contact, they tend to exchange their constituents until the composition of each phase attains a constant value; when that state is reached, we say that the phases are in equilibrium. The equilibrium compositions of two phases are often very different from one another, and it is precisely this difference that enables us to separate mixtures by distillation, extraction, and other phase-contacting operations.

The final, or equilibrium, phase compositions depend on several variables, such as temperature and pressure, and on the chemical nature and concentrations of the substances in the mixture. Phase-equilibrium thermodynamics seeks to establish the relations among the various properties (in particular, temperature, pressure, and composition) that ultimately prevail when two or more phases reach a state of equilibrium wherein all tendency for further change has ceased.

Because so much of life is concerned with the interaction between different phases, it is evident that phase-equilibrium thermodynamics is a subject of fundamental importance in many sciences, physical as well as biological. It is of special interest in chemistry and chemical engineering because so many operations in the manufacture of chemical products consist of phase contacting: Extraction, adsorption, distillation, leaching, and absorption are essential unit operations in chemical industry; an understanding of any one of them is based, at least in part, on the science of phase equilibrium.

Equilibrium properties are required for the design of separation operations; these, in turn, are essential parts of a typical chemical plant, as shown in Fig. 1-1. In this plant, the central part (stage II) is the chemical reactor and it has become frequent practice to call the reactor the heart of the plant. But, in addition, a plant needs a mouth (stage I) and a digestive system (stage III). Prior to reaction, the reactants must be prepared for reaction; because the raw materials provided by nature are usually mixtures, separation is often required to separate the desired reactants from other unwanted components that may interfere with the reaction. Downstream from the reactor, separation is necessary to separate desired from undesired products and, because reaction is rarely complete, it is also necessary to separate the unreacted reactants for recycle.

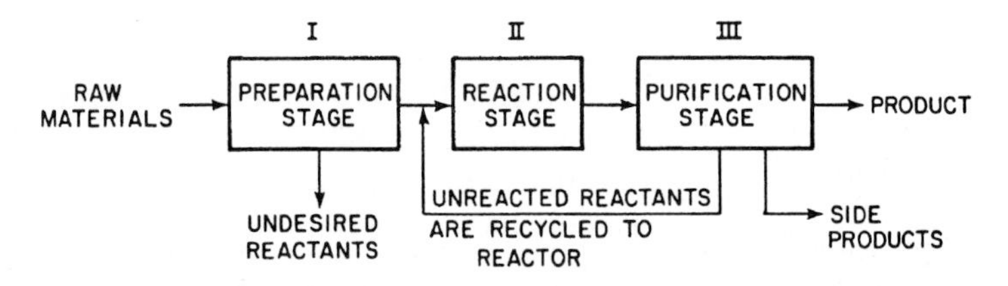

STAGES I AND III REQUIRE SEPARATION OPERATIONS (e.g., DISTILLATION, ABSORPTION, EXTRACTION). IN A TYPICAL CHEMICAL PLANT, 40–80% OF INVESTMENT IS FOR SEPARATION-OPERATION EQUIPMENT.

Figure 1-1 Schematic of a chemical plant.

Figure 1-1 illustrates why separations are so important in chemical engineering. In a typical large-scale chemical plant, the investment for separation operations is in the neighborhood of 50% and often it is appreciably more.

1.1 Essence of the Problem

We want to relate quantitatively the variables that describe the state of equilibrium of two or more homogeneous phases that are free to interchange energy and matter. By a *homogeneous phase* at equilibrium we mean any region in space where the intensive properties are everywhere the same.[1] *Intensive properties* are those that are independent of the mass, size, or shape of the phase; we are concerned primarily with the intensive properties temperature, density, pressure, and composition (often expressed in terms of mole fractions). We want to describe the state of two or more phases that are free to interact and that have reached a state of equilibrium. Then, given some of the equilibrium properties of the two phases, our task is to predict those that remain.

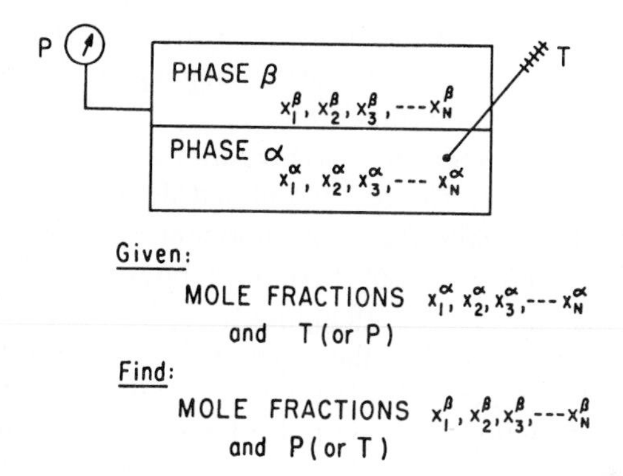

Figure 1-2 Statement of problem.

Figure 1-2 illustrates schematically the type of problem that phase-equilibrium thermodynamics seeks to solve. We suppose that two multicomponent phases, α and β, have reached an equilibrium state; we are given temperature T of the two phases and mole fractions $x_1^\alpha, x_2^\alpha, \cdots$, of phase α. Our task, then, is to find mole fractions $x_1^\beta, x_2^\beta, \cdots$, of phase β and pressure P of the system. Alternatively, we might know $x_1^\alpha, x_2^\alpha, \cdots$, and P and be asked to find $x_1^\beta, x_2^\beta, \cdots$, and T, or our problem might be characterized by other combinations of known and unknown variables. The number of intensive properties that must be specified to fix unambiguously the state of equilibrium is given by the *Gibbs phase rule*. In the absence of chemical reactions, the phase rule is:

$$\text{Number of independent intensive properties} = \text{Number of components} - \text{Number of phases} + 2$$

For example, for a two-component, two-phase system, the number of independent intensive properties is two. In such a system the intensive properties of interest usually

[1] We are here neglecting all special forces such as those due to gravitational, electric, or magnetic fields, surface forces, etc.

are x_1^α, x_1^β, T, and P.[2] Two of these, any two, must be specified before the remaining two can be found.

How shall we go about solving the problem illustrated in Fig. 1-2? What theoretical framework is available to give us a basis for finding a solution? When this question is raised, we turn to thermodynamics.

1.2 Application of Thermodynamics to Phase-Equilibrium Problems

One of the characteristics of modern science is abstraction. By describing a difficult, real problem in abstract, mathematical terms, it is sometimes possible to obtain a simple solution to the problem not in terms of immediate physical reality, but in terms of mathematical quantities that are suggested by an abstract description of the real problem. Thermodynamics provides the mathematical language that enables us to obtain an abstract solution of the phase-equilibrium problem.

Application of thermodynamics to phase equilibria in multicomponent systems is shown schematically in Fig. 1-3. The real world and the real problem are represented by the lower horizontal line, while the upper horizontal line represents the world of abstraction. The three-step application of thermodynamics to a real problem consists of an indirect mental process; instead of attempting to solve the real problem within the world of physically realistic variables, the indirect process first projects the problem into the abstract world, then seeks a solution within that world, and finally projects this solution back to physical reality. The solution of a phase-equilibrium problem using thermodynamics requires three steps:

I. The real problem is translated into an abstract, mathematical problem.

II. A solution is found to the mathematical problem.

III. The mathematical solution is translated back into physically meaningful terms.

The essential feature of step I is to define appropriate and useful mathematical functions to facilitate step II. The profound insight of Gibbs, who in 1875 defined such a function – the *chemical potential* – made it possible to achieve the goal of step II; the mathematical solution to the phase-equilibrium problem is given by the remarkably simple result that at equilibrium, the chemical potential of each component must be the same in every phase.

The difficult step is the last one, step III. Thanks to Gibbs, steps I and II present no further problems and essentially all work in this field, after Gibbs, has been

---

[2] Because $\sum_i x_i = 1$ for each phase, x_2^α and x_2^β are not additional variables in this case.

concerned with step III. From the viewpoint of a formal theoretical physicist, the phase-equilibrium problem has been solved completely by Gibbs' relation for the chemical potentials. A pure theoretician may require nothing further, but someone who is concerned with obtaining useful numerical answers to real problems must face the task of translating the abstract results of step II into the language of physical reality.

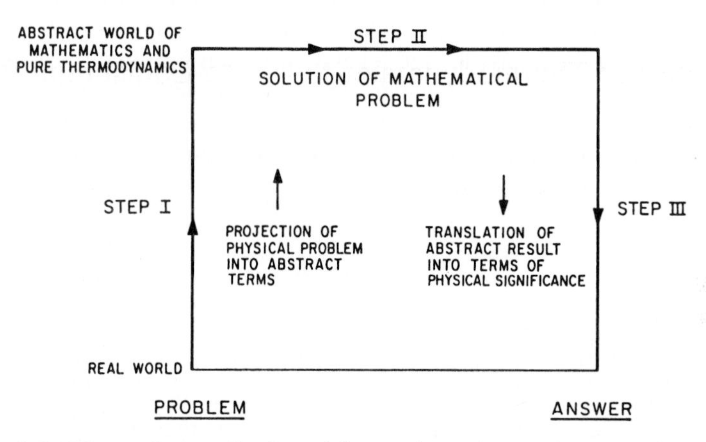

Figure 1-3 Three-step application of thermodynamics to phase-equilibrium problems.

Our concern in this book is concentrated almost exclusively on step III. In Chap. 2 we briefly review some of the important concepts that lead to Gibbs' equation, i.e., that for any component, the chemical potential must be the same in all equilibrated phases. In a sense, we may call Chap. 2 historical because it reproduces, in perhaps more modern terminology, work that was completed many years ago. However, in all the remaining chapters, we address the contemporary problem of how quantitatively to relate the chemical potential to the primary variables temperature, pressure, and composition. We should point out at once that this problem, designated by step III, is mostly outside the realm of classical thermodynamics and, therefore, much of the material in later chapters cannot be called thermodynamics in the strict sense. Classical thermodynamics by itself gives us important but also severely limited information on the relation between the abstract chemical potential and the real, experimentally accessible quantities temperature, pressure, and composition. For quantitative, numerical results, classical thermodynamics is bot sufficient. Step III must also utilize concepts from statistical thermodynamics, molecular physics, and physical chemistry.

To solve problems of the type illustrated in Fig. 1-2, we must make the transition from what we have, i.e., the abstract thermodynamic equation of equilibrium, toward what we want, i.e., quantitative information about temperature, pressure, and phase compositions. Thanks to Gibbs, the thermodynamic equation of equilibrium is now well known and we need not concern ourselves with it except as a place to start. In any problem concerning the equilibrium distribution of some component i between two phases α and β, we must always begin with the relation

$$\mu_i^\alpha = \mu_i^\beta \tag{1-1}$$

where μ is the chemical potential. It is then that our problem begins; we must now ask how μ_i^α is related to T, P, and $x_1^\alpha, x_2^\alpha, \cdots$, and similarly, how μ_i^β is related to T, P, and $x_1^\beta, x_2^\beta, \cdots$. To establish these relations, it is convenient to introduce some auxiliary functions such as *fugacity* and *activity*. These functions do not solve the problem for us, but they facilitate our efforts to find a solution because, in many cases, they make the problem somewhat easier to visualize; fugacity and activity are quantities much closer to our physical senses than the abstract concept chemical potential. Suppose, for example, that phase α is a vapor and phase β is a liquid. Then, as discussed in subsequent chapters, Eq. (1-1) can be rewritten

$$\varphi_i y_i P = \gamma_i x_i f_i^0 \tag{1-2}$$

where, in the vapor phase, y_i is the mole fraction and φ_i is the fugacity coefficient, and in the liquid phase, x_i is the mole fraction, γ_i is the activity coefficient, and f_i^0 is the fugacity of component i at some fixed condition known as the *standard state*.

The details of Eq. (1-2) are not important just now; they are covered later. What is important to note is the procedure whereby the highly abstract Eq. (1-1) has been transformed into the not-quite-so-abstract Eq. (1-2). Equation (1-2), unlike Eq. (1-1), at least contains explicitly three of the variables of interest, x_i, y_i, and P. Equation (1-2) is no more and no less fundamental than Eq. (1-1); one is merely a mathematical transformation of the other, and any claim Eq. (1-2) may have to being more useful is only a consequence of a fortunate choice of useful auxiliary functions in the transformation.

Much of this utility comes from the concept of ideality. If we define mixtures with certain properties as ideal mixtures, we then find, as a result of our choice of auxiliary functions, that the equation of equilibrium can be simplified further; for example, for a mixture of ideal gases $\varphi_i = 1$, and for ideal liquid mixtures at low pressures, $\gamma_i = 1$ when f_i^0 is given by the saturation pressure of pure liquid i at the temperature of interest. We thus find that some of the auxiliary functions (such as φ_i and γ_i) are useful because they are numerical factors, frequently of the order of unity, that establish the connection between real mixtures and those that, by judicious choice, have been defined as ideal mixtures.

From the viewpoint of formal thermodynamics, Eq. (1-2) is no better than Eq. (1-1); but from the viewpoint of experimental chemistry and chemical engineering, Eq. (1-2) is preferable because it provides a convenient frame of reference.

In the general case we cannot assume ideal behavior and we must then establish two relations, one for φ_i and γ_i:

$$\varphi_i = F_\varphi(T, P, y_1, y_2, \ldots) \tag{1-3}$$

$$\gamma_i = F_\gamma(T, P, x_1, x_2, \ldots) \tag{1-4}$$

In Chaps. 3, 5, and 12, we discuss in detail what we can say about function F_φ in Eq. (1-3). In Chap. 4, we digress with a brief discussion of the nature of intermolecular forces, because the functional relationships of both Eqs. (1-3) and (1-4) are determined by forces that operate between molecules. In Chaps. 6, 7 and 12, we are concerned with function F_γ in Eq. (1-4), and in Chaps. 10 and 11, primary attention is given toward determination of a useful f_i^0, such that activity coefficient γ is often close to unity. Chapter 8 discusses activity coefficients in systems containing polymers and Chap. 9 is devoted to liquid solutions of solutes that dissociate into ions.

Before discussing in detail various procedures for calculating fugacities and other useful auxiliary functions, we first give in Chap. 2 a brief survey of steps I and II indicated in Fig. 1-3.

CHAPTER 2

Classical Thermodynamics of Phase Equilibria

Thermodynamics as we know it today originated during the middle of the nineteenth century, and while the original thermodynamic formulas were applied to only a limited class of phenomena (such as heat engines), they have, as a result of suitable extensions, become applicable to a large number of problems in both physical and biological sciences. From its Greek root *(therme,* heat; *dynamis,* force), one might well wonder what "thermodynamics" has to do with the *distribution* of various components between various phases. Indeed, the early workers in thermodynamics were concerned only with systems of one component, and it was not until the monumental work of J. Willard Gibbs that thermodynamic methods were shown to be useful in the study of multicomponent systems. It was Gibbs who first saw the generality of thermodynamics. He was able to show that a thermodynamic procedure is possible for a wide variety of applications, including the behavior of chemical systems.

This chapter briefly reviews the essential concepts of the classical thermodynamic description of phase equilibria.[1] It begins with a combined statement of the first and second laws as applied to a closed, homogeneous system, and proceeds toward the laws of equilibrium for an open, heterogeneous system. For our purposes here, we exclude surface and tensile effects, acceleration, or change of position in an external field, such as a gravitational or electromagnetic field (other than along a surface of constant potential); for simplicity, we also rule out chemical and nuclear reactions.[2] We are then left with the classical problem of phase equilibrium that considers internal equilibrium with respect to three processes:

1. Heat transfer between any two phases within the heterogeneous system.
2. Displacement of a phase boundary.
3. Mass transfer of any component in the system across a phase boundary.

The governing potentials in the first two processes are temperature and pressure, respectively, and we assume prior knowledge of their existence; the governing potential for the third process, however, is considered not to be known a priori, and it is one of the prime responsibilities of classical thermodynamics of phase equilibria to "discover" and exploit the appropriate "chemical potential."[3] A heterogeneous system that is in a state of internal equilibrium is a system at equilibrium with respect to each of these three processes.

The chapter continues with some discussion of the nature of the chemical potential and the need for standard states, and then introduces the auxiliary functions fugacity and activity. The chapter concludes with a very simple example of how the thermodynamic equations of phase equilibrium may be applied to obtain a physically useful result.

2.1 Homogeneous Closed Systems

A *homogeneous system* is one with uniform properties throughout; i.e., a property such as density has the same value from point to point, in a macroscopic sense. A phase is a homogeneous system. A *closed system* is one that does not exchange matter with its surroundings, although it may exchange energy. Thus, in a closed system not undergoing chemical reaction, the number of moles of each component is constant. This constraint can be expressed as

$$dn_i = 0 \quad i = 1, 2, \ldots, m \tag{2-1}$$

[1] More complete discussions are given in references listed at the end of this chapter.
[2] However, see the final two paragraphs of App. A.
[3] This was first done by Gibbs in 1875.

where n_i is the number of moles of the ith component and m is the number of components present.

For a homogeneous, closed system, with the qualifications given previously, taking into account interactions of the system with its surroundings in the form of heat transfer and work of volumetric displacement, a combined statement of the first and second laws of thermodynamics is (Denbigh, 1981)

$$\boxed{dU \leq T_B dS - P_E dV} \tag{2-2}$$

where dU, dS, and dV are, respectively, small changes in energy, entropy, and volume of the *system* resulting from the interactions; each of these properties is a *state function* whose value in a prescribed state is independent of the previous history of the system. For our purposes here, the surroundings are considered to be two distinct bodies: a constant-volume heat bath, also at constant, uniform temperature T_B, in thermal contact only with the system; and another external body, at constant, uniform pressure P_E, in "volumetric" contact only with the system through a movable, thermally insulated piston.

Because Eq. (2-2) is our starting point, it is important to have a better understanding of its significance and range of validity, even though we do not attempt to develop or justify it here. However, before proceeding, we need to discuss briefly three important concepts: equilibrium state, reversible process, and state of internal equilibrium.

By an *equilibrium state* we mean one with no tendency to depart spontaneously, having in mind certain permissible changes or processes, i.e., heat transfer, work of volume displacement and, for open systems (next section), mass transfer across a phase boundary. In an equilibrium state, the properties are independent of time and of previous history of the system; further, they are stable, that is, not subject to "catastrophic" changes on slight variations of external conditions. We distinguish an equilibrium state from a steady state, insisting that in an equilibrium state there are no net fluxes of the kind under consideration (heat transfer, etc.) across a plane surface placed anywhere in the system.

In thermodynamics, we are normally concerned with a finite change in the equilibrium state of a system or a variation in an equilibrium state subject to specified constraints. A change in the equilibrium state of a system is called a *process.* A *reversible process* is one where the system is maintained in a state of virtual equilibrium throughout the process; a reversible process is sometimes referred to as one connecting a series of equilibrium states. This requires that the potential difference (between system and surroundings) causing the process to occur be only infinitesimal; then the direction of the process can be reversed by an infinitesimal increase or decrease, as the case may be, in the potential for the system or the surroundings. Any natural or actual process occurs irreversibly; we can think of a reversible process as a limit approached but never attained. The inequality in Eq. (2-2) refers to a natural (*irreversible*) process and the equality to a reversible process.

By a "single-phase system in a state of internal equilibrium" we mean one that is homogeneous (uniform properties) even though it may be undergoing an irreversible process as a result of an interaction with its surroundings. In practice, such a state may be impossible to achieve, but the concept is useful for a discussion of the significance of Eq. (2-2), to which we now return.

If the interaction of the system with its surroundings occurs reversibly (reversible heat transfer and reversible boundary displacement), the equality sign of Eq. (2-2) applies; in that event, $T_B = T$, the temperature of the system, and $P_E = P$, the pressure of the system. Hence, we may write

$$dU = TdS - PdV \tag{2-3}$$

The first term on the right is the heat absorbed by the system ($TdS = \delta Q_{rev}$), and the second term is the work done by the system ($\delta W_{rev} = PdV$). The form of this equation implies that the system is characterized by two independent variables or degrees of freedom, here represented by S and V.

If the interaction between system and surroundings occurs irreversibly, the inequality of Eq. (2-2) applies:

$$dU < T_B dS - P_E dV \tag{2-4}$$

In this case $\delta W = P_E\, dV$, but $\delta Q \neq T_B dS$. However, if the system is somehow maintained in a state of internal equilibrium during the irreversible interaction, that is, if it has uniform properties, then it is a system characterized by two independent variables and Eq. (2-3) applies. Hence, this equation may be applicable whether the process is externally reversible or irreversible. However, in the latter situation the terms TdS and PdV can no longer be identified with heat transfer and work, respectively.

To obtain the finite change in a thermodynamic property occurring in an actual process (from equilibrium state 1 to equilibrium state 2), the integration of an equation such as Eq. (2-3) must be done over a *reversible path* using the properties of the *system*. This results in an equation of the form

$$\Delta U = U_2 - U_1 = \int_{S_1}^{S_2} TdS - \int_{V_1}^{V_2} PdV \tag{2-5}$$

Because U is a *state function*, this result is independent of the path of integration, and it is independent of whether the system is maintained in a state of internal equilibrium or not during the actual process; it requires only that the initial and final states be equilibrium states. Hence the essence of classical (reversible) thermodynamics lies in the possibility of making such a calculation by constructing a convenient, reversible path to replace the actual or irreversible path of the process that is usually not amenable to an exact description.

Equation (2-3) represents the fundamental thermodynamic relation (Gibbs, 1961). If U is considered to be a function of S and V, and if this function U is known, then all other thermodynamic properties can be obtained by purely mathematical operations on this function. For example, $T = (\partial U/\partial S)_V$ and $P = -(\partial U/\partial V)_S$. While another pair of independent variables could be used to determine U, no other pair has this simple physical significance for the function U. We therefore call the group of variables U, S, V a *fundamental grouping*.

An important aspect of Eq. (2-2) is that it presents U as a potential function. If the variation dU is constrained to occur at constant S and V, then

$$dU_{S,V} \le 0 \tag{2-6}$$

Equation (2-6) says that at constant S and V, U tends toward a minimum in an actual or irreversible process in a closed system and remains constant in a reversible process. Because an actual process is one tending toward an equilibrium state, an approach to equilibrium at constant entropy and volume is accompanied by a decrease in internal energy. Equation (2-6), then, provides a criterion for equilibrium in a closed system; we shall make use of this criterion later.

Other extensive thermodynamic potentials for closed systems and other fundamental groupings can be obtained by using different pairs of the four variables P, V, T, and S as independent variables on the right-hand side of Eq. (2-3). Partial Legendre transformations (Callen, 1985) enable us to use three other pairs while retaining the important property of a fundamental equation. For example, suppose we wish to interchange the roles of P and V in Eq. (2-3) to have P as an independent variable. We then define a new function, viz. the original function, U, minus the product of the two quantities to be interchanged with due regard for the sign of the term in the original equation. That is, we define

$$H \equiv U - (-PV) = U + PV \tag{2-7}$$

where H, the *enthalpy* of the system, is a state function because it is defined in terms of state functions. Differentiation of Eq. (2-7) and substitution for dU in Eq. (2-3) gives

$$\boxed{dH = TdS + VdP} \tag{2-8}$$

and the independent variables are now S and P. The role of H as a potential for a closed system at constant S and P means that

$$dH_{S,P} \le 0 \tag{2-9}$$

Similarly, to interchange T and S (but not P and V) in Eq. (2-3), we define the *Helmholtz energy*

$$A \equiv U - TS \tag{2-10}$$

giving

$$\boxed{dA = -SdT - PdV} \tag{2-11}$$

and

$$dA_{T,V} \leq 0 \tag{2-12}$$

In this case, the independent variables or constraints are T and V. Finally, to interchange both T and S and P and V in Eq. (2-3) so as to use T and P as the independent variables, we define the *Gibbs energy*

$$G \equiv U - TS - (-PV) = H - TS \tag{2-13}$$

giving

$$\boxed{dG = -SdT + VdP} \tag{2-14}$$

and

$$dG_{T,P} \leq 0 \tag{2-15}$$

Table 2-1 gives a summary of the four fundamental equations and the roles of U, H, A, and G as thermodynamic potentials. Also included in the table are a set of identities resulting from the fundamental equations and the set of equations known as *Maxwell relations*. These relations are obtained from the fundamental equations by the application of Euler's reciprocity theorem that takes advantage of the fact that the order of differentiation in forming second partial derivatives is immaterial for continuous functions and their derivatives.

2.2 Homogeneous Open Systems

An *open system* can exchange matter as well as energy with its surroundings. We now consider how the laws of thermodynamics for a closed system can be extended to apply to an open system.

For a closed homogeneous system, we considered U to be a function only of S and V; that is,

$$U = U(S, V) \tag{2-16}$$

Table 2-1 Some important thermodynamic relations for a homogeneous closed system.

Definition of *H*, *A*, and *G*

$$H = U + PV$$
$$A = U - TS$$
$$G = U + PV - TS = H - TS = A + PV$$

Fundamental Equations

$$dU = TdS - PdV \qquad dA = -SdT - PdV$$
$$dH = TdS + VdP \qquad dG = -SdT + VdP$$

Extensive Functions as Thermodynamic Potentials

$$dU_{S,V} \le 0 \qquad dA_{T,V} \le 0$$
$$dH_{S,P} \le 0 \qquad dG_{T,P} \le 0$$

Maxwell Relations Resulting from the Fundamental Equations

$$\left(\frac{\partial T}{\partial V}\right)_S = -\left(\frac{\partial P}{\partial S}\right)_V \qquad \left(\frac{\partial S}{\partial V}\right)_T = \left(\frac{\partial P}{\partial T}\right)_V$$
$$\left(\frac{\partial T}{\partial P}\right)_S = \left(\frac{\partial V}{\partial S}\right)_P \qquad \left(\frac{\partial S}{\partial P}\right)_T = -\left(\frac{\partial V}{\partial T}\right)_P$$

Identities Resulting from the Fundamental Equations

$$\left(\frac{\partial U}{\partial V}\right)_T = T\left(\frac{\partial P}{\partial T}\right)_V - P \qquad \left(\frac{\partial H}{\partial P}\right)_T = V - T\left(\frac{\partial V}{\partial T}\right)_P$$
$$\left(\frac{\partial U}{\partial S}\right)_V = T = \left(\frac{\partial H}{\partial S}\right)_P \qquad \left(\frac{\partial H}{\partial P}\right)_S = V = \left(\frac{\partial G}{\partial P}\right)_T$$
$$\left(\frac{\partial U}{\partial V}\right)_S = -P = \left(\frac{\partial A}{\partial V}\right)_T \qquad \left(\frac{\partial A}{\partial T}\right)_V = -S = \left(\frac{\partial G}{\partial T}\right)_P$$

Heat Capacities

$$\left(\frac{\partial U}{\partial T}\right)_V = C_v \qquad \left(\frac{\partial S}{\partial T}\right)_V = \frac{C_v}{T}$$
$$\left(\frac{\partial H}{\partial T}\right)_P = C_p \qquad \left(\frac{\partial S}{\partial T}\right)_P = \frac{C_p}{T}$$

In an open system, however, there are additional independent variables. For these, we can use the mole numbers of the various components present. Hence, we must now consider U as the function

$$U = U(S, V, n_1, n_2, \ldots, n_m) \tag{2-17}$$

where m is the number of components. The total differential is then

$$dU = \left(\frac{\partial U}{\partial S}\right)_{V,n_i} dS + \left(\frac{\partial U}{\partial V}\right)_{S,n_i} dV + \sum_i \left(\frac{\partial U}{\partial n_i}\right)_{S,V,n_j} dn_i \tag{2-18}$$

where subscript n_i refers to all mole numbers and subscript n_j to all mole numbers other than the ith. Because the first two derivatives in Eq. (2-18) refer to a closed system, we may use the identities of Table 2-1. Further, we define the function μ_i as

$$\mu_i \equiv \left(\frac{\partial U}{\partial n_i}\right)_{S,V,n_j} \tag{2-19}$$

We may then rewrite Eq. (2-18) in the form

$$dU = TdS - PdV + \sum_i \mu_i dn_i \tag{2-20}$$

Equation (2-20) is the fundamental equation for an open system corresponding to Eq. (2-3) for a closed system. The function μ_i is an intensive quantity and we expect it to depend on temperature, pressure, and composition of the system. However, our primary task is to show that μ_i is a mass or *chemical potential*, as we might suspect from its position in Eq. (2-20) as a coefficient of dn_i, just as T (the coefficient of dS) is a thermal potential and P (the coefficient of dV) is a mechanical potential. Before doing this, however, we consider other definitions of μ_i and the corresponding fundamental equations for an open system in terms of H, A, and G. Using the defining equations for H, A, and G [Eqs. (2-7), (2-10), and (2-13)], we may substitute for dU in Eq. (2-20) in each case and arrive at the following further three fundamental equations for an open system:

$$dH = TdS + Vdp + \sum_i \mu_i dn_i \tag{2-21}$$

$$dA = -SdT - PdV + \sum_i \mu_i dn_i \tag{2-22}$$

$$dG = -SdT + VdP + \sum_i \mu_i dn_i \tag{2-23}$$

From the definition of μ_i given in Eq. (2-19) and from Eqs. (2-20) to (2-23), it follows that

$$\mu_i \equiv \left(\frac{\partial U}{\partial n_i}\right)_{S,V,n_j} = \left(\frac{\partial H}{\partial n_i}\right)_{S,P,n_j} = \left(\frac{\partial A}{\partial n_i}\right)_{T,V,n_j} = \left(\frac{\partial G}{\partial n_i}\right)_{T,P,n_j} \tag{2-24}$$

There are thus four expressions for μ_i where each is a derivative of an extensive property with respect to the amount of the component under consideration, and each uses a fundamental grouping of variables: U, S, V; H, S, P; A, T, V; and G, T, P. The quantity

μ_i is the partial molar Gibbs energy, but it is *not* the partial molar internal energy, enthalpy, or Helmholtz energy. This is because the independent variables T and P, chosen for defining partial molar quantities, are also the fundamental independent variables for the Gibbs energy G.

2.3 Equilibrium in a Heterogeneous Closed System

A *heterogeneous*, closed system is made up of two or more phases with each phase considered as an open system within the overall closed system. We now consider the conditions where the heterogeneous system is in a state of internal equilibrium with respect to the three processes of heat transfer, boundary displacement, and mass transfer.[4]

We already have four criteria with different sets of constraints for equilibrium in a closed system as given by the third set of equations in Table 2-1 with the equal sign in each case. However, these are in terms of the four extensive thermodynamic potentials U, H, A, and G. We can obtain more useful criteria in terms of the intensive quantities T, P, and μ_i. We expect that, to have thermal and mechanical equilibrium in the system, temperature and pressure must be uniform throughout the entire heterogeneous mass. If μ_i is the intensive potential governing mass transfer, we expect that μ_i must also have a uniform value throughout the whole heterogeneous system at equilibrium with respect to this process. Gibbs first gave the proof of this in 1875. He used the function U as a starting point rather than H, A, or G, probably because of the symmetry in the expression for dU in Eq. (2-20); each differential on the right is the differential of an extensive quantity and each coefficient is an intensive quantity. This means that the uniformity of all intensive potentials at equilibrium can be proved by consideration of only one function U. Details of this proof are given in App. A.

The general result for a closed, heterogeneous system consisting of π phases and m components is that at equilibrium with respect to the processes described earlier,

$$T^{(1)} = T^{(2)} = \cdots = T^{(\pi)} \tag{2-25}$$

$$P^{(1)} = P^{(2)} = \cdots = P^{(\pi)} \tag{2-26}$$

$$\begin{aligned} \mu_1^{(1)} &= \mu_1^{(2)} = \cdots = \mu_1^{(\pi)} \\ \mu_2^{(1)} &= \mu_2^{(2)} = \cdots = \mu_2^{(\pi)} \\ &\vdots \qquad \vdots \qquad\quad \vdots \\ \mu_m^{(1)} &= \mu_m^{(2)} = \cdots = \mu_m^{(\pi)} \end{aligned} \tag{2-27}$$

[4] We neglect here "special" effects such as surface forces; semipermeable membranes; and electric, magnetic or gravitational forces.

where the superscript in parentheses denotes the phase and the subscript denotes the component. This set of equations provides the basic criteria for phase equilibrium for our purposes. In the next two sections, we consider the number of independent variables (degrees of freedom) in systems of interest to us.

2.4 The Gibbs-Duhem Equation

We may characterize the intensive state of each phase present in a heterogeneous system at internal equilibrium by its temperature and pressure, and the chemical potential of each component present – a total of $m + 2$ variables. However, these are not all independently variable, and we now derive an important relation, known as the Gibbs-Duhem equation, that shows how the variables are related.

Consider a particular phase within the heterogeneous system as an open, homogeneous system. The fundamental equation in terms of U [Eq. (2-20)] is

$$dU = TdS - PdV + \sum_i \mu_i dn_i \tag{2-28}$$

We may integrate this equation from a state of zero mass ($U = S = V = n_1 = \ldots = n_m = 0$) to a state of finite mass ($U, S, V, n_1, \ldots, n_m$) at constant temperature, pressure, and composition; along this path of integration, all coefficients, including all μ_i in Eq. (2-28), are constant; integration gives

$$U = TS - PV + \sum_i \mu_i n_i \tag{2-29}$$

This equation may be regarded as expressing U as a function of T, P, composition, and the size of the system. The path of integration amounts to adding together little bits of the phase, each with the same temperature, pressure, and composition, to obtain a finite amount of phase. Because U is a state function, the result expressed by Eq. (2-29) is independent of the path of integration. Differentiation of this equation to obtain a general expression for dU comparable to that in Eq. (2-28) gives

$$dU = TdS + SdT - PdV - VdP + \sum_i \mu_i dn_i + \sum_i n_i d\mu_i \tag{2-30}$$

Comparing Eqs. (2-28) and (2-30), we have

$$\boxed{SdT - VdP + \sum_i n_i d\mu_i = 0} \tag{2-31}$$

Equation (2-31) is the *Gibbs-Duhem equation*, a fundamental equation in the thermodynamics of solutions used extensively in Chap. 6. For now we note that it places a restriction on the simultaneous variation of T, P, and the μ_i, for a single phase. Hence, of the $m + 2$ intensive variables that may be used to characterize a phase, only $m + 1$ are independently variable; a phase has $m + 1$ degrees of freedom.

2.5 The Phase Rule

When we consider the number of degrees of freedom in a heterogeneous system, we need to take into account the results of the preceding two sections. If the heterogeneous system is *not* in a state of internal equilibrium, but each phase is, the number of independent variables is $\pi(m + 1)$, because for each phase there are $m + 1$ degrees of freedom; a Gibbs-Duhem equation applies to each phase. However, if we stipulate that the entire system is in a state of internal equilibrium, then among the $\pi(m + 1)$ variables there are $(\pi - 1)(m + 2)$ equilibrium relations given by Eqs. (2-25) to (2-27). Thus the number of *degrees of freedom*, F, is the number of intensive variables used to characterize the system minus the number of relations or restrictions connecting them:

$$\begin{aligned} F &= \pi(m+1) - (\pi - 1)(m+2) \\ &= m + 2 - \pi \end{aligned} \tag{2-32}$$

In the type of system we have been considering, the number of components m is equal to the number of independently variable species in a chemical sense, because we have ruled out chemical reaction and all special restrictions.[5]

2.6 The Chemical Potential

The task of phase-equilibrium thermodynamics is to describe quantitatively the distribution at equilibrium of every component among all the phases present. For example, in distillation of a mixture of toluene and hexane we want to know how, at a certain temperature and pressure, the toluene (or hexane) is distributed between the liquid and the gaseous phases; or in extraction of acetic acid from an aqueous solution using benzene, we want to know how the acetic acid distributes itself between the two liquid phases. Gibbs obtained the thermodynamic solution to the phase-equilibrium problem many years ago when he introduced the abstract concept chemical potential. The goal of present work in phase-equilibrium thermodynamics is to relate the abstract chemical

[5] See the final paragraph of App. A.

potential of a substance to physically measurable quantities such as temperature, pressure, and composition.

To establish the desired relation, we must immediately face one apparent difficulty: We cannot compute an absolute value for the chemical potential but must content ourselves with computing changes in the chemical potential that accompany any arbitrary change in the independent variables temperature, pressure, and composition. This difficulty is apparent rather than fundamental; it is really no more than an inconvenience. It arises because the relations between chemical potential and physically measurable quantities are in the form of differential equations that, upon integration, give only differences. These relations are discussed in more detail in Chap. 3, but one example is useful here.

For a pure substance i, the chemical potential is related to the temperature and pressure by the differential equation

$$d\mu_i = -s_i\,dT + v_i\,dP \tag{2-33}$$

where s_i is the molar entropy and v_i the molar volume. Integrating and solving for μ_i at some temperature T and pressure P, we have

$$\mu_i(T,P) = \mu_i(T^r,P^r) - \int_{T^r}^{T} s_i\,dT + \int_{P^r}^{P} v_i\,dP \tag{2-34}$$

where superscript r refers to some arbitrary reference state.

In Eq. (2-34) the two integrals on the right side can be evaluated from thermal and volumetric data over the temperature range T^r to T and the pressure range P^r to P. However, the chemical potential $\mu_i(T^r,P^r)$ is unknown. Hence, the chemical potential at T and P can only be expressed relative to its value at the arbitrary reference state designated by T^r and P^r.

Our inability to compute an absolute value for the chemical potential complicates the use of thermodynamics in practical applications. This complication follows from a need for arbitrary reference states that are commonly called *standard states*. Successful application of thermodynamics to real systems frequently is based on a judicious choice of standard states, as shown by examples discussed in later chapters. For the present it is only necessary to recognize why standard states arise and to remember that they introduce a constant into our equation. This constant need not give us concern because it must always cancel out when we compute for some substance the change of chemical potential that results from a change of any, or all, of the independent variables.[6]

[6] Standard states are reference points; we use these frequently in daily life. For example, when traveling in Europe, the senior author is often asked "Where is Berkeley?" A possible reply could be "Berkeley is 2000 miles west of Podunk, Iowa." In that statement, Podunk is the standard state. But this standard state is not useful because no one in Europe knows where Podunk, Iowa might be. A more useful reply is "Berkeley is ten miles east

2.7 Fugacity and Activity

The chemical potential does not have an immediate equivalent in the physical world and it is therefore desirable to express the chemical potential in terms of some auxiliary function that might be more easily identified with physical reality. A useful auxiliary function is obtained by the concept *fugacity*.

In attempting to simplify the abstract equation of chemical equilibrium, G. N. Lewis first considered the chemical potential for a pure, ideal gas and then generalized to all systems the result he obtained for the ideal case. From Eq. (2-33),

$$\left(\frac{\partial \mu_i}{\partial P}\right)_T = v_i \tag{2-35}$$

Substituting the ideal-gas equation,

$$v_i = \frac{RT}{P} \tag{2-36}$$

and integrating at constant temperature,

$$\mu_i - \mu_i^0 = RT \ln \frac{P}{P^0} \tag{2-37}$$

Equation (2-37) says that for an ideal gas, the change in chemical potential, in isothermally going from pressure P^0 to pressure P, is equal to the product of RT and the logarithm of the pressure ratio P/P^0. Hence, at constant temperature, the change in the abstract thermodynamic quantity μ is a simple logarithmic function of the physically real quantity, pressure. The essential value of Eq. (2-37) is that it simply relates a mathematical abstraction to a common, intensive property of the real world. However, Eq. (2-37) is valid only for pure, ideal gases; to generalize it, Lewis defined a function f, called *fugacity*,[7] by writing for an isothermal change for any component in any system, solid, liquid, or gas, pure or mixed, ideal or not,

$$\mu_i - \mu_i^0 = RT \ln \frac{f_i}{f_i^0} \tag{2-38}$$

While either μ_i^0 or f_i^0 is arbitrary, both may not be chosen independently; when one is chosen, the other is fixed.

---

of San Francisco." Now San Francisco is the standard state. San Francisco is a better standard state not only because it is close to Berkeley but perhaps more important, essentially everyone in Europe knows where San Francisco is.

[7] From the Latin *fuga*, meaning flight or escape.

For a pure, ideal gas, the fugacity is equal to the pressure, and for a component i in a mixture of ideal gases, it is equal to its partial pressure y_iP. Because all systems, pure or mixed, approach ideal-gas behavior at very low pressures, the definition of fugacity is completed by the limit

$$\frac{f_i}{y_i P} \to 1 \quad \text{as} \quad P \to 0 \tag{2-39}$$

where y_i is the mole fraction of i.

Lewis called the ratio f / f^0 the *activity,* designated by symbol a. The activity of a substance gives an indication of how "active" a substance is relative to its standard state because it provides a measure of the difference between the substance's chemical potential at the state of interest and that at its standard state. Because Eq. (2-38) was obtained for an isothermal change, the temperature of the standard state must be the same as that of the state of interest. The compositions and pressures of the two states, however, need not be (and indeed usually are not) the same.

The relation between fugacity and chemical potential provides conceptual aid in performing the translation from thermodynamic to physical variables. It is difficult to visualize the chemical potential, but the concept of fugacity is less so. Fugacity is a "corrected pressure"; for a component in a mixture of ideal gases it is equal to the partial pressure of that component. The ideal gas is not only a limiting case for thermodynamic convenience but corresponds to a well-developed physical model based on the kinetic theory of matter. The concept of fugacity, therefore, helps to make the transition from pure thermodynamics to the theory of intermolecular forces; if the fugacity is a "corrected pressure", these corrections are due to nonidealities that can be interpreted by molecular considerations.

The fugacity provides a convenient transformation of the fundamental equation of phase equilibrium, Eq. (2-27). For phases α and β, respectively, Eq. (2-38) is

$$\mu_i^{\alpha} - \mu_i^{0\alpha} = RT \ln \frac{f_i^{\alpha}}{f_i^{0\alpha}} \tag{2-40}$$

and

$$\mu_i^{\beta} - \mu_i^{0\beta} = RT \ln \frac{f_i^{\beta}}{f_i^{0\beta}} \tag{2-41}$$

Substituting Eqs. (2-40) and (2-41) into the equilibrium relation, Eq. (2-27), yields

$$\mu_i^{0\alpha} + RT \ln \frac{f_i^{\alpha}}{f_i^{0\alpha}} = \mu_i^{0\beta} + RT \ln \frac{f_i^{\beta}}{f_i^{0\beta}} \tag{2-42}$$

We now consider two cases. First, suppose that the standard states for the two phases are the same; i.e., suppose

$$\mu_i^{0\alpha} = \mu_i^{0\beta} \tag{2-43}$$

In that case, it follows that

$$f_i^{0\alpha} = f_i^{0\beta} \tag{2-44}$$

Equations (2-42), (2-43), and (2-44) give a new form of the fundamental equation of phase equilibrium:

$$\boxed{f_i^{\alpha} = f_i^{\beta}} \tag{2-45}$$

Second, suppose that the standard states for the two phases are at the same temperature but not at the same pressure and composition. In that case, we use the exact relation between the two standard states:

$$\mu_i^{0\alpha} - \mu_i^{0\beta} = RT \ln \frac{f_i^{0\alpha}}{f_i^{0\beta}} \tag{2-46}$$

Substituting Eq. (2-46) into Eq. (2-42), we again have

$$\boxed{f_i^{\alpha} = f_i^{\beta}}$$

Equation (2-45) gives an useful result. It tells us that the equilibrium condition in terms of chemical potentials can be replaced without loss of generality by an equation that says, for any species i, the fugacities must be the same in all phases. (The condition that the activities must be equal holds only for the special case where the standard states in all phases are the same.) Equation (2-45) is equivalent to Eq. (2-27); from a strictly thermodynamic point of view, one is not preferable to the other. However, from the viewpoint of one who wishes to apply thermodynamics to physical problems, an equation that equates fugacities is often more convenient than one that equates chemical potentials. In much of our subsequent discussion, therefore, we regard Eqs. (2-25), (2-26), and (2-45) as the three fundamental equations of phase equilibrium.

Most of the chapters to follow present in detail relations between fugacity and independent variables temperature, pressure, and composition. However, before discussing the details of these relations, it is desirable to give a preview of where we are going, to present an illustration of how the various concepts in this chapter can, in at least one very simple case, lead to a relation possessing immediate physical utility.

2.8 A Simple Application: Raoult's Law

Consider the equilibrium distribution of a component in a binary system between a liquid phase and a vapor phase. We seek a simple relation describing the distribution of the components between the phases, i.e., an equation relating x, the mole fraction in the liquid phase, to y, the mole fraction in the vapor phase. We limit ourselves to a very simple system, whose behavior can be closely approximated by the assumption of several types of ideal behavior.

For component 1, the equilibrium equation says

$$f_1^V = f_1^L \tag{2-47}$$

where superscript V refers to the vapor and superscript L to the liquid. We now have the problem of relating the fugacities to the mole fractions. To solve this problem, we make two simplifying assumptions, one for each phase:

Assumption 1. The fugacity f_1^V, at constant temperature and pressure, is proportional to the mole fraction y_1. That is, we assume

$$f_1^V = y_1 f_{\text{pure 1}}^V \tag{2-48}$$

where $f_{\text{pure 1}}^V$ is the fugacity of pure component 1 as a vapor at the temperature and pressure of the mixture.

Assumption 2. The fugacity f_1^L, at constant temperature and pressure, is proportional to the mole fraction x_1. That is, we assume

$$f_1^L = x_1 f_{\text{pure 1}}^L \tag{2-49}$$

where $f_{\text{pure 1}}^L$ is the fugacity of pure component 1 as a liquid at the temperature and pressure of the mixture.

Assumptions 1 and 2 are equivalent to saying that both vapor-phase and liquid-phase solutions are ideal solutions; Eqs. (2-47) and (2-48) are statements of the *Lewis fugacity rule*. These assumptions are valid only for very limited conditions as discussed in later chapters. For mixtures of similar components, however, they provide reasonable approximations based on the naive but attractive supposition that the fugacity of a component in a given phase increases in proportion to its mole fraction in that phase.

Upon substituting Eqs. (2-48) and (2-49) into (2-47), the equilibrium relation now becomes

$$y_1 f_{\text{pure 1}}^V = x_1 f_{\text{pure 1}}^L \tag{2-50}$$

Equation (2-50) gives an ideal-solution relation using only mole fractions and pure-component fugacities. It is the basis of the original K charts $(K = y/x = f^L/f^V)$ used in the petroleum industry. Equation (2-50) can be simplified further by introducing two additional assumptions.

Assumption 3. Pure component 1 vapor at temperature T and pressure P is an ideal gas. It follows that

$$f_{\text{pure 1}}^{V} = P \tag{2-51}$$

Assumption 4. The effect of pressure on the fugacity of a condensed phase is negligible at moderate pressures. Further, we assume that the vapor in equilibrium with pure liquid 1 at temperature T is an ideal gas. It follows that

$$f_{\text{pure 1}}^{L} = P_1^s \tag{2-52}$$

where P_1^s is the saturation (vapor) pressure of pure liquid 1 at temperature T.

Substituting Eqs. (2-51) and (2-52) into (2-50) we obtain

$$\boxed{y_1 P = x_1 P_1^s} \tag{2-53}$$

Equation (2-53) is the desired, simple relation known as *Raoult's law.*

Equation (2-53) is of limited utility because it is based on severe simplifying assumptions. The derivation of Raoult's law has been given here only to illustrate the general procedure whereby relations in thermodynamic variables can, with the help of physical arguments, be translated into useful, physically significant, equations. In general, this procedure is considerably more complex but the essential problem is always the same: How is the fugacity of a component related to the measurable quantities temperature, pressure and composition? It is the task of molecular thermodynamics to provide useful answers to this question. All of the chapters to follow are concerned with techniques for establishing useful relations between the fugacity or chemical potential of a component in a phase and physicochemical properties of that phase. To establish such relations, we rely heavily on classical thermodynamics but we also utilize, when possible, concepts from statistical mechanics, molecular physics, and physical chemistry.

References

Bett, K. E., J. S. Rowlinson, and G. Saville, 1975, *Thermodynamics for Chemical Engineers.* Cambridge: The MIT Press.

Callen, H. B., 1985, *Thermodynamics and an Introduction to Termostatistics,* 2nd Ed. New York: John Wiley & Sons.

Denbigh, K. G., 1981, *The Principles of Chemical Equilibrium,* 4th Ed., Chaps. 1 and 2. Cambridge: Cambridge University Press.

Gibbs, J., 1961, *The Scientific Papers of J. Willard Gibbs,* Vol. I, pp. 55-100. New York: Dover Publications.

Guggenheim, E. A., 1967, *Thermodynamics,* 5th Ed., Chap. 1. Amsterdam: North-Holland.

Klotz, I. M. and R. M. Rosenberg, 1994, *Chemical Thermodynamics: Basic Theory and Methods,* 5th Ed. New York: John Wiley & Sons.

Kyle, B., 1991, *Chemical and Process Thermodynamics,* 2nd Ed. Englewood Cliffs: Prentice-Hall.

Pitzer, K. S., 1995, *Thermodynamics,* 3rd Ed. New York: McGraw-Hill.

Redlich, O., 1976, *Thermodynamics: Fundamentals, Applications.* Amsterdam: Elsevier.

Rowlinson, J. S. and F. L. Swinton, 1982, *Liquids and Liquid Mixtures,* 3rd Ed. London: Butterworths.

Prigogine, I. and R. Defay, 1954, *Chemical Thermodynamics* (Trans./Rev. by D. H. Everett). London: Longmans & Green.

Sandler, S. I., 1989, *Chemical and Engineering Thermodynamics,* 2nd Ed. New York: John Wiley & Sons.

Smith, J. M., H. C. Van Ness, and M. M. Abbott, 1996, *Introduction to Chemical Engineering Thermodynamics,* 5th Ed. New York: McGraw-Hill.

Tester, J. W. and M. Modell, 1996, *Thermodynamics and Its Applications,* 3rd Ed. Englewood Cliffs: Prentice-Hall.

Van Ness, H. C. and M. M. Abbott, 1982, *Classical Thermodynamics of Nonelectrolyte Solutions.* New York: McGraw-Hill.

Winnick, J., 1997, *Chemical Engineering Thermodynamics.* New York: John Wiley & Sons.

Problems[8]

1. The volume coefficient of expansion of mercury at 0°C is 18×10^{-5} (°C)$^{-1}$. The coefficient of compressibility κ_T is 5.32×10^{-6} (bar)$^{-1}$. If mercury were heated from 0°C to 1°C in a constant-volume system, what pressure would be developed?

$$\kappa_T = -\frac{1}{v}\left(\frac{\partial v}{\partial P}\right)_T$$

[8] Appendix J gives some fundamental constants and conversion factors to SI units.

2. Find expressions for $(\partial S/\partial V)_T$, $(\partial S/\partial P)_T$, $(\partial U/\partial V)_T$, $(\partial U/\partial P)_T$, and $(\partial H/\partial P)_T$ for a gas whose behavior can be described by the equation

$$P\left(\frac{V}{n}-b\right)=RT$$

Also find expressions for ΔS, ΔU, ΔH, ΔG, and ΔA for an isothermal change.

3. If the standard entropy of liquid water at 298.15 K is 69.96 J K^{-1} mol^{-1}, calculate the entropy of water vapor in its standard state (i.e., an ideal gas at 298.15 K and 1.01325 bar). The vapor pressure of water is 3168 Pa at 298.15 K and its enthalpy of vaporization is 2.436 kJ g^{-1}.

4. The residual volume α is the difference between the ideal-gas volume and the actual gas volume. It is defined by the equation

$$\alpha=\frac{RT}{P}-v$$

For a certain gas, α has been measured at 100°C and at different molar volumes; the results are expressed by the empirical equation $\alpha = 2 - (3/v^2)$, where v is in L mol^{-1}. The velocity of sound w is given by the formula

$$w^2=-gkv^2\left(\frac{\partial P}{\partial v}\right)_T$$

where g is the acceleration of gravity. Calculate the velocity of sound for this gas at 100° C when its molar volume is 2.3 liter, using $k = 1.4$. The molar mass is 100 g mol^{-1}.

5. A gas at 350°C and molar volume 600 cm^3 mol^{-1} is expanded in an isentropic turbine. The exhaust pressure is atmospheric. What is the exhaust temperature? The ideal-gas heat capacity at constant pressure is c_p^0 = 33.5 J K^{-1} mol^{-1}. The P-V-T properties of the gas are given by the van der Waals equation, with $a = 56\times10^5$ bar $(cm^3\ mol^{-1})^2$ and b = 45 cm^3 mol^{-1}.

6. Show that when the van der Waals equation of state is written in the virial form,

$$\frac{Pv}{RT}=1+\frac{B}{v}+\frac{C}{v^2}+\cdots$$

the second virial coefficient is given by

$$B=a-\frac{b}{RT}$$

7. The second virial coefficient B of a certain gas is given by

$$B = a - \frac{b}{T^2}$$

where a and b are constants.
Compute the change in internal energy for this gas in going, at temperature τ, from very low pressure to a pressure π. Use the equation

$$z = \frac{Pv}{RT} = 1 + \frac{BP}{RT}$$

8. Consider the equation of state

$$\left(P + \frac{n}{v^2 T^{1/2}}\right)(v - m) = RT$$

where n and m are constants for any gas. Assume that carbon dioxide follows this equation. Calculate the compressibility factor of carbon dioxide at 100°C and at a volume of 6.948 $dm^3\ kg^{-1}$.

9. The volumetric behavior of a gas is satisfactorily described by the equation of state

$$P = \frac{RT}{v - b}\left(1 - \frac{a}{RTv}\right)$$

where a and b are constants. To a very good approximation, the ideal-gas heat capacity of the gas, c_p^0, is temperature-independent. Derive an analytical expression for the molar internal energy of this gas in terms of temperature and molar volume. As a reference state, use one of temperature T_0 = 273 K and molar volume tending to infinity. Under reference state conditions the gas behaves practically as an ideal gas. Constants that may appear in the desired expression for internal energy are a, b, R, c_p^0, and T_0.

10. Consider an aqueous mixture of sugar at 25°C and 1 bar pressure. The activity coefficient of water is found to obey a relation of the form

$$\ln \gamma_w = A(1 - x_w)^2$$

where is normalized such that $\gamma_w \to 1$ as $x_w \to 1$ and A is an empirical constant dependent only on temperature. Find an expression for γ_s, the activity coefficient of sugar normalized such that $\gamma_s \to 1$ as $x_w \to 1$ (or as $x_s \to 0$). The mole fractions x_w and x_s refer to water and sugar, respectively.

11. Consider a binary liquid solution of components 1 and 2. At constant temperature (and low pressure) component 1 follows Henry's law for the mole fraction range $0 \le x_1 \le a$. Show that component 2 follows Raoult's law for the mole fraction range $(1 - a) \le x_2 \le 1$.

12. Using only data given in the steam tables, compute the fugacity of steam at 320°C and 70 bar.

13. The inversion temperature is the temperature where the Joule-Thomson coefficient changes sign and the Boyle temperature is the temperature where the second virial coefficient changes sign. Show that for a van der Waals gas the low-pressure inversion temperature is twice the Boyle temperature.

14. A gas, designated by subscript 1, is to be dissolved in a nonvolatile liquid. At a certain pressure and temperature the solubility of the gas in the liquid is x_1 (where x is the mole fraction). Assume that Henry's law holds. Show that the change in solubility with temperature is given by

$$\frac{d \ln x_1}{d(1/T)} = -\frac{\Delta \bar{h}_1}{R}$$

where

$$\Delta \bar{h}_1 = \bar{h}_{1\,\text{(in liquid solution)}} - h_{1\,\text{(pure gas)}}$$

at the same pressure and temperature. Based on physical reasoning alone, would you expect $\Delta \bar{h}_1$ to be positive or negative?

CHAPTER 3

Thermodynamic Properties from Volumetric Data

For any substance, regardless of whether it is pure or a mixture, most thermodynamic properties of interest in phase equilibria can be calculated from thermal and volumetric measurements. For a given phase (solid, liquid, or gas), thermal measurements (heat capacities) give information on how some thermodynamic properties vary with temperature, whereas volumetric measurements give information on how thermodynamic properties vary with pressure or density at constant temperature. Whenever there is a change of phase (e.g., fusion or vaporization), additional thermal and volumetric measurements are required to characterize that change.

Frequently, it is useful to express a selected thermodynamic function of a substance relative to that which the same substance has as an ideal gas at the same temperature and composition and at some specified pressure or density. This relative function is often called a *residual function*. The fugacity is a relative function because its numerical value is always relative to that of an ideal gas at unit fugacity; in other words, the standard-state fugacity f_i^0 in Eq. (2-38) is arbitrarily set equal to some fixed value, usually 1 bar.[1]

As indicated in Chap. 2, the thermodynamic function of primary interest is the fugacity that is directly related to the chemical potential; however, the chemical potential is directly related to the Gibbs energy, that, by definition, is found from the enthalpy and entropy. Therefore, a proper discussion of calculation of fugacities from volumetric properties must begin with the question of how enthalpy and entropy, at constant temperature and composition, are related to pressure. On the other hand, as indicated in Chap. 2, the chemical potential may also be expressed in terms of the Helmholtz energy; in that event the first question must be how entropy and energy, at constant temperature and composition, are related to volume. The answers to these questions may readily be found from Maxwell's relations. We can then obtain exact equations for the thermodynamic functions U, H, S, A, and G; from these we can derive the chemical potential and, finally, the fugacity.

If we consider a homogeneous mixture at some fixed composition, we must specify two additional variables. In practical phase-equilibrium problems, the common additional variables are temperature and pressure, and in Sec. 3.1 we give equations for the thermodynamic properties with T and P as independent variables. However, volumetric data are most commonly expressed by an *equation of state* that uses temperature and volume as independent variables, and therefore it is a matter of practical importance to have available equations for the thermodynamic properties in terms of T and V; these are given in Sec. 3.4. The equations in Secs. 3.1 and 3.4 contain no simplifying assumptions;[2] they are exact and are not restricted to the gas phase but, in principle, apply equally to all phases.

In Sec. 3.3 we discuss the fugacity of a pure liquid or solid, and in Secs. 3.2 and 3.5 we give examples based on the van der Waals equation. Finally, in Sec. 3.6 we consider briefly how the exact equations for the fugacity may, in principle, be used to solve fluid-phase equilibrium problems subject only to the condition that we have available a reliable equation of state, valid for fluid pure substances and their mixtures over a large density range.

3.1 Thermodynamic Properties with Independent Variables P and T

At constant temperature and composition, we can use one of Maxwell's relations to give the effect of pressure on enthalpy and entropy:

---

[1] Throughout this book we use the pressure unit bar, related to the SI pressure unit (pascal) by 1 bar = 10^5 pascal = 0.986923 atmosphere.

[2] The equations in Secs. 3.1 and 3.4 do, however, assume that surface effects and all body forces due to gravitational, electric, or magnetic fields, etc., can be neglected.

$$dH = \left[V - T\left(\frac{\partial V}{\partial T}\right)_{P,n_T} \right] dP \tag{3-1}$$

$$dS = -\left(\frac{\partial V}{\partial T}\right)_{P,n_T} dP \tag{3-2}$$

These two relations form the basis of the derivation for the desired equations. We will not present the derivation here; it requires only straightforward integrations clearly given in several publications by Beattie (1942, 1949, 1955). First, expressions for the enthalpy and entropy are found. The other properties are then calculated from the definitions of enthalpy, Helmholtz energy, and Gibbs energy:

$$U = H - PV \tag{3-3}$$

$$A = H - PV - TS \tag{3-4}$$

$$G = H - TS \tag{3-5}$$

$$\mu_i = \left(\frac{\partial G}{\partial n_i}\right)_{T,P,n_j} \tag{3-6}$$

$$RT \ln \frac{f_i}{f_i^0} = \mu_i - \mu_i^0 \tag{3-7}$$

The results are given in Eqs. (3-8) to (3-14). It is understood that all integrations are performed at constant temperature and constant composition.

The symbols have the following meanings:

h_i^0 = molar enthalpy of pure i as an ideal gas at temperature T;

s_i^0 = molar entropy of pure i as an ideal gas at temperature T and 1 bar;

$\mu_i^0 = h_i^0 - Ts_i^0$ and $f_i^0 = 1$ bar;

n_i = number of moles of i;

n_T = total number of moles;

$y_i = n_i / n_T$ = mole fraction of i.

All extensive properties denoted by capital letters (V, U, H, S, A, and G) represent the total property for n_T moles and therefore are *not* on a molar basis. Extensive properties on a molar basis are denoted by lowercase letters (v, u, h, s, a, and g). In Eqs. (3-10) to (3-13), pressure P is in bars.

$$U = \int_0^P \left[V - T\left(\frac{\partial V}{\partial T}\right)_{P,n_T} \right] dP - PV + \sum_i n_i h_i^0 \tag{3-8}$$

$$H = \int_0^P \left[V - T\left(\frac{\partial V}{\partial T}\right)_{P,n_T} \right] dP + \sum_i n_i h_i^0 \tag{3-9}$$

$$S = \int_0^P \left[\frac{n_T R}{P} - \left(\frac{\partial V}{\partial T}\right)_{P,n_T} \right] dP - R\sum_i n_i \ln y_i P + \sum_i n_i s_i^0 \tag{3-10}$$

$$A = \int_0^P \left(V - \frac{n_T RT}{P} \right) dP + RT\sum_i n_i \ln y_i P - PV + \sum_i n_i (h_i^0 - Ts_i^0) \tag{3-11}$$

$$G = \int_0^P \left(V - \frac{n_T RT}{P} \right) dP + RT\sum_i n_i \ln y_i P + \sum_i n_i (h_i^0 - Ts_i^0) \tag{3-12}$$

$$\mu_i = \int_0^P \left(\bar{v}_i - \frac{RT}{P} \right) dP + RT \ln y_i P + h_i^0 - Ts_i^0 \tag{3-13}$$

and finally

$$\boxed{RT \ln \varphi_i = RT \ln \frac{f_i}{y_i P} = \int_0^P \left(\bar{v}_i - \frac{RT}{P} \right) dP} \tag{3-14}$$

where $\bar{v}_i \equiv (\partial V / \partial n_i)_{T,P,n_j}$ is the *partial molar volume* of i. The dimensionless ratio $f_i / y_i P = \varphi_i$ is called the *fugacity coefficient*. For a mixture of ideal gases, $\varphi_i = 1$, as shown later.

Equations (3-8) to (3-14) enable us to compute all the desired thermodynamic properties for any substance relative to the ideal-gas state at 1 bar and at the same temperature and composition, provided that we have information on volumetric behavior in the form

$$V = F(T, P, n_1, n_2, \ldots) \tag{3-15}$$

To evaluate the integrals in Eqs. (3-8) to (3-14), the volumetric information required in function F must be available not only for pressure P, where the thermodynamic properties are desired, but for the entire pressure range 0 to P.

In Eqs. (3-8) and (3-11), the quantity V appearing in the PV product is the total volume at the system pressure P and at the temperature and composition used throughout. This volume V is found from the equation of state, Eq. (3-15).

For a pure component, $\bar{v}_i = v_i$, and Eq. (3-14) simplifies to

$$RT \ln\left(\frac{f}{P}\right)_{\text{pure } i} = \int_0^P \left(v_i - \frac{RT}{P}\right) dP \tag{3-16}$$

where v_i is the molar volume of pure i. Equation (3-16) is frequently expressed in the equivalent form

$$\boxed{\ln\left(\frac{f}{P}\right)_{\text{pure } i} = \int_0^P \frac{z-1}{P} dP} \tag{3-17}$$

where z, the *compressibility factor*, is defined by

$$z \equiv \frac{Pv}{RT} \tag{3-18}$$

The fugacity of any component i in a mixture is given by Eq. (3-14) that is not only general and exact but also remarkably simple. One might well wonder, then, why there are any problems at all in calculating fugacities and, subsequently, computing phase-equilibrium relations. The problem is not with Eq. (3-14) but with Eq. (3-15), where we have written the vague symbol F, meaning "some function". Herein lies the difficulty: What is F? Function F need not be an analytical function; sometimes we have available tabulated volumetric data that may then be differentiated and integrated numerically to yield the desired thermodynamic functions. But this is rarely the case, especially for mixtures. Usually, one must estimate volumetric behavior from limited experimental data. There is, unfortunately, no generally valid equation of state, applicable to a large number of pure substances and their mixtures over a wide range of conditions, including the liquid phase. There are some good equations of state useful for only a limited class of substances and for limited conditions; however, these equations are almost always pressure-explicit rather than volume-explicit. Therefore, it is necessary to express the derived thermodynamic functions in terms of the less convenient independent variables V and T as shown in Sec. 3.4. Before concluding this section, however, let us briefly discuss some of the features of Eq. (3-14).

First, we consider the fugacity of a component i in a mixture of ideal gases. In that case, the equation of state is

$$V = \frac{(n_1 + n_2 + \cdots)RT}{P} \tag{3-19}$$

and the partial molar volume of i is

$$\bar{v}_i \equiv \left(\frac{\partial V}{\partial n_i}\right)_{T,P,n_j} = \frac{RT}{P} \tag{3-20}$$

Substituting in Eq. (3-14) gives

$$f_i = y_i P \tag{3-21}$$

For a mixture of ideal gases, then, the fugacity of i is equal to its partial pressure, as expected.

Next, let us assume that the gas mixture follows Amagat's law at all pressures up to the pressure of interest. *Amagat's law* states that at fixed temperature and pressure, the volume of the mixture is a linear function of the mole numbers

$$V = \sum_i n_i v_i \tag{3-22}$$

where v_i is the molar volume of pure i at the same temperature and pressure and in the same phase.

Another way to state Amagat's law is to say that at constant temperature and pressure, the components mix isometrically, i.e., with no change in total volume. If there is no volume change, then the partial molar volume of each component must be equal to its molar volume in the pure state. It is this equality which is asserted by Amagat's law. Differentiating Eq. (3-22), we have

$$\bar{v}_i \equiv \left(\frac{\partial V}{\partial n_i}\right)_{T,P,n_j} = v_i \tag{3-23}$$

Substitution in Eq. (3-14) yields

$$RT \ln \frac{f_i}{y_i P} = \int_0^P \left(v_i - \frac{RT}{P}\right) dP \tag{3-24}$$

Upon comparing Eq. (3-24) with Eq. (3-16), we obtain

$$\boxed{f_i = y_i f_{\text{pure } i}} \tag{3-25}$$

Equation (3-25) is the *Lewis fugacity rule.* In Eq. (3-25), the fugacity of pure i is evaluated at the temperature and pressure of the mixture and for the same phase.

The Lewis fugacity rule is a particularly simple equation and is therefore widely used for evaluating fugacities of components in gas mixtures. However, it is not reliable because it is based on the severe simplification introduced by Amagat's law. The Lewis fugacity rule is discussed further in Chap. 5; for present purposes it is sufficient to understand clearly how Eq. (3-25) was obtained. The derivation assumes additivity of the volumes of all the components in the mixture at constant temperature and pressure; at high pressures, this is frequently a very good assumption because at liquid-like densities, fluids tend to mix with little or no change in volume. For example, volumetric data for the nitrogen/butane system at 171°C, shown in Fig. 3-1, indicate that at 690 bar the molar volume of the mixture is nearly a straight-line function of the mole fraction (Evans and Watson, 1956). At first glance, therefore, one might be tempted to conclude that for this system, at 690 bar and 171°C the Lewis fugacity rule should give an excellent approximation for the fugacities of the components in the mixture. A second look, however, shows that this conclusion is not justified because the Lewis fugacity rule assumes additivity of volumes not only at the pressure P of interest, but for the entire pressure range 0 to P. Figure 3-1 shows that at pressures lower than about 345 bar, the volumetric behavior deviates markedly from additivity. As indicated by Eq. (3-14), the partial molar volume $\bar{v}_i$ is part of an integral and, as a result, whatever assumption one wishes to make about $\bar{v}_i$ must hold not only at the upper limit but also for the entire range of integration.

3.2 Fugacity of a Component in a Mixture at Moderate Pressures

In the preceding section we first calculated the fugacity of a component in a mixture of ideal gases and then in an ideal mixture of real gases, i.e., one which obeys Amagat's law. To illustrate the use of Eq. (3-14) with a realistic example, we compute now the fugacity of a component in a binary mixture at moderate pressures. In this illustrative calculation, we use, for simplicity, a form of the van der Waals equation valid only to moderate pressures:

$$Pv = RT + \left(b - \frac{a}{RT}\right)P + \cdots \text{ terms in } P^2, P^3, \text{ etc.} \tag{3-26}$$

where a and b are the van der Waals constants for the mixture. To calculate the fugacity with Eq. (3-14), we must first find an expression for the partial molar volume; for this purpose, Eq. (3-26) is rewritten on a total (rather than molar) basis by substituting $V = n_T v$:

$$V = \frac{n_T RT}{P} + n_T b - \frac{n_T a}{RT} \tag{3-27}$$

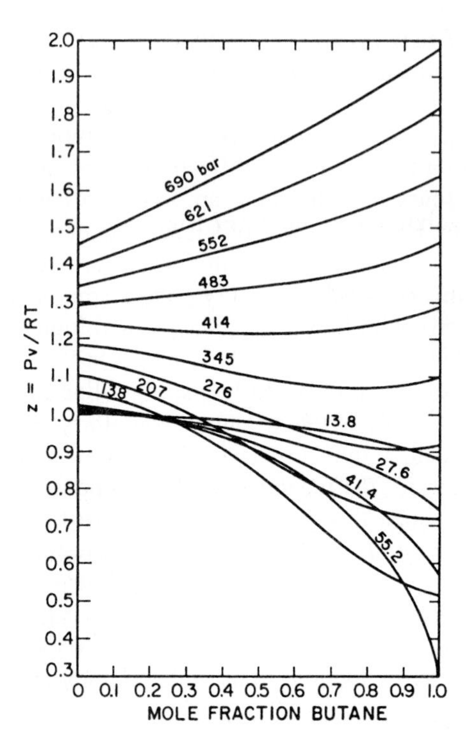

Figure 3-1 Compressibility factors for nitrogen/butane mixtures at 171°C (Evans and Watson, 1956).

We let subscripts 1 and 2 stand for the two components. Differentiating Eq. (3-27) with respect to n_1 gives

$$\bar{v}_1 = \left(\frac{\partial V}{\partial n_1}\right)_{T,P,n_2} = \frac{RT}{P} + \frac{\partial(n_T b)}{\partial n_1} - \frac{1}{RT}\frac{\partial(n_T a)}{\partial n_1} \tag{3-28}$$

We must now specify a *mixing rule*, i.e., a relation that states how constants a and b for the mixture depend on the composition. We use the mixing rules originally proposed by van der Waals:

$$a = y_1^2 a_1 + 2y_1 y_2\sqrt{a_1 a_2} + y_2^2 a_2 \tag{3-29}$$

$$b = y_1 b_1 + y_2 b_2 \tag{3-30}$$

To utilize these mixing rules in Eq. (3-28), we rewrite them

$$n_T a = \frac{n_1^2 a_1 + 2n_1 n_2 \sqrt{a_1 a_2} + n_2^2 a_2}{n_T} \tag{3-31}$$

$$n_T b = n_1 b_1 + n_2 b_2 \tag{3-32}$$

The partial molar volume for component 1 is

$$\bar{v}_1 = \left(\frac{\partial V}{\partial n_1}\right)_{T,P,n_2} = \frac{RT}{P} + b_1 - \frac{1}{RT}\frac{n_T(2n_1 a_1 + 2n_2\sqrt{a_1 a_2}) - n_T^2 a}{n_T^2} \tag{3-33}$$

In performing the differentiation, it is important to remember that n_2 is held constant and that therefore n_T cannot also be constant.

Algebraic rearrangement and subsequent substitution into Eq. (3-14) gives the desired result:

$$\varphi_1 = \frac{f_1}{y_1 P} = \exp\left[\left(b_1 - \frac{a_1}{RT}\right)\frac{P}{RT}\right]\exp\left[\frac{(a_1^{1/2} - a_2^{1/2})^2 y_2^2 P}{(RT)^2}\right] \tag{3-34}$$

Equation (3-34) contains two exponential factors to correct for nonideality. The first correction is independent of component 2 but the second is not, because it contains a_2 and y_2. We can therefore rewrite Eq. (3-34) by utilizing the boundary condition

$$\text{as } y_2 \to 0, \quad f_1 \to f_{\text{pure 1}} = P\exp\left[\left(b_1 - \frac{a_1}{RT}\right)\frac{P}{RT}\right] \tag{3-35}$$

Upon substitution, Eq. (3-34) becomes

$$f_1 = y_1 f_{\text{pure 1}} \exp\left[\frac{(a_1^{1/2} - a_2^{1/2})^2 y_2^2 P}{(RT)^2}\right] \tag{3-36}$$

When written in this form, we see that the exponential in Eq. (3-36) is a correction to the Lewis fugacity rule.

Figure 3-2 presents fugacity coefficients for several hydrocarbons in binary mixtures with nitrogen. In these calculations, Eq. (3-34) was used with $y_1 = 0.10$ and T = 343 K; in each case, component 2 is nitrogen. For comparison, we also show the fugacity coefficient of butane according to the Lewis fugacity rule; in that calculation, the second exponential in Eq. (3-34) was neglected. From Eq. (3-36) we see that the Lewis rule is poor for butane for two reasons: First, the mole fraction of butane is small (hence y_2^2 is near unity), and second, the difference in intermolecular forces between butane and nitrogen (as measured by $|a_1^{1/2} - a_2^{1/2}|$) is large. If the gas in excess were hydrogen or helium instead of nitrogen, the deviations from the Lewis rule for butane would be even larger.

For a component i, the Lewis fugacity rule is often poor when $y_i << 1$. However, the Lewis fugacity rule is usually good when y_i is close to unity because it becomes exact when $y_i = 1$.

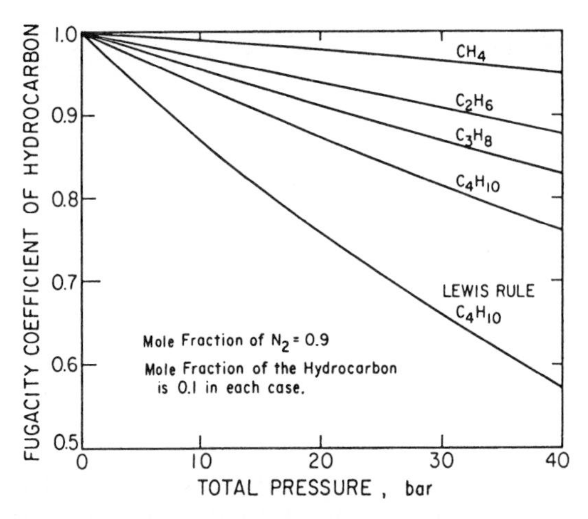

Figure 3-2 Fugacity coefficients of light hydrocarbons in binary mixtures with nitrogen at 343 K. Calculations based on simplified form of van der Waals' equation.

3.3 Fugacity of a Pure Liquid or Solid

The derivation of Eq. (3-16) is general and not limited to the vapor phase. It may be used to calculate the fugacity of a pure liquid or that of a pure solid. Such fugacities are of importance in phase-equilibrium thermodynamics because we frequently use a pure condensed phase as the standard state for activity coefficients.

To calculate the fugacity of a pure solid or liquid at given temperature T and pressure P, we separate the integral in Eq. (3-16) into two parts: The first part gives the fugacity of the saturated vapor at T and P^s (the saturation pressure), and the second part gives the correction due to the compression of the condensed phase to pressure P. At saturation pressure P^s, the fugacity of saturated vapor is equal to the fugacity of the saturated liquid (or solid) because the saturated phases are in equilibrium. Let superscript s refer to saturation and superscript c refer to condensed phase. Equation (3-16) for a pure component is now rewritten

$$RT\ln\frac{f_i^C}{P} = \int_0^{P_i^s}\left(v_i - \frac{RT}{P}\right)dP + \int_{P_i^s}^{P}\left(v_i^C - \frac{RT}{P}\right)dP \tag{3-37}$$

The first term on the right-hand side gives the fugacity of the saturated vapor, equal to that of the saturated condensed phase. Equation (3-37) becomes

$$RT\ln\frac{f_i^C}{P} = RT\ln\frac{f_i^S}{P_i^S} + \int_{P_i^s}^{P} v_i^C\,dP - RT\ln\frac{P}{P_i^S} \tag{3-38}$$

that can be rearranged to yield

$$\boxed{f_i^C = P_i^S\varphi_i^S\exp\left(\int_{P_i^s}^{P}\frac{v_i^C dP}{RT}\right)} \tag{3-39}$$

where $\varphi_i^S = f_i^S / P_i^S$.

Equation (3-39) gives the important result that the fugacity of a pure condensed component i at T and P is, to a first approximation, equal to P_i^S, the saturation (vapor) pressure at T. Two corrections must be applied. First, the fugacity coefficient φ_i^S corrects for deviations of the saturated vapor from ideal-gas behavior. Second, the exponential correction (often called the *Poynting correction*) takes into account that the liquid (or solid) is at a pressure P different from P_i^S. In general, the volume of a liquid (or solid) is a function of both temperature and pressure, but at conditions remote from critical, a condensed phase may often be regarded as incompressible and in that case the Poynting correction takes the simple form

$$\exp\left[\frac{v_i^C(P - P_i^S)}{RT}\right]$$

The two corrections are often, but not always, small and sometimes they are negligible. If temperature T is such that the saturation pressure P_i^s is low (say below 1 bar), then φ_i^S is very close to unity.[3]

Figure 3-3 gives fugacity coefficients for four liquids at saturation; these coefficients were calculated with Eq. (3-37) using vapor-phase volumetric data. Because the liquids are at saturation conditions, no Poynting correction is required.

[3] There are, however, a few exceptions. Substances that have a strong tendency to polymerize (e.g., acetic acid or hydrogen fluoride) may show significant deviations from ideal-gas behavior even at pressures near or below 1 bar. See Sec. 5.9.

We see that when plotted against reduced temperature, the results for the four liquids are almost (but not quite) superimposable; further, we note that φ_i^s differs considerably from unity as the critical temperature is approached. The correction φ_i^s always tends to decrease the fugacity f_i^c because for all pure, saturated substances $\varphi_i^s < 1$.

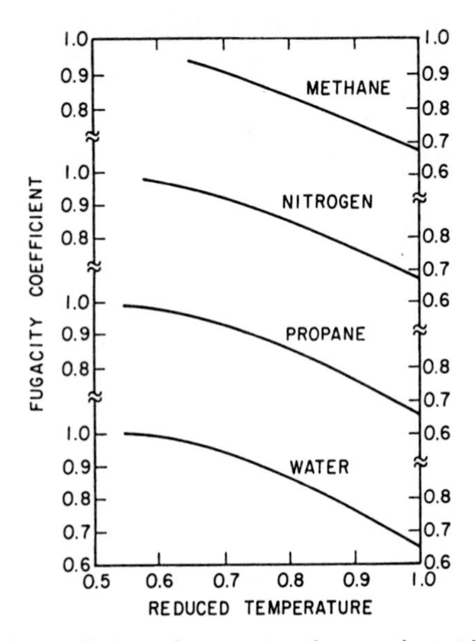

Figure 3-3 Fugacity coefficients from vapor-phase volumetric data for four saturated liquids.

The Poynting correction is an exponential function of the pressure; it is small at low pressure but may become large at high pressures or at low temperatures. To illustrate, Table 3-1 gives some numerical values of the Poynting correction for an incompressible component with v_i^c = 100 cm^3 mol^{-1} and T = 300 K.

Table 3-1 The Poynting correction: effect of pressure on fugacity of a pure, condensed and incompressible substance whose molar volume is 100 cm^3 mol^{-1} (T = 300 K).

Pressure in excess of saturation pressure (bar)	Poynting correction
1	1.00405
10	1.0405
100	1.499
1000	57.0

Figure 3-4 shows the fugacity of compressed liquid water as a function of pressure for three temperatures; the fugacities were calculated from thermodynamic properties reported by Keenan *et al.* (1978). The lowest temperature is much less, whereas the highest temperature is just slightly less than the critical temperature, 374°C. The saturation pressures are also indicated in Fig. 3-4. We see that, as indicated by Eq. (3-39), the fugacity of compressed liquid water is more nearly equal to the saturation pressure than to the total pressure. At the highest temperature, liquid water is no longer incompressible and its compressibility cannot be neglected in the Poynting correction.

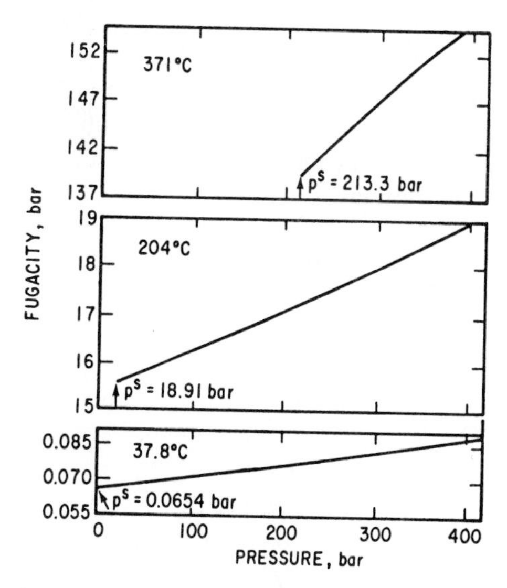

Figure 3-4 Fugacity of liquid water at three temperatures from saturation pressure to 414 bar. The critical temperature of water is 374°C.

3.4 Thermodynamic Properties with Independent Variables V and T

In Sec. 3.1 we gave expressions for the thermodynamic properties in terms of independent variables P and T. Because volumetric properties of fluids are usually (and more simply) expressed by equations of state that are pressure-explicit,[4] it is more convenient to calculate thermodynamic properties in terms of independent variables V and T.

At constant temperature and composition, we use one of Maxwell's relations to give the effect of volume on energy and entropy:

[4] If the equation of state is to hold for both vapor and liquid phases, it must necessarily be pressure-explicit. For example, water at 100°C and 1 atm has *two* equilibrium phases with much different volumes: saturated steam and liquid water. At a fixed P and T, a volume-explicit equation gives only one value for V. But a pressure-explicit equation, at fixed P and T, can give two or more values for V.

$$dU = \left[T\left(\frac{\partial P}{\partial T}\right)_{V,n_T} - P \right] dV \tag{3-40}$$

$$dS = \left(\frac{\partial P}{\partial T}\right)_{V,n_T} dV \tag{3-41}$$

These two equations form the basis of the derivation for the desired equations. Again, we do not present the derivation but refer to the publications of Beattie.

First, expressions are found for the energy and entropy. The other properties are then calculated from their definitions:

$$H = U + PV \tag{3-42}$$

$$A = U - TS \tag{3-43}$$

$$G = U + PV - TS \tag{3-44}$$

$$\mu_i = \left(\frac{\partial A}{\partial n_i}\right)_{T,V,n_j} \tag{3-45}$$

$$RT \ln \frac{f_i}{f_i^0} = \mu_i - \mu_i^0 \tag{3-46}$$

The results are given in Eqs. (3-47) to (3-53). As in Sec. 3.1, it is understood that all integrations are performed at constant temperature and composition. The symbols are the same as those defined after Eq. (3-7) with one addition:

$u_i^0 = h_i^0 - RT$ = molar energy of pure i as an ideal gas at temperature T.

In Eqs. (3-49) to (3-52), the units of V/n_iRT are bar^{-1}. Because the independent variable in Eqs. (3-47) to (3-49) is the volume, no additional terms for change of phase (e.g., enthalpy of vaporization) need be added to these equations when they are applied to a condensed phase.

$$U = \int_V^\infty \left[P - T\left(\frac{\partial P}{\partial T}\right)_{V,n_T} \right] dV + \sum_i n_i u_i^0 \tag{3-47}$$

$$H = \int_V^\infty \left[P - T\left(\frac{\partial P}{\partial T}\right)_{V,n_T} \right] dV + PV + \sum_i n_i u_i^0 \tag{3-48}$$

$$S = \int_V^\infty \left[\frac{n_T R}{V} - \left(\frac{\partial P}{\partial T}\right)_{V,n_T} \right] dV + R\sum_i n_i \ln \frac{V}{n_i RT} + \sum_i n_i s_i^0 \tag{3-49}$$

$$A = \int_V^\infty \left(P - \frac{n_T RT}{V} \right) dV - RT\sum_i n_i \ln \frac{V}{n_i RT} + \sum_i n_i (u_i^0 - Ts_i^0) \tag{3-50}$$

$$G = \int_V^\infty \left(P - \frac{n_T RT}{V} \right) dV - RT\sum_i n_i \ln \frac{V}{n_i RT} + PV + \sum_i n_i (u_i^0 - Ts_i^0) \tag{3-51}$$

$$\mu_i = \int_V^\infty \left[\left(\frac{\partial P}{\partial n_i}\right)_{T,V,n_j} - \frac{RT}{V} \right] dV - RT \ln \frac{V}{n_i RT} + RT + u_i^0 - Ts_i^0 \tag{3-52}$$

and finally,

$$\boxed{RT \ln \varphi_i = RT \ln \frac{f_i}{y_i P} = \int_V^\infty \left[\left(\frac{\partial P}{\partial n_i}\right)_{T,V,n_j} - \frac{RT}{V} \right] dV - RT \ln z} \tag{3-53}$$

where $z = Pv/RT$ is the compressibility factor of the mixture.

Equation (3-53) gives the fugacity of component i in terms of independent variables V and T; it is similar to Eq. (3-14) that gives the fugacity in terms of independent variables P and T. However, in addition to the difference in the choice of independent variables, there is another, less obvious difference: Whereas in Eq. (3-14) the key term is $\bar{v}_i$, the partial molar volume of i, in Eq. (3-53) the key term is $(\partial P / \partial n_i)_{T,V,n_j}$, and that is *not* a partial molar quantity.[5]

For a pure component Eq. (3-53) becomes

$$RT \ln\left(\frac{f}{P}\right)_{\text{pure } i} = \int_V^\infty \left(\frac{P}{n_i} - \frac{RT}{V} \right) dV - RT \ln z + RT(z-1) \tag{3-54}$$

[5] The definition of a partial molar property is applicable only to *extensive* properties differentiated at constant temperature and pressure. The total volume of a mixture is related to the partial molar volumes by a summation: $V = \sum_i n_i (\partial V / \partial n_i)_{T,P,n_j}$. The analogous equation for the total pressure is not valid: $P \neq \sum_i n_i (\partial P / \partial n_i)_{T,V,n_j}$. The derivative $(\partial P / \partial n_i)_{T,V,n_j}$ should not be regarded as a partial pressure; it does not have a significance analogous to that of the partial molar volume.

Equation (3-54) is not particularly useful; for a pure component, Eq. (3-17) is much more convenient. However, for mixtures, Eq. (3-53) is more useful than Eq. (3-14).

Equations (3-47) to (3-53) enable us to compute all thermodynamic properties relative to the properties of an ideal gas at 1 bar and at the same temperature and composition, provided that we have information on volumetric behavior in the form

$$P = F(T, V, n_1, n_2, \ldots) \tag{3-55}$$

Most equations of state are pressure-explicit [Eq. (3-55)] rather than volume-explicit [Eq. (3-15)]. Therefore, for phase-equilibrium problems, Eq. (3-53) is more useful than Eq. (3-14).

To use Eq. (3-53) for calculating the fugacity of a component in a mixture, volumetric data must be available, preferably in the form of an equation of state, at the temperature under consideration and as a function of composition and density, from zero density to the density of interest that corresponds to the lower limit V in the integral. The molar density of the mixture, n_T / V, corresponding to the lower integration limit, must often be found from the equation of state itself; because the specified conditions are usually composition, temperature, and pressure, the density is not ordinarily given. This calculation is tedious because it is usually of the trial-and-error type. However, regardless of the number of components in the mixture, the calculation need only be performed once for any given composition, temperature, and pressure; because the quantity V in Eq. (3-53) is for the entire mixture, it can be used in the calculation of the fugacities for all components. Only if the composition, temperature, or pressure changes must the trial-and-error calculation be repeated. It is probably because of the need for trial-and-error calculations that many authors of classic books on thermodynamics have paid little attention to Eq. (3-53). While trial-and-error calculations were highly undesirable before about 1960, in the age of computers they cause little concern.

For some applications like those requiring derived properties (e.g. composition derivatives of fugacity coefficients), one can achieve considerable computational simplification (Topliss *et al.*, 1988; Mollerup and Michelsen, 1992) in the calculation of thermodynamic properties when relations for the desired properties are defined in terms of (ρ, T, x_i) instead of (V, T, n_i), as in the equations above. Here, ρ is the molar density corresponding to pressure P, temperature T, and composition of the phase of interest, and $x_i = n_i / n_T$ is the mole fraction of component i.

For a given equation of state, the associated expression for the compressibility factor is

$$z = \frac{P(\rho, T, x_i)}{\rho RT} \tag{3-56}$$

The reduced molar residual Helmholtz energy, $\tilde{A} \equiv A^r / n_T RT$, is given by

$$\tilde{A} = \int_0^\rho \frac{z(\rho, T, x_i) - 1}{\rho} d\rho \tag{3-57}$$

The adjective "residual" means that the quantity is defined relative to a mixture of ideal gases at the same density, temperature, and composition as those of the mixture of interest.

The associated expression for the fugacity coefficient of component i becomes

$$\ln \varphi_i = \left[\frac{\partial(n_T \tilde{A})}{\partial n_i}\right]_{\rho, T, n_{j \neq i}} + (z-1) - \ln z \tag{3-58}$$

For a mixture with m components, the derivative of $\tilde{A}$ with respect to the number of moles of component i is

$$\left[\frac{\partial(n_T \tilde{A})}{\partial n_i}\right]_{\rho, T, n_{j \neq i}} = \tilde{A} + \left(\frac{D\tilde{A}}{Dx_i}\right)_{\rho, T, x_j} - \sum_{j=1}^{m} x_j \left(\frac{D\tilde{A}}{Dx_j}\right)_{\rho, T, x_i} \tag{3-59}$$

where the differential operator $(D / Dx_i)_{x_j}$ indicates differentiation with respect to x_i while all other x_j are held constant.

Similarly, we can derive equations of the partial derivatives of $\ln \varphi_i$ with respect to pressure (at constant temperature and composition), with respect to temperature (at constant pressure and composition), and with respect to composition (at constant pressure and temperature). Therefore, these equations relate the desired thermodynamic properties to $\tilde{A}$ and derivatives of $\tilde{A}$ with respect to its independent variables.

The main advantage of Eqs. (3-58) and (3-59) is that the fugacities are obtained from $\tilde{A}$ by differentiation [Eq. (3-58)] rather than by integration of terms obtained from the equation of state [Eq. (3-53)]. It is easier to differentiate than to integrate and not every expression is analytically integrable. To use Eqs. (3-58) and (3-59) it is necessary to have a model for $\tilde{A}$ as a function of temperature, density, and composition.

Topliss (1988) and Dimitrelis (1986) give efficient computation methods for phase equilibria using this formalism.

3.5 Fugacity of a Component in a Mixture According to van der Waals' Equation

To illustrate the applicability of Eq. (3-53), we consider a mixture whose volumetric properties are described by van der Waals' equation:

$$P = \frac{RT}{v-b} - \frac{a}{v^2} \tag{3-60}$$

In Eq. (3-60), v is the molar volume of the mixture and a and b are constants that depend on composition.

To substitute Eq. (3-60) into Eq. (3-53), we must first transform it from a molar basis to a total basis by substituting $v = V / n_T$, where n_T is the total number of moles. Equation (3-60) then becomes

$$P = \frac{n_T RT}{V - n_T b} - \frac{n_T^2 a}{V^2} \tag{3-61}$$

We want to calculate the fugacity of component i in the mixture at some given temperature, pressure, and composition. Differentiating Eq. (3-61) with respect to n_i, we have

$$\left(\frac{\partial P}{\partial n_i}\right)_{T,V,n_j} = \frac{RT}{V - n_T b} + \frac{n_T RT\left(\frac{\partial(n_T b)}{\partial n_i}\right)}{(V - n_T b)^2} - \frac{1}{V^2}\frac{\partial(n_T^2 a)}{\partial n_i} \tag{3-62}$$

Substituting into Eq. (3-53) and integrating, we obtain

$$RT \ln \frac{f_i}{y_i P} = RT \ln \frac{V - n_T b}{V}\Bigg]_V^\infty - n_T RT\left(\frac{\partial(n_T b)}{\partial n_i}\right)\frac{1}{(V - n_T b)}\Bigg]_V^\infty + \left(\frac{\partial(n_T^2 a)}{\partial n_i}\right)\frac{1}{V}\Bigg]_V^\infty - RT \ln z \tag{3-63}$$

At the upper limit of integration, as $V \to \infty$,

$$\ln \frac{V - n_T b}{V} \to 0 \qquad \frac{1}{V - n_T b} \to 0 \qquad \frac{1}{V} \to 0 \tag{3-64}$$

and Eq. (3-63) becomes

$$RT \ln \frac{f_i}{y_i P} = RT \ln \frac{V}{V - n_T b} + n_T RT\left[\frac{\partial(n_T b)}{\partial n_i}\right]\frac{1}{(V - n_T b)} - \left[\frac{\partial(n_T^2 a)}{\partial n_i}\right]\frac{1}{V} - RT \ln z \tag{3-65}$$

Equation (3-65) gives the desired result derived from rigorous thermodynamic relations. To proceed further, it is necessary to make assumptions concerning the composition dependence of constants a and b. These assumptions cannot be based on thermodynamic arguments but must be obtained from molecular considerations. Suppose that we have m components in the mixture. If we interpret constant b as a term proportional to the size of the molecules and if we assume that the molecules are spherical, then we might average the molecular diameters, giving

$$b^{1/3} = \sum_{i=1}^{m} y_i b_i^{1/3} \tag{3-66}$$

On the other hand, we may choose to average the molecular volumes directly and obtain the simpler relation

$$b = \sum_{i=1}^{m} y_i b_i \tag{3-67}$$

Neither Eq. (3-66) nor Eq. (3-67) is in any sense a "correct" mixing rule; both are based on arbitrary assumptions, and alternative mixing rules could easily be constructed based on different assumptions. Equation (3-67) is commonly used because of its mathematical simplicity.

At moderate densities, for mixtures whose molecules are not too dissimilar in size, the particular mixing rule used for b does not significantly affect the results. However, the fugacity of a component in a mixture is sensitive to the mixing rule used for the constant a. If we interpret a as a term that reflects the strength of attraction between two molecules, then for a mixture we may want to express a by averaging over all molecular pairs. Thus

$$a = \sum_{i=1}^{m} \sum_{j=1}^{m} y_i y_j a_{ij} \tag{3-68}$$

where a_{ij} is a measure of the strength of attraction between a molecule i and a molecule j. If i and j are the same chemical species, then a_{ij} is the van der Waals a for that substance. If i and j are chemically not identical and if we have no experimental data for the i-j mixture, we then need to express a_{ij} in terms of a_i and a_j. This need constitutes one of the key problems of phase-equilibrium thermodynamics. Given information on the intermolecular forces of each of two pure fluids, how can we predict the intermolecular forces in a mixture of these two fluids? There is no general answer to this question. An introduction to the study of intermolecular forces is given in Chap. 4 where it is shown that only under severe limiting conditions can the forces between molecule i and molecule j be related in a simple way to the forces between two molecules i and two molecules j. Our knowledge of molecular physics is, unfortunately, not sufficient to give generally reliable methods for predicting the properties of mixtures using only knowledge of the properties of pure components.

For $i \neq j$, it was suggested many years ago by Berthelot, on strictly empirical grounds, that

$$a_{ij} = (a_i a_j)^{1/2} \tag{3-69}$$

This relation, often called the *geometric-mean assumption*, was used extensively by van der Waals and his followers in their work on mixtures. Since van der Waals' time, the geometric-mean assumption has been used for other quantities in addition to van der Waals' a; it is commonly used for those parameters that are a measure of intermolecular attraction. Long after the time of van der Waals and Berthelot, it was shown by London (see Sec. 4.4) that under certain conditions there is some theoretical justification for the geometric-mean assumption.

If we adopt the mixing rules given by Eqs. (3-67), (3-68), and (3-69), the fugacity for component i, given by Eq. (3-65), becomes

$$\boxed{\ln \frac{f_i}{y_i P} = \ln \frac{v}{v-b} + \frac{b_i}{v-b} - \frac{2\sqrt{a_i}\sum\limits_{j=1}^{m} y_j \sqrt{a_j}}{vRT} - \ln z} \tag{3-70}$$

where v is the molar volume and z is the compressibility factor of the mixture.

Equation (3-70) indicates that, to calculate the fugacity of a component in a mixture at a given temperature, pressure, and composition, we must first compute constants a and b for the mixture using some mixing rules, for example, those given by Eqs. (3-67), (3-68), and (3-69). Using these constants and the equation of state, Eq. (3-60), we must then find the molar volume v of the mixture by Cardan's method or by trial and error; this step is the only tedious part of the calculation. Once the molar volume is known, the compressibility factor z is easily calculated and the fugacity is readily found from Eq. (3-70).

For a numerical example, consider the fugacity of hydrogen in a ternary mixture at 50°C and 303 bar containing 20 mol % hydrogen, 50 mol % methane, and 30 mol % ethane. Using Eq. (3-70), we find that the fugacity of hydrogen is 114.5 bar.[6] From the ideal-gas law, the fugacity is 60.8 bar, while the Lewis fugacity rule gives 72.3 bar. These three results are significantly different from one another. No reliable experimental data are available for this mixture at 50°C but in this particular case it is probable that 114.5 bar is much closer to the correct value than the results of either of the

[6] Constants a and b for each component were found from critical properties. The molar volume of the mixture, as calculated from van der Waals' equation, is 62.43 $cm^3\ mol^{-1}$.

two simpler calculations. However, such a conclusion cannot be generalized. The equation of van der Waals gives only an approximate description of gas-phase properties and sometimes, because of cancellation of errors, calculations based on simpler assumptions may give better results.

In deriving Eq. (3-70) we have shown how the rigorous expression, Eq. (3-53), can be used to calculate the fugacity of a component in a mixture once the equation of state for the mixture is given. In this particular calculation, the relatively simple van der Waals equation was used but the same procedure can be applied to any pressure-explicit equation of state, regardless of how complex it may be.

It is frequently said that the more complicated an equation of state, and the more constants it contains, the better representation it gives of volumetric properties. This statement is correct for a pure component if ample data are available to determine the constants with confidence and if the equation is used only under those conditions of temperature and pressure used to determine the constants. However, for predicting the properties of mixtures from pure-component data alone, the more constants one has, the more mixing rules are required and, because these rules are subject to much uncertainty, it frequently happens that a simple equation of state containing only two or three constants is better for predicting mixture properties than a complicated equation of state containing a large number of constants (Ackerman and Redlich, 1963; Shah and Thodos, 1965).

Chapter 5 discusses the use of equations of state for calculating fugacities in gas-phase mixtures and Chap. 12 discusses such use for calculating fugacities in both gas-phase and liquid-phase mixtures.

3.6 Phase Equilibria from Volumetric Properties

In Chap. 1 we indicated that the purpose of phase-equilibrium thermodynamics is to predict conditions (temperature, pressure, composition) which prevail when two or more phases are in equilibrium.

In Chap. 2 we discussed thermodynamic equations that determine the state of equilibrium between phases α and β. These are:[7]

Equality of temperatures: $T^{\alpha} = T^{\beta}$

Equality of pressures: $P^{\alpha} = P^{\beta}$

For each component i,
equality of fugacities: $f_i^{\alpha} = f_i^{\beta}$

To find the conditions that satisfy these equations, it is necessary to have a method for evaluating the fugacity of each component in phase α and in phase β. Such

[7] Note, however, the restrictions indicated in the footnote at the beginning of Sec. 2.3.

a method is supplied by Eq. (3-14) or (3-53); both are valid for any component in any phase. In principle, therefore, a solution to the phase-equilibrium problem is provided completely by either one of these equations together with an equation of state and the equations of phase equilibrium.

To illustrate these ideas, consider vapor-liquid equilibria[8] in a system with m components; suppose that we know pressure P and mole fractions x_1, x_2, ..., x_m, for the liquid phase. We want to find temperature T and vapor-phase mole fractions y_1, y_2, ..., y_m. We assume that we have available a pressure-explicit equation of state applicable to all components and to their mixtures over the entire density range, from zero density to the density of the liquid phase.
We can then compute the fugacity of each component, in either phase, by Eq. (3-53).

The number of unknowns is:

$y_1, y_2, \ldots, y_{m-1}$	$(m - 1)$ mole fractions.[9]
T	Temperature.
υ^V, υ^L	Molar volumes of vapor and liquid phases at equilibrium.

Total: $(m + 2)$ unknowns.

The number of independent equations is:

$f_i^V = f_i^L$	m equations, where f_i^V and f_i^L are found for each component i by Eq. (3-53).
$P = F(\upsilon^V, y_1, \ldots, T)$	Equation of state, applied once to the vapor phase and
$P = F(\upsilon^L, x_1, \ldots, T)$	once to the liquid phase.

Total: $(m + 2)$ independent equations.

Because the number of unknowns is equal to the number of independent equations, the unknown quantities can be found by simultaneous solution of all equations.[10] It is apparent, however, that the computational effort to do so is large, especially if the equation of state is not simple and if the number of components is high.

Calculation of vapor-liquid equilibria, along the lines outlined, was discussed many years ago by van der Waals who made use of the equation bearing his name, and in the period 1940-1952 extensive calculations for hydrocarbon mixtures were reported

[8] Because the equations relating chemical potential (or fugacity) to volumetric properties use integrals starting from zero density, evaluation of these integrals requires continuity from zero density to the density of interest. While it is possible to go continuously from vapor to liquid, it is not possible to go continuously from either fluid to a crystalline solid. Therefore, the discussion here is not applicable to fluid-solid equilibria.

[9] Because $\sum_i y_i = 1$, the mth mole fraction is fixed once $(m - 1)$ mole fractions are determined.

[10] We have considered here the case where P and x are known and T and y are unknown, but similar reasoning applies to other combinations of known and unknown quantities; the number of intensive variables that must be specified is given by the phase rule.

by Benedict *et al.* (1940, 1942, 1951) who used an eight-constant equation of state. Since that pioneering work, many others have made similar calculations using a variety of equations of state, as discussed in Chap. 12.

Calculation of phase equilibria from volumetric data alone requires a large computational effort but, thanks to modern computers, the effort does not by itself present a significant difficulty. The major disadvantage of this type of calculation is not computational; the more fundamental difficulty is that we do not have a satisfactory equation of state applicable to mixtures over a density range from zero density to liquid densities. Because of this crucial deficiency, phase-equilibrium calculations based on volumetric data alone are often doubtful. To determine the required volumetric data with the necessary degree of accuracy, a large amount of experimental work is required; rather than make all these volumetric measurements, it is usually more economical to measure the desired phase equilibria directly. For those mixtures where the components are chemically similar (e.g., mixtures of paraffinic and olefinic hydrocarbons) an equation-of-state calculation for vapor-liquid equilibria provides a reasonable possibility because many simplifying assumptions can be made concerning the effect of composition on volumetric behavior. But even in this relatively simple situation, there is much ambiguity when one attempts to predict properties of mixtures using equation-of-state constants determined from pure-component data. Benedict *et al.* used eight empirical constants to describe the volumetric behavior of each pure hydrocarbon; to fix *uniquely* eight constants, a very large amount of experimental data is required. Even with the generous amount of data that Benedict had at his disposal, his constants could not be determined without some ambiguity. But the essence of the difficulty arises when one must decide on mixing rules, i.e., on how these constants are to be combined for a mixture. Phase-equilibrium calculations are often sensitive to the mixing rules used, especially to the binary parameters that appear in these rules.

In summary, then, the equation-of-state method to attain a complete determination of phase equilibria is often not promising because we usually do not have sufficiently accurate knowledge of the volumetric properties of mixtures at high densities. Calculation of fugacities by Eq. (3-53) is practical for vapor mixtures but, special cases excepted, may not be practical for condensed mixtures. Even in vapor mixtures, the calculations are often not accurate because of our inadequate knowledge of volumetric properties. The accuracy of the fugacity calculated with Eq. (3-53) depends directly on the validity of the equation of state and, to determine the constants in a good equation of state, one must have either a large amount of reliable experimental data, or else some sound theoretical basis for predicting volumetric properties. For many mixtures we have little of either data or theory. Reliable volumetric data exist for the more common substances but these constitute only a small fraction of the number of listings in a chemical handbook. Reliable volumetric data are scarce for binary mixtures and they are rare for mixtures containing more than two components. Considering the nearly infinite number of ternary (and higher) mixtures possible, it is clear that there will never be sufficient experimental data to give an adequate empirical description of volumetric properties of mixed fluids. A strictly empirical approach to the phase-equilibrium problem is, therefore, subject to severe limitations. Progress can be achieved only by generalizing from limited, but reliable, experimental results, utilizing as much as possible techniques based on our theoretical knowledge of molecular behavior. Since the first edition of this book (1969) remarkable progress has been made; based on statistical mechanical derivations, promising equations of state are now available, for example, for polar and for hydrogen-bonded fluids, for electrolytes and colloidal particles in water, and for polymeric systems. Many of these advanced equations of state are beyond the scope of this book.

Significant progress in phase-equilibrium thermodynamics can be achieved only by increased use of the concepts of molecular physics. Therefore, before continuing our discussion of fugacities in Chap. 5, we turn in Chap. 4 to a brief survey of intermolecular forces.

References

Ackerman, F. J. and O. Redlich, 1963, *J. Chem. Phys.*, 38: 2740.

Beattie, J. A. and W. H. Stockmayer, 1942. In *Treatise on Physical Chemistry,* (H. S. Taylor and S. Glasstone, Eds.), Chap. 2. Princeton: Van Nostrand.

Beattie, J. A., 1949, *Chem. Rev.*, 44: 141.

Beattie, J. A., 1955. In *Thermodynamics and Physics of Matter,* (F. D. Rossini, Ed.), Chap. 3, Part C. Princeton: Princeton University Press.

Benedict, M., G. B. Webb, and L. C. Rubin, 1940, *J. Chem. Phys.*, 8: 334.

Benedict, M., G. B. Webb, and L. C. Rubin, 1942, *J. Chem. Phys.*, 10: 747.

Benedict, M., G. B. Webb, and L. C. Rubin, 1951, *Chem. Eng. Prog.*, 47: 419.

Dimitrelis, D. and J. M. Prausnitz, 1986, *Fluid Phase Equilibria,* 31: 1.

Evans, R. B. and G. M. Watson, 1956, *Chem. Eng. Data Series,* 1: 67.

Keenan, J. H., F. G. Keyes, P. G. Hill, and J. G. Moore, 1978, *Steam Tables: Thermodynamic Properties of Water Including Vapor, Liquid and Solid Phases.* New York: Wiley-Interscience.

Mollerup, J. M. and M. L. Michelsen, 1992, *Fluid Phase Equilibria,* 74: 1.

Shah, K. K. and G. Thodos, 1965, *Ind. Eng. Chem.*, 57: 30.

Topliss, R. J., D. Dimitrelis, and J. M. Prausnitz, 1988, *Comput. Chem. Engng.*, 12: 483.

Problems

1. Consider a mixture of m gases and assume that the Lewis fugacity rule is valid for this mixture. For this case, show that the fugacity of the mixture f_{mixt} is given by

$$f_{\text{mixt}} = \prod_{i=1}^{m} f_{\text{pure } i}^{y_i}$$

where y_i is the mole fraction of component i and $f_{\text{pure } i}$ is the fugacity of pure component i at the temperature and total pressure of the mixture.

2. A binary gas mixture contains 25 mol % A and 75 mol % B. At 50 bar total pressure and 100°C the fugacity coefficients of A and B in this mixture are, respectively, 0.65 and 0.90. What is the fugacity of the gaseous mixture?

3. At 25°C and 1 bar partial pressure, the solubility of ethane in water is very small; the equilibrium mole fraction is $x_{C_2H_6} = 0.33 \times 10^{-4}$. What is the solubility of ethane at 25°C when the partial pressure is 35 bar? At 25°C, the compressibility factor of ethane is given by the empirical relation

$$z = 1 - 7.63 \times 10^{-3} P - 7.22 \times 10^{-5} P^2$$

where P is in bar. At 25°C, the saturation pressure of ethane is 42.07 bar and that of water is 0.0316 bar.

4. Consider a binary mixture of components 1 and 2. The molar Helmholtz energy change Δa is given by

$$\frac{\Delta a}{RT} = \ln\frac{v}{v-b} - y_1 \ln\frac{v}{y_1 RT} - y_2 \ln\frac{v}{y_2 RT}$$

where v is the molar volume of the mixture; b is a constant for the mixture, depending only on composition; and Δa is the molar Helmholtz energy change in going isothermally from the standard state (pure, unmixed, ideal gases at 1 bar) to the molar volume v. The composition dependence of b is given by $b = y_1 b_1 + y_2 b_2$. Find an expression for the fugacity of component 1 in the mixture.

5. Derive Eq. (3-54). [Hint: Start with Eq. (3-51) or with Eq. (3-53).]

6. Oil reservoirs below ground frequently are in contact with underground water and, in connection with an oil drilling operation, you are asked to compute the solubility of water in a heavy oil at the underground conditions. These conditions are estimated to be 140°C and 410 bar. Experiments at 140°C and 1 bar indicate that the solubility of steam in the oil is $x_1 = 35 \times 10^{-4}$ (x_1 is the mole fraction of steam). Assume Henry's law in the form $f_1 = H(T)x_1$, where $H(T)$ is a constant, dependent only on the temperature, and f_1 is the fugacity of H_2O. Also assume that the vapor pressure of the oil is negligible at 140°C. Data for H_2O are given in the steam tables.

7. A gaseous mixture contains 50 mol % A and 50 mol % B. To separate the mixture, it is proposed to cool it sufficiently to condense the mixture; the condensed liquid mixture is then sent to a distillation column operating at 1 bar. Initial cooling to 200 K (without condensation) is to be achieved by Joule-Thomson throttling. If the temperature upstream of the throttling valve is 300 K, what is the required upstream pressure? The volumetric properties of this gaseous mixture are given by

$$v = \frac{RT}{P} + 50 - \frac{10^5}{T}$$

where v is the volume per mole of mixture in cm^3 mol^{-1}. The ideal-gas specific heats are:

	c_p^0 (J mol^{-1} K^{-1})
A	29.3
B	37.7

Some experimental data for the region from 0 to 50 bar indicate that the fugacity of a pure gas is given by the empirical relation

$$\ln\left(\frac{f}{P}\right) = -cP - dP^2$$

where P is the pressure (bar) and c and d are constants that depend only on temperature. For the region 60 to 100°C, the data indicate that

$$c = -0.067 + \frac{30.7}{T}$$

$$d = 0.0012 - \frac{0.416}{T}$$

where T is in kelvin. At 80°C and 30 bar, what is the molar enthalpy of the gas relative to that of the ideal gas at the same temperature?

8. A certain cryogenic process is concerned with an equimolar mixture of argon (1) and ethane (2) at 110 K. To design a separation process, a rough estimate is required for the enthalpy of mixing for this liquid mixture. Make this estimate using van der Waals' equation of state with the customary mixing rules, $b_{\text{mixt}} = \Sigma_i x_i b_i$ and $a_{\text{mixt}} = \Sigma_i \Sigma_j x_i x_j a_{ij}$ with $a_{ij} = (a_i a_j)^{1/2}(1 - k_{ij})$. Assume that $v_{\text{mixt}} = \Sigma_i x_i v_i$ and, because pressure is low, that $\Delta_{\text{mix}}U = \Delta_{\text{mix}}H$, where U is the internal energy and H is the enthalpy. Data are:

	v at 110 K (cm^3 mol^{-1})	a (bar cm^6 mol^{-2})	b (cm^3 mol^{-1})
Argon	32.2	1.04×10^6	25.0
Ethane	48.5	4.17×10^6	49.3[11]

Second-virial-coefficient data for this mixture indicate that k_{ij} $(i \neq j) = 0.1$.

[11] This b is obtained from critical volume. The inadequacy of the van der Waals equation is evident because, as shown here, $b > v$ at 110 K. Fortunately this unrealistic result does not affect the solution to Problem 8.

CHAPTER 4

Intermolecular Forces, Corresponding States and Osmotic Systems

Thermodynamic properties of any pure substance are determined by intermolecular forces that operate between the molecules of that substance. Similarly, thermodynamic properties of a mixture depend on intermolecular forces that operate between the molecules of the mixture. The case of a mixture, however, is necessarily more complicated because consideration must be given not only to interaction between molecules belonging to the same component, but also to interaction between dissimilar molecules. To interpret and correlate thermodynamic properties of solutions, it is therefore necessary to have some understanding of the nature of intermolecular forces. The purpose of this chapter is to give a brief introduction to the nature and variety of forces between molecules.

We must recognize at the outset that our understanding of intermolecular forces is far from complete and that quantitative results have been obtained for only simple and idealized models of real matter. Further, we must point out that analytic relations that link intermolecular forces to macroscopic properties (i.e., statistical mechanics) are also at present limited to relatively simple and idealized cases.[1] It follows, therefore, that, for most cases, we can use our knowledge of intermolecular forces in only an approximate manner to interpret and generalize phase-equilibrium data. Molecular physics is always concerned with models and we must beware whenever we are tempted to substitute models for nature. Frequently, the theory of intermolecular forces gives us no more than a qualitative, or perhaps semiquantitative, basis for understanding phase behavior, but even such a limited basis can be useful for understanding and correlating experimental results.

While the separation between molecular physics and practical problems in phase behavior is large, every year new results tend to reduce that separation. There can be no doubt that future developments in applied thermodynamics will increasingly utilize and rely on statistical mechanics and the theory of intermolecular forces.

When a molecule is in the proximity of another, forces of attraction and repulsion strongly influence its behavior. If there were no forces of attraction, gases would not condense to form liquids, and in the absence of repulsive forces, condensed matter would not show resistance to compression. The configurational properties of matter can be considered as a compromise between those forces that pull molecules together and those that push them apart; by *configurational properties* we mean those properties that depend on interactions between molecules rather than on the characteristics of isolated molecules. For example, the energy of vaporization for a liquid is a configurational property, but the specific heat of a gas at low pressure is not.

There are many different types of intermolecular forces, but for our limited purposes here, only a few important ones are considered. These forces may be classified under the following arbitrary but convenient headings:

- *Electrostatic forces* between charged particles (ions) and between permanent dipoles, quadrupoles, and higher multipoles.
- *Induction forces* between a permanent dipole (or quadrupole) and an induced dipole, that is, a dipole induced in a molecule with polarizable electrons.
- Forces of attraction (*dispersion forces*) and repulsion between nonpolar molecules.
- *Specific (chemical) forces* leading to association and solvation, i.e., to the formation of loose chemical bonds; hydrogen bonds and charge-transfer complexes are perhaps the best examples.

[1] However, thanks to advances in molecular simulation with powerful computers, it is now possible to calculate physical properties of some bulk substances provided that we have quantitative knowledge of their intermolecular forces. There is good reason to believe that molecular simulation will become increasingly useful as computers become even more powerful.

An introductory discussion of these forces is presented in most of this chapter, with special attention to those acting between nonpolar molecules and to the molecular theory of corresponding states. The chapter also includes some brief discussion of osmotic pressure and micelles.

4.1 Potential-Energy Functions

Molecules have kinetic energy as a result of their velocities relative to some fixed frame of reference; they also have potential energy as a result of their positions relative to one another. Consider two simple, spherically symmetric molecules separated by the distance r. The *potential energy* Γ shared by these two molecules is a function of r; the force F between the molecules is related to the potential energy by

$$\boxed{F = -\frac{d\Gamma}{dr}} \tag{4-1}$$

The negative of the potential energy, i.e., $-\Gamma(r)$, is the work which must be done to separate two molecules from the intermolecular distance r to infinite separation. Intermolecular forces are usually expressed in terms of potential-energy functions. The common convention is that a force of attraction is negative and one of repulsion is positive.

In the simplified discussion above, we have assumed that the force acting between two molecules depends on their relative position as specified by only one coordinate, r. For two spherically symmetric molecules, such as argon atoms, this assumption is valid; but for more complicated molecules, other coordinates, such as the angles of orientation, may be required as additional independent variables of the potential-energy function. A more general form of Eq. (4-1) is

$$F(r,\theta,\phi,\ldots) = -\nabla\Gamma(r,\theta,\phi,\ldots) \tag{4-2}$$

where ∇ is the gradient and θ, ϕ,... designate whatever additional coordinates may be needed to specify the potential energy.

In the next sections, for simplicity, we assume that ions, atoms, or molecules are in free space (vacuum). For electrostatic forces, extension to a medium other than vacuum is made through introduction of the *relative permittivity* (or *dielectric constant*) of the medium. While we use the SI unit system throughout, in some cases we also give units in other systems that remain in common use.

4.2 Electrostatic Forces

Of all intermolecular forces, those due to point charges are the easiest to understand and the simplest to describe quantitatively. If we regard two point electric charges of magnitudes q_i and q_j, respectively, separated from one another in vacuo by distance r, then the force F between them is given by Coulomb's relation, sometimes called the *inverse-square law:*[2]

$$F = \frac{q_i q_j}{4\pi\varepsilon_o r^2} \tag{4-3}$$

where F is in newtons, q in coulombs, r in meters, and ε_o, the dielectric permittivity of a vacuum, is $\varepsilon_o = 8.85419\times10^{-12}$ C^2 J^{-1} m^{-1}.

Upon integrating, the potential energy is

$$\Gamma_{ij} = \frac{q_i q_j}{4\pi\varepsilon_o r} + \text{constant of integration} \tag{4-4}$$

The usual convention is that the potential energy is zero at infinite separation. Substituting $\Gamma = 0$ when $r = \infty$, the constant of integration in Eq. (4-4) vanishes.

For charged molecules (i.e., ions), q_i and q_j are integral multiples of the unit charge e; therefore, the potential energy between two ions is

$$\boxed{\Gamma_{ij} = \frac{z_i z_j e^2}{4\pi\varepsilon_o r}} \tag{4-5}$$

where z_i and z_j are the ionic valences and $e = 1.60218\times10^{-19}$ C.[3]

For a medium other than vacuum, Eq. (4-5) becomes

$$\Gamma_{ij} = \frac{z_i z_j e^2}{4\pi\varepsilon r} \tag{4-5a}$$

---

[2] In the still-frequently-used cgs system of units, Coulomb's law (in vacuo) is written

$$F = \frac{q_i q_j}{r^2}$$

where F is in dynes, r is in centimeters, and q is in esu [or (erg cm)$^{1/2}$]. Note that in Eq. (4-3) the proportional factor $1/4\pi\varepsilon_o$ implies that SI units are being used; this factor does not appear when cgs (or other) units are used.

[3] In the cgs unit system, the unit charge is $e = 4.8024\times10^{-10}$ (erg cm)$^{1/2}$.

where ε, the absolute permittivity ($C^2\ J^{-1}\ m^{-1}$), is defined by $\varepsilon = \varepsilon_0 \varepsilon_r$; ε_0 is the permittivity of a vacuum, and ε_r is the (dimensionless) *dielectric constant* or permittivity relative[4] to that of a vacuum; ε_r is unity in vacuo but greater than unity otherwise (for water at 25°C, $\varepsilon_r = 78.41$).[5]

Compared to others physical intermolecular energies, the magnitude of the Coulomb energy as given by Eq. (4-5) is large and long range. To illustrate, consider the isolated ions Cl^- and Na^+ in contact. Distance r is given by the sum of the two ionic radii (r = 0.276 nm or 2.76 Å); the potential energy between these two ions in vacuo is

$$\Gamma_{ij} = \frac{(-1)(+1)(1.60218 \times 10^{-19})^2}{(4\pi)(8.8542 \times 10^{-12})(0.276 \times 10^{-9})} = -8.36 \times 10^{-19}\ \text{J}$$

This energy is about $200kT$, where k is Boltzmann's constant; kT is the thermal energy [at room temperature, $kT = (1.38 \times 10^{-23}\ \text{J K}^{-1}) \times (300\ \text{K}) = 0.0414 \times 10^{-19}\ \text{J}$]. $200kT$ is of the same order of magnitude as typical covalent bonds. Only when the two ions are about 560 Å far apart is the Coulomb interaction equal to kT.

Electrostatic forces between ions are inversely proportional to the square of the separation and therefore they have a much longer range than most other intermolecular forces that depend on higher powers of the reciprocal distance. These electrostatic forces make the dominant contribution to the configurational energy of salt crystals and are therefore responsible for the very high melting points of salts. In addition, the long-range nature of ionic forces is, at least in part, responsible for the difficulty in constructing a theory of electrolyte solutions.

Electrostatic forces can arise even for those particles that do not have a net electric charge. Consider a particle having two electric charges of the same magnitude e but of opposite sign, held a distance d apart. Such a particle has an electric couple or permanent *dipole moment* μ defined by

$$\mu = e\, d \tag{4-6}$$

Asymmetric molecules possess permanent dipoles resulting from an uneven spatial distribution of electronic charges about the positively charged nuclei. Symmetric molecules, like argon or methane, have zero dipole moment and those molecules having very little asymmetry generally have small dipole moments. Table 4-1 gives a rep-

[4] The relative permittivity of a substance is easily measured by comparing capacitance C of a capacitor with the sample to that without the sample (C_0) and using $\varepsilon_r = C/C_0$. Note that the magnitude of ε_r can have a significant effect on the magnitude of the interaction between ions in solution. For example, water at 25°C reduces the long-range interionic coulombic interaction energy by about two orders of magnitude from its value in vacuo. When the dielectric constant is used in Eq. (4-5a) and similar equations, the medium is considered a structureless continuum.

[5] According to the current international standard formulation (D. P. Fernández, A. R. H. Goodwin, E. W. Lemmon, J. M. H. Levelt-Sengers, and R. C. Williams, 1997, *J. Phys. Chem. Ref. Data*, 26: 1125).

resentative selection of molecules and their dipole moments. The common unit for dipole moment is the debye (D). The dipole moment of a pair of charges +*e* and -*e*, separated by 0.1 nm (or 1 Å) is $\mu = (1.60218\times10^{-19}\ C)\times(10^{-10}\ m) = 1.60218\times10^{-29}$ C m, corresponding to 4.8 D [1 D = 3.33569×10^{-30} C m = 10^{-18} esu = 10^{-18} (erg cm^3)$^{1/2}$].

Table 4-1 Permanent dipole moments.*

Molecule	μ (Debye)	Molecule	μ (Debye)
CO	0.10	CH_3I	1.64
C_3H_6	0.35	CH_3COOCH_3	1.67
$C_6H_5CH_3$	0.37	C_2H_5OH	1.70
PH_3	0.55	H_2O	1.84
HBr	0.80	HF	1.91
$CHCl_3$	1.05	C_2H_5F	1.92
$(C_2H_5)_2O$	1.16	$(CH_3)_2CO$	2.88
NH_3	1.47	$C_6H_5COCH_3$	3.00
$C_6H_5NH_2$	1.48	$C_2H_5NO_2$	3.70
C_6H_5Cl	1.55	CH_3CN	3.94
C_2H_5SH	1.56	$CO(NH_2)_2$	4.60
SO_2	1.61	KBr	9.07

* Taken from a more complete list of dipole moments given in Landolt-Börnstein, *Zahlenwerte und Funktionen,* 6th Ed., Vol. I, Part 3 (Berlin: Springer, 1951), and New Series, Group II, Vols. II and VI (Berlin: Springer, 1967, 1973); and in A. McClellan, *Tables of Experimental Dipole Moments,* Vol. 1 (San Francisco: W. H. Freeman, 1963), and Vol. 2 (El Cerrito: Rahara Enterprises, 1974).

The potential energy of two permanent dipoles i and j is obtained by considering the coulombic forces between the four charges. The energy of interaction depends on the distance between dipole centers and on the relative orientations of the dipoles, as illustrated in Fig. 4-1, where angles θ and ϕ give the orientation of the dipole axes.

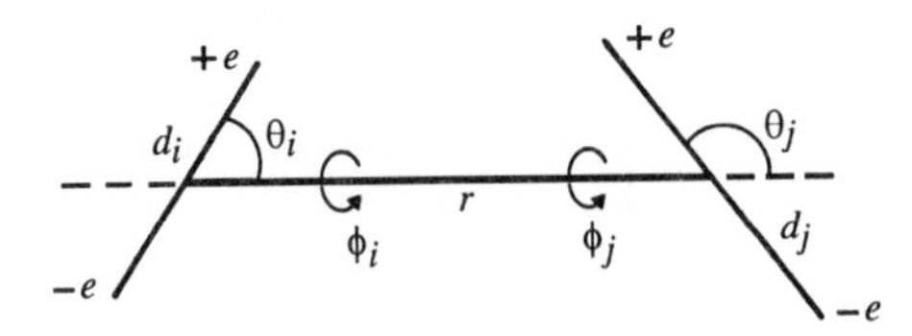

Figure 4-1 Orientation of two dipoles.

If the distance r between the dipoles is large compared to d_i and d_j, the potential energy is

$$\Gamma_{ij} = -\frac{\mu_i \mu_j}{4\pi\varepsilon_0 r^3}[2\cos\theta_i \cos\theta_j - \sin\theta_i \sin\theta_j \cos(\phi_i - \phi_j)] \tag{4-7}$$

The orientation making the potential energy a maximum is that corresponding to when the dipoles are in the same straight line, the positive end of one facing the positive end of the other; the energy is a minimum when the dipoles are in a straight line, the positive end of one facing the negative end of the other.

In an assembly of polar molecules, the relative orientations of these molecules depend on the interplay of two factors: The presence of an electric field set up by the polar molecules tends to line up the dipoles, whereas the kinetic (thermal) energy of the molecules tends to toss them about in a random manner. We expect, therefore, that as the temperature rises, the orientations become more random until in the limit of very high temperature, the average potential energy due to polarity becomes vanishingly small. This expectation is confirmed by experimental evidence; whereas at low and moderate temperatures the behavior of polar gases is markedly different from that of nonpolar gases, this difference tends to disappear as the temperature increases. It was shown by Keesom (1922) that at moderate and high temperatures, orientations leading to negative potential energies are preferred statistically. The average potential energy Γ_{ij} between two dipoles i and j in vacuum at a fixed separation r is found by averaging over all orientations with each orientation weighted according to its Boltzmann factor (Hirschfelder *et al.*, 1964). When the Boltzmann factors are expanded in powers of $1/kT$, Γ_{ij} becomes

$$\boxed{\overline{\Gamma}_{ij} = -\frac{2}{3}\frac{\mu_i^2 \mu_j^2}{(4\pi\varepsilon_0)^2 kT r^6} + \cdots} \tag{4-8}$$[6]

Equation (4-8) indicates that for a pure polar substance ($i = j$) the potential energy varies as the fourth power of the dipole moment. A small increase in the dipole

[6] This equation is based on a Boltzmann average

$$\overline{\Gamma} = \frac{\int \Gamma[\exp(-\Gamma / kT)]\, d\Omega}{\int [\exp(-\Gamma / kT)]\, d\Omega}$$

where $d\Omega$ is the element of the solid angle: $d\Omega = \sin\theta_1 \sin\theta_2\, d\theta_1\, d\theta_2\, d\phi$, where $\phi = \phi_1 - \phi_2$. However, it has been argued (J. S. Rowlinson, 1958, *Mol. Phys.*, 1: 414) that the proper average is

$$\overline{\Gamma} = -kT \ln \frac{\int [\exp(-\Gamma / kT)] d\Omega}{\int d\Omega}$$

When that average is used, the numerical coefficient in Eq. (4-8) is -1/3 instead of -2/3.

Equations (4-7) and (4-8) are for ideal dipoles, i.e. where $r >> d$. Therefore, these equations fail at small separations where r and d may be of similar magnitude. For real (non-ideal) dipoles, calculation of Γ and $\overline{\Gamma}$ becomes more complicated. See Cohen *et al.* , 1996, *J. Coll. Interface Sci.*, 177: 276.

moment can therefore produce a large change in the potential energy due to permanent dipole forces. Whereas the contribution of polar forces to the total potential energy is small for molecules having dipole moments of 1 debye or less, this contribution becomes increasingly significant for small molecules having larger dipole moments. Section 4.4 gives a quantitative comparison of the importance of polar forces relative to some other intermolecular forces.

In addition to dipole moments, it is possible for molecules to have quadrupole moments due to the concentration of electric charge at four separate points in the molecule. The difference between a molecule having a dipole moment and one having a linear quadrupole moment is shown schematically:[7]

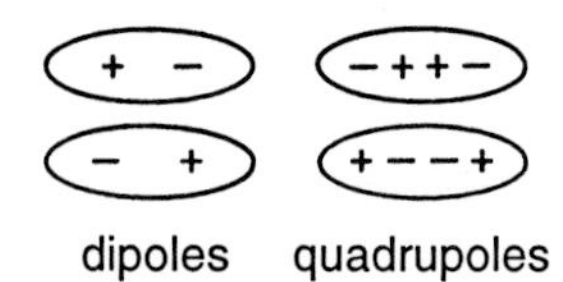

For example, carbon dioxide, a linear molecule, has no dipole moment but its quadrupole moment is sufficiently strong to affect its thermodynamic properties that are different from those of other nonpolar molecules of similar size and molecular weight. For the simplest case, a linear molecule, *quadrupole moment* Q is defined by the sum of the second moments of the charges:

$$Q = \sum_i e_i d_i^2 \tag{4-9}$$

where the charge e_i is located at a point a distance d_i away from some arbitrary origin and where all charges are on the same straight line. Quadrupole moment Q is independent of the position of the origin provided that the molecule has no net charge and no dipole moment. For nonlinear quadrupoles or for molecules having permanent dipoles, the definition of the quadrupole moment is more complicated (Buckingham, 1959; Buckingham and Disch, 1963).

Experimental determination of quadrupole moments is difficult and not many measurements have been made (Buckingham, 1959; Buckingham and Disch, 1963; Flygare and Benson, 1971; Sutter and Flygare, 1976). Table 4-2 gives quadrupole moments[8] for some molecules whose symmetry is sufficient to allow a single scalar to specify the moment. (To distinguish between the two quadrupole structures shown above, it is necessary to assign a direction – plus or minus – to the magnitude of the

[7] Note that two dipoles shown are identical because if the top molecule is rotated 180°, we obtain the bottom molecule. However, the two quadrupoles are not identical; we cannot obtain one from the other by rotation.

[8] The SI unit for quadrupole moment is C m^2; 1 C m^2 = 0.29979×10^{14} esu = 0.29979×10^{14} erg$^{1/2}$ cm$^{5/2}$. The unit buckingham (B) is also commonly used: 1 B = (1 debye)×(1 Å) = 10^{-26} esu .

quadrupole moment). Flygare (1971) and Sutter and Flygare (1976) have tabulated quadrupole moments for molecules of low symmetry.

Table 4-2 Quadrupole moments for selected molecules.*

Molecule	$Q \times 10^{40}$ (C m^2)
H_2	+2.2
C_2H_2	+10
C_2H_4	+5.0
C_2H_6	-2.2
C_6H_6	+12
N_2	-5.0
O_2	-1.3
N_2O	-10

* Taken from A. D. Buckingham, 1967, *Adv. Chem. Phys.*, 12; and from D. E. Stogryn and A. D. Stogryn, 1966, *Mol. Phys.*, 11: 371.

The potential energy between a quadrupole and a dipole, or between a quadrupole and another quadrupole, is a function of the distance of separation and the angles of mutual orientation. The average potential energy is found by averaging over all orientations; each orientation is weighted according to its Boltzmann factor (Hirschfelder *et al.*, 1964). Upon expanding in powers of $1/kT$, we obtain:

For dipole i-quadrupole j:

$$\bar{\Gamma}_{ij} = -\frac{\mu_i^2 Q_j^2}{(4\pi\varepsilon_0)^2 kT r^8} + \cdots \tag{4-10}$$

For quadrupole i-quadrupole j:

$$\bar{\Gamma}_{ij} = -\frac{7}{40}\frac{Q_i^2 Q_j^2}{(4\pi\varepsilon_0)^2 kT r^{10}} + \cdots \tag{4-11}$$

Whereas the scientific literature on dipole moments is extensive, considerably less is known about quadrupole moments and little work has been done on higher multipoles such as octapoles and hexadecapoles. The effect of quadrupole moments on thermodynamic properties is already much less than that of dipole moments and the effect of higher multipoles is usually negligible (Parsonage and Scott, 1962). This relative rank of importance follows because intermolecular forces due to multipoles higher than dipoles are extremely short range; for dipoles, the average potential energy is proportional to the sixth power of the inverse distance of separation and for quadrupoles the average potential energy depends on the tenth power of the reciprocal distance. For higher multipoles, the exponent is larger.

4.3 Polarizability and Induced Dipoles

A nonpolar molecule such as argon or methane has no permanent dipole moment but when such a molecule is subjected to an electric field, the electrons are displaced from their ordinary positions and a dipole is induced. In fields of moderate strength, the induced dipole moment μ^i is proportional to the field strength E:

$$\mu^i = \alpha E \tag{4-12}$$

where the proportionality factor α is a fundamental property of the substance. It is called the *polarizability*, that is, how easily the molecule's electrons can be displaced by an electric field. Polarizability can be calculated in several ways, most notably from dielectric properties and from index-of-refraction data. For asymmetric molecules, polarizability is not a constant but a function of the molecule's orientation relative to the direction of the field. Table 4-3 gives average polarizabilities of some representative molecules. In SI units (electric field in V m^{-1} and μ in C m), polarizability has the units C^2 J^{-1} m^2. However, it is common practice to present polarizabilities in units of volume[9] (such as cm^3), using the relation

$$\alpha' = \frac{\alpha}{4\pi\varepsilon_0}$$

The units of $4\pi\varepsilon_0$ are C^2 J^{-1} m^{-1}, and therefore α' (also called the *polarizability volume*) has dimensions of volume.

When a nonpolar molecule i is situated in an electric field set up by the presence of a nearby polar molecule j, the resultant force between the permanent dipole and the induced dipole is always attractive. The mean potential energy was first calculated by Debye and is usually associated with his name. It is given by

$$\overline{\Gamma}_{ij} = -\frac{\alpha_i \mu_j^2}{(4\pi\varepsilon_0)^2 r^6} \tag{4-13}$$

[9] $1\ C^2\ J^{-1}\ m^2 = 0.8988\times10^{16}\ cm^3$.

Table 4-3 Average polarizabilities.*

Molecule	$\alpha' \times 10^{24}$ (cm^3)	Molecule	$\alpha' \times 10^{24}$ (cm^3)
H_2	0.81	SO_2	3.89
H_2O	1.48	Xe	4.02
Ar	1.64	C_2H_6	4.50
N_2	1.74	Cl_2	4.61
CO	1.95	$(CH_3)_2O$	5.22
NH_3	2.22	HI	5.44
HCl	2.60	$(CH_3)_2CO$	6.42
CH_4	2.60	$CHCl_3$	8.50
CO_2	2.64	CCl_4	10.5
CH_3OH	3.23	C_6H_6	10.6
C_2H_2	3.36	Naphthalene	24.0
HBr	3.61	Anthracene	35.2

* C. G. Gray and K. E. Gubbins, 1984, *Theory of Molecular Fluids*, Vol. 1, Oxford: Clarendon Press.

Polar as well nonpolar molecules can have dipoles induced in an electric field. The general Debye formula, therefore, for the mean potential energy due to induction by permanent dipoles, is

$$\bar{\Gamma}_{ij} = -\frac{(\alpha_i \mu_j^2 + \alpha_j \mu_i^2)}{(4\pi\varepsilon_0)^2 r^6} \tag{4-14}$$

An electric field may also be caused by a permanent quadrupole moment. In that case, the average potential energy of induction between a quadrupole j and a nonpolar molecule i is again attractive; if both molecules i and j have permanent quadrupole moments,

$$\bar{\Gamma}_{ij} = -\frac{3}{2}\frac{(\alpha_i Q_j^2 + \alpha_j Q_i^2)}{(4\pi\varepsilon_0)^2 r^8} \tag{4-15}$$

For molecules with a permanent dipole moment, the potential energy due to induction is usually small when compared to the potential energy due to permanent dipoles; and similarly, for molecules with a permanent quadrupole moment, the induction energy is usually less than that due to quadrupole-quadrupole interactions.

4.4 Intermolecular Forces between Nonpolar Molecules

The concept of polarity has been known for a long time but until about 1930 there was no adequate explanation for the forces acting between nonpolar molecules. It was very puzzling, for example, why such an obviously nonpolar molecule as argon should nevertheless show serious deviations from the ideal-gas laws at moderate pressure. In 1930 it was shown by London that so-called nonpolar molecules are, in fact, nonpolar only when viewed over a period of time; if an instantaneous photograph of such a molecule were taken, it would show that, at a given instant, the oscillations of the electrons about the nucleus had resulted in distortion of the electron arrangement sufficient to cause a temporary dipole moment. This dipole moment, rapidly changing its magnitude and direction, averages zero over a short period of time; however, these quickly varying dipoles produce an electric field which then induces dipoles in the surrounding molecules. The result of this induction is an attractive force called the *induced dipole-induced dipole force*. Using quantum mechanics, London showed that, subject to certain simplifying assumptions, the potential energy between two simple, spherically symmetric molecules i and j at large distances is given by

$$\Gamma_{ij} = -\frac{3}{2}\frac{\alpha_i \alpha_j}{(4\pi\varepsilon_o)^2 r^6}\left(\frac{h\nu_{0i}\, h\nu_{0j}}{h\nu_{0i} + h\nu_{0j}}\right) \qquad \text{(4-16)[10]}$$

where h is Planck's constant and ν_0 is a characteristic electronic frequency for each molecule in its unexcited state. This frequency is related to the variation of the index of refraction n with light frequency ν by

$$n - 1 = \frac{c}{\nu_0^2 - \nu^2} \qquad \text{(4-17)}$$

where c is a constant. It is this relationship between index of refraction and characteristic frequency that is responsible for the name *dispersion* for the attractive force between nonpolar molecules.

For a molecule i, the product $h\nu_{0i}$ is very nearly equal to its first ionization potential I_i.[11] Equation (4-16) is therefore usually written in the form

---

[10] When molecules i and j are in a medium whose dielectric constant is ε, it is tempting to use Eq. (4-16) with ε_0 replaced by ε. However, it is incorrect to do so. The dielectric constant ε can be used in potentials arising from electrostatics (as in the previous section) but not for a potential arising from interactions between fluctuating induced dipoles. Equation (4-16) is the first term of a series in $(1/r)$. The next term is proportional to $(1/r)^8$. Therefore, Eq. (4-16) is not valid for very small r; it is inappropriate especially when $r \leq \sigma$, where σ is the molecular diameter.

$$\Gamma_{ij} = -\frac{3}{2}\frac{\alpha_i\alpha_j}{(4\pi\varepsilon_o)^2 r^6}\left(\frac{I_i I_j}{I_i + I_j}\right) \tag{4-18}$$

If molecules i and j are of the same species, Eq. (4-18) reduces to

$$\Gamma_{ii} = -\frac{3}{4}\frac{\alpha_i^2 I_i}{(4\pi\varepsilon_o)^2 r^6} \tag{4-19}$$

Equations (4-18) and (4-19) give the important result that the potential energy between nonpolar molecules is independent of temperature and varies inversely as the sixth power of the distance between them. The attractive force therefore varies as the reciprocal seventh power. This sharp decline in attractive force as distance increases explains why it is much easier to melt or vaporize a nonpolar substance than an ionic one where the dominant attractive force varies as the reciprocal second power of the distance of separation.[12]

London's formula is more sensitive to the polarizability than it is to the ionization potential, because, for typical molecules, α is (roughly) proportional to molecular size while I does not change much from one molecule to another. Table 4-4 gives a representative list of ionization potentials. Since the polarizabilities dominate, it can be shown that the attractive potential between two dissimilar molecules is approximately given by the geometric mean of the potentials between the like molecules at the same separation. We can rewrite Eqs. (4-18) and (4-19):

$$\Gamma_{ij} = k'\frac{\alpha_i\alpha_j}{r^6} \qquad \Gamma_{ii} = k'\frac{\alpha_i^2}{r^6} \qquad \Gamma_{jj} = k'\frac{\alpha_j^2}{r^6} \tag{4-20}$$

where k' is a constant that is approximately the same for the three types of interaction: i-i, j-j, and i-j. It then follows that

$$\Gamma_{ij} = (\Gamma_{ii}\Gamma_{jj})^{1/2} \tag{4-21}$$

[11] The first ionization potential is the work which must be done to remove one electron from an uncharged molecule M: $M \rightarrow e^- + M^+$. The second ionization potential is the work needed to remove the second electron according to: $M^+ \rightarrow e^- + M^{2+}$.

[12] Various authors, in addition to London, have derived expressions for the attractive portion of the potential function of two, spherically symmetric, nonpolar molecules. These expressions all agree on a distance dependence of r^{-6} but the coefficients differ considerably. A good review of this subject is presented in *Advances in Chemical Physics*, J. O. Hirschfelder, (Ed.), 1967, Vol. 12, New York: Wiley-Interscience. See also Stone (1996).

Table 4-4 First ionization potentials.*

Molecule	I (eV)[†]	Molecule	I (eV)[†]
1,3,5-$C_6H_3(CH_3)_3$	8.4	CCl_4	11.0
p-$C_6H_4(CH_3)_2$	8.5	C_3H_8	11.2
$C_6H_5CH(CH_3)_2$	8.7	C_2H_2	11.4
$C_6H_5CH_3$	8.9	$CHCl_3$	11.5
c-C_6H_{10}	9.0	NH_3	11.5
C_6H_6	9.2	H_2O	12.6
n-C_7H_{14}	9.5	HCl	12.8
C_5H_5N	9.8	CH_4	13.0
$(CH_3)_2CO$	10.1	Cl_2	13.2
$(C_2H_5)_2O$	10.2	CO_2	13.7
n-C_7H_{16}	10.4	CO	14.1
C_2H_4	10.5	H_2	15.4
C_2H_5OH	10.7	CF_4	17.8
C_2H_5Cl	10.8	He	24.5

* Taken from a more complete list given in Landolt-Börnstein, 1951, *Zahlenwerte und Funktionen*, 6th Ed., Vol. 1, Part 3, Berlin: Springer; and from W. A. Duncan, J. P. Sheridan, and F. L. Swinton, 1966, *Trans. Faraday Soc.*, 62: 1090.

[†] 1 eV = 1.60218×10^{-19} J.

Equation (4-21) gives some theoretical basis for the frequently applied *geometric-mean rule*, which is so often used in equations of state for gas mixtures and in theories of liquid solutions. We have already used this rule in Sec. 3.5 [compare Eq. (3-69)].

To show the relative magnitude of dipole, induction, and dispersion forces in some representative cases, London (1937) has presented calculated potential energies for a few simple molecules. His results are given in the form

$$\Gamma_{ii} = -\frac{B}{r^6} \tag{4-22}$$

where B is calculated separately for each contribution due to dipole, induction, and dispersion effects. In these calculations Eqs. (4-8), (4-14), and (4-19) were used; Table 4-5 gives some results similar to those given by London. The computed values of B indicate that the contribution of induction forces is small and that even for strongly polar substances, like ammonia, water, or acetone, the contribution of dispersion forces is far from negligible.

Table 4-5 Relative magnitudes of intermolecular forces between two identical molecules at 0°C.

Molecule	Dipole moment (debye)	$B \times 10^{79}$ (J m^6)		
		Dipole	Induction	Dispersion
CH_4	0	0	0	102
CCl_4	0	0	0	1460
$c\text{-}C_6H_{12}$	0	0	0	1560
CO	0.10	0.0018	0.0390	64.3
HI	0.42	0.550	1.92	380
HBr	0.80	7.24	4.62	188
HCl	1.08	24.1	6.14	107
NH_3	1.47	82.6	9.77	70.5
H_2O	1.84	203	10.8	38.1
$(CH_3)_2CO$	2.87	1200	104	486

Table 4-6 gives some calculated results for intermolecular forces between two molecules that are not alike. In these calculations Eqs. (4-8), (4-14), and (4-18) were used. Again, we notice that polar forces are not important when the dipole moment is less than about 1 debye and induction forces always tend to be much smaller than dispersion forces.

Table 4-6 Relative magnitudes of intermolecular forces between two different molecules at 0°C.

Molecules		Dipole moment (debye)		$B \times 10^{79}$ (J m^6)		
(1)	(2)	(1)	(2)	Dipole	Induction	Dispersion
CCl_4	$c\text{-}C_6H_{12}$	0	0	0	0	1510
CCl_4	NH_3	0	1.47	0	22.7	320
$(CH_3)_2CO$	$c\text{-}C_6H_{12}$	2.87	0	0	89.5	870
CO	HCl	0.10	1.08	0.206	2.30	82.7
H_2O	HCl	1.84	1.08	69.8	10.8	63.7
$(CH_3)_2CO$	NH_3	2.87	1.47	315	32.3	185
$(CH_3)_2CO$	H_2O	2.87	1.84	493	34.5	135

London's formula does not hold at very small separations where the electron clouds overlap and the forces between molecules are repulsive rather than attractive. Repulsive forces between nonpolar molecules at small distances are not understood as well as attractive forces at larger distances. Theoretical considerations suggest that the repulsive potential should be an exponential function of intermolecular separation, but it is more convenient (Amdur *et al.*, 1954) to represent the repulsive potential by an inverse-power law of the type

$$\Gamma = \frac{A}{r^n} \tag{4-23}$$

where A is a positive constant and n is a number usually taken to be between 8 and 16.

To take into account both repulsive and attractive forces between nonpolar molecules, it is customary to assume that the total potential energy is the sum of the two separate potentials:

$$\boxed{\Gamma_{\text{total}} = \Gamma_{\text{repulsive}} + \Gamma_{\text{attractive}} = \frac{A}{r^n} - \frac{B}{r^m}} \tag{4-24}$$

where A, B, n, and m are positive constants and where $n > m$. This equation was first proposed by Mie (1903) and was extensively investigated by Lennard-Jones. Equation (4-24) forms the basis of a variety of physicochemical calculations; it has been used especially to calculate thermodynamic and transport properties of dilute nonpolar gases (Hirschfelder *et al.*, 1964).

4.5 Mie's Potential-Energy Function for Nonpolar Molecules

Equation (4-24) gives the potential energy of two molecules as a function of their separation and it is apparent that at some distance r_{min}, Γ is a minimum; this minimum energy is designated by Γ_{min}. By algebraic rearrangement, *Mie's potential* can be rewritten

$$\Gamma = \frac{\varepsilon\left(n^n / m^m\right)^{1/(n-m)}}{n-m}\left[\left(\frac{\sigma}{r}\right)^n - \left(\frac{\sigma}{r}\right)^m\right] \qquad (4\text{-}25)^{13}$$

where $\varepsilon = -\Gamma_{\text{min}}$ and where σ is the intermolecular distance when $\Gamma = 0$.

London has shown from the theory of dispersion forces that $m = 6$ but we do not have a theoretical value for n. It is frequently convenient for calculation to let $n = 12$, in that case Eq. (4-25) becomes

[13] Parameter ε in Mie's equation should not be confused with dielectric constant ε.

$$\boxed{\Gamma = 4\varepsilon\left[\left(\frac{\sigma}{r}\right)^{12} - \left(\frac{\sigma}{r}\right)^{6}\right]} \tag{4-26}$$

Equation (4-26) is the *Lennard-Jones potential.*[14] It relates the potential energy of two molecules to their distance of separation in terms of two parameters: an energy parameter ε that, when multiplied by -1, gives the minimum energy corresponding to the equilibrium separation; and a distance parameter σ that is equal to the intermolecular separation when the potential energy is zero. An illustration of Eqs. (4-26) and (4-26a) is given in Fig. 4-2.

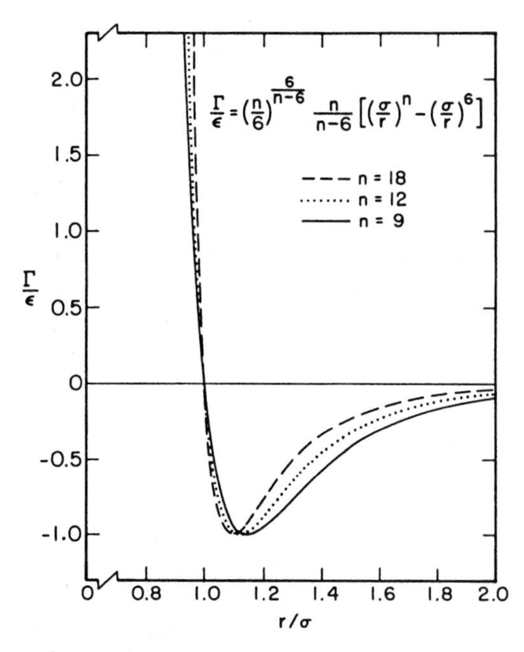

Figure 4-2 Three forms of Mie's potential for simple, nonpolar molecules.

Because of the steepness of the repulsion potential, the numerical values of r_{min} and σ are not far apart. For a Mie (*n*, 6) potential, we obtain

---

[14] Better agreement with experiment is often obtained by letting *n* be an adjustable parameter. In that case Eq. (4-25) becomes (Lichtenthaler and Schäfer, 1969):

$$\Gamma = \left(\frac{n}{6}\right)^{6/(n-6)} \frac{n}{n-6}\,\varepsilon\left[\left(\frac{\sigma}{r}\right)^{n} - \left(\frac{\sigma}{r}\right)^{6}\right] \tag{4-26a}$$

$$\sigma = \left(\frac{6}{n}\right)^{1/(n-6)} r_{\text{min}} \tag{4-27}$$

Constants ε, σ, and n can be estimated from a variety of physical properties as well as from spectroscopic and molecular-beam experiments.[15] Subject to several simplifying assumptions, they can be computed, for example, from the compressibility of solids at low temperatures or from specific-heat data of solids or liquids. More commonly, they are obtained from the variation of viscosity or self-diffusivity with temperature at low pressures, and, most commonly, from gas-phase volumetric properties as expressed by second virial coefficients (Moelwyn-Hughes, 1961).

Mie's potential applies to two nonpolar, spherically symmetric molecules that are completely isolated. In nondilute systems, and especially in condensed phases, two molecules are not isolated but have many other molecules in their vicinity. By introducing appropriate simplifying assumptions, it is possible to construct a simple theory of dense media using a form of Mie's two-body potential such as that of Lennard-Jones (Moelwyn-Hughes, 1961).

Consider a condensed system at conditions not far removed from those prevailing at the triple point. We assume that the total potential energy is due primarily to interactions between nearest neighbors. Let the number of nearest neighbors in a molecular arrangement be designated by z. In a system containing N molecules, the total potential energy Γ_t is then approximately given by

$$\Gamma_t = \frac{1}{2} Nz\Gamma \tag{4-28}$$

where Γ is the potential energy of an isolated pair. Factor 1/2 is needed to avoid counting each pair twice. Substituting Mie's equation into Eq. (4-28), we have

$$\Gamma_t = \frac{1}{2} Nz\left(\frac{A}{r^n} - \frac{B}{r^m}\right) \tag{4-29}$$ [16]

where r is the distance between two adjacent molecules.

Equation (4-29) considers only interactions between nearest-neighbor molecules. To account for additional potential energy resulting from interaction of a molecule with all of those outside its nearest-neighbor shell, numerical constants s_n and s_m (that are near unity) are introduced by rewriting the total potential energy:

---

[15] See G. Maitland, M. Rigby, and W. Wakeham, 1981, *Intermolecular Forces: Their Origin and Determination*, Oxford: Oxford University Press.

[16] Even if we neglect the effect of nonnearest neighbors, Eq. (4-28) is not exact because it assumes additivity of two-body potentials to give the potential energy of a multibody system

$$\Gamma_t = \frac{1}{2} N z \left(\frac{s_n A}{r^n} - \frac{s_m B}{r^m} \right) \tag{4-30}$$

When the condensed system is considered as a lattice such as that existing in a regularly spaced crystal, the constants s_n and s_m can be accurately determined from the lattice geometry. For example, a molecule in a crystal of the simple-cubic type has 6 nearest neighbors at a distance r, 12 at a distance $r\sqrt{2}$, 8 at a distance $r\sqrt{3}$, etc. The attractive energy of one molecule with respect to all of the others is then given by

$$\begin{aligned} \Gamma_{\text{attractive}} &= B\left[\frac{6}{r^m} + \frac{12}{\left(\sqrt{2}\,r\right)^m} + \frac{8}{\left(\sqrt{3}\,r\right)^m} + \ldots\right] \\ &= \frac{6B}{r^m}\left[1 + \frac{2}{\left(\sqrt{2}\right)^m} + \frac{4}{\left(\sqrt{3}\right)^{m+2}} + \ldots\right] \\ &= \frac{zBs_m}{r^m} \end{aligned} \tag{4-31}$$

and

$$s_m = 1 + \frac{2}{\left(\sqrt{2}\right)^m} + \frac{4}{\left(\sqrt{3}\right)^{m+2}} + \ldots \tag{4-32}$$

Similarly, s_n can be calculated for the repulsive potential. Table 4-7 shows summation constants for several geometrical arrangements.

Table 4-7 Summation constants s_n and s_m for cubic lattices (Moelwyn-Hughes, 1961).

n or m	Simple cubic $z = 6$	Body-centered cubic $z = 8$	Face-centered cubic $z = 12$
6	1.4003	1.5317	1.2045
9	1.1048	1.2368	1.0410
12	1.0337	1.1394	1.0110
15	1.0115	1.0854	1.0033

Having obtained numerical values for constants s_n and s_m, it is now possible to obtain a relation between the equilibrium distance of separation r_{min} of an isolated pair of molecules and the equilibrium distance of separation r_{min_t} between a molecule and its nearest neighbors in a condensed system. At equilibrium, the potential energy of the condensed system is a minimum; therefore,

$$\left(\frac{d\Gamma_t}{dr}\right)_{r=r_{\min_t}} = 0 \tag{4-33}$$

From Eq. (4-30) we obtain

$$(r_{\min_t})^{n-m} = \frac{s_n nA}{s_m mB} \tag{4-34}$$

and comparing this result with that obtained for an isolated pair of molecules, we have

$$\left(\frac{r_{\min}}{r_{\min_t}}\right)^{n-m} = \frac{s_m}{s_n} \tag{4-35}$$

Since m is equal to 6, and assuming that n is between 8 and 16, the values in Table 4-7 show that the equilibrium distance in an isolated pair is always a few percent larger than that in a condensed system. This leads to the interesting result that for a pair of adjacent molecules in a condensed system, the absolute value of the average potential energy is roughly 50% smaller than that corresponding to the equilibrium separation between a pair of isolated molecules.

Equation (4-35) can also be used to estimate $r_{\min}$ for an isolated pair of molecules from data for $r_{\min t}$ of the condensed system. At low temperatures, the condensed system is a crystal and the average distance between two neighboring molecules is approximately $r_{\min t}$. This approximation becomes increasingly better as the temperature falls and becomes exact at T = 0 K. Therefore, $r_{\min t}$ can be obtained from low-temperature molar-volume data if the lattice geometry is known. Another possibility is to determine $r_{\min t}$ directly from low-temperature X-ray scattering data.

By algebraic rearrangement of Eq. (4-30) it is possible to obtain a relation between the equilibrium potential energy ε of an isolated pair of molecules and the equilibrium potential energy $\Gamma_t(r = r_{\min t})$ of the condensed system:

$$\varepsilon = -\frac{1}{N}\Gamma_t(r = r_{\min_t})\frac{2}{zs_m}\left(\frac{s_n}{s_m}\right)^{m/(n-m)} \tag{4-36}$$

where $\Gamma_t(r = r_{\min t})$, the lattice energy at T = 0 K, is given by

$$\Gamma_t(r = r_{\min_t}) = \Delta_{\text{sub}}h_0 + \frac{9}{8}R\theta_D \tag{4-37}$$

In this equation, $\Delta_{sub}h_0$ is the enthalpy of sublimation at 0 K and the second term is the zero-point energy. The Debye temperature θ_D is obtained from the temperature dependence of the specific heat that, at very low temperatures, follows Debye's T law. Although an appreciable number of $\Delta_{sub}h_0$ values are tabulated (D'Ans-Lax, 1967), enthalpy-of-sublimation data are available only at temperatures above zero. However, $\Delta_{sub}h_0$ can be calculated from these data in connection with the temperature dependence of the specific heat. In this way $\Gamma_t(r = r_{min_t})$ can be estimated and Eq. (4-36) then yields an estimated value for ε (Mie, 1903). Since m is equal to 6, n must be estimated from at least one other physical property.

Mie's potential, as well as other similar potentials for nonpolar, spherically symmetrical molecules, contains one independent variable r and one dependent variable Γ. When these variables are nondimensionalized with characteristic molecular constants, the resulting potential function leads to a useful generalization known as the *molecular theory of corresponding states*, as discussed in Sec. 4.12.

4.6 Structural Effects

Intermolecular forces of nonspherical molecules depend not only on the center-to-center distance but also on the relative orientation of the molecules. The effect of molecular shape is most significant at low temperatures and when the intermolecular distances are small, i.e., especially in the condensed state. For example, there are significant differences among the boiling points of isomeric alkanes that have the same carbon number; a branched isomer has a lower boiling point than a straight chain, and the more numerous the branches, the lower the boiling point. To illustrate, Fig. 4-3 shows results for some pentanes and hexanes.

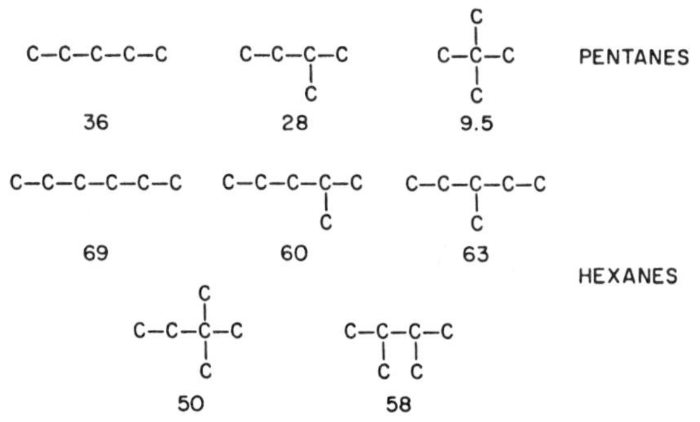

```
                                                C
                                                |
C—C—C—C—C          C—C—C—C                    C—C—C          PENTANES
                         |                      |
                         C                      C
    36                  28                     9.5

C—C—C—C—C—C      C—C—C—C—C        C—C—C—C—C
                         |              |
                         C              C
    69                  60             63
                                                             HEXANES
              C
              |
        C—C—C—C              C—C—C—C
              |                |  |
              C                C  C
             50                 58
```

Figure 4-3 Boiling points (in °C) of some alkane isomers.

A similar effect of branching on boiling point is observed within many families of organic compounds. It is reasonable that branching should lower the boiling point; with branching, the shape of a molecule tends to approach that of a sphere and the surface area per molecule decreases. Therefore, intermolecular attraction per pair of molecules becomes weaker and a lower kinetic energy kT is sufficient to overcome that attraction.

Boiling-point differences for isomers could also be due to different interactions between methyl and methylene groups. However, to explain differences in the boiling points of linear and branched alkanes only in terms of the methyl-methylene interaction would require an unreasonably large force-field difference between methyl and methylene groups. Differences in molecular shape provide a more likely explanation.

Similarly, other thermodynamic properties are affected by branching. Spectroscopic results (Te Lam *et al.*, 1974; Delmas and Purves, 1977; Tancrède *et al.*, 1977; Heintz and Lichtenthaler, 1977, 1984), for example, show the presence of orientational order among long chains of pure *n*-alkanes, which does not exist among more nearly spherical branched alkanes. Short-range orientational order in systems containing anisotropic molecules can be detected by studying the thermodynamic properties of a solution wherein the substance with orientational order is dissolved in a relatively inert solvent with little or no orientational order. Mixing liquids of different degrees of order usually brings about a net decrease of order, and hence positive contributions to the enthalpy $\Delta_{mix}h$ and entropy $\Delta_{mix}s$ of mixing. This effect is illustrated in Fig. 4-4, that shows the composition dependence of $\Delta_{mix}h$ for linear and branched decane mixed with cyclohexane at 25 and 40°C. The difference in $\Delta_{mix}h$ for the two isomers is surprisingly large. At mole fraction $x = 0.5$, $\Delta_{mix}h$ for the binary containing *n*-decane is nearly twice that for the binary containing isodecane. Further, the temperature dependence of $\Delta_{mix}h$ for the *n*-decane system is much stronger than that for the isodecane system. Upon mixing with cyclohexane (a globular molecule), short-range orientational order is destroyed and hence the mixing process requires more energy with the linear alkane than with the branched alkane. At higher temperatures, orientational order in the pure *n*-alkanes is already partially destroyed by thermal motion of the molecules; mixing isothermically at higher temperature requires less energy than at lower temperature. The branched alkanes show only little orientational order and therefore the temperature dependence of $\Delta_{mix}h$ for branched alkane/cyclohexane mixtures is weak.

4.7 Specific (Chemical) Forces

In addition to physical intermolecular forces briefly described in previous sections, there are specific forces of attraction which lead to the formation of new molecular species; such forces are called *chemical forces*. A good example of such a force is that between ammonia and hydrogen chloride; in this case, a new species, ammonium chloride, is formed. Such forces, in effect, constitute the basis of the entire science of chemistry and it is impossible to discuss them adequately in a few pages. However, it is important to recognize that chemical forces can, in many cases, be of major impor-

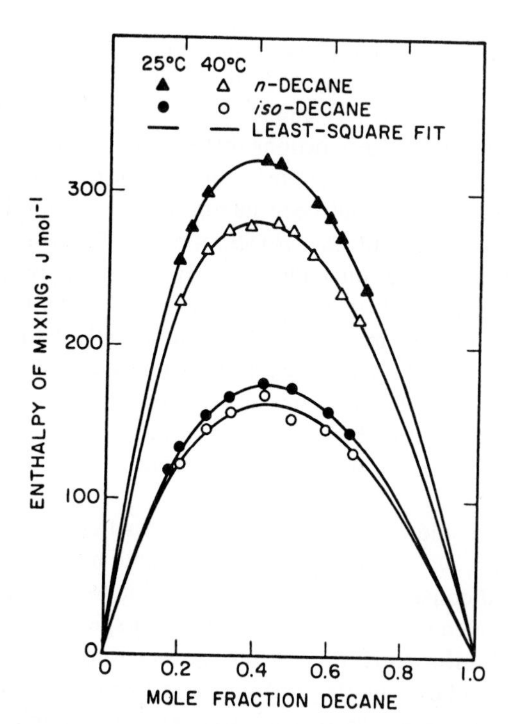

Figure 4-4 Effect of molecular structure on enthalpy of mixing: cyclohexane/*n*-decane and cyclohexane/isodecane (2,6-dimethyl octane).

tance in determining thermodynamic properties of solutions. Whereas in the previous sections we were able to write some simple formulas for the potential energies of physically interacting molecules, we cannot give simple quantitative relations that describe on a microscopic level interaction between chemically reactive molecules. Instead, we briefly discuss in a qualitative manner some relations between chemical forces and properties of solutions.

Numerous types of specific chemical effects are of importance in the thermodynamics of solutions. For example, the solubility of silver chloride in water is very small; however, if some ammonia is added to the solution, the solubility rises by several orders of magnitude due to the formation of a silver/ammonia complex. Acetone is a much better solvent for acetylene than for ethylene because acetylene, unlike ethylene, can form a hydrogen bond with the carbonyl oxygen in the solvent. Because of an electron donor-electron acceptor interaction, iodine is more soluble in aromatic solvents like toluene and xylene than in paraffinic solvents like heptane or octane. Finally, an example is provided by a well-known industrial process for the absorption of carbon dioxide in ethanol amine; carbon dioxide is readily soluble in this solvent because of specific chemical interaction between (acidic) solute and (basic) solvent.

The main difference between a physical and a chemical force lies in the criterion of saturation: Chemical forces are saturated but physical forces are not. The saturated nature of chemical forces is intimately connected with the theory of the covalent bond and also with the law of multiple proportions that says the ratio of atoms in a molecule is a small, integral number. If two hydrogen atoms meet, they have a strong tendency to form a hydrogen molecule H_2, but once having done so, they have no appreciable further tendency to form a molecule H_3. Hence, the attractive force between hydrogen atoms is "satisfied" (or saturated) once the stable H_2 molecule is formed. On the other hand, the purely physical force between, say, two argon atoms knows no such "satisfaction." Two argon atoms that are attracted to form a doublet still have a tendency to attract a third argon atom, and a triplet has a further tendency to attract a fourth. It is true that in the gaseous state, doublets are much more frequent than triplets but that is because in the dilute state, a two-body collision is much more probable than a three-body collision. In the condensed or highly concentrated state, there are aggregates of many argon atoms.

Chemical effects in solution are conveniently classified in terms of *association* or *solvation*. By the former, we mean the tendency of some molecules to form polymers; for example, acetic acid consists primarily of dimers due to hydrogen bonding. By solvation, we mean the tendency of molecules of different species to form complexes; for example, a solution of sulfur trioxide in water shows strong solvation by formation of sulfuric acid. This particular example illustrates a severe degree of solvation but there are many cases where the solvation is much weaker; for example, there is a tendency for chloroform to solvate with acetone due to hydrogen bonding between the primary hydrogen of the chloroform and the carbonyl oxygen of the acetone; this tendency has a profound effect on the properties of chloroform/acetone solutions. Chloroform also forms hydrogen bonds with diisobutyl ketone, but in this case the extent of complexing is much smaller because of steric hindrance and, as a result, mixtures of chloroform with diisobutyl ketone behave more ideally than do mixtures of chloroform with acetone. Solvation effects in solution are very common. When these effects are strong, they often produce negative deviations from Raoult's law since they necessarily decrease the volatilities of the original components.

It is easy to see that whenever solvation occurs in solution it has a marked effect on the thermodynamic properties of that solution. It is, perhaps, not as obvious that association effects are also of major importance. The reason is that the extent of association is a strong function of the composition, especially in the range dilute with respect to the associating component. Pure methanol, for example, exists primarily as dimer, trimer and tetramer, but when methanol is dissolved in a large excess of hexane, it exists primarily as a monomer. As the methanol concentration rises, more polymers are formed; the fraction of methanol molecules that exists in the associated form is strongly dependent on the number of methanol molecules present per unit volume of solution; as a result, the fugacity of methanol is a highly nonlinear function of its mole fraction.

The ability of a molecule to solvate or associate is closely related to its electronic structure. For example, if we want to compare the properties of aluminum trichloride and antimony trichloride, we note immediately an important difference in their electronic structures:

```
       ..                    ..
      :Cl:                  :Cl:
   ..  ..                ..  ..
  :Cl:Al                :Cl:Sb:
   ..  ..                ..  ..
      :Cl:                  :Cl:
       ..                    ..
```

In the trichloride, antimony has its octet of electrons, and therefore its chemical forces are saturated. Aluminum, however, has only six electrons and has a strong tendency to add two more. Consequently, aluminum trichloride solvates easily with any molecule that can act as an electron donor whereas antimony trichloride does not. This difference explains, at least in part, why aluminum trichloride, unlike antimony trichloride, is an excellent catalyst for some organic reactions, for example, the Friedel-Crafts reaction.

4.8 Hydrogen Bonds

The most common chemical effect encountered in the thermodynamics of solutions is that due to the *hydrogen bond*. While the "normal" valence of hydrogen is unity, many hydrogen-containing compounds behave as if hydrogen were bivalent. Studies of hydrogen fluoride vapor, for example, show that the correct formula is $(HF)_n$, where n depends on temperature and pressure, and may be as much as 6. The only reasonable explanation this is to write the structure of hydrogen fluoride in this manner:

--H—F----H—F----H—F--

where the solid line indicates the "normal" bond and the dashed line, an "auxiliary" bond. Similarly, studies on the crystal structure of ice show that each hydrogen is "normally" bonded to one oxygen atom and additionally attached to another oxygen atom:

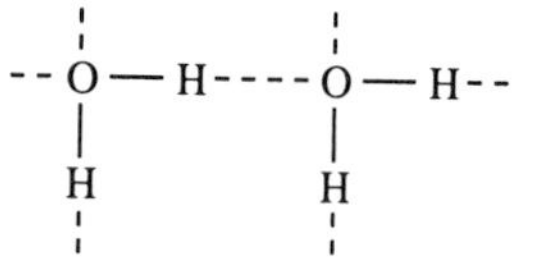

```
      ¦         ¦
   -- O — H---- O — H--
      |         |
      H         H
      ¦         ¦
```

It appears that two sufficiently negative atoms X and Y (that may be identical) may, in suitable circumstances, be united with hydrogen according to X—H---Y. Consequently, molecules containing hydrogen linked to an electronegative atom (as in acids, alcohols, and amines) show strong tendencies to associate with each other and to solvate with other molecules possessing accessible electronegative atoms.

The major difference between a hydrogen bond and a normal covalent bond is the former's relative weakness. The bond strength of most hydrogen bonds lies in the neighborhood 8 to 40 kJ mol^{-1}, whereas the usual covalent bond strength is in the region 200 to 400 kJ mol^{-1}. Therefore, the hydrogen bond is broken rather easily and it is for this reason that hydrogen-bonding effects usually decrease at higher temperatures where the kinetic energy of the molecules is sufficient to break these loose bonds.

Before we discuss the effect of hydrogen bonding on physical properties, we review a few characteristic properties of hydrogen bonds that have been observed experimentally (compare Fig. 4-5):

I. Distances between the neighboring atoms of the two functional groups (X—H---Y) in hydrogen bonds are substantially smaller than the sum of their van der Waals radii.

II. X—H stretching modes are shifted toward lower frequencies (lower wave numbers) upon hydrogen-bond formation.

III. Polarities of X—H bonds increase upon hydrogen-bond formation, often leading to complexes whose dipole moments are larger than those expected from vectorial addition.

IV. Nuclear-magnetic-resonance (NMR) chemical shifts of protons in hydrogen bonds are substantially smaller than those observed in the corresponding isolated molecules. The observed deshielding follows from reduced electron densities at protons participating in hydrogen bonding.

All effects (I to IV) are less pronounced in isolated hydrogen-bonded dimers than in hydrogen-bonded liquids or crystals, indicating the importance of long-range interactions through chains or networks of hydrogen bonds.

The thermodynamic constants for hydrogen-bonding reactions are generally dependent on the medium in which they occur (for a review see Christian and Lane, 1975). Tucker and Christian (1976), for example, report data for the 1:1 hydrogen-bonded complex of trifluoroethanol (TFE) with acetone in the vapor phase and in CCl_4 solution. To relate the vapor-phase results to the liquid-phase results, Tucker and Christian constructed the thermodynamic cycle shown in Fig. 4-6 for the TFE/acetone complex, including values of energy and Gibbs energy for both the vertical transfer reactions and the horizontal complex-formation reactions. Figure 4-6 shows that the transfer energies and Gibbs energies for each of the individual components are not small compared with the values for the association reaction. The energy of transfer of the complex into CCl_4 is 83% of that of the separated monomers; the corresponding Gibbs energy of transfer is 79% that of the monomers. The results indicate that the transfer energy and Gibbs energy of the complex are not even approximately canceled by the transfer energies and Gibbs energies of the constituent molecules. The complex

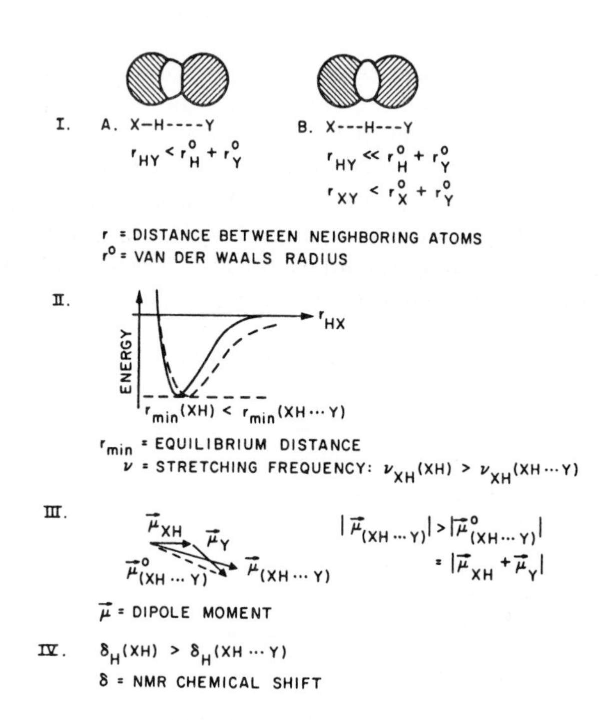

Figure 4-5 Characteristic properties of hydrogen-bonded systems: (I) intermolecular geometries: (A) ordinary, (B) strong hydrogen bonds; (II) vibrational spectra, XH stretching frequencies; (III) increase in polarity on complex formation; (IV) NMR deshielding effect observed at protons participating in hydrogen bonds. (P. Schuster, 1978, *The Fine Structure of the Hydrogen Bond in Intermolecular Interactions from Diatomics to Biopolymers*, B. Pullman, Ed., New York: Wiley & Sons).

is less stable in the relatively inert solvent than in the vapor phase, indicating an important solvent effect on hydrogen-bond formation. For most hydrogen-bonded complexes, stabilities decrease as the solvent changes from aliphatic hydrocarbon to chlorinated (or aromatic) hydrocarbon, to highly polar liquid.

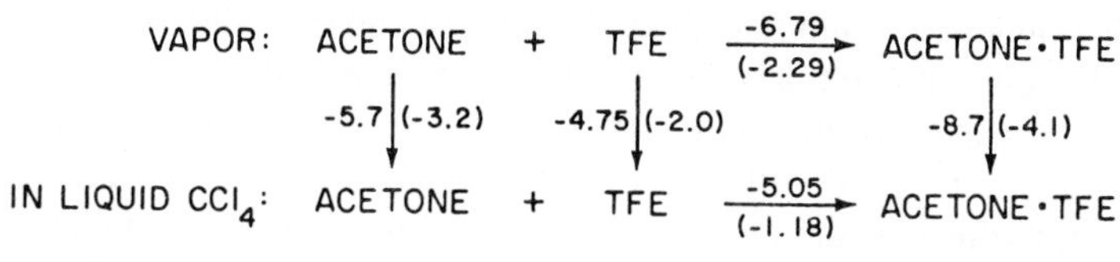

$$
\begin{array}{llcccc}
\text{VAPOR:} & \text{ACETONE} & + & \text{TFE} & \xrightarrow[(-2.29)]{-6.79} & \text{ACETONE·TFE} \\
 & \downarrow -5.7\ (-3.2) & & \downarrow -4.75\ (-2.0) & & \downarrow -8.7\ (-4.1) \\
\text{IN LIQUID } CCl_4\text{:} & \text{ACETONE} & + & \text{TFE} & \xrightarrow[(-1.18)]{-5.05} & \text{ACETONE·TFE}
\end{array}
$$

Figure 4-6 Thermodynamic data for complex-formation reactions and for transfer reactions: Trifluoroethanol (TFE)/acetone system at 25°C. Numbers in parentheses are Δg^0 (kJ mol^{-1}) based on the unit molarity, ideal-dilute-solution standard state. Remaining quantities are standard energies, Δu^0 (kJ mol^{-1}), for the various steps.

The strong effect of hydrogen bonding on physical properties is best illustrated by comparing some thermodynamic properties of two isomers: dimethyl ether and ethyl alcohol. These molecules both have the formula C_2H_6O but strong hydrogen bonding occurs only in the alcohol. Table 4-8 shows some of the properties. Due to the additional cohesive forces in the hydrogen-bonded alcohol, the boiling point, enthalpy of vaporization, and Trouton's constant are appreciably larger than those of the ether. Also, because ethanol can readily solvate with water, it is infinitely soluble in water whereas dimethyl ether is only partially soluble.

Table 4-8 Some properties of the isomers ethanol and dimethyl ether.

	Ethanol	Dimethyl ether
Normal boiling point (°C)	78	-25
Enthalpy of vaporization at normal boiling point (kJ mol^{-1})	42.6	18.6
Trouton's constant* (J mol^{-1} K^{-1})	121	74.9
Solubility in water at 18°C and 1 bar (g/100 g)	∞	7.12

* Trouton's constant is the entropy of vaporization at the normal boiling point.

Hydrogen bonding between molecules of the same component can frequently be detected by studying the thermodynamic properties of a solution wherein the hydrogen-bonded substance is dissolved in a nonpolar, relatively inert solvent.

When a strongly hydrogen-bonded substance such as ethanol is dissolved in an excess of a nonpolar solvent (such as hexane or cyclohexane), hydrogen bonds are broken until, in the limit of infinite dilution, all the alcohol molecules exist as monomers rather than as dimers, trimers, or higher aggregates. This follows simply from the law of mass action: In the equilibrium $n\text{A} \rightleftharpoons \text{A}_n$ (where n is an integer greater than one) the fraction of A molecules that are monomeric (i.e., not polymerized) increases with falling total concentration of all A molecules, polymerized or not. As the total concentration of A molecules in the solvent approaches zero, the fraction of all A molecules that are monomers approaches unity.

The strong dependence of the extent of polymerization on solute concentration results in characteristic thermodynamic behavior as shown in Figs. 4-7 and 4-8.

Figure 4-7 gives the enthalpy of mixing per mole of solute as a function of solute mole fraction at constant temperature and pressure. The behavior of strongly hydrogen-bonded ethanol is contrasted with that of nonpolar benzene; the qualitative difference indicates that the effect of the solvent on the ethanol molecules is markedly different from that on the benzene molecules. When benzene is mixed isothermally with a paraffinic or naphthenic solvent, there is a small absorption of heat and a small expansion due to physical (essentially dispersion) forces. However, when ethanol is dissolved in an "inert" solvent, hydrogen bonds are broken and, because such breaking requires energy, much heat is absorbed; further, because a hydrogen-bonded network of molecules tends to occupy somewhat less space than that corresponding to the sum of the individual nonbonded molecules, there is appreciable expansion in the volume of the mixture, as shown in Fig. 4-8.

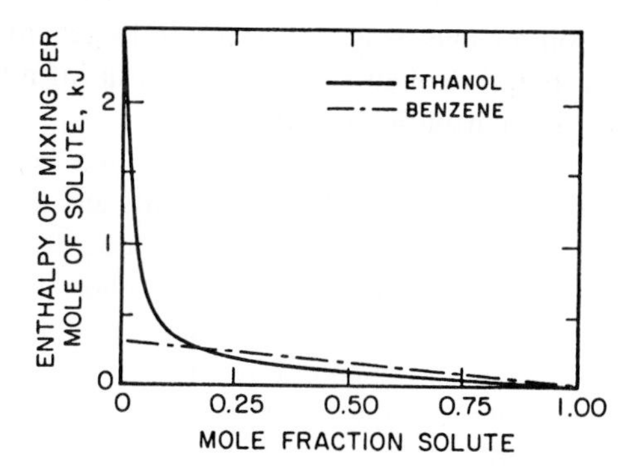

Figure 4-7 Hydrogen bonding in solution. Enthalpic effects for two chemically different solutes in cyclohexane at 20°C.

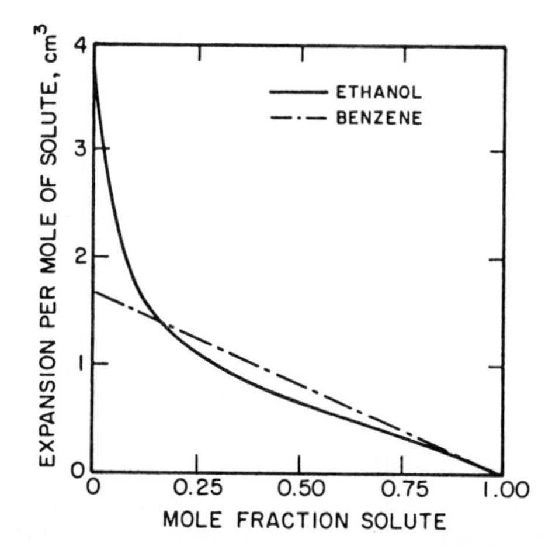

Figure 4-8 Hydrogen bonding in solution. Volumetric effects for two chemically different solutes in *n*-hexane at 6°C.

Evidence for hydrogen-bond formation between dissimilar molecules can be obtained in a variety of ways; good surveys are given by Pimentel and McClellan (1960), Schuster *et al.* (1976), and Huyskens *et al.* (1991). We shall not discuss them here but only discuss briefly two types of thermodynamic evidence best illustrated by an example, i.e., hydrogen bonding between acetone and chloroform.

Dolezalek (1908) observed that the partial pressures for liquid mixtures of acetone and chloroform were lower than those calculated from Raoult's law and he interpreted this negative deviation as a consequence of complex formation between the two dissimilar species. However, negative deviations from ideality can result from causes other than complex formation;[17] further, a binary liquid mixture that forms weak complexes between the two components may nevertheless have partial pressures slightly larger than those calculated from Raoult's law (Booth *et al.*, 1959). Thus Dolezalek's evidence, while pertinent, is not completely convincing.

More direct evidence for hydrogen-bond formation has been obtained by Campbell and Kartzmark (1960), who measured freezing points and enthalpies of mixing for mixtures of acetone with chloroform and with carbon tetrachloride. By comparing the properties of these two systems, it becomes evident that there is a large difference between the interaction of acetone with chloroform and the interaction of acetone with carbon tetrachloride, a molecule similar to chloroform except that the latter possesses an electron-accepting hydrogen atom that is capable of interacting with the electron-donating oxygen atom in acetone.

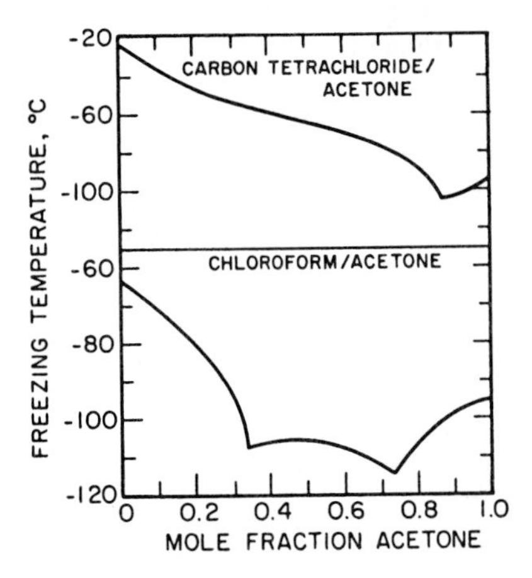

Figure 4-9 Evidence for hydrogen bonds: Freezing-point data.

Figure 4-9 gives freezing-point data for the two systems. Mixtures containing carbon tetrachloride exhibit simple behavior with a eutectic formed at -105°C and 87.5 mol % acetone. However, mixtures containing chloroform show more complicated behavior: Two eutectics are formed, one at -106°C and 31 mol % acetone and the other at -115°C and 74 mol % acetone, and there is a convex central section whose maximum is at 50 mol %. This maximum indicates that the compound $(CH_3)_2CO$---$HCCl_3$ exists in the solid state, although it is readily dissociated in the liquid state. The existence of

[17] For example, solutions of nonpolar polymers in nonpolar liquid solvents show strong negative deviations from Raoult's law, as discussed in Sec. 8.2.

such a compound is excellent evidence for strong interaction between the two dissimilar molecules. Since the maximum in the diagram occurs at the midpoint on the composition axis, we conclude that the complex has a 1:1 stoichiometric ratio, as we would expect from the structure of the molecules.

Figure 4-10 shows enthalpies of mixing for the two binary systems. Unfortunately, the original reference does not mention the temperature for these data but it is probably 25°C or some temperature nearby. The enthalpy of mixing of acetone with carbon tetrachloride is positive (heat is absorbed), whereas the enthalpy of mixing of acetone with chloroform is negative (heat is evolved), and it is almost one order of magnitude larger. These data provide strong support for a hydrogen bond formed between acetone and chloroform. The effect of physical intermolecular forces (dipole, induction, dispersion) causes a small amount of heat to be absorbed, as shown by the data for the carbon tetrachloride mixtures; in the chloroform mixtures, however, there is a chemical heat effect that not only cancels the physical contribution to the observed enthalpy of mixing but, because it is much larger, causes heat to be evolved. Because energy is needed to break hydrogen bonds, it necessarily follows that heat is liberated when hydrogen bonds are formed.

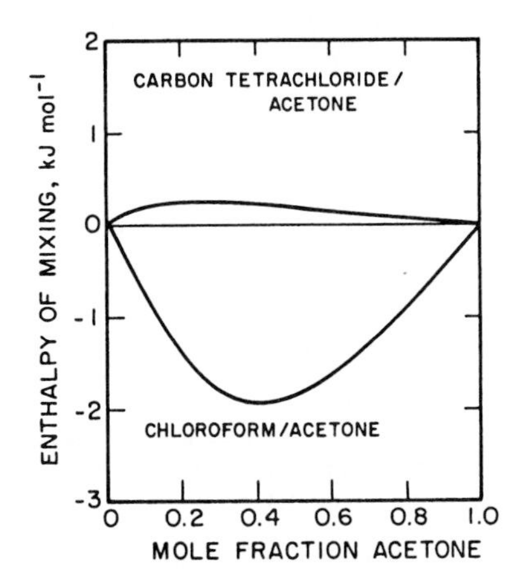

Figure 4-10 Evidence for hydrogen bonds: Enthalpy-of-mixing data.

The freezing-point data and the calorimetric results indicate that in the liquid phase an equilibrium exists:

$$(CH_3)_2C{=}O + HCCl_3 \rightleftharpoons (CH_3)_2C{=}O - - - HCCl_3$$

It can be shown from the law of mass action that when acetone is very dilute in chloroform, all of the acetone in solution is complexed with chloroform. For this reason the enthalpy data shown in Fig. 4-10 were replotted by Campbell and Kartzmark on coordinates of enthalpy of mixing per mole of acetone versus mole fraction of acetone. From the intercepts of this plot (0 mol % acetone), Δh = -8.75 kJ mol^{-1} for the system containing chloroform and Δh = 2.62 kJ mol^{-1} for the system containing carbon tetrachloride. From these results we can calculate that the enthalpy of complex formation, i.e., the enthalpy of the hydrogen bond, is (-8.75 – 2.62) = -11.37 kJ mol^{-1}.

In view of the experimental uncertainties that are magnified by this particular method of data reduction, Campbell and Kartzmark give Δh = (11.3 ± 0.4) kJ mol^{-1} as the enthalpy of hydrogen-bond formation. This result is in fair agreement with enthalpy data for similar hydrogen bonds determined by different measurements.

4.9 Electron Donor-Electron Acceptor Complexes

While the consequences of hydrogen bonding are probably the most common chemical effect in solution thermodynamics, chemical effects may also result from other kinds of bonding forces leading to loose complex formation between electron donors and electron acceptors (Andrews and Keefer, 1964; Foster 1973, 1974; Kuznetsov, 1995), sometimes called *charge-transfer complexes.*

The existence of donor-acceptor complexes can be established by a variety of experimental methods (Foster, 1973, 1974). Several types of data for donor-acceptor complexes are listed in Table 4-9 that also gives the designation "primary" or "secondary" for each type. Primary data are those derived from measurements by an interpretation requiring theoretical assumptions which are well established in fields other than donor-acceptor-complex studies. All data classified as secondary depend on a particular method for determining the concentration of complex in the sample.

Table 4-9 Sources of experimental data for donor-acceptor complexes (Gutman, 1978).

Data	Type*
1. Frequencies of charge-transfer absorption bonds	Primary
2. Geometry of solid complexes	Primary
3. NMR studies of motion in solid complexes	Primary (?)
4. Association constants	Secondary
5. Molar absorptivity or other measures of absorption intensity	Secondary
6. Enthalpy changes upon association	Secondary
7. Dipole moments	Secondary
8. Infrared frequency shifts	Secondary (?)
9. Infrared intensity changes	Secondary
10. NMR chemical shifts of magnetic nuclei in complexes	Secondary
11. Nuclear quadrupole resonance measurements on solid complexes	Primary

* "Primary" indicates that the data can be interpreted using well-established theoretical principles. "Secondary" indicates that data reduction requires simplifying assumptions that may be doubtful.

Ultraviolet spectroscopy is the method used most frequently to study donor-acceptor complexes; there are extensive tabulations of charge-transfer-frequency data in the literature (Briegleb, 1961; Andrews and Keefer, 1964; Rose, 1967; Mulliken and Person, 1969). These data provide a valuable source of information about complexes that exhibit a charge-transfer (CT) band. Spectroscopic data may be used to give a quantitative measure of complex stability (Rossotti and Rossotti, 1961).

The general basis of such measurements is given schematically in Fig. 4-11 showing two optical cells; in the upper cell we have in series two separate, dilute solutions of components A and B in some "inert" solvent. In the lower cell there is a single solution of A and B dissolved in the same "inert" solvent. We put equal numbers of A molecules in the top and bottom cells and we do the same with B. Monochromatic light of equal intensity is passed through both cells. If a complex is formed between A and B, and if the light frequency is in the absorption range of the complex, then light absorption is larger in the lower cell. On the other hand, if no complex is formed, light absorption is the same in both cells. The quantitative difference in light absorption provides a basis for subsequent calculations of complex stability and, when such spectroscopic measurements are performed at different temperatures, it is possible to calculate also the enthalpy and entropy of complex formation.

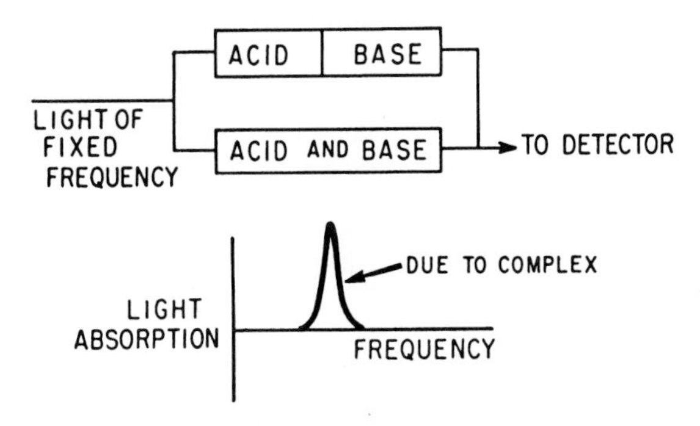

Figure 4-11 Spectroscopic measurement of acid-base complex.

Table 4-10 gives some results for complex formation between *s*-trinitrobenzene (an electron acceptor) and aromatic hydrocarbons (electron donors). The results show that complex stability rises with the number of methyl groups on the benzene ring, in agreement with various other measurements indicating that π-electrons on the aromatic ring become more easily displaced when methyl groups are added; for example, the ionization potentials of methyl-substituted benzenes fall with increasing methyl substitution.[18] For *s*-trinitrobenzene with aromatics, complex stability is strong. For complexes of aromatics with other, more common polar organic molecules, complex stability

[18] A lower ionization potential means that an electron can be removed more easily.

Table 4-10 Spectroscopic equilibrium constants and enthalpies of formation for *s*-trinitrobenzene/aromatic complexes dissolved in cyclohexane at 20°C.*

Aromatic	Equilibrium constant (L mol^{-1})	$-\Delta h$ (kJ mol^{-1})
Benzene	0.88	6.15
Mesitylene	3.51	9.63
Durene	6.02	11.39
Pentamethylbenzene	10.45	14.86
Hexamethylbenzene	17.50	18.30

* C. C. Thompson, Jr. and P. A. D. de Maine, 1965, *J. Phys. Chem.*, 69: 2766.

is much weaker but not negligible, as indicated by experimental results of Weimer (1966) shown in Table 4-11. For the polar solvents listed, Weimer found no complex formation with saturated hydrocarbons and as a result we may expect the thermodynamic properties of solutions of these polar solvents with aromatics to be significantly different from those of solutions of the same solvents with paraffins and naphthenes, as has been observed (Orye *et al.*, 1965, 1965a). The tendency of polar solvents to form complexes with unsaturated hydrocarbons, but not with saturated hydrocarbons, supplies a basis for various commercial separation processes in the petroleum industry, including the Edeleanu and Udex processes.

Table 4-11 Spectroscopic equilibrium constants of formation for polar solvent/*p*-xylene complexes dissolved in *n*-hexane at 25°C (Weimer, 1966).

Polar solvent	Equilibrium constant (L mol^{-1})
Acetone	0.25
Cyclohexanone	0.15
Triethyl phosphate	0.14
Methoxyacetone	0.12
Cyclopentanone	0.12
γ–Butyrolactone	0.09
n-Methylpyrrolidone	0.09
Propionitrile	0.07
Nitromethane	0.05
Nitroethane	0.05
2-Nitropropane	0 05
Citraconic anhydride	0.04
2-Nitro-2-methylpropane	0.03

Evidence for complex formation is often obtained from thermodynamic measurements. Agarwal (1978), for example, has measured volumes of mixing for 1,2,4-

trichlorobenzene with benzene, toluene, *p*-xylene, and mesitylene at 30°C as a function of composition. The results, shown in Fig. 4-12, are negative for the entire composition range for all four mixtures and the trend indicates that the interaction between unlike molecules rises with increasing electron-donating power of the hydrocarbon.

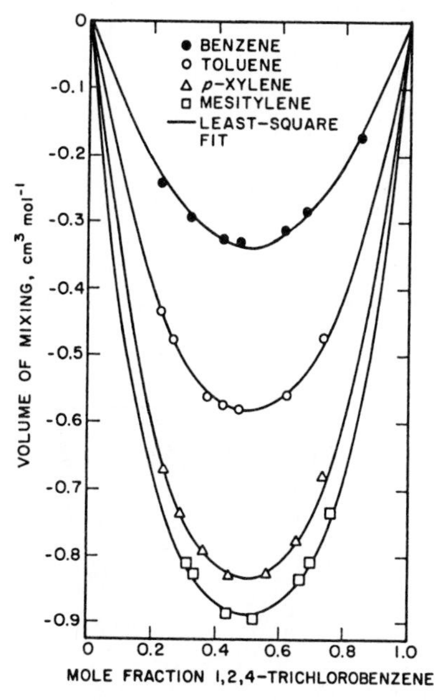

Figure 4-12 Effect of complex formation on volume of mixing. Binary systems containing 1,2,4-trichlorobenzene at 30°C.

A quantitative correlation exists between the volume of mixing and the ionization potential of the hydrocarbons, as shown in Fig. 4-13, where the volume of mixing (at equimolar composition) is a straight-line function of the ionization potential. The results in Fig. 4-13 give evidence for the existence of a donor-acceptor interaction between trichlorobenzene and aromatic hydrocarbons.

Sometimes, however, the interpretation of thermodynamic data does not provide such an obvious conclusion. Mahl *et al.* (1978) for example, have measured enthalpy and volume of mixing for tetrahydrofuran (THF) with benzene, toluene, and xylenes at 25°C as a function of composition. The results, given in Figs. 4-14 and 4-15, are negative for both properties over the entire composition range for all five mixtures, suggesting complex formation between unlike molecules. However, the magnitude of the enthalpy of mixing at equimolar composition follows the sequence benzene > toluene > xylenes. At first glance, this sequence is surprising because the ionization potential of benzene is larger than that of toluene, and that, in turn, is larger than that of xylene. One might, therefore, have expected a reverse sequence. However, the observed enthalpy of mixing is determined not only by chemical forces but also by physical forces.

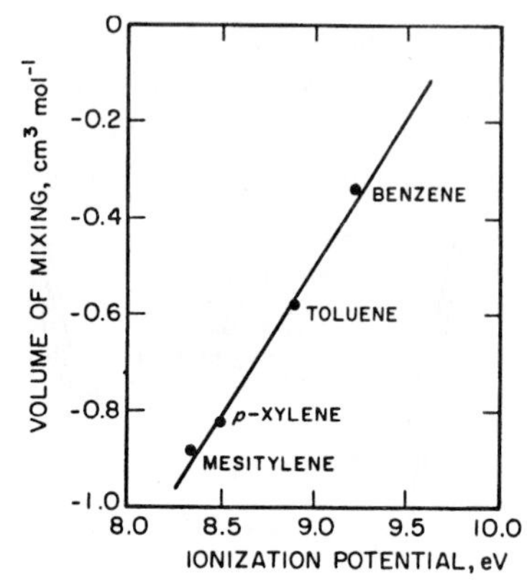

Figure 4-13 Correlation of volume of mixing with the ionization potential of the hydrocarbon. Equimolar 1,2,4-triclorobenzene/hydrocarbon mixtures at 30°C.

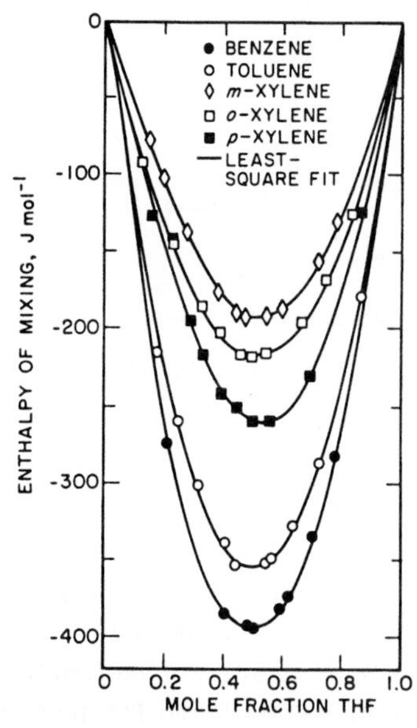

Figure 4-14 Effect of complex formation on enthalpy of mixing. Binary systems containing tetrahydrofuran (THF) at 25°C.

For these mixtures, chemical and physical forces are of the same order of magnitude, and therefore, the sequence of enthalpy effects is not necessarily the same as the sequence of complex stability. Exothermic mixing, together with volumetric contraction, provides good evidence of moderately strong interactions between unlike molecules in the liquid state. It seems that the lone pair of electrons on the oxygen atom of the THF forms charge-transfer complexes with benzene, toluene, and xylenes. However, precise quantitative information on complex stability cannot be determined from the enthalpic and volumetric data because of unknown solvent effects. For such information, spectroscopic data are more useful.

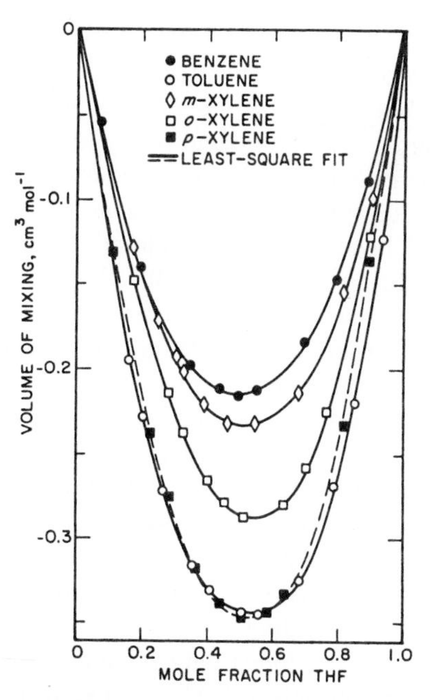

Figure 4-15 Effect of complex formation on volume of mixing. Binary systems containing tetrahydrofuran (THF) at 25°C.

Because most published data on molecular complexes are based on measurements in the liquid phase, solvent effects on molecular-complex properties must be taken into account in data interpretation. There is no solvent that may be regarded as completely inert; therefore, solute properties in solution are not identical to those in the vapor phase (Tamres, 1973). Reliable results on complex formation in the vapor phase are necessary for understanding the effect of a liquid solvent on solution equilibria, particularly since the theory (Mulliken, 1952) of charge-transfer complexes is based on the concept of the *isolated* donor, acceptor, and complex, a situation attained only in the low-pressure gas phase. A review of solvent effects is given by Davis (1975) and, in particular for biological systems, by Gutmann *et al.* (1997).

The examples shown above illustrate how specific (chemical) intermolecular forces can profoundly influence thermodynamic properties. At present, with rare exceptions, the theory of chemical bonding is not sufficiently developed to establish fundamental quantitative relations between specific intermolecular forces and thermodynamic properties, but in many cases it is possible to interpret thermodynamic behavior in terms of pertinent chemical phenomena. Such interpretation can then facilitate meaningful correlation of thermodynamic properties, based on molecular concepts.

4.10 Hydrophobic Interaction

It has been known for many years that some molecules are soluble in alcohol, ether, and many other solvents, but not in water; often these molecules have a dual nature: one part of molecule is soluble in water (the *hydrophilic*, i.e. water-loving, part), while another part is not water-soluble (the *hydrophobic*, i.e. water-fearing, part).

Molecules that are partly hydrophilic and partly hydrophobic[19] are forced by their dual nature to adopt unique orientations in an aqueous medium, that is, to form suitably organized structures. Such molecules, called *amphiphiles*, play important roles in living organisms at the cell level, and find potentially useful applications in biotechnology and in chemical industry (Tanford, 1980; Chen and Rajagoplan, 1990). The organized structures are called *micelles* (Hoffmann and Ulbricht, 1986). As schematically represented in Fig. 4-16(a) for a surfactant in aqueous solution, in micelles (spherical, ellipsoidal or in form of bilayers) the hydrophobic part (usually a long-chained hydrocarbon) is kept away from water, while the hydrophilic terminal groups (uncharged, cationic, or anionic) at the surface of the aggregates are water-solvated and keep the aggregates in solution. However, addition of a small amount of water to a surfactant-containing organic nonpolar phase may induce the formation of thermodynamically stable aggregates designated by *reverse micelles*, schematically represented in Fig. 4-16(b). In this case, the surfactant orients itself such that its terminal hydrophilic groups point inward, surrounding the water core, while the long hydrocarbon hydrophobic chains point outward into the bulk organic phase. An example is provided by the double-chained anionic surfactant AOT[20] [shown in Fig. 4-16(c)] in solutions of isooctane with small amounts of water. We do not discuss further this large subject but mention it only to indicate the tremendous importance of structural effects.[21]

The origin of the hydrophobic effect is drastically different from those described in the previous sections. It is mainly an entropic phenomenon often observed in nature. The hydrophobic effect arises mainly from the strong attractive forces (hydrogen bonds) between H_2O molecules in highly structured liquid water. These attractive forces must be disrupted or distorted when a solute is dissolved in water. Upon solubilization of a solute, hydrogen bonds in water are often not broken but they are maintained in distorted form. Water molecules reorient, or rearrange, themselves such that they can participate in hydrogen-bond formation, more or less as in bulk pure liquid water. In doing so, they create a higher degree of local order than that in pure liquid water, thereby producing a decrease in entropy. It is this loss of entropy (rather than

[19] Typical examples are surfactant molecules (the main components of detergents), such as sodium-*n*-dodecyl-1-sulfate, $C_{12}H_{25}SO_4Na$, that contains a long (hydrophobic) hydrocarbon chain and a terminal (hydrophilic) polar or ionic group.

[20] Surfactant AOT is sodium-di-2-ethylhexyl sulfosuccinate, with a molecular weight of 444 g mol^{-1}.

[21] For a review of applications of reversed micelles in biotechnology see M. J. Pires, M. R. Aires-Barros, and J. M. S. Cabral, 1996, *Biotechology Progress*, 12: 290; T. Ono and T. M. Goto, 1997, *Current Opinion in Colloid & Interface Science*, 2: 397.

enthalpy) that leads to an unfavorable Gibbs energy change for solubilization of non-polar solutes in water (Tanford, 1980). For example, hydrocarbons are only sparingly soluble in water, i.e. they have a highly unfavorable Gibbs energy of solubilization whose most important contribution is entropic.

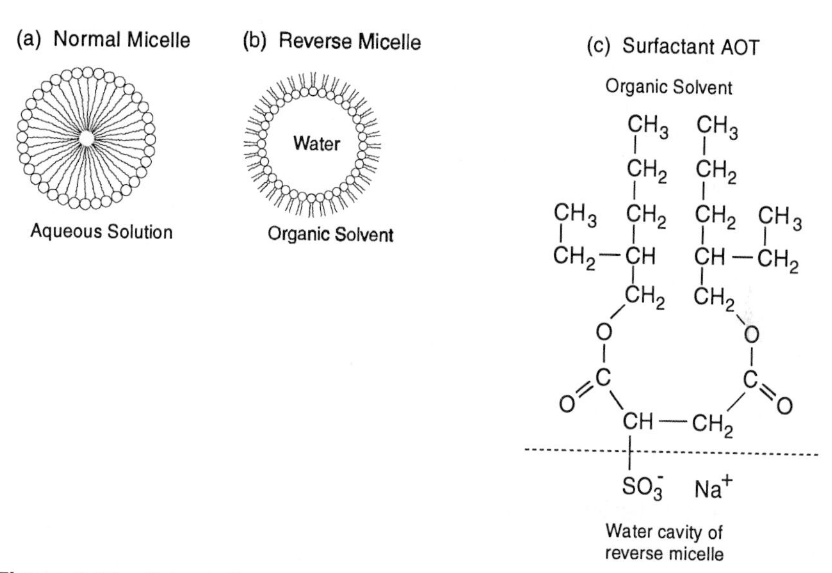

Figure 4-16 Schematic representation of a cross-section of micelles (a) and reverse micelles (b). In reverse micelles, the polar terminal group of the surfactant's molecules are directed towards the interior of the aggregate (forming an inner water core), and the hydrophobic chains are exposed to the organic solvent, as illustrated for the double-chained surfactant AOT (c).

For example, as shown in Table 4-12, at 25°C the standard Gibbs energy for the transfer (Δg^0) of n-butane from its pure liquid to water is about +24.7 kJ mol^{-1}. This quantity is the sum of an enthalpic contribution (Δh^0) and an entropic contribution ($T\Delta s^0$). For n-butane, these are, respectively, -3.3 kJ mol^{-1} and -28.0 kJ mol^{-1} at 25°C; the large decrease in entropy accounts for 85% of the Gibbs energy of solubilization. For other hydrocarbons (e.g. n-hexane), the entropic contribution to Δg^0 is even larger.

As summarized in Table 4-12, the standard entropy of transfer is strongly negative, due to the reorientation of the water molecules around the hydrocarbon. The poor solubility of hydrocarbons in water is not due to a large positive enthalpy of solution but rather to a large entropy decrease caused by what is called the *hydrophobic effect.* This effect is, in part, also responsible for the immiscibility of nonpolar substances (hydrocarbons, fluorocarbons, etc.) with water.

Closely related to the hydrophobic effect is the *hydrophobic interaction* (Israelachvili, 1992). This interaction is mainly entropic and refers to the unusually strong attraction between hydrophobic molecules (and surfaces) in water;[22] in many cases, this attraction is stronger than in vacuo. For example, from Table 4-5 and using Eq. (4-22), we can calculate that the energy of interaction of two contacting methane molecules (with a molecular diameter of 4 Å) in vacuo is -2.5×10^{-21} J. In water, the same interaction energy is -14×10^{-21} J.

Table 4-12 Change in standard molar Gibbs energy (Δg^0), enthalpy (Δh^0), and entropy ($T\Delta s^0$), all in kJ mol^{-1}, for the transfer of hydrocarbons from their pure liquids into water at 25°C (Tanford, 1980).

Hydrocarbon	Δg^0 (kJ mol^{-1})	Δh^0 (kJ mol^{-1})	$T\Delta s^0$ (kJ mol^{-1})
Ethane	16.3	-10.5	-26.8
Propane	20.5	-7.1	-27.6
n-Butane	24.7	-3.3	-28.0
n-Hexane	32.4	0	-32.4
Benzene	19.2	+2.1	-17.1
Toluene	22.6	+1.7	-20.9

4.11 Molecular Interactions in Dense Fluid Media

In previous sections we described the intermolecular forces that arise between molecules in the low-pressure gas phase. However, there is a significant difference in the physical interpretation of intermolecular forces between gaseous molecules and those between solute molecules in a liquid solvent. Molecules in the low-pressure gas phase interact in a "free" medium (i.e. a vacuum), but dissolved solutes interact in a solvent medium. Interactions between two molecules in a vacuum are described by a potential function (e.g. Lennard-Jones) but interactions between two molecules in a solvent medium are described by what is called the *potential of mean force* that plays a major role in colloid science and in the physical chemistry of protein solutions. The essential difference is that the interaction between two molecules in a solvent is influenced by the molecular nature of the solvent but there is no corresponding influence on the interaction of two molecules in (nearly) free space.

For example, for two solute molecules in a solvent, their intermolecular pair potential includes not only the direct solute-solute interaction energy, but also any changes in the solute-solvent and solvent-solvent interaction energies as the two solute molecules approach each other. A solute molecule can approach another solute molecule only by displacing solvent molecules from its path. Thus, at some fixed separation, while two molecules may attract each other in free space, they may repel each

[22] A molecular-thermodynamic model for hydrophobic hydration of small nonpolar molecules and of extended hydrophobic surfaces is given by N. A. M. Besseling and J. Lyklema, 1997, *J. Phys. Chem. B*, 101: 7604.

other in a solvent medium if the work that must be done to displace the solvent molecules exceeds that gained by the approaching solute molecules. Further, solute molecules often perturb the local ordering of solvent molecules. It the energy associated with this perturbation depends on the distance between the two dissolved molecules, it produces an additional *solvation* force between them.

The molecular nature of the solvent can produce potentials of mean force that are much different from the corresponding two-body potential in vacuo.[23] The potential of mean force is a measure of the intermolecular interaction of solute molecules in liquid solution. Solution theories, such as the *McMillan-Mayer theory* (1945), provide a direct quantitative relation between the potential of mean force and macroscopic thermodynamic properties (the osmotic virial coefficients) accessible to experiment. Osmotic virial coefficients are obtained through osmotic-pressure measurements.

Osmotic Pressure

Osmotic pressure is a phenomenon frequently encountered in nature, especially in biological systems. The first systematic quantitative studies were made in the late 19th century. The physical chemistry of osmotic pressure was developed by van't Hoff, one of the founders of physical chemistry (about 1890) and a pioneer in applying thermodynamics to the study of liquid solutions. We briefly summarize the main concepts of osmotic pressure.

Consider a system divided into two parts, α and β, by a semi-permeable membrane, as shown in Fig. 4-17.

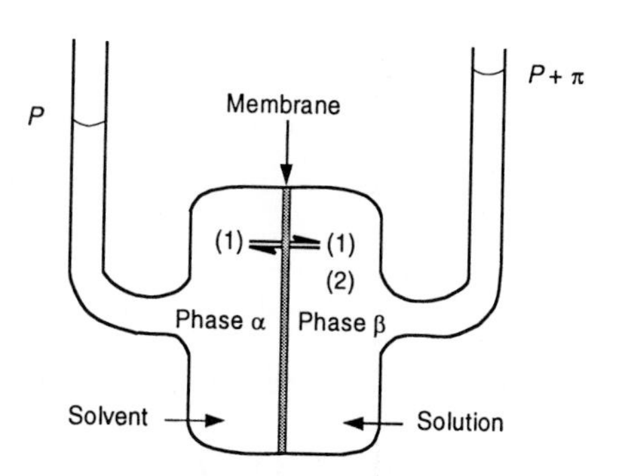

Figure 4-17 Schematic diagram of an osmotic-pressure measurement apparatus. The membrane is permeable to the solvent (1) but not to the solute (2).

[23] For a description of how solvent effects influence intermolecular and interparticle interaction potentials, see Israelachvili (1992).

The semi-permeable membrane is permeable to the solvent (1) but impermeable to the solute (2). Phases α and β are at the same temperature. The pressure on phase α is P, while that on phase β is $P + \pi$. The equation of chemical equilibrium is,

$$\mu_1^\alpha = \mu_1^\beta \tag{4-38}$$

where μ is the chemical potential, given by

$$\mu_1^\alpha = \mu_{\text{pure }1}^\alpha(T, P) \tag{4-38a}$$

$$\mu_1^\beta = \mu_{\text{pure }1}^\beta(T, P+\pi) + RT \ln a_1 \tag{4-39}$$

where a is the activity, related to composition through $a_1 = \gamma_1 x_1$, where γ is the activity coefficient and x is the mole fraction.

For a pure fluid, $(\partial\mu / \partial P)_T = \upsilon$. Assuming that the molar volume does not vary with pressure (incompressible fluid), we have

$$\mu_{\text{pure }1}(P+\pi) = \mu_{\text{pure }1}(P) + \pi \upsilon_{\text{pure }1} \tag{4-40}$$

Equation (4-38) can be rewritten as

$$-\ln a_1 = \frac{\pi \upsilon_{\text{pure }1}}{RT} \tag{4-41}$$

If the solution in phase β is dilute, x_1 is close to unity; in that event, γ_1 is also close to unity and Eq. (4-41) becomes

$$-\ln x_1 = \frac{\pi \upsilon_{\text{pure }1}}{RT} \tag{4-42}$$

When $x_2 << 1$, $\ln x_1 = \ln(1 - x_2) \approx -x_2$. Equation (4-42) becomes

$$\pi \upsilon_{\text{pure }1} = x_2 RT \tag{4-43}$$

Because $x_2 << 1$, $n_2 << n_1$ and $x_2 \approx n_2/n_1$. Equation (4-43) then becomes

$$\boxed{\pi V = n_2 RT} \tag{4-44}$$

where $V = n_1 \upsilon_{\text{pure }1}$ is the total volume available to n_2 moles of solute.

Equation (4-44) is the *van't Hoff equation* for osmotic pressure π, analogous to the ideal-gas equation.

The only important assumptions for Eq. (4-44) are:

- The solution is very dilute.
- The solution is incompressible.

An important application of Eq. (4-44) was recognized many years ago: If we measure π and T, and if we know the *mass* concentration of the solute (g/liter), we can then calculate the solute's molecular weight.

Osmometry provides a standard procedure for measuring molecular weights of large molecules (polymers or biomacromolecules such as proteins) whose molecular weights cannot be accurately determined from other colligative-property measurements such as boiling-point elevation or freezing-point depression.

Van't Hoff's equation is a limiting law; Eq. (4-44) is an asymptote that is approached as the concentration of solute goes to zero.

For finite concentrations, it is useful to write a series expansion in the *mass* concentration c_2, typically expressed in (g/liter):

$$\boxed{\frac{\pi}{c_2} = RT\left(\frac{1}{M_2} + B^* c_2 + C^* c_2^2 + \cdots\right)} \qquad \text{(4-45)}^{24}$$

where M_2 is the molar mass of solute. The *osmotic second virial coefficient* is designated by B^*; the third, by C^*, etc.

In Eq. (4-45), if we set $B^* = C^* = ... = 0$, we recover the van't Hoff equation.

For dilute solutions, we can neglect three-body (and higher) interactions in Eq. (4-44). Thus, a plot of π/c_2 against c_2 is linear (for small values of c_2), with intercept equal to RT/M_2 and slope equal to RTB^*_{22}. To illustrate, Fig. 4-18 shows osmotic-pressure data (McCarty and Adams, 1987; Haynes *et al.*, 1992) measured by *membrane*

[24] The osmotic virial expansion is analogous to the virial expansion for gases (to be presented in Chap. 5),

$$\frac{P}{\rho} = RT(1 + B\rho + C\rho^2 + \cdots) \qquad \text{(4-45a)}$$

where ρ is the *molar* density and B and C are the second and third virial coefficients. The molar concentration ρ is related to the mass solute concentration c_2 by $\rho = c_2/M_2$. Substituting in the above equation, we obtain

$$\frac{P}{c_2} = RT\left[\frac{1}{M_2} + \left(\frac{B}{M_2^2}\right)c_2 + \left(\frac{C}{M_2^3}\right)c_2^2 + \cdots\right] \qquad \text{(4-45b)}$$

Equations (4-45a) and (4-45b) are essentially the same but the dimensions of B and C are not the same as those of B^* and C^*.

osmometry[25] for aqueous protein solutions: α-chymotrypsin in a 0.1 M potassium sulfate buffer, at pH 5 and 25°C; and lysozyme and ovalbumin in a 0.06 M cacodylate buffer (aqueous solutions of dimethylarsinic acid sodium salt, $(CH_3)_2AsO_2Na \cdot 3H_2O$), at pH 5.8 and 37°C. Table 4-13 lists the proteins' osmotic second virial coefficients and number-average molecular weights, regressed from data shown in Fig. 4-18.

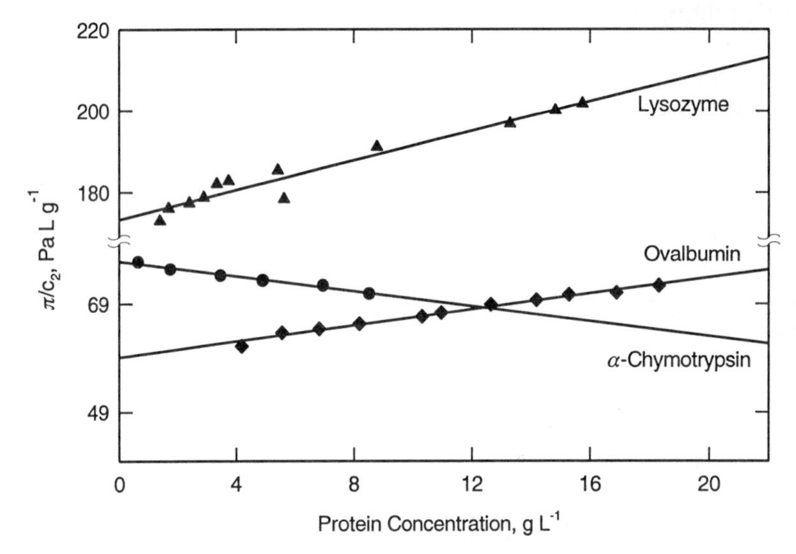

Figure 4-18 Osmotic-pressure data for α-chymotrypsin (●)(Haynes *et al.*, 1992) in 0.1 M potassium sulfate buffer, at pH 5 and 25°C, and for lysozyme (▲) and ovalbumin (◆) (McCarty and Adams, 1987) in 0.06 M cacodylate buffer, at pH 5.8 and 37°C.

Figure 4-18 shows that, contrary to what is observed for lysozyme and ovalbumin, π/c_2 for α-chymotrypsin falls with increasing c_2, giving a negative osmotic second virial coefficient. This negative value indicates that, at the prevailing experimental conditions, dilute α-chymotrypsin molecules are slightly attracted to each other.

Table 4-13 Osmotic second virial coefficients and number-average molecular weights for α-chymotrypsin, lysozyme, and ovalbumin in aqueous buffer solutions, regressed from the data shown in Fig. 4-18.

Protein	$B^*_{22} \times 10^7$ (L mol g^{-2})	M_2 (g mol^{-1})
α-Chymotrypsin	-2.72	32,200
Lysozyme	7.47	15,100
Ovalbumin	2.76	43,400

[25] A more accurate technique for experimentally determining osmotic second virial coefficients is provided by *low-angle laser-light scattering* (LALLS).

As discussed elsewhere (Hill, 1959, 1986), osmotic second virial coefficients (a macroscopic property) are related to (microscopic) intermolecular forces between two solute molecules. Quantitative data for B_{22}^* can provide useful information on interactions between, for example, polymer or protein molecules in solution.

Donnan Equilibria

The osmotic-pressure relation given by van't Hoff was derived for solutions of nonelectrolytes or else for solutions of electrolytes where the membrane's permeability did not distinguish between cations and anions. But now consider a chamber divided into two parts by a membrane that exhibits ion selectivity, i.e. some ions can flow through the membrane while others cannot. In this case, the equilibrium conditions become more complex because, in addition to the usual Gibbs equations for equality of chemical potentials, it is now also necessary to satisfy an additional criterion: electrical neutrality for each of the two phases in the chamber. The thermodynamics of equilibrium in systems containing an ion-selective membrane was first discussed by Donnan early in the 20th century.

To illustrate the essentials,[26] we consider an aqueous system containing three ionic species: Na^+, Cl^- and R^-, where R^- is some anion much larger than Cl^-. Water is in excess; all ionic concentrations are small. The chamber is divided into two equi-sized parts, phase α and phase β, by an ion-selective membrane as indicated in Fig. 4-19. This membrane is permeable to water, Na^+ and Cl^- but it is impermeable to R^-.

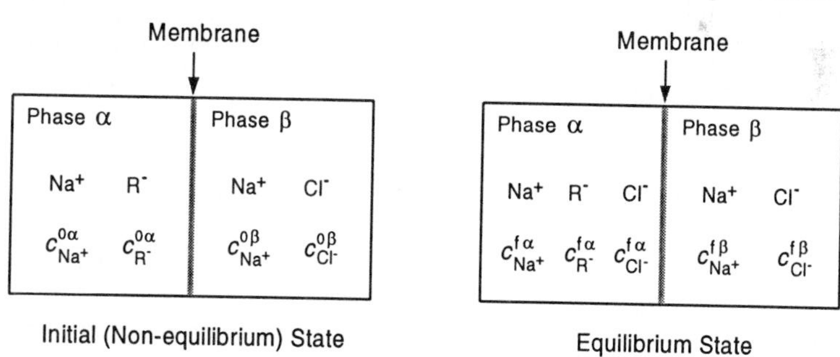

Figure 4-19 Schematic representation of Donnan equilibrium.

Initially (before equilibrium is attained), the left side contains only water, Na^+ and R^- at *molar* concentrations $c_{Na^+}^{0\alpha}$ and $c_{R^-}^{0\alpha}$. The right side initially contains only water and Na^+ and Cl^- at *molar* concentrations $c_{Na^+}^{0\beta}$ and $c_{Cl^-}^{0\beta}$. Electroneutrality requires that

[26] For a clear and simple discussion, see E. A. Moelwyn-Hughes, 1961, *Physical Chemistry,* 2nd Ed., New York: Pergamon Press. For a comprehensive discussion, see M. P. Tombs and A. R. Peacocke, 1974, *The Osmotic Pressure of Biological Macromolecules,* Oxford: Clarendon Press.

$$c_{\mathrm{Na+}}^{0\alpha} = c_{\mathrm{R-}}^{0\alpha} \qquad \text{and} \qquad c_{\mathrm{Na+}}^{0\beta} = c_{\mathrm{Cl-}}^{0\beta} \tag{4-46}$$

We now allow the initial state to attain equilibrium. Let δ represent the change in Na^+ concentration in α. Because R^- cannot move from one side to the other, the change in Cl^- concentration in β is $-\delta$.

At equilibrium, the final (f) concentrations are

In α:
$$c_{\mathrm{Na+}}^{f\alpha} = c_{\mathrm{Na+}}^{0\alpha} + \delta \qquad c_{\mathrm{Cl-}}^{f\alpha} = \delta \qquad c_{\mathrm{R-}}^{f\alpha} = c_{\mathrm{R-}}^{0\alpha} \tag{4-47}$$

In β:
$$c_{\mathrm{Na+}}^{f\beta} = c_{\mathrm{Na+}}^{0\beta} - \delta \qquad c_{\mathrm{Cl-}}^{f\beta} = c_{\mathrm{Cl-}}^{0\beta} - \delta \qquad c_{\mathrm{R-}}^{f\beta} = 0 \tag{4-48}$$

Our task now is to calculate δ from the known original concentrations.

For the solvent (s), we write

$$\mu_s^\alpha = \mu_s^\beta \tag{4-49}$$ [27]

We relate the chemical potential to the pressure and to the activity by

$$\mu_s^\alpha = \mu_s^* + P^\alpha v_s + RT \ln a_s^\alpha \tag{4-50}$$

and we use a similar relation for μ_s^β. Here v is the molar volume and a is the activity. The standard state (*) is pure liquid solvent at system temperature and at zero pressure. Substitution in Eq. (4-49) gives

$$\frac{RT}{v_s} \ln\left(\frac{a_s^\beta}{a_s^\alpha}\right) = P^\alpha - P^\beta = \pi \tag{4-51}$$

where π is osmotic pressure.

We also have

$$\mu_{\mathrm{NaCl}}^\alpha = \mu_{\mathrm{NaCl}}^\beta \tag{4-52}$$

Because sodium chloride is totally dissociated into sodium ions and chloride ions, Eq. (4-52) can be rewritten

$$\mu_{\mathrm{Na+}}^\alpha + \mu_{\mathrm{Cl-}}^\alpha = \mu_{\mathrm{Na+}}^\beta + \mu_{\mathrm{Cl-}}^\beta \tag{4-53}$$

[27] It is tempting to write the (erroneous) equilibrium equations $\mu_{\mathrm{Cl-}}^\alpha = \mu_{\mathrm{Cl-}}^\beta$ and $\mu_{\mathrm{Na+}}^\alpha = \mu_{\mathrm{Na+}}^\beta$. These equations are incorrect because the electric potential of phase α is not equal to that of phase β. See Chap. 9.

Substituting relations similar to Eq. (4-50) into Eq. (4-53), we obtain

$$\pi = \frac{RT}{\overline{v}_{Na+} + \overline{v}_{Cl-}} \ln\left(\frac{a^{\beta}_{Na+}\, a^{\beta}_{Cl-}}{a^{\alpha}_{Na+}\, a^{\alpha}_{Cl-}} \right) \tag{4-54}$$

where $\overline{v}_i$ is the partial molar volume of i when all solutes are at vanishingly low concentrations.

Upon equating π from Eq. (4-51) to π from Eq. (4-54), we have

$$\frac{a^{\beta}_{Na+}\, a^{\beta}_{Cl-}}{a^{\alpha}_{Na+}\, a^{\alpha}_{Cl-}} = \left(\frac{a^{\beta}_{s}}{a^{\alpha}_{s}} \right)^{\frac{\overline{v}_{Na+} + \overline{v}_{Cl-}}{v_s}} \tag{4-55}$$

In a very dilute solution, $a^{\alpha}_{s} = a^{\beta}_{s} = 1$ and the activity of a solute i is equal to its molar concentration, $a_i = c_i$. Equation (4-55) then becomes

$$c^{\alpha}_{Na+}\, c^{\alpha}_{Cl-} = c^{\beta}_{Na+}\, c^{\beta}_{Cl-} \tag{4-56}$$

From the definition of δ, Eq. (4-56) becomes

$$(c^{0\alpha}_{Na+} + \delta)\,\delta = (c^{0\beta}_{Na+} - \delta)^2 \tag{4-57}$$

giving

$$\boxed{\delta = \frac{(c^{0\beta}_{Na+})^2}{c^{0\alpha}_{Na+} + 2c^{0\beta}_{Na+}}} \tag{4-58}$$

When Eq. (4-58) is used to obtain δ, Eqs. (4-47) and (4-48) give the final equilibrium concentrations as a function of the initial concentrations.

Rearranging Eq. (4-58), we obtain the fraction of the original sodium chloride in β that has moved to α:

$$\frac{\delta}{c^{0\beta}_{Na+}} = \left(2 + \frac{c^{0\alpha}_{Na+}}{c^{0\beta}_{Na+}} \right)^{-1} \tag{4-59}$$

Also, the osmotic pressure is given by

$$\pi = 2RT(c^{0\alpha}_{\mathrm{Na}^+} - c^{0\beta}_{\mathrm{Cl}^-} + 2\delta) \tag{4-60}$$

Because the equilibrium concentration of Na^+ is not the same in both sides, we have a concentration cell (battery) with a difference in electric potential across the membrane. This difference in electric potential $\Delta\phi$ is given by the Nernst equation,

$$\Delta\phi = \frac{RT}{N_A e z_{\mathrm{Na}^+}} \ln \frac{a^{\beta}_{\mathrm{Na}^+}}{a^{\alpha}_{\mathrm{Na}^+}} \tag{4-61}$$

Upon setting activities equal to concentrations and using Eq. (4-58), Nernst's equation becomes

$$\Delta\phi = \frac{RT}{N_A e z_{\mathrm{Na}^+}} \ln \frac{c^{0\beta}_{\mathrm{Na}^+}}{c^{0\alpha}_{\mathrm{Na}^+} + c^{0\beta}_{\mathrm{Na}^+}} \tag{4-61a}$$

where z_{Na^+} is the valence of the sodium ion (+1), e is the charge of an electron, and N_A is Avogadro's constant. These results (and others based on similar arguments) are of much importance in biology and medicine because animals (and man) contain numerous semipermeable membranes that exhibit ion selectivity. Further, these results are of importance in industrial separation processes using charged membranes as, for example, electrodialysis and reverse osmosis, as discussed in Sec. 8.4.

4.12 Molecular Theory of Corresponding States

Classical or *macroscopic theory of corresponding states* was derived by van der Waals based on his well-known equation of state. It can be shown, however, that van der Waals' derivation is not tied to a particular equation but can be applied to any equation of state containing two arbitrary constants in addition to gas constant R.

From the principle of continuity of the gaseous and liquid phases, van der Waals showed that at the critical point

$$\left(\frac{\partial P}{\partial v}\right)_T = \left(\frac{\partial^2 P}{\partial v^2}\right)_T = 0 \tag{4-62}$$

These relations led van der Waals to the general result that for variables v (volume), T (temperature), and P (pressure) there exists a universal function such that[28]

---

[28] For example, van der Waals' equation in reduced variables is

$$F\left(\frac{v}{v_c}, \frac{T}{T_c}, \frac{P}{P_c}\right) = 0 \tag{4-63}$$

is valid for all substances; subscript c refers to the critical point. Another way of stating this result is to say that, if the equation of state for any one fluid is written in reduced coordinates (i.e., v/v_c, T/T_c, P/P_c), that equation is also valid for any other fluid.

Classical theory of corresponding states is based on mathematical properties of the macroscopic equation of state. *Molecular* or *microscopic theory of corresponding states*, however, is based on mathematical properties of the potential-energy function.

Intermolecular forces of a number of substances are closely approximated by the inverse-power potential function given by Eq. (4-24). The independent variable in this potential function is the distance between molecules. When this variable is made dimensionless, the potential function can be rewritten in a general way such that the dimensionless potential is a *universal function* F of the dimensionless distance of separation between molecules:

$$\frac{\Gamma_{ii}}{\varepsilon_i} = F\left(\frac{r}{\sigma_i}\right) \tag{4-64}$$

where ε_i is an energy parameter and σ_i is a distance parameter characteristic of the interaction between two molecules of species i. For example, if function F is given by the Lennard-Jones potential, then ε_i is the energy (times minus one) at the potential-energy minimum, and σ_i is the distance corresponding to zero potential energy. However, Eq. (4-64) is not restricted to the Lennard-Jones potential, nor is it restricted to an inverse-power function as given by Eq. (4-24). Equation (4-64) merely states that the reduced potential energy ($\Gamma_{ii}/\varepsilon_i$) is some universal function of the reduced distance (r/σ_i).

Once the potential-energy function of a substance is known, it is possible, at least in principle, to compute the macroscopic configurational properties of that substance by the techniques of statistical mechanics. Hence a universal potential-energy function, Eq. (4-64), leads to a universal equation of state and to universal values for all reduced configurational properties.

To obtain macroscopic thermodynamic properties from statistical mechanics, it is useful to calculate the *canonical partition function* of a system depending on temperature, volume, and number of molecules. For fluids containing small molecules, the partition function Q is expressed as a product of two factors,[29]

$$F = P_R - \frac{8T_R}{3v_R - 1} + \frac{3}{v_R^2} = 0$$

where $P_R = P/P_c$, $T_R = T/T_c$, and $v_R = v/v_c$.

[29] A more detailed discussion is given in App. B.

$$Q = Q_{\text{int}}(N,T)\, Q_{\text{trans}}(N,T,V) \tag{4-65}$$

where the translational contributions to the energy of the system are separated from all others, due to other degrees of freedom such as rotation and vibration. It is assumed that contributions from rotation and vibration depend only on temperature. These contributions are called *internal* because (by assumption) they are independent of the presence of other near-by molecules.

In the classical approximation, the translational partition function, Q_{trans}, splits into a product of two factors, one arising from the kinetic energy and the other from the potential energy. For a one-component system of N molecules, Q_{trans} is given by

$$Q_{\text{trans}} = \left(\frac{2\pi mkT}{h^2}\right)^{(3N/2)} \frac{1}{N!} \int \underset{V}{\cdots} \int \exp\left[-\frac{\Gamma_t(\mathbf{r}_1,\ldots,\mathbf{r}_N)}{kT}\right] d\mathbf{r}_1 \ldots d\mathbf{r}_N \tag{4-66}$$

where m is the molecular mass, k is Boltzmann's constant, h is Planck's constant, and $\Gamma_t(\mathbf{r}_1,\ldots,\mathbf{r}_N)$ is the potential energy of the entire system of N molecules whose positions are described by vectors $\mathbf{r}_1,\ldots,\mathbf{r}_N$. For a given number of molecules and known molecular mass, the first factor depends only on the temperature. The second factor, called the *configurational integral*, Z_N, depends on temperature and volume:

$$\boxed{Z_N = \int \underset{V}{\cdots} \int \exp\left[-\frac{\Gamma_t(\mathbf{r}_1,\ldots,\mathbf{r}_N)}{kT}\right] d\mathbf{r}_1 \ldots d\mathbf{r}_N} \tag{4-67}$$

Hence the configurational part provides the only contribution that depends on intermolecular forces. However, Z_N is *not* unity for an ideal gas ($\Gamma_t = 0$). For an ideal gas, $Z_N^{\text{id}} = V^N$.

The equation of state is obtained from Q (see App. B):

$$P = kT\left(\frac{\partial \ln Q}{\partial V}\right)_{T,N} = kT\left(\frac{\partial \ln Z_N}{\partial V}\right)_{T,N} \tag{4-68}$$

The equation of state depends only on Z_N when Eqs. (4-65) and (4-66) are valid. Therefore, the main problem in applying statistical mechanics to real fluids lies in the evaluation of the configurational partition function.

There are four assumptions that lead to the *molecular theorem of corresponding states*, clearly stated by Pitzer (1939) and Guggenheim (1945). They are:

1. The partition function is factored according to Eq. (4-65), where Q_{int} is independent of the volume per molecule.

2. The classical approximation Eq. (4-66) is used for Q_{trans}.[30]
3. The potential energy Γ_t is represented as the sum of the interactions $\Gamma_{ij}(r_{ij})$ of all possible pairs of molecules. For a given ij pair, Γ_{ij} depends only on the distance r_{ij} between them:

$$\Gamma_t = \sum_{i<j} \Gamma_{ij}(r_{ij}) \tag{4-69}$$

4. The potential energy of a pair of molecules, reduced by a characteristic energy, is represented as a universal function of the intermolecular distance, reduced by a characteristic length, i.e., Eq. (4-64).

Assumptions 3 and 4 are substituted into the configurational partition function. Further, we use reduced coordinates obtained by dividing the three-dimensional position vectors $\mathbf{r}_1,\ldots,\mathbf{r}_N$ by the scale factor σ^3. Thus

$$Z_N = \sigma^{3N} \int_{V/\sigma^3}^{\cdots} \int \exp\left[-\frac{\varepsilon}{kT} \sum_{i<j} F\left(\frac{r_{ij}}{\sigma}\right)\right] d\left(\frac{r_1}{\sigma^3}\right) \cdots d\left(\frac{r_N}{\sigma^3}\right) \tag{4-70}$$

Apart from the factor σ^{3N}, the configurational integral depends only on N, on the reduced temperature kT/ε, and on the reduced volume V/σ^3 (through the limits of the integral):

$$Z_N = \sigma^{3N} Z_N^* \left(\frac{kT}{\varepsilon}, \frac{V}{\sigma^3}, N\right) \tag{4-71}$$

where Z_N^* is a universal function.

Because configurational Helmholtz energy is given by

$$A^{\text{conf}} = -kT \ln Z_N \tag{4-72}$$

and because A^{conf} is an extensive property (proportional to N), we have

$$A^{\text{conf}} = N\Psi(T, \upsilon) \tag{4-73}$$

where the function ψ depends only on the intensive variables T and $\upsilon = V/N$.

[30] This assumption is not valid for low-molar-mass fluids like H_2, He, and Ne. For these fluids, quantum effects play an important role at low temperatures. However, these fluids can be included in corresponding-states correlations by using effective reduction parameters, as discussed in Chap. 5.

Equations (4-72) and (4-73) imply that the configurational integral must be of the form

$$Z_N = \sigma^{3N}\left[z^*\left(\frac{kT}{\varepsilon}, \frac{V}{N\sigma^3}\right)\right]^N \tag{4-74}$$

where function z^* depends only on intensive variables. Substitution of Eq. (4-74) into Eq. (4-68) gives the equation of state

$$\frac{P}{NkT} = \left(\frac{\partial \ln z^*}{\partial V}\right)_{T,N} \tag{4-75}$$

Introducing the reduced variables

$$\tilde{T} = \frac{kT}{\varepsilon} \qquad \tilde{v} = \frac{V}{N\sigma^3} \qquad \tilde{P} = \frac{P\sigma^3}{\varepsilon} \tag{4-76}$$

we find that z^* is a function of $\tilde{T}$ and $\tilde{v}$ [Eq. (4-74)]; the equation of state becomes

$$\tilde{P} = F^*(\tilde{T}, \tilde{v}) \tag{4-77}$$

where F^* is a universal function. The nature of this function depends only on the nature of the potential function Γ_{ij} in Eq. (4-69).

Equation (4-77) expresses the *molecular* (or *microscopic*) *theory of corresponding states.* This theory is analogous to the macroscopic theory of corresponding states expressed by Eq. (4-63); the difference lies in the reducing parameters.

The reduced quantities [Eq. (4-76)] are defined in terms of macroscopic variables T, V, P, N and molecular parameters ε and σ. The use of molecular parameters is important in the extension of the theorem of corresponding states to mixtures.

To relate the *macroscopic* and the *microscopic* theories of corresponding states, it is desirable to establish a connection between the parameters of one theory and those of the other. In the microscopic theory, there are two independent parameters: an energy parameter and a distance parameter. In the macroscopic theory, there appear to be three – v_c, T_c, and P_c – but only two of these are independent because, according to the theory, the compressibility factor at the critical point ($z_c = P_c v_c / RT_c$) is the same for all fluids.

The connection between the macroscopic and the microscopic theories of corresponding states can be established by substituting Eq. (4-77) into the relations given by Eq. (4-62). It then follows that the macroscopic critical properties v_c, T_c, and P_c are related to the molecular parameters ε and σ by

$$\frac{\varepsilon}{k} = c_1 T_c \tag{4-78}$$

$$\frac{2}{3}\pi N_A \sigma^3 = c_2 \upsilon_c \tag{4-79}$$

$$\frac{\varepsilon}{\sigma^3} = c_3 P_c \tag{4-80}$$

where N_A is Avogadro's constant (υ_c is per mole) and c_1, c_2, and c_3 are universal constants. For simple nonpolar molecules, i.e., those nonpolar molecules having a small number of atoms per molecule, these relations have been found empirically for the case where the generalized function F is replaced by the Lennard-Jones (12-6) potential (Hirschfelder *et al.*, 1964). For that particular case, we have, approximately,

$$c_1 = 0.77 \qquad c_2 = 0.75$$
$$c_3 = 7.42 \qquad z_c = \frac{2\pi}{3}\frac{c_1}{c_2 c_3} = 0.290 \tag{4-81}$$

Since the critical temperature is a measure of the kinetic energy of the fluid at a characteristic state (where the liquid and gaseous states become identical), the simple proportionality between energy parameter ε and critical temperature T_c is reasonable. Similarly, the critical volume reflects the size of the molecules; hence, the proportionality between distance parameter σ^3 and the critical volume is also reasonable. The proportionality of the critical pressure to the ratio ε/σ^3 follows because, according to the theory, the compressibility factor z_c is the same for all fluids.

While parameters ε_i and σ_i are directly related to the macroscopic properties of substance i, the macroscopic properties used to estimate ε_i and σ_i need not be critical properties. For example, ε_i can also be evaluated from the Boyle temperature and σ_i from the molar volume at the normal boiling point. Nor is it necessary, in principle, that the molecular parameters be determined from thermodynamic data since these parameters are also related to transport properties like viscosity and diffusivity (Hirschfelder *et al.*, 1964). However, potential parameters obtained from different properties of the same fluid tend to be different because the assumed potential function (e.g., Lennard-Jones) is not the "true" potential function but only an approximation. Further, the four assumptions cited above are not strictly true for most substances. If a particular potential-energy function is to be used for calculating equilibrium properties, it is best to evaluate the parameters from a property similar to the one being investigated.

An important advantage of the molecular theory, relative to the classical theory of corresponding states, is that the former permits calculation of other macroscopic properties (e.g. transport properties) in addition to those which may be calculated by classical thermodynamics from an equation of state. For the purposes of phase-equilibrium thermodynamics, however, the main advantage of the molecular theory is that it can be meaningfully extended to mixtures, thereby providing some aid in typical phase-equilibrium problems.

For mixtures, Eq. (4-67) can be written as

$$Z_N = \int_V \cdots \int \exp\left(-\frac{\Gamma_t}{kT}\right)(dr_A)^{N_A}(dr_B)^{N_B}\ldots \tag{4-82}$$

where $(dr_i)^{N_i}$ is an abbreviated notation for the coordinates of the centers of the N_i molecules of kind i. Again, Γ_t is given by Eq. (4-69), but in addition to the pair interactions between similar molecules, Eq. (4-69) now includes also those between dissimilar molecules. In view of the physical significance of the parameters ε_i and σ_i, it is possible to make reasonable predictions of what these parameters are for the interaction between dissimilar molecules. Thus, as a first approximation, the London theory suggests that, for the interaction of two unlike molecules i and j having nearly the same size and ionization potential,

$$\varepsilon_{ij} = (\varepsilon_i \varepsilon_j)^{1/2} \tag{4-83}$$

and on the basis of a hard-sphere model for molecular interaction,

$$\sigma_{ij} = \frac{1}{2}(\sigma_i + \sigma_j) \tag{4-84}$$

Equations (4-83) and (4-84) provide a basis for obtaining some properties of a variety of mixtures. Some applications of the molecular theory of corresponding states to phase-equilibrium calculations are discussed in later chapters.

4.13 Extension of Corresponding-States Theory to More Complicated Molecules

The theory of corresponding states as expressed by the generalized potential function [Eq. (4-64)] is a two-parameter theory and is therefore limited to those molecules whose pair-wise energies of interaction can be adequately described in terms of a function using only two parameters. Such molecules are called *simple* molecules; strictly speaking, only the heavier noble gases (argon, krypton, xenon) are "simple" but the properties of several others are closely approximated by Eq. (4-64). A simple molecule is one whose force field has a high degree of symmetry; that is equivalent to saying that the potential energy is determined only by the distance of separation and not by the relative orientation between two molecules. Nonpolar (or slightly polar) molecules like methane, oxygen, nitrogen, and carbon monoxide are therefore nearly simple molecules. For more complex molecules, however, it is necessary to introduce at least one additional parameter in the potential function and thereby to construct a three-parameter theory of corresponding states. This can be done in several ways but the most convenient for practical purposes is to divide molecules into different classes, each class corresponding to a particular extent of deviation from simple-molecule behavior (Pitzer *et al.*, 1955, 1957, 1958). This extension of the theory of corresponding states relaxes the fourth assumption of the four listed in the preceding section. Assumptions 1 and 2 are unaffected. Assumption 3 is relaxed somewhat in the sense that we now assume that it is sufficient to use an average potential function wherein we have averaged out all effects of asymmetry in the intermolecular forces. Extension of corresponding-states theory is mostly concerned with assumption 4.

In the three-parameter theory of corresponding states, Eq. (4-64) still applies but the generalized function F is now different for each class. The class must be designated by some third parameter; for practical purposes, a convenient parameter is one that is easily calculated from readily available data. Different parameters have been proposed but Pitzer's proposal is, perhaps, the most useful because his third parameter is calculated from experimental data that tend to be accurate as well as accessible. Pitzer defines an *acentric factor* ω that is a measure of the acentricity, i.e., the noncentral nature of intermolecular forces.

The definition of the acentric factor is arbitrary and chosen for convenience. According to two-parameter (simple-fluid) corresponding-states, the reduced saturation pressures of all liquids should be a universal function of their reduced temperatures; in fact, however, they are not, and Pitzer uses this empirical result as a measure of deviation from simple-fluid behavior. For simple fluids, it has been observed that at a temperature equal to 7/10 of the critical, the saturation pressure P^s divided by the critical pressure P_c is given very closely by

$$\frac{P^s}{P_c} = \frac{1}{10} \quad \text{at} \quad \frac{T}{T_c} = 0.7 \tag{4-85}$$

Pitzer therefore defines the acentric factor by

$$\boxed{\omega \equiv -\log\left(\frac{P^s}{P_c}\right)_{T/T_c=0.7} - 1.000} \tag{4-86}$$

For simple fluids, $\omega \approx 0$ and for more complex fluids, $\omega > 0$.[31] Acentric factors for some typical fluids are given in Table 4-14. (Appendix J presents a more extensive list). The acentric factor is easily determined from a minimum of experimental information; the data required are the critical temperature, the critical pressure, and the vapor pressure at a reduced temperature of 0.7. This reduced temperature is usually close to the normal boiling point where vapor-pressure data are most likely to be available.[32]

Table 4-14 Acentric factors.

Molecule	ω	Molecule	ω
CH_4	0.008	C_6H_6	0.212
O_2	0.021	CO_2	0.225
N_2	0.040	NH_3	0.250
CO	0.049	$CFCl_3$	0.188
C_2H_4	0.085	n-C_6H_{14}	0.296
C_2H_6	0.098	iso-C_8H_{18}	0.303
CF_4	0.191	H_2O	0.344
n-C_4H_{10}	0.193	n-C_8H_{18}	0.394

The *three-parameter corresponding-states theory* asserts that all fluids having the same acentric factor have the same reduced configurational properties at the same reduced temperature T_R and reduced pressure P_R. Pitzer and Brewer have tabulated these reduced configurational properties as determined from experimental data for representative fluids (Pitzer, 1995) as functions of T_R and P_R. Any property of a fluid, in reduced form, is assumed to be given by a function of the three variables: P_R, T_R, and ω.

For example, for the compressibility factor z, Pitzer used a truncated Taylor series in ω:

$$z(P_R,T_R,\omega) = z^{(0)}(P_R,T_R,\omega = 0) + \omega\, z^{(1)}(P_R,T_R) \tag{4-87}$$

where $z^{(0)}$ is the compressibility factor for a simple fluid (e.g., argon or methane for which $\omega = 0$) and $z^{(1)}$ represents the deviation of the real fluid from $z^{(0)}$. With this approach, Pitzer correlated volumetric and thermodynamic properties of normal fluids[33] and their mixtures, using functions $z^{(0)}$ and $z^{(1)}$ in tabular form, over the range of reduced temperatures, 0.8 to 4.0, and reduced pressures, 0 to 9.0.

[31] For quantum gases (He, H_2, Ne), ω is slightly negative. Since Pitzer's theory of corresponding states is not applicable to quantum gases, these negative acentric factors are not significant.

[32] Pitzer (1995) also discuss methods to determine ω in case vapor-pressure data are not available.

[33] Fluids that have strong hydrogen bonds, large dipole moments or quantum effects are excluded from the normal category.

However, the original three-parameter Pitzer correlation is inadequate for calculations performed in the critical region and for liquids at low temperatures. To overcome these limitations, several modifications as well as extensions of Pitzer's work to wider ranges of T_R and P_R have been published (Lu *et al.*, 1973; Schreiber and Pitzer, 1989), as briefly mentioned in Sec. 5.7. A comprehensive compilation was presented by Lee and Kesler (1975).

In the generalized three-parameter correlation of Lee and Kesler, the volumetric and thermodynamic functions (e.g., densities, fugacity coefficients, second virial coefficients, etc.) of fluids are analytically represented by a modified Benedict-Webb-Rubin (BWR) equation of state. This equation is written first for a simple fluid (referred to by superscript 0), such as argon or methane, and then for a reference fluid (superscript r), chosen to be n-octane. The compressibility factor z for a normal fluid at reduced temperature T_R, reduced volume v_R, and acentric factor ω is written in the form

$$z(T_R,v_R,\omega)=z^{(0)}(T_R,v_R,\omega=0)+\frac{\omega}{\omega^{(r)}}[z^{(r)}(T_R,v_R,\omega^{(r)})-z^{(0)}(T_R,v_R,\omega=0)] \quad (4\text{-}88)$$

In Eq. (4-88), the compressibility factors of both the simple fluid, $z^{(0)}$, and the reference fluid, $z^{(r)}$, are represented by the following reduced form of a modified BWR equation of state:

$$z=\frac{P_R v_R}{T_R}=1+\frac{b^*}{v_R}+\frac{c^*}{v_R^2}+\frac{d^*}{v_R^5}+\frac{c_4}{T_R^3 v_R^2}\left(\beta+\frac{\gamma}{v_R^2}\right)\exp\left(-\frac{\gamma}{v_R^2}\right) \quad (4\text{-}89)$$

where constants b^*, c^*, and d^* are functions of the reduced temperature. These constants, as well as constants c_4, β, and γ, are determined for the simple fluid from data for argon and methane, and for the reference fluid from data for n-octane. Lee and Kesler give the constants of Eq. (4-89) and also expressions for several thermodynamic functions derived from the same equation. To extend their correlation to mixtures, Lee and Kesler provide the necessary mixing rules.

Inclusion of a third parameter very much improves the accuracy of corresponding-states correlations. For normal fluids (other than those that are highly polar or those that associate strongly by hydrogen bonding), the accuracy of the gas-phase compressibility factors given in the tables of Lee and Kesler is 2% or better, and for many fluids it is much better. The tables are not applicable to strongly polar fluids, although they are often so used with surprising accuracy except at low temperatures near the saturated-vapor region.

4.14 Summary

Both *physical* and *chemical* forces play an important role in determining properties of solutions. In some cases chemical forces can be neglected, thereby leading to purely "physical" solutions, but in other cases chemical forces predominate. The dissolved condition, therefore, represents a state of wide versatility where in one extreme the solvent is merely a diluent with respect to the solute, while in the other extreme it is a chemical reactant.

The aim in applying thermodynamic methods to phase-equilibrium problems is to order, interpret, correlate, and finally, to predict properties of solutions. The extent to which this aim can be fulfilled depends in large measure on the degree of our understanding of intermolecular forces that are responsible for the molecules' behavior.

Intermolecular forces can be roughly classified into three categories. First, there are those that are purely *electrostatic* arising from the Coulomb force between charges. The interactions between charges, permanent dipoles, and quadrupoles, presented in Sec. 4.2, fall into this category. Second, there are *polarization forces* (Sec. 4.3) that arise from the dipole moments induced in the atoms and molecules by the electric fields of nearby charges of permanent dipoles. Finally, there are forces that are *quantum-mechanical* in nature. Such forces give rise to *covalent bonding* (including charge-transfer interactions) and to the *repulsive interactions* (due to the Pauli exclusion principle) that balance the attractive forces at very short distances.

These three categories are not rigid nor exhaustive, since for certain types of forces (e.g. van der Waals forces) an unambiguous classification is not possible, while other intermolecular interactions (such as magnetic forces) were not considered since they are always very weak for the systems of interest here.

Our quantitative knowledge of two-body intermolecular forces is limited to simple systems under ideal conditions, i.e. when the two molecules under consideration are isolated from all others. For non-simple systems, the effect of molecular structure and shape is often large but we have no adequate tools for describing such effects in a truly fundamental way; the best we can do is to estimate such effects by *molecular simulation* on computers.

Chemical forces (formation of weakly bonded dimers, trimers, etc.) are often dominant in determining thermodynamic properties. Numerous experimental methods (especially spectroscopy) can be used to augment and support thermodynamic measurements but as yet, our fundamental understanding of chemical forces is not satisfying.

For large molecules in solution, osmometry and light scattering provide information on intermolecular forces through the *potential of mean force*. Structural effects, as the hydrophobic effect, can exert a large influence.

For normal fluids, the theorem of corresponding states provides a powerful tool for estimating thermodynamic properties when experimental data are scarce.

Classical and *statistical* thermodynamics can define useful functions and derive relationships between them, but the numerical values of these functions cannot be determined by thermodynamics alone. Determination of numerical values, by either theory or experiment is, strictly speaking, outside the realm of thermodynamics; such values depend directly on the microscopic physics and chemistry of the molecules in the mixture. Future progress in phase-equilibrium thermodynamics depends, in part, on progress in statistical mechanics. However, progress in applications of phase-equilibrium thermodynamics is possible only with increased knowledge of intermolecular forces.

References

Agarwal, S., 1978, *J. Solution Chem.*, 7: 795.

Amdur, I. and A. L. Harkness, 1954, *J. Chem. Phys.*, 22: 664; Amdur, I. and E. A. Mason, *ibid.*, 670; Amdur, I., E. A. Mason, and A. L. Harkness, *ibid.*, 1071.

Andrews, L. J. and R. M. Keefer, 1964, *Molecular Complexes in Organic Chemistry.* San Francisco: Holden-Day.

Booth, D., F. S. Dainton, and K. J. Ivin, 1959, *Trans. Faraday Soc.*, 55: 1293.

Briegleb, G., 1961, *Elektronen-Donator-Acceptor Komplexe.* Berlin: Springer.

Buckingham, A. D., 1959, *Q. Rev. (Lond.)*, 13: 183.

Buckingham, A. D. and R. L. Disch, 1963, *Proc. R. Soc. (Lond.)*, A273: 275.

Campbell, A. N. and E. M. Kartzmark, 1960, *Can. J. Chem.*, 38: 652.

Chen, S. H. and R. Rajagoplan, 1990, *Micellar Solutions and Microemulsions.* Berlin: Springer.

Christian, S. D. and E. H. Lane, 1975. In *Solutions and Solubilities*, (M. R. J. Dack, Ed.), Chap. 6. New York: John Wiley & Sons.

D'Ans-Lax, 1967, *Taschenbuch für Chemiker und Physiker*, Vol. I. Berlin: Springer.

Davis, K. M. C., 1975. In *Molecular Association*, (R. Foster, Ed.). New York: Academic Press.

Delmas, G. and P. Purves, 1977, *J. Chem. Sac. Faraday Trans. II*, 73: 1828, 1838.

Dolezalek, F., 1908, *Z. Phys. Chem.*, 64: 727.

Flygare, W. H. and R. C. Benson, 1971, *Mol. Phys.*, 20: 225.

Foster, R., (Ed.), 1973, *Molecular Complexes*, Vol I. New York: Crane & Russak.

Foster, R., (Ed.), 1974, *Molecular Complexes*, Vol II. New York: Crane & Russak.

Guggenheim, E. A., 1945, *J. Chem. Phys.*, 13: 253.

Gutmann, V., 1978, *The Donor-Acceptor Approach to Molecular Interactions.* New York: Plenum Press.

Gutmann, F., C. Johnson, H. Keyzer, and L. Molnar, 1997, *Charge Transfer Complexes in Biological Systems.* New York: Marcel Dekker.

Haynes, C. A., K. Tamura, H. R. Körfer, H. W. Blanch, and J. M. Prausnitz, 1992, *J. Phys. Chem.*, 96: 505.

Heintz, A. and R. N. Lichtenthaler, 1977, *Ber. Bunsenges. Phys. Chem.*, 81: 921.

Heintz, A. and R. N. Lichtenthaler, 1984, *Angew. Chem. Int. Ed. Engl.*, 21: 184.

Hill, T. L., 1959, *J. Chem. Phys.*, 30: 93.

Hill, T. L., 1986, *An Introduction to Statistical Thermodynamics*. Reading: Addison-Wesley.

Hirschfelder, J. O., C. F. Curtiss, and R. B. Bird, 1964, *Molecular Theory of Gases and Liquids*. New York: John Wiley & Sons.

Hoffmann, H. and W. Ulbricht, 1986, *The Formation of Micelles*. In *Thermodynamic Data for Biochemistry and Biotechnology*, (H.-J. Hinz, Ed.). Berlin: Springer.

Huyskens, P. L., W. A. P. Luck, and T. Zeegers-Huyskens, (Eds.), 1991, *Intermolecular Forces – An Introduction to Modern Methods and Results*. Berlin: Springer.

Israelachvili, J. N., 1992, *Intermolecular and Surface Forces*, 2nd Ed. San Diego: Academic Press.

Keesom, W. H., 1922, *Comm. Leiden, Supl.* 24a, 24b; *Phys. Z.*, 22: 129.

Kuznetsov, A. M., 1995, *Charge Transfer in Physics, Chemistry and Biology: Physical Mechanisms of Elementary Processes and an Introduction to the Theory*. London: Gordon & Breach.

Lee, B. I. and M. G. Kesler, 1975, *AIChE J.*, 21: 510.

Lichtenthaler, R. N. and K. Schäfer, 1969, *Ber. Bunsenges. Phys. Chem.*, 73: 42.

London, F., 1937, *Trans. Faraday Soc.*, 33: 8.

Lu, B. C.-Y., J. A. Ruether, C. Hsi, and C.-H. Chiu, 1973, *J. Chem. Eng. Data.*, 18: 241.

Mahl, B. S., Z. S. Kooner, and J. R. Khurma, 1978, *J. Chem. Eng. Data*, 23: 150.

McCarty, B. W. and E. T. Adams, Jr., 1987, *Biophys. Chem.*, 28: 149.

McMillan, W. G. and J. E. Mayer, 1945, *J. Phys. Chem.*, 13: 276.

Mie, G., 1903, *Ann. Phys.*, 11: 657.

Moelwyn-Hughes, E. A., 1961, *Physical Chemistry*, 2nd Ed. Oxford: Pergamon Press.

Mulliken, R. S., 1952, *J. Am. Chem. Soc.*, 74: 811.

Mulliken, R. S. and W. B. Person, 1969, *Molecular Complexes: A Lecture and Reprint Volume*. New York: Wiley-Interscience.

Orye, R. V. and J. M. Prausnitz, 1965, *Trans. Faraday Soc.*, 61: 1338.

Orye, R. V., R. F. Weimer, and J. M. Prausnitz, 1965a, *Science*, 148: 74.

Parsonage, N. G. and R. L. Scott, 1962, *J. Chem. Phys.*, 37: 304.

Pimentel, G. C. and A. L. McClellan, 1960, *The Hydrogen Bond*. San Francisco: W. H. Freeman.

Pitzer, K. S., 1939, *J. Chem. Phys.*, 7: 583.

Pitzer, K. S., D. Z. Lippman, R. F. Curl, Jr., C. M. Huggins, and D. E. Petersen, 1955, *J. Am. Chem. Soc.*, 77: 3427, 3433; *ibid.*, 1957, *J. Am. Chem. Soc.*, 79: 2369; *ibid.*, 1958, *Ind. Eng. Chem.*, 50: 265.

Pitzer, K. S., 1995, *Thermodynamics*, 3rd Ed. New York: McGraw-Hill.

Rose, J., 1967, *Molecular Complexes*. Oxford: Pergamon Press.

Rossotti, F. J. C. and H. Rossotti, 1961, *The Determination of Stability Constants*. New York: McGraw-Hill.

Schreiber, D. R. and K. S. Pitzer, 1989, *Fluid Phase Equilibria*, 46: 113.

Schuster, P., G. Zundel, and C. Sandorfy, (Eds.), 1976, *The Hydrogen Bond – Recent Developments in Theory and Experiment*. Amsterdam: North-Holland.

Stone, A. J., 1996, *The Theory of Intermolecular Forces*. Oxford: Clarendon Press.

Sutter, D. H. and W. H. Flygare, 1976, *Top. Curr. Chem.*, 63: 89.

Tamres, M., 1973, in Foster (1973), p. 49.

Tancrède, P., P. Bothorel, P. de St. Romain, and D. Patterson, 1977, *J. Chem. Soc. Faraday Trans. II*, 73: 15, 29.

Tanford, C., 1980, *The Hydrophobic Effect: Formation of Micelles and Biological Membranes*, 2nd Ed. New York: John Wiley & Sons.

Te Lam, V., P. Picker, D. Patterson, and P. Tancrède, 1974, *J. Chem. Soc. Faraday Trans. II*, 70: 1465.

Tucker, E. E. and S. D. Christian, 1976, *J. Am. Chem. Soc.*, 98: 6109.

Weimer, R. F. and J. M. Prausnitz, 1966, *Spectrochim. Acta*, 22: 77.

Problems

1. Consider one molecule of nitrogen and one molecule of ammonia 25 Å apart and at a temperature well above room temperature. Compute the force acting between these molecules. Assume that the potential energies due to various causes are additive and use the simple, spherically symmetric formulas for these potentials.

2. Experimental studies show that when two molecules of methane are 1 nm apart, the force of attraction between them is 2×10^{-8} dyne. Using this result, we now want to estimate the force of attraction between two molecules of some substance B. We know little about substance B, but we know that its molecules are small and nonpolar. Further, we have critical data:

	T_c (K)	v_c (cm^3 mol^{-1})
Methane	191	99
Substance B	300	125

Using the theory of intermolecular forces and corresponding-states theory, estimate the force of attraction between two molecules of substance B when they are 2 nm apart.

3. Experimental studies for a simple, nonpolar gas A indicate that when two molecules of species A are 2 molecular diameters apart, the potential energy is -8×10^{-16} erg. Consider now two simple, nonpolar molecules of species B. When these two molecules are 2 molecular diameters apart, what is the potential energy?
Critical data for A and B are:

	T_c (K)	P_c (bar)	v_c (cm^3 mol^{-1})
A	120	31.9	90
B	180	36.1	120

4. The dipole moment of HCl is 3×10^{-30} C m and its (mean) polarizability is $(\alpha/4\pi\varepsilon_o) = 2.63\times10^{-30}$ m^3. For a center of mass separation of 0.5 nm, calculate the contributions to the intermolecular energy from dipole-dipole and from dipole-induced dipole interactions for the two relative orientations, $\rightarrow\rightarrow$ and $\rightarrow\uparrow$.

5. Consider a spherical, nonpolarizable molecule of radius 3 Å having at its center a dipole moment of 2 debye. This molecule is dissolved in a nonpolar liquid having a dielectric constant equal to 3.5. Calculate the energy that is required to remove this molecule from the solution.

(a) Consider a gaseous mixture of N_2 and CO and compare it with a gaseous mixture of N_2 and Ar. Which mixture is more likely to follow Amagat's law (approximately)? Why?

(b) Consider a diatomic molecule such as carbon monoxide. What is meant by the force constant of the C—O bond? How is it measured? Why is the heat capacity of CO significantly larger than that of argon?

6. What is meant by the terms "electron affinity" and "ionization potential"? How are these concepts related to the Lewis definition of acids and bases? In the separation of aromatics from paraffins by extraction, why is liquid SO_2 a better solvent than liquid NH_3?

7. Suppose that you want to measure the dipole moment of chlorobenzene in an inert solvent such as *n*-heptane. What experimental measurements would you make and how would you use them to compute the dipole moment?

8. Consider a binary solution containing components 1 and 2. The Lennard-Jones parameters ε_{11}, ε_{22}, and $\sigma_{11}$and σ_{22} are known. Assuming that $\sigma_{12} = 1/2(\sigma_{11} + \sigma_{22})$, express ε_{12} in terms of pure-component parameters. Also find under which conditions $\varepsilon_{12} = (\varepsilon_{11}\,\varepsilon_{22})^{1/2}$. (Consider the attractive part only.)

9. What is a hydrogen bond? Cite all the experimental evidence that supports the conclusion that phenol is a hydrogen-bonded substance.

10. Qualitatively compare the activity coefficient of acetone when dissolved in carbon tetrachloride with that when dissolved in chloroform.

11. The polymer polypropylene oxide, a polyether, is to be dissolved in a solvent at ambient temperature.

(a) Consider these solvents: chlorobenzene, cyclohexane, and chloroform. Which of these is likely to be the best solvent? The worst solvent? Explain.

(b) Would *n*-butanol be a good solvent? Would tertiary butanol be better? Explain. (The normal boiling point of *n*-butanol is 117.5°C; that of *t*-butanol is 82.9°C.)

(c) Consider a polymer like cellulose nitrate. Explain why (as observed) a mixture of two polar solvents is frequently more effective in dissolving this polymer than either polar solvent by itself.

(d) Suppose that you wished to evaporate hydrogen cyanide from a solution in carbon tetrachloride or in octane. Which case is likely to require more heat?

12. Sketch qualitatively a plot of compressibility factor z versus mole fraction (from zero to unity) at constant temperature and pressure, for the following mixtures at 25 bar:

(a) Dimethylamine/hydrogen at 170°C.

(b) Dimethylamine/hydrogen chloride at 170°C.

(c) Argon/hydrogen chloride at 80°C.

In the sketch indicate where $z = 1$.

13. Using your knowledge of intermolecular forces, explain the following observations:

(a) At 30°C, the solubilities of ethane and acetylene in *n*-octane are about the same. However, at the same temperature, the solubility of acetylene in dimethylformamide is very much larger than that of ethane.

(b) At 10°C and 40 bar total pressure, the K factor for benzene in the methane/benzene system is much larger than that in the hydrogen/benzene system, at the same temperature and pressure ($K = y/x$). However, at 10°C and 3 bar the two K factors are nearly the same.

(c) Hydrogen gas at 0°C contains 1 mol % carbon dioxide. When this gas is compressed isothermally to pressure P, carbon dioxide condenses. However, when methane gas, also at 0°C and also containing 1 mol % carbon dioxide, is isothermally compressed to the same pressure P, carbon dioxide does *not* condense.

(d) At 100°C and 50 bar, the compressibility factor z for ethane gas is less than unity. However, at the same temperature and pressure the compressibility factor for helium gas is larger than unity.

(e) Which among the following liquids would be the best solvent for poly(vinyl chloride), $-(-CH_2CHCl-)_n-$: i) *n*-Heptane; ii) Ethanol; iii) Cyclohexanone; iv) Chlorobenzene.

(f) State what is the lowest nonvanishing multipole moment (e.g. dipole, quadrupole, octopole, etc.) for each of the following molecules: i) Dichlorodifluoromethane; ii) Carbon tetrafluoride; iii) Carbon dioxide; iv) Mesitylene (1,3,5-trimethylbenzene).

(g) An absorber operating at 0°C and 600 psia uses heptane to absorb ethane and propane from natural gas. It is found that heptane losses due to evaporation constitute a significative economic cost and it is therefore decided to lower the operating temperature to -20°C. However, after this is done, it is found that heptane losses have increased rather than decreased. Explain.

14. What are the assumptions of the molecular theory of corresponding states?

(a) Does the Kihara potential (see Chap. 5) necessarily violate any of these assumptions? Explain.

(b) Which of these assumptions, if any, are *not* obeyed by hydrogen?
(c) Does the molecular theory of corresponding states have anything at all to say about c_p (specific heat at constant pressure)? If so, what?

15. A small drop of water at 25°C contains 0.01 molal potassium nitrate. The drop is completely surrounded by a semipermeable membrane; it is placed in an aqueous solution containing 0.01 M sodium nitrate and 2 g L^{-1} lysozyme at 25°C. The electric charge on lysozyme is -2. The semi-permeable membrane has a cut-off molecular weight of 8,000. The molar mass of lysozyme is 14,000 g mol^{-1}. What is the osmotic pressure in the drop?

16. A container at 300K is divided into two parts, α and β, separated by a semipermeable membrane which is permeable to the solvent (water) but not to the solute, a protein A at its isoelectric point. The molar mass of the protein is 5,000 g mol^{-1}.
Part β contains pure water. Part α contains protein A in aqueous solution; its concentration is 5 g/liter. Protein A has a tendency to dimerize according to

$$2A \rightleftharpoons A_2$$

The equilibrium constant for dimerization is 10^5.
What is the osmotic pressure in part α ?
Assume that at this low protein concentration, the liquid in part α behaves as an ideal dilute solution. In this case, ideal dilute solution is defined by relating the chemical potential of a solute i to its mole fraction x_i, according to

$$\mu_i = \text{constant} + RT \ln x_i$$

17. At 25°C osmotic-pressure data for aqueous solutions of the protein bovine serum albumin (BSA) in 0.15 M NaCl at pH 5.37 and 7.00 is given below. The isoelectric point of BSA is 5.37 in 0.15 M sodium chloride aqueous solutions.
(a) Estimate the molecular weight and the specific volume of BSA in the solution at pH 5.37.
(b) The specific volume of the unsolvated protein (which is considered spherical) is about 0.75 g cm^{-3}, and the excluded volume of the molecule is estimated to be 2.95×10^{-25} m^3 $molecule^{-1}$. At pH 5.37, does the solute appear to be solvated?
(c) For a charged particle in the presence of a low-molecular weight salt, the contribution of charge to the second osmotic coefficient is given by

$$B = \frac{1000z^2}{4M_2^2\rho_1 m_{MX}}$$

where z is the charge of the particle of molecular weight M_2, ρ_1 is the density of the solution, and m_{MX} is the concentration (molality) of the salt. From the osmotic-pressure data above, estimate the charge of BSA.

pH 5.37		pH 7.00	
BSA concentration (g/kg water)	π (mmHg)	BSA concentration (g/kg water)	π (mmHg)
8.95	2.51	16.76	5.19
17.69	5.07	29.46	9.90
27.28	8.35	50.15	19.27
56.20	19.33	55.92	21.61
58.50	20.12	56.10	24.04
61.30	22.30		

18. By membrane osmometry, osmotic pressures of aqueous solutions of the nonionic surfactant *n*-dodecylhexaoxyethylene monoether, $C_{12}H_{25}(OC_2H_4)_4OC_2H_5$, were measured at 25°C. At concentrations below $c_0 = 0.038$ g L^{-1}, no osmotic pressure develops, indicating membrane permeation by the micelar species. Above this concentration, an osmotic pressure is measured, indicating the presence of impermeable aggregate species. The following table gives osmotic pressure data for various $c - c_0$ values:

π (cm of solvent)	$c - c_0$ (g L^{-1})
4.90	29.72
6.53	38.12
7.62	43.90
10.58	58.46

a) Obtain the second virial osmotic coefficient and the molecular weight for the species responsible for the osmotic pressure.

b) Determine the number of molecules in the aggregate. Assuming they are spherical, estimate its molar volume and radius.

CHAPTER 5

Fugacities in Gas Mixtures

It was shown in Chap. 2 that the basic equation of equilibrium between two phases α and β, at the same temperature, is given by equality of fugacities for any component i in these phases:

$$f_i^{\alpha} = f_i^{\beta}$$

In many cases one of the phases is a gaseous mixture; in this chapter we discuss methods for calculating the fugacity of a component in such a mixture.

Formal thermodynamics for calculating fugacities from volumetric data was discussed in Chap. 3; the two key equations are:

$$\ln \varphi_i = \frac{1}{RT}\int_0^P \left[\left(\frac{\partial V}{\partial n_i}\right)_{T,P,n_j} - \frac{RT}{P}\right] dP \tag{5-1}$$

and

$$\ln \varphi_i = \frac{1}{RT} \int_V^{\infty} \left[\left(\frac{\partial P}{\partial n_i} \right)_{T,V,n_j} - \frac{RT}{V} \right] dV - \ln z \tag{5-2}$$

where the *fugacity coefficient* φ_i is defined by

$$\varphi_i \equiv \frac{f_i}{y_i P} \tag{5-3}$$

and z is the compressibility factor of the mixture.

Equation (5-l) is used whenever the volumetric data are given in volume explicit form; i.e., whenever

$$V = F_V(T, P, n_1, \ldots) \tag{5-4}$$

Equation (5-2) is used in the more common case when the volumetric data are expressed in pressure-explicit form, i.e., whenever

$$P = F_P(T, V, n_1, \ldots) \tag{5-5}$$

The mathematical relation between volume, pressure, temperature, and composition is called the *equation of state* and most forms of the equation of state are pressure-explicit. Therefore, Eq. (5-2) is frequently more useful than Eq. (5-1). At low or moderate densities, it is often possible to describe volumetric properties of a gaseous mixture in a volume-explicit form; in that case Eq. (5-1) can be used. At high densities, however, volumetric properties are much better represented in pressure-explicit form, requiring the use of Eq. (5-2).

Equations (5-l) and (5-2) are exact and if the information needed to evaluate the integrals is at hand, then the fugacity coefficient can be calculated exactly. The problem of calculating fugacities in the gas phase, therefore, is equivalent to the problem of estimating volumetric properties. Techniques for estimating such properties must come not from thermodynamics, but rather from molecular physics; it is for this reason that Chap. 4, concerned with intermolecular forces, precedes Chap. 5.

5.1 The Lewis Fugacity Rule

A particularly simple and popular approximation for calculating fugacities in gas-phase mixtures is given by the *Lewis rule*; the thermodynamic basis of the Lewis rule has already been given (Sec. 3.1). The assumption on which the rule rests states that at constant temperature and pressure, the molar volume of the mixture is a linear function of the mole fraction. This assumption (Amagat's law) must hold not only at the pressure of interest but for all pressures up to the pressure of interest.

As shown in Sec. 3.1, the fugacity of component i in a gas mixture can be related to the fugacity of pure gaseous i at the same temperature and pressure by the exact relation

$$RT \ln \frac{f_i}{y_i f_{\text{pure } i}} = \int_0^P (\overline{v}_i - v_i)\, dP \tag{5-6}$$

where $\overline{v}_i$ is the partial molar volume, defined shortly after Eq. (3-14). According to Amagat's law, $\overline{v}_i = v_i$, and assuming validity of this equality over the entire pressure range $0 \rightarrow P$, the Lewis fugacity rule follows directly from Eq. (5-6):

$$f_i = y_i f_{\text{pure } i} \tag{5-7}$$

or, in equivalent form,

$$\varphi_i = \varphi_{\text{pure } i} \tag{5-8}$$

where $f_{\text{pure } i}$ and $\varphi_{\text{pure } i}$ are evaluated for the pure gas at the same temperature and pressure as those of the mixture.

In effect, the Lewis rule assumes that at constant temperature and pressure, the fugacity coefficient of i is independent of the composition of the mixture and is independent of the nature of the other components in the mixture. These are drastic assumptions. On the basis of our knowledge of intermolecular forces, we recognize that for component i, deviations from ideal-gas behavior (as measured by φ_i) depend not only on temperature and pressure, but also on the relative amounts of component i and other components j, k, ...; further, we recognize that φ_i must depend on the chemical nature of these other components that interact with component i. The Lewis rule, however, precludes such dependence; according to it, φ_i is a function only of temperature and pressure but not of composition.

Nevertheless, the Lewis rule is frequently used because of its simplicity for practical calculations. We may expect that the partial molar volume of a component i is close to the molar volume of pure i at the same temperature and pressure whenever the intermolecular forces experienced by a molecule i in the mixture are similar to those that it experiences in the pure gaseous state. In more colloquial terms, a molecule i that feels "at home" while it is "with company", possesses properties in the mixture close to those it has in the pure state. It therefore follows that for a component i, the Lewis fugacity rule is:

- Always a good approximation at sufficiently low pressures where the gas phase is nearly ideal.

- Always a good approximation at any pressure whenever i is present in large excess (say, $y_i > 0.9$). The Lewis rule becomes exact in the limit as $y_i \rightarrow 1$.
- Often a fair approximation over a wide range of composition and pressure whenever the physical properties of all the components are nearly the same (e.g., nitrogen-carbon monoxide or benzene-toluene).
- Almost always a poor approximation at moderate and high pressures whenever the molecular properties of the other components are significantly different from those of i and when i is not present in excess. If y_i is small and the molecular properties of i differ much from those of the dominant component in the mixture, the error introduced by the Lewis rule may be extremely large (see Sec. 5.12).

One of the practical difficulties encountered in the use of the Lewis rule for vapor-liquid equilibria results from the frequent necessity of introducing a hypothetical state. At the temperature of the mixture, it often happens that P, the total pressure, exceeds P_i^s, the saturation pressure of pure component i. In that case $\varphi_{\text{pure } i}$, the fugacity coefficient of pure gas i at the temperature and pressure of the mixture, is fictitious because pure gas i cannot physically exist at these conditions. Calculation of $\varphi_{\text{pure } i}$ under such conditions requires assumptions about the nature of the hypothetical pure gas and consequently, further inaccuracies may result from use of the Lewis rule.

In summary, then, the Lewis fugacity rule is attractive because of convenience but it has no general validity. However, when applied in certain limiting situations, it frequently provides a good approximation.

5.2 The Virial Equation of State

As indicated at the beginning of this chapter, the problem of calculating fugacities for components in a gaseous mixture is equivalent to the problem of establishing a reliable equation of state for the mixture; once an equation of state exists, fugacities can be found by straightforward computation. Such computation presents no difficulties in principle although it may be tedious because it may require trial-and-error calculations.

Many equations of state have been proposed and each year additional ones appear in the literature, but most of them are either totally or at least partially empirical. All empirical equations of state are based on more or less arbitrary assumptions that are not generally valid. Because the constants in an empirical equation of state for a pure gas have at best only approximate physical significance, it is difficult (and frequently impossible) to justify mixing rules for expressing the constants of the mixture in terms of the constants of the pure components that comprise the mixture. As a result, because mixing rules introduce further arbitrary assumptions, for typical empirical equations of state, one set of mixing rules may work well for one or several mixtures but poorly for others (Cullen and Kobe, 1955).

To calculate with confidence fugacities in a gas mixture, it is advantageous to use an equation of state where the parameters have physical significance, i.e. where the parameters can be related directly to intermolecular forces. One equation of state that possesses this desirable ability is the *virial equation of state.*

Figure 5-1 shows a plot of the compressibility factor as a function of density for helium, methane, and three binary mixtures containing 10, 25 and 50 mole per cent water in methane. For these systems, the magnitudes of the intermolecular forces differ appreciably and depend strongly on density (or pressure). The compressibility factor for the methane mixture containing 10 mol % water deviates little from unity over a wide range of density, even when compared with the compressibility factors of pure methane or helium. However, as density increases, the compressibility factor plot for the same mixture shows a change in the slope from negative to positive. It is apparent that, to describe systems as those in Fig. 5-1, the parameters that appear in a gas-phase equation of state must account for a large variety of intermolecular forces that are responsible for the nonideality of the gas.

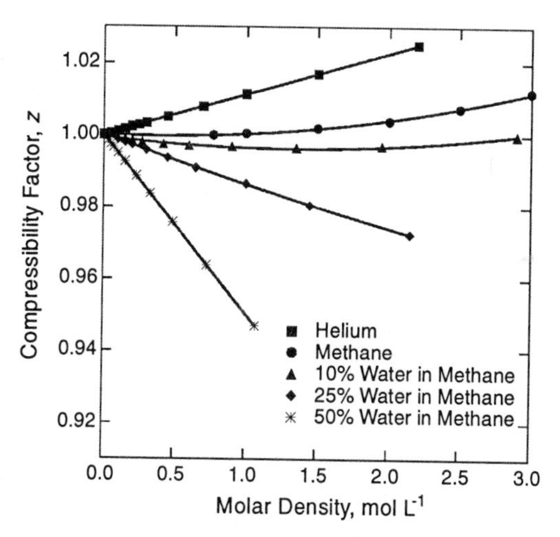

Figure 5-1 Compressibility factors for helium, methane and three water/methane mixtures as a function of density at 498.15 K (Joffrion and Eubank, 1988).

The virial equation of state for nonelectrolyte gases has a sound theoretical foundation, free of arbitrary assumptions (Mason and Spurling, 1969). The virial equation gives the compressibility factor as a power series in the reciprocal molar volume $1/v$:[1]

[1] Equation (5-9) is frequently written in the equivalent form

$$z = 1 + B\rho + C\rho^2 + D\rho^3 + \ldots,$$

where ρ, the molar density, is equal to $1/v$. It is most conveniently derived using the grand partition function (see App. B) as, for example, in Hill (1986). The derivation shows that the virial equation is remarkably general, provided that the intermolecular potential obeys certain well-defined restrictions.

$$z = \frac{Pv}{RT} = 1 + \frac{B}{v} + \frac{C}{v^2} + \frac{D}{v^3} + \ldots \qquad (5\text{-}9)$$

In Eq. (5-9), B is the second virial coefficient, C is the third virial coefficient, D is the fourth, and so on. All virial coefficients are independent of pressure or density; for pure components, they are functions only of the temperature. The unique advantage of the virial equation follows, as shown later, because there is a theoretical relation between the virial coefficients and the intermolecular potential. Further, in a gaseous mixture, virial coefficients depend on composition in an exact and simple manner.

The compressibility factor is sometimes written as a power series in the pressure:

$$z = \frac{Pv}{RT} = 1 + B'P + C'P^2 + D'P^3 + \ldots \qquad (5\text{-}10)$$

where coefficients B', C', D', . . . depend on temperature but are independent of pressure or density. For mixtures, however, these coefficients depend on composition in a more complicated way than do those appearing in Eq. (5-9). Relations between the coefficients in Eq. (5-9) and those in Eq. (5-10) are derived in App. C with the results

$$B' = \frac{B}{RT} \qquad (5\text{-}11)$$

$$C' = \frac{C - B^2}{(RT)^2} \qquad (5\text{-}12)$$

$$D' = \frac{D - 3BC + 2B^3}{(RT)^3} \qquad (5\text{-}13)$$

Equation (5-9) is usually superior to Eq. (5-10) in the sense that when the series is truncated after the third term, the experimental data are reproduced by Eq. (5-9) over a wider range of densities (or pressures) than by Eq. (5-10), provided that the virial coefficients are evaluated as physically significant parameters. In that case, the second virial coefficient B is properly evaluated from low-pressure P-V-T data by the definition

$$B = \lim_{\rho \to 0} \left(\frac{\partial z}{\partial \rho} \right)_T \qquad (5\text{-}14)$$

Similarly, the third virial coefficient must also be evaluated from *P-V-T* data at low pressures; it is defined by

$$C = \lim_{\rho \to 0} \frac{1}{2!}\left(\frac{\partial^2 z}{\partial \rho^2}\right)_T \tag{5-15}$$

Reduction of *P-V-T* data to yield second and third virial coefficients is illustrated in Fig. 5-2, taken from the work of Douslin (1962) on methane. In addition to Douslin's data, Fig. 5-2 also shows experimental results from several other investigators. The coordinates of Fig. 5-2 follow from rewriting the virial equation in the form

$$v\left(\frac{Pv}{RT} - 1\right) = B + \frac{C}{v} + \cdots \tag{5-16}$$

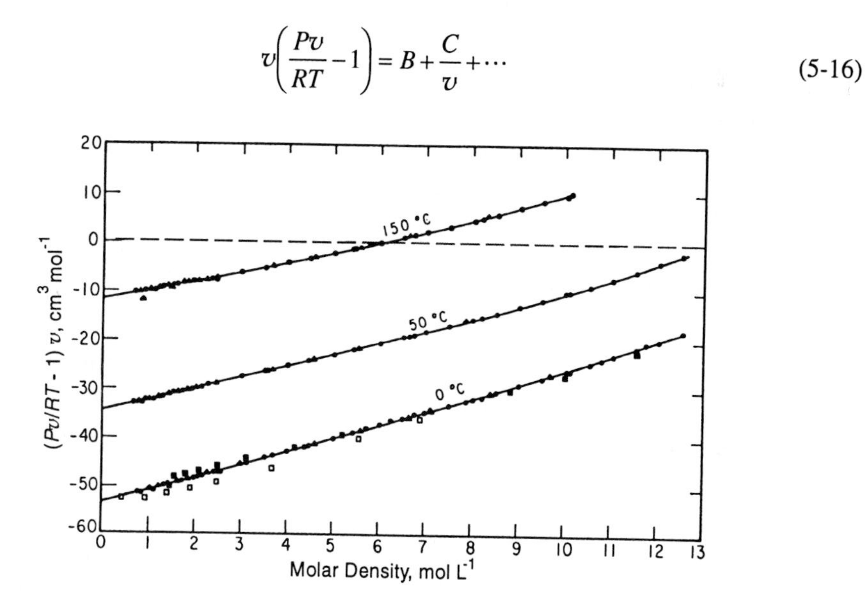

Figure 5-2 Reduction of *P-V-T* data for methane to yield second and third virial coefficients (data from various sources).

When isothermal data are plotted as shown, the intercept on the ordinate gives *B*, while *C* is found from the limiting slope as $1/v \to 0$. For mixtures, the same procedure is used, but in addition to isothermal conditions, each plot must also be at constant composition. An example is given in Fig. 5-3 for the mixture methanol/methyl acetate at several temperatures (Olf *et al.*, 1989).

An illustration of the applicability of Eqs. (5-9) and (5-10) at two different temperatures is given in Figs. 5-4 and 5-5, based on Michels' accurate volumetric data for argon (Michels *et al.*, 1960; Guggenheim and McGlashan, 1960; Munn, 1964; Munn *et al.*, 1965, 1965a). Using only low-pressure data along an isotherm, *B* and *C* were calculated as indicated by Eq. (5-16). These coefficients were then used to predict the compressibility factors at higher pressures (or densities). One prediction is based on Eq. (5-9) and the other on Eq. (5-10) together with Eqs. (5-11) and (5-12). Experimentally determined isotherms are also shown. In both cases, Eq. (5-9) is more successful than Eq. (5-10).[2]

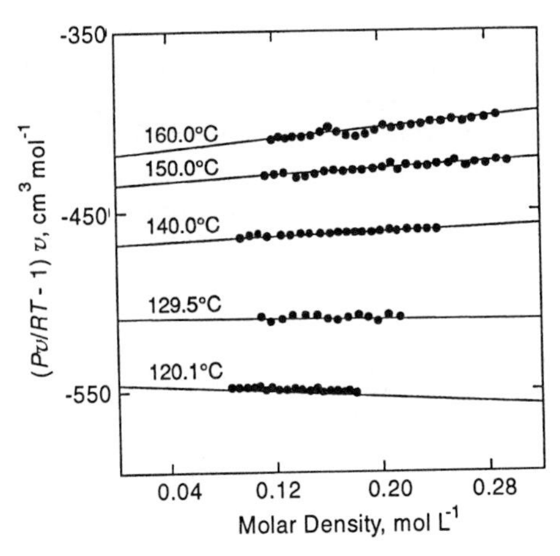

Figure 5-3 Reduction of *P-V-T* data for methanol/methyl acetate to yield second and third virial coefficients of approximately equimolar mixtures (Olf *et al.*, 1989).

For many gases, it has been observed that Eq. (5-9), when truncated after the third term (i.e., when D and all higher virial coefficients are neglected), gives a good representation of the compressibility factor to about one half the critical density and a fair representation to nearly the critical density.

For higher densities, the virial equation is of little practical interest. Experimental as well as theoretical methods are not sufficiently developed to obtain useful quantitative results for fourth and higher virial coefficients. The virial equation is, however, applicable to moderate densities as commonly encountered in many typical vapor-liquid and vapor-solid equilibria.

The significance of the virial coefficients lies in their direct relation to intermolecular forces. In an ideal gas, the molecules exert no forces on one another. In the real world, no ideal gas exists, but when the mean distance between molecules becomes very large (low density), all gases tend to behave as ideal gases. This is not surprising because intermolecular forces diminish rapidly with increasing intermolecular distance and therefore forces between molecules at low density are weak. However, as density rises, molecules come into closer proximity with one another and, as a result, interact more

---

[2] The comparisons shown in Figs. 5-4 and 5-5 are for the case where Eqs. (5-9) and (5-10) are truncated after the quadratic terms. When similar comparisons are made with these equations truncated after the linear terms, it often happens that, because of compensating errors, Eq. (5-10) provides a better approximation at higher densities than Eq. (5-9). See Chueh and Prausnitz (1967).

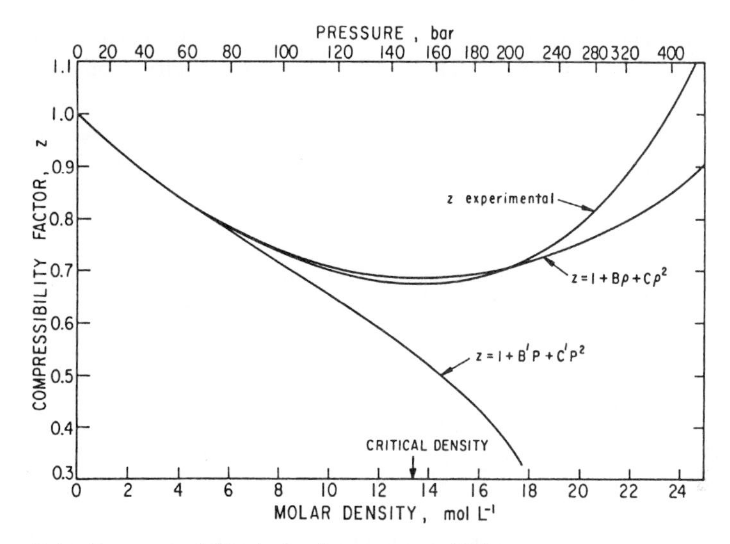

Figure 5-4 Compressibility factor for argon at -70°C.

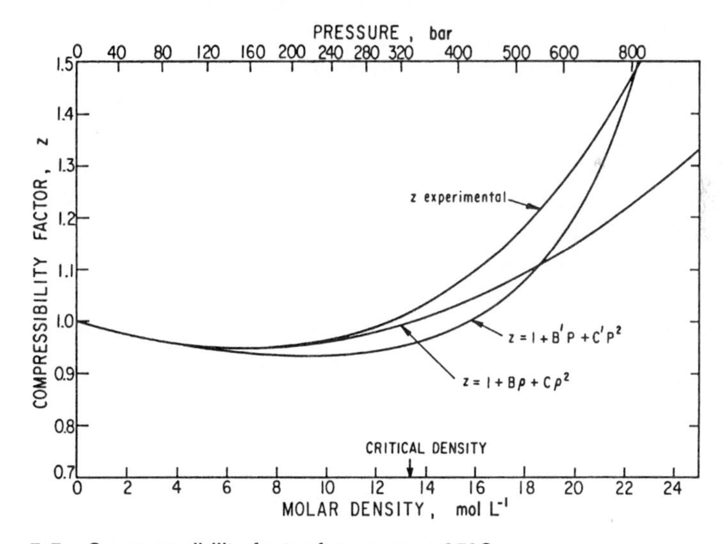

Figure 5-5 Compressibility factor for argon at 25°C.

frequently. The purpose of the virial coefficients is to take these interactions into account. The physical significance of the second virial coefficient is that it takes into account deviations from ideal behavior that result from interactions between two molecules. Similarly, the third virial coefficient takes into account deviations from ideal behavior that result from the interaction of three molecules. The physical significance of each higher virial coefficient follows in an analogous manner.

From statistical mechanics we can derive relations between virial coefficients and intermolecular potential functions (Hill, 1986). For simplicity, consider a gas composed of simple, spherically symmetric molecules such as methane or argon. The potential energy between two such molecules is designated by $\Gamma(r)$, where r is the distance between molecular centers. The second and third virial coefficients are given as functions of $\Gamma(r)$ and temperature by

$$\boxed{B = 2\pi N_A \int_0^{\infty}\left[1 - e^{-\Gamma(r)/kT}\right] r^2 dr} \tag{5-17}$$[3]

and

$$C = \frac{-8\pi^2 N_A^2}{3}\int_0^{\infty}\int_0^{\infty}\int_{|r_{12}-r_{13}|}^{r_{12}+r_{13}} f_{12} f_{13} f_{23} r_{12} r_{13} r_{23} dr_{12} dr_{13} dr_{23} \tag{5-18}$$

where $f_{ij} \equiv \exp(-\Gamma_{ij}/kT) - 1$, k is Boltzmann's constant and N_A is Avogadro's constant.[4]

Similar expressions can be written for the fourth and higher virial coefficients. While Eqs. (5-17) and (5-18) refer to simple, spherically symmetric molecules, we do not imply that the virial equation is applicable only to such molecules; rather, it is valid for essentially all stable, uncharged (electrically neutral) molecules, polar or nonpolar, including those with complex molecular structure. However, in a complex molecule the intermolecular potential depends not only on the distance between molecular centers but also on the spatial geometry of the separate molecules and on their relative orientation. In such cases, it is possible to relate the virial coefficients to the intermolecular potential, but the mathematical expressions corresponding to Eqs. (5-17) and (5-18) are necessarily more complicated.

Special care must be taken with "reactive" molecules, for example, molecules like acetic acid that dimerize, as discussed in Sec. 5.9.

---

[3] Similarly, the McMillan-Mayer solution theory provides the link between the osmotic second virial coefficient (B_{22}^*) of a solute (see Sec. 4.11) dilute in a solvent medium and the potential of mean force, w_{22}:

$$B_{22}^*(T,\mu_s) = 2\pi N_A \int_0^{\infty}[1 - e^{-w_{22}(r,\mu_s,T)/kT}]r^2 dr$$

where μ_s is the chemical potential of the solvent. Osmotic third virial coefficients can also be calculated from the potential of mean force. While in a gas B depends only on temperature, in a dilute solution B_{22}^* depends on temperature and chemical potential of the solvent.

[4] Equation (5-18) assumes, for convenience, that the potential energy of molecules 1, 2, and 3 is given by the sum of the three binary potential energies (additivity assumption): $\Gamma_{123}(r_{12}, r_{13}, r_{23}) = \Gamma_{12}(r_{12}) + \Gamma_{13}(r_{13}) + \Gamma_{23}(r_{23})$. This assumption is unfortunately not strictly correct, although in many cases, it provides a good approximation.

5.3 Extension to Mixtures

Perhaps the most important advantage of the virial equation of state for application to phase equilibrium problems lies in its direct extension to mixtures. This extension requires no arbitrary assumptions. The composition dependences of all virial coefficients are given by a generalization of the statistical-mechanical derivation used to derive the virial equation for pure gases.

First, consider the second virial coefficient that takes into account interactions between two molecules. In a pure gas, the chemical identity of each of the interacting molecules is always the same; in a mixture, however, there are various types of two-molecule interactions depending on the number of components present. In a binary mixture containing species i and j, there are three types of two-molecule interactions, designated i-i, j-j, and i-j. For each of these interactions there is a corresponding second virial coefficient that depends on the intermolecular potential between the molecules under consideration. Thus B_{ii} is the second virial coefficient of pure i that depends on Γ_{ii}; B_{jj} is the second virial coefficient of pure j that depends on Γ_{jj}; and B_{ij} is the second virial coefficient corresponding to the i-j interaction as determined by Γ_{ij}, the potential energy between molecules i and j. If i and j are spherically symmetric molecules, B_{ij} is given by the same expression as that in Eq. (5-17):

$$B_{ij} = 2\pi N_A \int_0^{\infty} \left[1 - e^{-\Gamma_{ij}(r)/kT}\right] r^2 dr \tag{5-19}$$

The three second virial coefficients B_{ii}, B_{jj}, and B_{ij} are functions only of the temperature; they are independent of density (or pressure) and, what is most important, they are independent of composition. Because the second virial coefficient is concerned with interactions between *two* molecules, it can be rigorously shown that the second virial coefficient of a mixture is a *quadratic* function of the mole fractions y_i and y_j. For a binary mixture of components i and j,

$$B_{\text{mixt}} = y_i^2 B_{ii} + 2 y_i y_j B_{ij} + y_j^2 B_{jj} \tag{5-20}$$

For a mixture of m components, the second virial coefficient is given by a rigorous generalization of Eq. (5-20):

$$B_{\text{mixt}} = \sum_{i=1}^{m} \sum_{j=1}^{m} y_i y_j B_{ij} \tag{5-21}$$

The third virial coefficient of a mixture is related to the various C_{ijk} coefficients that take into account interactions of three molecules i, j, and k; in a pure gas, the chemical identity of these three molecules is always the same but in a mixture, molecules i, j, and k may belong to different chemical species. In a binary mixture, for example, there are four C_{ijk} coefficients. Two of these correspond to the pure-component third virial coefficients and two of them are cross-coefficients. Because third virial coefficients take into account interactions of *three* molecules, it can be rigorously shown that the third virial coefficient of a mixture is a *cubic* function of the mole fractions. For a binary mixture of components i and j,

$$C_{\text{mixt}} = y_i^3 C_{iii} + 3y_i^2 y_j C_{iij} + 3y_i y_j^2 C_{ijj} + y_j^3 C_{jjj} \tag{5-22}$$

Equation (5-22) can be rigorously generalized for a mixture of m components:

$$\boxed{C_{\text{mixt}} = \sum_{i=1}^{m}\sum_{j=1}^{m}\sum_{k=1}^{m} y_i y_j y_k C_{ijk}} \tag{5-23}$$

Examining random-error propagation, Eubank and Hall (1990) found that there are optimum mixture compositions for calculation of cross-virial coefficients from binary gas-density experimental data. For determination of the cross-second virial coefficient B_{12} from B_{mixt} using Eq. (5-20), the equimolar mixture is optimum as expected. However, to determine C_{112} and C_{122} from (at least) two density measurements of the mixture third virial coefficients using Eq. (5-22), it is best to use density data at mole fractions 0.25 and 0.75.

Coefficient C_{ijk} is related to intermolecular potentials Γ_{ij}, Γ_{ik}, and Γ_{jk} by an equation of the same form as that of Eq. (5-18):

$$C_{ijk} = \frac{-8\pi^2 N_A^2}{3} \int_0^\infty \int_0^\infty \int_{|r_{ij}-r_{ik}|}^{r_{ij}+r_{ik}} f_{ij} f_{ik} f_{jk} r_{ij} r_{ik} r_{jk}\, dr_{ij}\, dr_{ik}\, dr_{jk} \tag{5-24}$$

where $f_{ij} \equiv \exp(-\Gamma_{ij}/kT) - 1$, $f_{ik} \equiv \exp(-\Gamma_{ik}/kT) - 1$, and $f_{jk} \equiv \exp(-\Gamma_{jk}/kT) - 1$.

The fourth, fifth, and higher virial coefficients of a gaseous mixture are related to the composition and to the various potential functions in an analogous manner: the nth virial coefficient of a mixture is a polynomial function of the mole fractions of degree n.

Equations (5-20) and (5-22) can be used to obtain, respectively, the second virial cross coefficients, B_{ij}, and the third virial cross coefficients, C_{iij} and C_{ijj}, from regression of the corresponding mixture virial coefficients.

If experimental data are available for several compositions, the cross coefficients can be obtained from

$$\lim_{y_j \to 0} \left(\frac{\partial B_{\text{mixt}}}{\partial y_j} \right)_T = 2(B_{ij} - B_{ii}) \tag{5-25}$$

and

$$\lim_{y_j \to 0} \left(\frac{\partial C_{\text{mixt}}}{\partial y_j} \right)_T = 3(C_{iij} - C_{iii}) \tag{5-26}$$

To illustrate, Figs. 5-6 and 5-7 show, respectively, experimental second and third virial coefficients for the carbon dioxide/water system as a function of temperature and mole fraction of water. As indicated in Fig. 5-7, uncertainties in estimating the limiting curvature makes it difficult to determine the third virial cross coefficients with high accuracy.

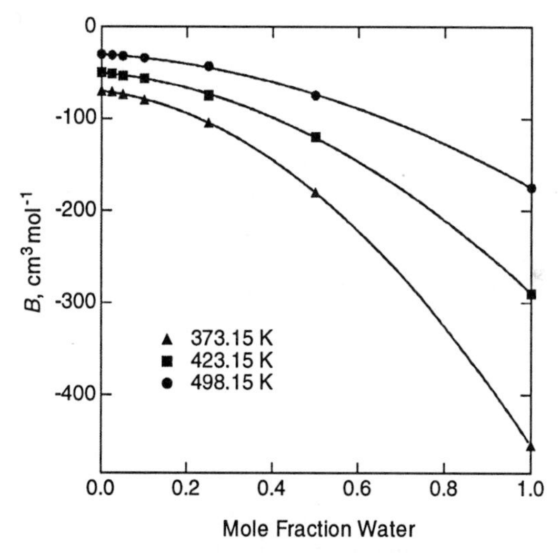

Figure 5-6 Experimental second virial coefficients for the CO_2/H_2O system as a function of the mole fraction of water, for various temperatures (Patel *et al.*, 1987).

Equations (5-21) and (5-23) are rigorous results from statistical mechanics and are not subject to any assumptions other than those upon which the virial equation itself is based. The proof for these equations is not elementary and no attempt is made to reproduce it here. Such a proof may be found in advanced texts.[5] However, the physi-

[5] See, for example, Hirschfelder *et al.* (1954), Hill (1986), or Mayer and Mayer (1977). It is shown in these, as well as in other references, that the *n*th virial coefficient is a polynomial in the mole fractions of degree *n*. These polynomial relations do not require the assumption of pairwise additivity of potential functions. That assumption

cal significance of Eqs. (5-21) and (5-23) is not difficult to understand because these equations are a logical consequence of the physical significance of the individual virial coefficients; each of the individual virial coefficients describes a particular interaction and the virial coefficient of the mixture is a summation of the individual virial coefficients, appropriately weighted with respect to composition.

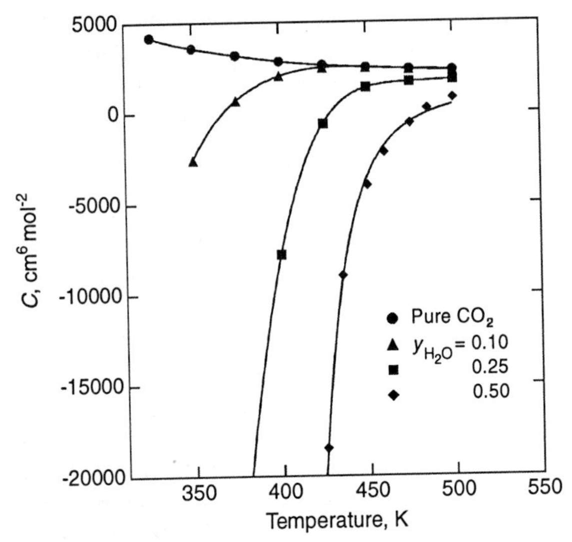

Figure 5-7 Experimental third virial coefficients for the CO_2/H_2O system as a function of temperature, for several mole fractions (Patel *et al.*, 1987).

Extension of the virial equation to mixtures is based on theoretical rather than empirical grounds and it is this feature of the virial equation that makes it useful for phase-equilibrium problems. Empirical equations of state that contain constants having only empirical significance are useful for pure components but cannot be extended to mixtures without the use of somewhat arbitrary mixing rules for combining the constants. Extension of the virial equation to mixtures, however, follows in a simple and rigorous way from the theoretical nature of that equation.

5.4 Fugacities from the Virial Equation

Once we have decided on a relation that gives the volumetric properties of a mixture as a function of temperature, pressure, and composition, we can readily compute fugacities.

---

is customarily used in evaluating the individual virial coefficients from potential functions but it is not required for establishing the composition dependence of the third (and higher) virial coefficients.

The virial equation for a mixture, truncated after the third term, is given by

$$z_{\text{mixt}} = \frac{Pv}{RT} = 1 + \frac{B_{\text{mixt}}}{v} + \frac{C_{\text{mixt}}}{v^2} \tag{5-27}$$

where z_{mixt} is the compressibility factor of the mixture, v is the molar volume of the mixture, and B_{mixt} and C_{mixt} are the virial coefficients of the mixture given by Eqs. (5-21) and (5-23). The fugacity coefficient for any component i in a mixture of m components is obtained by substitution in Eq. (3-53). When the indicated differentiations and integrations are performed, we obtain

$$\ln \varphi_i = \frac{2}{v} \sum_{j=1}^{m} y_j B_{ij} + \frac{3}{2v^2} \sum_{j=1}^{m} \sum_{k=1}^{m} y_j y_k C_{ijk} - \ln z_{\text{mixt}} \tag{5-28}$$

In Eq. (5-28) the summations are over all components, including component i. For example, for component 1 in a binary mixture, Eq. (5-28) becomes

$$\ln \varphi_1 = \frac{2}{v}(y_1 B_{11} + y_2 B_{12}) + \frac{3}{2}\frac{1}{v^2}(y_1^2 C_{111} + 2y_1 y_2 C_{112} + y_2^2 C_{122}) - \ln z_{\text{mixt}} \tag{5-29}$$

Similarly, for component 2,

$$\ln \varphi_2 = \frac{2}{v}(y_2 B_{22} + y_1 B_{12}) + \frac{3}{2}\frac{1}{v^2}(y_2^2 C_{222} + 2y_1 y_2 C_{122} + y_1^2 C_{112}) - \ln z_{\text{mixt}} \tag{5-30}$$

Equation (5-28) is one of the most useful equations for phase-equilibrium thermodynamics. It relates the fugacity of a component in the vapor phase to its partial pressure through the theoretically-derived virial equation of state. The major practical limitation of Eq. (5-28) lies in its restriction to moderate densities. It may be applied to any component in a gas mixture regardless of whether or not that component can exist as a pure vapor at the temperature and pressure of the mixture; no hypothetical states are introduced. Further, Eq. (5-28) is not limited to binaries but is applicable without further assumptions to mixtures containing any number of components. Finally, Eq. (5-28) is valid for many types of (nonionized) molecules, polar and nonpolar, although it is unfortunately true that for practical purposes, theoretical calculation of the various B and C coefficients from statistical mechanics is restricted to relatively simple substances. However, this limitation is not due to failure of the virial equation or of the thermodynamic equations in Sec. 3.4; rather, it is a result of our present inability to describe adequately the intermolecular potential between molecules of complex structure.

Because data for second virial coefficients[6] are much more plentiful than those for third virial coefficients, Eq. (5-28) is often truncated by omitting the quadratic term in density:

$$\boxed{\ln \varphi_i = \frac{2}{v} \sum_{j=1}^{m} y_j B_{ij} - \ln z_{\text{mixt}}} \tag{5-31}$$

where z_{mixt} is given by

$$z_{\text{mixt}} = 1 + \frac{B_{\text{mixt}}}{v} \tag{5-32}$$

When the volume-explicit form of the virial equation [Eq. (5-10)] is used instead of the pressure-explicit form [Eq. (5-9)], and when terms in the third virial coefficient C' (and higher) are omitted, we obtain [using Eq. (5-11)]

$$\boxed{\ln \varphi_i = \left(2 \sum_{j=1}^{m} y_j B_{ij} - B_{\text{mixt}} \right) \frac{P}{RT}} \tag{5-33}$$

Equation (5-33) is more convenient than Eq. (5-31) because it uses pressure, rather than volume, as the independent variable. Further, Eq. (5-33) is preferable because the assumption $C' = 0$ often provides a better approximation than the assumption $C = 0$ in Eq. (5-31). However, both Eqs. (5-31) and (5-33) are valid only at low or moderate densities, approximately equal to but not exceeding (about) one-half the critical density.

In the limit as component i becomes infinitely dilute in component j, Eq. (5-31) for a binary mixture reduces to

$$\ln \varphi_i^{\infty} = \frac{2B_{ij}}{v_{\text{pure } j}} - \ln z_{\text{pure } j}$$

Therefore, a reliable estimate of cross-coefficient B_{ij} is essential for the calculation of fugacity coefficients of dilute components.

[6] An extensive compilation of experimental virial coefficients is given by J. H. Dymond and E. B. Smith, 1980, *The Virial Coefficients of Pure Gases and Mixtures*, Oxford: Clarendon Press.

5.5 Calculation of Virial Coefficients from Potential Functions

In previous sections we discussed the nature of the virial equation of state and, in Eq. (5-28), we indicated the way it may be used to calculate the fugacity of a component in a gaseous mixture. We must now consider how to calculate the virial coefficients that appear in Eq. (5-28) and, to do so, we make use of our discussion of intermolecular forces in Chap. 4.

First, we must recognize that the first term on the right-hand side of Eq. (5-28) is frequently much more important than the second term; at low or moderate densities, the second term is sufficiently small to allow us to neglect it. This is fortunate because we can estimate B's with much more accuracy than we can estimate C's.

Equation (5-19) gives the relation between the second virial coefficient B_{ij} and the intermolecular potential function $\Gamma_{ij}(r)$ for spherically symmetrical molecules i and j, where i and j may, or may not, be chemically identical. If the potential function $\Gamma_{ij}(r)$ is known, then B_{ij} can be calculated by integration as indicated by Eq. (5-19) and similarly, if the necessary potentials are known, C_{ijk} can be found from Eq. (5-24). Such integrations have been performed for many types of potential functions corresponding to different molecular models. A few models are illustrated in Figs. 5-8 and 5-9. We now give a brief discussion of each of them with reference to second virial coefficients, followed by a short section on third virial coefficients.

Ideal-Gas Potential. The simplest (trivial) case is to assume that $\Gamma = 0$ for all values of the intermolecular distance r. In that case, the second, third, and higher virial coefficients are zero for all temperatures and the virial equation reduces to the ideal-gas law.

Hard-Sphere Potential. This model takes into account the nonzero size of the molecules but neglects attractive forces. It considers molecules to be like billiard balls; for hard-sphere molecules there are no forces between the molecules when their centers are separated by a distance larger than σ, the hard-sphere diameter, but the force of repulsion becomes infinitely large when they touch, at a separation equal to σ. The potential function $\Gamma(r)$ is given by

$$\Gamma = \begin{cases} 0 & \text{for} \quad r > \sigma \\ \infty & \text{for} \quad r \le \sigma \end{cases} \tag{5-34}$$

Substituting into Eqs. (5-19), we obtain for a pure component

$$B = \frac{2}{3}\pi N_A \sigma^3 \tag{5-35}$$

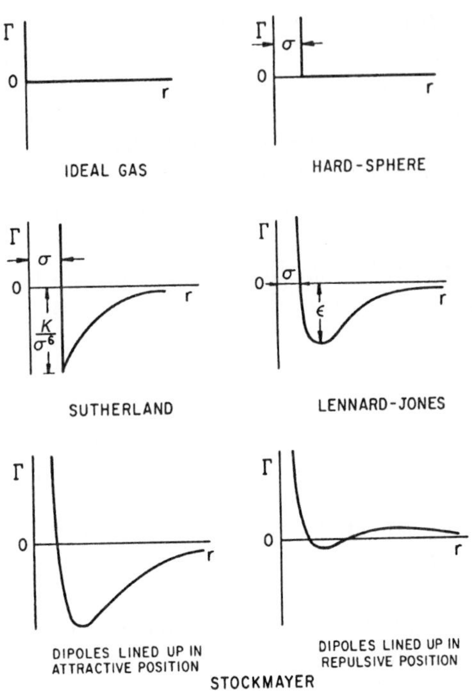

Figure 5-8 Potential functions with zero, one, or two adjustable parameters.

For mixtures, the second virial coefficient B_{ij} $(i \neq j)$ is

$$B_{ij} = \frac{2}{3}\pi N_A \left(\frac{\sigma_i + \sigma_j}{2} \right)^3 \tag{5-36}$$

The hard-sphere model gives a highly oversimplified picture of real molecules because, for a given gas, it predicts second virial coefficients that are independent of temperature. These results are in strong disagreement with experiment but give a rough approximation for the behavior of simple molecules at temperatures far above the critical. For example, helium or hydrogen have very small forces of attraction; near room temperature, where the kinetic energies of these molecules are much larger than their potential energies, the size of the molecules is the most significant factor that contributes to deviation from ideal-gas behavior. Therefore, at high-reduced temperatures, the hard-sphere model provides a reasonable but rough approximation. Because Eq. (5-34) requires only one characteristic constant, the hard-sphere model is a one-parameter model.

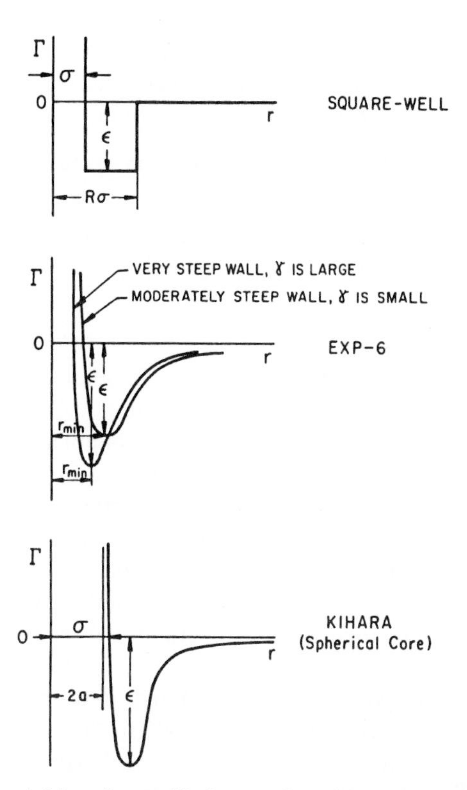

Figure 5-9 Potential functions with three adjustable parameters.

Sutherland Potential. According to London's theory of dispersion forces, the potential energy of attraction varies inversely as the sixth power of the distance of separation. When this result is combined with the hard-sphere model, the potential function becomes

$$\Gamma = \begin{cases} \infty & \text{for} \quad r \le \sigma \\ \dfrac{-K}{r^6} & \text{for} \quad r > \sigma \end{cases} \tag{5-37}$$

where K is a constant depending on the nature of the molecule. London's equation [Eq. (4-19)] suggests that K is proportional to the ionization potential and to the square of the polarizability. The *Sutherland model* provides a large improvement over the hard-sphere model and it is reasonably successful in fitting experimental second-virial-coefficient data with its two adjustable parameters. Like the hard-sphere model, however, it predicts that at high temperatures the second virial coefficient approaches a constant value, although the best available data show that it goes through a weak maximum at a temperature very much higher than the critical. This limitation is not serious in typical phase-equilibrium problems where such high-reduced temperatures are almost never encountered except, perhaps, for helium.

Lennard-Jones' Form of Mie's Potential. As discussed in Chap. 4, *Lennard-Jones*' form of Mie's equation is

$$\Gamma = 4\varepsilon\left[\left(\frac{\sigma}{r}\right)^{12} - \left(\frac{\sigma}{r}\right)^{6}\right] \tag{5-38}$$

where ε is the depth of the energy well (minimum potential energy) and σ is the collision diameter, i.e., the separation where $\Gamma = 0$. Equation (5-38) gives what is probably the best known two-parameter potential for small, nonpolar molecules. In Lennard-Jones' formula, the repulsive wall is not vertical but has a finite slope; this implies that if two molecules have very high kinetic energy, they may be able to interpenetrate to separations smaller than the collision diameter σ. Potential functions with this property are called *soft-sphere potentials*. The Lennard-Jones potential correctly predicts that, at a temperature very much larger than ε/k (k is Boltzmann's constant), the second virial coefficient goes through a maximum. The temperature where $B = 0$ is called the *Boyle temperature*.

When Lennard-Jones' potential is substituted into the statistical mechanical equation for the second virial coefficient [Eq. (5-17)], the required integration is not simple. However, numerical results have been obtained (Hirschfelder *et al.*, 1954) as shown in Fig. 5-10, where the reduced (dimensionless) virial coefficient is a function of the reduced (dimensionless) temperature. The reducing parameter for the virial coefficient is proportional to collision diameter σ raised to the third power and that for the temperature is proportional to characteristic energy ε.

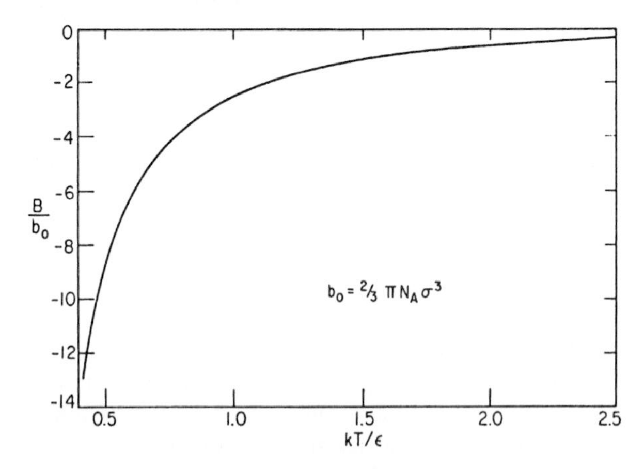

Figure 5-10 Second virial coefficients calculated from Lennard-Jones 6-12 potential.

For many gases, second virial coefficients, as well as other thermodynamic and transport properties, have been interpreted and correlated successfully with the Lennard-Jones potential. Unfortunately, however, it has frequently been observed that for a given gas, one set of parameters (ε and σ) is obtained from data reduction of one property (e.g., the second virial coefficient) while another set is obtained from data reduction of a different property (e.g., viscosity). If the Lennard-Jones potential were the *true* potential, then the parameters ε and σ should be the same for all properties of a given substance.

But even if attention is restricted to the second virial coefficient alone, there is good evidence that the Lennard-Jones potential is only an approximation, albeit a very good one in certain cases. It has been shown by Michels *et al.* (1958) that his highly accurate data for the second virial coefficient of argon over the temperature range -140 to +150°C cannot be fitted with the Lennard-Jones potential within the experimental error using only one set of parameters. This conclusion can be supported through a revealing series of calculations suggested by Michels *et al.* (1960). We take an experimental value of B corresponding to a certain temperature and then arbitrarily assume a value for ε. We now calculate the corresponding value of $b_0 = (2/3)N_A\sigma^3$ that is required to force agreement between the experimental B and that calculated from the Lennard-Jones function. Next, we repeat the calculation at the same temperature assuming some other value of ε. In this way we obtain a curve on a plot of b_0 versus ε. We now perform the same series of calculations for another experimental value of B at a different temperature and again obtain a curve; where the two curves intersect should be the "true" value of ε and b_0. However, we find that when we repeat these calculations for several different temperatures, all the curves do not intersect at one point, as they should if the Lennard-Jones potential were exactly correct.

Such a plot is shown in Fig. 5-11; instead of a point of intersection, the curves define an area that gives a region rather than a unique set of potential parameters. Therefore, we conclude that even for a spherically symmetric, nonpolar molecule such as argon, the Lennard-Jones potential is not completely satisfactory (Guggenheim and McGlashan, 1960; Munn, 1964; Munn *et al.*, 1965, 1965a). Such a conclusion, however, was reached only because Michels' data are of unusually high accuracy and were measured over a large temperature range. For many practical calculations the Lennard-Jones potential is adequate. Lennard-Jones parameters for some fluids are given in Table 5-1.

The Square-Well Potential. The Lennard-Jones potential is not a simple mathematical function. To simplify calculations, a crude potential was proposed having the general shape of the Lennard-Jones function. This crude potential is obviously an unrealistic simplification because it has discontinuities, but its mathematical simplicity and flexibility make it useful for practical calculations. The flexibility arises from the square-well potential's three adjustable parameters: the collision diameter, σ; the well depth (minimum potential energy), ε; and the reduced well width, R. The *square-well potential* function is

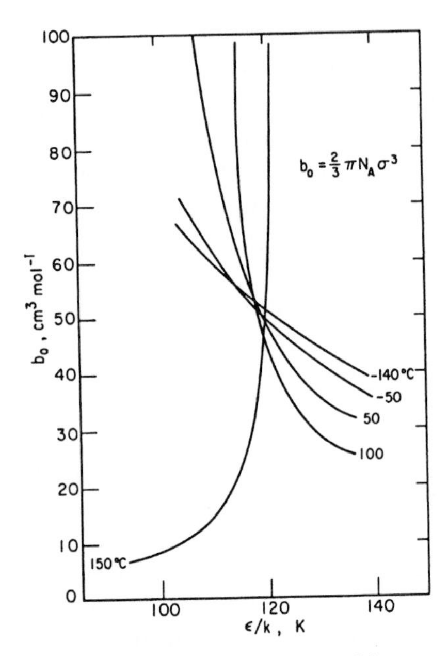

Figure 5-11 Lennard-Jones' parameters calculated from second-virial-coefficient data for argon. If perfect representation were given by Lennard-Jones' potential, all isotherms would intersect at one point.

Table 5-1 Parameters for the Lennard-Jones potential obtained from second-virial coefficient data.[§]

	σ (Å)	ε/*k* (K)
Ar	3.499	118.13
Kr	3.846	162.74
Xe	4.100	222.32
CH_4	4.010	142.87
N_2	3.694	96.26
C_2H_4	4.433	202.52
C_2H_6	5.220	194.14
C_3H_8	5.711	233.28
$C(CH_3)_4$	7.420	233.66
n-C_4H_{10}	7.152	223.74
C_6H_6	8.443	247.50
CO_2	4.416	192.25
n-C_5H_{12}	8.540	217.69

[§] L. S. Tee, S. Gotoh, and W. E. Stewart, 1966, *Ind. Eng. Chem. Fundam.*, 5: 356.

$$\Gamma = \begin{cases} \infty & \text{for} \quad r \leq \sigma \\ -\varepsilon & \text{for} \quad \sigma < r \leq R\sigma \\ 0 & \text{for} \quad r > R\sigma \end{cases} \tag{5-39}$$

leading to

$$B = b_0 R^3 \left(1 - \frac{R^3 - 1}{R^3} \exp \frac{\varepsilon}{kT} \right)$$

The square-well model has an infinitely steep repulsive wall and therefore, like the Sutherland model, it does not predict a maximum for the second virial coefficient. With its three adjustable parameters, good agreement can often be obtained between calculated and experimental second virial coefficients (Sherwood and Prausnitz, 1964).

The Exp-6 Potential. A potential function for nonpolar molecules should contain an attractive term of the London type in addition to a repulsive term; little is known about that term but it must depend strongly on the intermolecular distance. For the repulsive term Mie, and later Lennard-Jones, used a term inversely proportional to r, the intermolecular distance, raised to a large power. Theoretical calculations, however, have suggested that the repulsive potential is not an inverse-power function but rather an exponential function of r. A potential function that uses an exponential form for repulsion and an inverse sixth power for attraction is called an *exp-6 potential.* (It is also sometimes referred to as a *modified Buckingham potential.*) This potential function contains three adjustable parameters and is written[7]

$$\Gamma = \frac{\varepsilon}{1-(6/\gamma)} \left\{ \frac{6}{\gamma} \exp\left[\gamma \left(1 - \frac{r}{r_{\min}} \right) \right] - \left(\frac{r_{\min}}{r} \right)^6 \right\} \tag{5-40}$$

where $-\varepsilon$ is the minimum potential energy at intermolecular separation $r_{\min}$. The third parameter, γ, determines the steepness of the repulsive wall; in the limit, when $\gamma = \infty$, the exp-6 potential becomes the Sutherland potential that has a hard-sphere repulsive term.

The collision diameter σ (i.e., the intermolecular distance where $\Gamma = 0$) is only slightly less than the distance $r_{\min}$, but the exact relation depends on the value of γ, as shown in Table 5-2. Numerical results for the second virial coefficient, based on Eq. (5-40), are available (Sherwood and Prausnitz, 1964a). Good agreement can often be obtained between calculated and observed second virial coefficients.

[7] Equation (5-40) is valid only for $r > s$, where s (a very small distance) is that value for r where Γ goes through a (false) maximum. For completeness, therefore, it should be added that $\Gamma = \infty$ for $r < s$. The quantity s, however, is not an independent parameter and has no physical significance.

Table 5-2 Ratio σ/r_{min} for the exp-6 potential as a function of the repulsive steepness parameter γ.

γ	σ/r_{min}
15	0.894170
18	0.906096
20	0.912249
24	0.921911
30	0.932341
40	0.943914
100	0.970041
300	0.986692
∞	1.000000

The Kihara Potential. According to Lennard-Jones' potential, two molecules can interpenetrate completely provided that they have enough energy; according to this model, molecules consist of point centers surrounded by "soft" (i.e., penetrable) electron clouds. An alternative picture of molecules is to think of them as possessing impenetrable (hard) cores surrounded by penetrable (soft) electron clouds. This picture leads to a model proposed by Kihara. In crude mechanical terms, Kihara's model (for spherically symmetric molecules) considers a molecule to be a hard billiard ball with a foam-rubber coat; a Lennard-Jones molecule, by contrast, is a soft ball made exclusively of foam rubber.

Kihara (1953, 1958, 1963) writes a potential function identical to that of Lennard-Jones except that the intermolecular distance is taken not as that between molecular centers but rather as the distance between the surfaces of the molecules' cores.[8] For molecules with spherical cores, the *Kihara potential* is

$$\Gamma = \begin{cases} \infty & \text{for } r < 2a \\ 4\varepsilon\left[\left(\dfrac{\sigma - 2a}{r - 2a}\right)^{12} - \left(\dfrac{\sigma - 2a}{r - 2a}\right)^{6}\right] & \text{for } r \geq 2a \end{cases} \tag{5-41}$$

where a is the radius of the spherical molecular core, ε is the depth of the energy well, and σ is the collision diameter, i.e., the distance r between molecular centers when $\Gamma = 0$.

Equation (5-41) is for the special case of a spherical core, but a more general form has been presented by Kihara for cores having other convex shapes such as rods,

[8] When the cores are not spherical, the intermolecular distance is for the orientation that gives a minimum distance of separation.

tetrahedra, triangles, prisms, etc. (Connolly and Kandakic, 1960; Prausnitz and Keeler, 1961; Prausnitz and Myers, 1963). Numerical results, based on Kihara's potential, are available for second virial coefficients for several core geometries and, in particular, for reduced (spherical) core sizes a^*, where $a^* \equiv 2a/(\sigma - 2a)$. When $a^* = 0$, the results are identical to those obtained from Lennard-Jones' potential. Because it is a three-parameter function, Kihara's potential is successful in fitting thermodynamic data for a large number of nonpolar fluids, including some complex substances whose properties are represented poorly by the two-parameter Lennard-Jones potential. Figure 5-12 shows reduced second virial coefficients calculated from Kihara's potential.

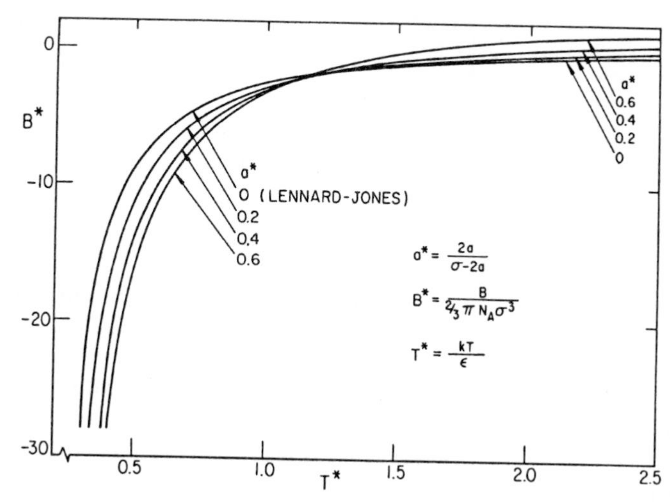

Figure 5-12 Second virial coefficients calculated from Kihara's potential with a spherical core of radius a.

In our discussion of Lennard-Jones' potential we indicated that Michels' highly accurate second-virial-coefficient data for argon could not be represented over a large temperature range by the Lennard-Jones potential using only one set of potential parameters. However, these same data can be represented within the very small experimental error by the Kihara potential using only one set of parameters (Myers and Prausnitz, 1962; Rossi and Danon, 1966; O'Connell and Prausnitz, 1968). The ability of Kihara's potential to do what Lennard-Jones' potential cannot do is hardly surprising because the former potential has three adjustable parameters whereas the latter has only two. In fitting data for argon, the three Kihara parameters were determined by trial and error until the deviation between experimental and theoretical second virial coefficients reached a minimum less than the experimental error. The magnitude of the core diameter obtained by this procedure is physically reasonable when compared to the diameter of the "impenetrable core" of argon as calculated from its electronic structure. Figure 5-13 shows results of a theoretical calculation of the electron density as a function of distance. While the results shown appear to justify confidence in the Kihara potential, the agreement indicated must not be considered "proof" of its validity. The "true" potential between two argon atoms is undoubtedly quite different from that given by Eq. (5-41) especially at very small separations. However, it appears that for practical calculation of common thermodynamic properties (except those at very high temperatures), Kihara's potential is one of the most useful potential functions now available.

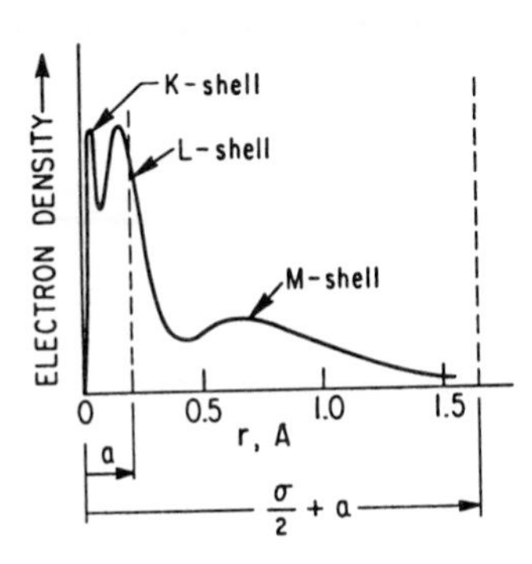

Figure 5-13 Charge distribution in argon (quoted by C. A. Coulson, 1962, *Valence,* 2nd Ed. London: Oxford University Press).

One practical application of Kihara's potential is for prediction of second virial coefficients at low temperatures where experimental data are scarce and difficult to obtain. To illustrate, Fig. 5-14 shows predicted and observed second virial coefficients for krypton at low temperatures; two sets of predictions were made, one with the Kihara potential and the other with the Lennard-Jones potential. In both cases potential parameters were obtained from experimental measurements made at room temperature and

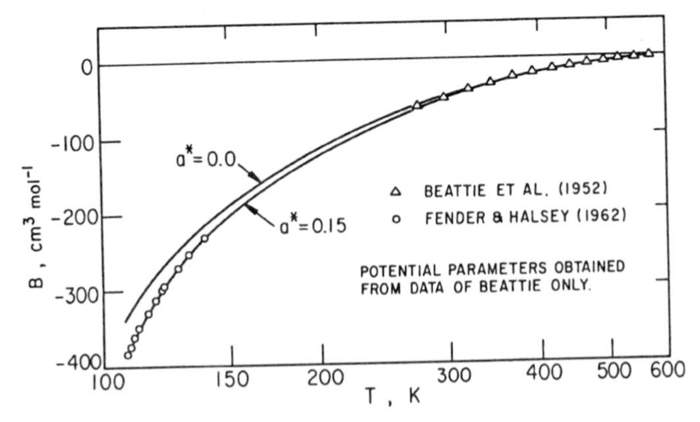

Figure 5-14 Second virial coefficients for krypton. Predictions at low temperatures based on Lennard-Jones potential ($a^* = 0$) and on Kihara potential.

above. It is evident from Fig. 5-14 that even for such a "simple" substance as krypton, the Kihara potential is significantly superior to the Lennard-Jones potential. Kihara parameters for some fluids are given in Table 5-3.

Table 5-3 Parameters for the Kihara potential (spherical core) obtained from second-virial-coefficient data.[§]

	a^*	σ (Å)	ε/k (K)
Ar	0.121	3.317	146.52
Kr	0.144	3.533	213.73
Xe	0.173	3.880	298.15
CH_4	0.283	3.565	227.13
N_2	0.250	3.526	139.2[†]
O_2	0.308	3.109	194.3[‡]
C_2H_6	0.359	3.504	496.69
C_3H_8	0.470	4.611	501.89
CF_4	0.500	4.319	289.7[‡]
$C(CH_3)_4$	0.551	5.762	557.75
n-C_4H_{10}	0.661	4.717	701.15
C_6H_6	0.750	5.335	832.0[†]
CO_2	0.615	3.760	424.16
n-C_5H_{12}	0.818	5.029	837.82

[§] L. S. Tee, S. Gotoh, and W. E. Stewart, 1966, *Ind. Eng. Chem. Fundam.*, 5: 363.
[†] A. E. Sherwood and J. M. Prausnitz, 1964, *J. Chem. Phys.*, 41: 429.
[‡] C. E. Hunt, unpublished results.

For mixtures, Kihara's potential gives B_{ij} $(i \neq j)$ when the pure-component core parameters and the unlike-pair potential parameters ε_{ij} and σ_{ij} are specified. The latter two are frequently related to the pure-component parameters by empirical combining rules. However, the core parameter for the i-j interaction can be derived exactly from the core parameters for the i-i and j-j interactions even for nonspherical cores (Kihara, 1953, 1958, 1963; Myers and Prausnitz, 1962).

The difficulty of determining "true" intermolecular potentials from second virial-coefficient data is illustrated in Figs. 5-15 and 5-16 that show several potential functions for argon and for neopentane. Each of these functions gives a good prediction of the second virial coefficient; the three-parameter potentials give somewhat better predictions than the two-parameter potentials, but all of them are in fairly good agreement with experiment. However, the various potential functions differ very much from one another, especially for neopentane.

Figures 5-15 and 5-16 give striking evidence that agreement between a particular set of experimental results and those calculated from a particular model should not be regarded as proof that the model is correct.[9] Models are useful in molecular thermodynamics but one must not confuse utility with truth. Figures 5-15 and 5-16 provide a powerful illustration of A. N. Whitehead's advice to scientists: "Seek simplicity but distrust it".

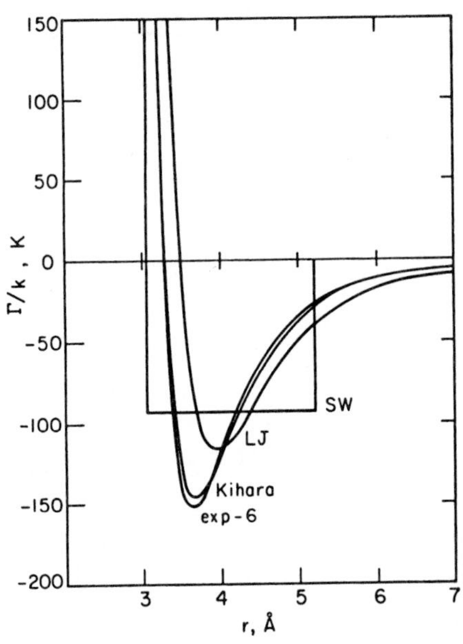

Figure 5-15 Potential functions for argon as determined from second-virial-coefficient data.

The Stockmayer Potential. All of the potential functions previously described are applicable only to nonpolar molecules. We now briefly consider molecules that have a permanent dipole moment; for such molecules Stockmayer proposed a potential that adds to the Lennard-Jones formula for nonpolar forces an additional term for the potential energy due to dipole-dipole interactions. Dipole-induced dipole interactions are not considered explicitly although it may be argued that because these forces, like London forces, are proportional to the inverse sixth power of the intermolecular separation, they are, in effect, included in the attractive term of the Lennard-Jones formula. For polar molecules, the potential energy is a function not only of intermolecular separation but also of relative orientation. *Stockmayer's potential* is

[9] A more nearly "true" potential can be obtained by simultaneous analysis of experimental data for a variety of properties: second virial coefficients, gas-phase viscosity, diffusivity, and thermal diffusivity. Such analysis has produced a "best" potential function for argon. See Dymond and Alder (1969) and Barker and Pompe (1968).

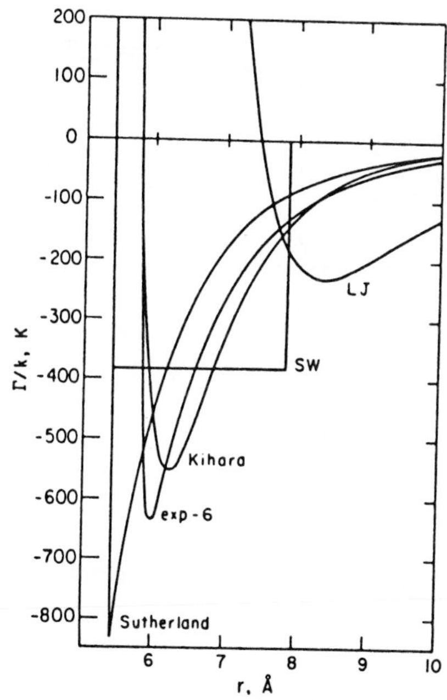

Figure 5-16 Potential functions for neopentane as determined from second-virial-coefficient data.

$$\Gamma = 4\varepsilon\left[\left(\frac{\sigma}{r}\right)^{12} - \left(\frac{\sigma}{r}\right)^{6}\right] + \frac{\mu^2}{r^3} F_\theta(\theta_1, \theta_2, \theta_3) \tag{5-42}$$

where F_θ is a known function of the angles θ_1, θ_2, and θ_3 that determine the relative orientation of the two dipoles [see Eq. (4-7)]. This potential function contains only two adjustable parameters because the dipole moment μ is an independently determined physical constant.

The collision diameter σ in Eq. (5-42) is the intermolecular distance where the potential energy due to forces other than dipole-dipole forces becomes equal to zero.

Numerical results, based on Stockmayer's potential, are available for the second virial coefficient (Rowlinson, 1949). Figure 5-17 shows reduced second virial coefficients calculated from Stockmayer's potential as a function of reduced temperature and reduced dipole moment. The top curve (zero dipole moment) is for nonpolar Lennard-Jones molecules and it is evident that the effect of polarity is to lower (algebraically) the second virial coefficient due to increased forces of attraction, especially at low temperatures, as suggested by Keesom's formula (see Sec. 4.2). Stockmayer's potential has been used successfully to fit experimental second-virial-coefficient data for a variety of polar molecules; Table 5-4 gives parameters for some polar fluids and Table 5-5 shows some illustrative results for trifluoromethane.

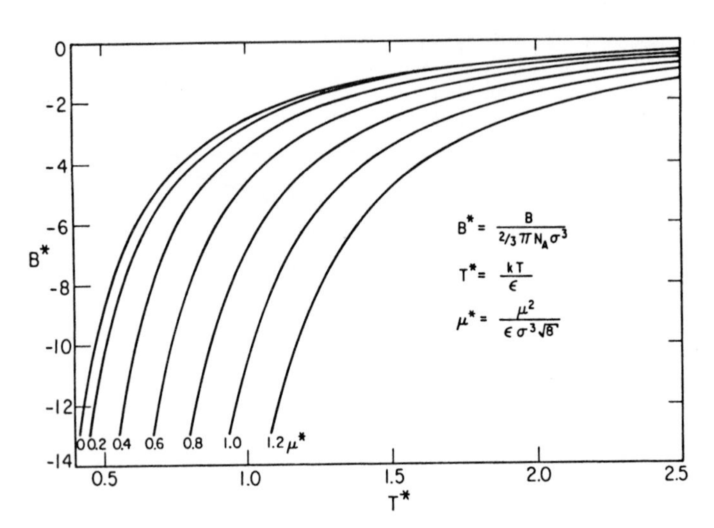

Figure 5-17 Second virial coefficients calculated from Stockmayer's potential for polar molecules.

Table 5-4 Parameters for Sockmayer's potential for polar fluids.*

	μ (debye)	σ (Å)	ε/k (K)
Acetonitrile	3.94	4.38	219
Nitromethane	3.54	4.16	290
Acetaldehyde	2.70	3.68	270
Acetone	2.88	3.67	479
Ethanol	1.70	2.45	620
Chloroform	1.05	2.98	1060
n-Butanol	1.66	2.47	1125
n-Butyl amine	0.85	1.58	1020
Methyl formate	1.77	2.90	684
n-Propyl formate	1.92	3.06	877
Methyl acetate	1.67	2.83	895
Ethyl acetate	1.76	2.99	956
Ethyl ether	1.16	3.10	935
Diethyl amine	1.01	2.99	1180

* R. F. Blanks and J. M. Prausnitz, 1962, *AIChE J.*, 8: 86.

The various potential models discussed above may be used to calculate B_{ij} as well as B_{ii}. The calculations for B_{ij} are exactly the same as those for B_{ii} when potential Γ_{ij} is used rather than potential Γ_{ii}. In the Stockmayer potential, when $i \neq j$, the reduced dipole moment μ^* in Fig. (5-17) becomes $\mu^* = \mu_i \mu_j (\varepsilon_{ij} \sigma_{ij}^3 \sqrt{8})^{-1}$.

Table 5-5 Second virial coefficients of trifluoromethane. Calculated values from Stockmayer potential with ε/k = 188 K, σ = 4.83 Å, and μ = 1.65 debye.

Temperature	$-B$ (cm^3 mol^{-1})	
(°C)	Experimental*	Calculated
0	233	215
25	187	185
50	154	150
75	127	127
100	107	108
150	76	76
200	53	53

* J. L. Belzile, S. Kaliaguine, and R. S. Ramalho, 1976, *Can. J. Chem. Eng.*, 54: 446.

5.6 Third Virial Coefficients

In the preceding section, attention was directed to the second virial coefficient. We now consider briefly our limited knowledge concerning third virial coefficients.

Equations (5-18) and (5-24) give expressions for the third virial coefficient in terms of three two-body intermolecular potentials. In the derivation of these equations, an important simplifying assumption was made; i.e., we assumed *pairwise additivity* of potentials. The third virial coefficient takes into account deviations from ideal-gas behavior due to three-molecule interactions; for a collision of three molecules i, j, and k, we need Γ_{ijk} the potential energy of the three-molecule assembly. However, in the derivation of Eqs. (5-18) and (5-24) it was assumed that

$$\Gamma_{ijk} = \Gamma_{ij} + \Gamma_{ik} + \Gamma_{jk} \tag{5-43}$$

Equation (5-43) says that the potential energy of the three molecules i, j, and k is equal to the sum of the potential energies of the three pairs i-j, i-k, and j-k. This assumption of pairwise additivity of intermolecular potentials is a common one in molecular physics because little is known about three-, four- (or higher) body forces. For an m-body assembly, the additivity assumption takes the form

$$\Gamma_{1,2,3,\ldots,m} = \sum_{\substack{\text{all possible} \\ ij \text{ pairs}}} \Gamma_{ij} \tag{5-44}$$

We also used this assumption in Sec. 4.5, where we briefly considered some properties of the condensed state. While there is no rigorous proof, it may well be that because of cancellation effects, the assumption of pairwise additivity becomes better as the number of particles increases. However, it is likely that the assumption is somewhat in error for a three-body assembly (Rowlinson, 1965). Therefore, calculations for the third virial coefficient using Eq. (5-18) must be considered as approximations.

For any realistic potential, the calculation of third virial coefficients is complicated and, to obtain numerical results, we require a computer. Numerical computations have been carried out for several potential functions and results for pure nonpolar components are available (Sherwood and Prausnitz, 1964a; Kihara, 1953, 1958, 1963; Graben and Present, 1962; Sherwood *et al.*, 1966). For example, Fig. 5-18 gives reduced third virial coefficients as calculated from Kihara's potential. In these calculations, a spherical core was used and pairwise additivity was assumed. The reduced third virial coefficient, reduced temperature, and reduced core are defined by

$$C^* = \frac{C}{\left(\frac{2}{3}\pi N_A \sigma^3\right)^2} \qquad T^* = \frac{kT}{\varepsilon} \qquad a^* = \frac{2a}{\sigma - 2a}$$

where $-\varepsilon$ is the minimum energy in the potential function, σ is the intermolecular distance when the potential is zero, and a is the core radius. For $a^* = 0$, the results shown are those obtained from Lennard-Jones' potential.

Some efforts have been made to include nonadditivity corrections in the calculation of third virial coefficients (Sherwood and Prausnitz, 1964a; Kihara, 1953, 1958, 1963). These corrections are based on a quantum-mechanical relation derived by Axilrod and Teller (1943) for the potential of three spherical, nonpolar molecules at separations where London dispersion forces dominate. The *nonadditive correction* is a function of the polarizability and at lower temperatures it is large; its overall effect is that it approximately doubles the calculated third virial coefficient at its maximum, steepens the slope near the peak value and shifts the maximum to a lower reduced temperature.

Calculated and observed third virial coefficients for argon are shown in Fig. 5-19. Calculated results are based on four potential functions; for each of these, the parameters were determined from second-virial-coefficient data. The solid lines include the nonadditivity correction but the dashed lines do not; it is clear that the nonadditivity correction is appreciable.

Barker and Henderson (1976) have presented a definitive study of the third virial coefficient of argon; their results are shown in Fig. 5-20. The lowest line shows calculations based on the assumption of pairwise additivity as given by Eq. (5-43). For the two-body potential for argon, Barker and Henderson used an expression obtained from data reduction using two-body experimental information: second virial coefficients and gas-phase transport properties at low densities. The middle line shows calculations based on a three-body potential (Γ_{ijk}) that includes first-order corrections to the additivity assumption. (This correction is called the *Axilrod-Teller correction*.) The top line shows calculations that include second- and third-order corrections. These calculations agree with experiment within experimental error.

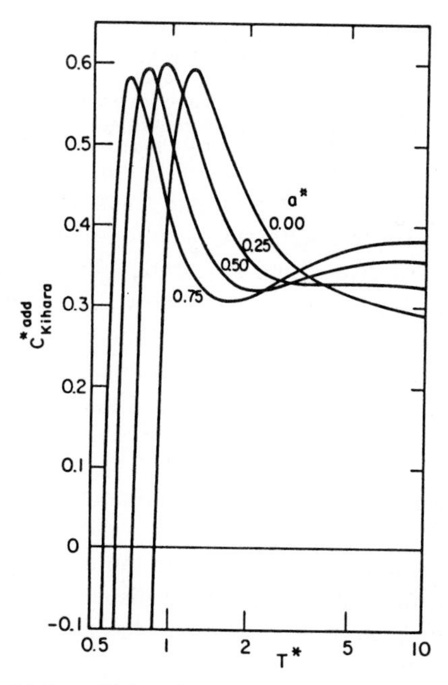

Figure 5-18 Third virial coefficient from Kihara potential assuming pairwise additivity.

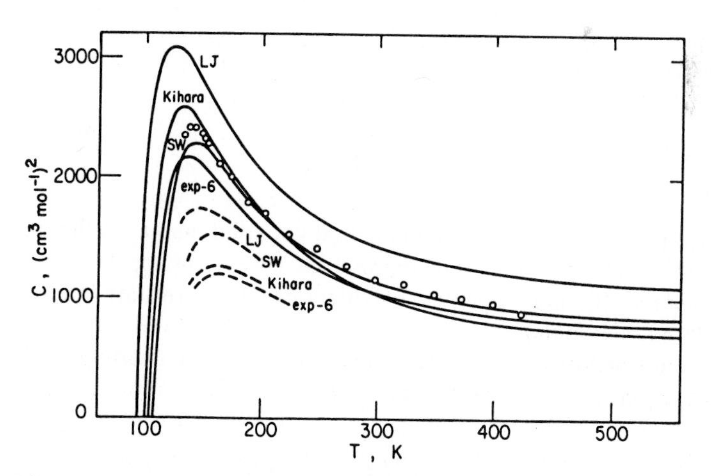

Figure 5-19 Calculated and observed third virial coefficients for argon. Solid lines include Axilrod-Teller nonadditivity corrections. Dashed lines show a portion of calculated results assuming additivity. Circles represent experimental data of Michels (1958).

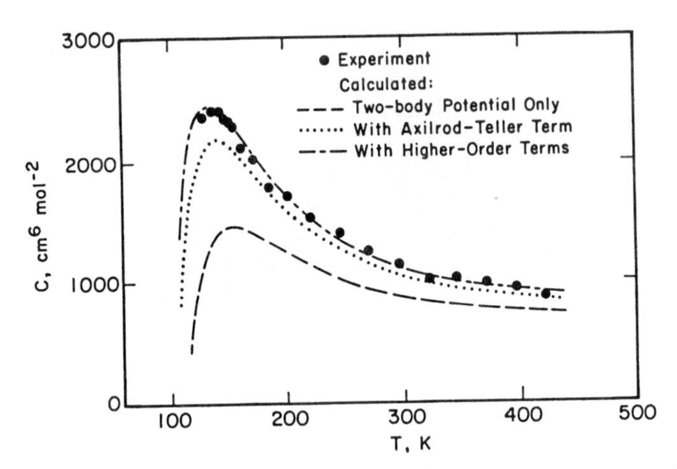

Figure 5-20 Third virial coefficients for argon (Barker and Henderson, 1976).

Without going into details, we can indicate the nature of the three-body corrections. First, we recall that in London's theory of dispersion forces, the potential is approximated by a series; for a two-body potential it has the form

$$\Gamma(r) = -\frac{C_6}{r^6} - \frac{C_8}{r^8} - \frac{C_{10}}{r^{10}} - \cdots \tag{5-45}$$

where r is the center-to-center distance between two molecules. The leading coefficient is

$$C_6 = \frac{3}{4}\frac{\alpha^2 I}{(4\pi\varepsilon_0)^2} \tag{5-46}$$

where α is the polarizability, I is the ionization potential, and ε_0 is the vacuum permittivity. Coefficients C_8 and C_{10} (not reproduced here) give the higher-order terms in London's potential. (In many practical calculations these terms are ignored.)

For a three-body potential, the first-order correction to the additivity assumption is obtained from London's theory, restricting attention to the leading (C_6/r^6) term. This first-order (Axilrod-Teller) correction to Γ_{ijk} is

$$\Delta\Gamma_{ijk}(r_{ij}, r_{jk}, r_{ik}) = \frac{9}{16}\frac{I\alpha^3(1+3\cos\theta_i\cos\theta_j\cos\theta_k)}{(4\pi\varepsilon_0)^2(r_{ij}r_{jk}r_{ik})^3} \tag{5-47}$$

where θ_i, θ_j, and θ_k are the three angles of a triangle whose sides are r_{ij}, r_{jk}, and r_{ik}.

Second-order corrections for nonadditivity are based on London's theory, restricting attention to the first two terms in the series; third-order corrections for nonadditivity are based on London's theory, restricting attention to the first three terms in the series. Still higher corrections appear to be negligible.

The important difference between Barker and Henderson's results (Fig. 5-20) and those of earlier workers (Fig. 5-19) lies in the two-body potential. The earlier work is based on a two-body potential that follows only from second virial-coefficient data and, as suggested by Fig. 5-15, these data alone do not yield a unique potential function. However, Barker and Henderson were able to use a *unique* two-body potential obtained from other dilute-gas data in addition to second-virial-coefficient data. The excellent agreement between theory and experiment, shown in Fig. 5-20, follows from an excellent two-body potential.

Due to experimental difficulties, there are few reliable values of third virial coefficients. It is therefore not possible to make a truly meaningful comparison between calculated and observed third virial coefficients; not only are experimental data not plentiful, but frequently they are of low accuracy. Even when calculated from very good P-V-T data, the accuracy of third virial coefficients is about one order of magnitude lower than that for second virial coefficients. To the extent that a comparison could be made, Sherwood (1964) found that the Lennard-Jones potential (with nonadditivity correction) generally predicted third virial coefficients that were too high, especially for larger molecules (such as pentane or benzene) where the predictions were very poor. The three-parameter potentials (square-well, exp-6, and Kihara) gave much better predictions; however, in view of the uncertainties in the data, and because corrections for nonadditive repulsive forces have been neglected, it is not possible to give a quantitative estimate of agreement between theory and experiment.

Little work has been done on the third virial coefficient of mixtures. The cross-coefficients, assuming additivity, can be calculated by Eq. (5-24) and the nonadditivity correction to these cross-coefficients, based on the formula of Axilrod and Teller, can also be computed as shown by Kihara (1953, 1958, 1963). However, the results of such calculations cannot be presented in a general manner; the coefficient C_{ijk} (for $i \neq j \neq k$) is a function of five independent variables for a two-parameter potential; for a three-parameter potential, eight independent variables must be specified.

An approximate method for calculating C_{ijk} was proposed by Orentlicher (1967), who showed that, subject to several simplifying assumptions, a reasonable estimate of C_{ijk} can be made based on Sherwood's numerical results for the third virial coefficient of pure gases.[10] Component i is chosen as a reference component; let $C_{ii}(T)$ stand for the third virial coefficient of pure i at the temperature T of interest. Let the potential function between molecules i and j be characterized by the collision diameter σ_{ij} and the energy parameter ε_{ij}. Similarly, the potential function for the i-k pair is characterized by σ_{ik} and ε_{ik}, and that for the j-k pair by σ_{jk} and ε_{jk}. Orentlicher's approximation is

[10] Orentlicher's approximation has been critically discussed by D. E. Stogryn, 1968, *J. Chem. Phys.*, 48: 4474.

$$\frac{C_{ijk}(T)}{C_{ii}(T)} = \left(\frac{\sigma_{ij}\sigma_{jk}\sigma_{ik}}{\sigma_{ii}^3} \right)^2 \frac{[C_{ij}^*(T_{ij}^*)C_{jk}^*(T_{jk}^*)C_{ik}^*(T_{ik}^*)]^{1/3}}{C_{ii}^*(T_{ii}^*)} \tag{5-48}$$

where $T_{ij}^* = kT/\varepsilon_{ij}$, etc., and where the individual reduced coefficients C_{ij}^*, etc., are obtained from available tables for pure components using any one of several popular potential functions.[11] Orentlicher's formula appears to give good results for mixtures where the components do not differ much in molecular size and characteristic energy. The accuracy of the approximation is difficult to assess, but it is probably useful for mixtures at those temperatures where the third virial coefficient of each component has already passed its maximum. Table 5-6 gives some observed and calculated third-virial cross coefficients for binary mixtures. Because the uncertainty in the experimental results is probably at least ±100 $(\text{cm}^3\ \text{mol}^{-1})^2$, agreement between calculated and experimental results is good for these particular mixtures.

Third virial coefficients for polar gases were calculated by Rowlinson (1951) using the Stockmayer potential and assuming additivity. Because experimental data are scarce and of low accuracy, it is difficult to make a meaningful comparison between calculated and experimental results.

Table 5-6 Experimental and calculated third-virial cross coefficients for some binary mixtures (Orentlicher, 1967).

Component			C_{112} $(\text{cm}^3\ \text{mol}^{-1})^2$	
1	2	Temperature (K)	Experimental	Calculated
Ar	N_2	273	1349	1510
		203	1706	1770
		163	2295	2420
N_2	Ar	273	1399	1340
		203	1780	1750
		163	2397	2330
CF_4	CH_4	273	4900	5250
		373	3400	3360
		473	2600	2700
		573	2400	2400
N_2	C_2H_4	323	2300	2300

[11] C_{ij}^* is a function of kT/ε_{ij} and, perhaps, of some additional parameter such as a_{ij}^* for the Kihara potential. This function, however, is the same as that for C_{ii}^*, that in turn depends on kT/ε_{ii} and, perhaps, on a_{ii}^*.

5.7 Virial Coefficients from Corresponding-States Correlations

Because there is a direct relation between virial coefficients and intermolecular potential, it follows from the molecular theory of corresponding states (Sec. 4.12) that virial coefficients can be correlated by data reduction with characteristic parameters such as critical constants. A few correlations are given in the following paragraphs.

The major part of this section is concerned with second virial coefficients for nonpolar gases. We cannot say much about third virial coefficients because of the scarcity of good experimental data and because of the nonadditivity problem mentioned in the preceding section. Further, our understanding of polar gases is not nearly as good as that of nonpolar gases, because again, good experimental data are not plentiful for polar gases, and because theoretical models, based on ideal dipoles, often provide poor approximations to the behavior of real polar molecules.

Equation (5-17) relates second virial coefficient B to intermolecular potential Γ. Following the procedure given in Sec. 4.12, we assume that the potential Γ can be written in dimensionless form by

$$\frac{\Gamma}{\varepsilon} = F\left(\frac{r}{\sigma}\right) \tag{5-49}$$

where ε is a characteristic energy parameter, σ is a characteristic size parameter, and F is a universal function of the reduced intermolecular separation. Upon substitution, Eq. (5-17) can then be rewritten in dimensionless form:

$$\frac{B}{2\pi N_A \sigma^3} = \int_0^\infty \left[1 - \exp\left(\frac{-\varepsilon F(r/\sigma)}{kT}\right)\right]\left(\frac{r}{\sigma}\right)^2 d\left(\frac{r}{\sigma}\right) \tag{5-50}$$

If we set σ^3 proportional to critical volume v_c, and ε/k proportional to critical temperature T_c, we obtain an equation of the form

$$\frac{B}{v_c} = F_B\left(\frac{T}{T_c}\right) \tag{5-51}$$

where F_B is a universal function of reduced temperature.

Equation (5-51) says that the reduced second virial coefficient is a generalized function of reduced temperature; this function can either be determined by specifying the universal potential function Γ/ε and integrating, as shown by Eq. (5-50), or by a direct correlation of experimental data for second virial coefficients.

For example, McGlashan and Potter (1962) plotted on reduced coordinates experimentally determined second virial coefficients for methane, argon, krypton, and xenon; the data for these four gases were well correlated by the empirical equation

$$\frac{B}{v_c} = 0.430 - 0.866\left(\frac{T}{T_c}\right)^{-1} - 0.694\left(\frac{T}{T_c}\right)^{-2} \tag{5-52}$$

Equation (5-52) was established from pure-component data but, utilizing the results of statistical mechanics and the molecular theory of corresponding states, it is readily extended to mixtures. To find B for a pure component, we use critical constants v_c and T_c for that component; for a mixture of m components we first recall that the second virial coefficient is a quadratic function of the mole fraction [Eq. (5-21)]:

$$B_{\text{mixt}} = \sum_{i=1}^{m}\sum_{j=1}^{m} y_i y_j B_{ij} \tag{5-53}$$

To calculate cross-coefficient B_{ij} $(i \neq j)$ we again use Eq. (5-52), but we must now specify parameters $v_{c_{ij}}$ and $T_{c_{ij}}$. If we use the common, semiempirical *combining rules* for the characteristic parameters of the potential function Γ_{ij}, i.e.,

$$\sigma_{ij} = \frac{1}{2}(\sigma_i + \sigma_j) \tag{5-54}$$

and

$$\varepsilon_{ij} = (\varepsilon_i \varepsilon_j)^{1/2} \tag{5-55}$$

it then follows from our previous assumptions that

$$v_{c_{ij}} = \frac{1}{8}\left(v_{c_i}^{1/3} + v_{c_j}^{1/3}\right)^3 \tag{5-56}$$

and

$$T_{c_{ij}} = (T_{c_i} T_{c_j})^{1/2} \tag{5-57}$$

The *geometric-mean approximation* for $T_{c_{12}}$ expressed by the above equation appears to be an upper limit; for asymmetric systems (i.e., mixtures whose components differ appreciably in molecular size), $T_{c_{12}}$ is usually smaller than the geometric mean of T_{c_1} and T_{c_2}. Subject to several simplifying assumptions, it can be shown from London's dispersion formula (Sec. 4.4) that

$$T_{c_{12}} = (T_{c_1} T_{c_2})^{1/2} \left[\frac{8(v_{c_1} v_{c_2})^{1/2}}{(v_{c_1}^{1/3} + v_{c_2}^{1/3})^3} \right]^q \left[\frac{2(I_1 I_2)^{1/2}}{I_1 + I_2} \right]$$

where I is the ionization potential and q is a positive exponent. Each of the bracketed quantities is equal to or less than unity and therefore the geometric mean for $T_{c_{12}}$ [Eq. (5-57)] represents an upper limit. While the London formula is useful for showing that the geometric mean for $T_{c_{12}}$ is a maximum, it is usually not reliable for a quantitative correction of the geometric-mean assumption.

To illustrate the applicability of Eq. (5-52) for a mixture, let us make the reasonable assumption that this equation holds also for nitrogen; we can then calculate the second virial coefficient of a mixture of argon and nitrogen using only the critical volume and critical temperature for each of these pure components. Table 5-7 shows results obtained at three temperatures for an equimolar mixture argon/nitrogen using Eqs. (5-52), (5-53), (5-56), and (5-57). Calculated results are in excellent agreement with those found experimentally by Crain and Sonntag (1965).

Table 5-7 Experimental and calculated second virial coefficients for an equimolar mixture of argon/nitrogen.

Temperature	B_{mixt} ($cm^3\ mol^{-1}$)	
(°C)	Calculated	Experimental
0	-16.5	-16.3
-70	-41.3	-40.4
-130	-89.4	-88.3

Equation (5-52) gives a good representation of the second virial coefficients of small, nonpolar molecules but for larger molecules Eq. (5-52) is no longer satisfactory. For example, McGlashan *et al.* (1962, 1964) measured second virial coefficients for normal alkanes and α-olefins containing up to eight carbon atoms; to represent their own data, as well as those of others, they used an amended form of Eq. (5-52), i.e.,

$$\frac{B}{v_c} = 0.430 - 0.886\left(\frac{T}{T_c}\right)^{-1} - 0.694\left(\frac{T}{T_c}\right)^{-2} - 0.0375(n-1)\left(\frac{T}{T_c}\right)^{-4.5} \qquad (5\text{-}58)$$

where n stands for the number of carbon atoms.[12] Clearly, for methane (n = 1) Eq. (5-58) reduces to Eq. (5-52). Because experimental critical volumes for α-olefins are not

[12] McGlashan and Potter's correlation applies equally well for some polymethyl compounds (e.g., tetramethylsilane, 2-methylbutane) taking the characteristic parameter n to be the number of methyl groups (J. M. Barbarín-Castillo and I. A. McLure, 1993, *J. Chem. Thermodynamics*, 25: 1521).

highly accurate and because critical volumes for α-olefins have not been measured for $n > 4$, McGlashan *et al.*, suggest that for these fluids the critical volumes be calculated from the expression

$$v_c(\text{cm}^3\ \text{mol}^{-1}) = 25.07 + 50.38n + 0.479n^2 \tag{5-59}$$

Equation (5-58) is plotted in Fig. 5-21 to illustrate the importance of the third parameter, n, especially at low reduced temperatures. The experimental data for hydrocarbons of different chain length clearly show the need for a third parameter in a corresponding-states correlation. This need indicates that a potential function containing only two characteristic constants is not sufficient for describing the thermodynamic properties of a series of compounds that, although chemically similar, differ appreciably in molecular size and shape.

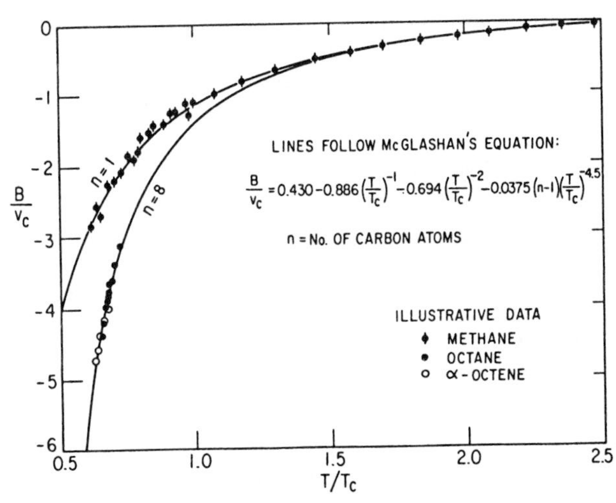

Figure 5-21 Corresponding-states correlation of McGlashan for second virial coefficients of normal paraffins and α-olefins.

While the addition of a third parameter significantly improves the accuracy of a corresponding-states correlation for the second virial coefficients of pure fluids, it unfortunately introduces the need for an additional combining rule when applied to the second virial coefficients of mixtures. Suppose, for example, that we want to use Eq. (5-58) for predicting the second virial coefficient of a mixture of propene and α-heptene. Let 1 stand for propene and 2 for heptene; how shall we compute the cross-coefficient B_{12}? For the characteristic volume and temperature we have the (at best) semitheoretical combining rules given by Eqs. (5-56) and (5-57). We must also decide on a value of n_{12} and, because B_{12} refers to the interaction of one propene molecule with one heptene molecule, we are led to the rule

$$n_{12} = \frac{1}{2}(n_1 + n_2) \tag{5-60}$$

that, in this case, gives $n_{12} = 5$. While this procedure appears reasonable, we must recognize that Eq. (5-60) has little theoretical basis; it is strictly an *ad hoc*, phenomenological equation that, ultimately, can be justified only empirically. McGlashan *et al.* measured volumetric properties of a nearly equimolar mixture of propene and α-heptene. The second virial coefficients of the mixture at several temperatures are shown in Fig. 5-22. Calculated results were obtained from Eq. (5-53); individual coefficients B_{11}, B_{22}, and B_{12} were determined from Eq. (5-58) and from the combining rules given by Eqs. (5-56), (5-57), and (5-60). In this system, agreement between calculated and experimental results is excellent; good agreement was also found for mixtures of propane and heptane and for mixtures of propane and octane (McGlashan and Potter, 1962).

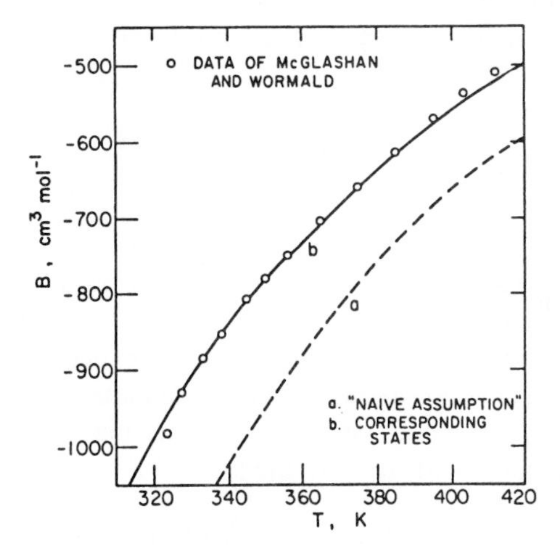

Figure 5-22 Second virial coefficients of a mixture containing 50.05 mole percent propene and 49.95 mole percent α-heptene.

For comparison, Fig. 5-22 also shows results calculated with the attractively simple assumption

$$B_{12} = \frac{1}{2}(B_{11} + B_{22}) \tag{5-61}$$

Guggenheim[13] appropriately calls Eq. (5-61) the "naive assumption". It can readily be shown that Eq. (5-61) follows from Amagat's law (or the Lewis rule); when Eq. (5-61) is substituted into the exact Eq. (5-53), we obtain the simple but *erroneous* result

[13] E. A. Guggenheim, 1952, *Mixtures,* Oxford: Clarendon Press.

$$B_{\text{mixt}} = y_1 B_{11} + y_2 B_{22} \tag{5-62}$$

Figure 5-22 shows that Eq. (5-62) is in significant disagreement with experimental data.

The Pitzer-Tsonopoulos Correlation. In Secs. 4.12 and 4.13 we pointed out that the applicability of corresponding-states correlations could be much extended if we distinguish between different classes of fluids and characterize these classes with an appropriate, experimentally accessible parameter. With this extension, the theory of corresponding states is applied separately to any one class containing a limited number of fluids rather than to a very large number of fluids belonging to different classes.

The classifying parameter suggested by Pitzer is ω, the *acentric factor*, a macroscopic measure of how much the force field around a molecule deviates from spherical symmetry. The acentric factor is (essentially) zero for spherical, nonpolar molecules such as the heavy noble gases and for small, highly symmetric molecules such as methane. The definition of ω was given earlier [Eq. (4-86)]; it is

$$\omega \equiv -\log\left(\frac{P^s}{P_c}\right)_{T/T_c=0.7} - 1.000 \tag{5-63}$$

where P^s is the saturation (vapor) pressure and P_c is the critical pressure. Some acentric factors are given in Table 4-14. Other classifying parameters have been proposed (Meissner and Seferian, 1951; Riedel, 1956; Rowlinson, 1955) but the acentric factor is the most practical because it is easily evaluated from experimental data that are both readily available and reasonably accurate for most common substances.

When applied to the second virial coefficient, the extended theory of corresponding states asserts that for all fluids in the same class,

$$\frac{B}{N_A \sigma^3} = F_\omega\left(\frac{kT}{\varepsilon}\right) \tag{5-64}$$

where, as before, σ is a characteristic molecular size and ε/k *is* a characteristic energy expressed in units of temperature. Function F_ω depends on the acentric factor ω and, for any given ω, it applies to all substances having that ω.

Upon replacing σ and ε/k with macroscopic parameters, Eq.(5-64) becomes

$$\frac{BP_c}{RT_c} = F_\omega\left(\frac{T}{T_c}\right) \tag{5-65}$$ [14]

---

[14] Corresponding states sets σ^3 proportional to v_c. But $v_c = z_c RT_c/P_c$. In Eq. (5-65), because z_c is a function only of ω, z_c has been absorbed in F_ω. By contrast, z_c appears explicitly in Eq. (5-66).

Schreiber and Pitzer (1989) have proposed a correlation of the form indicated by Eq. (5-65). They write

$$\frac{BP_c}{RT_c z_c} = c_1 + c_2 T_R^{-1} + c_3 T_R^{-2} + c_4 T_R^{-6} \tag{5-66}$$

Here, $T_R = T/T_c$ is the reduced temperature and z_c is the compressibility factor at the critical point approximated by $z_c = 0.291 - 0.08\omega$. Coefficients c_1, c_2, c_3, and c_4 depend on acentric factor according to

$$c_i = c_{i,0} + \omega\, c_{i,1} \tag{5-67}$$

Table 5-8 gives coefficients $c_{i,0}$ and $c_{i,1}$, determined from a large body of experimental data for nonpolar or slightly polar fluids. Highly polar fluids such as water, nitriles, ammonia, and alcohols were not included; also, the quantum gases helium, hydrogen and neon were omitted.

Table 5-8 Coefficients for Eq. (5-66).

i	$c_{i,0}$	$c_{i,1}$
1	0.442259	0.725650
2	-0.980970	0.218714
3	-0.611142	-1.249760
4	-0.00515624	-0.189187

Tsonopoulos (1974, 1975, 1978) gave another correlation for second virial coefficients in the form

$$\frac{BP_c}{RT_c} = F^{(0)}\left(\frac{T}{T_c}\right) + \omega F^{(1)}\left(\frac{T}{T_c}\right) \tag{5-68}$$

where

$$F^{(0)}\left(\frac{T}{T_c}\right) = 0.1445 - \frac{0.330}{T_R} - \frac{0.1385}{T_R^2} - \frac{0.0121}{T_R^3} - \frac{0.000607}{T_R^8} \tag{5-69}$$

$$F^{(1)}\left(\frac{T}{T_c}\right) = 0.0637 + \frac{0.331}{T_R^2} - \frac{0.423}{T_R^3} - \frac{0.008}{T_R^8} \tag{5-70}$$

Figure 5-23 shows the reduced second virial coefficient as a function of reciprocal reduced temperature for three ω. It is clear that the introduction of a third parameter has a pronounced effect, especially at low reduced temperatures.

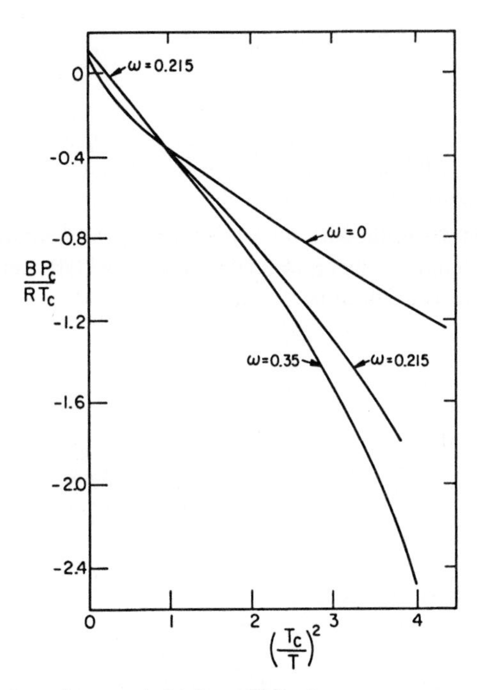

Figure 5-23 Reduced second virial coefficients.

Equations (5-68), (5-69), and (5-70) [and other similar equations like those given by Mak and Lielmezs (1989)] provide a good correlation for the second virial coefficients of normal fluids. A *normal fluid* is one whose molecules are of moderate size, are nonpolar or else slightly polar, and do not associate strongly (e.g., by hydrogen bonding); further, it is a fluid whose configurational properties can be evaluated to a sufficiently good approximation by classical, rather than quantum statistical mechanics. For the three quantum gases, helium, hydrogen, and neon, good experimental results are available (see App. C), and therefore there is no need to use a generalized correlation for calculating the second virial coefficients of these gases. However, effective critical constants for quantum gases are useful for calculating properties of mixtures that contain quantum fluids in addition to normal fluids, as discussed later.

For polar and hydrogen-bonded fluids, Tsonopoulos suggested that Eq. (5-68) be amended to read[15]

$$\frac{BP_c}{RT_c} = F^{(0)}\left(\frac{T}{T_c}\right) + \omega F^{(1)}\left(\frac{T}{T_c}\right) + F^{(2)}\left(\frac{T}{T_c}\right) \tag{5-71}$$

where

[15] Similar equations were proposed by Hayden and O'Connell (1975) and by Tarakad and Danner (1977).

$$F^{(2)}\left(\frac{T}{T_c}\right) = \frac{a}{T_R^6} - \frac{b}{T_R^8} \tag{5-72}$$

Constants a and b cannot easily be generalized, but for polar fluids that do not hydrogen-bond, Tsonopoulos found that $b = 0$.

Figure 5-24 shows constant a as a function of reduced dipole moment for a variety of non-hydrogen-bonded polar fluids. The equations given in the same figure present a compromise best fit for the polar fluids indicated.

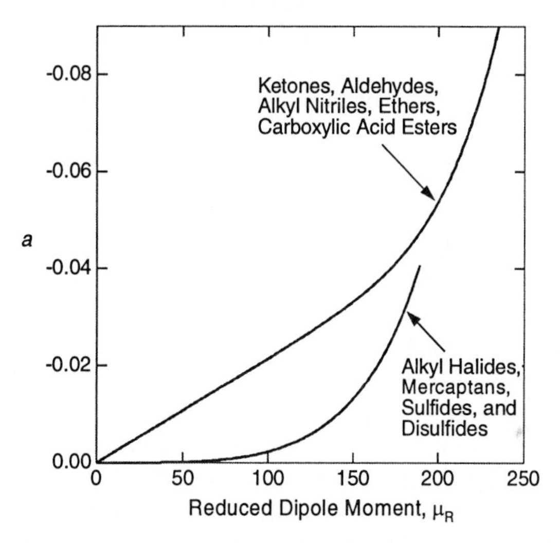

Figure 5-24 Correlating constant *a* for some polar fluids. Top curve for ketones, aldehydes, alkyl nitriles, ethers, and carboxylic acid esters; lower curve for alkyl halides, mercaptans, sulfides, and disulfides.

As shown in Fig. 5-24, Tsonopoulos (1990) suggests separate equations, one for each group of different chemical species. The top curve in Fig. 5.24 is recommended for ketones, aldehydes, alkyl nitriles, ethers, and carboxylic-acid esters,

$$a = -2.14 \times 10^{-4} \mu_R - 4.308 \times 10^{-21} (\mu_R)^8 \tag{5-73}$$

However, for alkyl halides, mercaptans, sulfides, and disulfides, the bottom curve of Fig. 5-24 should be used:

$$a = -2.188 \times 10^{-11} (\mu_R)^4 - 7.831 \times 10^{-21} (\mu_R)^8 \tag{5-74}$$

In Eqs. (5-73) and (5-74), μ_R is the reduced dipole moment defined by

$$\mu_R = 0.9869 \times 10^5 \frac{\mu^2 P_c}{T_c^2} \tag{5-75}$$

where the units are debye for the dipole moment μ, bar for P_c, and kelvin for T_c.

As Fig. 5-24 shows, Eqs. (5-73) and (5-74) approach closely at high μ_R suggesting a unique relation between a and μ_R for strongly polar, non-dimerizing organic compounds.

For water the recommended a is -0.0109.

It is difficult to correlate second virial coefficients of polar and hydrogen-bonded fluids because we do not adequately understand the pertinent intermolecular forces and also because reliable experimental second-virial-coefficient data for such fluids are relatively scarce. For strongly hydrogen-bonded fluids, the virial equation is not useful for describing vapor-phase imperfections. For such fluids a chemical theory (dimerization) is required, as discussed in Sec. 5.8.

Third Virial Coefficients. Reliable experimental data are rare for third virial coefficients of pure gases, and good third virial coefficients for gaseous mixtures are extremely rare. In addition to the scarcity of good experimental results, it is difficult to establish an accurate corresponding-states correlation because, as briefly discussed earlier (Secs. 5.2 and 5.6), third virial coefficients do not exactly satisfy the assumption of pairwise additivity of intermolecular potentials. Fortunately, however, as shown by Sherwood (1966), nonadditivity contributions to the third virial coefficient from the repulsive part of the potential tend to cancel, at least in part, nonadditivity contributions from the attractive part of the potential; further, these contributions appear to be a function primarily of the reduced polarizability and they are important only in part of the reduced temperature range, at temperatures near or below the critical temperature.

Several corresponding-states correlations are available for the third virial coefficients of gases and gas mixtures. Those of Chueh and Prausnitz (1967), Pope *et al.* (1973), De Santis and Grande (1979), and Orbey and Vera (1983) are suitable for nonpolar gases, whereas the correlation presented by Besher and Lielmezs (1992) also provides estimates for polar fluids. Typically these correlations give the reduced third virial coefficient through the generalized correlation,

$$C_R = \frac{CP_c^2}{(RT_c)^2} = F^{(0)}(T_R) + \omega F^{(1)}(T_R) \tag{5-76}$$

where $F^{(0)}$ and $F^{(1)}$ are regressed from experimental data.

For polar compounds, Besher and Lielmezs suggested that the above equation should take the form

$$C_R = F^{(0)}(T_R) + \omega F^{(1)}(T_R) + \mu_R^x F^{(2)}(T_R) \tag{5-77}$$

where μ_R is the reduced dipole moment [defined by Eq. (5-75)] and x is an empirical, substance-dependent constant.

For the correlation of Orbey and Vera, functions $F^{(0)}$ and $F^{(1)}$ in Eq. (5-76) are given by

$$F^{(0)}(T_R) = 0.01407 + \frac{0.02432}{T_R^{2.8}} - \frac{0.00313}{T_R^{10.5}} \tag{5-78}$$

$$F^{(1)}(T_R) = -0.02676 + \frac{0.01770}{T_R^{2.8}} + \frac{0.040}{T_R^{3.0}} - \frac{0.003}{T_R^{6.0}} - \frac{0.00228}{T_R^{10.5}} \tag{5-79}$$

Figure 5-25 compares experimental with calculated third virial coefficients for (nonpolar) sulfur hexafluoride obtained from the correlations of De Santis and Grande, Orbey and Vera, and Besher and Lielmezs. As Fig. 5-25 shows, all three methods compare well with experiment.

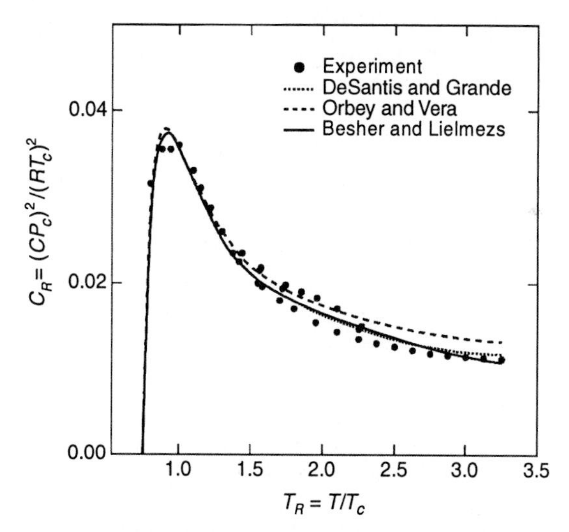

Figure 5-25 Reduced third virial coefficient of sulfur hexafluoride as a function of reduced temperature. Experiment: ● (from various sources). Calculated: ...... De Santis and Grande (1979); - - - - Orbey and Vera (1983); —— Besher and Lielmezs (1992).

For mixtures, it is difficult to predict the cross-coefficients C_{ijk}. Studies by Stogryn (1968, 1969, 1970) indicate that no simple approximation procedure is entirely satisfactory; however, the approximation suggested by Orentlicher is useful [see Eq. (5-48)]. In its simplest form, Orentlicher's approximation suggests that

$$C_{ijk} = (C_{ij}C_{jk}C_{ik})^{1/3} \tag{5-80}$$

Coefficient C_{ij} is the third virial coefficient for a hypothetical pure gas whose potential-energy function is characterized by distance parameter σ_{ij} and energy parameter ε_{ij} or, in corresponding-states terms,

$$C_{ij} = v_{c_{ij}}^2 F_c\left(\frac{T}{T_{c_{ij}}}, \omega_{ij}\right) \tag{5-81}$$

where F_c is the correlating equation for the third virial coefficient of a pure fluid.

A simple combining rule [e.g., Eq. (5-56)] is used for $v_{c_{ij}}$, while $T_{c_{ij}}$ is best estimated from second-virial coefficient (B_{ij}) data as discussed earlier in this section. For the third parameter ω_{ij} we use the arithmetic average.

Cross-virial-coefficient data are scarce and often not reliable. Measurements of cross virial coefficients are time-consuming and require meticulous care, especially for third virial coefficients. To reduce the number of cross parameters required to describe experimental data, McGregor *et al.* (1987) proposed a relation among the third cross virial coefficients that requires only binary data to predict multicomponent mixture behavior. The same relation also permits an accurate estimation of third cross virial coefficients from a single experimental isotherm at a single composition if the third virial coefficients are known for the pure components. McGregor used the simplifying assumption

$$3C_{iij} - 2C_{iii} - C_{jjj} = 3C_{ijj} - C_{iii} - 2C_{jjj} \equiv \delta C_{ij} \tag{5-82}$$

From Eq. (5-82), the mixture third virial coefficient is

$$C_{\text{mixt}} = \sum_{i=1}^{n} y_i C_{iii} + \sum_{i=1}^{n} \sum_{j=i+1}^{n} y_i y_j \delta C_{ij} \tag{5-83}$$

For a binary mixture,

$$C_{\text{mixt}} = y_1 C_{111} + y_1 y_2 \delta C_{12} + y_2 C_{222} \tag{5-84}$$

As a result, C_{mixt} given by Eq. (5-84) is formally identical to the quadratic mole-fraction expression for the mixture's second virial coefficient. Equation (5-84) requires only the same amount of data as that required for the calculation of second virial cross coefficients.

In Eq. (5-82), δC_{12} is a binary parameter whose evaluation requires data for only one binary mixture. A single composition for a binary mixture is sufficient to calculate B_{12} as well as C_{112} and C_{122}, provided that we know C_{111} and C_{222}.

Extension to multicomponent mixtures uses the assumption

$$3C_{ijk} - C_{iii} - C_{jjj} = \frac{1}{2}(\delta C_{ij} + \delta C_{ik} + \delta C_{jk}) \tag{5-85}$$

Pairwise additivity is not assumed in Eq. (5-85). Rather, it is assumed that certain relations exist among the magnitudes of three-body interactions. Therefore, the simplifying assumptions expressed by Eqs. (5-82) and (5-85) reduce the number of interaction parameters required to obtain, e.g., compressibility factor z from Eq. (5-9). Thus, to evaluate z for a ternary mixture, Eqs. (5-21) and (5-23) require three cross terms for the second virial coefficients and seven for the third. However, the simplified model of McGregor, briefly described above, reduces to three the number of cross terms needed for the third virial coefficient. In addition, McGregor's method can be used to predict third virial cross coefficients when fitted to experimental data for a single binary mixture, thereby permitting the description of multicomponent mixtures using only data for binary mixtures.

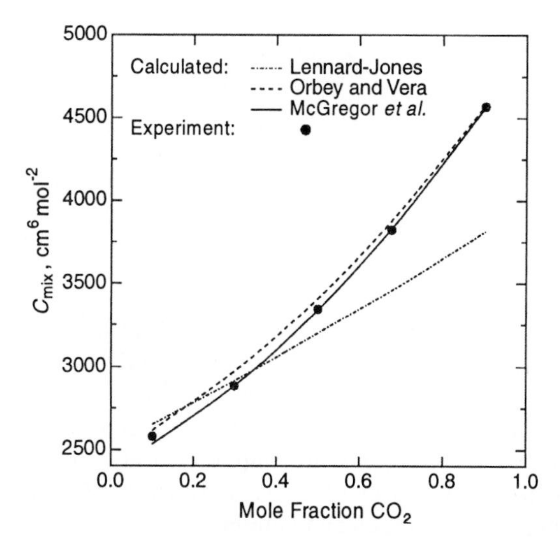

Figure 5-26 Third virial coefficients [from Eq. (5-22)] for the mixture CH_4/CO_2 at 300 K. ● Experiment; – · – · – Lennard-Jones; – – – – Orbey and Vera; —— McGregor.

A comparison between experimental and calculated mixture third virial coefficients obtained from several methods has been presented by Brugge *et al.* (1989). They calculated C_{mixt} from Eq. (5-22) using the Lennard-Jones 6-12 intermolecular potential $[C^* = C^*_{add} + \Delta C^*$, where $C^* = C/(B^*)^2 = C/(2/3\pi N_A\sigma^3)$; C^*_{add} is the pairwise contribution and ΔC^* is the non-additivity correction], the empirical correlation of Orbey and Vera, and the model of McGregor. Figure 5-26 shows a comparison for the system methane/carbon dioxide at 300 K. Large deviations (particularly for high CO_2 concentrations) result from the calculations using the Lennard-Jones potential. In general, the other two methods are able to represent third virial coefficients of the mixtures within experimental uncertainty.

Quantum Gases. Corresponding-states correlations for virial coefficients may also be applied to quantum gases (helium, hydrogen, and neon) by using *effective* critical constants and $\omega = 0$. Because the configurational properties of low-molecular-weight molecules are described by quantum, rather than classical mechanics, effective critical constants depend on temperature. Following the method proposed by Gunn *et al.* (1966), these correlations for quantum gases use temperature dependent effective critical constants given by

$$T_c = \frac{T_c^0}{1 + \frac{21.8}{MT}} \tag{5-86}$$

$$P_c = \frac{P_c^0}{1 + \frac{44.2}{MT}} \tag{5-87}$$

$$v_c = \frac{v_c^0}{1 - \frac{9.91}{MT}} \tag{5-88}$$

where M is the molar mass in grams per mole and T is in kelvin. Superscript ^{0} designates the "classical" critical constants that are the effective critical constants at high temperatures; as MT becomes large, $T_c \to T_c^0$, $P_c \to P_c^0$ and $v_c \to v_c^0$. Therefore, while Eqs. (5-86) to (5-88) are essentially empirical, they are consistent with the theoretical requirement that quantum corrections are important only for low-molecular weight compounds at low temperatures. Table 5-9 gives effective "classical" critical constants for neon and for isotopes of hydrogen and helium.

Table 5-9 "Classical" critical constants for quantum gases.

	T_c^0 (K)	P_c^0 (bar)	v_c^0 (cm³ mol⁻¹)
Ne	45.5	27.3	40.3
He^4	10.47	6.76	37.5
He^3	10.55	6.01	42.6
H_2	43.6	20.5	51.5
HD	42.9	19.9	52.3
HT	42.3	19.4	52.9
D_2	43.6	20.4	51.8
DT	43.5	20.6	51.2
T_2	43.8	20.8	51.0

For mixtures containing one or more quantum gases, binary characteristic constants $T_{c_{ij}}$ and $P_{c_{ij}}$ are obtained from combining rules

$$T_{c_{ij}} = \frac{(T^0_{c_i} T^0_{c_j})^{1/2}(1-k_{ij})}{1+\dfrac{21.8}{M_{ij}T}} \tag{5-89}$$

and

$$P_{c_{ij}} = \frac{P^0_{c_{ij}}}{1+\dfrac{44.2}{M_{ij}T}} \tag{5-90}$$

where

$$P^0_{c_{ij}} = \frac{z^0_{c_{ij}} R(T^0_{c_i} T^0_{c_j})^{1/2}(1-k_{ij})}{v^0_{c_{ij}}} \tag{5-91}$$

$$(v^0_{c_{ij}})^{1/3} = \frac{1}{2}\left[(v^0_{c_i})^{1/3} + (v^0_{c_j})^{1/3}\right] \tag{5-92}$$

$$z^0_{c_{ij}} = 0.291 - 0.08\left(\frac{\omega_i + \omega_j}{2}\right) \tag{5-93}$$

and

$$\frac{1}{M_{ij}} = \frac{1}{2}\left(\frac{1}{M_i} + \frac{1}{M_j}\right) \tag{5-94}$$

For a quantum gas, the effective $\omega = 0$. For non-quantum gases, T^0_c, P^0_c and v^0_c are the experimental critical temperature, pressure and volume. If no experimental value is available for the critical volume, it can be estimated by

$$v^0_c = \frac{RT^0_c}{P^0_c}(0.291 - 0.08\omega)$$

The "classical" critical volumes listed in Table 5-9 satisfy this relation with $\omega = 0$.

The combining rules listed above are useful for estimating cross-coefficient B_{ij} $(i \neq j)$ when either i or j (or both) is a quantum gas.

5.8 The "Chemical" Interpretation of Deviations from Gas-Phase Ideality

It was suggested many years ago that nonideal behavior of gases may be attributed to the formation of different chemical species. From the principle of Le Chatelier, one can readily show that, at low pressures, complex formation by association or solvation is negligible (resulting in ideal behavior); however, as pressure rises, significant chemical conversion may occur, sometimes leading to large deviations from ideal-gas behavior. For example, in pure gas A, various equilibria may be postulated:

$$2A \rightleftharpoons A_2, \qquad 3A \rightleftharpoons A_3, \qquad \text{etc.}$$

The equilibrium constant for dimerization reflects the interaction of two molecules at a time and therefore, a relation can be established between the dimerization equilibrium constant and the second virial coefficient (Mason and Spurling, 1969). Similarly, the trimerization equilibrium constant is related to the third virial coefficient, and so on. However, regardless of the degree of association, the "chemical" viewpoint considers the forces between molecules to be of chemical, rather than physical nature and it therefore attempts to explain nonideal behavior in terms of the formation of new chemical species. Polymerization reactions (dimerization, trimerization, etc.) result in negative deviations ($z < 1$) from ideal-gas behavior while dissociation reactions (primarily important at high temperatures) result in positive deviations ($z > 1$).

5.9 Strong Dimerization: Carboxylic Acids

The "chemical" viewpoint is particularly justified for systems with strong forces of attraction between molecules; good examples are organic acids or alcohols and other molecules capable of hydrogen bonding. To illustrate, consider the dimerization of acetic acid:

$$2\,CH_3{-}C(=O)OH \rightleftharpoons CH_3{-}C(=O{\cdots}H{-}O)(O{-}H{\cdots}O=)C{-}CH_3$$

We define an equilibrium constant in terms of the partial pressures rather than fugacities because we assume that the mixture of true species (i.e., monomers and dimers) behaves as an ideal gas.

$$K = \frac{p_{A_2}}{p_A^2} = \frac{y_{A_2} P^0}{y_A^2 P} \tag{5-95}$$

where P is the total pressure; y_A is the mole fraction of monomer; y_{A_2} is the mole fraction of dimer; and P^0 is the standard-state pressure equal to 1 bar. The equilibrium constant can be calculated from P-V-T data as follows: We measure V, the volume for one mole of acetic acid at total pressure P and temperature T. (In this connection, one mole means the formula weight of monomeric acetic acid.) Let n_A be the number of moles of monomer and n_{A_2} the number of moles of dimer; further, let α be the fraction of molecules that dimerize. Then n_T, the total number of "true" moles, is given by

$$n_T = n_A + n_{A_2} = (1-\alpha) + \frac{\alpha}{2} = 1 - \frac{\alpha}{2} \tag{5-96}$$

with the restraint that $n_A + 2n_{A_2} = 1$, as required by material balance. Because, by assumption, $PV = n_T RT$, we can compute α by substitution and obtain

$$\alpha = 2 - \frac{2PV}{RT} \tag{5-97}$$

Because

$$y_A = \frac{n_A}{n_T} \tag{5-98}$$

and

$$y_{A_2} = \frac{n_{A_2}}{n_T} \tag{5-99}$$

the equilibrium constant K is related to α by

$$K = \frac{\alpha(1-\alpha/2)P^0}{2(1-\alpha)^2 P} \tag{5-100}$$

Equation (5-100) shows that, because K is a function only of temperature, α must go to zero as P approaches zero; in other words, dimerization decreases as the pressure falls. This pressure dependence of the degree of dimerization is a direct consequence of Le Chatelier's principle.

Figure 5-27 gives dimerization constants for acetic acid and for propionic acid as a function of temperature based on experimental P-V-T measurements by McDougall (1936, 1941). In these gases, dimerization is very strong even at low pressures. For example, at 40°C and $P = 0.016$ bar, α for acetic acid is 0.8, and α for propionic acid is slightly larger, about 0.84. These gases show large deviations from ideal behavior at very

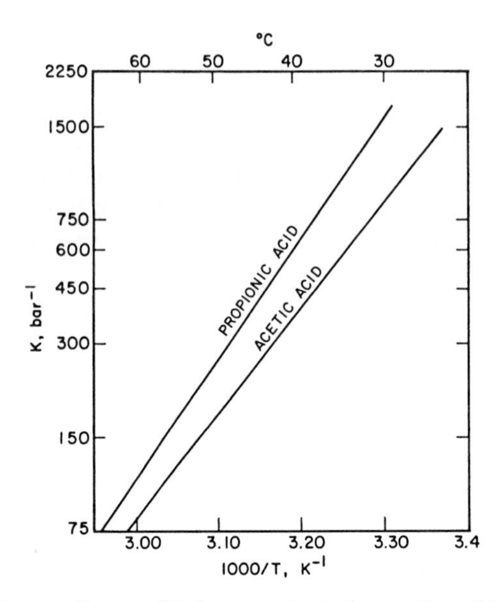

Figure 5-27 Dimerization equilibrium constants for acetic acid and propionic acid.

low pressures, well below 1 bar.[16] Tsonopoulos (1970) gives association constants for formic, acetic, propionic, butanoic, and trimethylacetic acids.

Because acetic and propionic acids are members of the same homologous series, it is reasonable to use the results shown in Fig. 5-27 to estimate the equilibrium constant for the solvation (or cross-dimerization) of these acids:

$$CH_3{-}C(=O){-}OH + CH_3{-}CH_2{-}C(=O){-}OH \rightleftharpoons CH_3{-}C\begin{matrix} =O\cdots H{-}O \\ {-}O{-}H\cdots O= \end{matrix}C{-}CH_2{-}CH_3$$

For this estimate, we assume that the enthalpy of the hydrogen bond is the same in the mixed dimer as that in either pure-component dimer. If A stands for acetic acid monomer and B stands for propionic acid monomer, we assume that there is zero enthalpy change for the reaction

[16] Because of strong dimerization in the vapor phase, the fugacity of a pure, saturated carboxylic acid is considerably lower than its saturation (vapor) pressure even when that pressure is small. Similarly, the fugacity coefficient of a carboxylic acid in a gaseous mixture is not close to unity even at low pressures. These deviations from ideal behavior are significant in vapor-liquid equilibria and failure to take them into account can lead to serious error. At low and moderate pressures, the fugacity of a strongly dimerized component, pure or in a mixture is set equal to the partial pressure of that component's monomer; this partial pressure can be calculated from the dimerization equilibrium constant. Details of such calculations have been presented by various authors, notably by Sebastiani and Lacquanti (1967), Tsonopoulos (1970), and Marek (1954, 1955).

$$1/2A_2 + 1/2B_2 \rightleftharpoons AB$$

However, the entropy change for this reaction is positive because distinguishability has been lost; whereas there are two different molecular species on the left side, there is only one molecular species on the right and therefore the reaction proceeds toward a more probable state. Details of the entropy calculation need not concern us here;[17] the change in entropy is given by $R\ln 2$. The standard Gibbs energy of forming AB dimers from A and B monomers can now be calculated from the following scheme:

$$2A \rightleftharpoons A_2 \qquad \Delta g^0_{A_2} = -RT\ln K_{A_2} \tag{5-101}$$

$$2B \rightleftharpoons B_2 \qquad \Delta g^0_{B_2} = -RT\ln K_{B_2} \tag{5-102}$$

$$A + B \rightleftharpoons AB \qquad \Delta g^0_{AB} = -\frac{1}{2}(\Delta g^0_{A_2} + \Delta g^0_{B_2}) - RT\ln 2 = -RT\ln K_{AB} \tag{5-103}$$

where Δg^0 is the molar Gibbs energy change in the standard state. Substitution gives $K_{AB} = 2\sqrt{K_{A_2}K_{B_2}}$.

From McDougall's data at 20°C, $K_{A_2} = 3.25\times10^{-3}$ and $K_{B_2} = 7.84\times10^{-3}$, giving $K_{AB} = 10.11\times10^{-3}$. Experimental measurements at 20°C by Christian (1957) indicate good agreement between the observed properties of acetic acid/propionic acid mixtures and those calculated from the three equilibrium constants. This favorable result is a fortunate consequence of the similar chemical nature of the mixture's components; in other mixtures, where the solvating components are chemically dissimilar, it is usually not possible to predict accurately the properties of the mixture from data for the pure components.

Much attention has been given to the system acetic acid/water. For example, Tsonopoulos (1970) suggests that when the total pressure is near 1 bar, vapor-phase nonidealities can be described by the following chemical equilibria:

$$2A \rightleftharpoons A_2 \tag{I}$$

$$3A \rightleftharpoons A_3 \tag{II}$$

$$2W \rightleftharpoons W_2 \tag{III}$$

$$A + W \rightleftharpoons AW \tag{IV}$$

[17] See R. Fowler and E. A. Guggenheim, 1952, *Statistical Thermodynamics*, p. 167, Cambridge: Cambridge University Press.

where A stands for acetic acid and W stands for water. Using *P-V-T* data for acetic acid vapor, equilibrium constants are evaluated for I and II; using *P-V-T* (second virial coefficient) data for water vapor, an equilibrium constant is evaluated for III. For the cross-dimer, Tsonopoulos assumes that the equilibrium constant for IV is equal to that for III.

From material balances and from the four simultaneous chemical equilibria, Tsonopoulos calculates p_A, the partial pressure of acetic acid monomer and p_W, that of water monomer. The fugacity coefficients of (stoichiometric) water (1) and (stoichiometric) acetic acid (2) are then given by

$$\varphi_1 = \frac{p_W}{p_1} \tag{5-104}$$

$$\varphi_2 = \frac{p_A}{p_2} \tag{5-105}$$

Equations (5-104) and (5-105) follow from the general relation (see App. G) stating that, in an associated and/or solvated mixture, the chemical potential of a (stoichiometric) component is equal to that of the component's monomer.

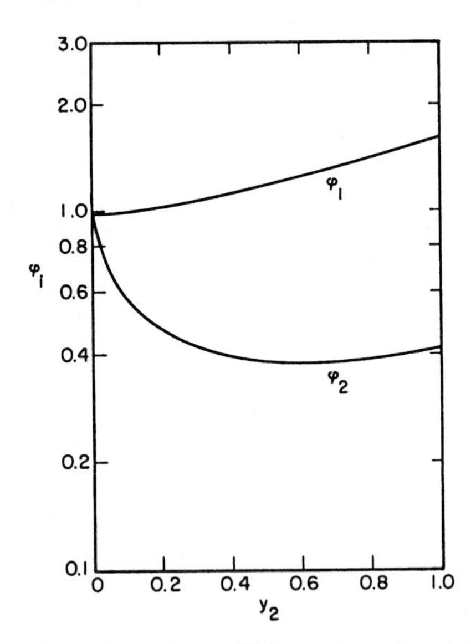

Figure 5-28 Vapor-phase fugacity coefficients for saturated mixtures of water (1)/acetic acid (2) at 1 bar.

Figure 5-28 shows calculated vapor-phase fugacity coefficients for the system water/acetic acid along the vapor-liquid saturation curve at 1 bar. The necessary dew-point (*T*, *y*) data were taken from Sebastiani and Lacquanti (1967). Although the total pressure is low, deviations from ideal-gas behavior ($\varphi_1 = \varphi_2 = 1$) are large. In this calculation at 1 bar, chemical equilibrium I is by far the most important. If equilibria II, III, and IV had been neglected, the results shown in Fig. 5-28 would not be seriously affected.

5.10 Weak Dimerizations and Second Virial Coefficients

The "chemical" theory of virial coefficients was proposed many years ago; it has been extensively developed by Lambert (1953) and collaborators in connection with their studies of organic polar gases and polar-gas mixtures. Following Lambert, we consider a pure polar gas A and assume that the forces between A molecules can be divided into two classes: The first class is concerned with "normal" intermolecular forces (such as dispersion forces) that exist between nonpolar as well as polar molecules, and the second is concerned with chemical association forces leading to the formation of new chemical species. At moderate densities, the first class of forces is responsible for a "normal" or "physical" second virial coefficient, whereas the second class is responsible for a dimerization equilibrium constant. The equation of state is written

$$PV = n_T(RT + BP) \tag{5-106}$$

with

$$B = B_{\text{nonpolar}} + B_{\text{polar}}$$

Lambert has shown that B_{polar} is directly proportional to the dimerization equilibrium constant:

$$B_{\text{polar}} = -RTK / P^0 \tag{5-107}$$

where P^0 is the standard-state pressure (1 bar).

Techniques for calculating B_{nonpolar} are somewhat arbitrary but, fortunately, this arbitrariness is not important for our purposes because, for any case where Lambert's procedure is of interest, B_{polar} is always much larger in magnitude than B_{nonpolar}. A reasonable method for estimating B_{nonpolar} is to use the reduced second-virial-coefficient equation of Schreiber and Pitzer [Eq. (5-66)] with the true critical constants but with a fictitious acentric factor as determined by the polar molecule's homomorph.[18]

[18] The homomorph (*homo* = same, *morph* = form) of a polar molecule is a nonpolar molecule having essentially the same size and shape as those of the polar molecule. For example, the homomorph for acetonitrile could be ethane.

However, dimerization can only account for negative second virial coefficients. While dissociation reactions can account for positive deviations from ideal-gas behavior, small positive deviations are commonly found for stable common gases (e.g. methane) at moderate pressures and at temperatures well above the critical; these deviations cannot realistically be ascribed to dissociation. For stable molecules at high reduced temperatures, positive second virial coefficients must be accounted for by the concept of excluded volume, i.e., the inaccessibility of some parts of space due to the finite size of the molecules.

In its simplest form, the excluded volume can be accounted for by an equation of state

$$P(V - n_T b_m) = n_T RT \tag{5-108}$$ [19]

where V is the total volume, n_T is the true total number of moles, and $n_T b_m$ is the excluded volume due to the finite size of the molecules.

Let us suppose that, in the absence of any dimerization, there is one mole of A. As before, let α be the fraction of molecules that associate. Then, the equilibrium constant for association is given by Eq. (5-100); for small degrees of association ($\alpha << 1$), Eq. (5-100) to a good approximation becomes

$$K = \frac{\alpha P^0}{2P} \tag{5-109}$$

The gas is a mixture of monomer A and dimer A_2. Because we assume that α is much less than unity, n_{A_2} is necessarily much smaller than n_A. Also, using Eqs. (5-96) and (5-109), the total number of moles present n_T is

$$n_T = 1 - \frac{\alpha}{2} = 1 - \frac{PK}{P^0} \tag{5-110}$$

Rewriting Eq. (5-108) gives

$$\frac{PV}{RT} = 1 + \frac{P}{RT}\left(n_T b_m - \frac{RTK}{P^0}\right) \tag{5-111}$$

At modest densities, it is reasonable to assume that the excluded volume for the dimer is the same as that for the monomer. Therefore,

$$n_T b_m = (n_A + 2n_{A_2})b \tag{5-112}$$

[19] Equation (5-108) is equivalent to assuming that $B_{\text{nonpolar}} = n_T b_m$.

where b is the excluded volume for the monomer. Because the above derivation is based on one mole of A in the absence of any dimerization (i.e., $n_A + 2n_{A_2} = 1$), then $n_T b_m = b$. This result and a comparison of Eq. (5-111) with Eq. (5-106) gives

$$B = b - \frac{RTK}{P^0} \tag{5-113}$$

Equation (5-113) was obtained for small degrees of dimerization ($\alpha << 1$). Therefore, the chemical theory of gas imperfections leads to an equation of state of the virial form in the limit $\alpha \to 0$. Through Eq. (5-113) it is possible to obtain dimerization equilibrium constants from experimental second virial coefficients, provided that these coefficients were obtained at low densities where α is much less than unity. For highly polar fluids, and especially for strongly hydrogen-bonded fluids, the condition $\alpha << 1$ is attained only at low pressures. To illustrate, consider acetic acid where $\alpha = 0.8$ at 1600 Pa (or 0.016 bar) and 40°C; at the same temperature for α to be small, say equal to 0.01, it is necessary to reduce the pressure to 1.3 Pa. While carboxylic acids are unusually strong in their tendency to dimerize even at very low pressures, other common fluids (e.g., alcohols, aldehydes, esters, etc.) begin to dimerize to an appreciable extent only at normal pressures (close to 1 bar).

Lambert's form of the virial equation of state is not useful for highly associating fluids because the valid pressure range for that equation is often extremely small. This limitation is not alleviated by introducing third (and higher) virial coefficients because these coefficients do not account for large α, but rather reflect the formation of trimers and higher aggregates.

In general, calculations based on the chemical theory of gas imperfections are not limited to small α; the theory holds for the entire range $0 \le \alpha \le 1$. In practice the chemical theory is limited to moderate pressures because, as pressure rises, trimer and higher aggregates are formed and few quantitative data are available on chemical equilibria beyond dimerization. For weakly interacting fluids (i.e. nonpolar or slightly polar components) there is little difference between results calculated from the chemical theory and those obtained from the "physical" virial equation. For some engineering applications, however, the chemical theory of gas imperfections presents an advantage over the virial equation because it is applicable to strongly polar and hydrogen-bonding fluids as well as to normal nonpolar fluids.

Experimental data for a large variety of polar gases have been reduced using the chemical theory described above, e.g. by Nothnagel *et al.* (1973) and by Olf *et al.* (1989a). When data are available over a range of temperature, a semilogarithmic plot of K versus $1/T$ gives information on the enthalpy and entropy for dimer formation according to

$$-\ln K = \frac{\Delta h^0}{RT} - \frac{\Delta s^0}{R} \tag{5-114}$$

The enthalpy Δh^0 and entropy Δs^0 of dimer formation in the standard state are obtained from experimental second virial coefficients using Eqs. (5-113) and (5-114). Excluded volume b in Eq. (5-113) is proportional to the volume of the molecules; for polyatomic molecules, b may be calculated from atomic radii and bond distances (Bondi, 1968). Some experimentally determined results are shown in Table 5-10 for several polar gases.

Table 5-10 Excluded volumes (obtained from the method of Bondi), standard enthalpies and entropies of dimerization (from second virial coefficients) for some polar gases.

	b (cm^3 mol^{-1})	$-\Delta h^0$ (kJ mol^{-1})	$-\Delta s^0$ (kJ mol^{-1} K^{-1})
Methyl acetate	170.2	10.04	0.0562
Acetonitrile	113.5	18.07	0.0703
Diethyl amine	223.4	12.00	0.0598
Diethyl ether	206.0	8.463	0.0526
Ammonia	54.3	8.995	0.0662
Nitromethane	121.9	15.13	0.0648
Water	45.9	13.91	0.0713

The "chemical" theory is readily extended to mixtures. A two-component mixture (say 1 and 2) contains five molecular species: A, A_2, B, B_2, and AB; their mole fractions are related by material balances and through three equilibrium constants, K_{A_2}, K_{B_2}, and K_{AB}; the first and second of these can be determined from volumetric data for the pure components, as discussed above. However K_{AB} requires volumetric data for the mixture.

When dimerizations are weak, cross term B_{12} of Eq. (5-20) is given by

$$B_{12} = \frac{1}{2}(b_1^{1/3} + b_2^{1/3})^3 - \frac{1}{2}\frac{RTK_{AB}}{P^0} \tag{5-115}$$

In the above equation we assumed that $b_{AB}^{1/3} = b_{12}^{1/3} = (1/2)(b_1^{1/3} + b_2^{1/3})$ [see Eq. (5-56)]. In the last term of Eq. (5-115), the factor 1/2 is a symmetry factor. When P-V-T data at low pressures are available for a binary mixture, it is then possible to find B_{12} and hence K_{AB}. Figure 5-29 compares the temperature dependence of experimental second cross virial coefficients of methanol/methyl acetate and acetonitrile/methyl acetate with results calculated from chemical theory.

The equations presented above assume that chemical equilibria are limited to a stoichiometry of two (dimers). In principle, this assumption is not necessary but, with few exceptions, in practice the chemical theory is limited to moderate pressures where trimers (and higher aggregates) are neglected.[20]

[20] A notable exception is hydrogen fluoride that forms hexamers. See, for example, M. Lencka and A. Anderko, 1993, *AIChE J.*, 39: 533, and Sec. 12.10.

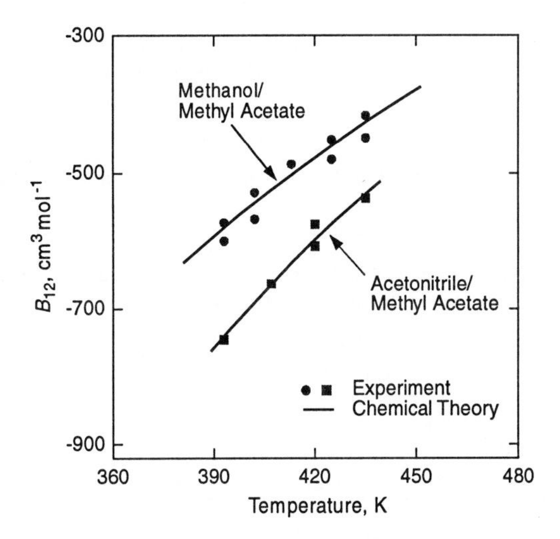

Figure 5-29 Second cross virial coefficients (Olf *et al.*, 1989a) of methanol/methyl acetate and acetonitrile/methyl acetate. ●, ■ Experiment; —— Calculated from chemical theory.

A few additional examples illustrate the "chemical" theory of gas nonideality for binary mixtures. First, we consider Carter's data (1963) for the acetonitrile/acetaldehyde system. Carter found that the experimental second-virial cross coefficient for the mixture was much more negative than that of either pure component, as shown in Fig. 5-30; the magnitude of the cross-coefficient is unexpectedly large, much larger than that predicted by Stockmayer's potential using pure-component parameters with common combining rules uncorrected for specific interactions. When Carter's experimental data are used to calculate chemical equilibrium constants as described above, and when the temperature dependence of these constants is investigated using Eq. (5-114), the enthalpy of formation for the acetonitrile/acetaldehyde complex is considerably larger in magnitude than that for either pure-component dimer. The observed enthalpies are compared with theoretically calculated energies in Table 5-11.[21]

The calculations are based on an electrostatic model with two structures, (a) and (b). In structure (a) the complex (or dimer) consists of two "touching" ideal spherical dipoles such that the dipoles are lined up in a direction perpendicular to the center-to-center separation, as shown at the bottom of Table 5-11. Structure (b) is similar, except that the dipoles are now lined up in a direction parallel to the center-to-center separation. From elementary electrostatic theory, the standard energy of formation of each of these structures is

$$\text{(a)}\quad \Delta u^0 = -\frac{\mu^2}{4\pi\varepsilon_0 r^3} \quad \left(\text{or } -\frac{\mu_i\mu_j}{4\pi\varepsilon_0 r^3} \text{ if } i \neq j\right) \tag{5-116}$$

[21] For our purposes here, no distinction need be made between energy and enthalpy.

$$\text{(b)}\quad \Delta u^0 = -\frac{2\mu^2}{4\pi\varepsilon_0 r^3} \quad \left(\text{or } -\frac{2\mu_i\mu_j}{4\pi\varepsilon_0 r^3} \text{ if } i \neq j\right) \tag{5-117}$$

Table 5-11 Standard energies of complex formation in the acetonitrile/acetaldehyde system (Carter, 1960).

Complex	Δu^0 (energy of formation, kJ mol^{-1})		
	Calculated		
	Parallel Structure (a)	End-to-end Structure (b)	Observed
Acetaldehyde/acetaldehyde	22.11	44.21	21.39
Acetonitrile/acetonitrile	14.15	28.30	16.58
Acetonitrile/acetaldehyde	17.27	34.54	30.54

Structure (a) Structure (b)

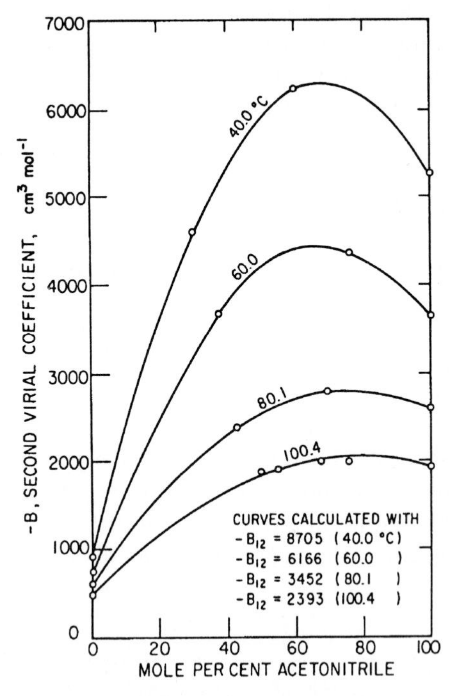

Figure 5-30 Second virial coefficients for the acetonitrile/acetaldehyde system.

where μ is the dipole moment of the monomer, ε_0 is the vacuum permittivity and r is the distance between the two centers of the two molecules that form the dimer. For acetonitrile and acetaldehyde the dipole moments are, respectively, 3.9 and 2.7 debye. Distances r were estimated from van der Waals radii reported by Pauling (1945).

The results in Table 5-11 suggest that the structure of the complex [structure (b), perhaps] is qualitatively different from that of the dimers that seem to have the properties of structure (a). In the absence of other experimental information, this interpretation of Carter's data should not be taken too seriously but it does provide a possible explanation concerning why the cross-virial coefficient for this system is so different from the virial coefficients of the pure components.

For a second example consider the system ammonia/acetylene studied by Cheh *et al.* (1966). This system is qualitatively different from that studied by Carter because, while acetaldehyde and acetonitrile are both strongly polar, in the ammonia/acetylene system only ammonia has a dipole moment. Nevertheless, in the interaction between unlike molecules, there are strong complexing forces that are not present in the interactions between like molecules; Fig. 5-31 gives B_{11}, B_{22}, and B_{12} and we see that B_{12} is considerably more negative than either B_{11} or B_{22}. Cheh *et al.*, using appropriate potential functions, reduced the experimental second virial coefficients of the pure gases by taking into account all physical dispersion, induction, dipole, and quadrupole forces. Then, using conventional combining rules, they calculated the (physical) cross-coefficient B_{12} and found that its magnitude was much too small, as shown in Fig. 5-32.

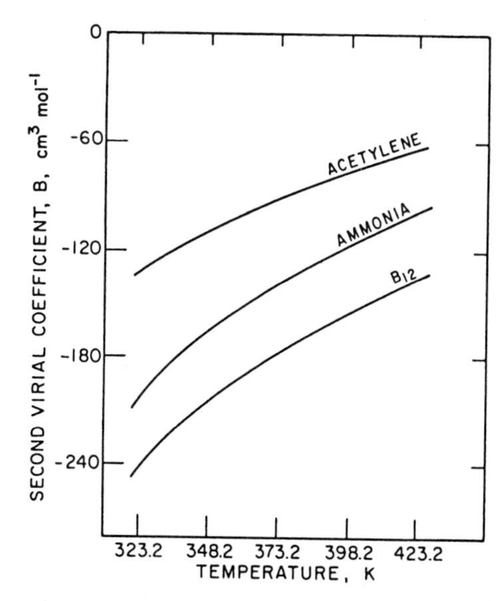

Figure 5-31 Experimental second virial coefficients for acetylene (1) and for ammonia (2). B_{12} is the observed cross coefficient for binary mixtures of (1) and (2).

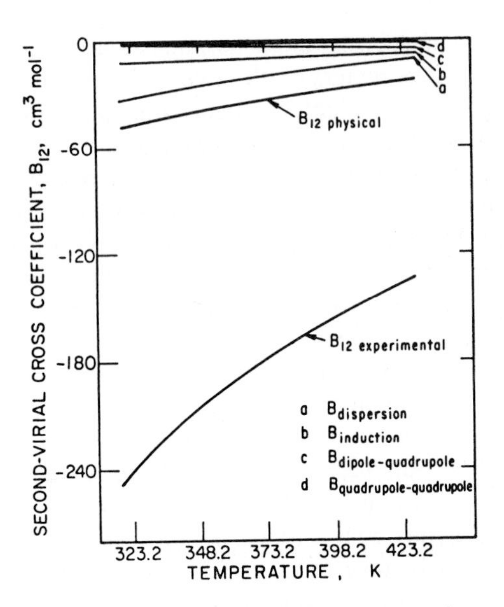

Figure 5-32 Second-virial cross coefficient for acetylene/ammonia mixtures. Calculated contributions from physical forces are insufficient to account for the observed large negative coefficient. $B_{12\,\mathrm{physical}}$ is the sum of contributions a, b, c and d.

The high negative B_{12} can be "explained" by the formation of a hydrogen bond between (acidic) acetylene and (basic) ammonia. This hydrogen-bond formation provides a reasonable explanation because there is much independent evidence for the basic properties of ammonia and there also is some independent evidence for the acidic properties of acetylene.[22]

Chemical equilibrium constant K for hydrogen-bond formation is found from

$$-\frac{1}{2}\frac{RTK}{P^0} = B_{12\,\mathrm{exp}} - B_{12\,\mathrm{physical}} \tag{5-118}$$

where P^0 is the standard-state pressure (1 bar).

From the variation of K with temperature, the standard enthalpy and entropy of complex formation are

$$\Delta h^0 = (-9250 \pm 1250)\ \mathrm{J\ mol^{-1}}$$

$$\Delta s^0 = (-75.3 \pm 3.8)\ \mathrm{J\ mol^{-1}\ K^{-1}}$$

[22] For example, acetylene is much more soluble in oxygen-containing solvents (such as acetone) than is ethane or ethylene.

These results are in good agreement with statistical-mechanical calculations for the formation of a loosely bonded complex of ammonia and acetylene.[23]

In some cases, specific chemical interaction between unlike molecules can lead to very large negative values for the second-virial cross coefficient B_{12}.

To illustrate, Fig. 5-33 shows some experimental results for the binary system trimethyl amine/methanol. The data of Millen and Mines (1974) indicate that there is strong interaction between the alcohol and the amine as suggested also by the shift in the (infrared) hydrogen-bond frequency of methanol (310 cm^{-1} expressed in wave numbers). The enthalpy and entropy of complex formation are, respectively, -28.9 kJ mol^{-1} and -98.3 J mol^{-1} K^{-1} at 25°C.

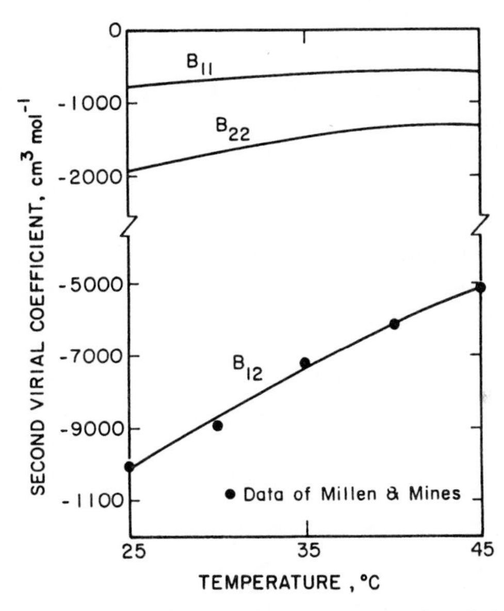

Figure 5-33 Second virial coefficients for the system trimethyl amine (1)/methanol (2).

The large negative B_{12} yields a remarkably low fugacity coefficient for methanol, infinitely dilute in trimethyl amine, as shown by Stein and Miller (1980). Figure 5-34 gives this fugacity coefficient at 45°C; even at low pressure, where gas-phase nonideality is usually negligible, the fugacity coefficient departs significantly from unity.

For calculating fugacity coefficients at moderate densities, it is desirable to have a correlation that includes both the "physical" contributions (second virial coefficients) and the "chemical" contributions (equilibrium constants) to vapor-phase nonideality. Hayden and O'Connell (1975), following earlier work by Nothnagel *et al.* (1973), presented

[23] The factor 1/2 in Eq. (5-118) was erroneously omitted in the work of Cheh *et al.* (1966) and therefore the entropy reported there is in error by $R \ln 2$.

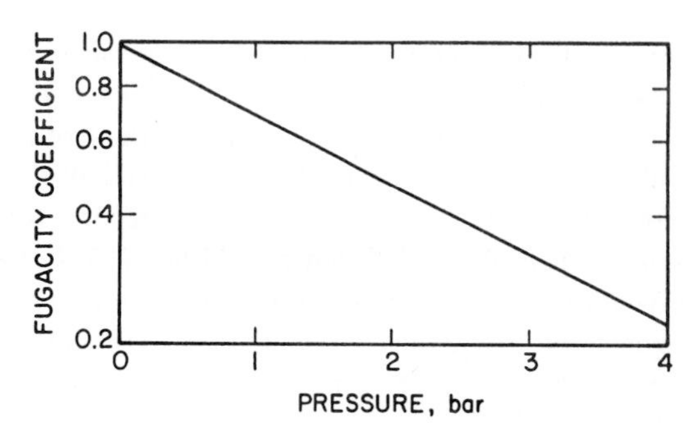

Figure 5-34 Fugacity coefficients for methanol at infinite dilution in trimethyl amine at 45°C (Stein and Miller, 1980).

a correlation of that form. Details of that correlation, including pertinent computer programs, are presented elsewhere.[24]

Figures 5-35 and 5-36 present fugacity coefficients for two binary systems, calculated with the Hayden-O'Connell correlation. In these calculations, application is directed at correlating isobaric vapor-liquid equilibria; the fugacity coefficients, therefore, are calculated at saturation, i.e., at dew-point conditions (Conti *et al.*, 1960; Wisniak and Tamir, 1976). Although the total pressure is only 1 bar, deviations from ideal-gas behavior are considerable in both systems.

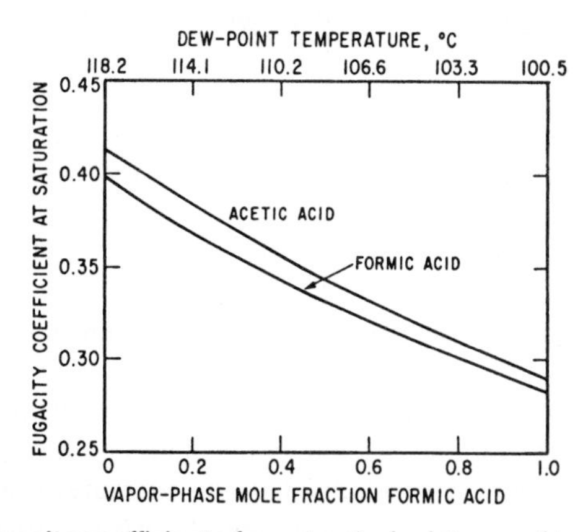

Figure 5-35 Fugacity coefficients for saturated mixtures of two carboxylic acids at 1 bar: formic acid (1)/acetic acid (2).

In the system formic acid/acetic acid, the fugacity coefficients are well below unity at all compositions because both components associate. In the system propionic acid/methyl isobutyl ketone, fugacity coefficients are close to unity only when the

[24] J. M. Prausnitz, T. F. Anderson, E. A. Grens, C. A. Eckert, R. Hsieh, and J. P. O'Connell, 1980, *Computer Calculations for Multicomponent Vapor-Liquid and Liquid-Liquid Equilibria*, Englewood Cliffs: Prentice Hall.

mole fraction of (associating) acid is small. As this mole fraction rises, deviations from ideal-gas behavior increase. The weak minimum in the fugacity coefficient for propionic acid is a consequence of two effects: As the concentration of acid rises, the tendency to associate increases but, because the temperature is also rising, the association constant decreases. At isobaric saturation, a change in composition is accompanied by a change in temperature. For propionic acid, an increase in mole fraction raises deviations from ideal-gas behavior but an increase in temperature lowers such deviations.

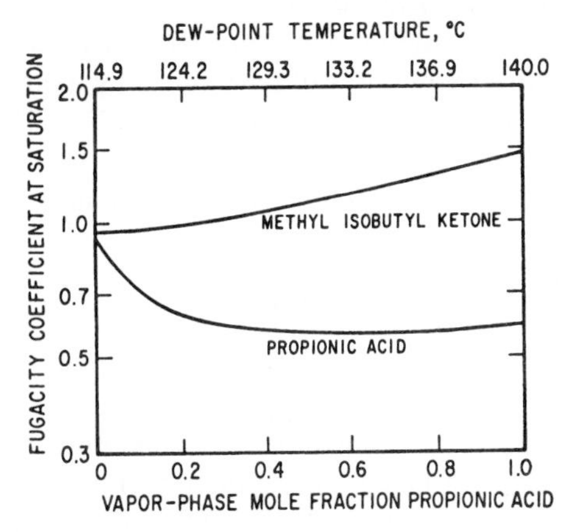

Figure 5-36 Fugacity coefficients for saturated mixtures of propionic acid (1)/methyl isobutyl ketone (2) at 1 bar.

These examples illustrate that forces between molecules are often highly specific and in such cases it is unfortunately not possible to predict even approximately the properties of a mixture from those of the pure components. This is not surprising if we consider a far-fetched analogy: Let us suppose that a sociologist in Russia has carefully studied the behavior of Russians and after years of observation knows all about them. He then goes to China and makes a similar, thorough study of the Chinese. With this knowledge, can he then predict the behavior of a society where Russians and Chinese intermingle freely? Probably not. This analogy is highly extreme but serves to remind us that molecules are not inert particles floating blindly through space; instead, they are complicated "individuals" whose "personalities" are sensitive to their environment.

5.11 Fugacities at High Densities

Previous sections pointed out advantages of the virial equation of state but, as indicated earlier, this equation has an important disadvantage, i.e., its inapplicability to gases at high densities. The density range for practical application of the virial equation varies somewhat with the temperature, but frequently the virial equation is not useful at densities larger than about 50% of the critical (when third and higher virial coefficients are neglected) or 75% of the critical (when fourth and higher virial coefficients are neglected). This disadvantage follows primarily from our limited knowledge regarding third and higher virial coefficients.

At present, gas-phase fugacities at high densities can best be calculated by using semi-empirical methods.

For example, the theorem of corresponding states can be extended to mixtures utilizing what is often called the *pseudocritical method* introduced by Kay (1936). According to this method, the configurational properties of a mixture may be calculated from the generalized, reduced properties of pure fluids by expressing the characteristic reducing parameters as a function of mixture composition. This idea is easily illustrated by using the two-parameter corresponding-states theory that expresses compressibility factor z as a universal function of reduced temperature and reduced pressure,

$$z = F_z\left(\frac{T}{T_c}, \frac{P}{P_c}\right) \tag{5-119}$$

The pseudocritical method assumes that the same universal function F_z applies to mixtures when T_c and P_c are taken, not as the (true) critical temperature and pressure of the mixture, but as pseudocritical constants, that is, characteristic parameters that depend on the composition of the mixture.

Kay assumed that T_c and P_c are linear functions of the mole fractions, although other functions relating T_c and P_c to composition are consistent with the pseudocritical idea (and numerous functions have been proposed by e.g. Reid and Leland, 1965; Leland and Chappelear, 1968). The pseudocritical and other empirical methods are reliable for many, but by no means all, cases. For example, for mixtures of small, nonpolar gases not near critical conditions, very good results can usually be obtained, while for mixtures containing polar components, reliability is frequently uncertain, and for all mixtures at or near critical conditions, regardless of molecular complexity, calculated fugacities are likely to be in error. These limitations follow directly from our incomplete knowledge of intermolecular forces between complex molecules and from our inability to describe in a simple way the behavior of fluids in the vicinity of the critical state.

The general and exact equations for the calculation of fugacities from volumetric properties have been derived and discussed in Chap. 3, and the two key relations are given again by Eqs. (5-1) and (5-2).

A more convenient method for calculating fugacities in gas mixtures is to use a phenomenological equation of state, i.e. an equation of state based on a molecular picture altough the details of the equation cannot be derived with theoretical rigor. Sometimes such equations are called semiempirical; most currently used phenomenological equations of state are based on a van der Waals-type picture of fluids.

While these equations of state are derived for pure fluids, they can be extended to mixtures upon making some simplifying assumptions. The key assumption is the *one-fluid theory* of mixtures. This theory assumes that the configurational properties of a mixture are the same as those of a hypothetical pure fluid whose characteristics (as expressed by constants in the equation of state) are some composition-dependent average of the characteristics of the pure components in the mixture.

The composition dependence of equation of state constants is given by *mixing rules* that are mostly empirical.

Given some phenomenological equation of state and using the one-fluid theory, we obtain a relation of the form

$$P = F[v, T, a(y), b(y), \ldots] \tag{5-120}$$

where v is the molar volume of the mixture and $a(y)$ and $b(y)$ designate constants a and b as functions of mole fraction y. The number of such constants is arbitrary but, to keep experimental-data requirements low, typical useful equations of state contain no more than two or three substance-specific constants. These constants are often expressed in terms of critical temperature, critical pressure and acentric factor.

As shown in Chap. 3, Eq. (5-120) is sufficient to calculate fugacities of all components in a fluid mixture. Chapter 12 presents details concerning calculation of fugacities based on Eq. (5-120).

5.12 Solubilities of Solids and Liquids in Compressed Gases

Vapor-phase fugacity coefficients are required in any phase-equilibrium calculation wherein one of the phases is a gas under high pressure. If the gas is at low pressure, vapor-phase fugacity coefficients are usually close to unity, although there are some exceptions (e.g. acetic acid). However, vapor-phase fugacity coefficients are particularly important for calculating the solubility of a solid or a high-boiling liquid in a dense gas because in such cases failure to include corrections for gas-phase nonideality can lead to serious errors. To illustrate how corrections to gas-phase nonideality can significantly influence phase behavior, we now briefly present some examples concerning the solubility of a condensed component (a solid and a liquid) in a dense gas.

First, we consider equilibrium between a compressed gas and a solid. This is a particularly simple case because the solubility of the gas in the solid is almost always negligible. The condensed phase may therefore be considered pure and thus all nonideal behavior in the system can be attributed entirely to the vapor phase. Gas-liquid equilibria, considered later, is a little more complicated because the solubility of the gas in the liquid, while it may be small, is usually not negligible.

Let subscript 1 stand for the light (gaseous) component and let subscript 2 stand for the heavy (solid) component. We want to calculate the solubility of the solid component in the gas phase at temperature T and pressure P. We first write the general equation of equilibrium for component 2:

$$f_2^{\mathcal{S}} = f_2^V \tag{5-121}$$

where superscript $\mathcal{S}$ stands for the solid phase.

Because the solid phase is pure, the fugacity of component 2 is given by

$$f_2^{\mathcal{S}} = P_2^s \varphi_2^s \exp\left(\int_{P_2^s}^{P} \frac{v_2^{\mathcal{S}}}{RT} dP\right) \tag{5-122}$$

as discussed in Sec. 3.3. In Eq. (5-122), P_2^s is the saturation (vapor) pressure of the pure solid, φ_2^s is the fugacity coefficient at saturation pressure P_2^s, and $v_2^{\mathcal{S}}$ is the solid molar volume, all at temperature T.

For the vapor-phase fugacity, we introduce fugacity coefficient φ_2 by recalling its definition:

$$\varphi_2 \equiv \frac{f_2^V}{y_2 P} \tag{5-123}$$

Substituting and solving for y_2, we obtain the desired solubility of the heavy component in the gas phase,

$$\boxed{y_2 = \frac{P_2^s}{P} E} \tag{5-124}$$

where

$$E \equiv \frac{\varphi_2^s}{\varphi_2} \exp\left(\int_{P_2^s}^{P} \frac{v_2^{\mathcal{S}}}{RT} dP\right) \tag{5-125}$$

The quantity E, nearly always greater than unity, is called the *enhancement factor*; that is, E is the correction factor that must be applied to the simple (ideal-gas) expression valid only at low pressure. The enhancement factor provides a measure of the extent that pressure enhances the solubility of the solid in the gas; as $P \to P_2^s$, $E \to 1$. It is a dimensionless measure of solvent power because it is the ratio of the observed solubility to the ideal-case solubility.

The enhancement factor contains three correction terms: φ_2^s takes into account nonideality of the pure saturated vapor; the Poynting correction gives the effect of pressure on the fugacity of the pure solid; and φ_2, the vapor-phase fugacity coefficient in the high-pressure gas mixture. Of these three correction terms, it is usually the last one that is by far the most important. In most practical cases, the saturation pressure P_2^s of the solid is small and thus φ_2^s is nearly equal to unity. The Poynting correction is not negligible, but it rarely accounts for an enhancement factor of more than 2 or 3 and frequently much less. However, the fugacity coefficient φ_2 can be so far removed from unity as to produce very large enhancement factors that, in some cases, can exceed 10^3 or more.

While enhancement factors in the range 10^4-10^6 are common, factors as high as 10^{12} have been reported for the solubility of solid oxygen in dense gaseous hydrogen (McKinley *et al.*, 1961).

Figure 5-37 shows the logarithm of the enhancement factor for solid carbon tetrachloride as a function of the density of supercritical carbon tetrafluoride at 249 K. The nearly linear function indicates that at high gaseous solvent densities (or pressures) very large deviations from nonideality are observed. The experimental data are fitted well using the Peng-Robinson equation of state (see Sec. 12.7).

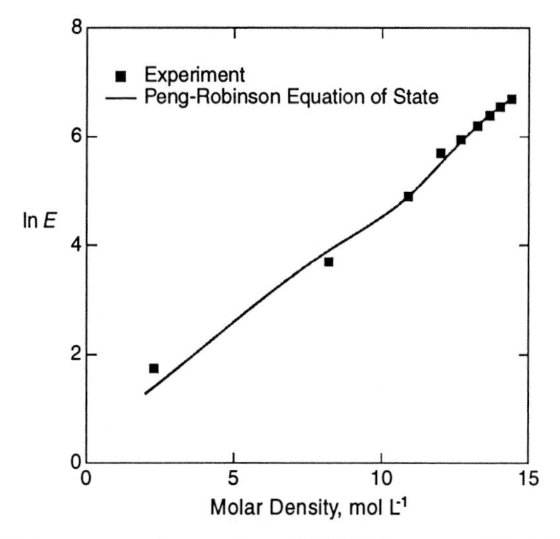

Figure 5-37 Enhancement factors for solid CCl_4 in supercritical CF_4 at 249 K.

Another example of enhanced solubility of a solid in a dense gas is given in Fig. 5-38 that shows the solubility of solid carbon dioxide in compressed air at 143 K; the dashed line gives the calculated solubility with $\varphi_2 = 1$. In this calculation the solubility was calculated with the Poynting correction but with the assumption of gas-phase ideality. The two lines at the top of the diagram were calculated from Eq. (5-28) based on the virial equation; one calculation included only second virial coefficients, while the other included second and third virial coefficients. In these calculations air was considered to be a mixture of oxygen and nitrogen and therefore the results shown are those for

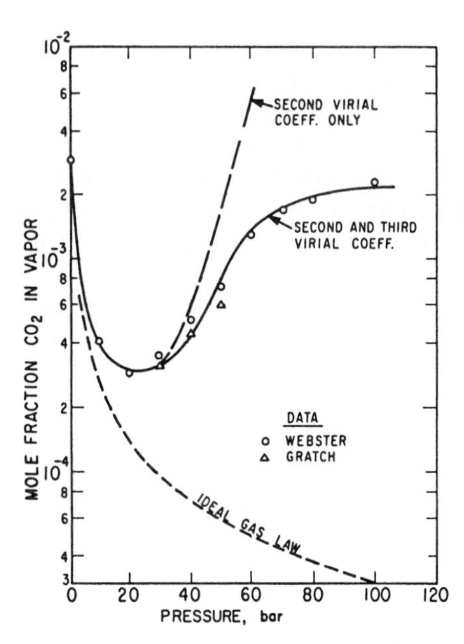

Figure 5-38 Solubility of solid carbon dioxide in air at 143 K.

a ternary system. The required virial coefficients were calculated from the Kihara potential using only pure-component parameters and customary combining rules. Calculated results are compared with experiment (Webster, 1952); when third virial coefficients are neglected, the calculations correctly predict the minimum in the curve of log y_2 versus P, but at pressures above that corresponding to the minimum mole fraction, third virial coefficients are required. In this particular system, agreement between calculated and experimental results at higher pressures is very good; unfortunately, such good agreement is not always obtained due to the difficulty of calculating third virial coefficients of mixtures. However, in the moderate-pressure region where third (and higher) virial coefficients can be neglected, it is almost always possible in nonpolar systems to make good estimates of the vapor-phase solubility of a solid (Ewald *et al.*, 1953). Such estimates are frequently useful for engineering purposes; for example, in a freezing-out process it is desirable to minimize the mole fraction of a heavy impurity in a gas stream or, in a gas-cooling process, it is necessary to know the solubility of a heavy component to prevent plugging of the flow lines by precipitation of solid.

At very high densities, the virial equation is not useful because almost no information is available on the fourth and higher virial coefficients; often, the second virial coefficient can be estimated well but the third virial coefficient can only be estimated approximately. Therefore, if the gas density is high (close to or exceeding the critical density), a mostly empirical equation of state must be used to relate the fugacity coefficient to the pressure, temperature, and gas-phase composition. To illustrate, Fig. 5-39 compares calculated (Kurnik and Reid, 1981) and experimental solubilities of naphthalene (melting point 80.2°C) in compressed ethylene as a function of temperature and pressure. The phase-equilibrium calculations shown in Fig. 5-39 were performed by Kurnik and Reid using the Peng-Robinson equation of state (see Sec. 12.3).

The results indicate S-shaped curves, each with a well-defined minimum at moderate pressure and a well-defined maximum at very high pressure. The maximum occurs at high densities when repulsive forces between ethylene and naphthalene become important; as the density (or pressure) rises further, naphthalene is "squeezed out" of the gaseous solution.

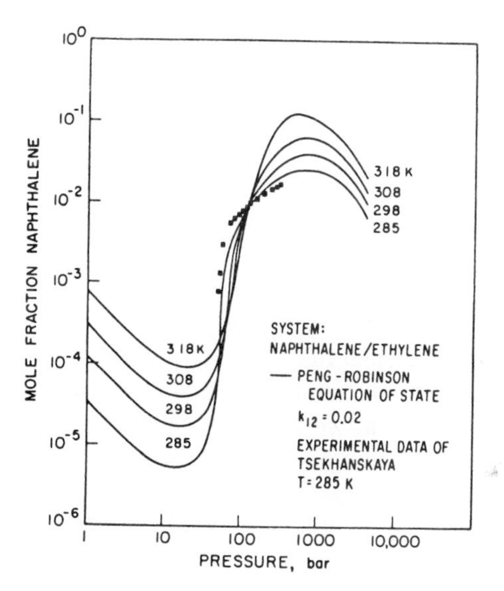

Figure 5-39 Solubility minima and maxima for naphthalene in compressed ethylene.

Kurnik and Reid have calculated maximum solubilities, shown by the dashed line in Fig. 5-40; the continuous lines indicate experimental data. The maximum solubility rises with increasing temperature because the vapor pressure of solid naphthalene rises with temperature. The pressure for maximum solubility falls with increasing temperature as we move away from the critical temperature of ethylene (282.4 K).

Finally, consider the solubility of a liquid in a compressed gas. This situation is somewhat more difficult than that discussed above because a gas is soluble to an appreciable extent in a liquid, whereas the solubility of a gas in a solid is usually negligible. A general discussion of high-pressure vapor-liquid equilibria is given in Chap. 12; here we restrict ourselves to equilibrium between a high-boiling liquid and a sparingly soluble gas at conditions remote from critical. We are interested in the solubility of the heavy (liquid) component in the vapor phase that contains the light (gaseous) component in excess; as before, let 1 stand for the (light) gaseous solvent and 2 for the (heavy) solute in the gas phase. For the heavy component, the equation of equilibrium is

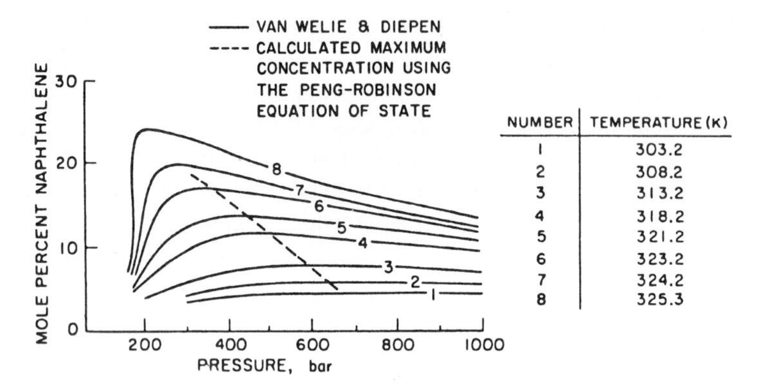

Figure 5-40 Solubility maxima for naphthalene in compressed ethylene.

$$f_2^L = f_2^V = \varphi_2 y_2 P \tag{5-126}$$

where superscripts V and L stand, respectively, for vapor and liquid and where φ_2 is the vapor-phase fugacity coefficient of component 2 in the gaseous mixture.

Fugacity coefficient φ_2 must again be calculated from an equation of state, but before we turn to that calculation, we must say something about the fugacity of component 2 in the liquid phase. If the gas is only sparingly soluble, the liquid-phase fugacity of component 2 can be calculated by assuming that the solubility of component 1 in the liquid is described by a pressure-corrected form of Henry's law (see Sec. 10.3): For component 1 we assume that, in the liquid phase, fugacity f_1^L is related to mole fraction x_1 by

$$f_1^L = H_{1,2} x_1 \exp\left(\int_{P_2^s}^{P} \frac{\overline{v}_1^\infty}{RT} dP \right) \tag{5-127}$$

where $H_{1,2}$ is Henry's constant, $\overline{v}_1^\infty$ is the partial molar volume of component 1 in the liquid phase at infinite dilution, and P_2^s is the saturation pressure of pure liquid component 2, all at system temperature T. Then, from the Gibbs-Duhem equation, it can be shown that

$$f_2^L = (1 - x_1) P_2^s \varphi_2^s \exp\left(\int_{P_2^s}^{P} \frac{v_2^L}{RT} dP \right) \tag{5-128}$$

where P_2^s is the saturation pressure, φ_2^s is the fugacity coefficient at saturation, and v_2^L is the molar volume, all of pure liquid 2, all at temperature T.

Substituting Eq. (5-128) into Eq. (5-126), and solving for the desired solubility y_2, we obtain

$$y_2 = \frac{(1-x_1)P_2^s\varphi_2^s \exp\left(\int_{P_2^s}^{P} \frac{v_2^L}{RT} dP\right)}{\varphi_2 P} \tag{5-129}$$

where x_1, the solubility of gas 1 in liquid 2, is calculated from

$$x_1 = \frac{y_1\varphi_1 P}{H_{1,2} \exp\left(\int_{P_2^s}^{P} \frac{\overline{v}_1^\infty}{RT} dP\right)} \tag{5-130}$$

Solution of Eq. (5-129) necessarily requires a trial-and-error calculation because x_1 as well as φ_1 and φ_2 depend on the composition of the vapor. For the conditions under consideration, $y_1 \approx 1$ and thus, as discussed in Sec. 5.1, the Lewis fugacity rule is a good approximation for the fugacity of component 1 (although it is a very bad approximation for the fugacity of component 2). For the first iteration, therefore, the calculations can be simplified by replacing φ_1 in Eq. (5-130) by $\varphi_{\text{pure 1}}$ (evaluated at the temperature and pressure of the system) and by setting $y_2 = 0$ in Eq. (5-28) for φ_2.

The fugacity of pure liquid 2 at temperature T and pressure P is readily calculated from the volumetric properties of pure component 2 as shown in Sec. 3.3. Henry's constant H must be determined experimentally or else estimated from a suitable correlation (see Sec. 10.5); it need be known only approximately because, under the conditions considered, $x_1 << 1$, and thus even a somewhat inaccurate estimate of x_1 does not seriously affect the value of y_2. A rough estimate of $\overline{v}_2^\infty$ is often sufficient here because the exponential correction to H may be small and often may be neglected.

In Eq. (5-129) it is the fugacity coefficient φ_2 that accounts for the nonideal behavior of liquid solubility in compressed gases. To illustrate, Fig. 5-41 shows the solubility of decane (normal b.p. 174°C) in compressed nitrogen at 50°C. Curve (a) gives calculated results assuming ideal-gas behavior and curve (c) shows results obtained when the Lewis fugacity rule is used for calculating the fugacity coefficient of decane (component 2). Curve (b) is based on the virial equation with B_{12} = -141 cm^3 mol^{-1}; it gives excellent agreement with experimental data (Prausnitz and Benson, 1959). Good agreement is perhaps not surprising because virial coefficient B_{12} used in the calculation was back-calculated from the experimental solubility data. However, a reasonable estimate of B_{12} could probably have been made without any data for this binary system by using one of the generalized equations discussed in Sec. 5.7.

Nonideal behavior of the decane/nitrogen system at 50°C is not large because in this case the fugacity coefficient of decane is of the order of magnitude of unity (or, in

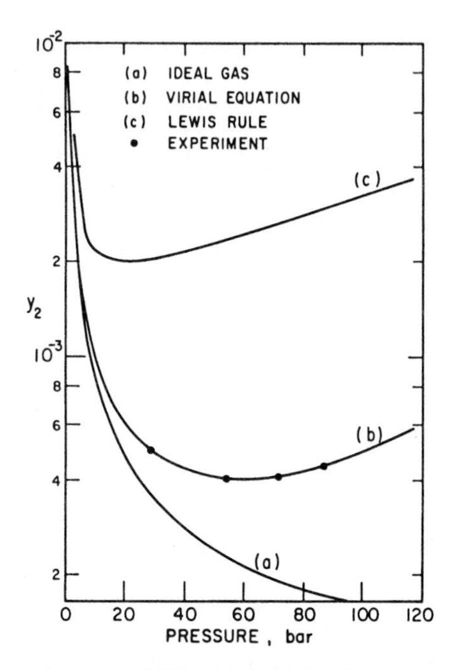

Figure 5-41 Vapor-phase solubility of decane in nitrogen at 50°C.

other words, the absolute value of B_{12} is not large). This result follows because nitrogen is a light gas with a potential energy much smaller than its kinetic energy at 50°C. Therefore, when a decane molecule escapes from the liquid into a space filled with gaseous nitrogen, it feels almost as if it were in a vacuum, or to put it more precisely, as if it were dissolved in an ideal gas. Thus the calculation based on the ideal-gas assumption is not seriously in error in this case; the solubility of decane in nitrogen is only a little larger than what it would be in an ideal-gas solvent because the attractive forces between solute and gaseous solvent are weak in this particular system at 50°C.

However, the ideal-gas assumption becomes increasingly bad for calculating the solubility of a liquid in a compressed gas when the potential energy of the gas is not small compared to its kinetic energy. To illustrate, Fig. 5-42 shows fugacity coefficients of decane at saturation in three gases. It is evident that the vapor-phase fugacity coefficient of decane (and hence its vapor-phase solubility) is strongly dependent on the nature of the gaseous component. Thus decane at 75°C and 100 bar is approximately 10 times more soluble in carbon dioxide than in nitrogen and approximately 40 times more soluble than in hydrogen. The relatively high solubility in carbon dioxide follows from the relatively large attractive forces between decane and carbon dioxide. A rough measure of the potential energy of a gaseous solvent is given by its critical temperature; therefore it is not surprising that carbon dioxide (T_c = 304.2 K) is a better solvent for decane than nitrogen (T_c = 126.2 K) that, in turn, is a better solvent than hydrogen (T_c = 33.3 K).

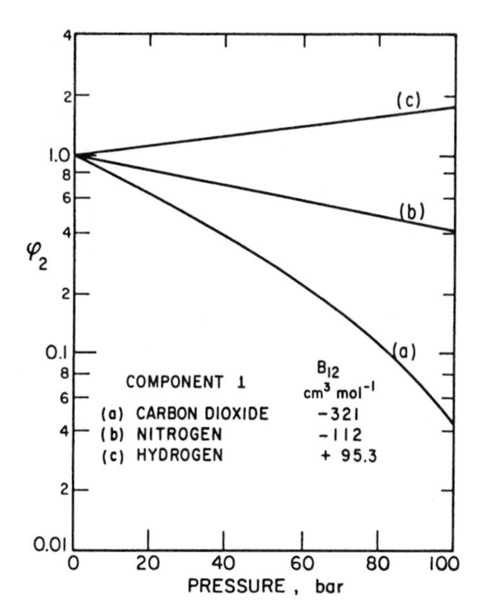

Figure 5-42 Fugacity coefficients of decane in hydrogen, nitrogen, and carbon dioxide at 75°C.

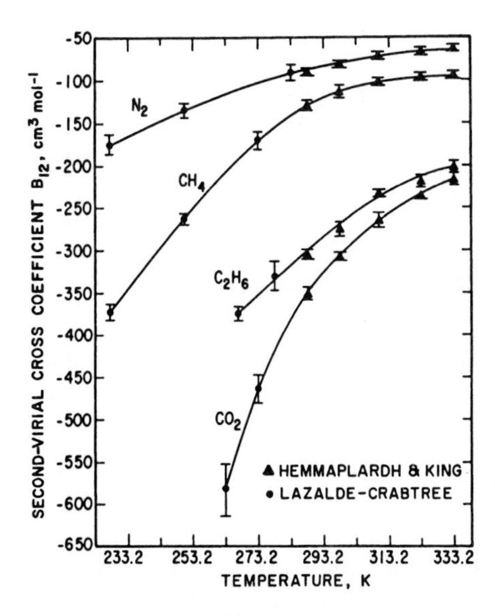

Figure 5-43 Second-virial cross coefficients for binary mixtures containing methanol.

As a final example, Fig. 5-43 shows some second-virial cross coefficients for binary mixtures of methanol, obtained from vapor-phase solubility measurements (Lazalde-Crabtree *et al.*, 1980). As the temperature falls, cross-coefficient B_{12} becomes increasingly negative, corresponding to the rising importance of attractive forces between the dissimilar molecules. Attractive forces between methanol and nitrogen are weak compared to those between methanol and carbon dioxide.

Methanol is an industrial solvent for sweetening natural gas by selective high-pressure absorption of the sour components carbon dioxide and hydrogen sulfide. Because of the solubility of methanol in the dense gas leaving the absorber, a high-pressure absorption process is necessarily accompanied by some solvent loss, that, at high flow rates, may represent an appreciable economic consideration. Figure 5-44 shows the calculated solubility of methanol in a typical effluent-gas mixture as a function of temperature and pressure. The calculations are based on equations of equilibrium similar to Eqs. (5-129) and (5-130) except that the necessary equations now are for a multicomponent, rather than a binary, mixture. For an industrial-scale natural-gas plant, the cost of solvent loss may be large, possibly a few million dollars per year.

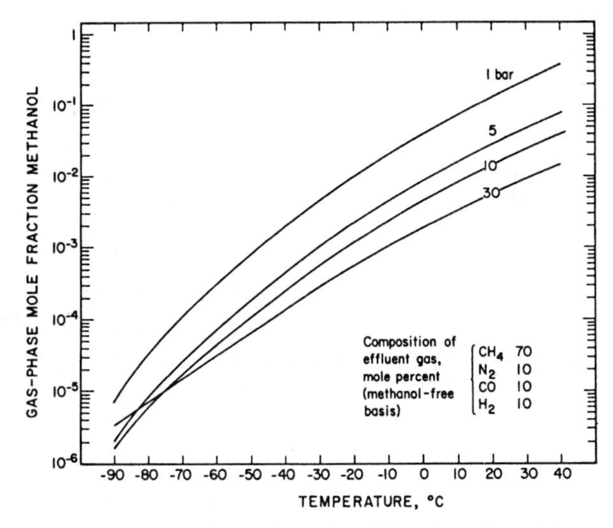

Figure 5-44 Solvent loss in absorption. Solubility of methanol in a natural gas effluent.

5.13 Summary

Efforts to describe quantitatively volumetric properties of real gases and gaseous mixtures extend over more than 200 years and the scientific literature abounds with articles on this subject. In this chapter we have given some of the main ideas with special reference to those contributions that appear to be most useful for technical applications. At the end of such a long chapter it is helpful to recapitulate and to state briefly the current status of efforts toward the calculation of gas-phase fugacities.

The *fugacity* of a component in a gas mixture at any temperature and density can be calculated exactly if volumetric data can be obtained for the mixture at the system temperature and over a density range from zero to the density of interest. However, because such data are rarely available, for practical purposes it is necessary to use approximations. A particularly simple approximation is provided by the *Lewis fugacity rule*; this rule is not generally reliable, although it gives good results for a component present in excess (Sec. 5.1).

There are many empirical *equations of state* and new ones continue to appear. These equations contain empirically determined constants and such constants have been reported for many common pure gases. To predict volumetric properties of a mixture, it is necessary to establish *mixing rules* that state how these constants depend on composition. For empirical equations of state, such mixing rules frequently cannot be established with confidence unless at least some data are available for the particular mixture of interest. Fugacity coefficients are usually sensitive to binary parameters in mixing rules.

A theoretically significant equation of state is the *virial equation* whose constants (virial coefficients) can be related to intermolecular potentials (Secs. 5.2, 5.3, and 5.4). The virial equation can readily be extended to *mixtures*; whereas for any pure component the virial coefficients depend only on temperature and on the intermolecular potential for that component, for a mixture they depend also on the potentials between molecules of those different components that comprise the mixture. The fundamental advantage of the virial equation is that it directly relates fugacities in mixtures to intermolecular forces. The practical disadvantage of the virial equation follows from insufficient understanding of intermolecular forces. As a result, we can at present apply the virial equation only to restricted cases: We can use it only for mixtures at moderate densities, and usually we can predict the required virial coefficients only for those mixtures whose components are nonpolar or weakly polar.

Many models for *intermolecular potential functions* have been proposed and from these, virial coefficients can be computed, provided that we can estimate the potential parameters (Secs. 5.5 and 5.6). For many practical purposes, however, it is simpler to calculate virial coefficients from correlations based on the theorem of corresponding states (Sec. 5.7).

For highly polar and hydrogen-bonded systems, virial coefficients can be interpreted "chemically" by relating them to the enthalpy and entropy of formation of chemically stable complexes (Secs. 5.8, 5.9, and 5.10). A "chemical" description of gas-phase nonideality is particularly useful for mixtures containing strongly hydrogen-bonded components like acetic acid.

Despite the limited practical use of the virial equation, we have devoted much attention to potential functions in this chapter because it is almost certain that in the future, fundamental advances in the description of volumetric properties will be based on research in statistical mechanics. Modern theories of fluids and molecular simulations require reliable potential functions.

At high densities, we must compute fugacity coefficients from mostly empirical equations of state or from *generalized correlations* such as those based on the theorem of corresponding states (Sec. 5.11). Research on the statistical mechanics of dense gases is progressing rapidly and we may expect practical results in the not-too-distant future; however, at present mostly empirical methods are more useful for most applications.

When applying an equation of state or a corresponding-states correlation to a gaseous mixture, flexibility in *mixing rules* is essential for good results. A mixing rule that is good for one system may not be good for another; mixing and combining rules should therefore contain one or two adjustable parameters determined, if possible, by a few data for the mixture under consideration. If no mixture data are available, they should be estimated by careful analysis of chemically similar systems where mixture data are at hand. Good estimates of fugacity coefficients can be made for constituents of dense gas mixtures that contain nonpolar (or slightly polar) components. However, for mixtures containing polar gases, or for any mixture near critical conditions, calculated fugacity coefficients are not likely to be highly accurate. Fugacity coefficients calculated from an equation of state are usually more sensitive to binary coefficients in mixing and combining rules than to any other detail in the equation of state, especially for those components that are dilute in the mixture.

In phase equilibria, the effect of gas-phase nonideality is particularly strong in those cases where a condensed component is sparingly soluble in a compressed gas. This solubility is strongly affected by the density and by the intermolecular forces between solute and gaseous solvent. In many cases, at high pressures, solubilities calculated with the assumption of ideal-gas behavior are in error by several orders of magnitude (Sec. 5.12).

We close this chapter by briefly repeating what we stressed at the beginning: Calculation of fugacities in gaseous mixtures is not a thermodynamic problem. The relations that express fugacity in terms of fundamental macroscopic properties are exact and well known; they easily lend themselves to numerical solution by computers. The difficulty we face lies in our inability to characterize and to predict with sufficient accuracy the configurational (essentially, volumetric) properties of pure fluids and even more, of mixed fluids; this inability, in turn, is a consequence of insufficient knowledge concerning intermolecular forces. To increase our knowledge, we require on the one hand new results from theoretical molecular physics and on the other, more accurate experimental data for equilibrium properties of dense mixtures, especially for those containing one or more polar components.

References

Axilrod, B. M. and E. Teller, 1943, *J. Chem. Phys.*, 11: 299.

Barker, J. A. and A. Pompe, 1968, *Aust. J. Chem.*, 21: 1683.

Barker, J. A. and D. Henderson, 1976, *Rev. Mod. Phys.,* 48: 587.

Besher, E. M. and J. Lielmezs, 1992, *Thermochim. Acta,* 200: 1.

Bondi, A., 1968, *Physical Properties of Molecular Crystals, Liquids, and Glasses.* New York: John Wiley & Sons.

Brugge, H. B., L. Yurttas, J. C. Holste, and K. R. Hall, 1989, *Fluid Phase Equilibria,* 51: 187.

Cheh, H. Y., J. P. O'Connell, and J. M. Prausnitz, 1966, *Can. J. Chem.,* 44: 429.

Christian, S. D., 1957, *J. Phys. Chem.,* 61: 1441.

Chueh, P. L. and J. M. Prausnitz, 1967, *AIChE J.,* 13: 896.

Connolly, J. F. and G. A. Kandalic, 1960, *Phys. Fluids,* 3: 463.

Conti, J. J., D. F. Othmer, and R. Gilmont, 1960, *J. Chem. Eng. Data,* 5: 301.

Crain, R. W. and R. E. Sonntag, 1965, *Adv. Cryog. Eng.,* 11: 379.

Cullen, E. J. and K. A. Kobe, 1955, *AIChE J.,* 1: 452.

De Santis, R. and B. Grande, 1979, *AIChE J.,* 25: 937.

Douslin, D. R., 1962. In *Progress in International Research on Thermodynamic and Transport Properties,* (J. F. Masi and D. H. Tsai, Eds.). New York: ASME/Academic Press.

Dymond, J. H. and B. J. Alder, 1969, *J. Chem. Phys.,* 51: 309.

Eubank, P. T. and K. R. Hall, 1990, *AIChE J.,* 36: 1661.

Ewald, A. H., W. B. Jepson, and J. S. Rowlinson, 1953, *Discuss. Faraday Soc.,* 15: 238.

Graben, H. W. and R. D. Present, 1962, *Phys. Rev. Lett.,* 9: 247.

Guggenheim, E. A. and M. L. McGlashan, 1960, *Proc. R. Soc. (Lond.),* A255: 456.

Gunn, R. D., P. L. Chueh, and J. M. Prausnitz, 1966, *AIChE J.,* 12: 937.

Hayden, J. G. and J. P. O'Connell, 1975, *Ind. Eng. Chem. Process Des. Dev.,* 14: 204.

Hill, T. L., 1986, *An Introduction to Statistical Thermodynamics.* Reading: Addison-Wesley.

Hirschfelder, J. O., C. F. Curtiss, and R. B. Bird, 1954, *Molecular Theory of Gases and Liquids.* New York: John Wiley & Sons.

Joffrion, L. and P. T. Eubank, 1988, *Fluid Phase Equilibria,* 43: 263.

Kay, W. B., 1936, *Ind. Eng. Chem.,* 28: 1014.

Kihara, T., 1953, *Rev. Mod. Phys.,* 25: 831.

Kihara, T., 1958, *Adv. Chem. Phys.,* 1: 276.

Kihara, T., 1963, *Adv. Chem. Phys.,* 5: 147.

Kurnik, R. T. and R. C. Reid, 1981, *AIChE J.,* 27: 861.

Lambert, J. D., 1953, *Discuss. Faraday Soc.,* 15: 226.

Lazalde-Crabtree, H., G. J. F. Breedveld, and J. M. Prausnitz, 1980, *AIChE J.,* 26: 462.

Leland, T. W. and P. S. Chappelear, 1968, *Ind. Eng. Chem.,* 60: 15.

Mak, P. C. N. and J. Lielmezs, 1989, *Ind. Eng. Chem. Res.* 28: 127.

Marek, J., 1954, *Collect. Czech. Chem. Comm.,* 19: 1074.

Marek, J., 1955, *Collect. Czech. Chem. Comm.,* 20: 1490.

Mason, E. A. and T. H. Spurling, 1969, *The Virial Equation of State.* In *The International Encyclopedia of Physical Chemistry and Chemical Physics.* Elmsford: Pergamon Press.

Mayer, J. E. and M. G. Mayer, 1977, *Statistical Mechanics,* 2nd Ed. New York: Wiley Interscience.

McDougall, F. H., 1936, *J. Am. Chem. Soc.,* 58: 2585.

McDougall, F. H., 1941, *J. Am. Chem. Soc.,* 63: 3420.

McGlashan, M. L. and D. J. B. Potter, 1962, *Proc. R. Soc. (Lond.),* A267: 478.

McGlashan, M. L. and C. J. Wormald, 1964, *Trans. Faraday Soc.,* 60: 646.

McGregor, D. R., J. C. Holste, P. T. Eubank, and K. R. Hall, 1987, *Fluid Phase Equilibria,* 35: 153.

McKinley, C., J. Brewer, and E. S. J. Wang, 1961, *Adv. Cryog. Eng.,* 7: 114.

Meissner, H. P., and R. Seferian, 1951, *Chem. Eng. Prog.,* 47: 579.

Michels, A., J. M. Levelt, and W. DeGraff, 1958, *Physica,* 24: 659.

Michels, A., W. DeGraaff, and C. A. Ten Seldam, 1960, *Physica,* 26: 393.

Millen, D. J. and G. M. Mines, 1974, *J. Chem. Soc. Faraday Trans. II,* 70: 693.

Munn, R. J., 1964, *J. Chem. Phys.,* 40: 1439.

Munn, R. J. and F. J. Smith, 1965, *J. Chem. Phys.,* 43: 3998.

Munn, R. J. and F. J. Smith, 1965a, *Discuss. Faraday Soc.,* 40.

Myers, A. L. and J. M. Prausnitz, 1962, *Physica,* 28: 303.

Nothnagel, K. H., D. S. Abrams, and J. M. Prausnitz, 1972, *Ind. Eng. Chem. Process Des. Dev.,* 12: 25.

O'Connell, J. P. and J. M. Prausnitz, 1968, *J. Phys. Chem.,* 72: 632.

Olf, G., A. Schnitzler, and J. Gaube, 1989, *Fluid Phase Equilibria,* 49: 49.

Olf, G., J. Spiske, and J. Gaube, 1989a, *Fluid Phase Equilibria,* 51: 209.

Orbey, H. and J. H. Vera, 1983, *AIChE J.,* 29: 107.

Orentlicher, M. and J. M. Prausnitz, 1967, *Can. J. Chem.,* 45: 373.

Patel, M. R., J. C. Holste, K. R. Hall, and P. T. Eubank, 1987, *Fluid Phase Equilibria,* 36: 279.

Pauling, L., 1945, *The Nature of the Chemical Bond.* Ithaca: Cornell University Press.

Pope, G. A., P. S. Chappelear, and R. Kobayashi, 1973, *J. Chem. Phys.,* 59: 423.

Prausnitz, J. M. and P. R. Benson, 1959, *AIChE J.,* 5: 161.

Prausnitz, J. M. and W. B. Carter, 1960, *AIChE J.,* 6: 611.[25] See also Renner, T. A. and M. Blander, 1977, *J. Phys. Chem.,* 81: 857.

Prausnitz, J. M. and R. N. Keeler, 1961, *AIChE J.,* 7: 399.

Prausnitz, J. M. and A. L. Myers, 1963, *AIChE J.,* 9: 5.

Reid, R. C. and T. W. Leland, 1965, *AIChE J.,* 11: 229.

Riedel, L., 1956, *Chem.-Ing.-Tech.,* 28: 557 (and earlier papers).

Rossi, J. C. and F. Danon, 1966, *J. Phys. Chem.,* 70: 942.

Rowlinson, J. S., 1949, *Trans. Faraday Soc.,* 45: 974.

Rowlinson, J. S., 1951, *J. Chem. Phys.,* 19: 827.

Rowlinson, J. S., 1955, *Trans. Faraday Soc.,* 51: 1317.

Rowlinson, J. S., 1965, *Discuss. Faraday Soc.,* 40: 19.

Schreiber, D. R. and K. S. Pitzer, 1989, *Fluid Phase Equilibria,* 46: 113.

[25] The equilibrium constants reported here are too large by a factor of 1.317. However, the enthalpies of complex-formation are not affected.

Sebastiani, E. and L. Lacquanti, 1967, *Chem. Eng. Sci.*, 22: 1155.
Sherwood, A. E. and J. M. Prausnitz, 1964, *J. Chem. Phys.*, 41: 429.
Sherwood, A. E. and J. M. Prausnitz, 1964a. *J. Chem. Phys.*, 41: 413.
Sherwood, A. E., A. G. DeRocco, and E. A. Mason, 1966, *J. Chem. Phys.*, 44: 2984.
Stein, F. P. and E. J. Miller, 1980, *Ind. Eng. Chem. Process Des. Dev.*, 19: 123.
Stogryn, D. E., 1968, *J. Chem. Phys.*, 48: 4474.
Stogryn, D. E., 1969, *J. Chem. Phys.*, 50: 4667.
Stogryn, D. E., 1970, *J. Chem. Phys.*, 52: 3671.
Tarakad, R. R. and R. P. Danner, 1977, *AIChE J.*, 23: 685.
Tsekhanskaya, Y. V., M. B. Iomtev, and E. V. Mushkina, 1964, *Zh. Fiz. Chim.*, 38: 2166.
Tsonopoulos, C. and J. M. Prausnitz, 1970, *Chem. Eng. J.*, 1: 273.
Tsonopoulos, C., 1974, *AIChE J.*, 20: 263.
Tsonopoulos, C., 1975, *AIChE J.*, 21: 827.
Tsonopoulos, C., 1978, *AIChE J.*, 24: 1112.
Tsonopoulos, C. and J. L. Heidman, 1990, *Fluid Phase Equilibria*, 57: 261.
Van Welie, G. S. A. and G. A. M. Diepen, 1961, *J. Rec. Trav. Chim.*, 80: 673.
Webster, T. J., 1952, *R. Soc. (Lond.)*, A214: 61.
Wisniak, J. and A. Tamir, 1976, *J. Chem. Eng. Data*, 21: 88.

Problems

1. Consider the following simple experiment: We have a container of fixed volume V; this container, kept at temperature T, has in it n_1 moles of gas 1. We now add isothermally to this container n_2 moles of gas 2 and we observe that the pressure rise is ΔP. Assume that the conditions are such that the volumetric properties of the gases and their mixture are adequately described by the virial equation neglecting the third and higher coefficients. The second virial coefficients of the pure gases are known. Find an expression that will permit calculation of B_{12}.

2. At -100°C, a gaseous mixture of one mole contains 1 mol % CO_2 and 99 mol % H_2. The mixture is compressed isothermally to 60 bar. Does any CO_2 precipitate? If so, approximately how much? [At -100°C, the saturation (vapor) pressure of pure solid CO_2 is 0.1392 bar and the molar volume of solid CO_2 is 27.6 cm^3 mol^{-1}.]

3. A gaseous mixture containing 30 mol % CO_2 and 70 mol % CH_4 is passed through a Joule-Thomson expansion valve. The gas mixture enters the valve at 70 bar and 40°C and leaves at 1 bar. Does any CO_2 condense? Assume that the heat capacities are independent of temperature. The data are as follows (1 = CH_4; 2 = CO_2):

$$B_{11} = 42.5 - 16.75\times10^3T^{-1} - 25.05\times 10^5T^{-2}$$

$$B_{22} = 40.4 - 25.39\times10^3T^{-1} - 68.7\times10^5T^{-2}$$

$$B_{12} = 41.4 - 19.50\times10^3T^{-1} - 37.3\times10^5T^{-2}$$

where B is in cm^3 mol^{-1} and T is in K. The c_p^0 for an ideal gas are: $c_{p,1}^0 = 35.8$ J mol^{-1} K^{-1} and $c_{p,2}^0 = 37.2$ J mol^{-1} K^{-1}. The vapor pressure equation for pure CO_2 is

$$\ln P^s\,(\text{bar}) = 10.807 - \frac{1980.24}{T} \qquad (243\,\text{K} < T < 303\,\text{K})$$

4. On the basis of the Stockmayer potential, devise a corresponding-states theorem for polar substances. How many reducing parameters are there, and what are they?

5. Using the generalized correlations of Pitzer (or Lee and Kesler), calculate:

(a) The energy of vaporization of acetylene at 0°C.

(b) The second virial coefficient of a mixture containing 80 mol % butane and 20 mol % nitrogen at 461 K. (The experimental value is -172 cm^3 mol^{-1}).

(c) The enthalpy of mixing (at 200 K and 100 bar) of one mole of methane, one mole of nitrogen, and one mole of hydrogen.

6. A gas stream at 25°C consists of 40 mol % nitrogen, 50 mol % hydrogen, and 10 mol % propane. The propane is to be removed from this stream by high-pressure absorption using a mineral oil of very low volatility. At the bottom of the column (where the gas enters), what is the driving force for the rate of absorption? The total pressure is 40 bar. The driving force for a component i is given by

$$\text{Driving force} = p_i - p_i^*$$

where p_i is the partial pressure of i in the gas bulk phase and p_i^* is the partial pressure of i in the gas phase in equilibrium with the liquid phase. In this column the liquid-to-gas flow ratio (L/G) is five moles of liquid per mole of gas. The column is to absorb 95% of the propane. Solubility of inert gases in the mineral oil is negligible. The data are as follows: The compressibility factor of the entering gas mixture is 0.95. The solubility of propane in the mineral oil is given by $f_{c_3} = Hx_{c_3}$ with $H = 53.3$ bar at 25°C.

7. A gas mixture contains 2 mol % ethane (1) and 98 mol % nitrogen (2). The ethane is to be recovered by absorption into heavy oil at 40°C and 50 bar. When this gas mixture is in equilibrium with the oil, what is the relative volatility $\alpha_{2,1}$ of nitrogen to ethane?

$$\alpha_{2,1} = \frac{y_2x_1}{x_2y_1}$$

Data, all at 40°C, are as follows:
Henry's constant for nitrogen in the oil: 1000 bar
Henry's constant for ethane in the oil: 100 bar
Second virial coefficients (cm^3 mol^{-1}): B_{11} = -172; B_{22} = 1.5; B_{12} = - 44.
Partial molar volumes in the oil (cm^3 mol^{-1}): nitrogen, 25; ethane, 60.

8. At 0°C and pressures to 20 bar, the solubility of methane (1) in methanol (2) follows Henry's law; Henry's constant is 1022 bar. The vapor pressure of methanol at 0°C is 0.0401 bar. Second virial coefficients (cm^3 mol^{-1}) at 0°C are B_{11} = -53.9, B_{12} = -166, and B_{22} = -4068. At 0°C and 20 bar, what is the solubility (mole fraction) of methanol in methane? Why is this solubility of interest in chemical engineering?

9. Acetic acid vapor has a tendency to dimerize. Show that in an equilibrium mixture of monomer and dimer the fugacity of the dimer is proportional to the square of the fugacity of the monomer.

10. Compare the systems:

hydrogen chloride/diisopropyl ether
hydrogen chloride/ethyl butyl ether

Second-virial cross coefficients B_{12} are available for both systems at 70°C. Compare the two coefficients. Are they likely to be positive or negative? Which one is more positive? Why?

11. At 143.5°C, the saturation (vapor) pressure of acetic acid is 2.026 bar. The dimerization constant for acetic acid at this temperature is 1.028 bar^{-1}. The molar liquid volume of acetic acid at 143.5°C is 57.2 cm^3 mol^{-1}. Calculate the fugacity of acetic acid at 143.5°C and 50 bar (see App. B).

12. The virial equation gives the compressibility factor as an expansion in the density

$$z = 1 + B\rho + C\rho^2 + \ldots \tag{1}$$

Because it is more convenient to use pressure as an independent variable, another form of the virial equation is given by

$$z = 1 + B'P + C'P^2 + \ldots \tag{2}$$

(a) Relate the virial coefficients B' and C' to the constants a and b that appear in the Redlich-Kwong equation.

(b) Consider a binary gas mixture of ethylene and nitrogen having mole fraction $y_{C_2H_4}$ = 0.2. Starting with Eq. (2), calculate the fugacity of ethylene at 50°C and 50 bar. Neglect virial coefficients beyond the second. For the second virial coefficient, use the relation found in (a) for the Redlich-Kwong equation with the customary mixing rules. (From experimental volumetric data, the fugacity of ethylene, under the conditions specified, is 8.87 bar.)

13. One mole of a binary gas mixture containing ethylene and argon at 40°C is confined to a constant-volume container. Compute the composition of this mixture that gives the maximum pressure. Assume that virial coefficients beyond the second may be neglected.

14. 100 moles of H_2 initially at 302.6 K and 20.7 bar is mixed adiabatically and isobarically with 2000 moles of ethylene initially at 259.5 K and 20.7 bar. What is the temperature of the final mixture?

15. A metallic sheet is to be coated with a thin layer of an organic solid A. To do so, it is proposed to vaporize the solid and to allow the vapor to condense on the sheet in an isolated chamber, as indicated below:

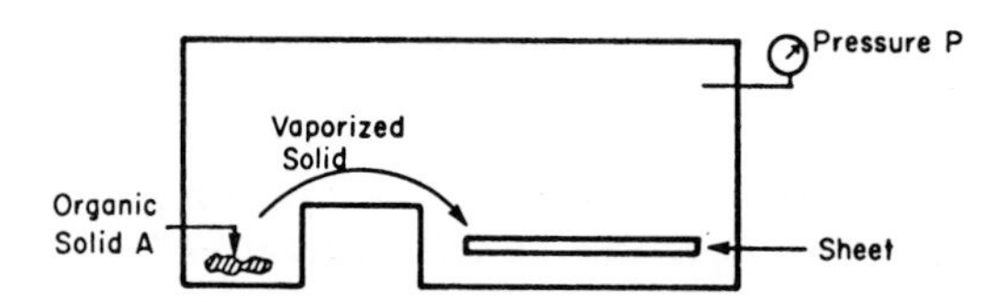

The condensation rate depends on the partial pressure of A. Since A is thermally unstable, it can be heated to no more than 350 K, where its vapor pressure is 1 torr. To increase the rate of condensation, it is proposed to fill the chamber with carbon dioxide gas at high pressure. What total pressure should the chamber have so that the partial pressure of A is 10 torr? Assume that the entire chamber is isothermal at 350 K and that the volumetric properties of CO_2-A mixtures are given by

$$v = \frac{RT}{P} + B$$

where B is the second virial coefficient and v is the volume per mole of mixture. At 350 K, for pure A, B = -500; for pure CO_2, B = -85 and the cross-coefficient $B_{A\text{-}CO_2}$ = -430, all in units of $cm^3\ mol^{-1}$. The molar volume of solid A at 350 K is 200 $cm^3\ mol^{-1}$.

16. At 35°C the vapor pressure of solid naphthalene is 2.80×10^{-4} bar and its density is 1.145 $g\ cm^{-3}$. Calculate the solubility of naphthalene in compressed ethylene at 35°C and at total pressure of 30 bar:

(a) Using the ideal-gas law.

(b) Using the virial equation of state truncated after the second term For the second virial coefficient B use the van der Waals result,

$$B = b - \frac{a}{RT}$$

Van der Waals constants a and b are related to the critical constants. For the gas mixture assume that a and b are a quadratic and a linear function of the mole fraction, respectively.

17. Moist air at 30 bar is cooled isobarically to -10°C. What is the maximum permissible moisture content (mole fraction gaseous water) that the air may have to be sure that no moisture condenses? Assume that the volumetric properties of moist air at 30 bar are given by the virial equation of state truncated after the second term. For this calculation, assume that dry air is 80 mol % nitrogen and 20 mol % oxygen.
Second virial coefficients at 10°C are (cm^3 mol^{-1}): B_{33} = -1500; B_{12} = -15; B_{11} = -25; B_{13} = -63, where 1 stands for nitrogen, 2 for oxygen, and 3 for water. At -10°C, the vapor pressure of ice is 1.95 torr and the density of ice is 0.92 g cm^{-3}.

18. At 300 K and modest pressures, the Joule-Thomson coefficient of hydrogen is negative (i.e., upon isenthalpic expansion, hydrogen becomes warmer). How much ethane must be added (per mole of hydrogen) to produce a Joule-Thomson coefficient of zero at 300 K and modest pressure? Use the truncated-virial equation with the correlation of McGlashan and Potter. Use the following reducing parameters (where 1 stands for hydrogen and 2 for ethane):

Species pair	v_c (cm^3 mol^{-1})	T_c (K)
11	51.5§	43.6§
22	148.0	305.4
12	91.4	111.9

§ Effective (not true) critical parameters

19. A subterranean natural-gas reservoir is at 60°C and 20 bar. The gas (primarily methane) is in contact with liquid water. The gas is pumped from the well to a pipeline operating at 40 bar and 25°C. Assuming that there is no entrainment of liquid water in the gas stream from the well, estimate the amount of liquid water [in moles of liquid water (2) removed per mole of methane (1) produced] that must be removed from the methane stream in the pipeline. Data:

	Vapor pressure of water (mmHg)	B_{11} (cm^3 mol^{-1})	B_{22} (cm^3 mol^{-1})	B_{12} (cm^3 mol^{-1})
25°C	24	-43	-1162	-62
60°C	149	-31	-790	-48

20. An equimolar mixture of CO_2 and CH_4 at 50 bar and 298 K is to be separated into pure CO_2 at 50 bar and 298 K an pure CH_4 at 50 bar and 298 K by a continuous flow process involving heat interactions with the environment at 298 K. Calculate the minimum possible amount of work per mole of mixture that would be required for this separation.

Assume that the gases follow the volume-explicit form of the virial equation of state, truncated after the second virial term, both in the pure state and in the mixtures. Second-virial coefficient data are (1 = CH_4; 2 = CO_2; B in cm^3 mol^{-1} and T in kelvin):

$$B_{11} = 42.5 - 16.75\times10^3\, T^{-1} - 25.05\times10^5\, T^{-2}$$

$$B_{12} = 40.4 - 25.39\times10^3\, T^{-1} - 68.7\times10^5\, T^{-2}$$

$$B_{22} = 41.4 - 19.50\times10^3\, T^{-1} - 37.3\times10^5\, T^{-2}$$

21. Using the square-well potential, find the fugacity of methyl chloride at 150°C and 30 atm. Potential parameters are: ε/k = 469 K; σ = 0.429 nm; R (well width) = 0.337σ.
Also calculate the enthalpy and entropy of dimerization of methyl chloride in the dilute-gas phase. For the entropy of dimerization, use an ideal-gas standard state of 1 mol L^{-1}.

22. Consider a gas for which the intermolecular potential $\Gamma(r)$ is given by

$$\Gamma = \begin{cases} \infty & \text{for} \quad r \le \sigma \\ -\dfrac{A}{r^n} & \text{for} \quad r > \sigma \end{cases}$$

where A and n are constants.

(a) Obtain an expression for the second virial coefficient of this gas at high temperature.
(b) Determine allowable values of n in this expression and the relationship of dB/dT with the attractive part of the potential.

23. Assume that argon at 273.15 K is represented by the square-well potential with ε/k = 141.06 K, σ = 0.2989 nm, and R = 1.55. Estimate the second virial coefficient for argon at 273.15 K and compare with the experimental value B = -22.08 cm^3 mol^{-1}.

24. The Sutherland potential describes the interaction introduced by van der Waals in his equation. Using the data below, obtain $B(T)$ for methane and n-pentane given by the Sutherland potential. Compare with experimental second-virial-coefficient data at 373 K.
For methane: ε/k = 434.8 K; σ = 0.3294 nm; B(373 K) = -20 cm^3 mol^{-1}.
For n-pentane: ε/k = 1096.8 K; σ = 0.4755 nm; B(373 K) = -621 cm^3 mol^{-1}.

25. A closed chamber, maintained at 40°C, consists of two equal parts, separated from each other by a membrane permeable only to helium. Each part has a volume of 300 cm^3. The pressure in the left part is 37.5 bar and the pressure in the right part is 84.5 bar.
The left part is filled with 0.99 moles ethane and 0.01 moles helium; the right part is filled with 0.99 moles nitrogen and 0.01 moles helium. After equilibrium is attained, what are the mole fractions of helium in each part of the chamber?

Assume that the virial equation truncated after the second term is applicable. The following second virial coefficients ($cm^3\ mol^{-1}$) are available (where 1 stands for helium, 2 for ethane, and 3 for nitrogen):

$$B_{11} = 17.17; \qquad B_{12} = 24.51$$

$$B_{22} = -169.23; \qquad B_{13} = 21.97$$

$$B_{33} = -1.55; \qquad B_{23} = -44.39$$

26. Hydrogen Fluoride associates in the gas phase; two possibilities are under consideration:

$$4(HF) \rightleftharpoons (HF)_4 \qquad \text{(a)}$$

$$6(HF) \rightleftharpoons (HF)_6 \qquad \text{(b)}$$

To test these possibilities, the vapor-phase densities of HF were measured at two pressures at 50°C. They are

Pressure (bar)	1.42	2.84
Density (g L^{-1})	1.40	5.45

In view of these measurements, which possibility, (a) or (b), is more likely? The molar mass of HF is 20 g mol^{-1}.

CHAPTER 6

Fugacities in Liquid Mixtures: Excess Functions

Calculation of fugacities from volumetric properties was discussed in Chap. 3; many of the relations derived there [in particular, Eqs. (3-4) and (3-53)] are general and may be applied to condensed phases as well as to the gas phase. However, often it is not practical to do so because the necessary integrations require that volumetric data be available at constant temperature and constant composition over the entire density range from the ideal-gas state (zero density) to the density of the condensed phase, including the two-phase region. It is a tedious task to obtain such data for fluid mixtures and very few data of this type have been reported. A more useful alternate method, therefore, is needed for calculation of fugacities in liquid solutions. Such a method is obtained by defining an ideal liquid solution and by describing deviations from ideal behavior in terms of excess functions; these functions yield the familiar activity coefficients that give a quantitative measure of departure from ideal behavior.

The fugacity of component i in a liquid solution is most conveniently related to the mole fraction x_i by

$$f_i^L = \gamma_i x_i f_i^0 \tag{6-1}$$

where γ_i is the activity coefficient and f_i^0 is the fugacity of i at some arbitrary condition known as the standard state. At any composition, the activity coefficient depends on the choice of standard state and the numerical value of γ_i has no significance unless the numerical value of f_i^0 is also specified.

Because the choice of standard state is arbitrary, it is convenient to choose f_i^0 such that γ_i assumes values close to unity and when, for a range of conditions, γ_i is exactly equal to unity, we say that the solution is ideal.[1] However, because of the intimate relation between the activity coefficient and the standard-state fugacity, the definition of solution ideality ($\gamma_i = 1$) is not complete unless the choice of standard state is clearly indicated. Either of two choices is frequently used. One of these leads to an ideal solution in the sense of Raoult's law and the other leads to an ideal solution in the sense of Henry's law.

6.1 The Ideal Solution

The history of modern science has shown repeatedly that a quantitative description of nature can often be achieved most successfully by first idealizing natural phenomena, i.e., by setting up a simplified model, either physical or mathematical, that crudely describes the essential behavior while neglecting details. (One of the outstanding characteristics of great contributors to modern science has been their ability to distinguish between what is essential and what is incidental.) The behavior of nature is then related to the idealized model by various correction terms that can be interpreted physically and that sometimes can be related quantitatively to those details in nature that were neglected in the process of idealization.

An ideal liquid solution is one where, at constant temperature and pressure, the fugacity of every component is proportional to some suitable measure of its concentration, usually the mole fraction. That is, at some constant temperature and pressure, for any component i in an ideal solution,

$$f_i^L = \Re_i x_i \tag{6-1a}$$

where $\Re_i$ is a proportionality constant dependent on temperature and pressure but independent of x_i.

[1] The two expressions "ideal solution" and "ideal mixture" are both in common usage. They are equivalent.

We notice at once from Eq. (6-1) that, if we let $f_i^0 = \Re_i$, then $\gamma_i = 1$. If Eq. (6-1a) holds for the entire range of composition (from $x_i = 0$ to $x_i = 1$), the solution is ideal in the sense of Raoult's law. For such a solution it follows from the boundary condition at $x_i = 1$ that the proportionality constant $\Re_i$ is equal to the fugacity of pure liquid i at the temperature of the solution.[2] For this case, if the fugacities are set equal to partial pressures, we then obtain the familiar relation known as Raoult's law.

In many cases, the simple proportionality between f_i^L and x_i holds only over a small range of composition. If x_i is near zero, it is still possible to have an ideal solution according to Eq. (6-1a), without, however, equating $\Re_i$ to the fugacity of pure liquid i. We call such a solution an ideal dilute solution leading to the familiar relation known as Henry's law.

The strict definition of an ideal solution requires that Eq. (6-1a) must hold not only at a special temperature and pressure of interest but also at temperatures and pressures in their immediate vicinity. This feature leads to an important conclusion concerning heat effects and volume changes of mixing for a solution ideal in the sense of Raoult's law. For such a solution, we have, at any T, P and x,

$$f_i(T,P,x) = f_{\text{pure } i}(T,P)x_i \tag{6-2}$$

where, for convenience, we have deleted superscript L. We now use two exact thermodynamic relations:[3]

$$\left(\frac{\partial \ln f_i}{\partial T}\right)_{P,x} = -\frac{\bar{h}_i - h_i^+}{RT^2} \qquad \left(\frac{\partial \ln f_{\text{pure } i}}{\partial T}\right)_P = -\frac{h_i - h_i^+}{RT^2} \tag{6-3}$$

and

$$\left(\frac{\partial \ln f_i}{\partial P}\right)_{T,x} = -\frac{\bar{v}_i}{RT} \qquad \left(\frac{\partial \ln f_{\text{pure } i}}{\partial P}\right)_T = -\frac{v_i}{RT} \tag{6-4}$$

where $\bar{h}_i$ is the partial molar enthalpy of component i in the liquid phase, h_i is the enthalpy of pure liquid i, h_i^+ is the enthalpy of pure i in the ideal-gas state, $\bar{v}_i$ is the partial molar volume of i, and v_i is the molar volume of pure i, both in the liquid phase, all at system temperature T and pressure P. Upon substitution of Eq. (6-2) we find that

[2] The standard-state fugacity of pure liquid i at system temperature is usually taken either at P_i^s, the saturation pressure of pure i, or else at P, the total pressure of the mixture. The latter choice is more common, especially at low or moderate pressures. At high pressures, special care must be taken in specifying the pressure of the standard state.

[3] Equation (6-3) is a form of the Gibbs-Helmholtz equation.

$$\bar{h}_i = h_i \tag{6-5}$$

and

$$\bar{v}_i = v_i \tag{6-6}$$

Because the partial molar enthalpy and partial molar volume of component i in an ideal solution are, respectively, the same as the molar enthalpy and molar volume of pure i at the same temperature and pressure, it follows that the formation of an ideal solution occurs without evolution or absorption of heat and without change of volume.

Mixtures of real fluids do not form ideal solutions, although mixtures of similar liquids often exhibit behavior close to ideality. However, all solutions of chemically stable nonelectrolytes behave as ideal dilute solutions in the limit of very large dilution. The correction terms that relate the properties of real solutions to those of ideal solutions are called excess functions.

6.2 Fundamental Relations of Excess Functions

Excess functions are thermodynamic properties of solutions that are in excess of those of an ideal (or ideal dilute) solution at the same conditions of temperature, pressure, and composition.[4] For an ideal solution all excess functions are zero. For example, G^E, the *excess Gibbs energy*, is defined by

$$\boxed{G^E \equiv G_{(\text{actual solution at } T,P \text{ and } x)} - G_{(\text{ideal solution at same } T,P, \text{ and } x)}} \tag{6-7}$$

Similar definitions hold for excess volume V^E, excess entropy S^E, excess enthalpy H^E, excess internal energy U^E, and excess Helmholtz energy A^E. Relations between these excess functions are exactly the same as those between the total functions:

$$H^E = U^E + PV^E \tag{6-8}$$

$$G^E = H^E - TS^E \tag{6-9}$$

$$A^E = U^E - TS^E \tag{6-10}$$

Also, partial derivatives of extensive excess functions are analogous to those of the total functions. For example,

[4] Most excess functions, but not all, are extensive. However, it follows from the definition that we cannot have an excess pressure, temperature, or composition. See R. Missen, 1969, *Ind. Eng. Chem. Fundam.*, 8: 81.

$$\left(\frac{\partial G^E}{\partial T}\right)_{P,x} = -S^E \tag{6-11}$$

$$\left(\frac{\partial G^E/T}{\partial T}\right)_{P,x} = -\frac{H^E}{T^2} \tag{6-12}$$

$$\left(\frac{\partial G^E}{\partial P}\right)_{T,x} = V^E \tag{6-13}$$

Excess functions may be positive or negative; when the excess Gibbs energy of a solution is greater than zero the solution is said to exhibit positive deviations from ideality, whereas if it is less than zero the deviations from ideality are said to be negative.

Partial molar excess functions are defined in a manner analogous to that used for partial molar thermodynamic properties. If M is an extensive thermodynamic property, then $\overline{m}_i$, the partial molar M of component i, is defined by

$$\boxed{\overline{m}_i \equiv \left(\frac{\partial M}{\partial n_i}\right)_{T,P,n_j}} \tag{6-14}$$

where n_i is the number of moles of i and where subscript n_j designates that the number of moles of all components other than i are kept constant. Similarly,

$$\overline{m}_i^E \equiv \left(\frac{\partial M^E}{\partial n_i}\right)_{T,P,n_j} \tag{6-15}$$

Also, from Euler's theorem, we have that

$$M = \sum_i n_i \overline{m}_i \tag{6-16}$$

It then follows that

$$\boxed{M^E = \sum_i n_i \overline{m}_i^E} \tag{6-17}$$

For our purposes, an extensive excess property is a homogeneous function of the first degree in the mole numbers.[5]

For phase-equilibrium thermodynamics, the most useful partial excess property is the partial molar excess Gibbs energy that is directly related to the activity coefficient. The partial molar excess enthalpy and partial molar excess volume are related, respectively, to the temperature and pressure derivatives of the activity coefficient. These relations are summarized in the next section.

6.3 Activity and Activity Coefficients

The activity of component i at some temperature, pressure, and composition is defined as the ratio of the fugacity of i at these conditions to the fugacity of i in the standard state, that is a state at the same temperature as that of the mixture and at some specified condition of pressure and composition:

$$a_i(T,P,x) \equiv \frac{f_i(T,P,x)}{f_i(T,P^0,x^0)} \tag{6-18}$$

where P^0 and x^0 are, respectively, an arbitrary but specified pressure and composition.

The activity coefficient γ_i is the ratio of the activity of i to some convenient measure of the concentration of i, usually the mole fraction[6]

$$\gamma_i \equiv \frac{a_i}{x_i} \tag{6-19}$$

The relation between partial molar excess Gibbs energy and activity coefficient is obtained by first recalling the definition of fugacity. At constant temperature and pressure, for a component i in solution,

$$\bar{g}_{i(\text{real})} - \bar{g}_{i(\text{ideal})} = RT[\ln f_{i(\text{real})} - \ln f_{i(\text{ideal})}] \tag{6-20}$$

Next, we introduce the partial molar excess function $\bar{g}_i^E$ by differentiation of Eq. (6-7) at constant T, P, and n_j:

---

[5] Many, but not all, extensive excess properties can be defined in this way. See Redlich, 1968, "Fundamental Thermodynamics since Caratheodory," *Rev. Mod. Phys.*, 40: 556; and O. Redlich, 1976, *Thermodynamics: Fundamentals, Applications*, (Amsterdam: Elsevier).

[6] For electrolyte solutions, it *is* often more convenient to use molality instead of mole fraction (see Sec. 9.1). For polymer solutions, mole fractions are not useful; instead, weight fractions or volume fractions are more appropriate.

$$\bar{g}_i^E = \bar{g}_{i(\text{real})} - \bar{g}_{i(\text{ideal})} \tag{6-21}$$

Substitution then gives

$$\bar{g}_i^E = RT \ln \frac{f_{i(\text{real})}}{f_{i(\text{ideal})}} \tag{6-22}$$

and substituting Eq. (6-1a), we obtain

$$\bar{g}_i^E = RT \ln \frac{f_i}{\Re x_i} \tag{6-23}$$

It follows from Eq. (6-1a) that an ideal solution is one where the activity is equal to the mole fraction; if we set the standard-state fugacity f_i^0 equal to $\Re_i$ we then have

$$a_i = \gamma_i x_i = \frac{f_i}{\Re_i} \tag{6-24}$$

But for an ideal solution [Eq. (6-1a)], f_i is equal to $\Re_i x_i$ and therefore, for an ideal solution, $\gamma_i = 1$ and $a_i = x_i$. Substitution of Eq. (6-24) into Eq. (6-23) gives the important and useful result[7]

$$\boxed{\bar{g}_i^E = RT \ln \gamma_i} \tag{6-25}$$

Substitution into Eq. (6-17) gives the equally important relation[8]

$$\boxed{g^E = RT \sum_i x_i \ln \gamma_i} \tag{6-26}$$

---

[7] A shorter but equivalent derivation of Eq. (6-25) follows from writing

$$\bar{g}_i^E \equiv \bar{g}_i - \bar{g}_{i(\text{ideal})} \quad \text{and} \quad \bar{g}_i^E = RT \ln[f_i \,/\, f_{i(\text{ideal})}]$$

Because $f_i = \gamma_i x_i f_i^0$ and $f_{i(\text{ideal})} = x_i f_i^0$, we obtain $\bar{g}_i^E = RT \ln \gamma_i$.

[8] Equations (6-25) and (6-26) are related. Equation (6-26) can be rewritten as $n_T g^E = RT \sum n_i \ln \gamma_i$, where n_i is the number of moles of component i and n_T is the total number of moles. Differentiating with respect to n_i at constant T, P, and all other mole numbers n_j,

$$\left(\frac{\partial n_T g^E}{\partial n_i}\right)_{T,P,n_j} = \bar{g}_i^E = RT\left[\ln \gamma_i + \sum_k n_k \left(\frac{\partial \ln \gamma_k}{\partial n_i}\right)_{T,P,n_j}\right]$$

where the summation is over all components, including i. The Gibbs-Duhem equation asserts that this summation is zero, yielding Eq. (6-25).

where g^E is the molar excess Gibbs energy. Equations (6-25) and (6-26) are used repeatedly in the remainder of this chapter as well as in later chapters.

We now want to consider the temperature and pressure derivatives of the activity coefficient. Let us first discuss the case where the excess Gibbs energy is defined relative to an ideal solution that is ideal over the entire range of composition in the sense of Raoult's law. In this case,

$$\Re_i = f_i \text{(pure liquid } i \text{ at } T \text{ and } P \text{ of solution)} \tag{6-27}$$

and

$$\ln \gamma_i = \ln f_i - \ln x_i - \ln f_{\text{pure } i} \tag{6-28}$$

Using Eq. (6-3), differentiation with respect to temperature at constant P and x gives

$$\left(\frac{\partial \ln \gamma_i}{\partial T} \right)_{P,x} = \frac{h_{\text{pure } i} - \bar{h}_i}{RT^2} = -\frac{\bar{h}_i^E}{RT^2} \tag{6-29}$$

where $\bar{h}_i^E$ is the partial molar enthalpy of i minus the molar enthalpy of pure liquid i at the same temperature and pressure. Differentiation with respect to pressure at constant T and x gives[9]

$$\left(\frac{\partial \ln \gamma_i}{\partial P} \right)_{T,x} = \frac{\bar{v}_i - v_{\text{pure } i}}{RT} = -\frac{\bar{v}_i^E}{RT} \tag{6-30}$$

where $\bar{v}_i^E$ is the partial molar volume of i minus the molar volume of pure liquid i at the same temperature and pressure.

Now let us consider the case where the excess Gibbs energy is defined relative to an ideal dilute solution. It is useful to define excess functions relative to an ideal dilute solution whenever the liquid mixture cannot exist over the entire composition range, as happens, for example, in a liquid mixture containing a gaseous solute. If the critical temperature of solute 2 is lower than the temperature of the mixture, then a liquid phase cannot exist as $x_2 \to 1$, and relations based on an ideal mixture in the sense of Raoult's law can be used only by introducing a hypothetical standard state for solute 2.

---

[9] In some cases $\Re_i$ is set equal to the fugacity of pure liquid i at temperature T and at its own saturation pressure P_i^s (rather than the total pressure P, that may vary with the composition in an isothermal mixture). In that case Eq. (6-29) is not affected significantly because the enthalpy of a pure liquid is usually a weak function of pressure. However, Eq. (6-30) becomes

$$\left(\frac{\partial \ln \gamma_i}{\partial P} \right)_{T,x} = \frac{\bar{v}_i}{RT} \tag{6-30a}$$

However, relations based on an ideal dilute solution avoid this difficulty. The proportionality constant $\Re_2$ is not determined from the pure-component boundary condition $x_2 = 1$, but rather from the boundary condition of the infinitely dilute solution, i.e., $x_2 \to 0$. For an ideal dilute solution, we have for the solute 2,[10]

$$\Re_2 = \lim_{x_2 \to 0} \frac{f_2}{x_2} = H_{2,1} \tag{6-31}$$

where $H_{2,1}$ is Henry's constant for solute 2 in solvent 1.

However, for the solvent (component 1, present in excess) we obtain the same result as before:

$$\Re_1 = \lim_{x_2 \to 0} \frac{f_1}{x_1} = f_{\text{pure liquid 1}} \tag{6-32}$$

For the solute, the activity coefficient is given by

$$\gamma_2 = \frac{f_2}{x_2 H_{2,1}} \tag{6-33}$$

Substituting into Eq. (6-25) and differentiating with respect to temperature we obtain, as before,

$$\left(\frac{\partial \ln \gamma_2}{\partial T}\right)_{P,x} = -\frac{\bar{h}_2^E}{RT^2} \tag{6-34}$$

However, $\bar{h}_2^E$ now has a different meaning; it is given by

$$\bar{h}_2^E = \bar{h}_2 - \bar{h}_2^\infty \tag{6-35}$$

where $\bar{h}_2^\infty$ is the partial molar enthalpy of solute 2 in an infinitely dilute solution.

The effect of pressure on the activity coefficient of the solute is given by

$$\left(\frac{\partial \ln \gamma_2}{\partial P}\right)_{T,x} = \frac{\bar{v}_2}{RT} \tag{6-36}$$

---

[10] For any binary system, Henry's constant $H_{2,1}$ depends on both temperature and pressure. At temperatures remote from the solvent's critical temperature, unless the pressure is large, the effect of pressure on $H_{2,1}$ is often negligible (see Sec. 10.3).

where $\overline{v}_2$ is the partial molar volume of solute 2.[11] The derivatives with respect to temperature and pressure of the activity coefficient for component 1 (the solvent) are the same as those given by Eqs. (6-29) and (6-30).

For all components in a mixture, the partial molar excess Gibbs energies (and the activity coefficients) are related to one another by a fundamental relation known as the *Gibbs-Duhem equation*, discussed in Sec. 6.6.

6.4 Normalization of Activity Coefficients

It is convenient to define activity in such a way that for an ideal solution, activity is equal to the mole fraction or, equivalently, that the activity coefficient is equal to unity. Since we have distinguished between two types of ideality (one leading to Raoult's law and the other leading to Henry's law), it follows that activity coefficients may be normalized (that is, become unity) in two different ways.

If activity coefficients are defined with reference to an ideal solution in the sense of Raoult's law [Eq. (6-27)], then for each component i the normalization is

$$\gamma_i \to 1 \quad \text{as} \quad x_i \to 1 \tag{6-37}$$

Because this normalization holds for both solute and solvent, Eq. (6-37) is called the *symmetric* convention for normalization.

However, if activity coefficients are defined with reference to an ideal dilute solution [Eq. (6-31)], then

$$\begin{aligned} \gamma_1 \to 1 \quad &\text{as} \quad x_1 \to 1 \quad \text{(solvent)} \\ \gamma_2 \to 1 \quad &\text{as} \quad x_2 \to 0 \quad \text{(solute)} \end{aligned} \tag{6-38}$$

Because solute and solvent are not normalized in the same way, Eq. (6-38) gives the *unsymmetric* convention for normalization. To distinguish between symmetrically and unsymmetrically normalized activity coefficients, it is useful to denote with an

---

[11] The derivation of Eq. (6-36) requires that Henry's constant in Eq. (6-31) be evaluated at the temperature of the solution and at some pressure that usually is the saturation pressure of pure 1. The relationship between Henry's constant at system pressure P and Henry's constant at pressure P_i^s is given by

$$H_{2,1}(P) = H_{2,1}(P_1^s)\exp\left(\int_{P_1^s}^{P} \frac{\overline{v}_2^\infty}{RT} dP\right)$$

If, in Eq. (6-31), $H_{2,1}(P)$ is used rather than $H_{2,1}(P_1^s)$, then Eq. (6-36) must be replaced by

$$\left(\frac{\partial \ln \gamma_2}{\partial P}\right)_{T,x} = \frac{\overline{v}_2 - \overline{v}_2^\infty}{RT} \tag{6-36a}$$

where $\overline{v}_2^\infty$ is the partial molar volume of solute 2 at infinite dilution.

asterisk (*) the activity coefficient of a component that approaches unity as its mole fraction goes to zero. With this notation, Eq. (6-38) becomes

$$\begin{aligned} &\gamma_1 \to 1 \quad \text{as} \quad x_1 \to 1 \quad \text{(solvent)} \\ &\gamma_2^* \to 1 \quad \text{as} \quad x_2 \to 0 \quad \text{(solute)} \end{aligned} \tag{6-38a}$$

The two methods of normalization are illustrated in Fig. 6-1. In the dilute region ($x_2 \ll 1$), $\gamma_2^* = 1$ and the solution is ideal;[12] however, $\gamma_2 \neq 1$ and therefore, while the dilute solution is ideal in the sense of Henry's law, it is not ideal in the sense of Raoult's law.

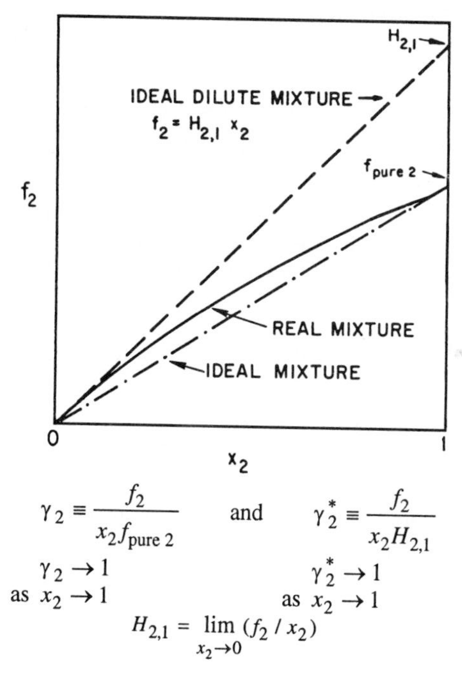

Figure 6-1 Normalization of activity coefficients.

Figure 6-2 shows experimental activity coefficient data for typical aqueous solutions that follow the symmetric convention ($\gamma_{methanol} \to 1$ as $x_{methanol} \to 1$) and the unsymmetric convention ($\gamma_{NaCl}^* \to 1$ as $x_{NaCl} \to 0$) for normalization of activity coefficients. Methanol activity coefficients become essentially constant for low solute concentrations. (The activity coefficient of methanol at infinite dilution in water at 25°C is

[12] It can readily be shown from the Gibbs-Duhem equation that for the region where $\gamma_2^* = 1$, $\gamma_1 = 1$.

$\gamma^{\infty} = 1.74$). In contrast, ideal behavior of aqueous NaCl solutions is approached as the solution becomes infinitely dilute.

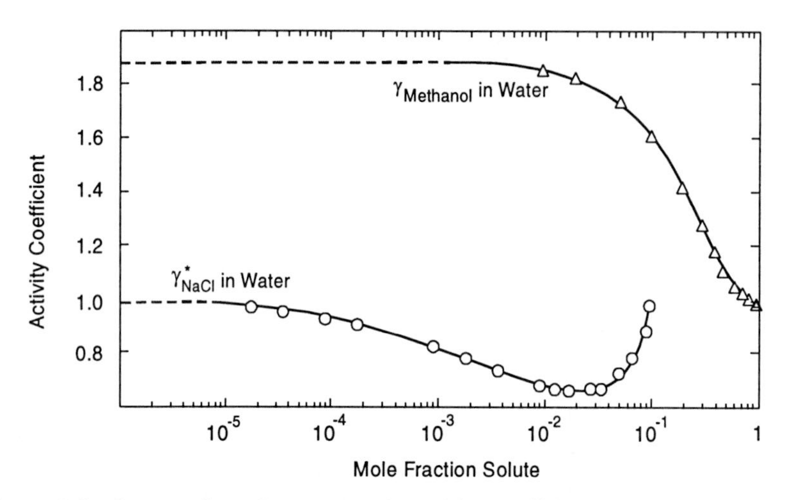

Figure 6-2 Symmetric and unsymmetric activity coefficient conventions. Experimental data at 25°C for the activity coefficients of methanol in water and sodium chloride in water (Ragal *et al.*, 1994). Solid lines are smooth data and dashed lines are extrapolations.

Symmetric normalization of activity coefficients is easily extended to solutions containing more than two components; the activity coefficient of any component approaches unity as its mole fraction goes to unity. However, as discussed elsewhere,[13] extension of unsymmetric normalization of activity coefficients to multicomponent solutions requires care.

In binary mixtures, activity coefficients that are normalized symmetrically are easily related to activity coefficients that are normalized unsymmetrically. The definitions of γ_2 and γ_2^* are

$$\gamma_2 = \frac{f_2}{x_2 f_{\text{pure 2}}} \tag{6-39}$$

$$\gamma_2^* = \frac{f_2}{x_2 H_{2,1}} \tag{6-40}$$

Therefore,

$$\frac{\gamma_2}{\gamma_2^*} = \frac{H_{2,1}}{f_{\text{pure 2}}} \tag{6-41}$$

[13] H. C. Van Ness and M. M Abbott, 1979, *AIChE J.*, 25: 645.

Because

$$\lim_{x_2 \to 0} \gamma_2^* = 1 \tag{6-42}$$

we obtain

$$\lim_{x_2 \to 0} \gamma_2 = \frac{H_{2,1}}{f_{\text{pure }2}} \tag{6-43}$$

Substitution in Eq. (6-41) gives

$$\frac{\gamma_2}{\gamma_2^*} = \lim_{x_2 \to 0} \gamma_2 \tag{6-44}$$

By a similar argument, we can also show that

$$\frac{\gamma_2^*}{\gamma_2} = \lim_{x_2 \to 1} \gamma_2^* \tag{6-45}$$

Both Eqs. (6-44) and (6-45) relate to each other the two activity coefficients of the solute, one normalized by the symmetric convention and the other by the unsymmetric convention. However, Eq. (6-44) is much more useful than Eq. (6-45) because the limit given on the right side of Eq. (6-44) corresponds to a real physical situation, whereas the limit on the right side of Eq. (6-45) corresponds to a situation that is hypothetical (physically unreal) whenever component 2 cannot exist as a pure liquid at the temperature of the solution.

6.5 Activity Coefficients from Excess Functions in Binary Mixtures

At a fixed temperature, the molar excess Gibbs energy g^E of a mixture depends on the composition of the mixture and, to a smaller extent, on pressure. At low or moderate pressures, well removed from critical conditions, the effect of pressure is negligible; it is therefore not considered in this section.[14]

We now consider a binary mixture where the excess properties are taken with reference to an ideal solution wherein the standard state for each component is the pure liquid at the temperature and pressure of the mixture. In that case, any expression for the molar excess Gibbs energy must obey the two boundary conditions:

[14] The effect of pressure on g^E is important only if the pressure is large, as discussed in Sec. 12.4.

$$g^E = 0 \quad \text{when} \quad x_1 = 0$$
$$g^E = 0 \quad \text{when} \quad x_2 = 0$$

Two-Suffix Margules Equation. The simplest nontrivial expression that obeys these boundary conditions is

$$g^E = Ax_1x_2 \tag{6-46}$$

where A is an empirical constant with units of energy, characteristic of components 1 and 2, that depends on the temperature but not on composition.

Equation (6-46) immediately gives expressions for activity coefficients γ_1 and γ_2 by substitution in the relation between activity coefficient and excess Gibbs energy [Eq. (6-25)]:

$$RT \ln \gamma_i = \bar{g}_i^E = \left(\frac{\partial n_T g^E}{\partial n_i} \right)_{T,P,n_j} \tag{6-47}$$

where n_i is the number of moles of i and n_T is the total number of moles. Remembering that $x_1 = n_1 / n_T$ and $x_2 = n_2 / n_T$, we obtain

$$\ln \gamma_1 = \frac{A}{RT} x_2^2 \tag{6-48}$$

$$\ln \gamma_2 = \frac{A}{RT} x_1^2 \tag{6-49}$$

Equations (6-48) and (6-49), often called the *two-suffix Margules equations*, provide a good representation for many simple liquid mixtures, i.e., for mixtures of molecules that are similar in size, shape, and chemical nature. The two equations are symmetric: When $\ln \gamma_1$ and $\ln \gamma_2$ are plotted against x_2 (or x_1), the two curves are mirror images. At infinite dilution, the activity coefficients of both components are equal:

$$\gamma_1^\infty \equiv \lim_{x_1 \to 0} \gamma_1 = \exp\left(\frac{A}{RT} \right) \tag{6-50}$$

$$\gamma_2^\infty \equiv \lim_{x_2 \to 0} \gamma_2 = \exp\left(\frac{A}{RT} \right) \tag{6-51}$$

Coefficient A may be positive or negative, and while it is in general a function of temperature, it frequently happens that for simple systems over a small temperature range, A is nearly constant. For example, vapor-liquid equilibrium data of Pool *et al.* (1962), for argon/oxygen are well represented by the two-suffix Margules equations as shown in Fig. 6-3. At 83.8 K, A is 148.1 J mol^{-1} and at 89.6 K, A is 141.0 J mol^{-1} For this simple system, A is a weak function of temperature.

Another binary system whose excess Gibbs energy is well represented by the two-suffix Margules equation is the benzene/cyclohexane system studied by Scatchard *et al.* (1939). Their results are also shown in Fig. 6-3. The variation of A with temperature is again not large, although not negligible; at 30, 40, and 50°C, the values of A are, respectively, 1268, 1185, and 1114 J mol^{-1}. In nonpolar solutions, A frequently falls with rising temperature.

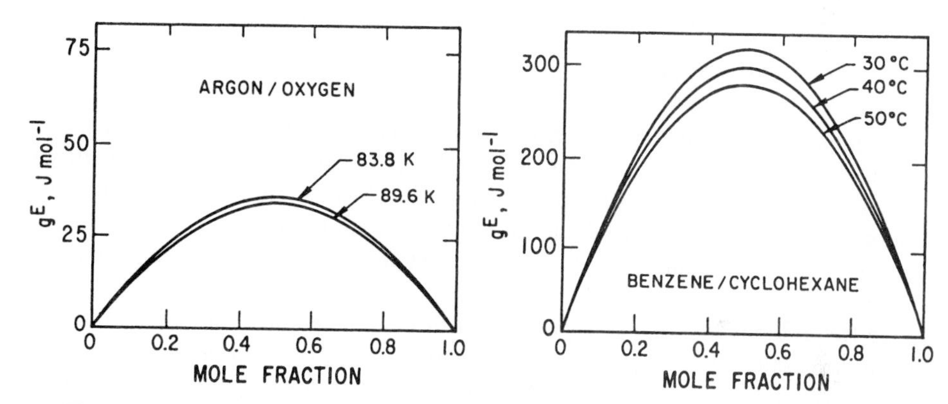

Figure 6-3 Applicability of two-suffix Margules equation to simple binary mixtures.

Equation (6-46) is a very simple relation; in the general case, a more complex equation is needed to represent adequately the excess Gibbs energy of a binary solution. Because the boundary conditions given just before Eq. (6-46) must be obeyed regardless of the complexity of the solution, one convenient extension of Eq. (6-46) is to write a series expansion:

$$g^E = x_1 x_2 [A + B(x_1 - x_2) + C(x_1 - x_2)^2 + D(x_1 - x_2)^3 + \ldots] \qquad (6\text{-}52)^{15}$$

[15] For binary mixtures containing one highly associating component, often better representation of the composition dependence of g^E (or any other excess function) is obtained by mutltiplying the Redlich-Kister expansion by a skewing factor:

$$g^E = x_1 x_2 \sum_j A_j (x_1 - x_2)^j \left[\frac{1}{1 + K(x_1 - x_2)} \right]$$

where K is a constant to be determined from experimental data. If $K = 0$, Eq. (6-52) is recovered.

where B, C, D, ... are additional, temperature-dependent parameters that must be determined from experimental data. Equation (6-55) is the *Redlich-Kister expansion* and, upon using Eq. (6-47), we obtain these expressions for the activity coefficients

$$RT\ln\gamma_1 = a^{(1)}x_2^2 + b^{(1)}x_2^3 + c^{(1)}x_2^4 + d^{(1)}x_2^5 + \ldots \tag{6-53}$$

$$RT\ln\gamma_2 = a^{(2)}x_1^2 + b^{(2)}x_1^3 + c^{(2)}x_1^4 + d^{(2)}x_1^5 + \ldots \tag{6-54}$$

where

$$a^{(1)} = A + 3B + 5C + 7D \qquad a^{(2)} = A - 3B - 5C - 7D$$

$$b^{(1)} = -4(B + 4C + 9D) \qquad b^{(2)} = 4(B - 4C + 9D)$$

$$c^{(1)} = 12(C + 5D) \qquad c^{(2)} = 12(C - 5D)$$

$$d^{(1)} = -32D \qquad d^{(2)} = 32D$$

The number of parameters (A, B, C, ...) that should be used to represent the experimental data depends on the molecular complexity of the solution, on the quality of the data, and on the number of data points available. Typical vapor-liquid equilibrium data reported in the literature justify no more than two or at most three constants; very accurate and extensive data are needed to warrant the use of four or more empirical parameters.

The Redlich-Kister expansion provides a flexible algebraic expression for representing the excess Gibbs energy of a liquid mixture. The first term in the expansion is symmetric in x and gives a parabola when g^E is plotted against x. The odd-powered correction terms [first (B), third (D), ...] are asymmetric in x and therefore tend to skew the parabola either to the left or right. The even-powered correction terms [second (C), fourth (E), ...] are symmetric in x and tend to flatten or sharpen the parabola. To illustrate, we show in Fig. 6-4 the first three terms of the Redlich-Kister expansion for unit value of the coefficients A, B, and C.

Redlich-Kister equations provide not only a convenient method for representing liquid-phase activity coefficients, but also for classifying different types of liquid solutions. From experimental data, γ_1 and γ_2 are calculated and these are then plotted as $\log(\gamma_1/\gamma_2)$ versus x_1, as shown in Figs. 6-5, 6-6, and 6-7 taken from Redlich *et al.* (1952). Strictly, Eqs. (6-52), (6-53), and (6-54) apply to isothermal data, but they are often applied to isobaric data; provided that the temperature does not change much with composition, this practice does not introduce large errors. From Eqs. (6-53) and (6-54) we obtain, after some rearrangement,

$$RT\ln\frac{\gamma_1}{\gamma_2} = A(x_2 - x_1) + B(6x_1x_2 - 1) + C(x_1 - x_2)(8x_1x_2 - 1) + D(x_1 - x_2)^2(10x_1x_2 - 1) + \ldots \tag{6-55}$$

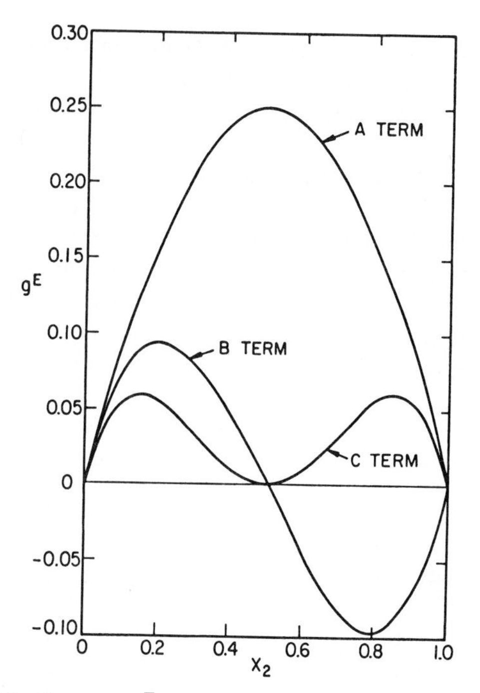

Figure 6-4 Contributions to g^E in Redlich-Kister equation (for $A = B = C = 1$).

For simple solutions, $B = C = D = \ldots = 0$, and a plot of $\log(\gamma_1/\gamma_2)$ versus x_1 gives a straight line, as shown in Fig. 6-5 for the system *n*-hexane/toluene. In this case a good representation of the data is obtained with $A/RT = 0.352$.

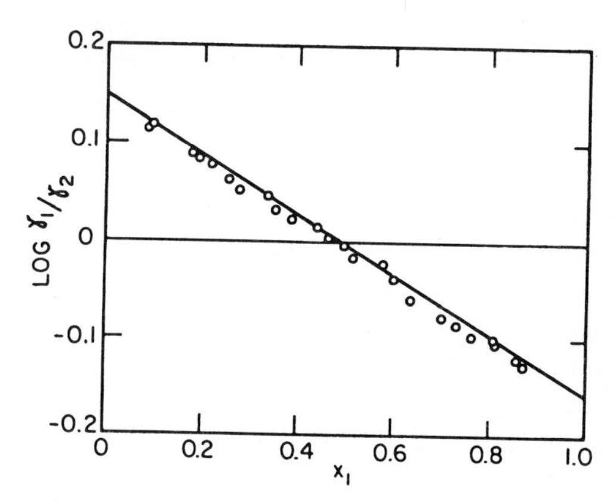

Figure 6-5 Activity-coefficient ratio for a simple mixture. Experimental data for *n*-hexane (1)/toluene (2) at 1.013 bar. The line is drawn to satisfy the area (consistency) test given by Eq. (6-92).

Data for a somewhat more complicated solution are shown in Fig. 6-6. In this case the plot has some curvature and two parameters are required to represent it adequately; they are A/RT = 0.433 and B/RT = 0.104. Although the systems in Figs. 6-5 and 6-6 are similar (both are mixtures of a paraffinic and an aromatic hydrocarbon), it may appear surprising that two parameters are needed for the second system, while only one is required for the first. It is likely that two parameters are required for the second system because the difference in the molecular sizes of the two components is much larger in the isooctane/benzene system than it is in the *n*-hexane/toluene system. At 25°C, the ratio of molar volumes (paraffin to aromatic) in the system containing benzene is 1.86, while it is only 1.23 in the system containing toluene. The effect of molecular size on the representation of activity coefficients is shown more clearly by Wohl's expansion discussed in Sec. 6.10.

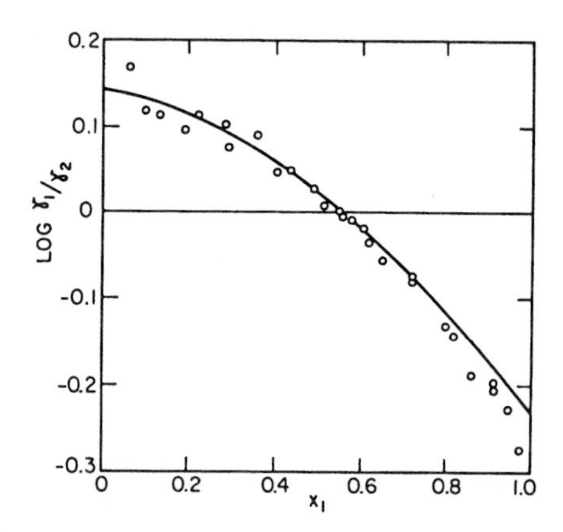

Figure 6-6 Activity-coefficient ratio for a mixture of intermediate complexity. Experimental data for benzene (1)/isooctane (2) at total pressures ranging from 0.981 to 1.013 bar.

Finally, Fig. 6-7 shows data for a highly complex solution. The plot is not only curved but has a point of inflection; four parameters are required to give an adequate representation. This mixture, containing an alcohol and a saturated hydrocarbon, is complex because the degree of hydrogen bonding of the alcohol is strongly dependent on the composition, especially in the region dilute with respect to alcohol.

The number of parameters in Eq. (6-55) required to represent activity coefficients of a binary mixture gives an indication of the apparent complexity of the mixture, thereby providing a means for classification. If the number of required parameters is large (four or more parameters), the mixture is classified as a complex solution, and if it is small (one parameter), the mixture is classified as a simple solution. Most solutions of nonelectrolytes commonly encountered in chemical engineering are of intermediate complexity, requiring two or three parameters in the Redlich-Kister expansion.

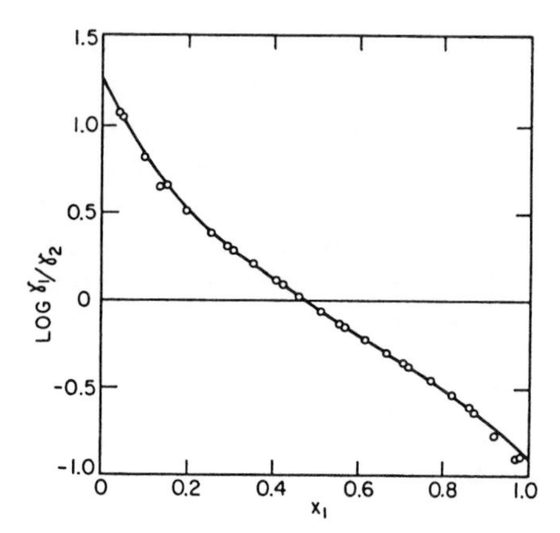

Figure 6-7 Activity-coefficient ratio for a highly complex mixture. Experimental data for ethanol (1)/methylcyclohexane (2) in the region 30-35°C.

Classification of solutions as determined by the number of required parameters is necessarily arbitrary. When a good fit of experimental data is obtained with one or two parameters, we cannot necessarily conclude that the mixture is truly simple in a molecular sense; in many cases a one- or two-parameter equation may fortuitously provide an adequate representation. For example, binary mixtures of two alcohols (such as methanol/ethanol) are certainly highly complex because of the various types of hydrogen bonds that can exist in such mixtures; nevertheless, it is frequently possible to represent the excess Gibbs energies of such solutions with a one- or two-parameter equation.

Similarly, the system acetic acid/water is surely complex. However, as shown in Fig. 6-8, the excess Gibbs energy is nearly a parabolic function of the composition. This apparent simplicity is due to cancellation; when we look at excess enthalpy and excess entropy, the mixture's complexity becomes apparent.

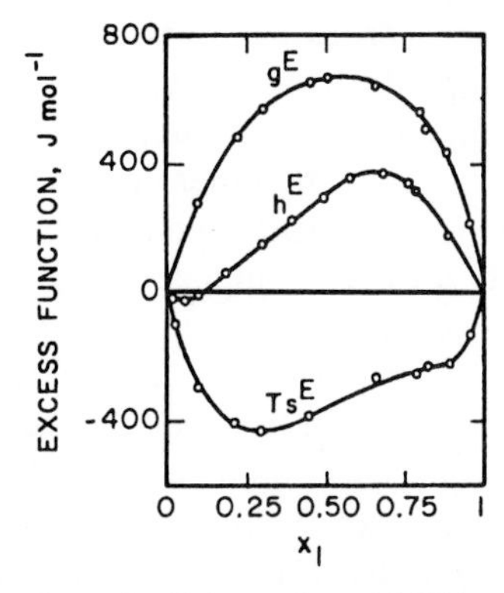

Figure 6-8 Excess functions for the acetic acid (1)/water (2) system at 25°C (R. Haase *et al.*, 1973, *Z. Naturforschung,* 28a: 1740).

Numerous equations have been proposed for expressing analytically the composition dependence of the excess Gibbs energies of binary mixtures. Most of these equations are empirical and we mention several of the better known ones in Sec. 6.10 but some have at least a little theoretical basis and we discuss a few of these in Chap. 7. However, before continuing our presentation of equations for representing excess Gibbs energies of nonideal liquid mixtures, we first want to discuss how the Gibbs-Duhem equation can be used to reduce experimental effort and to test experimental data for thermodynamic consistency.

6.6 Activity Coefficients for One Component from Those of the Other Components

In a mixture, the partial molar properties of the components are related to one another by one of the most useful equations in thermodynamics, the *Gibbs-Duhem equation*. This equation says that at constant temperature and pressure,

$$\sum_i x_i d\overline{m}_i = 0 \tag{6-56}$$

where $\overline{m}_i$ is any partial molar property. Equation (6-56) holds for ideal as well as real solutions and can be rewritten in terms of excess partial properties:

$$\sum_i x_i d\overline{m}_i^E = 0 \tag{6-57}$$

A derivation and more detailed discussion of this equation is given in App. D.

There are two important applications of Eq. (6-57). First, in the absence of complete experimental data on the properties of a mixture, Eq. (6-57) frequently may be used to calculate additional properties; for example, in a binary solution, if experimental measurements over a range of concentration yield activity coefficients of only one component, activity coefficients of the other component can be computed for the same concentration range. Second, if experimental data are available for a directly measured partial molar property for each component over a range of composition, it is then possible to check the data for thermodynamic consistency; if the data satisfy the Gibbs-Duhem equation, they are thermodynamically consistent and it is likely that they are reliable, but if they do not, it is certain that they are incorrect.

While the Gibbs-Duhem equation is applicable to all partial excess properties, it is most useful for the partial molar excess Gibbs energy that is directly related to the activity coefficient by Eq. (6-25). In terms of activity coefficients, Eq. (6-57) is

$$\sum_i x_i d\ln\gamma_i = 0 \qquad \text{(constant } T \text{ and } P) \tag{6-58}$$

Equation (6-58) is a differential relation between the activity coefficients of all the components in the solution. Hence, in a solution containing m components, data for activity coefficients of $m-1$ components may be used to compute the activity coefficient of the mth component. To illustrate, consider the simplest case, i.e., a binary solution for which isothermal data are available for one component and where the pressure is sufficiently low to permit neglect of the effect of pressure on the liquid-phase activity coefficient. For this case, the Gibbs-Duhem equation may be written

$$x_1 \frac{d\ln\gamma_1}{dx_1} = x_2 \frac{d\ln\gamma_2}{dx_2} \tag{6-59}$$

Assume now that data have been obtained for γ_1 for various x_1. By graphical integration of Eq. (6-59) one may then obtain values of γ_2. A simpler procedure, less exact in principle but much easier to use, is to curve-fit the data for γ_1 to an algebraic expression using x_1 (or x_2) as the independent variable. Once such an expression has been obtained, the integration of Eq. (6-59) may be performed analytically. For this purpose it is convenient to rewrite Eq. (6-59):

$$\frac{d\ln(\gamma_1/\gamma_2)}{dx_2} = \frac{1}{x_2}\frac{d\ln\gamma_1}{dx_2} \tag{6-60}$$

Assume now, as first suggested in 1895 by Margules, that the data for γ_1 can be represented by an empirical equation of the form[16]

$$\ln\gamma_1 = \sum_k \alpha_k x_2^{\beta_k} \qquad (\beta_k > 1) \tag{6-61}$$

where α_k and β_k are empirical constants to be determined from the data. Substituting (6-61) into (6-60) yields

$$\frac{d\ln(\gamma_1/\gamma_2)}{dx_2} = \sum_k \alpha_k \beta_k x_2^{\beta_k - 2} \tag{6-62}$$

Integrating Eq. (6-62) gives

[16] We need $\beta_k > 1$ to avoid singularities in $\ln\gamma_2$ when $x_2 = 1$.

$$\ln\gamma_2 = \ln\gamma_1 - \sum_k \frac{\alpha_k\beta_k}{\beta_k - 1} x_2^{\beta_k - 1} - I \tag{6-63}$$

where I is a constant of integration. To eliminate $\ln\gamma_1$ in Eq. (6-63), we substitute Eq. (6-61):

$$\ln\gamma_2 = \sum_k \alpha_k x_2^{\beta_k} - \sum_k \frac{\alpha_k\beta_k}{\beta_k - 1} x_2^{\beta_k - 1} - I \tag{6-64}$$

To evaluate I it is necessary to impose a suitable boundary condition. If component 2 can exist as a pure liquid at the temperature of the solution, then it is common to use pure liquid component 2 at that temperature as the standard state for γ_2; in that case,

$$\gamma_2 = 1 \quad \text{when} \quad x_2 = 1.$$

The constant of integration then is

$$I = \sum_k \alpha_k - \sum_k \frac{\alpha_k\beta_k}{\beta_k - 1} \tag{6-65}$$

and the expression for $\ln\gamma_2$ becomes

$$\ln\gamma_2 = \sum_k \alpha_k x_2^{\beta_k} - \sum_k \frac{\alpha_k}{\beta_k - 1}(\beta_k x_2^{\beta_k - 1} - 1) \tag{6-66}$$

Equation (6-66) is a general relation and it is strictly for convenience that one customarily uses only positive integers for β_k. To illustrate the use of Eq. (6-66), suppose that the data for γ_1 can be adequately represented by Eq. (6-61) terminated after the fourth term (k = 2, 3, 4) with $\beta_k = k$. Equation (6-61) becomes

$$\ln\gamma_1 = \alpha_2 x_2^2 + \alpha_3 x_2^3 + \alpha_4 x_2^4 \tag{6-67}$$

known as the *four-suffix Margules equation*.[17] Coefficients α_2, α_3, and α_4 must be found from the experimental data that give γ_1 as a function of the mole fraction. When this four-suffix Margules equation for $\ln\gamma_1$ is substituted into Eq. (6-66), the result for γ_2 is

$$\ln\gamma_2 = (\alpha_2 + \tfrac{3}{2}\alpha_3 + 2\alpha_4)x_1^2 - (\alpha_3 + \tfrac{8}{3}\alpha_4)x_1^3 + \alpha_4 x_1^4 \tag{6-68}$$

[17] The n-suffix Margules equation gives $\ln\gamma_1$ as a polynomial in x_2 of degree n.

The important feature of Eq. (6.68) is that γ_2 is given in terms of constants that are determined exclusively from data for γ_1.

In a binary system, calculating the activity coefficient of one component from data for the other is a common practice whenever the two components in the solution differ markedly in volatility. In that event, the measurements frequently give the activity coefficient of only the more volatile component and the activity of the less volatile component is found from the Gibbs-Duhem equation. For example, if one wished to have information on the thermodynamic properties of some high-boiling liquid (such as a polymer) dissolved in, say, benzene, near room temperature, then the easiest procedure would be to measure the activity (partial pressure) of the benzene in the solution and to compute the activity of the other component from the Gibbs-Duhem equation as outlined above; in this case it would not be practical to measure the extremely small partial pressure of the high-boiling component over the solution.

While the binary case is the simplest, the method just discussed can be extended to systems of any number of components. In these more complicated cases the computational work is greater but the theoretical principles are the same.

In carrying out numerical work, it is important that activity coefficients be calculated as rigorously as possible from the equilibrium data. We recall the definition of activity coefficient:

$$\gamma_i \equiv \frac{f_i}{x_i f_i^0} \tag{6-69}$$

Quantities f_i and f_i^0 must be computed with care; in vapor-liquid equilibrium, gas-phase corrections for both of these quantities are frequently important. For solutions of liquids, it is useful to use the pure component as the standard state and, to simplify calculations, it is common practice to evaluate γ_i from the data by the simplified expression

$$\gamma_i = \frac{y_i P}{x_i P_i^s} \tag{6-70}$$

where P_i^s is the saturation (vapor) pressure of pure i. Equation (6-70), however, is only an approximate form of Eq. (6-69) because it neglects gas-phase corrections and the Poynting factor. In some cases, the approximation is justified but before Eq. (6-70) is used, it is important to inquire if the simplifying assumptions apply to the case under consideration. For mixtures of strongly polar or hydrogen-bonding components, or for mixtures at cryogenic temperatures, gas-phase corrections may be significant even at pressures near or below 1 bar.

The Gibbs-Duhem equation interrelates the activity coefficients of all the components in a mixture. In a binary mixture, $\ln\gamma_1$ and $\ln\gamma_2$ are not independent, as indicated by Eq. (6-59) and the subsequent relations, Eqs. (6-67) and (6-68). The interrelation of the activity coefficients can be used to reduce experimental effort, as indicated in the next two sections, and to test experimental data for thermodynamic consistency, as indicated in Sec. 6.9.

6.7 Partial Pressures from Isothermal Total-Pressure Data

Complete description of vapor-liquid equilibrium for a system gives equilibrium compositions of both phases as well as temperature and total pressure. In a typical experimental investigation, temperature or total pressure is held constant; in a system of m components, complete measurements require that for each equilibrium state, data must be obtained for either the temperature or pressure and for $2(m - 1)$ mole fractions. Even in a binary system this represents a significant experimental effort; it is advantageous to reduce this effort by utilizing the Gibbs-Duhem equation for calculation of at least some of the desired information.

We consider a procedure for calculating partial pressures from isothermal total-pressure data for binary systems. According to this procedure, total pressures are measured as a function of the composition of one of the phases (usually the liquid phase) and no measurements are made of the composition of the other phase. Instead, the composition of the other phase is calculated from the total-pressure data with the help of the Gibbs-Duhem equation. The necessary experimental work is thereby much reduced.

Numerous techniques have been proposed for making this kind of calculation and it is not necessary to review all of them here. However, to indicate the essentials of these techniques, one representative and useful procedure is given below.

Barker's Numerical Method.[18] The total pressure for a binary system is written

$$P = \gamma_1 x_1 P_1^{s'} + \gamma_2 x_2 P_2^{s'} \tag{6-71}$$

where $P_i^{s'}$ is the "corrected" vapor pressure of component i:

$$P_1^{s'} \equiv P_1^s \exp\left[\frac{(v_1^L - B_{11})(P - P_1^s) - P\delta_{12}y_2^2}{RT}\right] \tag{6-72}$$

$$P_2^{s'} \equiv P_2^s \exp\left[\frac{(v_2^L - B_{22})(P - P_2^s) - P\delta_{12}y_1^2}{RT}\right] \tag{6-73}$$

[18] Barker (1953); Abbott and Van Ness (1977).

where δ_{12} is related to the second virial coefficients by

$$\delta_{12} \equiv 2B_{12} - B_{11} - B_{22} \tag{6-74}$$

At constant temperature, activity coefficients γ_1 and γ_2 are functions only of composition.

Equation (6-71) is rigorous provided we assume that the vapor phase of the mixture, as well as the vapors in equilibrium with the pure components, are adequately described by the volume-explicit virial equation terminated after the second virial coefficient; that the pure-component liquid volumes are incompressible over the pressure range in question; and that the liquid partial molar volume of each component is invariant with composition. The standard states for the activity coefficients in Eq. (6-71) are the pure components at the same temperature and pressure as those of the mixture.[19]

Barker's method is used to reduce experimental data that give the variation of total pressure with liquid composition at constant temperature. One further relation is needed in addition to Eqs. (6-71) to (6-74), and that is an equation relating the activity coefficients to mole fractions. This relation may contain any desired number of undetermined numerical coefficients that are then found from the total pressure data as shown below. For example, suppose we assume that

$$\ln \gamma_1 = \alpha x_2^2 + \beta x_2^3 \tag{6-77}$$

where α and β are unknown constants. Then, from the Gibbs-Duhem equation [Eq. (6-68)] it follows that

$$\ln \gamma_2 = \left(\alpha + \frac{3}{2}\beta\right) x_1^2 - \beta x_1^3 \tag{6-78}$$

Equations (6-71) to (6-78) contain only two unknowns, α and β. (It is assumed that values for the quantities v^L, B, P^s, and δ_{12} are available. It is true that y is also unknown but once α and β are known, y can be determined.)

In principle, Eqs. (6-71) to (6-78) could yield α and β using only two points on the experimental P-x curve. In practice, however, more than two points are required; we prefer to utilize all reliable experimental points and then optimize the values of α

---

[19] From these assumptions it follows that

$$\ln \gamma_1 = \ln \frac{y_1 P}{x_1 P_1^s} + \frac{(B_{11} - v_1^L)(P - P_1^s)}{RT} + \frac{P y_2^2 \delta_{12}}{RT} \tag{6-75}$$

and

$$\ln \gamma_2 = \ln \frac{y_2 P}{x_2 P_2^s} + \frac{(B_{22} - v_2^L)(P - P_2^s)}{RT} + \frac{P y_1^2 \delta_{12}}{RT} \tag{6-76}$$

and β to give the best agreement between the observed total pressure curve and that calculated with parameters α and β.

The calculations are iterative because y_1 and y_2 can only be calculated after α and β have been determined; the method of successive approximations must be used. In the first approximation, y_1 and y_2 are set equal to zero in Eqs. (6-72) and (6-73). Then α and β are found, and immediately thereafter y_1 and y_2 are computed (from the first approximation of the parameters α and β) using Eqs. (6-75) to (6-78). The entire calculation is then repeated except that the new values of y_1 and y_2 are now used in Eqs. (6-72) and (6-73). We proceed in this way until the assumed and calculated values of y_1 and y_2 are in agreement; usually three or four successive approximations are sufficient. The form of Eqs. (6-77) and (6-78) is arbitrary; we may use any desired set of equations with any desired number of constants, provided that the two equations satisfy the Gibbs-Duhem equation.

Although Barker's numerical method is too complicated for manual calculation, it can easily be programmed for a computer capable of rapidly transforming isothermal P-x data to isothermal y-x data. An illustration of Barker's method is provided by Hermsen (1963) who measured isothermal total vapor pressures for the benzene/cyclopentane system at 25, 35, and 45°C. Hermsen assumed that the excess Gibbs energy of this system is described by a two-parameter expansion of the Redlich-Kister type:[20]

$$\frac{g^E}{RT} = x_1 x_2 [A' + B'(x_1 - x_2)] \tag{6-79}$$

where subscript 1 refers to benzene and subscript 2 to cyclopentane. Activity coefficients are obtained from Eq. (6-79) by differentiation according to Eq. (6-25). They are

$$\ln \gamma_1 = (A' + 3B')x_2^2 - 4B' x_2^3 \tag{6-80}$$

and

$$\ln \gamma_2 = (A' - 3B')x_1^2 + 4B' x_1^3 \tag{6-81}$$

As indicated by Eqs. (6-72) and (6-73), Barker's method requires the molar liquid volumes of the pure liquids and the three second virial coefficients B_{11}, B_{22}, and B_{12}. For the benzene/cyclopentane system these are given in Table 6-1.

Hermsen's calculated and experimental results are given in Table 6-2a. At a given temperature, the total pressure and the liquid composition were determined. Vapor-phase compositions were not measured. From the experimental measurements, optimum values of A' and B' were found such that the calculated total pressures [Eq. (6-71)] reproduce as closely as possible the experimental ones. The optimum values are given in Table 6-2b.

[20] Coefficients A' and B' are dimensionless.

Table 6-1 Second virial coefficients and liquid molar volumes* for benzene (1) and cyclopentane (2) (cm^3 mol^{-1}).

Temp. (°C)	v_1^L	v_2^L	B_{11}	B_{22}	B_{12}	δ_{12}
25	89.39	94.71	-1314	-1054	-1176	16
35	90.49	95.98	-1224	-983	-1096	15
45	91.65	97.29	-1143	-919	-1024	14

* Hermsen (1963).

Table 6-2a Experimental and calculated results for the system benzene (1)/cyclopentane (2).*

		Pressure (bar)				
x_1	y_1 (calc.)	Exp.	Calc.	γ_1	γ_2	g^E (J mol^{-1})
			25°C			
0.1417	0.0655	0.3921	0.3921	1.408	1.010	142
0.2945	0.1324	0.3578	0.3580	1.253	1.043	239
0.4362	0.1984	0.3244	0.3246	1.151	1.095	280
0.5166	0.2410	0.3044	0.3055	1.108	1.135	282
0.5625	0.2682	0.2920	0.2918	1.087	1.160	277
0.8465	0.5510	0.1974	0.1976	1.010	1.380	143
			35°C			
0.1417	0.0684	0.5740	0.5739	1.375	1.009	136
0.2945	0.1391	0.5253	0.5250	1.234	1.040	230
0.4362	0.2091	0.4767	0.4769	1.140	1.088	270
0.5166	0.2543	0.4473	0.4475	1.100	1.125	272
0.5625	0.2829	0.4299	0.4298	1.080	1.148	267
0.8465	0.5732	0.2961	0.2962	1.009	1.350	138
			45°C			
0.1417	0.0697	0.8161	0.8164	1.353	1.009	134
0.2945	0.1421	0.7464	0.7471	1.219	1.039	226
0.4362	0.2142	0.6783	0.6788	1.130	1.085	263
0.5166	0.2607	0.6374	0.6371	1.092	1.119	265
0.5625	0.2903	0.6131	0.6120	1.074	1.141	260
0.8465	0.5862	0.4239	0.4244	1.008	1.325	133

* Hermsen (1963).

Table 6-2b Constants in Eq. (6-79) for the system benzene (1)/cyclopentane (2).*

	25°C	35°C	45°C
A'	0.45598	0.42463	0.40085
B'	-0.01815	-0.01627	0.02186

* Hermsen (1963).

In the benzene/cyclopentane system, deviations from ideality are not large and only weakly asymmetric; therefore, a two-parameter expression for the excess Gibbs energy is sufficient. For other systems, where the excess Gibbs energy is large or strongly asymmetric, it is advantageous to use a more flexible expression for the excess Gibbs energy, or, for example, to include higher terms in the Redlich-Kister expansion. To illustrate, Orye (1965) has measured total pressures for five binary systems containing a hydrocarbon and a polar solvent and he has reduced these data with Barker's method, using a three-parameter Redlich-Kister expansion for the excess Gibbs energy. Orye's results at 45°C are given in Table 6-3, and a typical plot of total and partial pressures is shown for one of the systems in Fig. 6-9. To justify three parameters, the number of experimental determinations of total pressure must necessarily be larger than that required to justify only two parameters. Whereas measurements for six compositions at any one temperature were sufficient for Hermsen to fix two parameters for the moderately nonideal benzene/cyclopentane system, Orye's measurements included about 15 compositions to specify three parameters for the more strongly nonideal hydrocarbon/polar-solvent systems.

Table 6-3 Excess Gibbs energies of five binary systems obtained from total-pressure measurements at 45°C.*

$$\frac{g^E}{RT} = x_1 x_2 [A' + B'(x_1 - x_2) + C'(x_1 - x_2)^2]$$

System (1)/(2)	A'	B'	C'
Toluene/acetonitrile	1.17975	-0.05992	0.12786
Toluene/2,3-butanedione	0.79810	0.01763	-0.01023
Toluene/acetone	0.66365	-0.00477	0.00227
Toluene /nitroethane	0.76366	0.07025	0.06190
Methylcyclohexane/acetone	1.69070	-0.00010	0.18324

* Orye, 1965.

Total-pressure measurements at constant temperature are particularly convenient for obtaining excess Gibbs energies of those binary liquid mixtures whose components have similar volatilities. Such measurements can be made rapidly; with care and experience, they can be very accurate.

A detailed discussion of data reduction is given by Van Ness and coworkers (Byer *et al.*, 1973; Abbott and Van Ness, 1975) who emphasize that isothermal P-x measurements often provide the best source of experimental data. For many typical mixtures, it is not necessary to measure y.

For data reduction, Van Ness and coworkers used a four-suffix Margules equation written in the form

$$\frac{g^E}{RTx_1x_2} = A'x_2 + B'x_1 - D'x_1x_2 \tag{6-82}$$

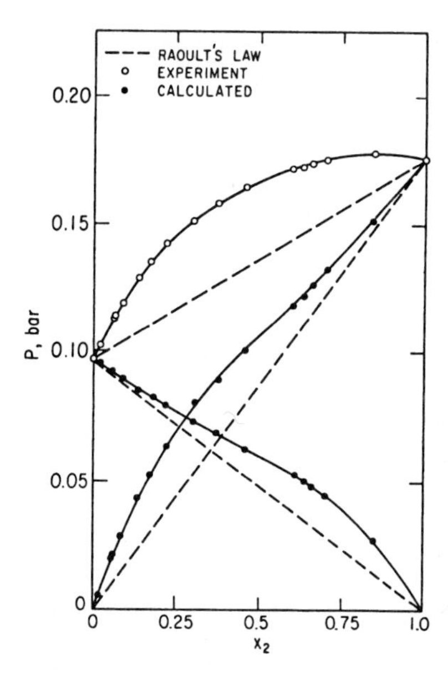

Figure 6-9 Total and partial pressures of toluene (1)/2,3-butanedione (2) at 45°C.

From *P*-*x* measurements for 14 binary systems at 30°C, constants A', B', and D' were obtained; results are shown in Table 6-4. The 14 systems represent a wide variety with some systems showing strong positive deviations from Raoult's law ($g^E > 0$) and others showing strong negative deviations ($g^E < 0$). One system (No. 6) exhibits ideal behavior.

Table 6-4 Excess Gibbs energy from *P*-*x* measurements at 30°C.* [Constants in Eq. (6-82)].

System (1)/(2)	A'	B'	D'
1. Carbon tetrachloride/THF†	-0.25704	-0.18188	0.04760
2. Chloroform/THF	-1.39352	-1.58092	0.58606
3. Dichloromethane/THF	-0.93341	0.87287	0.22232
4. Carbon tetrachloride/furan	0.28639	0.27034	0.01189
5. Chloroform/furan	-0.08350	-0.11890	0.02847
6. Dichloromethane/furan	0	0	0
7. THF/furan	-0.39970	-0.37125	-0.06410
8. Dichloromethane/methyl acetate	-0.42260	-0.63028	0.27851
9. Dichloromethane/acetone	0.58905	-0.76638	-0.11940
10. Dichloromethane/1,4-dioxane	-0.63128	-0.95516	-0.06863
11. Chloroform/1,4-dioxane	-0.75571	-1.58181	0.12739
12. Pyridine/acetone	0.19441	0.20447	0.02998
13. Pyridine/chloroform	-1.16104	-0.70714	0.37199
14. Pyridine/dichloromethane	-0.57919	-0.44873	0.03523

* Byer *et al.* (1973); † THF = tetrahydrofuran.

Figure 6-10 indicates a particularly convenient way to illustrate the nonidealities of the 14 systems; the coordinates of the plot are suggested by the Redlich-Kister expansion [Eq. (6-52)]: for simple mixtures, the plot gives a straight horizontal line; for slightly complex mixtures, the line is straight but not horizontal; and for complex mixtures, the line is curved. These three cases correspond, respectively, to those shown in Figs. 6-5, 6-6, and 6-7.

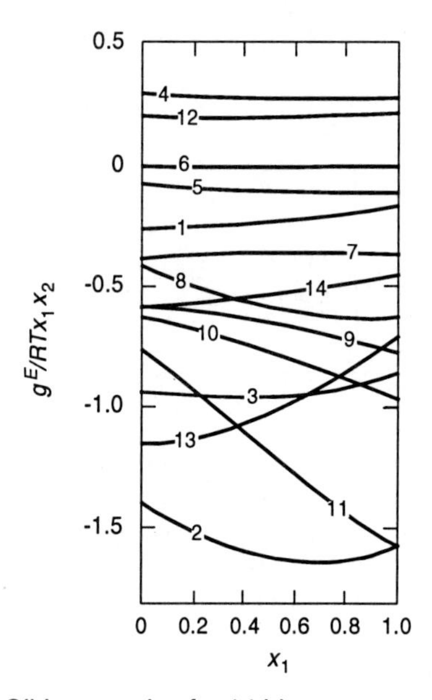

Figure 6-10 Excess Gibbs energies for 14 binary systems listed in Table 6-4.

6.8 Partial Pressures from Isobaric Boiling-Point Data

As indicated in the preceding section, the Gibbs-Duhem equation can be used to convert isothermal P-x data for a binary system into y-x data; similarly, it can be used to convert isobaric T-x data into y-x data. However, the latter calculation is often less useful because the Gibbs-Duhem equation for an isobaric, nonisothermal system (see App. D) contains a correction term proportional to the enthalpy of mixing and this correction term is not always negligible. The isothermal, nonisobaric Gibbs-Duhem equation also contains a correction term (that is proportional to the volume change on mixing), but in mixtures of two liquids at low pressures this term may safely be neglected.

Rigorous reduction of isobaric T-x data, therefore, requires data for the enthalpy of mixing at the boiling point of the solution. Such data are almost never at hand and if the object of a particular study is to obtain accurate isobaric y-x data, then it is usually easier to measure the y-x data directly in an equilibrium still than to obtain enthalpy-of-mixing data in addition to the *T-x* data. However, for approximate results, sufficient for some practical applications, boiling-point determinations may be useful because of experimental simplicity; it is a simple matter to place a liquid mixture in a manostated flask and to measure the boiling temperature. We now discuss briefly how isobaric *T-x* data may be reduced to yield an isobaric *y-x* diagram.

We assume that in the Gibbs-Duhem equation the correction term for nonisothermal conditions may be neglected. Further, we assume for simplicity that the gas phase is ideal and that the two-suffix Margules equation is adequate for the relation between activity coefficient and mole fraction:

$$RT \ln \gamma_1 = A x_2^2 \tag{6-83}$$

We assume that A is a constant independent of temperature, pressure, and composition. The Gibbs-Duhem equation then gives

$$RT \ln \gamma_2 = A x_1^2 \tag{6-84}$$

The problem now is to find parameter A from the *T-x* data. Once A is known, it is a simple matter to calculate the *y-x* diagram. To be consistent with the approximate nature of this calculation, we here use the simplified definition of the activity coefficient as given by Eq. (6-70).

To find A, we write

$$P = \text{constant} = x_1 P_1^s \exp\left(\frac{A}{RT} x_2^2\right) + x_2 P_2^s \exp\left(\frac{A}{RT} x_1^2\right) \tag{6-85}$$

From the *T-x* data and from the vapor-pressure curves of the pure components, everything in Eq. (6-85) is known except A. Unfortunately, Eq. (6-85) is not explicit in A but for any point on the *T-x* diagram, a value of A may be found by trial and error. Thus, in principle, the boiling point for one particular mixture of known composition is sufficient to determine A. However, to obtain a more representative value, it is preferable to measure boiling points for several compositions of the mixture, to calculate a value for A for each boiling point, and then either to use an optimum average value in the subsequent calculations or, if the data warrant doing so, to reject Eqs. (6-83) and (6-84) and, instead, to use a two- (or three-) parameter equation for relating activity coefficients to composition.

To illustrate, we consider boiling-point data for the diisopropyl ether/2-propanol system obtained at atmospheric pressure. Table 6-5 gives experimental boiling points. From Eq. (6-85) we find an average value of $A = (3.18 \pm 0.13)$ kJ mol^{-1}. When this average value is used, a *y-x* diagram is obtained, as shown by the line in Fig. 6-11. The points represent the experimental *y-x* data of Miller and Bliss (1940). In this case, agreement between the observed and calculated y-x diagram is good, but one should not assume that this will always be the case.

Table 6-5 Boiling points of diisopropyl ether/2-propanol mixtures at 1.013 bar.*

Mol % ether in liquid	Temperature (°C)	Mol % ether in liquid	Temperature (°C)
0	82.30	58.4	66.77
8.4	76.02	73.2	66.20
18.0	72.48	75.4	66.18
28.2	69.93	84.6	66.31
38.5	68.18	89.1	66.33
43.6	67.79	91.8	66.77
47.7	67.56	98.9	67.73
52.0	67.19	100.0	68.50

* Miller and Bliss (1940).

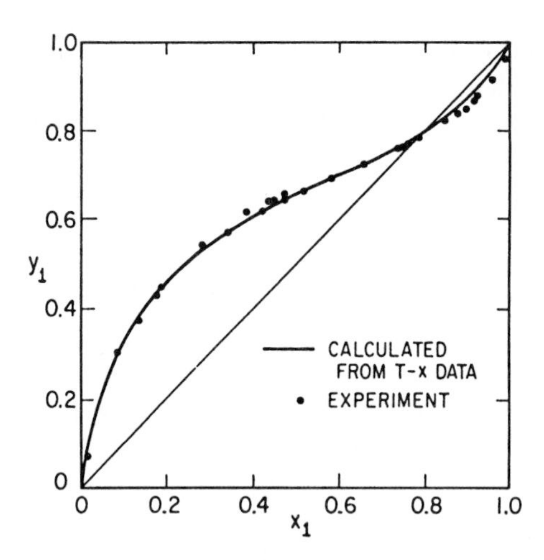

Figure 6-11 Vapor-liquid equilibrium for the diisopropyl ether (1)/2-propanol (2) system at 1.013 bar.

The method outlined above is rough and we can make many refinements; for example, we may use additional coefficients in the expressions for the activity coefficients or we may correct for vapor-phase nonideality (Adler and Adler, 1973). But these refinements are frequently not worthwhile when the temperature correction in the Gibbs-Duhem equation is neglected. It appears to be unavoidable that, unless the enthalpy of mixing can be estimated with at least fair accuracy, the boiling-point method for obtaining the y-x diagram is necessarily an approximation.

Techniques for estimating the isothermal or isobaric y-x diagram with the help of the Gibbs-Duhem equation have received a large amount of attention in the literature and many variations on this theme have been proposed. One particularly popular topic is concerned with the use of azeotropic data; if a binary system has an azeotrope, and if we know its composition, temperature, and pressure, we can compute the two constants of any two-constant equation for the activity coefficients and thereby calculate the entire x-y diagram (Miller and Bliss, 1940). This method assumes validity of the isothermal, isobaric Gibbs-Duhem equation. Good results are often obtained provided that the azeotropic mole fraction is in the interval 0.3 to 0.7.

While Van Ness (Byer *et al.*, 1973; Abbott and Van Ness, 1975) has stressed the relation between experimental P-x data and the Gibbs-Duhem equation, an interesting variation has been proposed by Christian *et al.* (1960), who describes apparatus and calculations for measuring P-y rather than P-x data to obtain the y-x diagram. A method for computing equilibrium phase compositions from dew point (T-y) data has been given by Bellemans (1959), and calculations using (P-x) data for ternary systems have been described by McDermott and Ellis (1965). All of these techniques have but one aim: To reduce the experimental effort needed to characterize liquid-mixture properties. Such techniques are useful but they have one serious limitation: Results obtained by data reduction with the Gibbs-Duhem equation cannot be checked for thermodynamic consistency because the method of calculation already forces the results to be thermodynamically consistent. Thus the Gibbs-Duhem equation may be used either to extend limited data or to test more complete data, but not both. In the next section we briefly discuss the basic principles for performing such tests.

6.9 Testing Equilibrium Data for Thermodynamic Consistency

The Gibbs-Duhem equation interrelates activity coefficients of all components in a mixture. Therefore, if data are available for all activity coefficients, these data should obey the Gibbs-Duhem equation; if they do not, the data cannot be correct. If they do obey the Gibbs-Duhem equation, the data are probably, although not necessarily, correct; it is conceivable that a given set of incorrect data may fortuitously satisfy the Gibbs-Duhem equation, but this is not likely. Unfortunately, there are many phase-equilibrium data in the literature that do not satisfy the Gibbs-Duhem equation and therefore must be incorrect.

To illustrate, we consider the simplest case: a binary solution of two liquids at low pressure for which isothermal activity-coefficient data have been obtained. For this case, the Gibbs-Duhem equation is Eq. (6-59).

A theoretically simple technique is to test the data directly with Eq. (6-59); that is, plots are prepared of $\ln\gamma_1$ versus x_1 and $\ln\gamma_2$ versus x_2 and slopes are measured. These slopes are then substituted into Eq. (6-59) at various compositions to see if the Gibbs-Duhem equation is satisfied. While this test appears to be both simple and exact, it is of little practical value; because experimental data inevitably exhibit some scatter, it is difficult to measure slopes with sufficient accuracy. Hence the "slope method" provides at best a rough measure of thermodynamic consistency that can only be applied in a semiquantitative manner. For example, if, at a given composition, $d\ln\gamma_1/dx_1$ is positive, then $d\ln\gamma_2/dx_2$ must also be positive, and if $d\ln\gamma_1/dx_1$ is zero, $d\ln\gamma_2/dx_2$ must also be zero. The slope method can therefore be used easily to detect serious errors in the equilibrium data.

For quantitative purposes it is much easier to use an integral rather than a differential (slope) test. While integral tests are popular and used often, unfortunately, they do not provide a stringent criterion for thermodynamic consistency. The most widely used integral test was proposed by Redlich and Kister (1948) and also by Herington (1947) and is derived below.

The molar excess Gibbs energy is related to activity coefficients by

$$\frac{g^E}{RT} = x_1 \ln\gamma_1 + x_2 \ln\gamma_2 \tag{6-86}$$

Differentiating with respect to x_1 at constant temperature and pressure gives

$$\frac{d(g^E/RT)}{dx_1} = x_1\frac{\partial\ln\gamma_1}{\partial x_1} + \ln\gamma_1 + x_2\frac{\partial\ln\gamma_2}{\partial x_1} + \ln\gamma_2\frac{dx_2}{dx_1} \tag{6-87}$$

Noting that $dx_1 = -dx_2$ and substituting the Gibbs-Duhem equation [Eq. (6-59)], we obtain

$$\frac{d(g^E/RT)}{dx_1} = \ln\frac{\gamma_1}{\gamma_2} \tag{6-88}$$

Integration with respect to x_1 gives

$$\int_0^1 \frac{d(g^E/RT)}{dx_1}dx_1 = \int_0^1 \ln\frac{\gamma_1}{\gamma_2}dx_1 = \frac{g^E}{RT}(\text{at } x_1 = 1) - \frac{g^E}{RT}(\text{at } x_1 = 0) \tag{6-89}$$

If the pure liquids at the temperature of the mixture are used as the standard states,

$$\begin{aligned} \ln\gamma_1 &\to 0 \quad \text{as} \quad x_1 \to 1 \\ \ln\gamma_2 &\to 0 \quad \text{as} \quad x_1 \to 0 \end{aligned} \tag{6-90}$$

and

$$\begin{aligned} \frac{g^E}{RT}(\text{at } x_1 = 1) &= 0 \\ \frac{g^E}{RT}(\text{at } x_1 = 0) &= 0 \end{aligned} \tag{6-91}$$

Equation (6-89) therefore becomes

$$\boxed{\int_0^1 \ln \frac{\gamma_1}{\gamma_2} dx_1 = 0} \tag{6-92}$$

Equation (6-92) provides what is called an *area test* of phase-equilibrium data. A plot of $\ln(\gamma_1 / \gamma_2)$ versus x_1 is prepared; a typical plot of this type is shown in Fig. 6-7. Because the integral on the left-hand side of Eq. (6-92) is given by the area under the curve shown in the figure, the requirement of thermodynamic consistency is met if that area is zero, i.e., if the area above the x-axis is equal to that below the x-axis. These areas can be measured easily and accurately with a planimeter and thus the area test is a particularly simple one to carry out.

Unfortunately, however, the area test has little value for deciding whether or not a set of activity-coefficient data is, or is not, thermodynamically consistent. As indicated by Eq. (6-92), the area test uses the *ratio* of γ_1 to γ_2; when this ratio is calculated, the pressure cancels out:

$$\frac{\gamma_1}{\gamma_2} = \frac{\varphi_1 y_1 / x_1 f_1^0}{\varphi_2 y_2 / x_2 f_2^0} \tag{6-93}$$

Therefore, the area test does not utilize what is probably the most valuable (and usually the most accurate) measurement, total pressure P.[21] As pointed out by Van Ness (1995), the area test is severely limited because (except for minor corrections – see footnote 21) the only data needed to construct the plot are x and y data and the ratio of the two pure-component vapor pressures, P_1^s / P_2^s. Therefore, for isothermal data, the area test does little more than determine whether or not the vapor pressure ratio P_1^s / P_2^s is appropriate to the set of measured x-y values. A plot of $\ln(\gamma_1 / \gamma_2)$ versus x_1 may be sensitive to scatter in the x-y data but it tells us nothing about the internal consistency of such data.

The only meaningful way to check thermodynamic consistency of experimental data consists of three steps: First, measure all three quantities P, x, and y at constant T; next, select any two of these measured quantities and predict the third using the Gibbs-Duhem equation; finally, compare the predicted third quantity with the measured quantity.

Van Ness and coworkers (Byer *et al.*, 1973; Abbott and Van Ness, 1975) have illustrated such a procedure by examining experimental P-x-y data (Fried *et al.*, 1967) for the system pyridine (1)/tetrachloroethylene (2) at 60°C; a plot of these data is shown

[21] A minor qualification is necessary because P does enter into the calculation of fugacity coefficients φ_i and, sometimes, standard-state fugacity f_i^0. However, these are secondary effects that become negligible at low pressure. The primary effect of pressure is not included in Eq. (6-92).

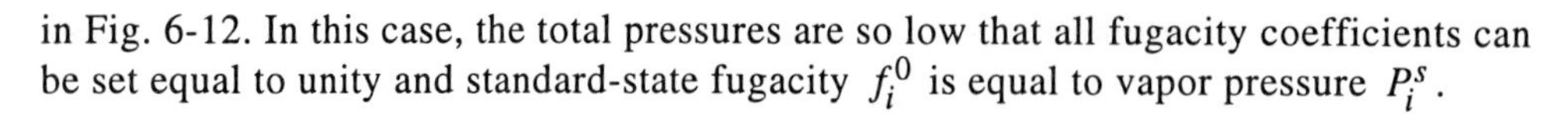

in Fig. 6-12. In this case, the total pressures are so low that all fugacity coefficients can be set equal to unity and standard-state fugacity f_i^0 is equal to vapor pressure P_i^s.

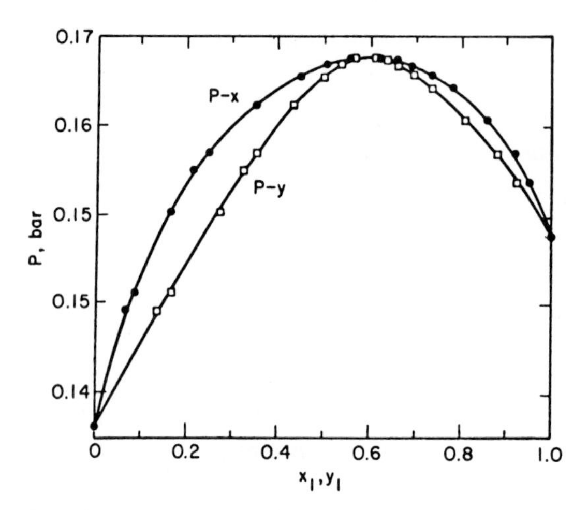

Figure 6-12 Experimental vapor-liquid equilibrium data for the system pyridine (1)/tetrachloroethylene (2) at 60°C.

Van Ness and coworkers fit the data to Eq. (6-82) in three ways, using

1. *P-x-y* data.
2. *x-y* data only.
3. *P-x* data only.

Results of data reduction are shown in Tables 6-6 and 6-7. Table 6-6 indicates that the coefficients in the expression for g^E depend significantly on the choice of data[22] used but, more important for testing data, deviations between calculated and measured quantities also differ markedly, as indicated in Table 6-7.

Table 6-6 Constants for Eq. (6-82) for the system pyridine/tetrachloroethylene at 60°C (Byer *et al.*, 1973).

	From *P-x-y* data	From *x-y* data	From *P-x* data
A'	0.93432	0.77882	0.82030
B'	0.84874	0.68925	0.77826
D'	0.48897	0.03721	0.09045

[22] Nicolaides and Eckert (1978).

The first horizontal row in Table 6-7 tells us nothing about thermodynamic consistency because all three measured quantities were used to obtain the coefficients in Eq. (6-82). The deviations merely tell us to what extent that equation can fit the total experimental data. In the second horizontal row, we must look at the deviations in pressure, because that quantity was *not* used in determining the coefficients of Eq. (6-82); similarly, in the third horizontal row, we must look at deviations in y.

Table 6-7 Deviations between calculated and measured quantities for the system pyridine/tetrachloroethylene at 60°C (Byer *et al.*, 1973).

	Δ = calculated quantity – measured quantity			
	Δy		$\Delta P \times 10^3$ bar	
Data used	RMS	Max.	RMS	Max
P-x-y	0.0058	0.0119	0.57	1.32
x-y	0.0018	0.0036	1.80	2.49
P-x	0.0054	0.0092	0.33	0.67

Experimentally, it is much easier to measure pressure with high accuracy; accurate measurements of y are usually more difficult. When P-x data are used, the root-mean-square (RMS) error in y is probably within the experimental error of y; but when x-y data are used, the RMS error in P is probably larger than the experimental error. The thermodynamic consistency of these data, therefore, is reasonably good but not as good as it might be, probably because of inaccuracies in y.

While RMS values are useful for assessing thermodynamic consistency, a more useful procedure is to plot ΔP (or Δy) versus x to determine if the deviations are primarily of one sign or if they scatter uniformly about zero. If the deviations show a clear trend, they may well be suspect. However if the deviations scatter about zero, without bias, and if the deviations are small, the data are likely to be of good quality.

It is not easy to devise a truly significant test for thermodynamic consistency because it does not appear to be possible to define an unambiguous quantitative criterion of consistency; all experimental data have some uncertainty and the judgment of "good" or "bad" depends on the system, the experimental method, the standards set by the judge, and most important, by the ultimate use of the data; for some applications, rough approximations are sufficient, while for others very high accuracy is essential. The problem is further complicated by the question of how different experimental measurements are weighted; some measurements are worth more than others because experimental conditions vary; for example, in typical binary vapor-liquid experiments, data at the two ends of the composition scale are likely to have the lowest accuracy (unless special precautions are observed), but it is precisely these measurements that give the most valuable information for determining the excess Gibbs energy.

The literature is rich with articles on testing for thermodynamic consistency because it is much easier to test someone else's data than to obtain one's own in the laboratory. Much (but by no means all) of this literature is obscured by excessive use of statistics. It has been said that "Patriotism is the last refuge of a scoundrel." Similarly, we might say that "Statistics is the last refuge of a poor experimentalist" or, in a more positive way, that a gram of good data is worth more than a ton of consistency tests.

6.10 Wohl's Expansion for the Excess Gibbs Energy

In Sec. 6.5 we discussed briefly some expressions for the excess Gibbs energy of binary solutions. We now continue this discussion with a general method for expressing excess Gibbs energies as proposed by Wohl (1946). One of the main advantages of this method is that some rough physical significance can be assigned to the parameters that appear in the equations; as a result, and as shown later in Sec. 6.14, *Wohl's expansion* can be extended systematically to multicomponent solutions.

Wohl expresses the excess Gibbs energy of a binary solution as a power series in z_1 and z_2, the effective volume fractions of the two components:

$$\frac{g^E}{RT(x_1q_1+x_2q_2)} = 2a_{12}z_1z_2 + 3a_{112}z_1^2z_2 + 3a_{122}z_1z_2^2 + 4a_{1112}z_1^3z_2 + 4a_{1222}z_1z_2^3 + 6a_{1122}z_1^2z_2^2 + \ldots \tag{6-94}$$

where

$$z_1 \equiv \frac{x_1q_1}{x_1q_1+x_2q_2} \quad \text{and} \quad z_2 \equiv \frac{x_2q_2}{x_1q_1+x_2q_2}$$

Wohl's equation contains two types of parameters, q's and a's. The q's are effective volumes, or cross sections, of the molecules; q_i is a measure of the size of molecule i, or of its "sphere of influence" in the solution. A large molecule has a larger q than a small one and, in solutions of nonpolar molecules of similar shape, it is often a good simplifying assumption that the ratio of the q's is the same as the ratio of the pure-component liquid molar volumes. The a's are interaction parameters whose physical significance, while not precise, is in a rough way similar to that of virial coefficients. Parameter a_{12} is a constant characteristic of the interaction between molecule 1 and molecule 2; parameter a_{112} is a constant characteristic of the interaction between three molecules, two of component 1 and one of component 2, and so on. The probability that any nearest-neighbor pair of two molecules consists of one molecule of component 1 and one molecule of component 2 is assumed to be $2z_1z_2$; similarly, the probability that a triplet of three nearest-neighbor molecules consists of molecules 1, 1, and 2 is assumed to be $3z_1^2z_2$, and so forth. Thus, there is a crude analogy between

Wohl's equation and the virial equation of state, but it is no more than an analogy because, while the virial equation has an exact theoretical basis, Wohl's equation cannot be derived from any rigorous theory without drastic simplifying assumptions.

When, as in Eq. (6-94), the excess Gibbs energy is taken with reference to an ideal solution in the sense of Raoult's law, only interactions involving at least two dissimilar molecules are contained in Eq. (6-94); that is, terms of the type z_1^2, z_1^3, ... and z_2^2, z_2^3, ... do not explicitly appear in the expansion. This is a necessary consequence of the boundary condition that g^E must vanish as x_1 or x_2 becomes zero.

However, if g^E is taken relative to an ideal dilute solution that is dilute in, say, component 2, then Wohl's expansion takes the form

$$\frac{g^{E*}}{RT(x_1q_1 + x_2q_2)} = -a_{22}z_2^2 - a_{222}z_2^3 - a_{2222}z_2^4 - \dots \qquad \text{(6-95)}[23]$$

In this case a_{22} is the self-interaction coefficient characteristic of the interaction between two molecules of component 2, a_{222} is the self-interaction coefficient characteristic of the interaction between three molecules of component 2, and so on. Because g^{E*} in Eq. (6-95) refers to a solution very dilute in component 2, it is not interaction between molecules 2 and molecules 1 but rather interaction between molecules of component 2 that cause deviation from ideal behavior and hence a nonvanishing g^{E*}.

Equation (6-94) is a well-known expression for mixtures whose components can exist as pure liquids at the solution temperature. Equation (6-95) is not known well but it is sometimes useful for solutions of gases or solids in liquids (see Chaps. 10 and 11).

van Laar Equation. To illustrate the generality of Eq. (6-94), we consider first the case of a binary solution of two components that are not strongly dissimilar chemically but that have different molecular sizes. An example is a solution of benzene (molar volume 89 cm^3 mol^{-1} at 25°C) and isooctane (molar volume 166 cm^3 mol^{-1} at 25°C). We make the simplifying assumption that interaction coefficients a_{112}, a_{122}, ... and higher may be neglected; i.e., Wohl's expression is truncated after the first term. In that case, Eq. (6-94) becomes

$$\frac{g^E}{RT} = \frac{2a_{12}x_1x_2q_1q_2}{x_1q_1 + x_2q_2} \qquad \text{(6-96)}$$

that is the *van Laar equation*. From Eq. (6-25), expressions for the activity coefficients can be found. They are[24]

[23] The minus signs before the coefficients are arbitrarily introduced for convenience. For most mixtures a_{22} in Eq. (6-95) is a positive number. The asterisk on g^{E*} indicates that g^{E*} is taken relative to an *ideal dilute* solution.

[24] A' and B' are dimensionless.

$$\ln\gamma_1 = \frac{A'}{\left(1+\dfrac{A'}{B'}\dfrac{x_1}{x_2}\right)^2} \tag{6-97}$$

and

$$\ln\gamma_2 = \frac{B'}{\left(1+\dfrac{B'}{A'}\dfrac{x_2}{x_1}\right)^2} \tag{6-98}$$

where $A' = 2q_1 a_{12}$ and $B' = 2q_2 a_{12}$.

Equations (6-97) and (6-98) are the familiar van Laar equations commonly used to represent activity-coefficient data. These equations include two empirical constants, A' and B'; the ratio of A' to B' is the same as the ratio of the effective volumes q_1 and q_2 and it is also equal to the ratio of $\ln\gamma_1^\infty$ to $\ln\gamma_2^\infty$. Whereas Eqs. (6-97) and (6-98) contain only two parameters, Eq. (6-96) appears to be a three-parameter equation. However, from the empirically determined values of A' and B' it is not possible to find a value of the interaction coefficient a_{12} unless some independent assumption is made concerning the value of q_1 or q_2. For practical purposes it is not necessary to know the values of q_1 and q_2 separately because it is only their ratio that is important.

Figure 6-13 gives activity coefficients for the benzene/isooctane system at 45°C. The data of Weissman and Wood (1960) are well represented by the van Laar equations with $A' = 0.419$ and $B' = 0.745$.

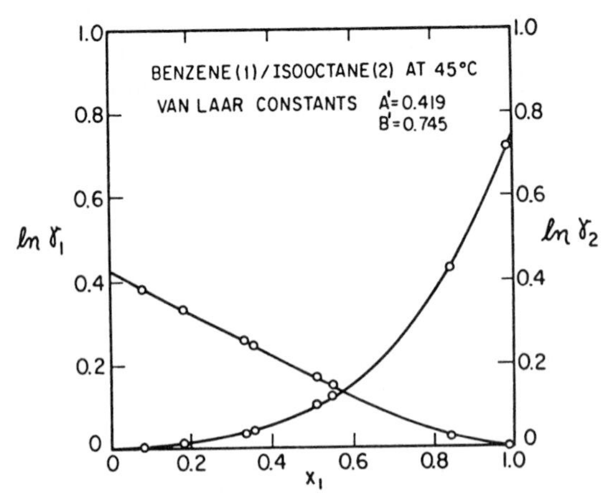

Figure 6-13 Application of van Laar's equations to a mixture whose components differ appreciably in molecular size.

The derivation of the van Laar equations suggests that they should be used for solutions of relatively simple, preferably nonpolar liquids but empirically it has been found that these equations are frequently able to represent activity coefficients of more complex mixtures. In such mixtures, the physical significance of the van Laar constants is more obscure; the constants must be regarded as essentially little more than empirical parameters in a thermodynamically consistent equation. The van Laar equations are widely used; they have become popular for applied work because of their flexibility and because of their mathematical simplicity relative to the many other equations that have been proposed.[25] In the special case where van Laar constants A' and B' are equal, the van Laar equations are identical to the two-suffix Margules equations [Eqs. (6-48) and (6-49)].

Whenever Wohl's expansion is used as the basis for an expression giving the activity coefficient as a function of composition, the resulting equation holds for a fixed temperature and pressure; that is, the constants in the equation are not dependent on composition but instead are functions of temperature and pressure. Because the effect of pressure on liquid-phase properties is usually small (except at high pressures and at conditions near critical), the pressure dependence of the constants can usually be neglected; however, the temperature dependence is often not negligible. Although many industrial operations (e.g., distillation) are conducted at constant pressure rather than constant temperature, there is nevertheless a strong temptation to assume that the constants in equations such as those of van Laar are temperature independent. Physically, this assumption is most unreasonable; as indicated by Eq. (6-29), the activity coefficient of a component in solution is independent of temperature only in an athermal solution, i.e., one where the components mix isothermally and isobarically without evolution or absorption of heat. For practical applications, however, this assumption is often tolerable provided that the temperature range in question is not large. For example, Fig. 6-14 shows activity coefficients for the propanol/water system calculated from the data of Gadwa (1936) at a constant pressure of 1.013 bar. These activity coefficients are well represented by the van Laar equations with $A' = 2.60$ and $B' = 1.13$. In this particular case the assumption of temperature-invariant constants appears to be a good approximation because at a constant pressure of 1.013 bar, the boiling temperature varies only from 87.8 to 100°C.

When, as frequently happens, experimental data are insufficient to specify the temperature dependence of the activity coefficient, either one of two simplifying approximations is usually made. The first, mentioned in the preceding paragraph, is to assume that at constant composition the activity coefficient is invariant with temperature; the second is to assume that at constant composition $\ln\gamma$ is proportional to the reciprocal of the absolute temperature. The first assumption is equivalent to assuming that the solution is athermal ($h^E = 0$), and the second is equivalent to assuming that the solution is regular ($s^E = 0$). Real solutions are neither athermal nor regular but more

[25] As pointed out by Abbott and Van Ness (1982), van Laar equation implies that when the experimental data are plotted in the form x_1x_2/g^E versus x_1, a straight line should be obtained.

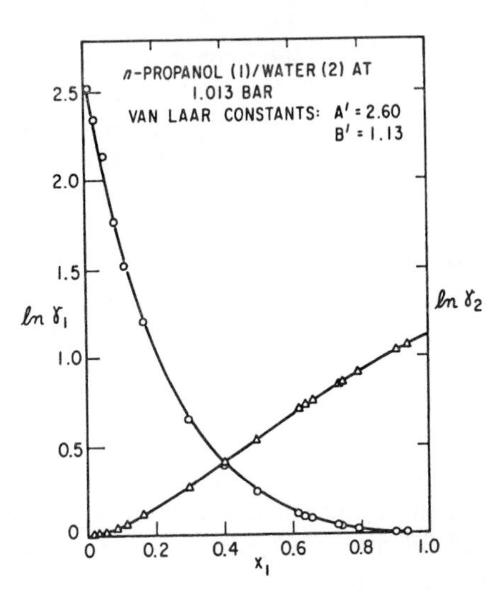

Figure 6-14 Application of van Laar's equations to an isobaric system. In this system the temperature varies only from 87.8 to 100°C.

often than not, the assumption of regularity provides a better approximation to the effect of temperature on the activity coefficient than does the assumption of athermal behavior.

A general empirical procedure is to assume that at constant x,

$$\ln \gamma_i = c + dT^{-1} \tag{6-99}$$

where c and d are empirical constants that depend on composition. When $c = 0$, we recover regular-solution behavior and when $d = 0$, we recover athermal-solution behavior. Unfortunately, in typical cases, the magnitude of c is similar to that of dT^{-1}.

In vapor-liquid equilibrium, the effect of temperature on γ is often not of major importance because the effect of temperature on the standard-state fugacity (essentially, the vapor pressure in most cases) is usually much stronger than that on the activity coefficient. Therefore, vapor-liquid equilibria are often insensitive to changes in γ with T. However, in liquid-liquid equilibria, the standard-state fugacity plays no role and for such equilibria, small effects of temperature on the activity coefficients can have a large influence on the liquid-liquid phase diagram.

Margules Equations. Next we consider a binary solution of two components whose molecular sizes are not much different. In that case we assume that $q_1 = q_2$ in Wohl's expansion. Neglecting terms higher than the fourth power in the mole fraction, and again using Eq. (6-25) to obtain expressions for the activity coefficients, we now obtain

$$\ln\gamma_1 = A'x_2^2 + B'x_2^3 + C'x_2^4 \tag{6-100}$$

$$\ln\gamma_2 = (A' + \tfrac{3}{2}B' + 2C')x_1^2 - (B' + \tfrac{8}{3}C')x_1^3 + C'x_1^4 \tag{6-101}$$

where

$$A' = q(2a_{12} + 6a_{112} - 3a_{122} + 12a_{1112} - 6a_{1122})$$

$$B' = q(6a_{122} - 6a_{112} - 24a_{1112} - 8a_{1222} + 24a_{1122})$$

$$C' = q(12a_{1112} + 12a_{1222} - 18a_{1122})$$

To simplify matters, and because experimental data are usually limited, it is common to truncate the expansions after the cubic terms, i.e., to set $C' = 0$; in that case the equation is called the *three-suffix Margules equation* that has two parameters. Only in those cases where the data are sufficiently precise and plentiful is the expansion truncated after the quartic terms and we then have a *four-suffix Margules equation* with three parameters as given in Eqs. (6-100) and (6-101). On the other hand, if the mixture is a simple one, containing similar components, it is sometimes sufficient to retain only the quadratic term (*two-suffix Margules equation*).

Although the assumption $q_1 = q_2$ suggests that Margules equations should be used only for mixtures whose components have similar molar volumes, it is nevertheless used frequently for all sorts of liquid mixtures, regardless of the relative sizes of the different molecules. The primary value of the Margules and van Laar equations lies in their ability to serve as simple empirical equations for representing experimentally determined activity coefficients with only a few constants. When, as is often the case, experimental data are scattered or scarce, these equations can be used to smooth the data and, more important, they serve as an efficient tool for interpolation and extrapolation with respect to composition.[26]

The three-suffix Margules equations have been used to reduce experimental vapor-liquid equilibrium data for many systems; to illustrate, Fig. 6-15 shows results for three binaries at 50°C: acetone/methanol, acetone/chloroform, and chloroform/methanol (Severns *et al.*, 1955). Each of these binaries has an azeotrope at 50°C.

Figure 6-15 shows that the thermodynamic properties of these three systems differ markedly from one another; in the acetone/methanol system there are strong positive deviations from ideality, while in the acetone/chloroform system there are equally strong negative deviations; in the chloroform/methanol system there are very large positive deviations at the chloroform-rich side, and at the methanol-rich side the activity coefficient of chloroform exhibits unusual behavior because it goes through a

[26] As pointed out by Abbott and Van Ness (1982), the three-suffix Margules equation implies that when the experimental data are plotted in the form g^E / x_1x_2 versus x_1, a straight line should be obtained.

maximum.[27] Despite these large differences, the three-suffix Margules equations give a good representation of the data for all three systems. The Margules constants,[28] as obtained from the experimental data at 50°C, are

	A'	B'
Acetone (1)/chloroform (2)	-0.553	-0.276
Acetone (1)/methanol (2)	0.334	0.368
Chloroform (1)/methanol (2)	2.89	-2.17

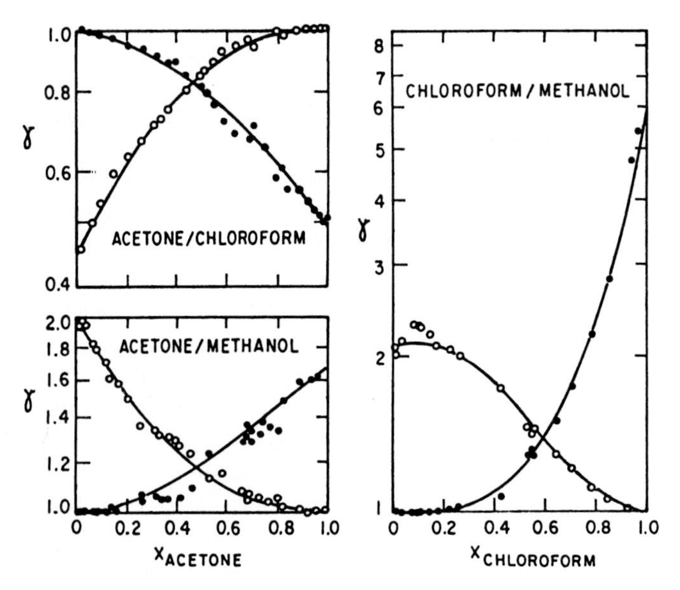

Figure 6-15 Activity coefficients for three binary systems at 50°C. Lines calculated from three-suffix Margules equations.

Scatchard-Hamer Equation. For one additional example of the flexibility of Wohl's expansion we consider the case where the series is truncated after the third-order terms but where, instead of assuming that $q_1 = q_2$, we assume that

$$\frac{q_1}{q_2} = \frac{v_1}{v_2}$$

[27] One of the advantages of the three-suffix Margules equations is that they are capable of representing maxima or minima; two-suffix Margules equations and van Laar equations cannot do so. However, maxima and minima are only rarely observed in isothermal plots of activity coefficient versus mole fraction.

[28] These constants pertain to Eqs. (6-100) and (6-101) with $C' = 0$.

where v_1 and v_2 are, respectively, the molar volumes of the pure liquids at the temperature of the solution. Truncating Wohl's expansion after the cubic terms, we then obtain expressions for the activity coefficients first proposed by Scatchard and Hamer (1935). They are

$$\ln \gamma_1 = A' z_2^2 + B' z_2^3 \quad (6\text{-}102)$$

$$\ln \gamma_2 = \left(A' + \frac{3}{2} B' \right) \left(\frac{v_2}{v_1} \right) z_1^2 - B' \left(\frac{v_2}{v_1} \right) z_1^3 \quad (6\text{-}103)$$

where $A' = v_1(2a_{12} + 6a_{112} - 3a_{122})$ and $B' = v_1(6a_{122} - 6a_{112})$.

The *Scatchard-Hamer equations* use only two adjustable parameters but they have not received much attention in the extensive literature on vapor-liquid equilibria. They are only slightly more complex than the popular van Laar equations and the three-suffix Margules equations. This is unfortunate because in the general case the assumptions of Scatchard and Hamer appear more reasonable than those of van Laar or those in the Margules equation truncated after the cubic terms.

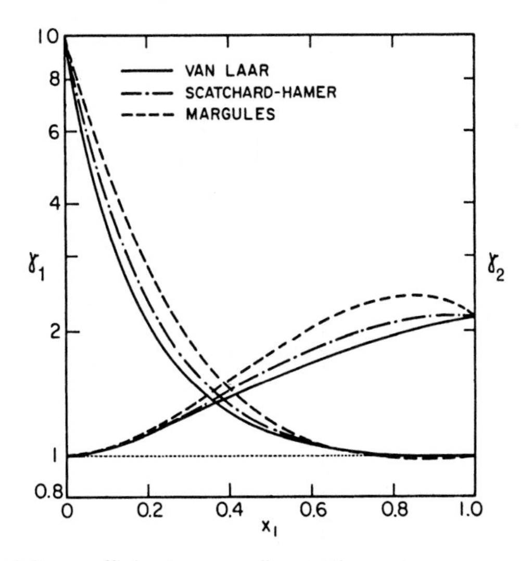

Figure 6-16 Activity coefficients according to three, two-parameter equations with $\gamma_1^\infty = 10$ and $\gamma_2^\infty = 2.15$. For the Scatchard-Hamer equation, $v_2/v_1 = 2/3$.

The behavior of the Scatchard-Hamer equations may be considered as intermediate between that of the van Laar equations and that of the three-suffix Margules equations. All three equations contain two adjustable parameters, and if for each of these equations we arbitrarily determine these parameters from values of γ_1^∞ and γ_2^∞, we can then compare the three equations as shown in Fig. 6-16.

6.11 Wilson, NRTL, and UNIQUAC Equations

Many equations have been proposed for the relation between activity coefficients and mole fractions and new ones appear every year.[29] Some, but not all, of these can be derived from Wohl's general method. In particular, three equations that are useful for many practical calculations cannot be obtained by Wohl's formulation.

Wilson Equation. Based on molecular considerations, Wilson (1964) presented the following expression for the excess Gibbs energy of a binary solution:

$$\frac{g^E}{RT} = -x_1 \ln(x_1 + \Lambda_{12}x_2) - x_2 \ln(x_2 + \Lambda_{21}x_1) \tag{6-104}$$

The activity coefficients derived from this equation are

$$\ln\gamma_1 = -\ln(x_1 + \Lambda_{12}x_2) + x_2\left(\frac{\Lambda_{12}}{x_1 + \Lambda_{12}x_2} - \frac{\Lambda_{21}}{\Lambda_{21}x_1 + x_2}\right) \tag{6-105}$$

$$\ln\gamma_2 = -\ln(x_2 + \Lambda_{21}x_1) - x_1\left(\frac{\Lambda_{12}}{x_1 + \Lambda_{12}x_2} - \frac{\Lambda_{21}}{\Lambda_{21}x_1 + x_2}\right) \tag{6-106}$$

In Eq. (6-104) the excess Gibbs energy is defined with reference to an ideal solution in the sense of Raoult's law; Eq. (6-104) obeys the boundary condition that g^E vanishes as either x_1 or x_2 becomes zero.

Wilson's equation has two adjustable parameters, Λ_{12} and Λ_{21}. In Wilson's derivation, these are related to the pure-component molar volumes and to characteristic energy differences by

$$\Lambda_{12} \equiv \frac{v_2}{v_1}\exp\left(-\frac{\lambda_{12} - \lambda_{11}}{RT}\right) \tag{6-107}$$

$$\Lambda_{21} \equiv \frac{v_1}{v_2}\exp\left(-\frac{\lambda_{21} - \lambda_{22}}{RT}\right) \tag{6-108}$$

---

[29] See, for example, E. Hala, J. Pick, V. Fried and O. Vilim, 1967, *Vapor-Liquid Equilibrium*, 2nd Ed., trans. G. Standart, Part 1, (Oxford: Pergamon Press); C. Black, 1959, *AIChE J.* 5: 249; M. Hiranuma and K. Honma, 1975, *Ind. Eng. Chem. Process Des. Dev.*, 14: 221. A particularly simple but unusual equation has been proposed by H. Mauser, 1958, *Z. Elektrochem.*, 62: 895. There are many, many others, too many to list here.

where v_i is the molar liquid volume of pure component i and the λ's are energies of interaction between the molecules designated in the subscripts. To a fair approximation, the differences in the characteristic energies are independent of temperature, at least over modest temperature intervals. Therefore, Wilson's equation gives not only an expression for the activity coefficients as a function of composition but also an estimate of the variation of the activity coefficients with temperature. This may provide a practical advantage in isobaric calculations where the temperature varies as the composition changes. For accurate work, $(\lambda_{12}-\lambda_{11})$ and $(\lambda_{21}-\lambda_{22})$ should be considered temperature-dependent but in many cases this dependence can be neglected without serious error.

Wilson's equation appears to provide a good representation of excess Gibbs energies for a variety of miscible mixtures. It is particularly useful for solutions of polar or associating components (e.g., alcohols) in nonpolar solvents. The three-suffix Margules equation and the van Laar equation are usually not adequate for such solutions. For a good data fit, an equation of the Margules type or a modification of van Laar's equation (Black, 1959) may be used, but such equations require at least three parameters and, more important, these equations are not readily generalized to multicomponent solutions without further assumptions or ternary parameters.

A study of Wilson's equation by Orye (1965a) shows that for approximately 100 miscible binary mixtures of various chemical types, activity coefficients were well represented by Wilson equation; in essentially all cases this representation was as good as, and in many cases better than, the representation given by the three-suffix (two-constant) Margules equation and by the van Laar equation. Similar conclusions were obtained by Gmehling *et al.*,[30] who report Wilson parameters for many binary systems.

To illustrate, Table 6-8 gives calculated and experimental vapor compositions for the nitromethane/carbon tetrachloride system. The calculations were made twice, once using the van Laar equation and once using Wilson's equation; in both cases, required parameters were found from a least-squares computation using experimental *P-x* data at 45°C given by Brown and Smith (1957). In both calculations the average error in the predicted vapor compositions is not high, but for the calculation based on van Laar's equation it is almost three times as large as that based on Wilson's equation.

A similar calculation is shown in Fig. 6-17 for the ethanol/isooctane system. Wilson and van Laar parameters were calculated from the isothermal vapor- pressure data of Kretschmer (1948). In this case, the Wilson equation is much superior to the van Laar equation that erroneously predicts an immiscible region for this system at 50°C.

For isothermal solutions that do not exhibit large or highly asymmetric deviations from ideality, Wilson's equation does not offer any particular advantages over the more familiar three-suffix Margules or van Laar equations, although it appears to be as good as these. For example, isothermal vapor-liquid equilibrium data for the cryogenic systems argon/nitrogen and nitrogen/oxygen are represented equally well by all three equations.

[30] J. Gmehling, U. Onken, and W. Arlt, DECHEMA Chemistry Data Series, starting in 1977.

Table 6-8 Calculated vapor compositions from fit of *P-x* data at 45°C [nitromethane (1)/ carbon tetrachloride (2)].

Experimental*				Calculated y_1	
x_1	P (bar)	y_1		Wilson	van Laar
0	0.3348	0		0	0
0.0459	0.3832	0.130		0.147	0.117
0.0918	0.3962	0.178		0.191	0.183
0.1954	0.4039	0.222		0.225	0.247
0.2829	0.4034	0.237		0.236	0.262
0.3656	0.4019	0.246		0.243	0.264
0.4659	0.3984	0.253		0.251	0.261
0.5366	0.3958	0.260		0.258	0.259
0.6065	0.3910	0.266		0.266	0.259
0.6835	0.3828	0.277		0.279	0.266
0.8043	0.3528	0.314		0.318	0.304
0.9039	0.2861	0.408		0.410	0.411
0.9488	0.2279	0.528		0.524	0.540
1	0.2256	1		1	1
			Error:	±0.004	±0.011
				$\Lambda_{12} = 0.1156$	$A' = 2.230$
				$\Lambda_{21} = 0.2879$	$B' = 1.959$

* Brown and Smith (1957).

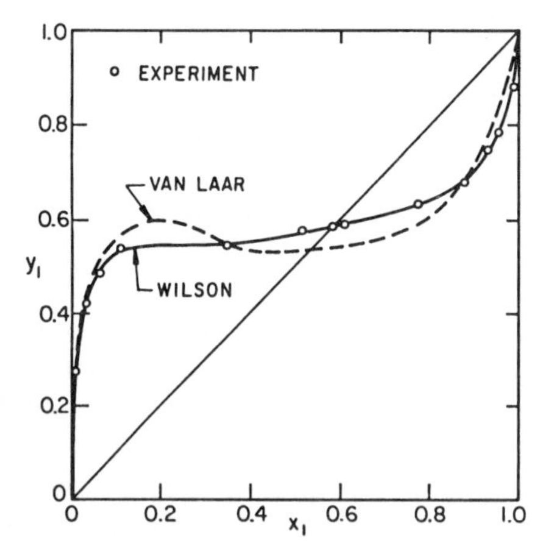

Figure 6-17 Vapor-liquid equilibrium for the ethanol (1)/ isooctane (2) system at 50°C. Lines calculated from *P-x* data. The van Laar equations erroneously predict partial immiscibility.

Wilson's equation has two disadvantages that are not serious for many applications. First, Eqs. (6-105) and (6-106) are not useful for systems where the logarithms of the activity coefficients, when plotted against x, exhibit maxima or minima. (Van Laar equations are also not useful for this case). Such systems, however, are not common. The second and more serious disadvantage of Wilson's equation lies in its inability to predict limited miscibility. When Wilson's equation is substituted into the equations of thermodynamic stability for a binary system (see next section), no parameters Λ_{12} and Λ_{21} can be found that indicate the existence of two stable liquid phases.[31] Wilson's equation, therefore, should be used only for liquid systems that are completely miscible or else for those limited regions of partially miscible systems where only one liquid phase is present.

NRTL Equation. The basic idea in Wilson's derivation of Eq. (6-104) follows from the concept of local composition that is discussed further in Sec. 7.7. This concept was also used by Renon (1968) in his derivation of the NRTL (*n*on*r*andom, *t*wo-*l*iquid) equation; however, Renon's equation, unlike Wilson's, is applicable to partially miscible as well as completely miscible systems. The NRTL equation for the excess Gibbs energy is

$$\frac{g^E}{RT} = x_1 x_2 \left(\frac{\tau_{21} G_{21}}{x_1 + x_2 G_{21}} + \frac{\tau_{12} G_{12}}{x_2 + x_1 G_{12}} \right) \tag{6-109}$$

where

$$\tau_{12} = \frac{g_{12} - g_{22}}{RT} \qquad \tau_{21} = \frac{g_{21} - g_{11}}{RT} \tag{6-110}$$

$$G_{12} = \exp(-\alpha_{12}\tau_{12}) \qquad G_{21} = \exp(-\alpha_{12}\tau_{21}) \tag{6-111}$$

The significance of g_{ij} is similar to that of λ_{ij} in Wilson's equation; g_{ij} is an energy parameter characteristic of the i-j interaction. Parameter α_{12} is related to the nonrandomness in the mixture; when α_{12} is zero, the mixture is completely random and Eq. (6-109) reduces to the two-suffix Margules equation. The NRTL equation contains three parameters, but reduction of experimental data for a large number of binary systems indicates that α_{12} varies from about 0.20 to 0.47; when experimental data are scarce, the value of α_{12} can often be set arbitrarily; a typical choice is α_{12} = 0.3. From Eq. (6-109), the activity coefficients are

[31] For partially miscible systems, Wilson (1964) suggested that the right-hand side of Eq. (6-104) be multiplied by a constant greater than unity. This suggestion not only introduces a third parameter but, more important, creates difficulties when the equation is applied to ternary (or higher) systems (see Sec. 6.15).

$$\ln\gamma_1 = x_2^2\left[\tau_{21}\left(\frac{G_{21}}{x_1 + x_2G_{21}}\right)^2 + \frac{\tau_{12}G_{12}}{(x_2 + x_1G_{12})^2}\right] \tag{6-112}$$

$$\ln\gamma_2 = x_1^2\left[\tau_{12}\left(\frac{G_{12}}{x_2 + x_1G_{12}}\right)^2 + \frac{\tau_{21}G_{21}}{(x_1 + x_2G_{21})^2}\right] \tag{6-113}$$

For moderately nonideal systems, the NRTL equation offers no advantages over the simpler van Laar and three-suffix Margules equations. However, for strongly nonideal mixtures, and especially for partially immiscible systems,[32] the NRTL equation often provides a good representation of experimental data if care is exercised in data reduction to obtain the adjustable parameters. For example, consider the nitroethane/isooctane system studied by Renon; below 30°C, this system has a miscibility gap. Reduction of liquid-liquid equilibrium data below 30°C and vapor-liquid equilibrium data at 25 and 45°C gave the results shown in Fig. 6-18. The parameters $(g_{12} - g_{22})$ and $(g_{21} - g_{11})$ appear to be linear functions of temperature, showing no discontinuities in the region of the critical solution temperature.

Renon's and Wilson's equations are readily generalized to multicomponent mixtures as discussed in Sec. 6.15.

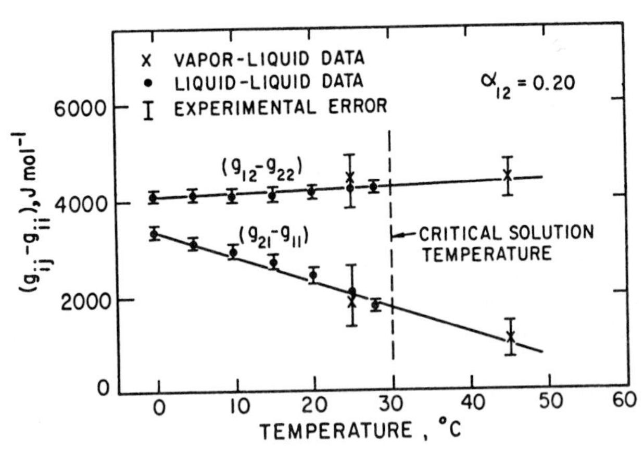

Figure 6-18 Parameters in NRTL equation for the nitroethane (1)/isooctane (2) system calculated from vapor-liquid and liquid-liquid equilibrium data.

[32] See App. E.

UNIQUAC Equation. A critical examination of the derivation of the NRTL equation shows that this equation, like those obtained from Wohl's expansion, is more suitable for h^E than g^E (Renon and Prausnitz, 1969). Further, because experimental data for typical binary mixtures are usually not sufficiently plentiful or precise to yield three meaningful binary parameters, attempts were made (Abrams, 1975; Maurer, 1978; Anderson, 1978; Kemeny and Rasmussen, 1981) to derive a two-parameter equation for g^E that retains at least some of the advantages of the equation of Wilson without restriction to completely miscible mixtures. Abrams derived an equation that, in a sense, extends the quasichemical theory of Guggenheim (see Sec. 7.6) for nonrandom mixtures to solutions containing molecules of different size. This extension was therefore called the *uni*versal *qua*si-*c*hemical theory or, in short, UNIQUAC. As discussed in Sec. 7.7, the UNIQUAC equation for g^E consists of two parts, a *combinatorial part* that attempts to describe the dominant entropic contribution, and a *residual part* that is due primarily to intermolecular forces that are responsible for the enthalpy of mixing. The combinatorial part is determined only by the composition and by the sizes and shapes of the molecules; it requires only pure-component data. The residual part, however, depends also on intermolecular forces; the two adjustable binary parameters, therefore, appear only in the residual part. The UNIQUAC equation is

$$\frac{g^E}{RT} = \left(\frac{g^E}{RT}\right)_{\text{combinatorial}} + \left(\frac{g^E}{RT}\right)_{\text{residual}} \tag{6-114}$$

For a binary mixture,

$$\left(\frac{g^E}{RT}\right)_{\text{combinatorial}} = x_1 \ln\frac{\Phi_1^*}{x_1} + x_2 \ln\frac{\Phi_2^*}{x_2} + \frac{z}{2}\left(x_1 q_1 \ln\frac{\theta_1}{\Phi_1^*} + x_2 q_2 \ln\frac{\theta_2}{\Phi_2^*}\right) \tag{6-115}$$

$$\left(\frac{g^E}{RT}\right)_{\text{residual}} = -x_1 q_1' \ln(\theta_1' + \theta_2'\tau_{21}) - x_2 q_2' \ln(\theta_2' + \theta_1'\tau_{12}) \tag{6-116}$$

where the coordination number z is set equal to 10. Segment fraction, Φ^*, and area fractions, θ and θ', are given by

$$\Phi_1^* = \frac{x_1 r_1}{x_1 r_1 + x_2 r_2} \qquad \Phi_2^* = \frac{x_2 r_2}{x_1 r_1 + x_2 r_2} \tag{6-117}$$

$$\theta_1 = \frac{x_1 q_1}{x_1 q_1 + x_2 q_2} \qquad \theta_2 = \frac{x_2 q_2}{x_1 q_1 + x_2 q_2} \tag{6-118}$$

$$\theta'_1 = \frac{x_1 q'_1}{x_1 q'_1 + x_2 q'_2} \qquad \theta'_2 = \frac{x_2 q'_2}{x_1 q'_1 + x_2 q'_2} \tag{6-119}$$

Parameters r, q, and q' are pure-component molecular-structure constants depending on molecular size and external surface areas. In the original formulation, $q = q'$. To obtain better agreement for systems containing water or lower alcohols, q' values for water and alcohols were adjusted empirically by Anderson (1978) to give an optimum fit to a variety of systems containing these components. For alcohols, the surface of interaction q' is smaller than the geometric external surface q, suggesting that intermolecular attraction is dominated by the OH group (hydrogen bonding). Table 6-9 presents a few values of these structural parameters. For fluids other than water or lower alcohols, $q = q'$.

For each binary mixture, there are two adjustable parameters, τ_{12} and τ_{21}. These, in turn, are given in terms of characteristic energies Δu_{12} and Δu_{21}, by

$$\tau_{12} = \exp\left(-\frac{\Delta u_{12}}{RT}\right) \equiv \exp\left(-\frac{a_{12}}{T}\right) \tag{6-120}$$

$$\tau_{21} = \exp\left(-\frac{\Delta u_{21}}{RT}\right) \equiv \exp\left(-\frac{a_{21}}{T}\right) \tag{6-121}$$

For many cases, Eqs. (6-120) and (6-121) give the primary effect of temperature on τ_{12} and τ_{21}. Characteristic energies Δu_{12} and Δu_{21} are often only weakly dependent on temperature. Table 6-10 gives some binary parameters a_{12} and a_{21}.

Activity coefficients γ_1 and γ_2 are given by

$$\begin{aligned}\ln\gamma_1 = {} & \ln\frac{\Phi_1^*}{x_1} + \frac{z}{2} q_1 \ln\frac{\theta_1}{\Phi_1^*} + \Phi_2^*\left(l_1 - \frac{r_1}{r_2} l_2\right) \\ & - q'_1 \ln(\theta'_1 + \theta'_2\tau_{21}) + \theta'_2 q'_1\left(\frac{\tau_{21}}{\theta'_1 + \theta'_2\tau_{21}} - \frac{\tau_{12}}{\theta'_2 + \theta'_1\tau_{12}}\right)\end{aligned} \tag{6-122}$$

$$\begin{aligned}\ln\gamma_2 = {} & \ln\frac{\Phi_2^*}{x_2} + \frac{z}{2} q_2 \ln\frac{\theta_2}{\Phi_2^*} + \Phi_1^*\left(l_2 - \frac{r_2}{r_1} l_1\right) \\ & - q'_2 \ln(\theta'_2 + \theta'_1\tau_{12}) + \theta'_1 q'_2\left(\frac{\tau_{12}}{\theta'_2 + \theta'_1\tau_{12}} - \frac{\tau_{21}}{\theta'_1 + \theta'_2\tau_{21}}\right)\end{aligned} \tag{6-123}$$

Table 6-9 Some structural parameters for UNIQUAC equation.*

Component	r	q
Carbon tetrachloride	3.33	2.82
Chloroform	2.70	2.34
Formic acid	1.54	1.48
Methanol	1.43	1.43
Acetonitrile	1.87	1.72
Acetic acid	1.90	1.80
Nitroethane	2.68	2.41
Ethanol	2.11	1.97
Acetone	2.57	2.34
Ethyl acetate	3.48	3.12
Methyl ethyl ketone	3.25	2.88
Diethylamine	3.68	3.17
Benzene	3.19	2.40
Methylcyclopentane	3.97	3.01
Methyl isobutyl ketone	4.60	4.03
n-Hexane	4.50	3.86
Toluene	3.92	2.97
n-Heptane	5.17	4.40
n-Octane	5.85	4.94
Water	0.92	1.40

Component	q'	Component	q'
Water	1.00	C_4-alcohols	0.88
CH_3OH	0.96	C_5-alcohols	1.15
C_2H_5OH	0.92	C_6-alcohols	1.78
C_3-alcohols	0.89	C_7-alcohols	2.71

* These parameters are dimensionless because they are (arbitrarily) taken relative to the size and surface area of a $-CH_2-$ unit in a high-molecular-weight paraffin.

where

$$l_1 = \frac{z}{2}(r_1 - q_1) - (r_1 - 1) \tag{6-124}$$

$$l_2 = \frac{z}{2}(r_2 - q_2) - (r_2 - 1) \tag{6-125}$$

The UNIQUAC equation is applicable to a wide variety of nonelectrolyte liquid mixtures containing nonpolar or polar fluids such as hydrocarbons, alcohols, nitriles, ketones,

Table 6-10 Some binary parameters for UNIQUAC equation.*

		Energy parameters (K)	
System (1)/(2)	T (K)	a_{12}	a_{21}
Acetonitrile/benzene	318	-40.70	299.79
n-Hexane/nitromethane	318	230.64	-5.86
Acetone/chloroform	323	-171.71	93.93
Ethanol/*n*-octane	348	-123.57	1354.92
Formic acid/acetic acid	374-387	-144.58	241.64
Propionic acid/methyl isobutyl ketone	390-411	-78.49	136.46
Acetone/water	331-368	530.99	-100.71
Acetonitrile/water	350-364	294.10	61.92
Acetic acid/water	373-389	530.94	-299.90
Formic acid/water	374-380	924.01	-525.85
Methylcyclopentane/ethanol	333-349	1383.93	-118.27
Methylcyclopentane/benzene	344-352	56.47	-6.47
Ethanol/carbon tetrachloride	340-351	-138.90	947.20
Ethanol/benzene	350-369	-75.13	242.53
Methyl ethyl ketone/*n*-heptane	328	-29.64	1127.95
Methanol/benzene	528	-56.35	972.09
Chloroform/ethanol	323	934.23	-208.50
Chloroform/*n*-heptane	323	-19.26	88.40
Ethanol/*n*-heptane	323	-105.23	1380.30
Acetone/methanol	323	379.31	-108.42
Methanol/ethyl acetate	335-347	-107.54	579.61

* Data sources are given by Anderson (1978).

aldehydes, organic acids, etc. and water, including partially miscible mixtures. With only two adjustable binary parameters, it cannot always represent high-quality data with high accuracy, but for many typical mixtures encountered in chemical practice, UNIQUAC provides a satisfactory description.[33]

The main advantages of UNIQUAC follow first, from its (relative) simplicity, using only two adjustable parameters, and second, from its wide range of applicability. To illustrate, some results are shown in Figs. 6-19 to 6-24.

Figure 6-19 shows experimental and calculated phase equilibria for the acetonitrile/benzene system at 45°C. This system exhibits moderate, positive deviations from Raoult's law. The high-quality data of Brown and Smith (1955) are very well represented by the UNIQUAC equation.

---

[33] UNIQUAC parameters are given for many binary systems by J. Gmehling, U. Onken, and W. Arlt, DECHEMA Chemistry Data Series, since 1977. See also Prausnitz *et al.*, 1980.

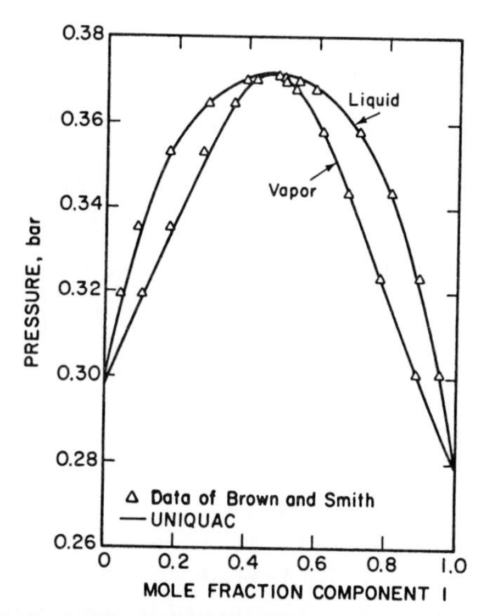

Figure 6-19 Moderate positive deviations from ideality. Vapor-liquid equilibria for the acetonitrile (1)/benzene (2) system at 45°C.

Figure 6-20 shows the isothermal data of Edwards (1962) for *n*-hexane and nitroethane. This system also exhibits positive deviations from Raoult's law; however, these deviations are much larger than those shown in Fig. 6-19. At 45°C, the mixture shown in Fig. 6-20 is only 15°C above its critical solution temperature (see Sec. 6.13).

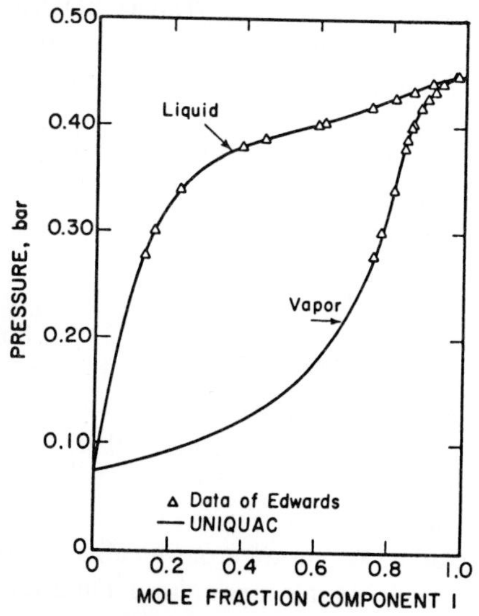

Figure 6-20 Strong positive deviations from ideality. Vapor-liquid equilibria for the *n*-hexane (1)/ nitroethane (2) system at 45°C.

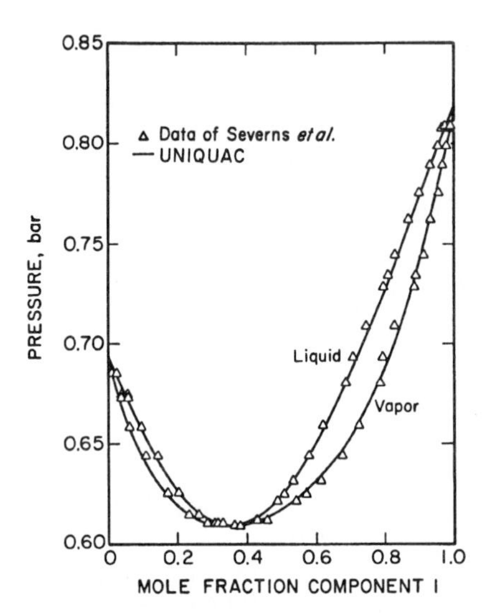

Figure 6-21 Negative deviations from ideality. Vapor-liquid equilibria for the acetone (1)/chloroform (2) system at 50°C.

The acetone/chloroform system, shown in Fig. 6-21, exhibits strong negative deviations from Raoult's law because of hydrogen bonding between the single hydrogen atom of chloroform and the carbonyl oxygen of acetone.

Figure 6-22 shows a fit of the UNIQUAC equation to the isobaric data of Nakanishi *et al.* (1967) for the methanol/diethylamine system; this system also exhibits strong negative deviations from Raoult's law. The UNIQUAC equation correctly reproduces a weak minimum in the activity coefficient of methanol. Agreement with experiment is not as good as that in previous examples because the data are somewhat scattered, particularly near the azeotrope.

At moderate pressures, vapor-phase nonideality is usually small in comparison to liquid-phase nonideality. However, when associating carboxylic acids are present, vapor-phase nonideality may dominate. These acids dimerize appreciably in the vapor phase even at low pressures; fugacity coefficients are well removed from unity. To illustrate, Fig. 6-23 shows observed and calculated vapor-liquid equilibria for two systems containing an associating component. In Fig. 6-23(a), both components strongly associate with themselves and with each other. In Fig. 6-23(b), only one of the components associates strongly. For both systems, representation of the data is very good. However, the interesting feature of these systems is that, whereas the fugacity coefficients are significantly remote from unity, the activity coefficients show only minor deviations from ideal-solution behavior. Figures 5-35 and 5-36 indicate that the fugacity coefficients show marked departure from ideality. In these systems, the major contribution to nonideality occurs in the vapor phase. Failure to take into account these strong vapor-phase nonidealities would result in erroneous activity-coefficient parameters, a_{12} and a_{21}, leading to poor prediction of multicomponent equilibria.

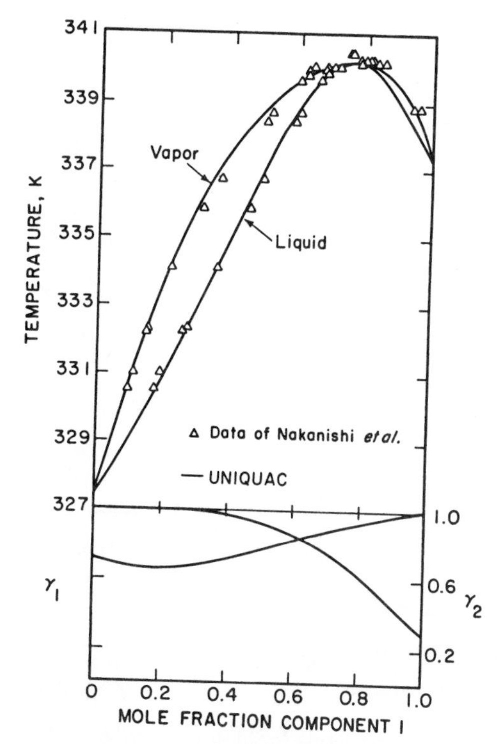

Figure 6-22 Vapor-liquid equilibria for a mixture containing two hydrogen-bonding components. Temperature-composition diagram for the system methanol (1)/diethylamine (2) at 0.973 bar.

6.12 Excess Functions and Partial Miscibility

In preceding sections, we have been concerned with mixtures of liquids that are completely miscible. We now consider briefly the thermodynamics of binary liquid systems wherein the components are only partially miscible.[34]

At a fixed temperature and pressure, a stable state is that which has a minimum Gibbs energy. Thermodynamic stability analysis tells us that a liquid mixture splits into two separate liquid phases if upon doing so, it can lower its Gibbs energy. To fix ideas, consider a mixture of two liquids, 1 and 2, whose calculated Gibbs energy of mixing at constant temperature and pressure is given by the heavy line (1) in Fig. 6-24. If the composition of the mixture is that corresponding to point *a*, then the molar Gibbs energy of that mixture is

[34] See App. E.

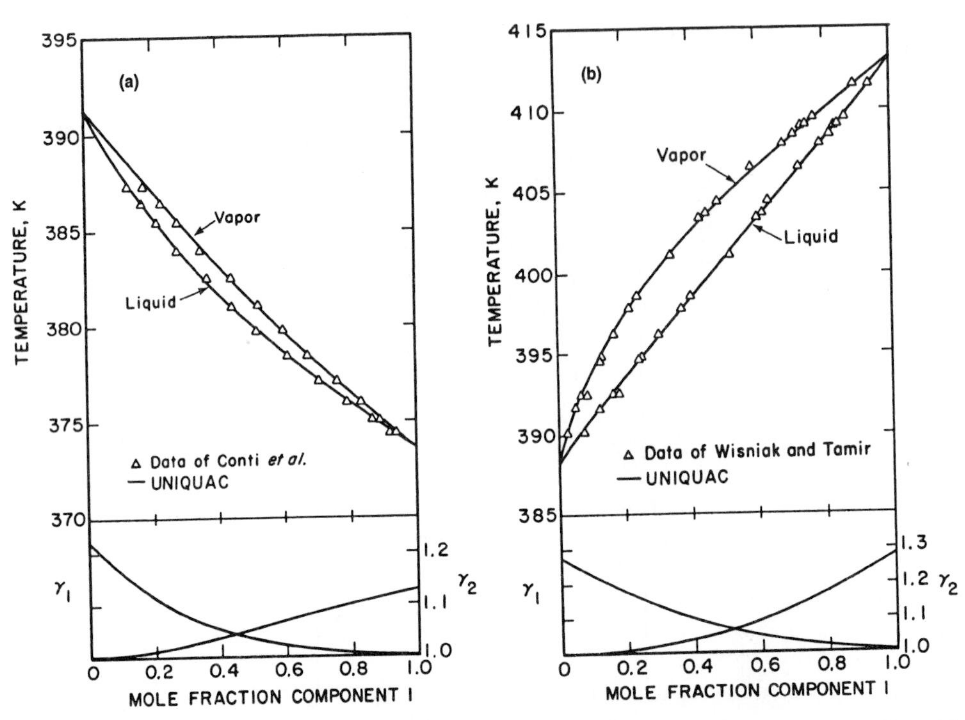

Figure 6-23 Vapor-liquid equilibria for binary systems containing two carboxylic acids or one carboxylic acid and one ketone. Temperature-composition diagrams and activity coefficients at 1.013 bar for the systems: (a) formic acid (1)/acetic acid (2); (b) propionic acid (1)/methyl isobutyl ketone (2).

$$\underset{(\text{at } a)}{g_{\text{mixt}}} = x_1 g_{\text{pure }1} + x_2 g_{\text{pure }2} + \Delta g_a \tag{6-126}$$

However, if the mixture splits into two separate liquid phases, one having mole fraction x_1' and the other having mole fraction x_1'', then the Gibbs energy change upon mixing is given by point b and the molar Gibbs energy of the two-phase mixture is

$$\underset{(\text{at } b)}{g_{\text{mixt}}} = x_1 g_{\text{pure }1} + x_2 g_{\text{pure }2} + \Delta g_b \tag{6-127}$$

Mole fractions x_1' and x_1'' in Eq. (6-127) represent the overall composition and they are the same as those in Eq. (6-126).

It is evident from Fig. 6-24 that point b represents a lower Gibbs energy of the mixture than does point a. Therefore, at temperature T_1 the liquid mixture having overall composition x_1 splits into two liquid phases having mole fractions x_1' and x_1''. Point b represents the lowest possible Gibbs energy that the mixture may attain subject to the restraints of fixed temperature, pressure, and overall composition x_1.

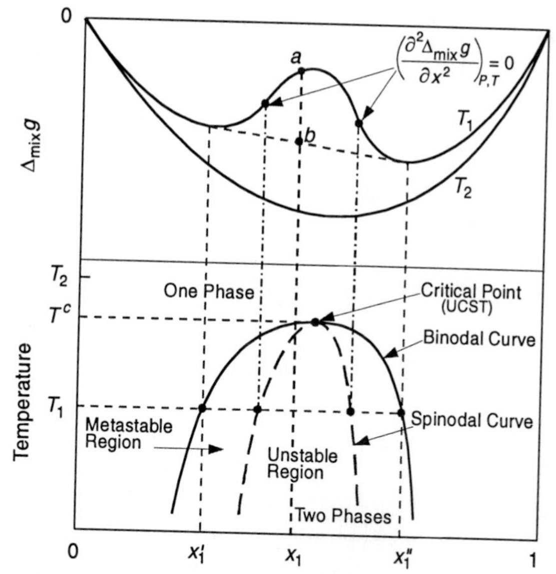

Figure 6-24 Molar Gibbs energy of mixing and *T*-*x* diagram of a mixture at constant pressure: T_1, partially miscible; T_2, totally miscible.

A decrease in the Gibbs energy of a binary liquid mixture due to the formation of another liquid phase can occur only if a plot of the Gibbs energy change of mixing against mole fraction is, in part, concave downward. Therefore, the condition for instability of a binary liquid mixture is

$$\left(\frac{\partial^2 g_{\text{mixt}}}{\partial x^2}\right)_{T,P} < 0 \qquad \text{(6-128)}^{35}$$

or

$$\left(\frac{\partial^2 \Delta_{\text{mix}} g}{\partial x^2}\right)_{T,P} < 0 \qquad \text{(6-129)}$$

In the *T*-*x* diagram shown in the lower part of Fig. 6-24, T^c is the *critical solution temperature* (see Sec. 6.13). At temperatures $T > T^c$, the mixture is completely miscible because for all mole fractions $(\partial^2 \Delta_{\text{mix}} g / \partial x^2)_{T,P} > 0$. At $T < T^c$, the mixture is partially miscible because in part of the mole fraction range $(\partial^2 \Delta_{\text{mix}} g / \partial x^2)_{T,P} < 0$. The *binodal curve* is the boundary between the one-phase region and the two-phase region. Within the two-phase region, the *spinodal curve* $[(\partial^2 \Delta_{\text{mix}} g / \partial x^2)_{T,P} = 0]$ distinguishes

[35] In Eqs. (6-128) and (6-129) x stands for either x_1 or x_2.

the unstable region $[(\partial^2\Delta_{\text{mix}}g/\partial x^2)_{T,P} < 0]$ from the metastable region $[(\partial^2\Delta_{\text{mix}}g/\partial x^2)_{T,P} > 0]$. If the overall mole fraction of the mixture falls within the unstable region, spontaneous demixing occurs when going from the one-phase to the two-phase region.

Let us now introduce an excess function into Eq. (6-128). As before, we define the excess Gibbs energy of a mixture relative to the Gibbs energy of an ideal mixture in the sense of Raoult's law:

$$g^E \equiv g_{\text{mixt}} - RT(x_1 \ln x_1 + x_2 \ln x_2) - x_1 g_{\text{pure 1}} - x_2 g_{\text{pure 2}} \tag{6-130}$$

Substituting into Eq. (6-128), we obtain for instability

$$\boxed{\left(\frac{\partial^2 g^E}{\partial x_1^2}\right)_{T,P} + RT\left(\frac{1}{x_1} + \frac{1}{x_2}\right) < 0} \tag{6-131}$$

For an ideal solution, $g^E = 0$ for all x and in that event the inequality is never obeyed for any values of x_1 and x_2 in the interval zero to one. Therefore, we conclude that an ideal solution is always stable and cannot exhibit phase splitting.

Suppose now that the excess Gibbs energy is not zero but is given by the simple expression

$$g^E = Ax_1x_2 \tag{6-132}$$

where A is a constant dependent only on temperature. Then

$$\left(\frac{\partial^2 g^E}{\partial x_1^2}\right)_{T,P} = -2A \tag{6-133}$$

and substitution in Eq. (6-131) gives

$$-2A < -RT\left(\frac{1}{x_1} + \frac{1}{x_2}\right) \tag{6-134}$$

Multiplying both sides by -1 inverts the inequality sign, and the condition for instability becomes

$$2A > RT\left(\frac{1}{x_1} + \frac{1}{x_2}\right) = \frac{RT}{x_1x_2} \tag{6-135}$$

The smallest value of A that satisfies inequality (6-135) is

$$A = 2RT \tag{6-136}$$

and, therefore, instability occurs whenever

$$\boxed{\frac{A}{RT} > 2} \tag{6-137}$$

The borderline between stability and instability of a liquid mixture is called *incipient instability*. This condition corresponds to a critical state and it occurs when the two points of inflection shown in Fig. 6-24 merge into a single point. Incipient instability, therefore, is characterized by the two equations

$$\left(\frac{\partial^2 g_{\text{mixt}}}{\partial x^2}\right)_{T,P} = 0 \tag{6-138}$$

and

$$\left(\frac{\partial^3 g_{\text{mixt}}}{\partial x^3}\right)_{T,P} = 0 \tag{6-139}$$

An equivalent, but more useful characterization of incipient instability is provided by introducing into Eqs. (6-168) and (6-139) the activity function given by

$$g_{\text{mixt}} = RT(x_1 \ln a_1 + x_2 \ln a_2) + x_1 g_{\text{pure 1}} + x_2 g_{\text{pure 2}} \tag{6-140}$$

We then obtain for incipient instability

$$\left(\frac{\partial \ln a_1}{\partial x_1}\right)_{T,P} = 0 \tag{6-141}$$

and

$$\left(\frac{\partial^2 \ln a_1}{\partial x_1^2}\right)_{T,P} = 0 \tag{6-142}$$

We now want to illustrate graphically instability, incipient instability, and stability in a binary liquid mixture. Figure 6-25 gives a plot of activity versus mole fraction as calculated from the simple excess Gibbs energy expression given by Eq. (6-132); the activity is given by

$$\ln a_1 = \ln \gamma_1 + \ln x_1 = \frac{A}{RT} x_2^2 + \ln x_1 \tag{6-143}$$

When $A/RT > 2$, the curve has a maximum and a minimum; for this case there are two stable liquid phases whose compositions are given by x_1' and x_1'' as shown schematically in Fig. 6-25.[36] When $A/RT = 2$, the maximum and minimum points coincide and we have incipient instability. For $A/RT < 2$, only one liquid phase is stable.

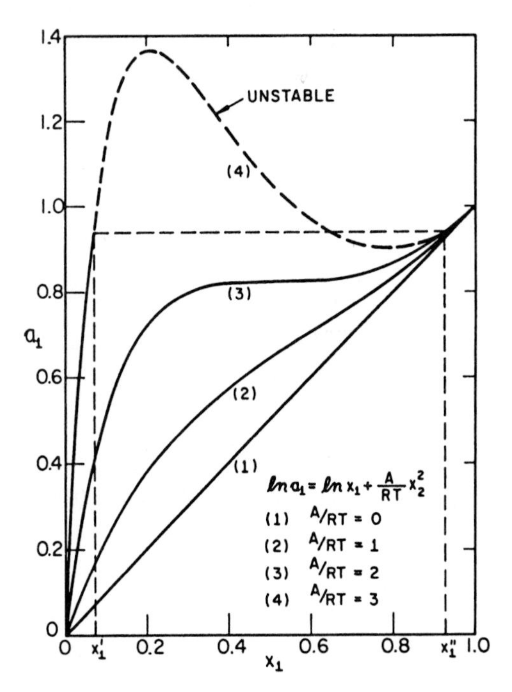

Figure 6-25 Activity of component 1 in a binary liquid solution for different values of *A/RT*. Curve (3) shows incipient instability.

A plot similar to Fig. 6-25 is shown in Fig. 6-26 that presents experimental results reported by Butler (1937) for four binary aqueous systems containing methyl, ethyl, *n*-propyl, or *n*-butyl alcohol at 25°C. Methyl and ethyl alcohols are completely miscible with water and when their activities are plotted against their mole fractions, there is no point of inflection. Propyl alcohol is also completely miscible with water at

---

[36] Compositions x_1' and x_1'' are found from simultaneous solution of two equilibrium relations: $(\gamma_1 x_1)' = (\gamma_1 x_1)''$ and $(\gamma_2 x_2)' = (\gamma_2 x_2)''$ plus material balances $x_1' + x_2' = 1$ and $x_1'' + x_2'' = 1$. See App. E.

this temperature but just barely so; the plot of activity versus mole fraction almost shows a point of inflection. Butyl alcohol, however, is miscible with water over only small ranges of concentration. If a continuous line were drawn through the experimental points shown in Fig. 6-26, it would necessarily have to go through a maximum and a minimum, analogous to an equation-of-state isotherm on P-V coordinates in the two-phase region.

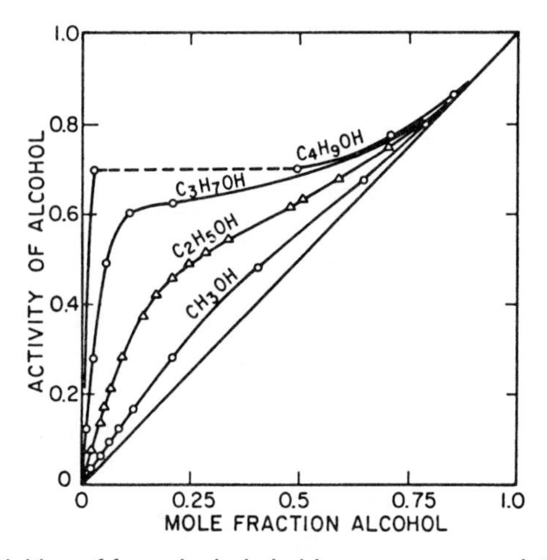

Figure 6-26 Activities of four alcohols in binary aqueous solution at 25°C. Data from Butler (1937).

6.13 Upper and Lower Consolute Temperatures

As indicated in the preceding section, the condition for instability of a binary liquid mixture depends on the nonideality of the solution and on the temperature. In the simplest case, when the excess Gibbs energy is given by a one-parameter equation such as Eq. (6-132), the temperature T^c for incipient instability is given by

$$T^c = \frac{A}{2R} \tag{6-144}$$

Temperature T^c is the *consolute temperature*;[37] when the excess Gibbs energy is given by a temperature-independent one-parameter Margules equation, T^c is always a maximum, but in general it may be a maximum (upper) or a minimum (lower) temperature on a T-x diagram, as shown in Fig. 6-27.

[37] Or, alternatively, the *critical solution temperature*.

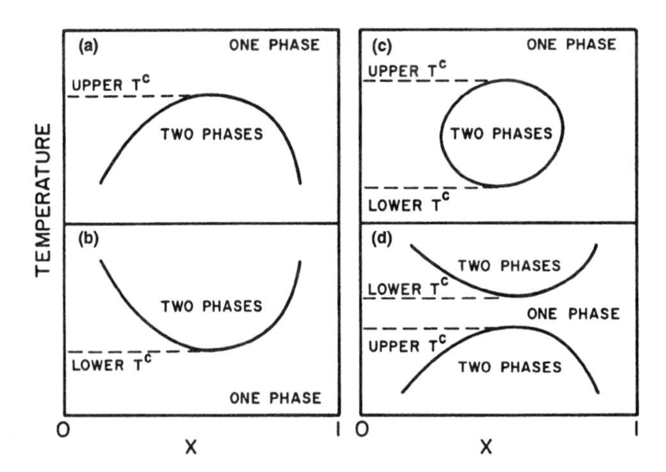

Figure 6-27 Phase stability in four binary liquid mixtures.

Some binary systems have both upper and lower consolute temperatures. Upper critical solution temperatures are more common than lower critical solution temperatures, although the latter are sometimes observed in mixtures of components that form hydrogen bonds with one another (e.g., aqueous mixtures of amines). Such phase behavior is often observed in polymer solutions, as discussed in Sec. 8.2. In many simple liquid mixtures, parameter A is a weak function of temperature and therefore vapor-liquid-equilibrium measurements obtained at some temperature not far from T^c may be used to estimate the consolute temperature. However, it sometimes happens that A depends on temperature in a complicated way and, unless data are taken very near T^c, Eq. (6-144) provides only an approximation even in those cases where the simple one-parameter equation is adequate for the excess Gibbs energy. Figure 6-27 illustrates four possible cases of phase stability corresponding to the four types of temperature dependence for parameter A shown in Fig. 6-28. In Fig. 6-27, in case (a), T^c is a maximum and in case (b) it is a minimum; in case (c), as temperature rises, there is first a minimum and then a maximum, and in case (d), as temperature rises, there is first a maximum and then a minimum.

Figure 6-28 shows that when A/RT falls with increasing temperature, we may obtain an upper T^c; if A/RT rises with increasing temperature, we may obtain a lower T^c. If A/RT goes through a maximum, we may obtain first, a lower T^c followed by an upper T^c and we may obtain the reverse when A/RT goes through a minimum.

When the excess Gibbs energy is given by Eq. (6-132), we always find that the composition corresponding to the consolute temperature is $x_1 = x_2 = 1/2$. However, when the excess Gibbs energy is given by a function that is not symmetric with respect to x_1 and x_2, the coordinates of the consolute point are not at the composition midpoint.

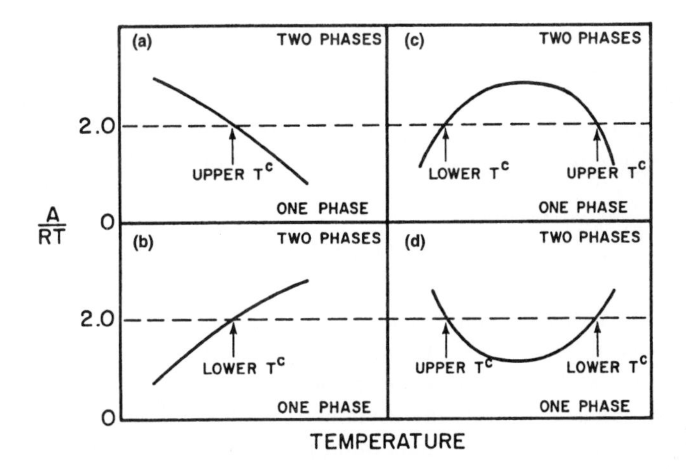

Figure 6-28 Phase stability in four binary liquid mixtures. For each, the excess Gibbs energy is given by a two-suffix Margules equation.

For example, if the excess Gibbs energy is given by van Laar's equation in the form[38]

$$g^E = \frac{Ax_1x_2}{\frac{A}{B}x_1 + x_2} \tag{6-145}$$

we find, upon substitution into Eqs. (6-141) and (6-142) that the coordinates of the consolute point are

$$T^c = \frac{2x_1x_2\frac{A^2}{B}}{R\left(\frac{A}{B}x_1 + x_2\right)^3}$$

$$x_1^c = \frac{\left[\left(\frac{A}{B}\right)^2 + 1 - \frac{A}{B}\right]^{1/2} - \frac{A}{B}}{1 - \frac{A}{B}} \tag{6-146}$$

[38] For a discussion of how different models for g^E affect critical mixing, see J. Wisniak, 1983, *Chem. Eng. Sci.*, 38: 969. The same author also analyses the influence of a third component on the mutual solubilities of two liquids (1984, *Chem. Eng. Sci.*, 39: 111).

In Sec. 6.12 we indicated that when the excess Gibbs energy is assumed to follow a two-suffix Margules expression as given by Eq. (6-132), incipient instability occurs when $A = 2RT$. A two-suffix Margules equation, however, is only a rough approximation for many real mixtures and a more accurate description is given by including higher terms. If we write the excess Gibbs energy in the three-parameter Redlich-Kister series,

$$g^E = x_1 x_2 [A + B(x_1 - x_2) + C(x_1 - x_2)^2] \tag{6-147}$$

and substitute this series into the equations of incipient instability, we obtain the results given in Figs. 6-29 and 6-30, taken from Shain (1963). Figure 6-29 gives the effect of the coefficient B (when $C = 0$) on the maximum value of A/RT for complete miscibility. Coefficient B reflects the asymmetry of the excess Gibbs energy function (see Sec. 6.5), and we see that the maximum permissible value of A/RT decreases below 2 as the asymmetry rises. Figure 6-30 gives the effect of both coefficients B and C on the maximum value of A/RT for complete miscibility. As discussed in Sec. 6.5, the third term in the Redlich-Kister expansion is symmetric in the mole fraction; it affects the flatness or steepness of the excess Gibbs energy curve. Positive values of C make the excess Gibbs energy curve flatter, and from Fig. 6-30 we see that for small values of B, positive values of C increase the maximum permissible value of A/RT beyond 2 to 2.4.

Shain's calculations show that whereas large positive values of the constant A favor limited miscibility, values of the constant B of either sign increase the tendency toward limited miscibility for a given value of A; however, small positive values of the constant C tend to decrease the tendency for phase separation. These calculations offer a useful guide but they must not be taken too seriously because they are based on a twofold differentiation of the excess Gibbs energy function. Extremely accurate data are required to assign precise quantitative significance to a function based on the second derivative of such data.

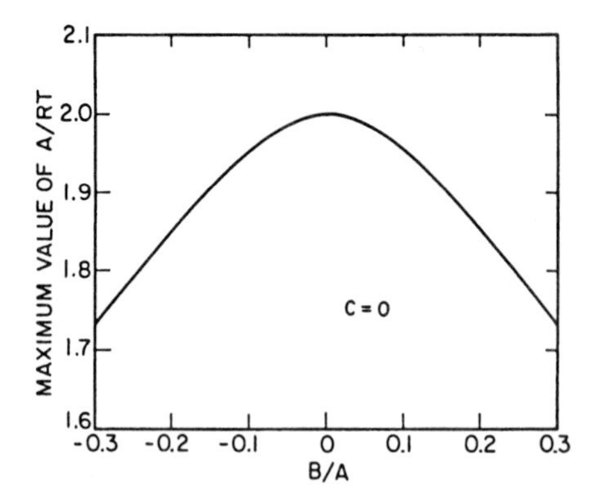

Figure 6-29 Effect of Redlich-Kister coefficient *B* on maximum values of *A/RT* for complete miscibility.

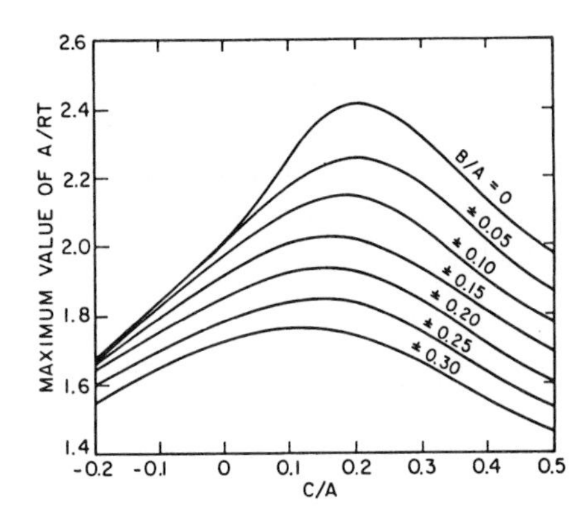

Figure 6-30 Effect of Redlich-Kister coefficient C on maximum values of A/RT for complete miscibility.

In concluding this section, we note that the thermodynamics of phase stability in binary liquid systems does not require that instability can occur only if the excess Gibbs energy is positive. In principle, a binary liquid mixture may be only partially miscible even though it has a negative excess Gibbs energy. However, such behavior is unlikely because the composition dependence of a negative excess Gibbs energy would have to be unusual to satisfy the condition of instability, Eq. (6-131). A good review of this subject is given by Sørensen *et al.* (1979, 1979a, 1979b 1980). A compilation of liquid-liquid equilibrium data is given by Sørensen and Arlt (1979).

Appendix E gives some further discussion on liquid-liquid equilibrium.

6.14 Excess Functions for Multicomponent Mixtures

So far in this chapter, we have primarily considered binary mixtures. We now turn to a discussion of mixtures containing more than two components, with particular attention to their excess Gibbs energies.

One of the main uses of excess functions for describing the thermodynamic properties of liquid mixtures lies in establishing thermodynamically consistent relations for multicomponent mixtures containing any desired number of components; from these relations we can then calculate activity coefficients needed to find liquid-phase fugacities. Expressions for the excess functions require a number of constants and many of these can be evaluated from binary data alone; in some cases, it is possible to obtain all the required constants from binary data. Application of excess functions to mixtures of more than two components, therefore, provides a significant laborsaving device that minimizes the experimental effort required to describe a mixture of many components.

Wohl's Equation. To illustrate the utility of excess functions, we consider first a ternary mixture as described by Wohl's method (see Sec. 6.10). Extension to systems containing more than three components will then be evident.

As for a binary solution, the Gibbs energy is again a summation of two-body, three-body, etc., interactions. When the excess Gibbs energy is relative to an ideal solution in the sense of Raoult's law, we have

$$\frac{g^E}{RT(x_1q_1+x_2q_2+x_3q_3)} = 2a_{12}z_1z_2 + 2a_{13}z_1z_3 + 2a_{23}z_2z_3 + 3a_{112}z_1^2z_2 + 3a_{122}z_1z_2^2 + 3a_{113}z_1^2z_3 + 3a_{133}z_1z_3^2 + 3a_{223}z_2^2z_3 + 3a_{233}z_2z_3^2 + 6a_{123}z_1z_2z_3 + \ldots \tag{6-148}$$

Two-Suffix Margules Equation. First, consider a simple case: Suppose that components 1, 2, and 3 are chemically similar and of approximately the same size. We assume that $q_1 = q_2 = q_3$ and that all three-body terms (and higher) may be neglected. Equation (6-148) simplifies to

$$\frac{g^E}{RT} = 2qa_{12}x_1x_2 + 2qa_{13}x_1x_3 + 2qa_{23}x_2x_3 \tag{6-149}$$

The activity coefficients follow from differentiation as given by Eq. (6-25). They are

$$\ln\gamma_1 = A'_{12}x_2^2 + A'_{13}x_3^2 + (A'_{12} + A'_{13} - A'_{23})x_2x_3 \tag{6-150}$$

$$\ln\gamma_2 = A'_{12}x_1^2 + A'_{23}x_3^2 + (A'_{12} + A'_{23} - A'_{13})x_1x_3 \tag{6-151}$$

$$\ln\gamma_3 = A'_{13}x_1^2 + A'_{23}x_2^2 + (A'_{13} + A'_{23} - A'_{12})x_1x_2 \tag{6-152}$$

where $A'_{12} = 2qa_{12}$, $A'_{13} = 2qa_{13}$, and $A'_{23} = 2qa_{23}$.

Equations (6-150), (6-151), and (6-152) possess a great advantage: All the constants may be obtained from binary data without further assumptions. Constant $2qa_{12}$ is given by data for the 1-2 binary; constants $2qa_{13}$ and $2qa_{23}$ are given, respectively, by data for the 1-3 and 2-3 binaries. Thus, by assumption, no ternary data are required to calculate activity coefficients for the ternary mixture.

Van Laar Equation. A somewhat more realistic, but still simplified, model of a ternary solution is provided by again assuming that all three-body terms (and higher) in Eq. (6-148) may be neglected, but this time we do not assume equality of all q terms. These assumptions lead to the van Laar equations for a ternary mixture. The molar excess Gibbs energy is given by

$$\frac{g^E}{RT} = \frac{2q_2a_{12}x_1x_2 + 2q_3a_{13}x_1x_3 + \dfrac{2q_2q_3}{q_1}a_{23}x_2x_3}{x_1 + \dfrac{q_2}{q_1}x_2 + \dfrac{q_3}{q_1}x_3} \tag{6-153}$$

To simplify notation, let

$$A'_{12} = 2q_1a_{12} \qquad A'_{21} = 2q_2a_{12}$$

$$A'_{13} = 2q_1a_{13} \qquad A'_{31} = 2q_3a_{13}$$

$$A'_{23} = 2q_2a_{23} \qquad A'_{32} = 2q_3a_{23}$$

Upon differentiation according to Eq. (6-25), the activity coefficient of component 1 is

$$\ln\gamma_1 = \frac{x_2^2A'_{12}\left(\dfrac{A'_{21}}{A'_{12}}\right)^2 + x_3^2A'_{13}\left(\dfrac{A'_{31}}{A'_{13}}\right)^2 + x_2x_3\dfrac{A'_{21}}{A'_{12}}\dfrac{A'_{31}}{A'_{13}}\left(A'_{12} + A'_{13} - A'_{32}\left(\dfrac{A'_{13}}{A'_{31}}\right)\right)}{\left(x_1 + x_2\dfrac{A'_{21}}{A'_{12}} + x_3\dfrac{A'_{31}}{A'_{13}}\right)^2} \tag{6-154}$$

Expressions for γ_2 and γ_3 are of exactly the same form as that for γ_1. To obtain γ_2, Eq. (6-154) should be used with this change of all subscripts on the right-hand side: Replace 1 with 2; replace 2 with 3; and replace 3 with 1. To obtain γ_3, replace 1 with 3; replace 2 with 1; and replace 3 with 2. Notice that if $q_1 = q_2 = q_3$, then Eq. (6-154) reduces to Eq. (6-150).

All parameters in Eq. (6-154) may be obtained from binary data, as indicated by Eqs. (6-97) and (6-98).[39]

[39] Wohl's expansion requires that parameter q_i depend only on pure-component properties of component i; therefore, the three sets of binary parameters ($A'_{12}, A'_{21}; A'_{13}, A'_{31}; A'_{23}, A'_{32}$) are not independent because, according to the definitions given immediately after Eq. (6-153),

$$\frac{A'_{13}/A'_{31}}{A'_{12}/A'_{21}} = \frac{q_2}{q_3} = \frac{A'_{23}}{A'_{32}}$$

Three-Suffix Margules Equation. The three-suffix Margules equations can also be extended to a ternary mixture by Wohl's method, but now all the constants cannot be found from binary data alone unless an additional assumption is made. We assume that the q's are equal to one another but we retain three-body terms in the expansion for excess Gibbs energy; higher-body terms are neglected. Equation (6-148) becomes:

$$\frac{g^E}{RT} = 2qa_{12}x_1x_2 + 2qa_{13}x_1x_3 + 2qa_{23}x_2x_3 + 3qa_{112}x_1^2x_2$$
$$+3qa_{122}x_1x_2^2 + 3qa_{113}x_1^2x_3 + 3qa_{133}x_1x_3^2 + 3qa_{223}x_2^2x_3 \qquad (6\text{-}155)$$
$$+3qa_{233}x_2x_3^2 + 6qa_{123}x_1x_2x_3$$

All constants appearing in Eq. (6-155) can be obtained from binary data except the last one, qa_{123}. This last constant is characteristic of the interaction between three different molecules: one of component 1, one of component 2, and one of component 3. It is a ternary constant and, in principle, can be obtained only from ternary data.

To simplify notation, let

$$A'_{12} = q(2a_{12} + 3a_{122}) \qquad A'_{21} = q(2a_{12} + 3a_{112})$$

$$A'_{13} = q(2a_{13} + 3a_{133}) \qquad A'_{31} = q(2a_{13} + 3a_{113})$$

$$A'_{23} = q(2a_{23} + 3a_{233}) \qquad A'_{32} = q(a_{23} + 3a_{223})$$

and

$$Q' = \frac{3q}{2}(a_{122} + a_{112} + a_{133} + a_{113} + a_{233} + a_{223} - 4a_{123})$$

All constants of type A'_{ij} can be determined from binary data alone.[40] However, constant Q' requires information on the ternary mixture because it is a function of a_{123}.

Consider now the binary van Laar equations [Eqs. (6-97) and (6-98)] at infinite dilution. Algebraic substitution gives

$$\frac{(\ln\gamma_1^\infty/\ln\gamma_3^\infty)_{\text{binary 13}}}{(\ln\gamma_1^\infty/\ln\gamma_2^\infty)_{\text{binary 12}}} = \left(\frac{\ln\gamma_2^\infty}{\ln\gamma_3^\infty}\right)_{\text{binary 23}}$$

This interdependence of parameters suggests that experimental data for the 1-3 binary and for the 1-2 binary can be used to *predict* (in part) behavior of the 2-3 binary. Unfortunately, such prediction is rarely reliable.

[40] Notice that constants A'_{12} and A'_{21} are simply related to constants A' and B' as defined after Eqs. (6-100) and (6-101) for the case where four-body (and higher) interactions are neglected. The relations are: $A'_{12} = A' + B'$ and $A'_{12} = A' + B'/2$. Similar relations can be written for the 1-3 and 2-3 mixtures.

The activity coefficient for component 1 is given by

$$\begin{aligned}\ln\gamma_1 = {} & A'_{12}x_2^2(1-2x_1)+2A'_{21}x_1x_2(1-x_1)+A'_{13}x_3^2(1-2x_1) \\ & +2A'_{31}x_1x_3(1-x_1)-2A'_{23}x_2x_3^2-2A'_{32}x_2^2x_3 \\ & +[\tfrac{1}{2}(A'_{12}+A'_{21}+A'_{13}+A'_{23}+A'_{32})-Q'](x_2x_3-2x_1x_2x_3)\end{aligned} \quad (6\text{-}156)$$

Expressions for γ_2 and γ_3 can be obtained from Eq. (6-156) by a change of all subscripts on the right-hand side. For γ_2 replace 1 with 2; replace 2 with 3; and 3 with 1. For γ_3 replace 1 with 3; 2 with 1; and 3 with 2.

If no ternary data are available, it is possible to estimate a_{123} by a suitable assumption. A reasonable but essentially arbitrary assumption is to set $Q' = 0$. An extensive study of Eq. (6-156) has been made by Adler *et al.* (1966).

In principle, only one experimental ternary point is required to determine qa_{123}. In practice, however, it is not advisable to base a parameter on one point only. For accurate work it is best to measure vapor-liquid equilibria for several ternary compositions that, in addition to the binary data, can then be used to evaluate a truly representative ternary constant.

The paragraphs above have shown how Wohl's method may be used to derive expressions for activity coefficients in a ternary mixture; exactly the same principles apply for obtaining expressions for activity coefficients containing four, five, or more components. The generalization of Eqs. (6-149) and (6-153) to mixtures containing any number of components shows that all the constants may be calculated from binary data alone. However, the generalization of Eq. (6-155) to solutions containing any number of components shows that the constants appearing in the expressions for the activity coefficients must be found from data on all possible constituent ternaries as well as binaries. For the generalization of Eq. (6-155), data on quaternary, quinternary, etc. mixtures are not needed.

To illustrate the applicability of the three-suffix Margules equation to ternary systems, we consider three strongly nonideal ternary systems at 50°C studied by Severns *et al.* (1955). They are:

I. Acetone/methyl acetate/methanol.

II. Acetone/chloroform/methanol.

III. Acetone/carbon tetrachloride/methanol.

Margules constants for the three ternary systems are given in Table 6-11. For system I a good representation of the ternary data was obtained by using binary data only and setting $Q' = 0$ in Eq. (6-156). In system II the ternary data required a small but significant ternary constant $Q' = -0.368$ that, if it had been neglected, would introduce some, but not serious, error. However, in system III the ternary data required an appreciable ternary constant $Q' = 1.15$ that cannot be neglected because it is of the same order of magnitude as the various A'_{ij} constants for this system.

Table 6-11 Three-suffix Margules constants for three ternary systems at 50°C.*

System	Margules constants	
Acetone (1)/methyl acetate (2)/methanol (3)	$A'_{12} = 0.149$	$A'_{21} = 0.115$
	$A'_{13} = 0.701$	$A'_{31} = 0.519$
	$A'_{23} = 1.07$	$A'_{32} = 1.02$
	$Q' = 0$	
Acetone (1)/chloroform (2)/methanol (3)	$A'_{12} = 0.83$	$A'_{21} = -0.69$
	$A'_{13} = 0.701$	$A'_{31} = 0.519$
	$A'_{23} = 0.715$	$A'_{32} = 1.80$
	$Q' = -0.368$	
Acetone (1)/carbon tetrachloride (2)/methanol (3)	$A'_{12} = 0.715$	$A'_{21} = 0.945$
	$A'_{13} = 0.701$	$A'_{31} = 0.519$
	$A'_{23} = 1.76$	$A'_{32} = 2.52$
	$Q' = 1.15$	

* Severns *et al.* (1955)

Kohler Equation. Kohler's model permits prediction of thermodynamic properties of multicomponent liquid solutions from binary data only. As opposed to the Wilson, NRTL, and UNIQUAC equations, Kohler's model does not impose restrictions on the functional form of the binary excess Gibbs energy expressions nor does it limit the number of adjustable binary parameters. A general polynomial expansion permits all types of binary data (such as vapor-liquid and liquid-liquid equilibria, excess enthalpy, etc.) to be optimized simultaneously to obtain one self-consistent expression for the binary excess Gibbs energy at all compositions and temperatures.

For a ternary system of components 1, 2, and 3, Kohler relates g^E of the ternary mixture to the g^E's of the three constituent binaries by:

$$g_{123}^E = (x_1 + x_2)^2 g_{12}^E + (x_1 + x_3)^2 g_{13}^E + (x_2 + x_3)^2 g_{23}^E \tag{6-157}$$

where x_i is the liquid-phase mole fraction of component i. Any equation (not necessarily the same for each binary) can be used to represent g_{ij}^E of the binary ij system.

Suppose that g^E of the binary liquid mixtures is given by the following Legendre polynomial expansion:

$$g_{12}^E = x_1 x_2 \sum_{i=0}^{N} (a_i + b_i T + c_i T \ln T) F_i(x_2 - x_1) \tag{6-158}$$

where a_i, b_i, and c_i are adjustable parameters; upper limit N is chosen as required to fit the experimental data available for the binary mixture; and $F_i(x_2 - x_1)$ is a Legendre polynomial of order i.[41] Using standard thermodynamics, we obtain from Eq. (6-157) expressions for several excess properties:

$$h^E = x_1 x_2 \sum_{i=0}^{N} (a_i - c_i T) F_i(x_2 - x_1) \tag{6-159}$$

$$c_P^E = -x_1 x_2 \sum_{i=0}^{N} c_i F_i(x_2 - x_1) \tag{6-160}$$

$$\bar{g}_1^E = RT \ln \gamma_1 = x_2^2 \sum_{i=0}^{N} (a_i + b_i T + c_i T \ln T)[F_i(x_2 - x_1) - 2x_1 F_i'(x_2 - x_1)] \tag{6-161}$$

In Eq. (6-161), $F_i'(x_2 - x_1)$ is the first derivative of $F_i(x_2 - x_1)$ with respect to $(x_2 - x_1)$. Therefore, using *one* simultaneous least-square regression analysis for all data available for each binary (VLE and LLE data, h^E, c_P^E, γ^∞, etc.), one self-consistent set of parameters a_i, b_i, and c_i is obtained. Because the number of parameters in Eq. (6-157) is unconstrained, the data can be fitted as precisely as justified by the extent and quality of the data.

An application of Kohler's model to ternary and quaternary systems is provided by Mier *et al.* (1994, 1994a) and by Talley *et al.* (1993). Figure 6-31 compares predictions of Kohler's method (with the coefficients for Eq. (6-157) in Table 6-12) with experimental liquid-liquid equilibria at 318.15 K for the ternary system benzene/acetonitrile/*n*-heptane. In the calculations shown in Fig. 6-31 no ternary data were used in the optimization procedure. The predicted curve is good for ternary compositions near the partially miscible binary but becomes increasingly poor as the plait point is approached. As with other models (Wilson, NRTL, UNIQUAC), the predicted plait point is in error because Kohler's model also neglects the contribution of concentration fluctuations that are important near critical conditions. A reasonable (but not rigorous) attempt to take such fluctuations into account has been presented by de Pablo (1988, 1989, 1990).

---

[41] A polynomial expansion in $(x_2 - x_1)^i$ such as that of Redlich-Kister [Eq. (6-52)] can also be used. However, Legendre polynomials offer several advantages (e.g. their terms are independent and therefore may be truncated from the series to yield reasonable approximations) as pointed out by A. D. Pelton and C. W. Bale, 1986, *Metall. Trans.*, 17A: 1057.

Table 6-12 Coefficients of Eq. (6-157) for the constituent binaries of the ternary system benzene/acetonitrile/*n*-heptane.*

Binary	i	a_i (J mol^{-1})	b_i (J K^{-1} mol^{-1})	c_i (J K^{-1} mol^{-1})
n-Heptane/benzene	0	8126.8	-106.21	14.694
	1	2121.4	-29.426	4.1280
	2	836.13	-15.927	2.3391
n-Heptane/acetonitrile	0	-63237	1585.9	-236.48
	1	46486	-1010.1	150.13
	2	31281	-649.74	96.114
	3	16283	-327.87	48.126
Benzene/acetonitrile	0	232.58	40.371	-5.5903
	1	2577.0	-46.445	0.58260
	2	200.52	-0.54606	0.10603

* Talley *et al.* (1993).

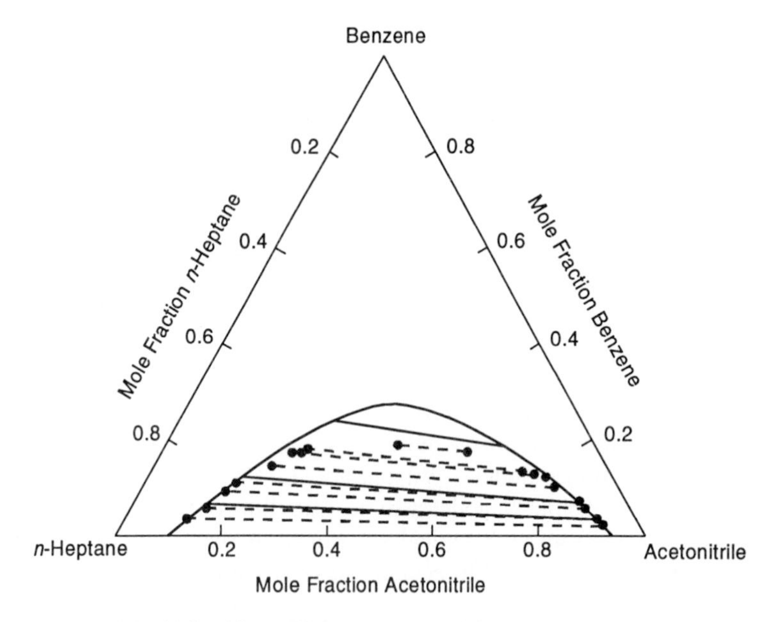

Figure 6-31 Liquid-liquid equilibria at 318.15 K for the ternary benzene/acetonitrile/*n*-heptane. • - - - • Experimental tie line. —— Calculated from Kohler's model with the binary parameters in Table 6-12.

6.15 Wilson, NRTL, and UNIQUAC Equations for Multicomponent Mixtures

The equations discussed in Sec. 6.11 are readily extended to as many components as desired without any additional assumptions and without introducing any constants other than those obtained from binary data.

Wilson Equation. For a solution of m components, Wilson's equation is

$$\frac{g^E}{RT} = -\sum_{i=1}^{m} x_i \ln\left(\sum_{j=1}^{m} x_j \Lambda_{ij}\right) \tag{6-162}$$

where

$$\Lambda_{ij} \equiv \frac{v_j}{v_i} \exp\left(-\frac{\lambda_{ij} - \lambda_{ii}}{RT}\right) \tag{6-163}$$

$$\Lambda_{ji} \equiv \frac{v_i}{v_j} \exp\left(-\frac{\lambda_{ji} - \lambda_{jj}}{RT}\right) \tag{6-164}$$

The activity coefficient for any component k is given by

$$\ln \gamma_k = -\ln\left(\sum_{j=1}^{m} x_j \Lambda_{kj}\right) + 1 - \sum_{i=1}^{m} \frac{x_i \Lambda_{ik}}{\sum_{j=1}^{m} x_j \Lambda_{ij}} \tag{6-165}$$

Equation (6-165) requires only parameters that can be obtained from binary data; for each possible binary pair in the multicomponent solution, two parameters are needed.

Orye (1965a) has tested Eq. (6-165) for a variety of ternary systems, using only binary data, and finds that for most cases good results are obtained. For example, Fig. 6-32 compares calculated and observed vapor compositions for the acetone/methyl acetate/methanol system at 50°C. A similar comparison is also shown for calculations based on the van Laar equation. No ternary data were used in either calculation; the binary constants used are given in Table 6-13. For this ternary system, the Wilson equations give a much better prediction than the van Laar equations but, as indicated in Table 6-11, the three-suffix Margules equations (using binary data only) can also give a good prediction because no ternary constant is required.

Table 6-13 Parameters for Wilson and van Laar equations for the system acetone (1)/methyl acetate (2)/methanol (3) at 50°C (Orye, 1965a).

Wilson equation	van Laar equation*
$\Lambda_{12} = 0.5781$	$A'_{12} = 0.1839$
$\Lambda_{13} = 0.6917$	$A'_{13} = 0.5965$
$\Lambda_{21} = 1.3654$	$A'_{21} = 0.1106$
$\Lambda_{23} = 0.6370$	$A'_{23} = 0.9446$
$\Lambda_{31} = 0.7681$	$A'_{31} = 0.5677$
$\Lambda_{32} = 0.4871$	$A'_{32} = 1.0560$

* Eq. (6-154)

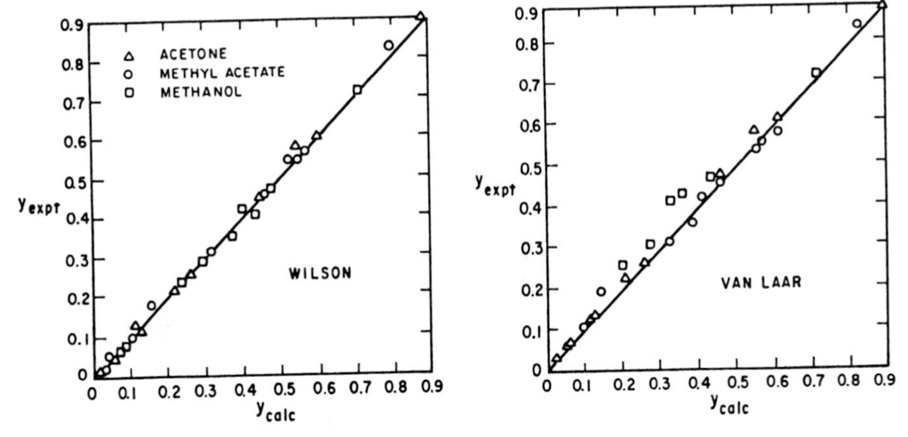

Figure 6-32 Experimental and calculated vapor compositions for the ternary system acetone/methyl acetate/methanol at 50°C. Calculations use only binary data.

Similar calculations for the ternary system acetone/methanol/chloroform are shown in Fig. 6-33, and again Wilson's equations, based on binary data only, give a better prediction than van Laar's equations. However, for this system the three-suffix Margules equations cannot give as good a prediction when only binary data are used because, as shown in Table 6-11, a significant ternary constant is required.

A final example of the applicability of Wilson's equation is provided by Orye's calculations for the system ethanol/methylcyclopentane/benzene at 1.013 bar. Wilson parameters were found from experimental data for the three binary systems (Myers, 1956; Sinor and Weber, 1960; Wehe and Coates, 1955); vapor compositions in the ternary system were then calculated for six cases and compared with experimental results as shown in Table 6-14. Wilson's equation again provides a good description for this ternary that has large deviations from ideal behavior.

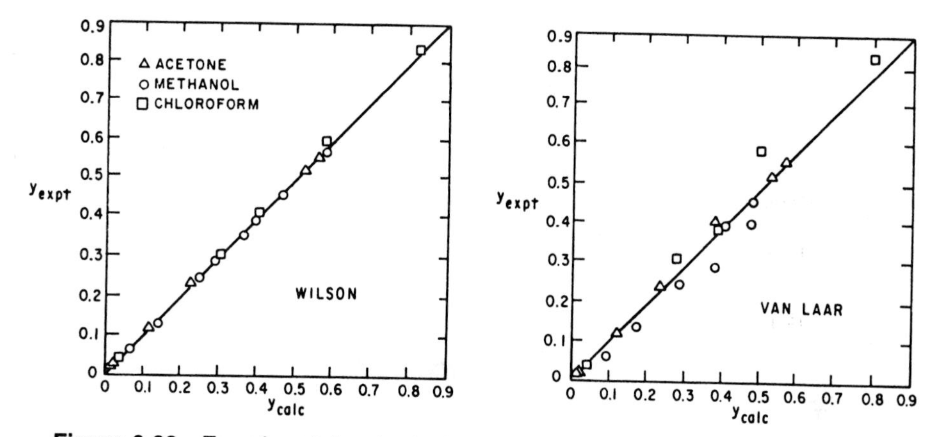

Figure 6-33 Experimental and calculated vapor compositions for the ternary system acetone/methanol/chloroform at 50°C. Calculations use only binary data.

Table 6-14 Calculated vapor compositions for the system ethanol (1)/methylcyclopentane (2)/benzene (3) at 1.013 bar using Wilson parameters obtained from binary data only.

T (K)	Component	Experimental* x	Experimental* y	Calculated y
336.15	1	0.047	0.258	0.258
	2	0.845	0.657	0.660
	3	0.107	0.084	0.081
338.85	1	0.746	0.497	0.502
	2	0.075	0.232	0.223
	3	0.178	0.271	0.275
335.85	1	0.690	0.432	0.434
	2	0.182	0.403	0.401
	3	0.128	0.165	0.165
340.85	1	0.878	0.594	0.603
	2	0.068	0.296	0.283
	3	0.053	0.110	0.114
337.15	1	0.124	0.290	0.300
	2	0.370	0.365	0.365
	3	0.505	0.345	0.335
334.05	1	0.569	0.386	0.383
	2	0.359	0.538	0.542
	3	0.071	0.076	0.075

Wilson parameters (kJ mol^{-1}):

$\lambda_{12} - \lambda_{11} = 9.2315$ $\quad \lambda_{31} - \lambda_{33} = 0.5246$

$\lambda_{21} - \lambda_{22} = 1.0266$ $\quad \lambda_{23} - \lambda_{22} = 0.0557$

$\lambda_{13} - \lambda_{11} = 5.8163$ $\quad \lambda_{32} - \lambda_{33} = 1.0413$

* J. E. Sinor and J. H. Weber (1960).

NRTL Equation. For a solution of m components, the NRTL equation is

$$\frac{g^E}{RT} = \sum_{i=1}^{m} x_i \frac{\sum_{j=1}^{m} \tau_{ji} G_{ji} x_j}{\sum_{l=1}^{m} G_{li} x_l} \tag{6-166}$$

where

$$\tau_{ji} = \frac{g_{ji} - g_{ii}}{RT} \tag{6-167}$$

$$G_{ji} = \exp(-\alpha_{ji} \tau_{ji}) \qquad (\alpha_{ji} = \alpha_{ij}) \tag{6-168}$$

The activity coefficient for any component i is given by

$$\ln \gamma_i = \frac{\sum_{j=1}^{m} \tau_{ji} G_{ji} x_j}{\sum_{l=1}^{m} G_{li} x_l} + \sum_{j=1}^{m} \frac{x_j G_{ij}}{\sum_{l=1}^{m} G_{lj} x_l} \left(\tau_{ij} - \frac{\sum_{r=1}^{m} x_r \tau_{rj} G_{rj}}{\sum_{l=1}^{m} G_{lj} x_l} \right) \tag{6-169}$$

Equations (6-166) and (6-169) contain only parameters obtained from binary data.

For nine ternary systems shown in Table 6-15, Renon (1968) predicted ternary vapor-liquid equilibria with Eq. (6-166) using binary data only. He also calculated ternary equilibria with Wohl's equation [Eq. (6-156)] both with and without a ternary constant. Table 6-15 indicates that Eq. (6-166) gives a good prediction of multicomponent equilibrium from binary equilibrium data alone.

UNIQUAC Equation. For a multicomponent system, the UNIQUAC equation for the molar excess Gibbs energy is given by the sum of

$$\frac{g^E\,\text{(combinatorial)}}{RT} = \sum_{i=1}^{m} x_i \ln \frac{\Phi_i^*}{x_i} + \frac{z}{2} \sum_{i=1}^{m} q_i x_i \ln \frac{\theta_i}{\Phi_i^*} \tag{6-170}$$

and

$$\frac{g^E\,\text{(residual)}}{RT} = -\sum_{i=1}^{m} q'_i x_i \ln\left(\sum_{j=1}^{m} \theta'_j \tau_{ji} \right) \tag{6-171}$$

Table 6-15 Comparison of NRTL and Wohl's equations for prediction of ternary vapor-liquid equilibria.

System	Mean arithmetic deviation in individual component's vapor mole fraction×10^3		95% confidence limits in vapor mole fraction×10^3	
	NRTL (with no ternary constant)	Wohl* (with best ternary constant)	NRTL (with no ternary constant)	Wohl* (with best ternary constant)
n-Heptane	3	3	2	8
Toluene	2	-4	1	5
Methyl ethyl ketone	-5	1	2	8
n-Heptane	4	0	3	4
Benzene	2	8	4	7
Ethanol (1.013 bar)	-6	-8	6	8
n-Heptane	5	0	7	4
Benzene	3	-5	4	7
Ethanol (0.533 bar)	-7	5	9	8
n-Heptane	-5	8	4	14
Toluene	-3	-2	5	8
Methanol	8	-6	8	19
Benzene	-1	13	5	22
Carbon tetrachloride	-3	3	4	20
Methanol (35°C)	4	10	7	39
Benzene	-3	-15	3	21
Carbon tetrachloride	-2	7	4	13
Methanol (55°C)	5	8	7	29
Acetone	-5	-11	4	18
Chloroform	-3	11	4	8
Methanol	8	0	3	12
Acetone	-4	-9	3	12
Methanol	1	8	7	15
Methyl acetate	3	1	5	8
Ethanol	-4	-6	7	22
Ethyl acetate	5	1	22	57
Water	1	5	17	49

* Eq. (6-156)

where segment fraction Φ^* and area fractions θ and θ' are given by

$$\Phi_i^* = \frac{r_i x_i}{\sum_{j=1}^{m} r_j x_j} \qquad \theta_i = \frac{q_i x_i}{\sum_{j=1}^{m} q_j x_j} \qquad \theta_i' = \frac{q_i' x_i}{\sum_{j=1}^{m} q_j' x_j}$$

and

$$\tau_{ij} = \exp\left(-\frac{a_{ij}}{T}\right) \quad \text{and} \quad \tau_{ji} = \exp\left(-\frac{a_{ji}}{T}\right)$$

The coordination number z is set equal to 10. For any component i, the activity coefficient is given by

$$\ln\gamma_i = \ln\frac{\Phi_i^*}{x_i} + \frac{z}{2}q_i \ln\frac{\theta_i}{\Phi_i^*} + l_i - \frac{\Phi_i^*}{x_i}\sum_{j=1}^{m} x_j l_j$$
$$-q_i' \ln\left(\sum_{j=1}^{m}\theta_j'\tau_{ji}\right) + q_i' - q_i'\sum_{j=1}^{m}\frac{\theta_j'\tau_{ij}}{\sum_{k=1}^{m}\theta_k'\tau_{kj}} \qquad (6\text{-}172)$$

where

$$l_j = \frac{z}{2}(r_j - q_j) - (r_j - 1) \qquad (6\text{-}173)$$

Equation (6-172) requires only pure-component and binary parameters.

Using UNIQUAC, Table 6-16 summarizes vapor-liquid equilibria predictions for several representative ternary systems and one quaternary system. Calculated results agree well with experimental pressures (or temperatures) and vapor-phase compositions.

The largest errors in predicted compositions occur for the systems acetic acid/formic acid/water and acetone/acetonitrile/water, where experimental uncertainties are significantly greater than those for other systems.

Moderate errors in the total pressure calculations are evident for the systems chloroform/ethanol/*n*-heptane and chloroform/acetone/methanol. Here strong hydrogen bonding between chloroform and alcohol creates unusual deviations from ideality; for both alcohol/chloroform systems, the activity coefficients show well-defined extrema. Because extrema are often not well reproduced by the UNIQUAC equation, these binaries are not represented as well as the others. The overall ternary deviations are similar to those for the worst-fitting binaries, methanol/chloroform and ethanol/chloroform. In spite of the relatively large deviations in calculated pressure, predicted vapor compositions agree well with the experimental data of Severns *et al.* (1955). Fortunately, extrema in activity coefficients are rare in binary systems.

Predictions for the other isobaric systems show good agreement. Excellent agreement is obtained for the system carbon tetrachloride/methanol/benzene, where the binary data are of superior quality.

The results shown in Table 6-16 suggest that UNIQUAC can be used with confidence for typical multicomponent systems of nonelectrolytes, provided that good experimental binary data are available to determine reliable binary parameters.

Table 6-16 Prediction of multicomponent vapor-liquid equilibrium with UNIQUAC equation using binary data only.*

System†	Number of data points	Pressure (bar); Temperature (°C)	Deviation in temperature or percent deviation in pressure Avg. (max)	Deviation in vapor composition (mol%) Avg. (max.)
MCP Ethanol Benzene	48	1.013 60-71	0.25 (0.31)°C	0.51 (-3.03) 0.55 (2.99) 0.35 (-1.25)
Acetic acid Formic acid Water	40	1.013 102-110	0.55 (-1.80)°C	1.00 (-2.08) 1.60 (3.77) 2.18 (-5.36)
Acetone Acetonitrile Water	30	1.016 63-92	1.13 (-3.67)°C	1.22 (3.24) 1.27 (-3.45) 1.53 (-4.68)
Methanol CTC Benzene	8	0.8866-0.9559 55	0.11 (-0.27)%	0.44 (0.99) 0.39 (-0.89) 0.09 (017)
MEK n-Heptane Toluene	39	1.013 77-103	0.17 (-0.63)°C	0.79 (2.00) 0.52 (-1.31) 0.38 (-1.18)
Chloroform Ethanol n-Heptane	92	0.3493-0.6679 50	1.57 (-3.30)%	‡
Chloroform Acetone Methanol	29	0.6173-0.8599 50	1.10 (-3.12)%	0.86 (1.03) 0.77 (2.68) 0.81 (1.03)
Chloroform Methanol Ethyl acetate	72	1.013 56-72	0.36 (1.77)°C	0.74 (2.06) 1.11 (2.40) 0.80 (2.47)
n-Hexane MCP Ethanol Benzene	10	1.013 60-65	0.38 (-0.45)°C	0.31(0.60) 0.44 (0.95) 0.55 (-1.13) 0.44 (0.96)

* Data references are given in Anderson (1978).

† MCP = methylcyclopentane; CTC = carbon tetrachloride; MEK = methyl ethyl ketone.

‡ P-T-x data only.

While Wilson's equation is not applicable to liquid mixtures with miscibility gaps, the NRTL equation and the UNIQUAC equation may be used to describe such mixtures. However, when these equations are applied to ternary (or higher) systems, it is often not possible to predict multicomponent liquid-liquid equilibria using only experimental binary data.

Reduction of typical binary vapor-liquid or liquid-liquid experimental data does not yield a unique set of binary parameters; several sets may reproduce the data equally well within experimental error. Multicomponent vapor-liquid equilibrium calculations are not highly sensitive to the choice of binary parameters, but multicomponent liquid-liquid equilibrium calculations depend strongly on that choice. As discussed in App. E, only a few ternary data are needed to guide the selection of "best" binary parameters. When that selection is made with care, the NRTL equation or the UNIQUAC equation can often represent ternary (or higher) liquid-liquid equilibria with good accuracy.

To obtain improved correlation of ternary liquid-liquid equilibria, Nagata (1989) introduced adjustable ternary parameters in the residual contribution of the UNIQUAC equation:

$$\frac{g^E\,(\text{residual})}{RT} = -\sum_{i=1}^{m} q_i' x_i \ln\left(\sum_{j=1}^{m} \theta_j' \tau_{ji} + \sum_{j<k}^{m}\sum^{m} \theta_j' \theta_k' \tau_{jki}\right)$$

with $\tau_{jki} \neq 0$ only for $i \neq j \neq k$, i.e. three adjustable ternary parameters for each ternary system.

Figure 6-34 gives an example of Nagata's modification of the UNIQUAC equation applied to liquid-liquid equilibria of the three ternaries forming the quaternary system acetonitrile/aniline/*n*-heptane/benzene (Nagata and Tamura, 1998). Calculated results with the binary and ternary parameters in Table 6-17 agree well with experiment. However, if calculations were carried out using only the binary energy parameters a_{12} and a_{21}, large deviations from experiment would be observed, particularly in the critical region. Good agreement with experiment is obtained only if ternary parameters are used, determined from optimization of ternary LLE data; however, UNIQUAC here is not a predictive method.

6.16 Summary

This chapter is concerned with activity coefficients to calculate fugacities of components in a liquid mixture. In nonideal mixtures, these *activity coefficients* depend strongly on composition. At conditions remote from critical, they often depend only weakly on temperature and very weakly on pressure. Unless the pressure is high, we can usually neglect the effect of pressure on activity coefficients.

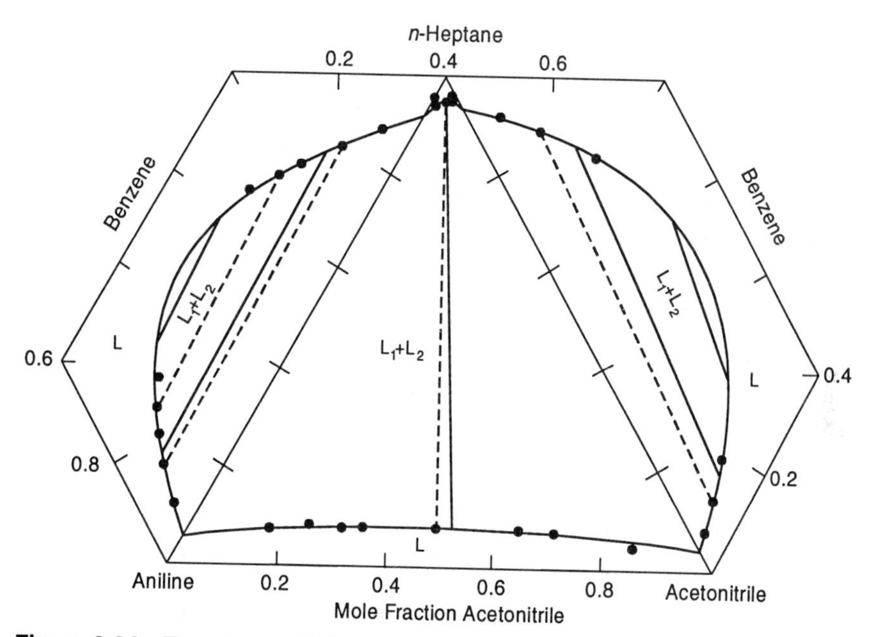

Figure 6-34 Experimental (●) and calculated (UNIQUAC) liquid-liquid equilibria at 298 K for the ternary systems aniline/acetonitrile/*n*-heptane, aniline/*n*-heptane /benzene, and n-heptane/acetonitrile/benzene. • – – – – • Experimental tie line. ——— Calculated from UNIQUAC equation with binary and ternary parameters obtained from binary VLE data and ternary LLE data in Table 6-17.

Table 6-17 Binary and ternary interaction parameters for modified UNIQUAC equation (Nagata, 1989) obtained, respectively, from fitting of binary vapor-liquid equilibrium data and ternary liquid-liquid equilibrium data (Nagata and Tamura, 1998).

System	T (K)	a_{12} (K)	a_{21} (K)	τ_{123}	τ_{132}	τ_{231}
Acetonitrile/aniline	293.15	189.51	-95.66			
Acetonitrile/benzene	293.15	37.28	251.01			
Acetonitrile/*n*-heptane	298.15	439.86	990.57			
Aniline/benzene	298.15	-26.62	260.03			
Aniline/*n*-heptane	298.15	383.90	632.43			
Benzene/*n*-heptane	318.15	61.74	69.44			
Acetonitrile/aniline/ *n*-heptane	298.15			1.9672	-0.2053	-0.9188
Acetonitrile/*n*-heptane/ benzene	298.15			0.1037	0.0591	-0.1512
Aniline/*n*-heptane/ benzene	298.15			-0.6050	0.4504	0.7545

* RMS = Root Mean Square

The numerical value of an activity coefficient is meaningful only when the standard-state fugacity to which it refers is clearly specified. The *excess Gibbs energy* g^E for a mixture is defined in terms of all activity coefficients in that mixture [Eq. (6-26)]; using the *Gibbs-Duhem relation*, individual activity coefficients are then related to g^E through Eq. (6-25).

It is advantageous to construct an analytical function for g^E because that function provides a useful method for interpolating and extrapolating activity-coefficient data. For a binary mixture, some physical or mathematical model is used to express g^E as a function of composition, using a few (typically two) adjustable binary parameters that may be temperature dependent. When these binary parameters are evaluated from *limited* binary activity-coefficient data, the physical or mathematical model *predicts* liquid-phase activity coefficients at new compositions (and to a lesser extent, at new temperatures) where experimental data are not available.

Numerous models for g^E have been proposed. Only a few of these are reviewed in this chapter.

For those binary mixtures where the pure components do not differ much in volatility, the best method for obtaining activity-coefficient data is to measure the total pressure as a function of liquid-phase composition at constant temperature. It is not necessary to measure the vapor composition; that composition can be calculated from the total-pressure data using the gibbs-Duhem equation.

When a function for g^E has been established for a binary mixture and the parameters are known, that function may be used to determine if the binary system has a *miscibility gap*; if it does, the function can be used to calculate mutual solubilities. Inverting, experimental mutual-solubility data can be used to fix the binary parameters (see App. E).

For practical applications, the most important use of the g^E function follows from extension to *multicomponent* (ternary or higher) systems. Such extension requires primarily binary parameters; in some cases a few ternary parameters are desirable but in many cases binary parameters alone are sufficient for vapor-liquid equilibria. Therefore, the g^E function is a "scale-up" tool, in the sense that fugacities in ternary (and higher) mixtures can be calculated using only experimental information for binary systems. This "scale-up" provides a very large reduction in the experimental effort required to characterize a multicomponent liquid mixture.

The methods discussed in Chaps. 5 and 6 can be used directly to establish a computerized technique for calculating multicomponent *vapor-liquid equilibria* as required, for example, in the design of separation equipment such as distillation columns. Details, including computer programs, are given elsewhere (Prausnitz *et al.*, 1980); the essential steps, however, are easily stated.

To fix ideas, suppose we have a liquid mixture containing m components at pressure P. Mole fractions $x_1, x_2, \ldots, x_m$ are known. We want to find vapor-phase mole fractions $y_1, y_2, \ldots, y_m$ and the equilibrium temperature T. To find these $m+1$ unknowns, we need $m+1$ independent equations:

One material balance:

$$\sum_{i}^{m} y_i = 1$$

m equations of equilibrium:

$$\varphi_i y_i P = \gamma_i x_i f_i^0 \qquad \text{for every component } i.$$

For the standard-state fugacity f_i^0 we usually choose the fugacity of pure liquid i at system temperature and pressure (see Chap. 3). For the vapor-phase fugacity coefficient φ_i we choose some vapor-phase equation of state (as discussed in Chap. 5) and then use Eq. (3-53). For the liquid-phase activity coefficient we use some equation for the excess Gibbs energy (as discussed in this chapter) and then use Eq. (6-25).

Simultaneous solution of the $m+1$ equations is, in principle, straightforward, using iterative techniques with a computer. Since we do not know mole fractions y_i or temperature T, we must *iterate* with respect to several quantities. However, iteration with respect to T is our primary concern; calculated results are strongly sensitive to T because, while φ_i and γ_i depend only weakly on temperature, f_i^0 is usually a very strong function of temperature. On the other hand, the only quantity that depends on y_i is φ_i, and that dependence is usually weak.

Similar considerations are used to calculate *liquid-liquid equilibria*. Unlike vapor-liquid equilibria, liquid-liquid equilibria are highly sensitive to small changes in the constants that appear in the equation for molar excess Gibbs energy g^E. In many cases it is not possible to predict g^E for a ternary system (using only binary data) with sufficient accuracy for liquid-liquid equilibrium calculations; typically some ternary data are required.

Detailed numerical methods are given in Prausnitz *et al.* (1980) and are not of concern here. However, it is important to recognize that the methods discussed in Chaps. 5 and 6 are essentially sufficient to solve many practical problems in multicomponent phase equilibrium.

References

Abbott, M. M. and H. C. Van Ness, 1975, *AIChE J.*, 21: 62.

Abbott, M. M. and H. C. Van Ness, 1977, *Fluid Phase Equilibria*, 1: 3.

Abrams, D. and J. M. Prausnitz, 1975, *AIChE J.*, 21: 116.

Adler, H. E. and S. B. Adler, 1973, *Ind. Eng. Chem. Process Des. Dev.*, 12: 71.

Adler, S. B., L. Friend, and R. L. Pigford, 1966, *AIChE J.*, 12: 629.

Anderson, T. F. and J. M. Prausnitz, 1978, *Ind. Eng. Chem. Process Des. Dev.*, 17: 552, 561.

Barker, J. A., 1953, *Aust. J. Chem.*, 6: 207.

Bellemans, A., 1959, *Bull. Soc. Chim. Belg.*, 68: 355.

Black, C., 1959, *AIChE J.*, 5: 249.

Brown, I. and F. Smith, 1955, *Aust. J. Chem.*, 8: 62.

Brown, I. and F. Smith, 1957, *Aust. J. Chem.*, 10: 423.

Butler, J. A. V., 1937, *Trans. Faraday Soc.*, 33: 229.

Byer, S. M., R. E. Gibbs, and H. C. Van Ness, 1973, *AIChE J.*, 19: 238.

Christian, S. D., E. Neparko, and H. E. Affsprung, 1960, *J. Phys. Chem.*, 64: 442.

de Pablo, J. J. and J. M. Prausnitz, 1988, *AIChE J.*, 34: 1595.

de Pablo, J. J. and J. M. Prausnitz, 1989, *Fluid Phase Equilibria*, 50: 101.

de Pablo, J. J. and J. M. Prausnitz, 1990, *Fluid Phase Equilibria*, 59: 1.

Edwards, J. B., 1962, Ph.D. Dissertation, Georgia Institute of Technology.

Fried, V., P. Gallant, and G. B. Schneier, 1967, *J. Chem. Eng. Data*, 12: 504.

Gadwa, T. A., 1936, Dissertation, Massachusetts Institute of Technology. Quoted by Carlson, H. C., and A. P. Colburn, 1942, *Ind. Eng. Chem.*, 34: 581.

Herington, E. F. G., 1947, *Nature*, 160: 610.

Hermsen, R. W. and J. M. Prausnitz, 1963, *Chem. Eng. Sci.*, 18: 485.

Kemeny, S. and P. Rasmussen, 1981, *Fluid Phase Equilibria*, 7: 197.

Kretschmer, C. B., 1948, *J. Am. Chem. Soc.*, 70: 1785.

Maurer, G. and J. M. Prausnitz, 1978, *Fluid Phase Equilibria*, 2: 91.

McDermott, C. and S. R. M. Ellis, 1965, *Chem. Eng. Sci.*, 20: 545.

Mier, W., R. N. Lichtenthaler, A. H. Roux, and J.-P. E. Grolier, 1994, *J. Chem. Thermodynamics*, 26: 1323.

Mier, W., R. N. Lichtenthaler, A. H. Roux, and J.-P. E. Grolier, 1994a, *J. Chem. Thermodynamics*, 26: 1335.

Miller, H. C. and H. Bliss, 1940, *Ind. Eng. Chem.*, 32: 123.

Myers, H. S., 1956, *Ind. Eng. Chem.*, 48: 1104.

Nagata, I., 1989, *Fluid Phase Equilibria*, 51: 53.

Nagata, I. and K. Tamura, 1998, *J. Chem. Thermodynamics*, 30: 179.

Nakanishi, K., H. Shirai, and T. Minamiyama, 1967, *J. Chem. Eng. Data*, 12: 591.

Nicolaides, G. L. and C. A. Eckert, 1978, *Ind. Eng. Chem. Fundam.*, 17: 331.

Orye, R. V. and J. M. Prausnitz, 1965, *Trans. Faraday Soc.*, 61: 1338.

Orye, R. V. and J. M. Prausnitz, 1965a, *Ind. Eng. Chem.*, 57: 19.

Pool, R. A. H., G. Saville, T. M. Herrington, B. D. C. Shields, and L. A. K. Staveley, 1962, *Trans. Faraday Soc.*, 58: 1692.

Prausnitz, J. M., E. A. Grens, T. F. Anderson, C. A. Eckert, R. Hsieh, and J. P. O'Connell, 1980, *Computer Calculations for Multicomponent Vapor-Liquid and Liquid-Liquid Equilibria*. Englewood Cliffs: Prentice-Hall.

Rafal, M., J. W. Berthold, N. C. Scrivner, and S. L. Grise, 1994, *Models for Electrolyte Solutions*. In *Models for Thermodynamic and Phase-Equilibria Calculations*, (S. I. Sandler, Ed.). New York: Marcel Dekker.

Redlich, O. and A. T. Kister, 1948, *Ind. Eng. Chem.*, 40: 345.

Redlich, O., A. T. Kister, and C. E. Turnquist, 1952, *Chem. Eng. Prog. Symp. Ser.*, 48: 49.

Renon, H. and J. M. Prausnitz, 1968, *AIChE J.*, 14: 135.

Renon, H. and J. M. Prausnitz, 1969, *AIChE J.*, 15: 785.

Scatchard, G. and W. J. Hamer, 1935, *J. Am. Chem. Soc.*, 57: 1805.

Scatchard, G., S. E. Wood, and J. M. Mochel, 1939, *J. Phys. Chem.*, 43: 119.

Severns, W. H., A. Sesonske, R. H. Perry, and R. L. Pigford, 1955, *AIChE J.*, 1: 401.

Shain, S. A. and J. M. Prausnitz, 1963, *Chem. Eng. Sci.*, 18: 244.

Sinor, J. E. and J. H. Weber, 1960, *J. Chem. Eng. Data*, 5: 243.

Sørensen, J. M. and W. Arlt, 1979, *Liquid-Liquid Equilibrium Data Collection*, DECHEMA Chemistry Data Series, Vol. V/1-3. Frankfurt: DECHEMA.

Sørensen, J. M., T. Magnussen, and Aa. Fredenslund, 1979a, *Fluid Phase Equilibria*, 2: 297.

Sørensen, J. M., T. Magnussen, and Aa. Fredenslund, 1979b, *Fluid Phase Equilibria*, 3: 47.

Sørensen, J. M., T. Magnussen, and Aa. Fredenslund, 1980, *Fluid Phase Equilibria*, 4: 151.

Talley, P. K., J. Sangster, C. W. Bale, and A. D. Pelton, 1993, *Fluid Phase Equilibria*, 85: 101.

Van Ness, H. C. and M. M. Abbott, 1982, *Classical Thermodynamics of Nonelectrolyte Solutions*, p. 225. New York: McGraw-Hill.

Van Ness, H. C., 1995, *Pure & Appl. Chem.*, 67: 859.

Wehe, A. H. and J. Coates, 1955, *AIChE J.*, 1: 241.

Weissman, S. and S. E. Wood, 1960, *J. Chem. Phys.*, 32: 1153.

Wilson, G. M., 1964, *J. Am. Chem. Soc.*, 86: 127.

Wohl, K., 1946, *Trans. AIChE*, 42: 215.

Problems

1. Experimental studies have been made on the isothermal vapor-liquid equilibrium of a ternary mixture. The measured quantities are liquid mole fractions x_1, x_2, x_3; vapor mole fractions y_1, y_2, y_3; total pressure P; and absolute temperature T. From these measurements indicate how to calculate the activity coefficient of component 1. All components are liquids at temperature T; the saturation pressure of component 1 is designated by P_1^s . The molar liquid volume of component 1 is v_1^L .

For the vapor phase, assume that the volume-explicit, truncated virial equation of state holds:

$$z = 1 + \frac{B_{\text{mixt}} P}{RT}$$

All necessary virial coefficients B_{ij} (i = 1, 2, 3; j = 1, 2, 3) are available.

2. Consider a solution of two similar liquids which are miscible in all proportions over a wide range of temperature. The excess Gibbs energy of this solution is adequately represented by the equation

$$g^E = Ax_1x_2$$

where A is a constant depending only on temperature.
Over a wide range of temperature the ratio of the vapor pressures of the pure components is constant and equal to 1.649. Over this same range of temperature the vapor phase may be considered ideal.
We want to find out whether or not this solution has an azeotrope. Find the range of values A may have for azeotropy to occur.

3. There is something unusual about the system hexafluorobenzene (1)/benzene (2). The following data were obtained at 343 K (x and y refer to the liquid- and vapor-phase mole fractions). Plot isothermal x-y and P-x-y diagrams. What is unusual about this binary system?
Data are as follows:

x_1	y_1	P (kPa)	x_1	y_1	P (kPa)
0.0000	0.0000	73.408	0.5267	0.5035	72.451
0.0938	0.0991	74.168	0.6014	0.5826	71.966
0.1847	0.1831	74.318	0.7852	0.7834	71.294
0.2740	0.2624	74.110	0.8959	0.8995	71.374
0.3468	0.3447	73.627	1.0000	1.0000	71.851
0.4539	0.4299	73.002			

4. Total-pressure data are available for the entire concentration range of a binary solution at constant temperature. At the composition $x_1 = a$, the total pressure is a maximum. Show that at the composition $x_1 = a$, this solution has an azeotrope, i.e., that the relative volatility at this composition is unity. Assume that the vapor phase is ideal.

5. Consider a liquid mixture of components 1, 2, 3, and 4. The excess Gibbs energies of all the binaries formed by these components obey relations of the form

$$g_{ij}^E = A_{ij}x_ix_j$$

where A_{ij} is the constant characteristic of the i-j binary.
Derive an expression for the activity coefficient of component 1 in the quaternary solution.

6. Construct T-x_1 and y_1-x_1 diagrams for the cyclohexanone (1)/phenol (2) system at 30 kPa. Available data are:

$$\frac{g^E}{RT} = -2.1x_1x_2$$

$$\ln P_1^s\,(\text{kPa}) = 15.0886 - \frac{4093.3}{t(°\text{C}) + 236.12}$$

$$\ln P_2^s\,(\text{kPa}) = 14.4130 - \frac{3490.885}{t(°\text{C}) + 174.569}$$

Vapor-phase nonidealities may be neglected.

7. Limited vapor-liquid equilibrium data have been obtained for a solution of two slightly dissimilar liquids A and B over the temperature range 20 to 100°C. From these data it is found that the variation of the limiting activity coefficients (symmetric convention) with temperature can be represented by the empirical equation

$$\ln \gamma_A^\infty = \ln \gamma_B^\infty = 0.15 + \frac{10.0}{t(°\text{C})}$$

where γ^∞ is the activity coefficient at infinite dilution. Estimate the enthalpy of mixing of an equimolar mixture of A and B at 60°C.

8. The partial molar enthalpy of water in concentarted sulfuric acid solutions containing less than 20 moles H_2O per mol H_2SO_4, at 293 K and 1 bar, is given by

$$\overline{H}_w\,(\text{kJ mol}^{-1}) = -\frac{134x_A^2}{(1 + 0.7983x_A)^2}$$

where subscripts w and A stand for water and acid, respectively, x_A is the mole fraction of acid, and the referance state for expressing $\overline{H}_w$ is pure liquid water at the temperature and pressure of the mixture. For a certain application, 2 mol H_2O and 1 mol H_2SO_4 are mixed isothermally in an open vessel, which is equipped with cooling coils. The flow of the cooling medium is controlled so that the entire mixing process occurs isothermally at 293 K.

(a) Calculate the infinite dilution partial molar enthalpy of H_2O in H_2SO_4 at 293 K.
(b) Calculate how much heat must be removed during the mixing process.

9. (a) Vapor-phase spectroscopic data clearly show that sulfur dioxide and normal butene-2 form a complex. However, thermodynamic data at 0°C show that liquid mixtures of these components exhibit slight positive deviations from Raoult's law. Is this possible? Or do you suspect that there may be experimental error?

(b) Qualitatively compare the excess Gibbs energies of mixtures containing sulfur dioxide and butene-2 with those containing sulfur dioxide and isobutene. Explain.

10. From the total-pressure data below compute the y-x diagram for ethyl alcohol (1)/chloroform (2) system at 45°C. Assume ideal gas behavior.

Data are as follows:

x_2	P (kPa)	x_1	P (kPa)
0	23.038	0	57.795
0.05	26.664	0.01	58.462
0.10	31.064	0.02	58.928
0.15	35.463	0.03	59.315
0.20	39.730	0.04	59.648
0.25	43.463	0.05	59.902
0.30	46.796	0.06	60.115
0.35	49.596	0.07	60.302
0.40	52.129	0.08	60.462
0.45	53.796	0.09	60.595
0.50	55.462	0.10	60.688
0.55	56.862		

Compare your computed results with the experimental data of Scatchard and Raymond (1938, *J. Am. Chem. Soc.*, 60: 1275).

11. For the system 2-propanol (1)/water (2) at 45°C, experimental data indicate that at infinite dilution the activity coefficients are $\gamma_1^\infty = 12.0$ and $\gamma_2^\infty = 3.89$. Using Wilson's equation, and assuming ideal-gas behavior, construct the pressure-composition P-x-y diagram for this system at 45°C (see App. E). Compare with the experimental data of Sada and Morisue (1975, *J. Chem. Eng. Jap.*, 8: 191).

Data are as follows:

x_1	y_1	P (kPa)
0.0462	0.3936	15.252
0.0957	0.4818	17.412
0.1751	0.5211	18.505
0.2815	0.5455	19.132
0.4778	0.5981	19.838
0.6046	0.6411	20.078
0.7694	0.7242	19.985
0.8589	0.8026	19.585

12. You want to estimate the *y-x* diagram for the liquid mixture carbon tetrabromide (1)/nitroethane (2) at 0.5 bar total pressure. You have available boiling-point *T-x* data for this system at 0.5 bar as well as pure-component vapor-pressure data as a function of temperature. Explain how you would use the available information to construct the desired diagram. Assume that a computer is available. Set up the necessary equations and define all symbols used. State all assumptions made. Are there any advantages in using the Wilson equation for the solution of this problem? Explain.

13. For the binary mixture 2-butanone/cyclohexane:
(a) Identify UNIFAC groups for each component.
(b) Using UNIFAC calculate the activity of each component for an equimolar mixture of 2-butanone/cyclohexane at 75°C.
(c) Plot the *y-x* diagram for this binary at 75°C.
At 75°C and 1 bar pressure, the activity coefficients for 2-butanone (1) and cyclohexane (2) are:

x_1	γ_1	γ_2
0	4.377	1.00
0.2	2.268	1.071
0.8	1.040	2.186
1.0	1.00	3.105

Antoine's constants for vapor pressure correlation are available in Reid *et al.*, 1979, *Properties of Gases and Liquids*, New York: McGraw-Hill.

14. Because of fire hazards, large quantities of flammable liquids are often stored in outdoor tanks. Consider a mixture of *n*-hexane and nitroethane, stored outdoors, in a chemical plant located in a northern climate. It is undesirable to have phase separation in the storage tank because, when pumped back into the plant for participation in a chemical process, it would be necessary to remix the two liquid phases to obtain the original composition.
The lowest outside temperature is estimated to be -40°C. Is it likely that an equimolar mixture will separate into two liquid phases at this temperature? To answer this question, use the UNIQUAC equation. Consider possible advantages of a graphical solution.
UNIQUAC parameters are as follows:

	r	q
n-Hexane (1)	4.5	3.86
Nitroethane (2)	2.68	2.41
	a_{12} = 231 K	a_{21} = -5.86 K

15. If one found data in the literature for the enthalpy of mixing $\Delta_{mix}h$ and the entropy change of mixing $\Delta_{mix}s$ (not excess entropy) at a particular T and P for a pair of liquids miscible in all proportions and upon plotting $\Delta_{mix}h/RT$ and $\Delta_{mix}s/R$ versus x_1 on the same graph found that the curves crossed, would one have reason to question the validity of the data? Explain.

16. At 300 K, some experimental data are available for dilute liquid mixtures of components 1 and 2. When 1 is dilute in an excess of 2, Henry's constant $H_{1,2} = 2$ bar. When 2 is dilute in an excess of 1, Henry's constant $H_{2,1} = 1.60$ bar. Estimate the vapor composition which is in equilibrium with an equimolar liquid mixture of 1 and 2 at 300 K. Assume that the vapor is an ideal gas. At 300 K, the pure-component vapor pressures are:

	Vapor pressure bar)
Pure liquid 1	1.07
Pure liquid 2	1.33

17. A ternary liquid mixture at 300 K contains components 1, 2, and 3; all liquid-phase mole fractions are equal to 1/3. The vapor pressures of the pure components (in kPa) at 300 K are $P_1^s = 53.3$, $P_2^s = 40$, and $P_3^s = 53.3$ Estimate the composition of the vapor in equilibrium with this mixture.
The following binary data are available:
1-2 binary: γ_1^∞ (infinite dilution) = 1.3 at 320 K.
1-3 binary: This binary forms an azeotrope at 300 K when $x_1 = x_3 = 1/2$ and $P = 60$ kPa.
2-3 binary: At 270 K, this binary has an upper consolute temperature when $x_3 = x_2 = 1/2$.
Clearly state all assumptions made.

18. At 25°C, a binary liquid mixture contains nonpolar components 1 and 2. Data for the dilute regions of this mixture indicate that $\gamma_1^\infty = 9.3$ and $\gamma_2^\infty = 4.7$ At 25°C, are liquids 1 and 2 miscible in all proportions or is there a miscibility gap?

19. It is reported in the literature that the excess Gibbs function for a binary system of A and B as determined from vapor-liquid equilibria is given by the following relations:

At 30°C: $g^E/RT = 0.500x_Ax_B$.

At 50°C: $g^E/RT = 0.415\, x_Ax_B$.

At 70°C: $g^E/RT = 0.330\, x_Ax_B$.

The vapor pressures of pure A and pure B are given by

$$\ln P_A^s = 11.92 - \frac{4050}{T} \qquad \text{and} \qquad \ln P_B^s = 12.12 - \frac{4050}{T}$$

where P^s is in bar and T in K. Making reasonable assumptions, determine:

(a) Whether this system forms an azeotrope at any of the listed temperatures, and if so, the azeotropic compositions.

(b) Whether this system forms an azeotrope at 760 mmHg. Make your reasoning clear.

(c) Another literature source gives the enthalpy of mixing for this system at 50°C:

$$\frac{\Delta_{\text{mix}} h}{RT} = (1.020 + 0.112 x_A) x_A x_B$$

Is this equation completely consistent with the data given for g^E/RT? If not, give some indication of the degree of inconsistency.

CHAPTER 7

Fugacities in Liquid Mixtures: Models and Theories of Solutions

When two or more pure liquids are mixed to form a liquid solution, it is the aim of solution theory to express the properties of the liquid mixture in terms of intermolecular forces and fundamental liquid structure. To minimize the amount of experimental information required to describe a solution, it is desirable to express the properties of a solution in terms that can be calculated completely from the properties of the pure components. Present theoretical knowledge has not yet reached a stage of development where this can be done with any degree of generality, although some results of limited utility have been obtained. Most current work in the theory of solutions utilizes the powerful methods of statistical mechanics that relate macroscopic (bulk) properties to microscopic (molecular) phenomena.[1]

[1] For an introduction see, e.g., T. M. Reed and K. E. Gubbins, 1973, *Applied Statistical Mechanics,* (New York: McGraw-Hill); D. A. McQuarrie, 1985, *Statistical Thermodynamics,* (Mill Valley: University Science Books); T. L. Hill, 1986, *An Introduction to Statistical Thermodynamics,* (Reading: Addison-Wesley); and K. Lucas, 1991,

In this chapter, we introduce some of the theoretical concepts that have been used to describe and to interpret solution properties. We cannot give a complete treatment; we attempt, however, to give a brief survey of those ideas that bear promise for practical applications.

The simplest theory of liquid solutions is that due to Raoult, who set the partial pressure of any component equal to the product of its vapor pressure and its mole fraction in the liquid phase; at modest pressures, this simple relation often provides a reasonable approximation for those liquid solutions whose components are chemically similar. However, Raoult's relation becomes exact only as the components of the mixture become identical, and its failure to represent the behavior of real solutions is due to differences in molecular size, shape, and intermolecular forces of the pure components. It appears logical, therefore, to use Raoult's relation as a reference and to express observed behavior of real solutions as deviations from behavior calculated by Raoult's law. This treatment of solution properties was formalized by Lewis in the early twentieth century, and since then it has become customary to express the behavior of real solutions in terms of activity coefficients. Another way of stating the aim of solution theory, then, is to say that it aims to predict numerical values of activity coefficients in terms of properties (or constants) that have molecular significance and that, hopefully, may be calculated primarily from the properties of the pure components.

One of the first systematic attempts to describe quantitatively the properties of fluid mixtures was made by van der Waals and his coworkers early in the twentieth century, shortly before the work of Lewis. Therefore, most of van der Waals' work on fluid mixtures appears in a form that today strikes us as awkward. However, no one can deny that he and his colleagues at Amsterdam were the first great pioneers in a field that, since about 1890, has attracted the serious attention of a large number of outstanding physical scientists.[2] One of van der Waals' students and later collaborators was van Laar, and it was primarily through van Laar's work that the basic ideas of the Amsterdam school became well known. It is convenient, therefore, to begin by discussing van Laar's theory of solutions and then to show how this simple but inadequate theory led to the more useful theory of regular solutions advanced by Scatchard and Hildebrand.

Applied Statistical Thermodynamics, (Berlin: Springer). For more specialized discussions see T. Boublik, I. Nezbeda, and H. Hlavaty, 1980, *Statistical Thermodynamics of Simple Liquids and Their Mixtures,* (New York: Elsevier); K. Singer (Ed.), 1973, *Statistical Mechanics,* (London: The Royal Society of Chemistry); D. Chandler, 1987, *Introduction to Modern Statistical Mechanics,* (New York: Oxford University Press).

[2] Van der Waals' thesis is translated by J. S. Rowlinson, 1988, *Van der Waals: On the Continuity of the Gaseous and Liquid States,* (Amsterdam: North-Holland). This book also contains insightful comments on the van der Waals theory of fluids from a modern point of view.

7.1 The Theory of van Laar

One of the essential requirements for a successful theory in physical science is judicious simplification. If one wishes to do justice to all the aspects of a problem, one very soon finds oneself in a hopelessly complicated situation. To make progress, it is necessary to ignore certain aspects of a physical situation and to retain others; the wise execution of this choice often makes the difference between a result that is realistic and one that is merely academic. Van Laar's essential contribution was that he chose good simplifying assumptions that made the problem tractable and yet did not greatly violate physical reality.

Van Laar considered a mixture of two liquids: x_1 moles of liquid 1 and x_2 moles of liquid 2. He assumed that the two liquids mix at constant temperature and pressure in such a manner that:

1. There is no volume change, i.e., $v^E = 0$.
2. The entropy of mixing is given by that corresponding to an ideal solution, i.e., $s^E = 0$,

where superscript E stands for excess. Since, at constant pressure,

$$g^E = u^E + Pv^E - Ts^E \tag{7-1}$$

it follows from van Laar's simplifying assumptions that

$$g^E = u^E \tag{7-2}$$

To calculate the energy change of mixing, van Laar constructed a three-step, isothermal, thermodynamic cycle wherein the pure liquids are first vaporized to some arbitrarily low pressure, mixed at this low pressure, and then recompressed to the original pressure, as illustrated in Fig. 7-1. The energy change is calculated for each step and, since energy is a state function independent of path, the energy change of mixing, Δu, is given by the sum of the three energy changes. That is,

$$\Delta u = u^E = \Delta u_{\mathrm{I}} + \Delta u_{\mathrm{II}} + \Delta u_{\mathrm{III}} \tag{7-3}$$

Step I. The two pure liquids are vaporized isothermally to the ideal-gas state. The energy change accompanying this process is calculated by the thermodynamic equation

$$\left(\frac{\partial u}{\partial v}\right)_T = T\left(\frac{\partial P}{\partial T}\right)_v - P \tag{7-4}$$

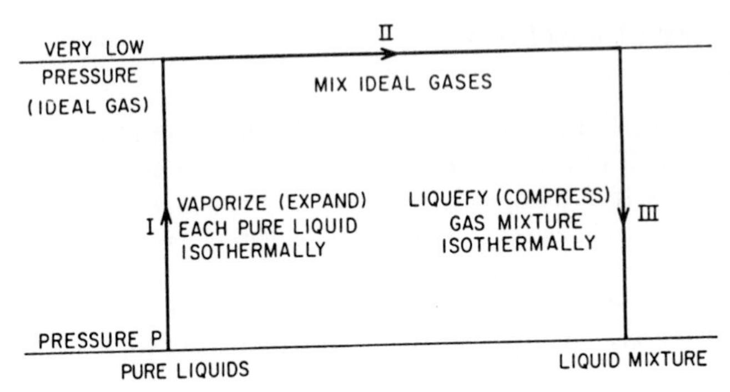

Figure 7-1 Thermodynamic cycle for forming a liquid mixture from the pure liquids at constant temperature.

Van Laar then (unfortunately) assumed that the volumetric properties of the pure fluids are given by the van der Waals equation. In that case,

$$\left(\frac{\partial u}{\partial v}\right)_T = \frac{a}{v^2} \tag{7-5}$$

where a is the constant appearing in the van der Waals equation. With x_1 moles of liquid 1 and x_2 moles of liquid 2, we obtain exactly one mole of mixture. Then

$$x_1(u_{\text{ideal}} - u)_1 = \int_{v_1^L}^{\infty} \frac{a_1 x_1}{v^2} dv - \frac{a_1 x_1}{v_1^L} \tag{7-6}$$

and

$$x_2(u_{\text{ideal}} - u)_2 = \int_{v_2^L}^{\infty} \frac{a_2 x_2}{v^2} dv - \frac{a_2 x_2}{v_2^L} \tag{7-7}$$

where u_{ideal} is the energy of the ideal gas and v^L is the molar volume of the pure liquid. Now, according to van der Waals' theory, the molar volume of a liquid well below its critical temperature can be replaced approximately by the constant b. Thus

$$\Delta u_{\text{I}} = \frac{a_1 x_1}{b_1} + \frac{a_2 x_2}{b_2} \tag{7-8}$$

Step II. Isothermal mixing of gases at very low pressure (i.e., ideal gases) proceeds with no change in energy. Thus

$$\Delta u_{\text{II}} = 0 \tag{7-9}$$

Step III. The ideal-gas mixture is now compressed isothermally and condensed at the original pressure. The thermodynamic equation (7-4) also holds for a mixture, and van Laar assumed that the volumetric properties of the mixture are also given by the van der Waals equation. Thus

$$\Delta u_{\text{III}} = -\frac{a_{\text{mixt}}}{b_{\text{mixt}}} \tag{7-10}$$

It is now necessary to express constants a and b for the mixture in terms of the constants for the pure components. Van Laar used the expressions

$$a_{\text{mixt}} = x_1^2 a_1 + x_2^2 a_2 + 2x_1 x_2 \sqrt{a_1 a_2} \tag{7-11}$$

$$b_{\text{mixt}} = x_1 b_1 + x_2 b_2 \tag{7-12}$$

Equation (7-11) follows from the assumption that only interactions between two molecules are important and that a_{12}, the constant characteristic of the interaction between two dissimilar molecules, is given by the geometric-mean law. Equation (7-12) follows from the assumption that there is no volume change upon mixing the two liquids.

Equations (7-8) to (7-12) are now substituted in Eq. (7-3). Algebraic rearrangement gives

$$g^E = \frac{x_1 x_2 b_1 b_2}{x_1 b_1 + x_2 b_2}\left(\frac{\sqrt{a_1}}{b_1} - \frac{\sqrt{a_2}}{b_2}\right)^2 \tag{7-13}$$

The activity coefficients are obtained by differentiation as discussed in Sec. 6.3 and we obtain

$$\boxed{\ln \gamma_1 = \frac{A'}{\left(1 + \dfrac{A'}{B'}\dfrac{x_1}{x_2}\right)^2}} \tag{7-14}$$

and

$$\boxed{\ln \gamma_2 = \frac{B'}{\left(1 + \dfrac{B'}{A'}\dfrac{x_2}{x_1}\right)^2}} \tag{7-15}$$

where

$$A' = \frac{b_1}{RT}\left(\frac{\sqrt{a_1}}{b_1} - \frac{\sqrt{a_2}}{b_2}\right)^2 \tag{7-16}$$

and

$$B' = \frac{b_2}{RT}\left(\frac{\sqrt{a_1}}{b_1} - \frac{\sqrt{a_2}}{b_2}\right)^2 \tag{7-17}$$

Equations (7-14) and (7-15) are the *van Laar equations* that relate the activity coefficients to temperature, composition, and to the properties of the pure components, i.e., (a_1, b_1) and (a_2, b_2).

Two important features of van Laar equations should be noted. One is that the logarithms of the activity coefficients are inversely proportional to the absolute temperature. This result, however, is independent of van Laar's thermodynamic cycle and follows directly from the assumption that $s^E = 0$.[3] The other important feature is that according to van Laar's theory, the activity coefficients of both components are never less than unity; hence, this theory always predicts positive deviations from Raoult's law. This result follows from Eq. (7-11), which says that

$$a_{\text{mixt}} < x_1 a_1 + x_2 a_2 \tag{7-18}$$

whenever $a_1 \neq a_2$.

Because constant a is proportional to the forces of attraction between molecules, Eq. (7-11) [or (7-18)] implies that the forces of attraction between the molecules in the mixture *are less* than what they would be if they were additive on a molar basis. If van Laar had assumed a rule where

$$a_{\text{mixt}} > x_1 a_1 + x_2 a_2 \tag{7-19}$$

he would have obtained

$$\gamma_i < 1 \qquad \text{for all } x < 1. \tag{7-20}$$

On the other hand, had he assumed that

$$a_{\text{mixt}} = x_1 a_1 + x_2 a_2 \tag{7-21}$$

he would have obtained that

[3] At constant pressure and composition, the derivative of g^E with respect to temperature is $-s^E$. When $s^E = 0$, it follows that $\ln\gamma$ is proportinal to T^{-1} at constant pressure and composition.

$$\gamma_1 = \gamma_2 = 1 \qquad \text{for all } x \tag{7-22}$$

Thus, we can see that the rules that one uses to express the constants for a mixture in terms of the constants for the pure components have a large influence on the predicted results.

As one might expect, quantitative agreement between van Laar's equations and experimental results is not good. However, this poor agreement is not due as much to van Laar's simplifications as it is to his adherence to the van der Waals equation and to the mixing rules used by van der Waals to extend that equation to mixtures.

One of the implications of van Laar's theory is the relation between solution nonideality and the critical pressures of the pure components. According to van der Waals' equation of state, the square root of the critical pressure of a pure fluid is proportional to $\sqrt{a}/b$. Therefore, van Laar's theory predicts that the nonideality of a solution rises with increasing difference in the critical pressures of the components; for a solution whose components have identical critical pressures, van Laar's theory predicts ideal behavior. These predictions, unfortunately, are contrary to experiment.

If we regard A' and B' as adjustable parameters, van Laar equations are useful empirical relations that have been used successfully to correlate experimental activity coefficients for many binary systems, including some that show large deviations from ideal behavior (see Sec. 6.10).

7.2 The Scatchard-Hildebrand Theory

Van Laar had recognized that a simple theory of solutions could be constructed if we restrict attention to those cases where the excess entropy and the excess volume of mixing could be neglected. Several years later, Hildebrand found that the experimental thermodynamic properties of iodine solutions in various nonpolar solvents appeared to be substantially in agreement with these simplifying assumptions. Hildebrand (1929) called these solutions regular and later defined a *regular solution* as one where the components mix with no excess entropy provided that there is no volume change upon mixing. Another way of saying this is to define a regular solution as one that has vanishing excess entropy of mixing at constant temperature and constant volume.

Both Hildebrand and Scatchard, working independently and a continent apart, realized that van Laar's theory could be greatly improved if it could be freed from the limitations of van der Waals' equation of state. This can be done by defining a parameter c according to

$$\boxed{c \equiv \frac{\Delta_{\text{vap}} u}{v^L}} \tag{7-23}$$

where $\Delta_{vap}u$ is the energy of complete vaporization, that is, the energy change upon isothermal vaporization of the saturated liquid to the ideal-gas state (infinite volume). Parameter c is the *cohesive-energy density.*

Having defined c, the key step made by Hildebrand and Scatchard consisted in generalizing Eq. (7-23) to a binary liquid mixture by writing, per mole of mixture,

$$-(u_{\text{liquid}} - u_{\text{ideal gas}})_{\text{mixt}} = \frac{c_{11}v_1^2x_1^2 + 2c_{12}v_1v_2x_1x_2 + c_{22}v_2^2x_2^2}{x_1v_1 + x_2v_2} \tag{7-24}$$

where superscript L has been dropped from the v's. Equation (7-24) assumes that the energy of a binary liquid mixture (relative to the ideal gas at the same temperature and composition) can be expressed as a quadratic function of the volume fraction and it also implies that the volume of a binary liquid mixture is given by the mole-fraction average of the pure-component volumes (i.e., $v^E = 0$). Constant c_{11} refers to interactions between molecules of species 1; c_{22} refers to interactions between molecules of species 2, and c_{12} refers to interactions between unlike molecules. For saturated liquids, c_{11} and c_{22} are functions only of temperature.

To simplify notation, we introduce symbols Φ_1 and Φ_2 that designate *volume fractions* of components 1 and 2, defined by

$$\Phi_1 \equiv \frac{x_1v_1}{x_1v_1 + x_2v_2} \tag{7-25}$$

$$\Phi_2 \equiv \frac{x_2v_2}{x_1v_1 + x_2v_2} \tag{7-26}$$

Equation (7-24) now becomes

$$-(u_{\text{liquid}} - u_{\text{ideal gas}})_{\text{mixt}} = (x_1v_1 + x_2v_2)[c_{11}\Phi_1^2 + 2c_{12}\Phi_1\Phi_2 + c_{22}\Phi_2^2] \tag{7-27}$$

The molar energy change of mixing (that is also the excess energy of mixing) is defined by

$$\Delta_{\text{mix}}u = u^E \equiv u_{\text{mixt}} - x_1u_1 - x_2u_2 \tag{7-28}$$

Equation (7-23) (for each component) and Eq. (7-27) are now substituted into Eq. (7-28); also, we utilize the relation for ideal gases,

$$\Delta_{\text{mix}}u = u_{\text{ideal}}^E = 0 \tag{7-29}$$

Algebraic rearrangement then gives

$$u^E = (c_{11} + c_{22} - 2c_{12})\Phi_1\Phi_2(x_1v_1 + x_2v_2) \tag{7-30}$$

Scatchard and Hildebrand now make what is probably the most important assumption in their theory. They assume that for molecules whose forces of attraction are due primarily to dispersion forces, there is a simple relation between c_{11}, c_{22}, and c_{12} as suggested by London's formula (see Sec. 4.4), i.e.,

$$c_{12} = (c_{11}c_{22})^{1/2} \tag{7-31}$$

Substituting Eq. (7-31) into Eq. (7-30) gives

$$u^E = (x_1v_1 + x_2v_2)\Phi_1\Phi_2(\delta_1 - \delta_2)^2 \tag{7-32}$$

where

$$\boxed{\delta_1 \equiv c_{11}^{1/2} = \left(\frac{\Delta_{\text{vap}}u}{v}\right)_1^{1/2}} \tag{7-33}$$

and

$$\boxed{\delta_2 \equiv c_{22}^{1/2} = \left(\frac{\Delta_{\text{vap}}u}{v}\right)_2^{1/2}} \tag{7-34}$$

The positive square root of c is given the symbol δ, called the *solubility parameter*.

To complete their theory of solutions, Scatchard and Hildebrand make one additional assumption, i.e., that at constant temperature and pressure the excess entropy of mixing vanishes. This assumption is consistent with Hildebrand's definition of regular solutions because in the treatment outlined above we had already assumed that there is no excess volume. With the elimination of excess entropy and excess volume at constant pressure, we have

$$g^E = u^E \tag{7-35}$$

The activity coefficients follow upon using Eq. (6-25). They are

$$\boxed{RT\ln\gamma_1 = v_1\Phi_2^2(\delta_1 - \delta_2)^2} \tag{7-36}$$

and

$$\boxed{RT \ln \gamma_2 = v_2 \Phi_1^2 (\delta_1 - \delta_2)^2} \tag{7-37}$$

Equations (7-36) and (7-37) are the *regular-solution equations*, and they have much in common with the van Laar relations [Eqs. (7-14) and (7-15)]. The regular-solution equations can easily be rearranged into the van Laar form by writing for parameters A' and B',

$$A' = \frac{v_1}{RT}(\delta_1 - \delta_2)^2 \tag{7-38}$$

and

$$B' = \frac{v_2}{RT}(\delta_1 - \delta_2)^2 \tag{7-39}$$

The regular-solution equations always predict $\gamma_i \geq 1$; i.e., a regular solution can exhibit only positive deviations from Raoult's law. This result is again a direct consequence of the geometric-mean assumption; it follows from Eq. (7-31), wherein the cohesive-energy density corresponding to the interaction between dissimilar molecules is given by the geometric mean of the cohesive-energy densities corresponding to interaction between similar molecules.

Solubility parameters δ_1 and δ_2 are functions of temperature, but the difference between these solubility parameters, $\delta_1 - \delta_2$, is often nearly independent of temperature. Since the regular-solution model assumes that the excess entropy is zero, it follows that at constant composition the logarithm of each activity coefficient must be inversely proportional to the absolute temperature. Hence, the model assumes that, as the temperature is varied at constant composition,

$$v_1 \Phi_2^2 (\delta_1 - \delta_2)^2 = \text{constant} \tag{7-40}$$

and

$$v_2 \Phi_1^2 (\delta_1 - \delta_2)^2 = \text{constant} \tag{7-41}$$

For many solutions of nonpolar liquids Eqs. (7-40) and (7-41) are reasonable approximations provided that the temperature range is not large and that the solution is remote from critical conditions.

Table 7-1 gives liquid molar volumes and solubility parameters for some typical nonpolar liquids at 25°C and for a few liquefied gases at 90 K. By inspection of the solubility parameters of different liquids, it is easily possible to make some qualitative statements about deviations from ideality of certain mixtures. Remembering that the logarithm of the activity coefficient varies directly as the square of the difference in solubility parameters, we can see, for instance, that a mixture of carbon disulfide with n-hexane exhibits large positive deviations from Raoult's law, whereas a mixture of carbon tetrachloride and cyclohexane is nearly ideal. The difference in solubility parameters of mixture components provides a measure of solution nonideality. For example, the solubility parameters shown in Table 7-1 bear out the well-known observation that whereas mixtures of aliphatic hydrocarbons are nearly ideal, mixtures of aliphatic hydrocarbons with aromatics show appreciable nonideality.

Table 7-1 Molar liquid volumes and solubility parameters of some nonpolar liquids.*

	v (cm^3 mol^{-1})	δ (J cm^{-3})$^{1/2}$
Liquefied gases at 90 K		
Nitrogen	38.1	10.8
Carbon monoxide	37.1	11.7
Argon	29.0	13.9
Oxygen	28.0	14.7
Methane	35.3	15.1
Carbon tetrafluoride	46.0	17.0
Ethane	45.7	19.4
Liquid solvents at 25 °C		
Perfluoro-n-heptane	226	12.3
Neopentane	122	12.7
Isopentane	117	13.9
n-Pentane	116	14.5
n-Hexane	132	14.9
1-Hexene	126	14.9
n-Octane	164	15.3
n-Hexadecane	294	16.3
Cyclohexane	109	16.8
Carbon tetrachloride	97	17.6
Ethyl benzene	123	18.0
Toluene	107	18.2
Benzene	89	18.8
Styrene	116	19.0
Tetrachloroethylene	103	19.0
Carbon disulfide	61	20.5
Bromine	51	23.5

* More complete tables are given by Barton (1991).

Regular-solution equations give a good semiquantitative representation of activity coefficients for many solutions containing nonpolar components. Because of various simplifying assumptions that have been made in the derivation, we cannot expect complete quantitative agreement between calculated and experimental results, but for approximate work, i.e., for reasonable estimates of (nonpolar) equilibria in the absence of any mixture data, the regular-solution equations provide useful results.

Figures 7-2, 7-3, and 7-4 show *y-x* diagrams for three representative nonpolar systems. Vapor-liquid equilibria were calculated first using Raoult's law and then using the regular-solution equations; experimentally observed equilibria are also shown and it is evident that for two systems, results based on the regular-solution theory provide a considerable improvement over those calculated by Raoult's law; for the third system, neopentane/carbon tetrachloride, the regular-solution equations overcorrect. For mixtures of nonpolar liquids, it is fair to say that whereas Raoult's law gives a zeroth approximation, the regular-solution equations usually give a first approximation to vapor-liquid equilibria. While regular-solution results are not always good, for nonpolar systems they are usually reasonable, and whenever an estimate of phase equilibria is required, the theory of regular solutions provides a valuable guide. [It must again be emphasized that Eqs. (7-36) and (7-37) are not valid for solutions containing polar components]. The only major failure of the theory of regular solutions for nonpolar fluids appears to be when it is applied to certain solutions containing fluorocarbons (Scott, 1958); the reasons for this failure are only partly understood.

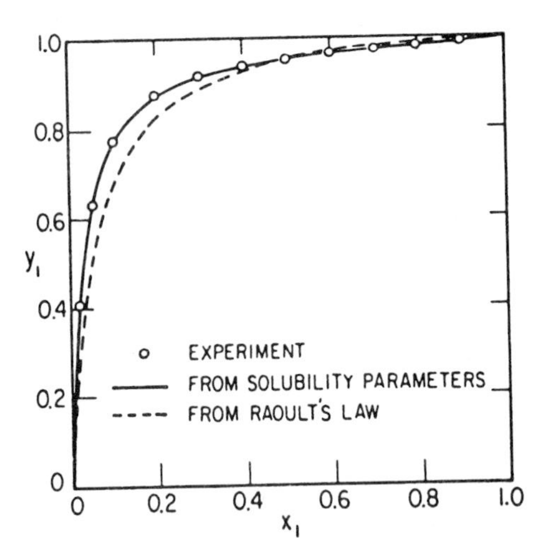

Figure 7-2 Vapor-liquid equilibria for CO (1)/CH_4 (2) mixtures at 90.7 K.

For mixtures that are nearly ideal, the regular-solution equations are often poor in the sense that predicted and observed excess Gibbs energies differ appreciably; however, for nearly ideal mixtures such errors necessarily have only a small effect on calculated vapor-liquid equilibria. For practical applications, the regular-solution equations are most useful for nonpolar mixtures having appreciable nonideality. Solubility-parameter theory provides a fairly good estimate of the excess Gibbs energies of most mixtures of common nonpolar liquids, especially when the excess Gibbs energy is large.

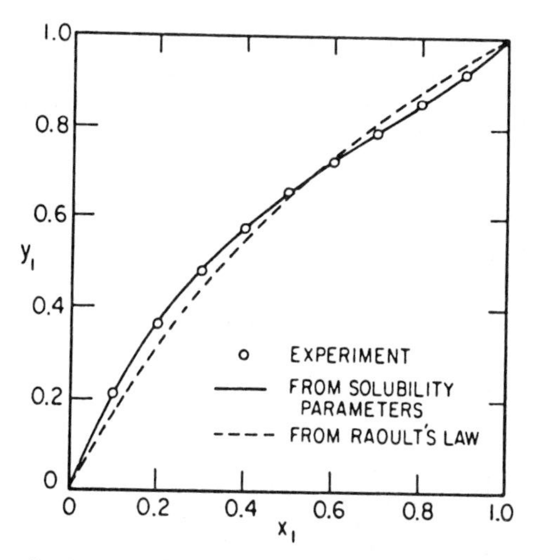

Figure 7-3 Vapor-liquid equilibria for C_6H_6 (1)/n-C_7H_{16} (2) at 70°C.

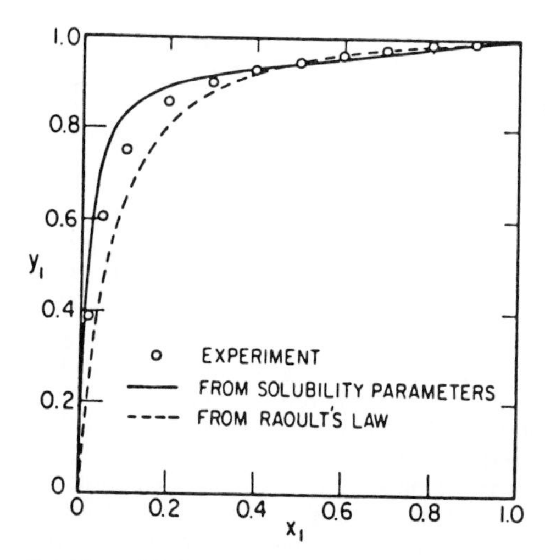

Figure 7-4 Vapor-liquid equilibria for neo-C_5H_{12} (1)/CCl_4 (2) at 0°C.

For small deviations from ideality, Eqs. (7-36) and (7-37) are less reliable because small errors in the geometric-mean assumption and in the solubility parameters become relatively more serious when δ_1 and δ_2 are close to one another.

Scott (1956) has shown that solubility-parameter theory fits excess Gibbs energies of most binary systems of nonpolar liquids to within 10 to 20% of the thermal energy RT. (At room temperature RT is nearly 2500 J mol^{-1}). McGlashan (1962) bears this out as indicated in Fig. 7-5, where a comparison is made between calculated and observed excess Gibbs energies for 21 binary systems near room temperature at the composition midpoint $x_1 = x_2 = 1/2$. The dashed lines were drawn 370 J mol^{-1} ($\approx 0.15RT$) above and below the solid line that corresponds to perfect agreement between theory and experiment.

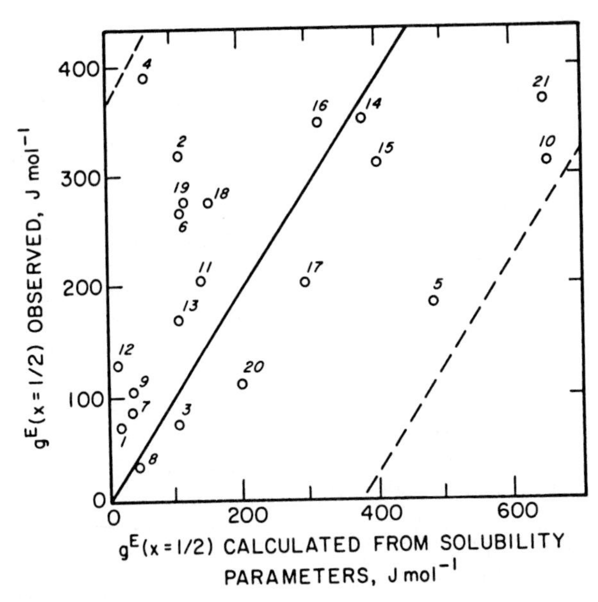

Figure 7-5 Excess Gibbs energies from the regular-solution equation. Binary systems shown are: 1. *c*-C_6H_{12}/CCl_4; 2. *c*-C_6H_{12}/C_6H_6; 3. *c*-C_6H_{12}/*n*-C_6H_{14}; 4. *c*-$C_6H_{12}/C_6H_5CH_3$; 5. *c*-$C_6H_{12}/C(CH_3)_4$; 6. C_6H_6/*c*-C_5H_{10}; 7. C_6H_6/CCl_4; 8. $C_6H_6/C_2H_4Cl_2$; 9. $CCl_4/CHCl_3$; 10. $CCl_4/C(CH_3)_4$; 11. CCl_4/CH_3I; 12. $TiCl_4/CCl_4$; 13. $SiCl_4/CCl_4$; 14. C_6H_6/*n*-C_7H_{16}; 15. C_6H_6/*n*-C_6H_{14}; 16. $C_6H_5CH_3$/*n*-C_6H_{14}; 17. $C_6H_5CH_3$/*n*-C_7H_{16}; 18. $C_6H_5CH_3$/*c*-$C_6H_{11}CH_3$; 19. CCl_4/CH_2Cl_2; 20. *n*-C_6H_{14}/CCl_4; 21. C_6H_6/*i*-C_8H_{18}. Systems 8, 9, 11, and 19 each contain one component whose polarity is not negligible and, strictly speaking, they should not be included in this list. However, since there are no specific effects (e.g., hydrogen bonding) in these systems, regular-solution theory still gives the right order of magnitude for g^E for these particular mixtures. Most of the data are at 25°C. The lowest temperature (0°C) is for system 10 and the highest (65°C) for system 18. According to regular-solution theory, the excess Gibbs energy is independent of temperature to a first approximation. Dashed lines indicate $\pm 0.15RT$, here taken as ± 370 J mol^{-1}.

Figure 7-5 suggests that solubility parameters are primarily useful for semiquantitative estimates of activity coefficients in liquid mixtures. Solubility parameters can tell us readily the magnitude of nonideality that is to be expected in a mixture of two nonpolar liquids. In addition, solubility parameters can form a basis for a more quantitative application when modified empirically. One example of such an application is provided by Chao and Seader (1961), who used solubility parameters to correlate phase equilibria for hydrocarbon mixtures over a wide range of conditions. Two other applications, one concerned with gas solubility and the other with solubility of solid carbon dioxide at low temperatures, are discussed in later chapters.

The theory of Scatchard and Hildebrand is essentially the same as that of van Laar but it is liberated from the narrow confines of the van der Waals equation or of any other equation of state. We know that the assumptions of regularity ($s^E = 0$) and isometric mixing ($v^E = 0$) at constant temperature and pressure are not correct even for simple mixtures but, due to cancellation of errors, these assumptions frequently do not seriously affect calculations of the excess Gibbs energy. (When regular-solution theory is used to calculate excess enthalpies, the results are usually much worse.) However, the most serious defect of the theory is the geometric-mean assumption. This assumption can be relaxed by writing instead of Eq. (7-31) the more general relation[4]

$$c_{12} = (1 - l_{12})(c_{11}c_{22})^{1/2} \tag{7-42}$$

where l_{12} is a constant, small compared to unity, characteristic of the 1-2 interaction. From London's theory of dispersion forces, an expression can be obtained for l_{12} in terms of molecular parameters but such an expression has little quantitative value.

In mixtures of chemically similar components (e.g., cyclohexane/*n*-hexane), deviations from the geometric mean are primarily a result of differences in molecular shape and subsequent differences in molecular packing. Our limited ability to describe properly the geometric arrangement of polyatomic molecules in the liquid phase is one of the main reasons for the inadequacy of currently existing theories of solution.

When Eq. (7-42) is used in place of Eq. (7-31), the activity coefficients are given by

$$\ln \gamma_1 = \frac{v_1 \Phi_2^2}{RT}\left[(\delta_1 - \delta_2)^2 + 2l_{12}\delta_1\delta_2\right] \tag{7-43}$$

and

$$\ln \gamma_2 = \frac{v_2 \Phi_1^2}{RT}\left[(\delta_1 - \delta_2)^2 + 2l_{12}\delta_1\delta_2\right] \tag{7-44}$$

Equations (7-43) and (7-44) show immediately that if δ_1 and δ_2 are close to each other, even a small value of l_{12} can significantly affect the activity coefficients. For example, suppose $T = 300$ K, $v_1 = 100$ cm^3 mol^{-1}, and δ_1 and δ_2 are 14.3 and 15.3 (J cm^{-3})$^{1/2}$, respectively. Then, at infinite dilution, we find that for $l_{12} = 0$, $\gamma_1^\infty = 1.04$. However, if $l_{12} = 0.03$, we obtain $\gamma_1^\infty = 1.77$. Even if l_{12} is as small as 0.01, we obtain $\gamma_1^\infty = 1.24$. These illustrative results show why the solubility-parameter theory is not quantitatively reliable for components whose solubility parameters are very nearly the

[4] The l_{12} used here is related to, but different from, k_{12} used in Sec. 5.7.

same. As the difference between δ_1 and δ_2 becomes larger, the effect of deviation from the geometric mean becomes less serious. However, it is apparent that even small deviations from the geometric mean, 1 or 2%, can have an appreciable effect on calculated activity coefficients and that much improvement in predicted results can often be achieved when only one (reliable) binary datum is available for evaluating l_{12}.

Efforts to correlate l_{12} have met with little success. In his study of binary cryogenic mixtures, Bazúa (1971) found no satisfactory variation of l_{12} with pure-component properties, although some rough trends were found by Cheung and Zander (1968) and by Preston (1970). In many typical cases l_{12} is positive and becomes larger as the differences in molecular size and chemical nature of the components increase. For example, for carbon dioxide/paraffin mixtures at low temperatures, Preston found that l_{12} = -0.02 (methane), +0.08 (ethane), +0.08 (propane), and +0.09 (butane).

Since l_{12} is an essentially empirical parameter, it depends on temperature. However, for typical nonpolar mixtures over a modest range of temperature, that dependence is usually small.

For mixtures of aromatic and saturated hydrocarbons, Funk (1970) found a systematic variation of l_{12} with the structure of the saturated component, as shown in Fig. 7-6. In this case, a good correlation could be established because experimental data are relatively plentiful and because the correlation is restricted to a narrow class of mixtures. Figure 7-7 shows the effect of l_{12} on calculating relative volatility in a typical binary system.

Our inability to correlate l_{12} for a wide variety of mixtures follows from our lack of understanding of intermolecular forces, especially between molecules at short separations.

One of the early improvements in the regular-solution theory was to replace the ideal entropy of mixing with the Flory-Huggins equations for mixing molecules appreciably different in size (Sec. 8.2). Another improvement was proposed by Gonsalves and Leland (1978) using some theoretical knowledge about the structure (molecular packing) of a fluid mixture. The results for their modified regular-solution theory show an improvement in the calculated excess Gibbs energy and excess enthalpy when the molecules in the mixture differ appreciably in size and shape. For molecules of approximately the same size, the modified theory gives essentially the same results as those from the original regular-solution theory.

The most important assumption in the calculation of excess functions is the one that concerns the unlike-pair interaction. Small errors in predicting this interaction can often offset completely any improvement derived from a better description of liquid structure.

Several authors have tried to extend regular-solution theory to mixtures containing polar components, but unless the classes of components considered are restricted, such extension has only semiquantitative significance. In establishing these extensions, the cohesive energy density is divided into separate contributions from nonpolar (dispersion) forces and from polar forces:

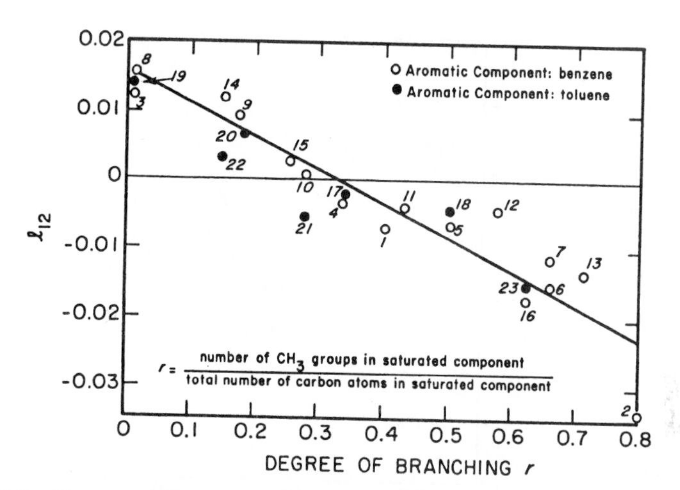

Figure 7-6 Binary parameter l_{12} for aromatic-saturated hydrocarbon mixtures at 50°C. Binary systems shown are: 1. Benzene (2)/Pentane (1); 2. Benzene (2)/Neopentane (1); 3. Benzene (2)/Cyclopentane (1); 4. Benzene (2)/Hexane (1); 5. Benzene (2)/2-Methylpentane (1); 6. Benzene (2)/2,2-Dimethylbutane (1); 7. Benzene (2)/2,3-Dimethylbutane (1); 8. Benzene (2)/ Cyclohexane (1); 9. Benzene (2)/Methylcyclopentane (1); 10. Benzene (2)/ Heptane (1); 11. Benzene (2)/3-Methylhexane (1); 12. Benzene (2)/2,4-Dimethylpentane (1); 13. Benzene (2)/2,2,3-Trimethylbutane (1); 14. Benzene (2)/Methylcyclohexane (1); 15. Benzene (2)/Octane (1); 16. Benzene (2)/2,2,4-Trimethylpentane (1); 17. Toluene (2)/Hexane (1); 18. Toluene (2)/3-Methylpentane (1); 19. Toluene (2)/Cyclohexane (1); 20. Toluene (2)/Methylcyclopentane (1); 21. Toluene (2)/Heptane (1); 22. Toluene (2)/Methylcyclohexane (1); 23. Toluene (2)/2,2,4-Trimethylpentane (1).

$$\left(\frac{\Delta_{\text{vap}}u}{v}\right)_{\text{total}} = \left(\frac{\Delta_{\text{vap}}u}{v}\right)_{\text{nonpolar}} + \left(\frac{\Delta_{\text{vap}}u}{v}\right)_{\text{polar}} \tag{7-45}$$

Equations (7-43) and (7-44) are used with the substitutions

$$\delta_1^2 = \tau_1^2 + \lambda_1^2 \tag{7-46}$$

$$\delta_2^2 = \tau_2^2 + \lambda_2^2 \tag{7-47}$$

$$l_{12}\delta_1\delta_2 = (\tau_1^2 + \lambda_1^2)^{1/2}(\tau_2^2 + \lambda_2^2)^{1/2} - (\lambda_1\lambda_2 + \tau_1\tau_2 + \psi_{12}) \tag{7-48}$$

where λ_i is the *nonpolar solubility parameter* [$\lambda_i^2 = (\Delta_{\text{vap}}u/v)_{\text{nonpolar}}$] and τ_i is the *polar solubility parameter* [$\tau_i^2 = (\Delta_{\text{vap}}u/v)_{\text{polar}}$]. Binary parameter ψ_{12} is not negligible, as shown by Weimer (1965) in his correlation of activity coefficients at infinite dilution for hydrocarbons in polar non-hydrogen-bonding solvents.

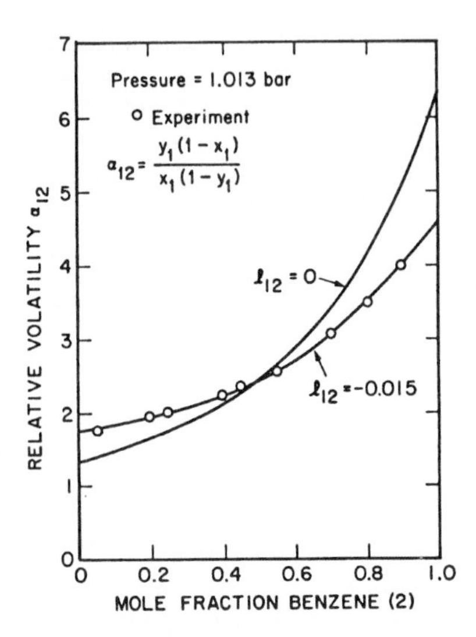

Figure 7-7 Comparison of experimental volatilities with volatilities calculated by Scatchard-Hildebrand theory for 2,2-dimethylbutane (1)/benzene (2).

Further extension of the Scatchard-Hildebrand equation to include hydrogen-bonded components makes little sense theoretically, because the assumptions of regular-solution theory are seriously in error for mixtures containing such components. Nevertheless, some semiquantitative success has been achieved by Hansen *et al.* (1967, 1967a, 1971) and others (Burrell, 1968; Gardon, 1966; Nelson *et al.*, 1970; Mark *et al.*, 1969; Barton, 1991) interested in establishing criteria for formulating solvents for paints and other surface coatings. Also, Null and Palmer (1969) and Null (1970) have used extended solubility parameters for establishing an empirical correlation of activity coefficients. Barton (1991) has given a comprehensive review of extended solubility parameters and their applications. Panayiotou (1997) has developed an equation of state model that provides analytical expressions to estimate solubility parameters as functions of temperature, pressure and mixture composition. This model is general, also applicable to complex systems containing molecules forming hydrogen bonds.

One of the main advantages of the regular-solution equations is their simplicity, and this simplicity is retained when the regular-solution model is extended to solutions containing more than two components. The derivation for the multicomponent case is analogous to that given for the binary case. The molar energy of a liquid mixture containing m components is written

$$-(u_{\text{liquid}} - u_{\text{ideal gas}})_{\text{mixt}} = \frac{\sum_i^m \sum_j^m v_i v_j x_i x_j c_{ij}}{\sum_i^m x_i v_i} \tag{7-49}$$

The volume fraction of component j is now defined by

$$\Phi_j \equiv \frac{x_j v_j}{\sum_i^m x_i v_i} \tag{7-50}$$

and the excess energy of mixing is defined by

$$u^E \equiv u_{\text{mixt}} - \sum_i^m x_i u_i \tag{7-51}$$

By assumption, the cohesive-energy density c_{ij} is given by the geometric mean,

$$c_{ij} = (c_{ii} c_{jj})^{1/2} \tag{7-52}$$

Again assuming that

$$s^E = v^E = 0 \tag{7-53}$$

we again have

$$g^E = u^E \tag{7-54}$$

Substitution and algebraic rearrangement, coupled with Eq. (6-25), gives a remarkably simple result for the activity coefficient of component j in a multicomponent solution:

$$\boxed{RT \ln \gamma_j = v_j (\delta_j - \bar{\delta})^2} \tag{7-55}$$

where

$$\bar{\delta} = \sum_i^m \Phi_i \delta_i \tag{7-56}$$

Parameter $\bar{\delta}$ is a volume-fraction average of the solubility parameters of all the components in the solution; the summation in Eq. (7-56) is over all components, including component j.

Equation (7-55) has the same advantages and disadvantages as Eqs. (7-36) and (7-37). It is useful for providing estimates of equilibria in nonpolar solutions and, with empirical modifications, it can serve as a basis for quantitative correlations.

Equations (7-43) and (7-44) can also be generalized for mixtures containing more than two components; the general expression for the activity coefficient, however, is no longer as simple as that given by Eq. (7-55). For a mixture of m components, it is

$$RT\ln\gamma_k = v_k \sum_i^m \sum_j^m \Phi_i \Phi_j \left(D_{ik} - \frac{1}{2} D_{ij} \right) \tag{7-57}$$

where

$$D_{ij} \equiv (\delta_i - \delta_j)^2 + 2l_{ij}\delta_i\delta_j \tag{7-58}$$

For every component i, $l_{ii} = D_{ii} = 0$. Equation (7-57) reduces to Eq. (7-55) only if $l_{ij} = 0$ for every ij pair.

Regular-solution theory is attractive because of its simplicity. For many liquid mixtures that contain nonpolar molecules, this theory can predict equilibria with fair accuracy and for many more, it can correlate liquid-phase activity coefficients using only one adjustable parameter to correct for deviations from the geometric-mean assumption.

For mixtures that contain large molecules (polymers) or for those that contain strongly polar or hydrogen-bonding molecules, the theory of regular solutions is inadequate; for such mixtures other theories are better, as described later in this chapter and in Chap. 8. However, before turning to such mixtures, it is useful to discuss briefly an alternate procedure applicable, in principle, to all fluid mixtures, although in practice it is usually applied only to relatively simple mixtures. This procedure is based on an equation of state applied to both the vapor phase and the liquid phase, following equations given in Sec. 3.4. This procedure is also discussed in Chapter 12.

7.3 Excess Functions from an Equation of State

For a liquid mixture, we can calculate the conventional molar excess Gibbs energy g^E provided that we have available an equation of state that is valid for the entire density range from zero to liquid density.

Because a realistic equation of state is inevitably pressure-explicit, it is more convenient to calculate the molar excess Helmholtz energy $\boldsymbol{a}^E$. As shown elsewhere,[5] at low pressures, we can use the excellent approximation

$$\boldsymbol{a}_v^E = g_P^E \tag{7-59}$$

[5] J. H. Hildebrand, J. M. Prausnitz, and R. L. Scott, 1970, *Regular and Related Solutions,* New York: Van Nostrand Reinhold.

where subscript v indicates constant volume and subscript P indicates constant pressure. The relation between molar Helmholtz energy, $\boldsymbol{a}$, and the equation of state is discussed in Chap. 3; the fundamental working equation is

$$P = -\left(\frac{\partial \boldsymbol{a}}{\partial v}\right)_{T,x} \tag{7-60}$$[6]

where subscript x indicates constant composition and v is the molar volume.

To illustrate, we use the Frisch-van der Waals equation of state,

$$\frac{Pv}{RT} = \frac{1+\xi+\xi^2}{(1-\xi)^2} - \frac{a}{RTv} \tag{7-61}$$

where reduced density $\xi = b/4v$ and where, for a binary mixture, constants a and b are given by customary mixing rules quadratic in mole fraction x:

$$\begin{aligned} a &= x_1^2 a_1 + 2x_1 x_2 a_{12} + x_2^2 a_2 \\ b &= x_1^2 b_1 + 2x_1 x_2 b_{12} + x_2^2 b_2 \end{aligned} \tag{7-62}$$

For binary parameter b_{12} we write

$$b_{12}^{1/3} = \frac{1}{2}(b_1^{1/3} + b_2^{1/3}) \tag{7-63}$$

and for the other binary parameter,

$$a_{12} = b_{12}\left(\frac{a_1 a_2}{b_1 b_2}\right)^{1/2}(1-k_{12}) \tag{7-64}$$

where, for simple mixtures, $|k_{12}| << 1$.

Equation (3-50) is used to find Helmholtz energy A for the mixture, for pure liquid 1 and for pure liquid 2. The molar excess Helmholtz energy $\boldsymbol{a}^E$ is given by

$$\begin{aligned} \boldsymbol{a}^E(T,x) = {} & \boldsymbol{a}_{\text{mixt}}(T,x,v_{\text{mixt}}^*) - x_1\boldsymbol{a}(T,x_1=1,v_1) - x_2\boldsymbol{a}(T,x_2=1,v_2) \\ & - RT(x_1 \ln x_1 + x_2 \ln x_2) \end{aligned} \tag{7-65}$$

[6] To avoid confusion with equation-of-state parameter a, we use in this section $\boldsymbol{a}$ for molar Helmholtz energy.

where $\boldsymbol{a} = A/n_T$, n_T is the total number of moles, and $v^*_{\text{mixt}} = x_1 v_1 + x_2 v_2$. Here v_1 is the molar volume of pure liquid 1 and v_2 is the molar volume of pure liquid 2, it being understood that system temperature T is well below T_{c1} and T_{c2}, where T_c is the critical temperature. Upon substituting Eq. (3-50) into Eq. (7-65), constants u_i^0 and s_i^0 cancel out.

Using $\boldsymbol{a}_v^E = g_P^E$, we can find the molar excess enthalpy h^E by differentiating

$$h^E = \left[\frac{\partial(g^E/T)}{\partial(1/T)}\right]_{P,x} \tag{7-66}$$

where it is understood that both g^E and h^E refer to mixing at constant temperature and pressure.

The molar excess volume v^E is found by solving the equation of state three times, once for the mixture, once for pure liquid 1, and once for pure liquid 2:

$$v^E = v_{\text{mixt}}(T,P,x) - x_1 v(T,P,x_1 = 1) - x_2 v(T,P,x_2 = 1) \tag{7-67}$$

Calculations to obtain excess functions have been performed by Marsh (1980) for nine binary systems. For each pure liquid, constants a and b were found from critical data. Binary parameter k_{12} was found from the experimental g^E at $x_1 = x_2 = 0.5$.

Table 7-2 shows experimental and predicted h^E and v^E for equimolar mixtures. Agreement is fair for mixtures of nearly spherical molecules (first six systems) but it is poor for C_6F_6/C_6H_6, where there is an enhanced interaction between unlike molecules (negative k_{12}), and for the last two systems where the molecules no longer have even approximately spherical shape.

Table 7-2 Experimental and predicted molar excess enthalpies and molar excess volumes for nine equimolar liquid mixtures (Marsh, 1980).

			h^E (J mol^{-1})		v^E (cm^3 mol^{-1})	
System	k_{12}x10^3	Temp. (°C)	Exp.	Calc.	Exp.	Calc.
C_6H_6/CCl_4	8	25	116	130	0	0.16
$C_6H_6/c\text{-}C_6H_{12}$	30	25	799	514	0.65	0.64
$c\text{-}C_6H_{12}/CCl_4$	7	25	166	115	0.17	0.13
$C_6F_6/c\text{-}C_6H_{12}$	81	40	1534	1325	2.57	2.83
$CCl_4/c\text{-}C_5H_{10}$	3	25	79	42	-0.04	-0.03
$C_2H_4Cl_2/C_6H_6$	4	20	60	65	0.25	0
C_6F_6/C_6H_6	-11	40	-435	-76	0.80	-0.93
$n\text{-}C_6H_{14}/c\text{-}C_6H_{12}$	2	20	216	110	0.10	-0.37
$n\text{-}C_6H_{14}/n\text{-}C_{16}H_{34}$	58	25	112	-81	-0.54	-3.10

Some promising efforts have been made to construct equations of state for nonspherical molecules and it is likely that these will become increasingly useful for liquid mixtures containing such molecules, as discussed later in this chapter and in Chap. 8. However, for practical calculations, it is often convenient to abandon the equation-of-state approach and to use, instead, approximate theories of solutions based on the idea that in the condensed state, molecules arrange themselves in a lattice-like structure where each molecule (or molecular segment) occupies one lattice point. These ideas are discussed in the next sections.

7.4 The Lattice Model

Since the liquid state is in some sense intermediate between the crystalline state and the gaseous state, it follows that there are two types of approach to a theory of liquids. The first considers liquids to be gas-like; a liquid is pictured as a dense and highly nonideal gas whose properties can be described by some equation of state; that of van der Waals' is the best known example. An equation-of-state description of pure liquids can readily be extended to liquid mixtures as was done by van der Waals and by some of his disciples like van Laar and later by many others.

The second approach considers a liquid to be solid-like, in a quasicrystalline state, where the molecules do not translate fully in a chaotic manner as in a gas, but where each molecule tends to stay in a small region, a more or less fixed position in space about which it vibrates back and forth. The quasicrystalline picture of the liquid state supposes molecules to sit in a regular array in space, called a *lattice,* and therefore liquid and liquid mixture models based on this simplified picture are called *lattice models.*[7] These theories are described in detail elsewhere (Barker, 1963; Guggenheim, 1966) and their proper study requires familiarity with the methods of statistical mechanics. We give here only a brief introduction to the lattice theory of solutions.

Since the lattice theory of liquids assumes that molecules are confined to lattice positions (sometimes called cages), calculated entropies (disorder) are low by what is called the "communal entropy". While this is a serious deficiency, it tends to cancel when lattice theory is used to calculate excess properties of liquid mixtures.

Molecular considerations suggest that deviations from ideal behavior in liquid solutions are due primarily to the following effects: First, forces of attraction between unlike molecules are quantitatively different from those between like molecules, giving rise to a nonvanishing enthalpy of mixing; second, if the unlike molecules differ significantly in size or shape, the molecular arrangement in the mixture may be appreciably different from that for the pure liquids, giving rise to a nonideal entropy of mixing;

[7] An exhaustive attempt has been made by Eyring and coworkers to describe liquids as consisting of gas-like and solid-like molecules (H. Eyring and M. S. John, 1969, *Significant Liquid Structures,* New York: John Wiley & Sons). While this attempt has had some empirical success, its main ideas and assumptions are in direct conflict with many physicochemical data for liquid structure.

and finally, in a binary mixture, if the forces of attraction between one of three possible pair interactions are very much stronger (or very much weaker) than those of the other two, there are certain preferred orientations of the molecules in the mixture that, in extreme cases, may induce thermodynamic instability and demixing (incomplete miscibility).

We consider a mixture of two simple liquids 1 and 2. Molecules of types 1 and 2 are small and spherically symmetric and the ratio of their sizes is close to unity. We suppose that the arrangement of the molecules in each pure liquid is that of a regular array as indicated in Fig. 7-8; all the molecules are situated on lattice points that are equidistant from one another. Molecular motion is limited to vibrations about the equilibrium positions and is not affected by the mixing process. We suppose further that for a fixed temperature, the lattice spacings for the two pure liquids and for the mixture are the same, independent of composition (i.e., $v^E = 0$).

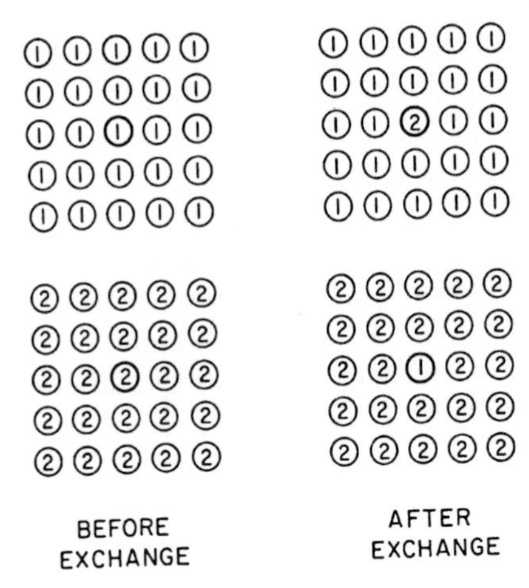

Figure 7-8 Physical significance of interchange energy. The energy absorbed in the process above is $2w$. [See Eq. (7-71)].

To derive an expression for the potential energy of a liquid, pure or mixed, we assume that the potential energy is pairwise additive for all molecular pairs and that only nearest neighbors need be considered in the summation. This means that the potential energy of a large number of molecules sitting on a lattice is given by the sum of the potential energies of all pairs of molecules that are situated immediately next to one another. For uncharged nonpolar molecules, intermolecular forces are short-range and therefore we assume in this simplified discussion that we can neglect contributions to the total potential energy from pairs that are not nearest neighbors.

Consider that each of N_1 molecules of type 1 and N_2 molecules of type 2 has z nearest (touching) neighbors. (z is the *coordination number* and may have a value between 6 and 12 depending on the type of packing, i.e., the way in which the molecules are arranged in three-dimensional space; empirically, for typical liquids at ordinary conditions, z is close to 10.) The total number of nearest neighbors is $(z/2)(N_1 + N_2)$ and there are three types of nearest neighbors: 1-1, 2-2, and 1-2. Let N_{11} be the number of nearest-neighbor pairs of type 11, N_{22} that of type 22, and N_{12} that of type 12. These three numbers are not independent; they are restricted by the following conservation equations:

$$\begin{aligned} zN_1 &= 2N_{11} + N_{12} \\ zN_2 &= 2N_{22} + N_{12} \end{aligned} \tag{7-68}$$

The total potential energy of the lattice U_t is then given by

$$U_t = N_{11}\Gamma_{11} + N_{22}\Gamma_{22} + N_{12}\Gamma_{12} \tag{7-69}$$

where, as in Chap. 4, Γ_{11} is the potential energy of a 1-1 pair, Γ_{22} that of a 2-2 pair, and Γ_{12} that of a 1-2 pair. Substitution of N_{11} and N_{22} from Eq. (7-68) gives

$$U_t = \frac{z}{2}N_1\Gamma_{11} + \frac{z}{2}N_2\Gamma_{22} + \frac{w}{2}N_{12} \tag{7-70}$$

where w, the *interchange energy*, is defined by

$$w \equiv z\left[\Gamma_{12} - \frac{1}{2}(\Gamma_{11} + \Gamma_{22})\right] \tag{7-71}$$

Equation (7-70) gives the potential energy of a binary mixture and also that of a pure liquid; in the latter case, N_{12} and either N_1 or N_2 are set equal to zero. In Eq. (7-70), the last term is the energy of mixing.

The physical significance of w is illustrated in Fig. 7-8; z pairs of type 1 and z pairs of type 2 are separated to form $2z$ dissimilar (1-2) pairs. Therefore, the change in energy that accompanies the interchange process shown in Fig. 7-8 is equal to $2w$.

To obtain thermodynamic properties, it is convenient to calculate first the canonical partition function (see App. B) of the lattice, given by

$$Q_{\text{lattice}} = \sum_{N_{12}} g(N_1, N_2, N_{12}) \exp\left(-\frac{U_t}{kT}\right) \tag{7-72}$$

where g is the combinatorial factor (degeneracy),[8] equal to the number of ways of arranging N_1 molecules of type 1 and N_2 molecules of type 2 on a lattice with a total of

[8] Combinational factor g should not be confused with molar Gibbs energy g.

$(N_1 + N_2)$ sites. For a pure component, whose molecules are of the type discussed here, $g = 1$. The summation over all N_{12} that give the same U_t can be replaced by retaining only the maximum term (see App. B).

For the Helmholtz energy change of mixing we have

$$\Delta_{\text{mix}} A = A_{\text{mixt}} - (A_{\text{pure }1} + A_{\text{pure }2}) \tag{7-73}$$

Using the relation between Helmholtz energy and the canonical partition function given in App. B (Table B-l), we obtain

$$\boxed{\Delta_{\text{mix}} A = -kT \ln\left[g(N_1, N_2, N_{12}) \exp\left(-\frac{wN_{12}}{kTz} \right) \right]} \tag{7-74}$$

Our task now is to say something about N_{12}. In view of the similarity between the two types of molecules, we assume that a mixture of 1 and 2 is *completely random*, i.e., a mixture where all possible arrangements of the molecules on the lattice are equally probable. For that case let $N_{12} = N_{12}^*$. By simple statistical arguments it can be shown (Guggenheim, 1952) that for a completely random arrangement,

$$N_{12}^* = \frac{zN_1N_2}{N_1 + N_2} \tag{7-75}$$

and

$$g(N_1, N_2, N_{12}^*) = \frac{(N_1 + N_2)!}{N_1!N_2!} \tag{7-76}$$

Substitution of Eqs. (7-75) and (7-76) into Eq. (7-74) and Stirling's approximation[9] gives

$$\frac{\Delta_{\text{mix}} A}{kT} = N_1 \ln \frac{N_1}{N_1 + N_2} + N_2 \ln \frac{N_2}{N_1 + N_2} + \frac{w}{kT} \frac{N_1 N_2}{N_1 + N_2} \tag{7-77}$$

or, in molar units,

$$\frac{\Delta_{\text{mix}} a}{RT} = x_1 \ln x_1 + x_2 \ln x_2 + \frac{w}{kT} x_1 x_2 \tag{7-78}$$

[9] $\ln N! = N \ln N - N$, when N is large (see App. B).

Notice that Eq. (7-78) is symmetric with respect to mole fraction x. For $w = 0$ we obtain an ideal solution; therefore, the molar excess Helmholtz energy is

$$\frac{a^E}{RT} = \frac{w}{kT} x_1 x_2 \tag{7-79}$$

Since w is independent of temperature (by assumption), we obtain for the entropy of mixing

$$\boxed{\frac{\Delta_{\text{mix}} s}{R} = -\frac{1}{R}\left(\frac{\partial \Delta_{\text{mix}} a}{\partial T}\right)_v = -x_1 \ln x_1 - x_2 \ln x_2} \tag{7-80}$$

This is the entropy of mixing for an ideal solution ($w = 0$); for this type of mixture, the excess entropy is zero ($s^E = 0$).

In view of the assumptions made about the lattice spacing for pure liquids and the mixture, we assume that the mixing process at constant pressure and temperature produces no changes in volume ($v^E = 0$).

As indicated in Secs. 7.1 and 7.2 (theories of van Laar and of Scatchard and Hildebrand), a solution for which $s^E = v^E = 0$ is called a *regular solution.*

For a regular solution the excess Gibbs energy, the excess Helmholtz energy, the excess enthalpy (or enthalpy of mixing) and the excess energy (or energy of mixing) are all equal:

$$g^E = a^E = h^E = u^E = N_A w x_1 x_2 \tag{7-81}$$

where N_A is Avogadro's constant. The activity coefficients follow from Eq. (6-25):

$$\boxed{\ln \gamma_1 = \frac{w}{kT} x_2^2} \tag{7-82}$$

$$\boxed{\ln \gamma_2 = \frac{w}{kT} x_1^2} \tag{7-83}$$

These results are of the same form as that of the two-suffix Margules equations. However, in Eqs. (7-82) and (7-83) parameter w has a well-defined physical significance.

From Eq. (7-71) we see that if the potential energy for a 1-2 pair is equal to the arithmetic mean of the potentials for the 1-1 and 2-2 pairs, then $w = 0$ and $\gamma_1 = \gamma_2 = 1$ for all x; we then have an ideal solution. However, as discussed in Chap. 4, for simple, nonpolar molecules Γ_{12} is more nearly equal to the geometric mean than to the arithmetic mean of Γ_{11} and Γ_{22}. Because the magnitude of the geometric mean is always less than that of the arithmetic mean, and since Γ_{12}, Γ_{11}, and Γ_{22} are negative in sign, it follows that for mixtures of simple, nonpolar molecules, Eqs. (7-82) and (7-83) predict positive deviations from ideal-solution behavior, in agreement with experiment.

7.5 Calculation of the Interchange Energy from Molecular Properties

Because the interchange energy w is related to the potential energies, it should be possible to obtain a numerical value for w from information on potential functions. Various attempts to do this have been reported and one of them, due to Kohler (1957), is particularly simple.

The potential function Γ depends on r, the distance between molecules. Kohler assumes that for the pure liquids,[10]

$$r_{11} = \left(\frac{v_1}{N_A}\right)^{1/3} \tag{7-84}$$

and

$$r_{22} = \left(\frac{v_2}{N_A}\right)^{1/3} \tag{7-85}$$

where v stands for the molar liquid volume and N_A is Avogadro's constant.

In the mixture, Kohler assumes

$$r_{12} = \frac{r_{11} + r_{22}}{2} \tag{7-86}$$

Basing his calculations on London's theory of dispersion forces (see Chap. 4), Kohler then writes[11]

$$\Gamma_{11} = -\frac{\alpha_1^2}{r_{11}^6}\xi_1 \tag{7-87}$$

[10] These assumptions are not completely consistent with the assumptions of the lattice theory, where $r_{11} = r_{22} = r_{12}$. Strictly, the lattice theory requires that $v_1 = v_2$, and that very much limits its applicability. A certain degree of inconsistency frequently results when an idealized theory is applied to real phenomena.

[11] ξ is closely related to the ionization potential (see Sec. 4.4).

$$\Gamma_{22} = -\frac{\alpha_2^2}{r_{22}^6}\xi_2 \tag{7-88}$$

$$\Gamma_{12} = -\frac{2^6 \alpha_1 \alpha_2}{(r_{11}+r_{22})^6}\frac{2\xi_1\xi_2}{\xi_1+\xi_2} \tag{7-89}$$

where α is the polarizability and ξ_i is calculated from $\Delta_{\text{vap}}h_i$, the molar enthalpy of vaporization, by

$$\xi_i = \frac{2r_{ii}^6}{z\alpha_i^2}\left(\frac{\Delta_{\text{vap}}h_i - RT}{N_A}\right) \tag{7-90}$$

When these expressions are substituted into Eq. (7-71), it is possible to obtain the interchange energy w as needed in the calculation of activity coefficients, Eqs. (7-82) and (7-83). One of the advantages of Kohler's method is that, because of cancellation, no separate estimate of the coordination number z is required; further, the three potential energies Γ_{11}, Γ_{22}, and Γ_{12} are calculated separately and it is not necessary to assume that Γ_{12} is the geometric mean of the other two.

Using Kohler's method, calculations have been made for the excess Gibbs energies of four simple binary systems, each at the composition midpoint where $x_1 = x_2 = 0.5$. Calculated results are compared in Table 7-3 with experiment and agreement is fairly good. However, we must remember that the applicability of this type of calculation is limited to mixtures where the molecules of the two components are not only nonpolar but also essentially spherical and similar in size. As a result, the equations that we have described are useful only for a small class of mixtures; when calculations based on Kohler's method are made for systems outside of this small class, agreement with experiment is usually poor.

Table 7-3 Excess Gibbs energies for equimolar, binary mixtures. Calculations based on lattice theory and Kohler's method for evaluating the interchange energy.

System	T (K)	g^E ($x_1 = x_2$) (J mol^{-1})	
		Lattice theory	Experimental
Argon/methane	90.7	67.0	71.2
Nitrogen/methane	90.7	247	134
Benzene/cyclohexane	298	176	322
Carbon tetrachloride/cyclohexane	298	54.4	71.2

Numerous efforts have been made to extend calculations similar to those of Kohler to more complex systems. In general, they are not successful because of our inadequate understanding of intermolecular forces. With few exceptions, we cannot predict forces between dissimilar species, using only experimental data for similar species. At present, a reasonable procedure for testing a theory is to fit that theory to one binary experimental property and then to see if that theory can predict other binary properties. This was the procedure used by Marsh, shown in Table 7-2.

7.6 Nonrandom Mixtures of Simple Molecules

One of the important assumptions made in the previous sections was that when the molecules of two components are mixed, the arrangement of the molecules is completely random; i.e., the molecules have no tendency to segregate either with their own kind or with the other kind of molecule. In a completely random mixture, a given molecule shows no preference in the choice of its neighbors.

Because intermolecular forces operate between molecules, a completely random mixture in a two-component system of equisized molecules can only result if these forces are the same for all three possible molecular pairs 1-1, 2-2, and 1-2.[12] In that event, however, there would also be no energy change upon mixing. Strictly, then, only an ideal mixture can be completely random.

In a mixture where the pair energies Γ_{11}, Γ_{22}, and Γ_{12} are not the same, some ordering (nonrandomness) must result. For example, suppose that the magnitude of the attractive energy between a 1-2 pair is much larger than that between a 1-1 and 2-2 pair; in that case, there is a strong tendency to form as many 1-2 pairs as possible. An example of such a situation is provided by the system chloroform/acetone, where hydrogen bonds can form between unlike molecules but not between like molecules. Or, suppose that the attractive forces between a 1-1 pair are much larger than those between a 1-2 or a 2-2 pair; in that event, a molecule of type 1 prefers to surround itself with other molecules of type 1 and more 1-1 pairs exists in the mixture than would exist in a purely random mixture having the same composition. An example of such a situation is the diethyl ether/pentane system; because diethyl ether has a large dipole moment whereas pentane is nonpolar, ether molecules interact by dipole-dipole forces that, on the average, are attractive; but between ether and pentane and between pentane and pentane there are no dipole-dipole forces.

In the lattice theory (for w independent of T), entropy is a measure of randomness; the entropy of mixing for a completely random mixture [Eq. (7-80)] is always larger than that of a mixture that is incompletely random, regardless of whether nonrandomness is due to preferential formation of 1-2 or 1-1 (or 2-2) pairs. Excess entropy due to ordering (i.e., nonrandomness) is always negative.

---

[12] The model based on the lattice theory is not so restrictive. For molecules of the same size, there is no energy of mixing and no departure from randomness when the interchange energy is zero, i.e., when $\Gamma_{12} = 1/2(\Gamma_{11} + \Gamma_{22})$.

Guggenheim (1952) has constructed a lattice theory for molecules of equal size that form mixtures that are not necessarily random. This theory is not rigorous but utilizes a simplification known as the *quasichemical approximation.* The essential ideas of this theory are summarized below.

For a completely random mixture, we set $N_{12} = N_{12}^*$, given by Eq. (7-75), where * designates complete randomness. If $w < 0$, we expect $N_{12} > N_{12}^*$ (e.g. chloroform/acetone), and if $w > 0$, we expect $N_{12} < N_{12}^*$ (e.g. diethyl ether/pentane). Now consider the "reaction"

$$(1-1)+(2-2) \rightleftharpoons 2\cdot(1-2) \tag{7-91}$$

For this "reaction" a chemical equilibrium constant K is defined by

$$K = \frac{(N_{12})^2}{N_{11}N_{22}} \tag{7-92}$$

According to Eq. (7-71), the energy change for this "reaction" is $2w/z$. From thermodynamics, the temperature derivative of ln K is

$$\left[\frac{\partial \ln K}{\partial(1/T)}\right]_v = -\frac{\Delta_r u}{R} \tag{7-93}$$

where $\Delta_r u$ is the molar energy change of the "reaction". We assume that $\Delta_r u$ is independent of temperature and that

$$\frac{\Delta_r u}{R} = \frac{2w}{kz} \tag{7-94}$$

Integration gives

$$K = C\exp\left(-\frac{2w}{zkT}\right) \tag{7-95}$$

where C is a constant, independent of w and T. We can find C from the limiting case $w = 0$, i.e., when mixing is completely random:

$$K^* = C = \frac{(N_{12}^*)^2}{N_{11}^* N_{22}^*} \tag{7-96}$$

where * designates random mixing. Using N_{12}^* given by Eq. (7-75) and the two conservation Eqs. (7-68), we obtain from Eq. (7-96) $C = 4$.

Combination of Eqs. (7-92) and (7-95) gives the key relation between N_{12}, N_{11}, and N_{22}:

$$\frac{(N_{12})^2}{N_{11}N_{22}} = 4\exp\left(-\frac{2w}{zkT}\right) = \frac{4}{\eta^2} \tag{7-97}$$

where $\eta \equiv \exp(w/zkT)$.

We now relate N_{12} to N_{12}^* by introducing a parameter β according to

$$N_{12} = N_{12}^*\left(\frac{2}{\beta+1}\right) = \frac{zN_1N_2}{N_1+N_2}\left(\frac{2}{\beta+1}\right) \tag{7-98}$$

For the random case, β = 1. Using Eq. (7-97) and the conservation equations, Eq. (7-68), we find that β is given by

$$\beta = [1+4x_1x_2(\eta^2-1)]^{1/2} \tag{7-99}$$

where x is the mole fraction. When $w = 0$ ($\eta^2 = 1$), $\beta = 1$, as expected. As used here, β depends not only on w, but also on the composition (x_1, x_2). For $w > 0$, we obtain $\eta^2 > 1$ and $\beta > 1$ and therefore, $N_{12} < N_{12}^*$; for $w < 0$, we obtain $\eta^2 < 1$ and $\beta < 1$; therefore, $N_{12} > N_{12}^*$. As $w/kT \to \infty$, $\eta^2 \to \infty$ and $\beta \to \infty$; therefore, $N_{12} \to 0$ (no mixing at all). As $w/kT \to -\infty$, $\eta^2 \to 0$, and $\beta \to 0$ (at $x_1 = x_2 = 0.5$) and, therefore, $N_{12} \to 2N_{12}^*$. This case corresponds to formation of a stable 1-2 complex.

From Eqs. (7-98) and (7-70), the excess energy of mixing is

$$u^E = u^{E^*}\left(\frac{2}{\beta+1}\right) \tag{7-100}$$

where u^{E^*} is the excess energy for the completely random mixture given by Eq. (7-81). The excess Helmholtz energy is obtained by integrating the thermodynamic equation

$$\left[\frac{\partial(a^E/T)}{\partial(1/T)}\right]_{v,x} = u^E \tag{7-101}$$

The boundary condition is $\beta \to 1$ as $T \to \infty$ (complete randomness). The integration again assumes that w is independent of temperature. However, as indicated by Eq. (7-99), β is temperature dependent. The result for the molar excess Helmholtz energy is

$$\boxed{\frac{a^E}{RT} = \frac{z}{2}\left[x_1\ln\frac{\beta-1+2x_1}{x_1(\beta+1)} + x_2\ln\frac{\beta-1+2x_2}{x_2(\beta+1)}\right]} \tag{7-102}$$

If $\beta = 1$, $a^E = 0$, as expected. However, $\beta = 1$ only when $w/kT = 0$, corresponding to an ideal solution. Equation (7-102) shows that a^E depends on z, whereas in the simpler approximation [Eq. (7-79)], a^E is independent of z.

Equation (7-102) can be simplified if we restrict attention to moderate values of (w/zkT). For most cases, w/zkT is smaller than unity and therefore we can expand the exponential $\exp(2w/zkT)$ that appears in Eqs. (7-99) and (7-102). Neglecting higher terms, the molar excess functions are

$$\frac{g^E}{RT} = \left(\frac{w}{kT}\right) x_1 x_2 \left[1 - \frac{1}{2}\left(\frac{2w}{zkT}\right) x_1 x_2 + \ldots\right] \tag{7-103}$$

$$\frac{h^E}{RT} = \left(\frac{w}{kT}\right) x_1 x_2 \left[1 - \left(\frac{2w}{zkT}\right) x_1 x_2 + \ldots\right] \tag{7-104}$$

$$\frac{s^E}{R} = -\left(\frac{w}{kT}\right) x_1 x_2 \left[\frac{1}{2}\left(\frac{2w}{zkT}\right) x_1 x_2 + \ldots\right] \tag{7-105}$$

All excess functions are symmetric in x. However, excess Gibbs energy is no longer equal to excess enthalpy and excess entropy is no longer zero. Only in the limit, as $(2w/zkT) \to 0$, we obtain

$$g^E \to h^E \quad \text{and} \quad s^E \to 0$$

as expected. In other words, the earlier results based on the assumption of complete randomness become a satisfactory approximation as the interchange energy per pair of molecules becomes small relative to the thermal energy kT. For a given mixture, randomness increases as the temperature rises or, at a fixed temperature, randomness increases as the interchange energy falls.

The excess entropy given by Eq. (7-105) is never positive; for any nonvanishing value of w, positive or negative, s^E is always negative. For this particular model, therefore, the entropy of mixing is a maximum for the completely random mixture.[13] However, the contribution of nonrandomness to the excess Gibbs energy and to the excess enthalpy may be positive or negative, depending on the sign of the interchange energy.

The excess Gibbs energy, given by Eq. (7-103) based on the quasichemical approximation, is not very different from Eq. (7-81) based on the assumption of random mixing. Figure 7-9 compares excess Gibbs energies calculated by the two equations and

[13] For many nonpolar mixtures of nearly equi-sized molecules, positive excess entropies have been observed experimentally. These observations are a result of other effects (neglected by the lattice theory) such as changes in volume and changes in excitation of internal degrees of freedom (rotation, vibration) that may result from the mixing process.

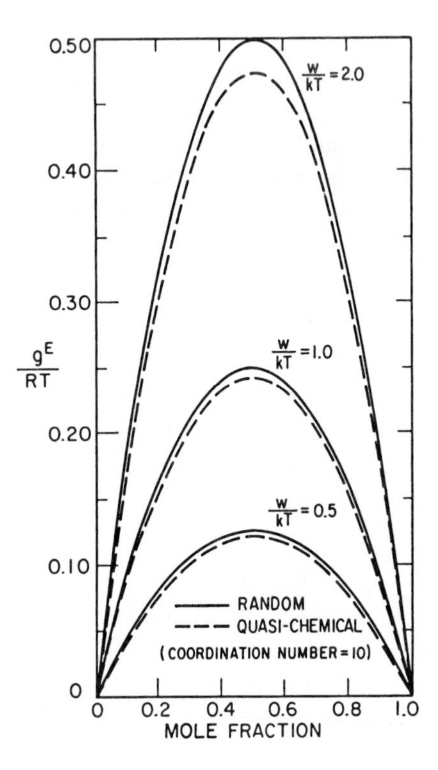

Figure 7-9 Effect of nonrandomness on excess Gibbs energies of binary mixtures.

it is evident that for totally miscible mixtures the correction for nonrandom mixing is not large.

However, deviations from random mixing become significant when w/kT is large enough to induce limited miscibility of the two components. The criteria for incipient demixing (instability) are[14]

$$\frac{\partial^2 \Delta_{\text{mix}} g}{\partial x^2} = \frac{\partial^3 \Delta_{\text{mix}} g}{\partial x^3} = 0 \tag{7-106}$$

where $\Delta_{\text{mix}} g$ is the change in the total (not excess) molar Gibbs energy upon mixing:

$$\Delta_{\text{mix}} g = RT(x_1 \ln x_1 + x_2 \ln x_2) + g^E \tag{7-107}$$

When Eq. (7-81) is substituted into Eq. (7-106), we find that T^c, the *upper consolute temperature*, is given by

[14] See Sec. 6.12.

$$\boxed{T^c = \frac{w}{2k}} \tag{7-108}$$

The upper consolute temperature is the maximum temperature for limited miscibility: for $T > T^c$ there is only one stable liquid phase (complete miscibility), whereas for $T < T^c$ there are two stable liquid phases.

In contrast to Eq. (7-108), when results based on the quasichemical approximation are substituted into Eq. (7-106), we obtain

$$T^c = \frac{w}{kz[\ln z - \ln(z-2)]} \tag{7-109}$$

When $z = 10$, we find

$$\boxed{T^c = \frac{w}{2.23k}} \tag{7-110}$$

Equation (7-110) shows that the consolute temperature according to the quasichemical approximation, is about 10% lower than that computed from the assumption of random mixing. This is a significant change, although when compared to experiment, it is not sufficiently large. However, a large effect becomes noticeable when we compute the coexistence curve, the locus of mutual solubilities of the two components at temperatures below the upper consolute temperature.

Figure 7-10 shows calculated results for the change in Gibbs energy due to mixing for four values of w/kT; calculations were performed first, assuming random mixing and second, assuming the quasichemical approximation. When $w/kT = 1.8$, both theories predict complete miscibility. When $w/kT = 2.0$, the random-mixing theory predicts incipient instability, whereas the quasichemical theory predicts complete miscibility. When $w/kT = 2.23$, the random theory indicates the existence of two liquid phases whose compositions are given by the two minima in the curves; the more refined theory merely predicts incipient demixing. When $w/kT = 2.5$, both theories indicate the existence of two liquid phases, but the compositions of the two phases as given by one theory are different from those given by the other. These compositions are given by the minima in the curves and we see that the mutual solubilities predicted by the quasichemical approximation are about twice those predicted by the random-mixing assumption.These illustrative calculations show that the effect of ordering (i.e., nonrandomness) is not important except when the components are near or below their consolute temperature.

As a further illustration, we show in Figs. 7-11 and 7-12 some calculations of Eckert (1964) for the methane/carbon tetrafluoride system. From second-virial-coefficient

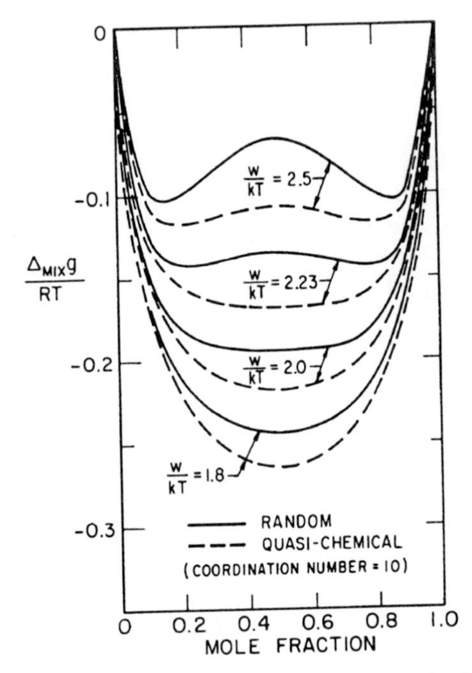

Figure 7-10 Effect of nonrandomness on Gibbs energy of mixing.

data for mixtures of the two gases near room temperature, Eckert estimates the interchange energy [Eq. (7-71)] and then calculates the excess Gibbs energy at 105.5 K for the liquid mixture. The results are shown in Fig. 7-11 along with the experimental data of Thorp and Scott (1956); agreement with experiment is good and there is not much difference between calculations based on random mixing and those based on the quasi-chemical approximation.

Croll and Scott (1958) have observed that methane and carbon tetrafluoride are not completely miscible below about 94 K; Eckert therefore calculated the coexistence curve and the results are shown in Fig. 7-12. The random-mixing theory predicts a consolute temperature that is too high by about 15 K; the consolute temperature predicted by the quasichemical theory is also too high but considerably less so. It is clear from Fig. 7-12 that for calculation of mutual solubilities in a pair of incompletely miscible liquids, the quasichemical theory provides a significant improvement over the random mixing theory. However, even in the improved theory there are still many features that are known to be incorrect. While the quasichemical theory is a step in the right direction, it provides no more than an approximation that is still far from a satisfactory theory of liquid mixtures.

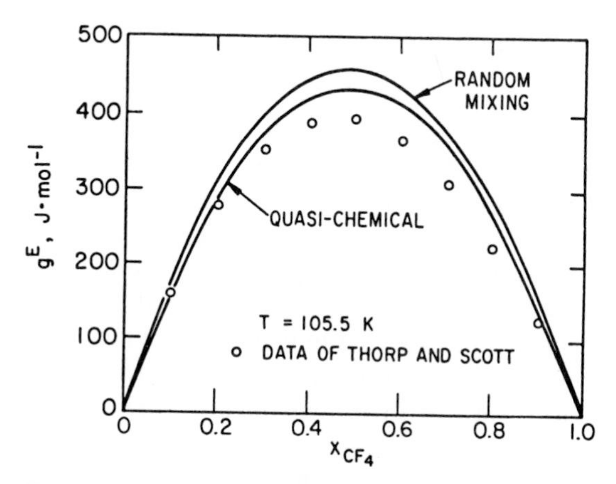

Figure 7-11 Excess Gibbs energy of methane/carbon tetrafluoride system.

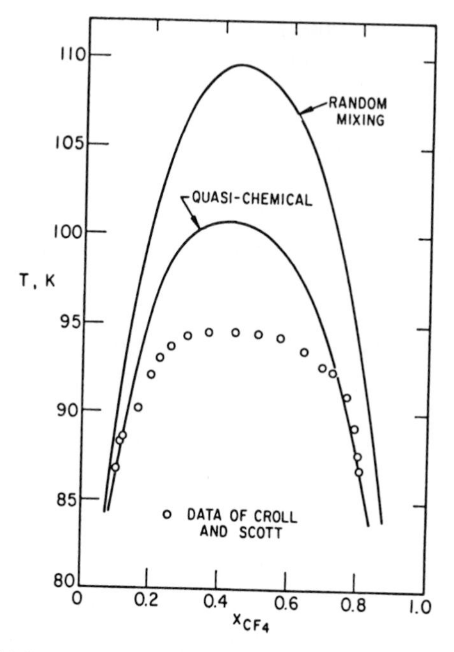

Figure 7-12 Liquid-liquid coexistence curve for the methane/carbon tetrafluoride system.

7.7 The Two-Liquid Theory

Extension of a corresponding-states theory to mixtures is based on the fundamental idea that a mixture can be considered to be a hypothetical pure fluid whose characteristic molecular size and potential energy are composition averages of the characteristic sizes and energies of the mixture's components (*one-fluid theory*). In macroscopic terms, effective critical properties (*pseudocriticals*) are composition averages of the component critical properties. However, this fundamental idea is not limited to one hypothetical pure fluid; it can be extended to include more than one hypothetical fluid, leading to m-fluid theories. These theories use as a reference a suitable (usually mole-fraction) average of the properties of m hypothetical pure fluids (Hicks, 1976). For example, *two-fluid theories*, as discussed by Scott (1956) and by Leland *et al.* (1969), use two pure reference fluids. For simple mixtures, one-fluid and two-fluid theories give similar results when compared with experiment (Henderson and Leonard, 1971). Watson and Rowlinson (1969), for example, have obtained good agreement between experimental and calculated bubble points for the ternary system argon/nitrogen/oxygen and the three corresponding binary systems.[15] Table 7-4 shows results for the nitrogen/oxygen system. In the entire pressure range, the differences between one-fluid and two-fluid models are small.

Table 7-4 Calculation of bubble temperatures of the system nitrogen/oxygen.

	Experiment		One-fluid model		Two-fluid model	
x_{N_2}	T (K)	y_{N_2}	T (K)	y_{N_2}	T (K)	y_{N_2}
			$P = 1.013$ bar			
0.075	88.3	0.252	87.6	0.269	87.8	0.263
0.496	81.6	0.805	81.0	0.802	81.1	0.802
0.986	77.8	0.996	77.3	0.995	77.3	0.996
			$P = 6.079$ bar			
0.121	108.8	0.259	108.2	0.270	108.4	0.268
0.507	102.1	0.720	101.6	0.719	101.8	0.722
0.951	97.1	0.979	96.9	0.977	96.9	0.978
			$P = 18.238$ bar			
0.087	128.8	0.147	127.9	0.154	128.1	0.150
0.532	120.4	0.665	119.6	0.666	119.9	0.667
0.928	114.9	0.957	114.4	0.953	114.4	0.956

Because one-fluid theory is easier to use, it is usually preferred. However, two-fluid theory provides a useful point of departure for deriving semiempirical equations

[15] Vera and Prausnitz (1971, *Chem. Eng. Sci.*, 26: 1772), have presented similar calculations for these systems using a reduced equation of state.

to represent thermodynamic excess functions for highly nonideal mixtures. To illustrate, we present first a brief discussion of two-fluid theory for simple mixtures, and second, we show a derivation of the UNIQUAC equation for complex mixtures given by Maurer (1978).

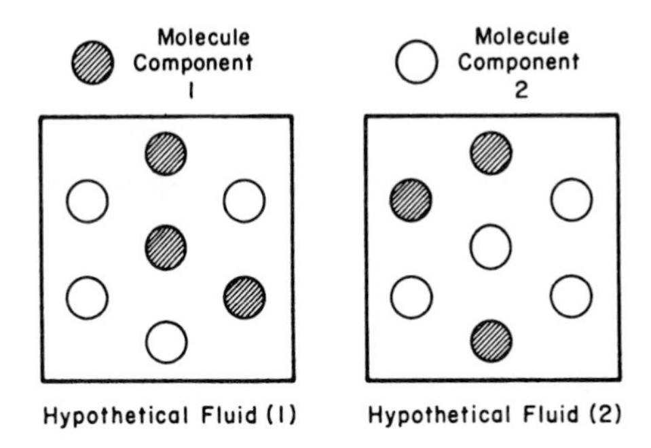

Figure 7-13 Essential idea of the two-fluid theory of binary mixtures. Hypothetical fluid (1) has a molecule 1 at the center. Hypothetical fluid (2) has a molecule 2 at the center.

To fix ideas, consider a binary mixture as shown in Fig. 7-13. Each molecule is closely surrounded by other molecules; we refer to the immediate region around any central molecule as that molecule's cell. In a binary mixture of components 1 and 2, we have two types of cells: One type contains molecule 1 at its center and the other contains molecule 2 at its center. The chemical nature (1 or 2) of the molecules surrounding a central molecule depends on the mole fractions x_1 and x_2. Let $M^{(1)}$ be some extensive configurational property M of a hypothetical fluid consisting only of cells of type 1; similarly, let $M^{(2)}$ be that same configurational property of a hypothetical fluid all of whose cells are of type 2. The two-fluid theory assumes that the extensive configurational property M of the mixture is given by

$$\boxed{M_{\text{mixt}} = x_1 M^{(1)} + x_2 M^{(2)}} \tag{7-111}$$

For mixtures of nonpolar components, $M^{(1)}$ and $M^{(2)}$ can, perhaps, be calculated from corresponding-states correlations for pure fluids by suitably averaging characteristic molecular (or critical) properties. For example, let M stand for the molar configurational enthalpy h^{conf}. We assume that

$$\frac{h^{\text{conf}}}{RT} = F\left(\frac{kT}{\varepsilon}, \frac{P\sigma^3}{\varepsilon}\right) \tag{7-112}$$

where σ and ε are characteristic size and energy parameters and where F is a function determined from experimental pure-component properties. To find $h^{\text{conf}(1)}$ we assume

$$\frac{h^{\text{conf}(1)}}{RT} = F\left(\frac{kT}{\varepsilon^{(1)}}, \frac{P\sigma^{(1)^3}}{\varepsilon^{(1)}}\right) \tag{7-113}$$

where F is the same function as in Eq. (7-112) and $\varepsilon^{(1)}$ and $\sigma^{(1)}$ denote composition averages for cells of type 1. For example, we might assume that

$$\varepsilon^{(1)} = x_1\varepsilon_{11} + x_2\varepsilon_{12} \tag{7-114}$$

$$\sigma^{(1)} = x_1\sigma_{11} + x_2\sigma_{12} \tag{7-115}$$

where ε_{ij} and σ_{ij} are constants characteristic of the i-j interaction. Similarly,

$$\frac{h^{\text{conf}(2)}}{RT} = F\left(\frac{kT}{\varepsilon^{(2)}}, \frac{P\sigma^{(2)^3}}{\varepsilon^{(2)}}\right) \tag{7-116}$$

and, using the same assumption,

$$\varepsilon^{(2)} = x_2\varepsilon_{22} + x_1\varepsilon_{12} \tag{7-117}$$

$$\sigma^{(2)} = x_2\sigma_{22} + x_1\sigma_{12} \tag{7-118}$$

The molar configurational enthalpy of the mixture is given by Eq. (7-111); it is

$$h^{\text{conf}}_{\text{mixt}} = x_1 h^{\text{conf}(1)} + x_2 h^{\text{conf}(2)} \tag{7-119}$$

The two-fluid theory for a binary mixture can be extended to mixtures containing any number of components. If there are m components, then there are m types of cells and the two-fluid binary theory becomes an m-fluid theory for an m-component mixture.

The UNIQUAC equation, proposed by Abrams (1975), has been derived by Maurer (1978) using phenomenological arguments based on a two-fluid theory. This method is similar to that used by Renon (1969) in the derivation of the three-parameter Wilson equation. Maurer's derivation avoids those inconsistencies that arise when a lattice one-fluid theory is used to derive UNIQUAC or any similar local-composition equation. The essential step in Maurer's derivation is the adoption of Wilson's assumption that local compositions can be related to overall compositions through Boltzmann factors.

We consider a binary mixture of molecules of components 1 and 2, where molecules 1 and 2 have arbitrary size and shape. As discussed in Sec. 8.3, molecules of component 1 consist of r_1 segments and each molecule has an external surface area proportional to q_1. Similar parameters are defined for molecules of component 2. For unisegmental (small, spherical) molecules, $r = q = 1$; for chain molecules, $q/r < 1$, but as the number of chain segments becomes large, q/r approaches a constant near 2/3.

For a polysegmented molecule i, the number of neighboring segments (belonging to other molecules) is zq_i, where z is the coordination number.

To fix ideas, consider first a unisegmental molecule ($r = q = 1$). Suppose one molecule of component 1 is isothermally vaporized from its pure liquid denoted by superscript (0) and then condensed into the center of a cell as shown in the left side of Fig. 7-13 [hypothetical fluid (1)]. In this case, $r = q = 1$ also for molecules of component 2. A molecule 1 in the pure liquid has $z^{(0)}$ nearest neighbors. Since intermolecular forces are short range, we assume pairwise additivity neglecting all except nearest neighbors; the energy of vaporization per molecule is $(1/2)z^{(0)}U_{11}^{(0)}$ where $U_{11}^{(0)}$ characterizes the potential energy of two nearest neighbors in pure liquid 1.

The central molecule in hypothetical fluid (1) is surrounded by $z^{(1)}\theta_{11}$ molecules of species 1 and $z^{(1)}\theta_{21}$ molecules of species 2, where θ_{11} is the local surface fraction of component 1, about central molecule 1, and θ_{21} is the local surface fraction of component 2, about central molecule 1 (note that $\theta_{11} + \theta_{21} = 1$). We now assume that $z^{(1)}$ is the same as $z^{(0)}$. The energy released by the condensation process is $(1/2)z(\theta_{11}U_{11}^{(1)} + \theta_{21}U_{21}^{(1)})$, where we have dropped the superscript on z. We make a similar transfer for a molecule 2 from the pure liquid, denoted by superscript (0), to a hypothetical fluid, denoted by superscript (2).

If we consider a mixture consisting of x_1 moles of hypothetical fluid (1) and x_2 moles of hypothetical fluid (2), the configurational part of an extensive property M of that mixture is given by Eq. (7-111). In particular, the total change in energy in transferring x_1 moles of species 1 from pure liquid 1 and x_2 moles of species 2 from pure liquid 2 into the "two-liquid" mixture, i.e., the molar excess energy u^E, is given by

$$u^E = \frac{1}{2}zx_1N_A[q_1(\theta_{11}U_{11}^{(1)} + \theta_{21}U_{21}^{(1)} - U_{11}^{(0)})] + \frac{1}{2}zx_2N_A[q_2(\theta_{22}U_{22}^{(2)} + \theta_{12}U_{12}^{(2)} - U_{22}^{(0)})] \tag{7-120}$$

where N_A is Avogadro's constant. Because the local surface fractions must obey the conservation equations

$$\theta_{21} + \theta_{11} = 1 \qquad \theta_{12} + \theta_{22} = 1 \tag{7-121}$$

and assuming that $U_{11}^{(1)} = U_{11}^{(0)}$ and $U_{22}^{(2)} = U_{22}^{(0)}$, Eq. (7-120) simplifies to

$$u^E = \frac{1}{2} z N_A [x_1 \theta_{21} q_1 (U_{21} - U_{11}) + x_2 \theta_{12} q_2 (U_{12} - U_{22})] \tag{7-122}$$

where we have now dropped the superscripts. Following Wilson (1964), we now assume

$$\frac{\theta_{21}}{\theta_{11}} = \frac{\theta_2}{\theta_1} \exp\left[\frac{-\frac{1}{2} z (U_{21} - U_{11})}{kT}\right] \tag{7-123}$$

and

$$\frac{\theta_{12}}{\theta_{22}} = \frac{\theta_1}{\theta_2} \exp\left[\frac{-\frac{1}{2} z (U_{12} - U_{22})}{kT}\right] \tag{7-124}$$

where θ is the surface fraction:

$$\theta_1 = \frac{x_1 q_1}{x_1 q_1 + x_2 q_2} \qquad \theta_2 = \frac{x_2 q_2}{x_1 q_1 + x_2 q_2} \tag{7-125}$$

When these assumptions are coupled with Eq. (7-121), we obtain

$$u^E = x_1 q_1 \theta_{21} \Delta u_{21} + x_2 q_2 \theta_{12} \Delta u_{12} \tag{7-126}$$

and

$$\theta_{21} = \frac{\theta_2 \exp(-\Delta u_{21} / RT)}{\theta_1 + \theta_2 \exp(-\Delta u_{21} / RT)} \tag{7-127}$$

$$\theta_{12} = \frac{\theta_1 \exp(-\Delta u_{12} / RT)}{\theta_2 + \theta_1 \exp(-\Delta u_{12} / RT)} \tag{7-128}$$

where

$$\Delta u_{21} = \frac{1}{2} z (U_{21} - U_{11}) N_A \qquad \Delta u_{12} = \frac{1}{2} z (U_{12} - U_{22}) N_A \tag{7-129}$$

Equation (7-126) is the fundamental relation based on two-fluid theory, utilizing the notion of local composition.

To obtain an expression for the molar excess Helmholtz energy, we use (at constant volume and composition),

$$\frac{d(a^E / T)}{d(1/T)} = u^E \tag{7-130}$$

where a^E is the excess Helmholtz energy per mole of mixture.

Integrating from $1/T_0$ to $1/T$, we have

$$\frac{a^E}{T} = \int_{1/T_0}^{1/T} u^E d(1/T) + \text{constant of integration} \tag{7-131}$$

We evaluate the constant of integration by letting $1/T_0 \to 0$. At very high temperature, we assume that components 1 and 2 form an athermal mixture (compare Sec. 8.2). As our boundary condition, we use the equation of Guggenheim (1952) for athermal mixtures of molecules of arbitrary size and shape,

$$\begin{aligned}\left(\frac{a^E}{RT}\right)_{\text{athermal}} &= -\left(\frac{s^E}{R}\right)_{\text{combinatorial}} \\ &= x_1 \ln\frac{\Phi_1^*}{x_1} + x_2 \ln\frac{\Phi_2^*}{x_2} + \frac{1}{2}z\left(q_1 x_1 \ln\frac{\theta_1}{\Phi_1^*} + q_2 x_2 \ln\frac{\theta_2}{\Phi_2^*}\right)\end{aligned} \tag{7-132}$$

where

$$\Phi_1^* = \frac{x_1 r_1}{x_1 r_1 + x_2 r_2} \quad \text{and} \quad \Phi_2^* = \frac{x_2 r_2}{x_1 r_1 + x_2 r_2} \tag{7-133}$$

Assuming that Δu_{21} and Δu_{12} are independent of temperature and that, as shown by Hildebrand and Scott (1950), at low pressures $(a^E)_{T,V} \approx (g^E)_{T,P}$, Eq. (7-131) gives

$$\left(\frac{a^E}{RT}\right)_{T,V} \approx \left(\frac{g^E}{RT}\right)_{T,P} = \left(\frac{g^E}{RT}\right)_{\text{combinatorial}} + \left(\frac{g^E}{RT}\right)_{\text{residual}} \tag{7-134}$$

where

$$\left(\frac{g^E}{RT}\right)_{\text{combinatorial}} = x_1 \ln\frac{\Phi_1^*}{x_1} + x_2 \ln\frac{\Phi_2^*}{x_2} + \frac{1}{2}z\left(x_1 q_1 \ln\frac{\theta_1}{\Phi_1^*} + x_2 q_2 \ln\frac{\theta_2}{\Phi_2^*}\right) \tag{7-135}$$

$$\left(\frac{g^E}{RT}\right)_{\text{residual}} = -x_1 q_1 \ln\left[\theta_1 + \theta_2 \exp\left(\frac{-\Delta u_{21}}{RT}\right)\right] - x_2 q_2 \ln\left[\theta_2 + \theta_1 \exp\left(\frac{-\Delta u_{12}}{RT}\right)\right] \tag{7-136}$$

that is the UNIQUAC equation.

As discussed in Sec. 6.11, Eq. (7-134) gives a good empirical representation of excess functions for a large variety of liquid mixtures. For mixtures containing more than two components, Eq. (7-134) can readily be generalized (Abrams and Prausnitz, 1975) without additional assumptions; the general result contains only pure-component and binary parameters.

While UNIQUAC has considerable empirical success, molecular-dynamic calculations suggest that the nonrandomness assumption [Eqs. (7-123) and (7-124)] is too strong; the magnitudes of the arguments of the Boltzmann factors are too large. In other words, UNIQUAC over-corrects for deviations from random mixing. While the basic ideas of UNIQUAC are useful, it is clear that significant modifications in the details are required to provide UNIQUAC with a sound molecular basis.[16]

7.8 Activity Coefficients from Group-Contribution Methods

For engineering purposes, it is often necessary to make some estimate of activity coefficients for mixtures where only fragmentary data, or no data at all, are available. For vapor-liquid equilibria, such estimates can be made using a *group-contribution* method as illustrated in Fig. 7-14. A molecule is divided (somewhat arbitrarily) into functional groups. Molecule-molecule interactions are considered to be properly weighted sums of group-group interactions. Therefore, for a multifunctional component in a multicomponent system, group-contribution methods assume that each functional group behaves in a manner independent of the molecule in which it appears. Once quantitative information on the necessary group-group interactions is obtained from reduction of experimental data for binary systems, it is then possible to calculate molecule-molecule interactions (and therefore phase equilibria) for molecular pairs where no experimental data are available. The fundamental advantage of this procedure is that when attention is directed to typical mixtures of nonelectrolytes, the number of possible distinct functional groups is much smaller than the number of distinct molecules or, more directly, the number of distinct group-group interactions is very much smaller than the number of possible distinct molecule-molecule interactions.

Calculation of activity coefficients from group contributions was suggested in 1925 by Langmuir, but this suggestion was not practical until a large database and readily accessible computers became available. A systematic development known as the ASOG[17] (*a*nalytical *s*olution *o*f *g*roups) method was established by Derr and Deal, (1969, 1973).

---

[16] J. Fischer and F. Kohler, 1983, *Fluid Phase Equilibria,* 14: 177; Y. Hu, E. G. Azevedo, and J. M. Prausnitz, 1983, *ibid.*, 13: 351; K. Nakanishi and H. Tanaka, 1983, *ibid.*, 13: 371; and D. J. Phillips and J. F. Brennecke, 1993, *Ind. Eng. Chem. Res.*, 32: 943.

[17] Parameters for the ASOG method are listed by K. Tochigi, D. Tiegs, J. Gmehling, and K. Kojima, 1990, *J. Chem. Eng. Japan*, 23: 453.

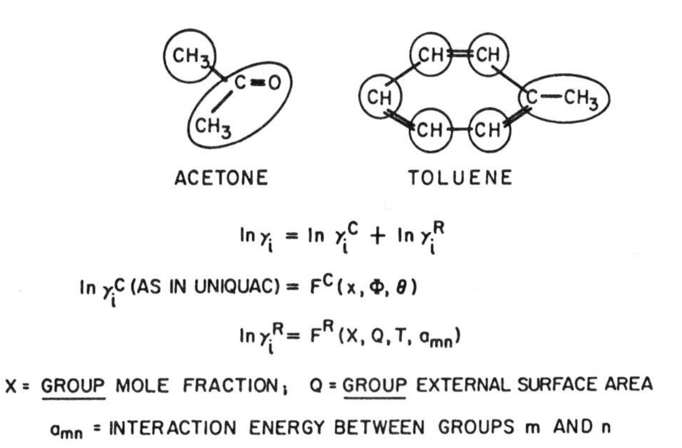

Figure 7-14 Activity coefficients from group contribution illustrated for a mixture of acetone and toluene. Acetone has two groups and toluene has six, as shown. For a component *i*, activity coefficient γ_i consists of two contributions, γ_i^C and γ_i^R where superscript *C* stands for configurational and superscript *R* stands for residual. Here F^C is a specified function of *molecular* composition and structure: mole fraction *x*, volume fraction Φ and surface fraction θ; F^R is a specified function of *group* composition, structure and interaction energies: *X*, *Q* and a_{mn}. Both functions F^C and F^R are obtained from the UNIQUAC model. The key parameters are the group-group interaction parameters for all pairs of groups (*n*,*m*) in the solution. In UNIFAC, for each pair, we use two parameters: a_{mn} and a_{nm}.

A similar but more convenient method,[18] based on the UNIQUAC equation, was developed by Fredenslund, Jones and Prausnitz (1975) and discussed by Fredenslund *et al.* (1977); this method, called UNIFAC (*uni*versal *f*unctional *a*ctivity *c*oefficient), is described in a monograph by Fredenslund *et al.* (1977a), but since its publication, numerous modifications and extensions have appeared (Fredenslund and Rasmussen, 1985; Gmehling, 1986; Larsen *et al.*, 1987). A large number of group-group interaction parameters is available (Skold-Jørgensen *et al.* 1979; Macedo *et al.* 1983; Hansen *et al.* 1991; Gmehling *et al.*, 1993; Fredenslund and Sørensen, 1994); as new experimental data are reported, this number will rise. UNIFAC has been successfully used for the design of distillation columns (including azeotropic and extractive distillation) where the required multicomponent activity coefficients were estimated because of a lack of experimental information.

Separate UNIFAC correlations have been proposed for liquid-liquid equilibria (Magnussen *et al.*, 1981; Gupte and Danner, 1987; Hooper *et al.*, 1988), but these tend to be less accurate than those for vapor-liquid equilibria and, therefore, are not used widely. Some details concerning group-contribution methods (including correlations for polymer-solvent systems) and some other methods for estimating activity coefficients are briefly discussed in App. F.

[18] A comprehensive comparison between the predictive capabilities of ASOG and UNIFAC methods is presented by J. Gmehling, D. Tiegs, and U. Knipp, 1990, *Fluid Phase Equilibria*, 54:147.

For engineering design, correlations (in particular, group-contribution methods) are attractive because it is almost always easier and faster to make a calculation than to perform an experiment. However, because UNIFAC (and similar methods) frequently provide only rough approximations, it is often necessary to obtain at least a few *reliable* experimental results, either from the literature or from the laboratory. On the other hand, many experimental results reported in the literature are of poor quality and in that event, a calculated activity coefficient *may* be more reliable. There is no substitute for judgment. Users of correlations are privileged but they must also be cautious.

7.9 Chemical Theory

The models described in the previous sections attempt to explain solution nonideality in terms of physical intermolecular forces. These models relate the activity coefficients to physical quantities that reflect the size of the molecules and the physical forces (primarily London dispersion forces) operating between them. An alternate approach to the study of solution properties is based on a different premise, viz., that molecules in a liquid solution interact with each other to form new chemical species and that solution nonideality, therefore, is a consequence of chemical reactions.

We can distinguish between two types of reactions, association and solvation. Association refers to formation of chemical aggregates[19] or dimers, trimers, etc. consisting of identical monomers. An association can be represented by reactions of the type

$$n\text{B} \rightleftharpoons \text{B}_n$$

where B is the monomer and n is the degree of association (or polymerization). A common case of association is dimerization ($n = 2$); a well-known example is dimerization of acetic acid:

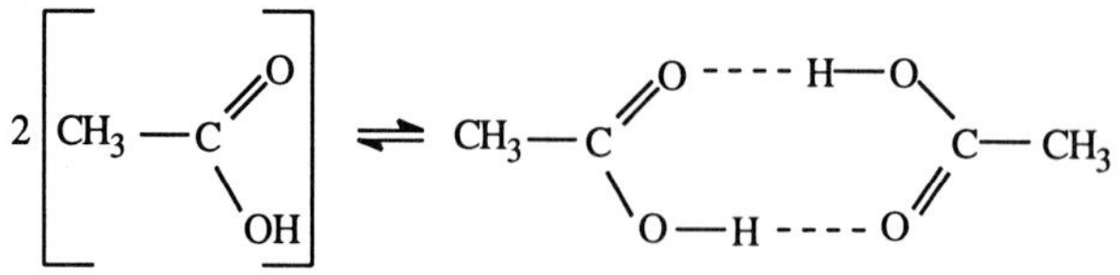

$$2\left[\text{CH}_3\text{—C}\begin{matrix}\text{//O}\\ \text{\\OH}\end{matrix}\right] \rightleftharpoons \text{CH}_3\text{—C}\begin{matrix}\text{//O}\cdots\text{H—O}\backslash\\ \backslash\text{O—H}\cdots\text{O//}\end{matrix}\text{C—CH}_3$$

In this case, dimerization is due to hydrogen bonding that is responsible for the most common form of association in liquid solutions.

[19] While aggregates are loosely bonded chemical compounds, this does not imply that they can necessarily be separated and exist by themselves. Such separations are possible only for stable compounds. The aggregates discussed here are often insufficiently stable for isolated existence.

Solvation refers to formation of chemical aggregates of two or more molecules that are not all identical, represented by the general equation

$$n\text{A} + m\text{B} \rightleftharpoons \text{A}_n\text{B}_m$$

A well-known example ($n = m = 1$) is solvation of chloroform and diethyl ether:

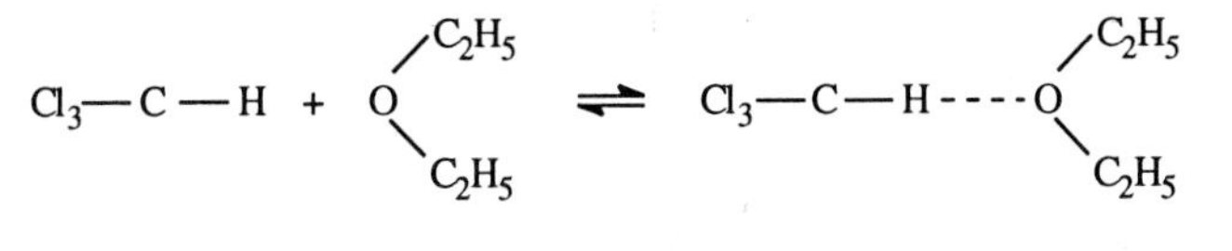

$$Cl_3\text{—C—H} + O(C_2H_5)_2 \rightleftharpoons Cl_3\text{—C—H}\cdots O(C_2H_5)_2$$

In this case, formation of the new species is again due to hydrogen bonding.

Another example of solvation is given by a charge-transfer complex between nitrobenzene and mesitylene:

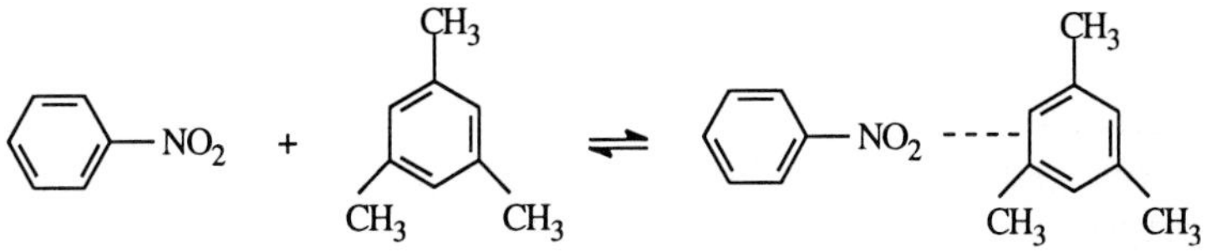

$$C_6H_5\text{—}NO_2 + C_6H_3(CH_3)_3 \rightleftharpoons C_6H_5\text{—}NO_2 \cdots C_6H_3(CH_3)_3$$

In this case, a weak chemical bond is formed because mesitylene is a good electron donor (Lewis base) and nitrobenzene is a good electron acceptor (Lewis acid).

The chemical theory of solutions postulates existence of chemically distinct species in solution that are assumed to be in chemical equilibrium. In its original form, the theory then assumes that these chemically distinct substances form an ideal solution. According to these assumptions, the observed nonideality of a solution is only an apparent one because it is based on an apparent, rather than a true, account of the solution's composition.

The chemical theory of solutions was first developed by Dolezalek (1908) at about the same time van Laar (1910) was publishing his work on solutions. The different points of view represented by these early workers in solution theory caused them to be hard enemies and some of their publications contain much bitter polemic.

Dolezalek's theory has the advantage that it can readily account for both positive and negative deviations from ideality for molecules of similar size; also, unlike the van Laar-Scatchard-Hildebrand theory of regular solutions, it is applicable to mixtures containing polar and hydrogen-bonded liquids. Its great disadvantage lies in its arbitrariness in deciding what "true" chemical species are present in the solution, and in our inability to assign equilibrium constants to the postulated equilibria without experimental data on the solution under consideration. The chemical theory of solutions, therefore, has little predictive value; it can almost never give quantitative predictions of solution behavior from pure-component data alone. However, for those solutions where chemical forces are dominant, the chemical theory has much qualitative and interpretative value; if some data on such a solution are available, they can often be interpreted along reasonable chemical lines and, therefore, the chemical theory can serve as a tool for interpolation and cautious extrapolation of limited data. A chemical rather than a physical view is frequently useful for correlation of solution nonidealities in a class of chemically similar mixtures (e.g., alcohols in paraffinic solvents). In the next two sections, we discuss the properties of associated solutions. Solvated solutions are discussed in Secs. 7.12 and 7.13.

7.10 Activity Coefficients in Associated Solutions

Suppose we have a liquid mixture of two components, 1 and 2. Component 1 is a nonpolar substance that we designate A, but component 2 is a polar substance that we designate B and that, we assume, can dimerize according to

$$2\mathrm{B} \rightleftharpoons \mathrm{B}_2$$

The equilibrium constant for this dimerization is given by

$$K = \frac{a_{\mathrm{B}_2}}{a_{\mathrm{B}}^2} \tag{7-137}$$

where a_{B} is the activity of monomer B molecules and a_{B_2} is the activity of dimer B_2 molecules.

We assume that species B and B_2 are in equilibrium with one another and we also assume that species A, B, and B_2 form an ideal liquid solution. In an ideal solution the activities can be replaced by the mole fractions and Eq. (7-137) becomes

$$K = \frac{\mathfrak{z}_{\mathrm{B}_2}}{\mathfrak{z}_{\mathrm{B}}^2} \tag{7-138}$$

where $\mathfrak{z}$ stands for the "true" mole fraction.

If there are n_1 moles of component 1 and n_2 moles of component 2, then

$$n_1 = n_{\mathrm{A}} \tag{7-139}$$

and

$$n_2 = n_{\mathrm{B}} + 2n_{\mathrm{B}_2} \tag{7-140}$$

where n_B is the number of moles of monomer B and n_{B_2} is the number of moles of dimer B_2. The "true" total number of moles is $n_A + n_B + n_{B_2}$, equal to $n_1 + n_2 - n_{B_2}$. The three "true" mole fractions are

$$z_A = \frac{n_1}{n_1 + n_2 - n_{B_2}} \tag{7-141}$$

$$z_B = \frac{n_B}{n_1 + n_2 - n_{B_2}} = \frac{n_2 - 2n_{B_2}}{n_1 + n_2 - n_{B_2}} \tag{7-142}$$

$$z_{B_2} = \frac{n_{B_2}}{n_1 + n_2 - n_{B_2}} \tag{7-143}$$

If we combine Eq. (7-138) with Eqs. (7-142) and (7-143) and then eliminate n_{B_2} with Eq. (7-141), we obtain an expression that we can solve for z_A and from this we can obtain the desired result for the activity coefficient of component 1. The algebra is straightforward but involved, and is not reproduced here. Remembering that $x_1 = n_1/(n_1 + n_2)$ and that $x_2 = n_2/(n_1 + n_2)$, we obtain

$$\gamma_1 = \frac{a_1}{x_1} = \frac{2k}{(2k-1)x_1 + kx_2 + (x_1^2 + 2kx_1x_2 + kx_2^2)^{1/2}} \tag{7-144}$$

where $k \equiv 4K + 1$.

By similar stoichiometric considerations, the activity coefficient for the dimerizing component can be shown to be[20]

$$\gamma_2 = \frac{a_2}{x_2} = \left(\frac{k^{1/2} + 1}{x_2}\right)\left[\frac{-x_1 + (x_1^2 + 2kx_1x_2 + kx_2^2)^{1/2}}{(2k-1)x_1 + kx_2 + (x_1^2 + 2kx_1x_2 + kx_2^2)^{1/2}}\right] \tag{7-145}$$

Figure 7-15 shows plots of Eqs. (7-144) and (7-145) for three dimerization equilibrium constants. It is a property of both Eqs. (7-144) and (7-145) that for all $K > 0$, positive deviations from Raoult's law result; thus $\gamma_1 \geq 1$ and $\gamma_2 \geq 1$, as shown. For $K = \infty$ all molecules of component 2 are dimerized; in this limiting case we have

$$\lim_{K\to\infty} \gamma_1 = \frac{1}{x_1 + (x_2/2)} \quad \text{and} \quad \lim_{K\to\infty} \gamma_2 = \frac{1}{(2x_1x_2 + x_2^2)^{1/2}} \tag{7-146}$$

[20] Appendix G presents the general relation for associated and solvated mixtures. It is shown there that the chemical potential of a (stoichiometric) component is equal to that of the component's monomer.

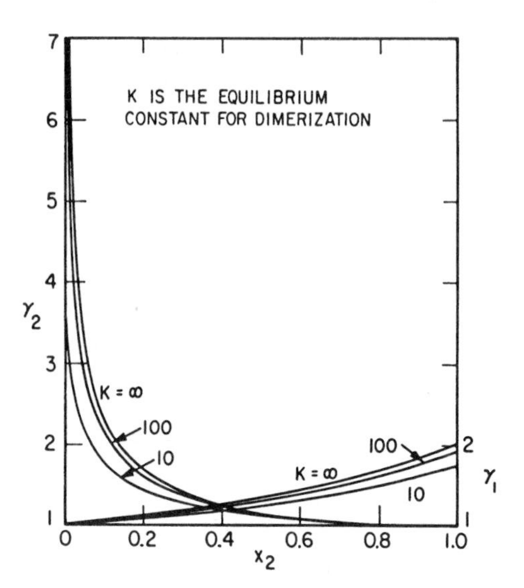

Figure 7-15 Activity coefficients of a dimerizing component (2) and an "inert" solvent (1).

At infinite dilution,

$$\lim_{\substack{K\to\infty \\ x_1\to 0}} \gamma_1 = 2 \qquad \text{and} \qquad \lim_{\substack{K\to\infty \\ x_2\to 0}} \gamma_2 = \infty \tag{7-147}$$

For the nondimerizing component, the largest possible value of γ_1 is 2.

To test the dimerization model represented by Eqs. (7-144) and (7-145), we can compare calculated and observed activity coefficients for a binary system containing an organic acid in a nonpolar "inert" solvent because spectroscopic, cryoscopic, and distribution data indicate that organic acids have a strong tendency to dimerize. Figure 7-16 shows a comparison for propionic acid dissolved in *n*-octane. Because the experimental data (Johnson *et al.*, 1954) were unfortunately obtained at constant pressure rather than constant temperature, the comparison is not completely straightforward because the dimerization equilibrium constant depends on temperature.

The experimental data shown in Fig. 7-16 follow the general trend predicted by Eq. (7-145), but even with $K = \infty$, the observed deviations from ideality are larger than those calculated. Because dimerization of an organic acid is exothermic, dimerization constant K falls with rising temperature but it is apparent that the variable temperature of the experimental data is not responsible for the lack of agreement between theoretical and experimental results. In this case, the chemical theory is evidently able to account qualitatively for the observed activity coefficients but, because physical effects are neglected by the chemical theory, quantitative agreement is not obtained.

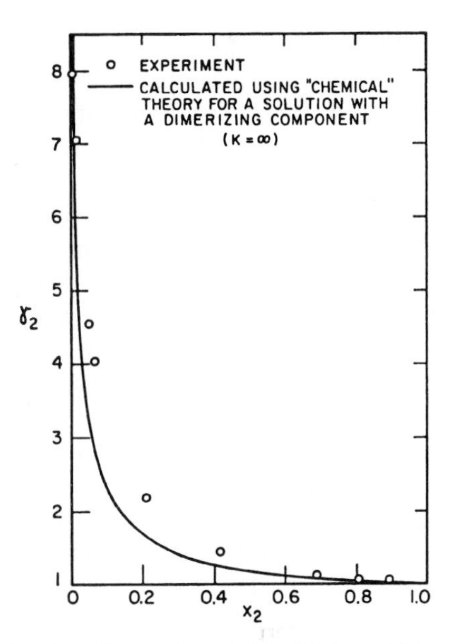

Figure 7-16 Calculated and experimental activity coefficients for propionic acid (2) dissolved in *n*-octane (1) at 1.013 bar (temperature between 121.3 and 141.1°C).

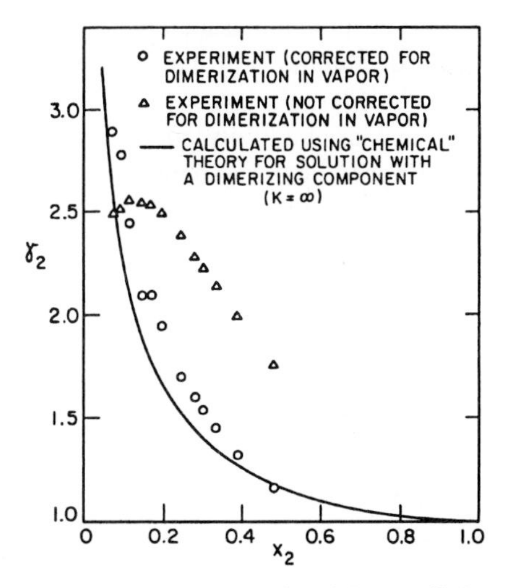

Figure 7-17 Calculated and experimental activity coefficients for acetic acid (2) dissolved in benzene (1) at 25°C.

Another comparison between calculated and experimental results, given in Fig. 7-17, shows activity coefficients for acetic acid in the benzene/acetic acid system at 25°C. Experimental data were reported by Hovorka and Dreisbach (1934) and, again, qualitative agreement is obtained between experimental and theoretical results.[21] The activity coefficients for acetic acid in benzene are somewhat lower than those for propionic acid in octane; because of the higher polarizability of π electrons in benzene, it is likely that the forces of attraction between benzene and acid are stronger than those between octane and acid.

The two examples given in Figs. 7-16 and 7-17 show only fair agreement between theory and experiment. But even if the agreement were good, it should not by itself be considered proof of the validity of the chemical theory. Good agreement would have shown only that the assumptions of the chemical theory, with the help of one adjustable parameter, are consistent with the experimental facts, but the assumptions of some other theory, again with one adjustable parameter, may be equally consistent with these facts. In other words, when agreement is good, we may say that the chemical theory offers a possible, but by no means unique, explanation of the observed thermodynamic properties. A chemical theory of solution behavior must always be viewed with suspicion unless there is independent evidence to support it. In the case of propionic (or acetic) acid, dissolved in octane or some other relatively "inert" solvent, we believe that the chemical explanation is reasonable because of independent (e.g., spectroscopic and cryoscopic) evidence That evidence strongly supports the assumption that organic acids do, in fact, dimerize in nonpolar solvents.

In Eq. (7-144), even, a very large values of the dimerization equilibrium constant K cannot produce high activity coefficients for the "inert" component; yet, such activity coefficients have been observed in some associated solutions, notably for nonpolar solvents dissolved in an excess of alcohol. Further, the activity coefficients of the associating component are often considerably larger than those calculated by Eq. (7-145) with $K = \infty$. To explain these very large activity coefficients, it has been proposed that the associating component undergoes chain association; i.e., it forms not only dimers but also trimers, tetramers, etc., according to

$$2B \rightleftharpoons B_2$$
$$B + B_2 \rightleftharpoons B_3$$
$$B + B_3 \rightleftharpoons B_4 \qquad \text{etc.}$$

For example, phenol is known from a variety of physicochemical data to form a multiple chain according to the structure

[21] Figure 7-17 presents two sets of results, each based on a particular method of data reduction. Because total pressure at 25°C is much less than 1 bar, activity coefficients represented by triangles were calculated from the experimental data without any vapor-phase correction; neither the fugacity of the acid in the vapor-phase mixture nor the (standard- state) fugacity of pure acetic acid were corrected for nonideal behavior. However, as discussed in Sec. 5.9, vapor-phase corrections are important for carboxylic acids even at pressures of the order of 10^{-2} bar. When such corrections are included in data reduction, the results obtained are those represented by circles.

Because the tendency of phenol to form chains is a strong function of phenol concentration (especially in the dilute region), it follows that the activity of phenol, when dissolved in some solvent, shows large deviations from ideal behavior (Tucker and Christian, 1978).

The thermodynamics of solutions that contain a component capable of multiple association has been considered by many authors; good discussions of this subject are given by Kortüm and Buchholz-Meisenheimer (1952), by Prigogine and Defay (1954), and by Tucker and Lippert (1976). We discuss here only the main concepts and present some typical results.

To reduce the number of adjustable parameters, it is common to assume that the equilibrium constant for the formation of a chain is independent of the chain length. That is, if we write the general association equilibrium

$$\mathrm{B} + \mathrm{B}_{n-1} \rightleftharpoons \mathrm{B}_n$$

then the equilibrium constant K_n is

$$K_n = \frac{a_{\mathrm{B}_n}}{a_{\mathrm{B}}\, a_{\mathrm{B}_{n-1}}} \tag{7-148}$$

where a is the activity. The simplifying assumption is that $K_2 = K_3 = \ldots = K_n = K$.[22]

If, as before, we assume that the solution of "true" species is an ideal solution, then we can replace activity a by "true" mole fraction $\mathfrak{z}$. The mathematical details are tedious and not reproduced here, but it can be shown that the activity coefficient of component 2, the associating component, is given by

$$\gamma_2 = \frac{(1+K)(1+KZ^2)}{(1+KZ)^2} \tag{7-149}$$

where

[22] Studies have been reported where this assumption is not made (Tucker and Lippert, 1976). However, the algebraic complexity and the amount of data required for meaningful data reduction are much larger when the equilibrium constant is allowed to depend on the degree of association and frequently the extra labor is not justified. One fine example, however, of such a detailed study is that by H. Wolff and A. Höpfner (1962, *Z. Elektrochem.*, 66: 149) for solutions of methylamine and hexane. These authors, using extensive data for the range -55 to 20°C, report equilibrium constants and enthalpies of formation for dimers, trimers, and tetramers of methylamine. At a given temperature, the authors found that the three equilibrium constants are close but not identical.

$$Z \equiv \frac{[1+4Kx_2(1-x_2)]^{1/2}-1}{2K(1-x_2)} \tag{7-150}$$

The activity coefficient of the nonassociating component 1 is given by

$$\gamma_1 = 1 + KZ^2 \tag{7-151}$$

Figure 7-18 shows activity coefficients for the multiply-associated component for several values of equilibrium constant K. At small mole fractions, the activity coefficients are now much larger than those shown in Fig. 7-15 that considered only dimerization. The activity coefficients for the associated component are large at the dilute end but fall rapidly as the concentration rises; this characteristic behavior is in excellent agreement with experimental results for solutions of alcohols in nonpolar solvents. For example, Prigogine *et al.* (1949, 1951) have shown that observed excess Gibbs energies of solutions of alcohols in carbon tetrachloride can be closely approximated by ascribing the nonideality of these solutions to multiple association. The small difference between calculated and observed excess Gibbs energies is probably due to physical forces that, by assumption, have been neglected in this purely chemical treatment. We shall return to this point in the next section, but first we want to consider some of the implications of the chemical theory of associated solutions.

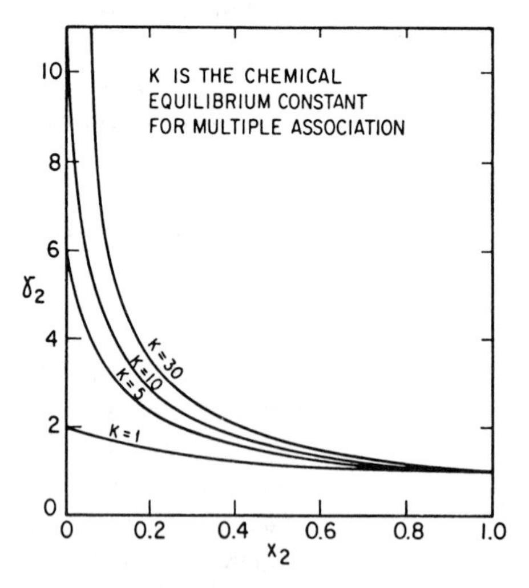

Figure 7-18 Activity coefficients of a multiply-associated component in an "inert" solvent.

When considering the validity of any theory of solutions, it is important to inquire whether or not it can be supported by evidence based on several physicochemical properties. The validity of any theory of solution is much enhanced if it can be shown to be in agreement with observed physical properties other than those used to obtain activity coefficients. For the theory of associated solutions, where association is due to hydrogen bonding, such support is provided by infrared spectroscopy as shown by many workers, notably by Hoffmann, Errera, Sack, and others (see Tucker and Lippert, 1976; Pimentel and McClellan, 1960). For example, consider a solution of methanol in a nonpolar solvent such as carbon tetrachloride. The frequency of vibration of the OH group is in the infrared spectrum and this frequency is strongly affected by whether or not it is "free" (i.e., attached only to a carbon atom by a normal covalent bond) or whether it is also attached to another OH group through hydrogen bonding. Therefore, by measuring the intensity of absorption at the frequency corresponding to this "free" vibration, it is possible to determine the concentration of alcohol that is in the monomeric, nonassociated state. Spectroscopic measurements thus provide an independent check on the theory of associated solutions.

If we let α_{B_1} stand for the fraction of alcohol molecules in the monomeric form, then, as shown by Prigogine, we can relate α_{B_1} to the unsymmetrically normalized activity coefficients of alcohol (2) and carbon tetrachloride (1) by the remarkably simple equation

$$\alpha_{B_1} = \frac{\gamma_2^*}{\gamma_1} \tag{7-152}$$

where γ_2^* and γ_1 are activity coefficients normalized by the unsymmetric convention (see Sec. 6.4):

$$\gamma_2^* \to 1 \quad \text{as} \quad x_2 \to 0$$

$$\gamma_1 \to 1 \quad \text{as} \quad x_1 \to 1$$

Figure 7-19 compares values of α_{B_1} determined spectroscopically, with those determined by standard thermodynamic measurements for the system methanol/carbon tetrachloride. The good agreement lends support to the essential ideas of the theory of multiple association for mixtures of alcohols and nonpolar solvents; however, studies by Van Ness *et al.* (1967) and Nagata (1977, 1978) indicate that the structure of alcohol solutions is probably considerably more complex than that assumed by Prigogine.

The simple theory of multiply-associated solutions (Kortüm and Buchholz-Meisenheimer, 1952), also establishes a relation between the activity coefficient of the nonassociating solvent and $\bar{n}$, the average chain length of the associating component:

$$\bar{n} = \frac{\gamma_1 x_2}{1 - \gamma_1 x_1} \tag{7-153}$$

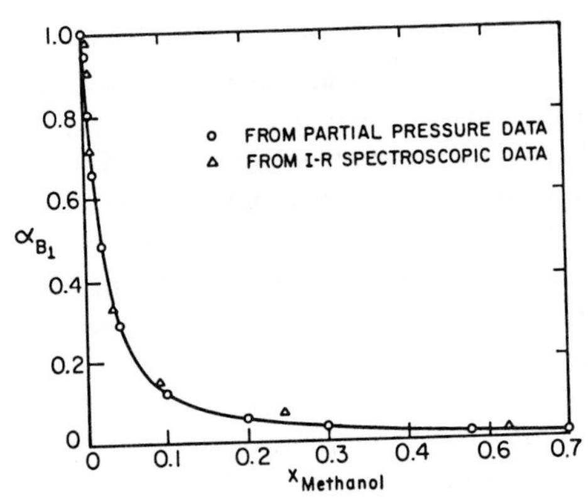

Figure 7-19 Fraction of monomeric methanol molecules in carbon tetrachloride solution at 20°C.

A comparison can then be made between average chain lengths calculated from activity coefficients and from other physicochemical measurements such as spectroscopy and cryoscopy. Again, good agreement is often obtained when results based on different methods of measurement are compared with one another.

Figure 7-20 shows average chain lengths as a function of alcohol concentration for several systems (Mecke, 1948). The degree of association rises rapidly in the dilute region; this is consistent with the experimental observation that the activity coefficient of an alcohol in a nonpolar solvent is large at infinite dilution but then falls quickly as the mole fraction of alcohol rises.

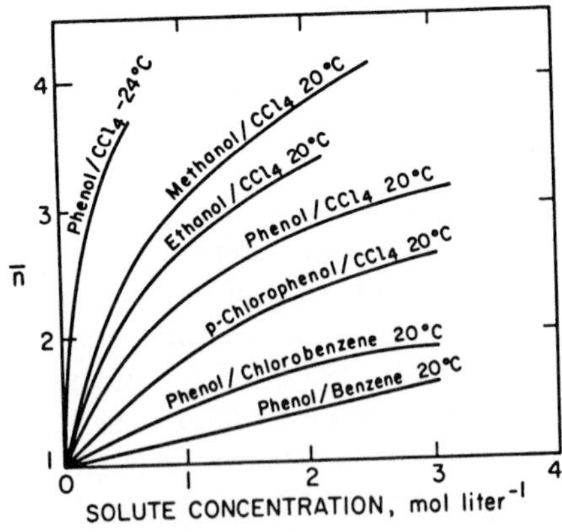

Figure 7-20 Average chain length of multiply-associated solutes in various solvents, as determined by infra-red spectroscopy (Mecke, 1948).

Figure 7-20 shows that for phenol in carbon tetrachloride, the extent of association falls with rising temperature, as expected, because the association reaction (hydrogen bonding) is exothermic. However, Fig. 7-20 also shows that for phenol at a constant temperature of 20°C, the average chain length depends not only on the alcohol concentration but also on the nature of the solvent.[23] This observation is not consistent with the assumptions of the chemical theory that postulates that the role of any solvent is merely that of an inert dispersing agent for the associated component; in other words, the chemical theory assumes that solvent-solute interactions make no contribution to solution nonideality. According to the theory of multiple association, at a fixed temperature, the activity coefficient of an associating component should be a function only of its mole fraction regardless of the nature of the "inert" solvent. However, experimental data for many solutions of alcohols in nonpolar solvents show that this is not the case, indicating that the concept of an inert solvent is a convenient, but unreal oversimplification.

7.11 Associated Solutions with Physical Interactions

The chemical solution theory of Dolezalek assumes that all deviations from ideal behavior are due to formation and decomposition of chemical species and that once the mole fractions of the true species are known, the properties of the solution can be calculated without further consideration to interactions between the true species. By contrast, the physical theory states that the true species are the same as the apparent species and that while there are physical (van der Waals) intermolecular forces, there are no chemical reactions in the solution.

The chemical and the physical theories of solutions are extreme, one-sided statements of what we now believe to be the correct situation. In certain limiting cases, each theory provides a satisfactory approximation: When forces between molecules are weak, no new stable chemical species are formed and the physical theory applies; on the other hand, when forces between molecules are strong, these forces result in the formation of chemical bonds; because the energies for chemical bond formation are significantly larger than those corresponding to van der Waals forces,[24] the chemical theory for such cases provides a reasonable description. In general, both physical and chemical forces should be taken into account. A comprehensive theory of solutions should allow for a smooth transition from one limit of a "physical" solution to the other limit of a "chemical" solution.

It is difficult to formulate a theory that takes into account both physical and chemical effects without thereby introducing involved algebra and, what is worse, a large number of adjustable parameters. Nevertheless, a few attempts have been made

[23] Notice that $\bar{n}$ for phenol is significantly lower in chlorobenzene and benzene than in carbon tetrachloride. This result is probably due to solvation between the alcohol and the π-electrons of the aromatic solvents.

[24] Roughly speaking, weak forces between molecules are those that have energies less than RT, while strong forces between molecules are those whose energies are considerably larger than RT. At room temperature RT is about 2500 J mol^{-1}.

and one of the more successful is based on the theory of polymer mixtures by Flory (1944). This theory has been applied to mixtures containing aliphatic alcohols and paraffinic hydrocarbons (Scatchard, 1949; Kretschmer and Wiebe, 1954; Smith and Brown, 1973); in such mixtures, alcohol polymerizes to form chains, but these chains interact with the paraffin only through van der Waals forces. We now briefly outline the essential ideas of the theory of associated solutions following the discussion of Renon (1967).

Let A stand for the hydrocarbon and B for the alcohol. We assume that:

1. The alcohol exists in the solution in the form of linear, hydrogen-bonded polymers $B_1, B_2, \ldots, B_n, \ldots$, formed by successive reactions of the type

$$B + B_{n-1} \rightleftharpoons B_n$$

2. The association constant for the reaction above is independent of n.
3. The molar volume of an n-mer is given by the molar volume of the monomer multiplied by n.
4. There are physical interactions between all molecules. These lead to contributions to liquid-mixture nonideality that can be described by expressions of the van Laar form.
5. The temperature dependence of the association constant K is such that the enthalpy of formation of a hydrogen bond is independent of temperature and degree of association.

On the basis of his lattice model, Flory (1944) derived expressions for the thermodynamic properties of solutions of polymers differing only in molecular weight; these expressions are especially suited for chemical equilibria between linear, polymeric species. We can use Flory's results for the entropy of mixing to obtain the proper expression for the equilibrium constant. It is

$$K(T) = \frac{\Phi_{B_{n+1}}}{\Phi_{B_n}\Phi_{B_1}} \frac{n}{n+1} \tag{7-154}$$

where Φ_{B_n} is the volume fraction of species B_n.

The derivation of Eq. (7-154) is given elsewhere (Flory, 1944). Flory's result clearly shows that the equilibrium constant should not be expressed in terms of mole fractions as in Sec. 7.10.

As shown by several authors (Scatchard, 1949; Kretschmer and Wiebe, 1954; Smith and Brown, 1973; Renon and Prausnitz, 1967; Redlich and Kister, 1947), the excess Gibbs energy, taken relative to an ideal solution of alcohol and hydrocarbon, can be separated into two contributions, one chemical and the other physical:

$$g^E = g_c^E + g_p^E \tag{7-155}$$

The chemical contribution g_c^E results from the dependence of the "true" composition of the solution on the chemical equilibria indicated by assumption 1. The excess entropy introduced by mixing of polymeric species is taken into account in the calculation of the chemical contribution. From Flory's theory we have

$$\frac{g_c^E}{RT} = x_A \ln\frac{\Phi_A}{x_A} + x_B \ln\frac{\Phi_{B_1}}{\Phi_{B_1}^* x_B} + Kx_B(\Phi_{B_1} - \Phi_{B_1}^*) \tag{7-156}$$

where x_A and x_B are the overall (stoichiometric) mole fractions, Φ_{B_1} is the (true) volume fraction of molecular species B_1, the alcohol monomer, and

$$\Phi_{B_1}^* = \lim_{x_A \to 0} \Phi_{B_1}$$

We can obtain Φ_{B_1} from the equilibrium constant:

$$\Phi_{B_1} = \frac{1 + 2K\Phi_B - \sqrt{1 + 4K\Phi_B}}{2K^2\Phi_B} \tag{7-157}$$

where Φ_B is the overall volume fraction of alcohol. The volume fraction of alcohol monomer in pure alcohol then becomes

$$\Phi_{B_1}^* = \frac{1 + 2K - \sqrt{1 + 4K}}{2K^2} \tag{7-158}$$

The physical contribution g_p^E is given by a one-parameter equation as suggested by Scatchard (1949):

$$g_p^E = \beta\Phi_A\Phi_B(x_A v_A + x_B v_B) \tag{7-159}$$

where β is a physical interaction parameter related to the hydrocarbon-alcohol monomer interaction, and v_A and v_B are liquid molar volumes.

Activity coefficients and the enthalpy are found by appropriate differentiation. The results are:

$$\ln \gamma_{\mathrm{A}} = \ln \frac{\Phi_{\mathrm{A}}}{x_{\mathrm{A}}} + \Phi_{\mathrm{B}}\left(1 - \frac{\upsilon_{\mathrm{A}}}{\upsilon_{\mathrm{B}}}\right) + K\frac{\upsilon_{\mathrm{A}}}{\upsilon_{\mathrm{B}}}\Phi_{\mathrm{B}}\Phi_{\mathrm{B}_1} + \frac{\beta}{RT}\upsilon_{\mathrm{A}}\Phi_{\mathrm{B}}^2 \tag{7-160}$$

$$\ln \gamma_{\mathrm{B}} = \ln \frac{\Phi_{\mathrm{B}_1}}{\Phi^*_{\mathrm{B}_1} x_{\mathrm{B}}} + \Phi_{\mathrm{A}}\left(1 - \frac{\upsilon_{\mathrm{B}}}{\upsilon_{\mathrm{A}}}\right) + K(\Phi_{\mathrm{B}}\Phi_{\mathrm{B}_1} - \Phi^*_{\mathrm{B}_1}) + \frac{\beta}{RT}\upsilon_{\mathrm{B}}\Phi_{\mathrm{A}}^2 \tag{7-161}$$

$$h^E = h_c^E + h_p^E \tag{7-162}$$

$$h_c^E = -K\Delta h^0\left[x_{\mathrm{B}}\frac{\partial \ln(\Phi_{\mathrm{B}_1}/\Phi^*_{\mathrm{B}_1})}{\partial K} + x_{\mathrm{B}}(\Phi_{\mathrm{B}_1} + \Phi^*_{\mathrm{B}_1}) + Kx_{\mathrm{B}}\left(\frac{\partial \Phi_{\mathrm{B}_1}}{\partial K} - \frac{\partial \Phi^*_{\mathrm{B}_1}}{\partial K}\right)\right] \tag{7-163}$$

$$h_p^E = \beta' \Phi_{\mathrm{A}}\Phi_{\mathrm{B}}(x_{\mathrm{A}}\upsilon_{\mathrm{A}} + x_{\mathrm{B}}\upsilon_{\mathrm{B}}) \tag{7-164}$$

where

$$\beta' = \beta - T\frac{d\beta}{dT} \tag{7-165}$$

and Δh^0 is the molar enthalpy of hydrogen-bond formation.

For the Gibbs energy of any alcohol/hydrocarbon system at a fixed temperature, the theory requires only one physical interaction parameter β and one equilibrium constant K. The equilibrium constant, however, depends only on the alcohol and is independent of the hydrocarbon solvent.

Renon's reduction for 11 binary alcohol/hydrocarbon systems is typically represented by Figs. 7-21 and 7-22. Considering the totality of the data, Renon chose only one value for Δh^0, i.e., -25.1 kJ mol^{-1}. This value fixes the temperature dependence of K for all alcohols according to

$$\frac{d \ln K}{d(1/T)} = -\frac{\Delta h^0}{R} \tag{7-166}$$

Each alcohol is further characterized by the value of K at one temperature. At 50°C, K is 450 for methanol, 190 for ethanol, 60 for isopropanol, and 90 for n-propanol.[25] These constants were obtained upon considering the totality of the data for each alcohol, but giving more weight to the more sensitive data (enthalpies) in the region where the model is physically most reasonable, i.e., at low temperatures and at high alcohol concentrations.

[25] Other values for K have been suggested. See, for example, A. Nath and E. Bender, 1981, *Fluid Phase Equilibria*, 7: 275, 289.

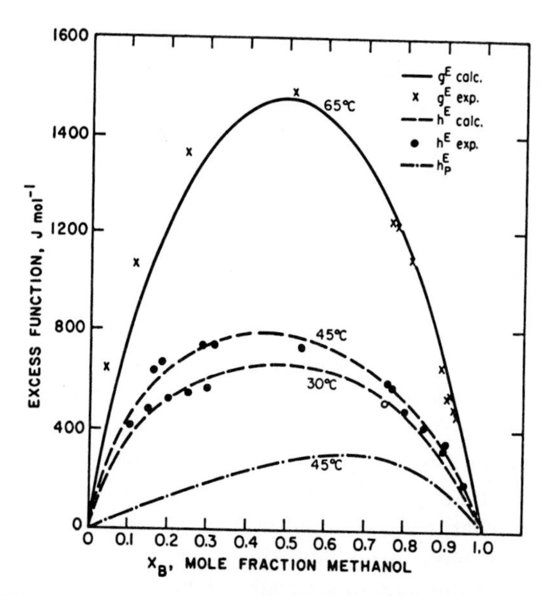

Figure 7-21 Excess functions for methanol/*n*-hexane mixtures. Here h_p^E stands for the physical (as opposed to chemical) contribution to the excess enthalpy.

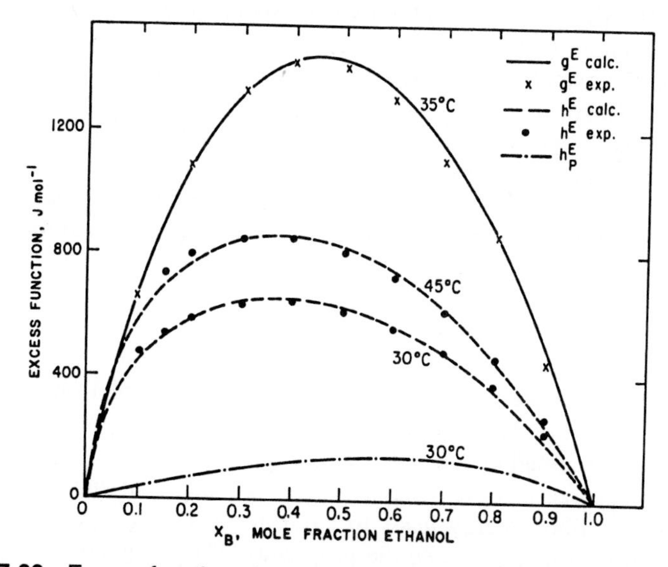

Figure 7-22 Excess functions for ethanol/*n*-hexane mixtures. Here h_p^E stands for the physical (as opposed to chemical) contribution to the excess enthalpy.

Renon's data reduction indicates that the theory described gives a good representation of the experimental data in accord with the physical meaning of the model. Discrepancies become large only where the degree of alcohol polymerization is small, i.e., at low alcohol concentrations and at higher temperatures. A particularly sensitive test of the theory is provided by comparison of calculated and experimental results for excess enthalpies at several temperatures.

In spite of its simplifying assumptions, the Flory-Scatchard model of associated solutions gives a good representation of the properties of concentrated solutions of alcohols in saturated hydrocarbons. It probably takes into account the major effects, but perhaps neglects others, such as formation of cyclic polymers.

The theory just described can be extended to include other chemical effects such as, for example, solvation between solute and solvent, or even association of both components, each with itself and with the other. For each assumed chemical equilibrium, a characteristic equilibrium constant must be introduced and thus a more general treatment, including various types of chemical equilibria, results in complicated algebraic expressions and, what is worse, requires a large number of empirical parameters.

Numerous models have been developed for solutions containing associating molecules making use of the idea to separate thermodynamic excess properties into a chemical and a physical contribution. They differ from each other with respect to the particular expressions used for the two contributions. Using the same basic ideas for the chemical contribution as discussed above, such models are, e. g., the *dis*persive-*qua*si-*c*hemical (DISQUAC) model proposed by Kehiaian (1983, 1985), the *e*xtended-*r*eal-*a*ssociated-*s*olution (ERAS) model developed by Heintz (1985), the *l*attice-*f*luid-*as*sociation (LFAS) model of Panayiotou (1988, 1991), and the model proposed by Nagata *et al.* (1997). All these models have been used to describe phase equilibria and thermodynamic excess properties. For example, recent variations include the DISQUAC model by González (1997, 1997a), the LFAS model by Ormanoudis and Panayiotou (1993), and the ERAS model by Heintz and Papaioannou (1998) and Kammerer *et al.* (1998). However, with these models, phase equilibria and excess properties usually cannot be described simultaneously using the same model parameters.

A more fundamental method to account for association and/or solvation is the *s*tatistical *a*ssociated-*f*luid *t*heory (SAFT) developed by Chapman *et al.* (1989, 1990). The SAFT equation of state has been successfully used to model phase behavior and thermodynamic properties for a large variety of simple and complex fluid and fluid mixtures. SAFT is discussed in Sec. 7.17.

We must recognize that the distinction between chemical and physical contributions to the excess Gibbs energy leads to an arbitrary and, perhaps, artificial model based on a simplified picture of solution properties. The designation of molecular interactions as either chemical or physical is only a convenience that probably cannot be justified by a sophisticated modern theory of intermolecular forces. Nevertheless, a joint chemical and physical description of equilibrium properties of mixtures, as exemplified by Eqs. (7-160) and (7-161), provides a reasonable and useful description for highly nonideal solutions that is a considerable improvement over the ideal chemical theory of Dolezalek on the one hand and the purely physical theory of van Laar on the other.[26]

7.12 Activity Coefficients in Solvated Solutions

The chemical theory of solutions has frequently been used to describe thermodynamic properties of binary solutions where two components form complexes. There are many experimental studies of such solutions that, if the complex is stable enough, are characterized by negative deviations from Raoult's law. To illustrate, we consider first a simple case, a binary solution where complexes form according to

$$A + B \rightleftharpoons AB \tag{7-167}$$

The equilibrium constant K is related to the activities of the three species by

$$K = \frac{a_{AB}}{a_A a_B} \tag{7-168}$$

If the solution is formed from n_1 moles of A and n_2 moles of B, and if at equilibrium, n_{AB} moles of complex are formed, the true mole fractions z of A, B, and AB are

$$z_A = \frac{n_1 - n_{AB}}{n_1 + n_2 - n_{AB}} \tag{7-169}$$

$$z_B = \frac{n_2 - n_{AB}}{n_1 + n_2 - n_{AB}} \tag{7-170}$$

$$z_{AB} = \frac{n_{AB}}{n_1 + n_2 - n_{AB}} \tag{7-171}$$

Following Dolezalek, we assume that the true species form an ideal solution and therefore the activity of each species is equal to its true mole fraction. Equations (7-168) to (7-171) may then be used to eliminate n_{AB}. The apparent mole fractions of the two components are x_1 (for A) and x_2 (for B). They are given by

---

[26] A fine study incorporating both chemical and physical effects has been presented by Calado and Staveley, 1979, *Fluid Phase Equilibria,* 3: 153, who correlated vapor-liquid equilibrium data for the system NO/Kr at 115.76 K. In this system there is appreciable dimerization of NO to $(NO)_2$.

$$x_1 = \frac{n_1}{n_1 + n_2} \quad \text{and} \quad x_2 = \frac{n_2}{n_1 + n_2}$$

Algebraic rearrangement then gives for the activity coefficients:

$$\gamma_1 = \frac{a_A}{x_1} = \frac{kx_1 - 2 + 2(1 - kx_1 x_2)^{1/2}}{kx_1^2} \tag{7-172}$$

$$\gamma_2 = \frac{a_B}{x_2} = \frac{kx_2 - 2 + 2(1 - kx_1 x_2)^{1/2}}{kx_2^2} \tag{7-173}$$

where $k \equiv 4K/(K+1)$. Because of symmetry in Eq. (7-167), γ_1 depends on x_1 in exactly the same way as γ_2 depends on x_2.

Figure 7-23 shows γ_1 as a function of x_1 for several values of equilibrium constant K. When $K = 0$, γ_1 for all x_1 as expected, because in that case no complex is formed, and therefore, by assumption, there is no deviation from ideal behavior. At the other extreme, when $K = \infty$, activity coefficients of both components go to zero at the midpoint ($x_1 = x_2 = 1/2$) because at this particular composition all molecules are complexed and no uncomplexed molecules A or B remain.

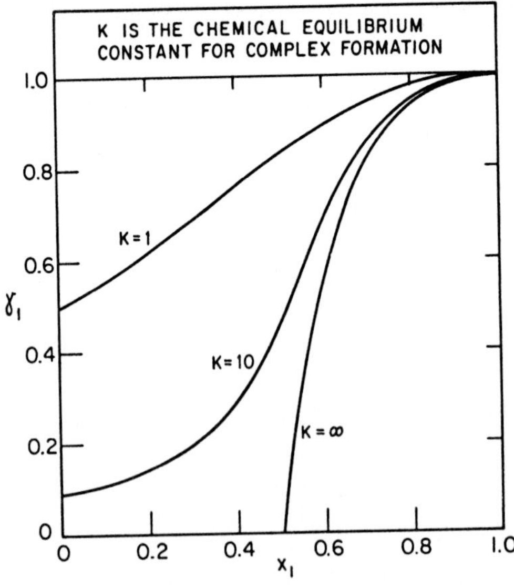

Figure 7-23 Activity coefficients of a solvating component that forms a bimolecular complex.

Equations (7-172) and (7-173) predict negative deviations from Raoult's law for $K > 0$ and as a result, it has unfortunately become all too common immediately to ascribe observed negative deviations from Raoult's law to solvation effects. It is true that strong negative deviations usually result from complex formation and conversely, if strong complexing is known to occur, negative deviations usually result. However, these conclusions are not always valid because they are based on a strictly chemical theory that neglects all physical effects. In systems where complexing is weak, physical effects are by no means negligible and as a result, weakly solvated solutions sometimes exhibit positive deviations from Raoult's law. Also, small negative deviations from Raoult's law may occur in the absence of complexing as has been observed, for example, in mixtures of normal paraffins, where differences in molecular size give positive excess entropy.

Equations (7-172) and (7-173) have been applied to a large number of solutions; one of these is the diethyl ether/chloroform system. Vapor-liquid equilibrium data for this system at several temperatures can be reduced using Eqs. (7-172) and (7-173) and the following equilibrium constants are then obtained:

Temperature (°C)	20	60	80	100
Equilibrium constant, K	2.96	1.00	0.80	0.70

Because the formation of a hydrogen bond between ether and chloroform is an exothermic reaction, we expect the equilibrium constant to fall with rising temperature, as indeed it does. From the equilibrium constants, we can calculate Δh^0, the enthalpy of hydrogen-bond formation, and we find that for ether/chloroform, Δh^0 is close to -12.5 kJ mol^{-1}, a reasonable value, in approximate agreement with those obtained by other experimental measurements.

Equations (7-172) and (7-173) have been derived for the case where the two components form a 1:1 complex. Similar equations can be derived for cases where the stoichiometry of the complex is 2:1, or 3:1, or 3:2, etc. This flexibility is both a strength and a weakness of the chemical theory of solutions. It is a strength because it can "explain" solution behavior for any sort of chemical interaction and therefore it has, potentially, a wide range of applicability. It is a weakness because, unless other information is available, the stoichiometry of the complex is another adjustable parameter, in addition to the equilibrium constant. Thus, if a particular assumed stoichiometry does not fit the experimental data, one can try another, and so on, and eventually a fit is obtained. Such a fit, however, has no physical significance unless there is independent evidence from the molecular structure of the components to verify the assumed stoichiometry. For diethyl ether/chloroform mixtures, it would be difficult to justify any complex other than one having a 1:1 stoichiometry.

As shown by Harris (1969), we can relax Dolezalek's assumption that the "true" chemical species form an ideal solution. Harris assumed that a mixture of "true" species is described by an equation of the van Laar type. For example, we again consider a mixture of molecules A and B that interact strongly to form complex AB:

$$\text{A} + \text{B} \rightleftharpoons \text{AB} \tag{7-174}$$

Let K be the equilibrium constant for this chemical equilibrium and let z_A, z_B, and z_{AB} stand for the true mole fractions. Then

$$K = \frac{z_{AB}}{z_A z_B} \frac{\gamma'_{AB}}{\gamma'_A \gamma'_B} \tag{7-175}$$

where γ' stands for the true activity coefficient. Dolezalek assumed that all γ' are equal to unity. Harris, however, makes the more reasonable assumption that for any true component k,[27]

$$RT \ln \gamma'_k = v_k \left(\sum_j \alpha_{k_j} \Phi_j - \frac{1}{2} \sum_i \sum_j \alpha_{ij} \Phi_i \Phi_j \right) \tag{7-176}$$

where Φ is the volume fraction:

$$\Phi_k = \frac{z_k v_k}{\sum_j z_j v_j}$$

and where v_i is the liquid molar volume of i and α_{ij} is a (van Laar) parameter for physical interaction of molecules i and j. Subscripts i, j, and k are understood in this case to range over the three possible species A, B, and AB. Equation (7-176) contains three physical parameters: $\alpha_{A\text{-}B}$, $\alpha_{A\text{-}AB}$, and $\alpha_{B\text{-}AB}$. To limit the number of adjustable parameters to two (one chemical parameter K and one physical parameter α), it is necessary to use plausible physical arguments for relating $\alpha_{A\text{-}AB}$ and $\alpha_{B\text{-}AB}$ to $\alpha_{A\text{-}B}$, as described by Harris (1969).

To reduce experimental vapor-liquid equilibrium data with Eqs. (7-175) and (7-176), we use a powerful theorem discussed in detail by Prigogine and Defay (1954) and summarized in App. G. It can be rigorously shown that for a mixture of components 1 (species B) and 2 (species A), the *apparent* activity coefficients and *apparent* mole fractions are related to the *true* activity coefficients and *true* mole fractions by

$$\gamma_2 = \frac{z_A \gamma'_A}{x_2} \tag{7-177}$$

$$\gamma_1 = \frac{z_B \gamma'_B}{x_1} \tag{7-178}$$

[27] For mixtures of components of greatly different size, improved representation of the true activity coefficients may be obtained by adding to the right-hand side of Eq. (7-176) the Flory-Huggins term: $RT[\ln(\Phi_k / z_k) + 1 - \Phi_k / z_k]$.

Equations (7-177) and (7-178) are independent of any physical model. They follow directly from the assumption that the "true" species A, B, and AB are in equilibrium.

For data reduction, Eqs. (7-175) to (7-178) must be combined with material balances relating true mole fractions to apparent mole fractions. This is done most conveniently in terms of the normalized extent of complex formation ξ $(0 \leq \xi \leq 1/2)$:

$$z_{AB} = \frac{\xi}{1-\xi} \tag{7-179}$$

$$z_{A} = \frac{x_2 - \xi}{1-\xi} \tag{7-180}$$

$$z_{B} = \frac{x_1 - \xi}{1-\xi} \tag{7-181}$$

To illustrate Harris' extension of Dolezalek's theory, Fig. 7-24 gives results of data reduction for solutions of acetylene in three organic solvents. Acetylene forms hydrogen bonds with butyrolactone and N-methylpyrrolidone but not with hexane. To represent the experimental data for the two polar solvents, two parameters (K and α) are

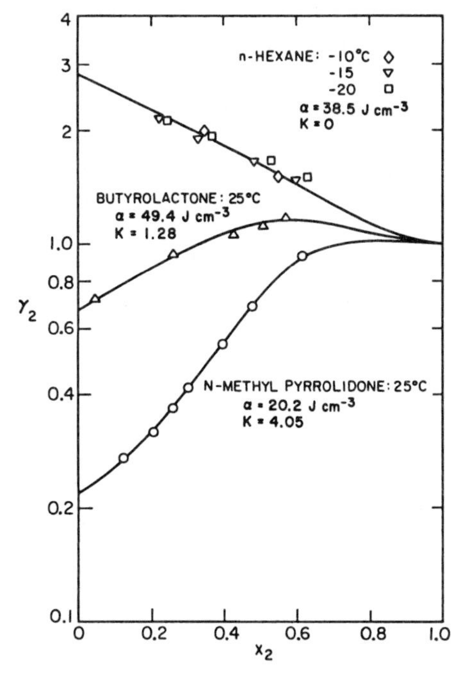

Figure 7-24 Calculated and observed activity coefficients for acetylene (2) in three organic solvents.

required, whereas the experimental data for hexane are represented with only one parameter, because $K = 0$. In hexane, a "physical" solvent, acetylene exhibits positive deviations from Raoult's law over the entire composition range. In chemical solvents, however, acetylene exhibits negative deviations from Raoult's law at the solvent-rich end and positive deviations at the acetylene-rich end. Dolezalek's theory (with one chemical equilibrium constant) cannot account for such behavior. The modification of the chemical theory proposed by Harris combines into one comprehensive model the ideas of both Dolezalek and van Laar.[28]

7.13 Solutions Containing Two (or More) Complexes

The chemical theory can be extended in a straightforward way to the case where several complexes of different stoichiometry are formed by the two components. For the same reasons as those stated in the preceding section, this possibility for extension carries with it an advantage and a danger. The advantage lies in the gain in flexibility; the danger arises because if one postulates enough complexes, one necessarily introduces a sufficient number of adjustable parameters to fit any data for any system.

An excellent study of a binary solution that contains two types of complexes has been reported by McGlashan and Rastogi (1958) who investigated thermodynamic properties of the *p*-dioxan/chloroform system; *p*-dioxan is a symmetric cyclic ether containing two oxygen atoms and therefore McGlashan and Rastogi postulated the existence in solution of two complexes having the structures

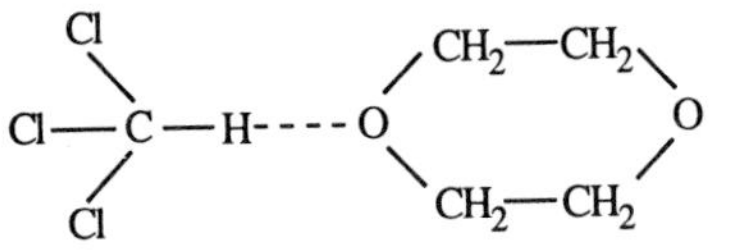

```
     Cl          CH2—CH2
       \        /       \
 Cl—C—H----O             O
       /        \       /
     Cl          CH2—CH2
```

and

```
     Cl          CH2—CH2                    Cl
       \        /       \                  /
 Cl—C—H----O             O----H—C—Cl
       /        \       /                  \
     Cl          CH2—CH2                    Cl
```

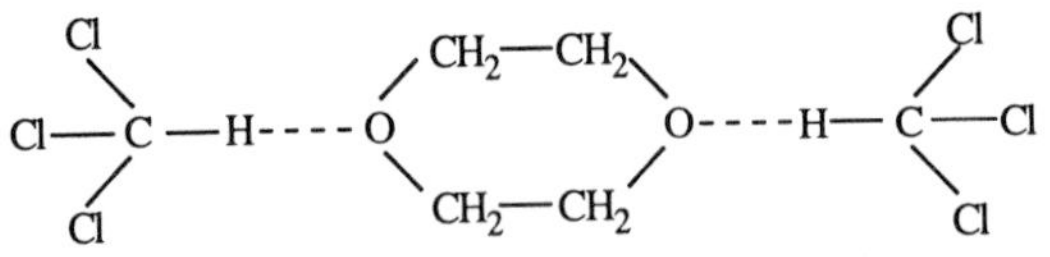

Let A stand for dioxan and B for chloroform. Two equilibria are postulated:

---

[28] An engineering-oriented study incorporating both solvation equilibria and physical effects was presented by Rivas, 1979, *AIChE J.*, 25: 975, who correlated solubility data for CO_2 and H_2S in mixed solvents for sweetening of sour natural (or synthetic) gases.

$$A + B \rightleftharpoons AB \tag{7-182}$$

$$A + 2B \rightleftharpoons AB_2 \tag{7-183}$$

Two equilibrium constants are defined by

$$K_1 = \frac{a_{AB}}{a_A a_B} \tag{7-184}$$

$$K_2 = \frac{a_{AB_2}}{a_A a_B^2} \tag{7-185}$$

We assume that this (apparent) binary, nonideal solution is an ideal solution of four true components (A, B, AB, and AB_2). The activity of each true component is then equal to its true mole fraction $\mathfrak{z}$ and Eqs. (7-184) and (7-185) become

$$K_1 = \frac{\mathfrak{z}_{AB}}{\mathfrak{z}_A \mathfrak{z}_B} \tag{7-186}$$

$$K_2 = \frac{\mathfrak{z}_{AB_2}}{\mathfrak{z}_A \mathfrak{z}_B^2} \tag{7-187}$$

By material balance,

$$\mathfrak{z}_A + \mathfrak{z}_B + \mathfrak{z}_{AB} + \mathfrak{z}_{AB_2} = 1 \tag{7-188}$$

Eliminating $\mathfrak{z}_{AB}$ and $\mathfrak{z}_{AB_2}$ with Eqs. (7-186) and (7-187), Eq. (7-188) becomes, after rearrangement,

$$\frac{1 - \mathfrak{z}_A - \mathfrak{z}_B}{\mathfrak{z}_A \mathfrak{z}_B} = K_1 + K_2 \mathfrak{z}_B \tag{7-189}$$

or

$$\frac{1 - a_A - a_B}{a_A a_B} = K_1 + K_2 a_B \tag{7-190}$$

Activities a_A and a_B are obtained from experimental vapor-liquid equilibrium data.

A plot of the left side of Eq. (7-190) versus a_B should give a straight line whose intercept and slope yield the two equilibrium constants K_1 and K_2. Such a plot was constructed by McGlashan and Rastogi (1958) who found that at 50°C, $K_1 = 1.11$ and $K_2 = 1.24$.

Once numerical values are given for the two equilibrium constants, activity coefficients can be calculated as shown by McGlashan and Rastogi. The activity coefficient of B (chloroform) is given by

$$\gamma_B = \frac{z_B}{x_B} \tag{7-191}$$

and the true mole fraction of B is related by material balances to x_B, the apparent mole fraction, by

$$x_B = \frac{(1+K_1)z_B + K_2 z_B^2(2-z_B)}{1+K_1 z_B(2-z_B)+K_2 z_B^2(3-2z_B)} \tag{7-192}$$

The activity coefficient of A (dioxan) is given by

$$\gamma_A = \frac{1-z_B}{(1+K_1 z_B + K_2 z_B^2)(1-x_B)} \tag{7-193}$$

Figure 7-25 gives a plot of calculated and observed activity coefficients. The excellent agreement shows that for this system calculations based on the assumption of the existence of two justifiable complexes can account for the system's thermodynamic behavior.

McGlashan and Rastogi also measured calorimetrically the enthalpy of mixing for this system. Using the chemical solution theory just described, the enthalpy of mixing can be related to the enthalpy of complex formation.

Because

$$\frac{g^E}{RT} = x_A \ln\gamma_A + x_B \ln\gamma_B \tag{7-194}$$

and because

$$h^E = -RT^2\left[\frac{\partial(g^E/RT)}{\partial T}\right]_{P,x} \tag{7-195}$$

substitution of Eqs. (7-191) and (7-193) for γ_A and γ_B gives

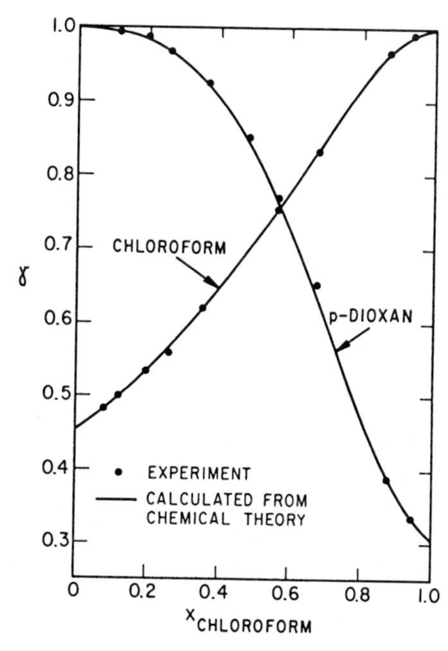

Figure 7-25 Activity coefficients for a solvated mixture: *p*-dioxan/chloroform at 50°C (McGlashan and Rastogi, 1958).

$$h^E = \left(\frac{x_A \breve{\gamma}_B}{1 + K_1 \breve{\gamma}_B + K_2 \breve{\gamma}_B^2}\right)(K_1 \Delta h_1^0 + \breve{\gamma}_B K_2 \Delta h_2^0) \tag{7-196}$$

where Δh_1^0 is the enthalpy of formation of complex AB and Δh_2^0 is the enthalpy of formation of complex AB_2:

$$\Delta h_1^0 = RT^2 \frac{d \ln K_1}{dT} \tag{7-197}$$

$$\Delta h_2^0 = RT^2 \frac{d \ln K_2}{dT} \tag{7-198}$$

By fitting experimentally determined enthalpies of mixing to Eq. (7-196), the enthalpies of complex formation are

$$\Delta h_1^0 = -8.4 \text{ kJ mol}^{-1}$$

$$\Delta h_2^0 = -15.1 \text{ kJ mol}^{-1}$$

These are reasonable values; Δh_1^0 is the enthalpy of formation of one ether-oxygen hydrogen bond and it agrees well with results determined by spectroscopic, cryoscopic, and other methods. The enthalpy of formation for the complex containing two hydrogen bonds is not quite twice that for the formation of one hydrogen bond. This is not surprising because the two oxygen atoms in dioxan are only separated by two carbon atoms; therefore, the effect of hydrogen bonding one oxygen atom has an appreciable effect on the other oxygen atom. If the two oxygen atoms were farther apart, one might expect that Δh_2^0 would be more nearly equal to $2\Delta h_1^0$.

7.14 Distribution of a Solute between Two Immiscible Solvents

The chemical theory of solution attempts to explain thermodynamic properties in terms of actual (true) chemical species present in solution. An explanation of this sort can sometimes be applied toward interpreting and extending data on partition coefficients for a solute between two immiscible liquid solvents. We present next one example illustrating such application (Moelwyn-Hughes, 1940).

Consider two liquid phases α and β; a solute, designated by subscript 1, is distributed between these two phases. First, we consider a simple case. Suppose that the mole fraction of solute in either phase is very small and that we can therefore assume the two solutions to be ideal dilute solutions. We then have

$$f_1^{\alpha} = H_{1,\alpha} x_1^{\alpha} \tag{7-199}$$

$$f_1^{\beta} = H_{1,\beta} x_1^{\beta} \tag{7-200}$$

where $H_{1,\alpha}$ is Henry's constant for solute 1 in phase α and $H_{1,\beta}$ is Henry's constant for solute 1 in phase β. Equating fugacities of component 1 in the two phases, we obtain the *partition coefficient K*,

$$\boxed{K = \frac{x_1^{\alpha}}{x_1^{\beta}} = \frac{H_{1,\beta}}{H_{1,\alpha}}} \tag{7-201}$$

At constant temperature and pressure, for sufficiently dilute solutions, the partition coefficient in Eq. (7-201) is a constant, independent of composition. Equation (7-201) is frequently called the *Nernst distribution law.*

Because mole fractions x_1^{α} and x_1^{β} are very small, they are, respectively, proportional to the concentrations of solute 1 in phase α and in phase β; it is therefore customary to use a somewhat different partition coefficient, K', expressed in terms of concentrations c rather than mole fractions x:

$$K' = \frac{c_1^\alpha}{c_1^\beta} = \frac{\rho^\alpha x_1^\alpha}{\rho^\beta x_1^\beta} \tag{7-202}$$

where ρ^α and ρ^β are, respectively, the molar densities of phases α and β. For very small x_1, ρ^α and ρ^β are the densities of the pure solvents.

Many cases are known where the Nernst distribution law is not consistent with experiment, even though the solute mole fractions are small; in other words, in these cases, the equations for ideal dilute solutions [Eqs. (7-199) and (7-200)] are not obeyed in either (or both) of the liquid phases at the particular concentrations investigated.[29] In many cases, departure from Nernst's law may be ascribed to chemical effects. We now consider such a case: the distribution of benzoic acid between the two (essentially) immiscible solvents water and benzene near room temperature. The explanation for the failure of Nernst's law can, in this case, be found by taking into account the tendency of organic acids to dimerize in a nonpolar solvent.

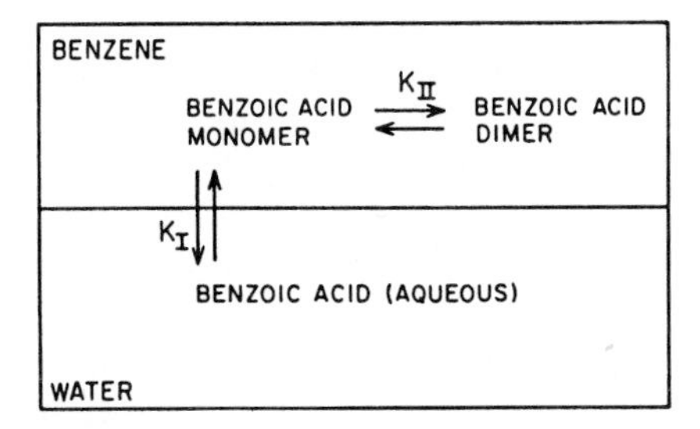

Figure 7-26 Distribution of benzoic acid between benzene and water.

We postulate two equilibria as shown in Fig. 7-26:

1. Phase-distribution equilibrium between the two phases:

   Acid in water $\rightleftharpoons$ monomer acid in benzene

2. Chemical equilibrium in the benzene phase:

   Monomer acid $\rightleftharpoons$ dimer acid

For each equilibrium there is an equilibrium constant:

[29] The ideal-dilute-solution equation is always approached for any nonelectrolyte when the mole fraction is sufficiently small, but just how small depends on the system. For a solute that associates in solution, it may be very small indeed, sometimes smaller than can be measured by common analytical methods.

$$K_{\text{I}} = \frac{c_{\text{M}}^{\text{B}}}{c^{\text{W}}} \tag{7-203}$$

$$K_{\text{II}} = \frac{c_{\text{D}}^{\text{B}}}{(c_{\text{M}}^{\text{B}})^2} \tag{7-204}$$

In Eqs. (7-203) and (7-204), c stands for concentration of benzoic acid; superscript B stands for the benzene phase and superscript W for the water phase. Subscript M stands for monomer and subscript D for dimer.[30]

Let c_{T}^{B} stand for the total concentration of benzoic acid in benzene. By material balance,

$$c_{\text{T}}^{\text{B}} = c_{\text{M}}^{\text{B}} + 2c_{\text{D}}^{\text{B}} \tag{7-205}$$

Substitution of Eqs. (7-203) and (7-204) into Eq. (7-205) gives the distribution law

$$\boxed{\frac{c_{\text{T}}^{\text{B}}}{c^{\text{W}}} = K_{\text{I}} + 2K_{\text{I}}^2 K_{\text{II}} c^{\text{W}}} \tag{7-206}$$

In this case, then, the distribution coefficient (i.e., the ratio of c_{T}^{B} to c^{W}) is not constant, as it would be according to Nernst's law, but varies linearly with the concentration of benzoic acid in water. Experimental data for this system (Nernst, 1891) are plotted in Fig. 7-27 in the form suggested by Eq. (7-206). The straight line obtained confirms the prediction based on the chemical theory of solutions; from the slope and intercept $K_{\text{I}} = 1.80$ (dimensionless) and $K_{\text{II}} = 176$ dm^3 mol^{-1}. By plotting the data in this way, one can, using only a few experimental points, interpolate and slightly extrapolate with confidence.

In some systems, where the tendency of the solute to dimerize is strong, K_{II} is very large. If, in the system just discussed, benzoic acid dimerized strongly, we would have

$$2K_{\text{I}}^2 K_{\text{II}} c^{\text{W}} >> K_{\text{I}} \tag{7-207}$$

giving the distribution law

$$\boxed{\frac{\sqrt{c_{\text{T}}^{\text{B}}}}{c^{\text{W}}} = (2K_{\text{I}}^2 K_{\text{II}})^{1/2} = \text{constant}} \tag{7-208}$$

Various examples of this distribution law have been found. Whenever a system behaves according to Eq. (7-208), it is considered good evidence that the solute molecules

[30] In the dilute aqueous phase, benzoic acid is probably completely solvated by hydrogen bonding with water.

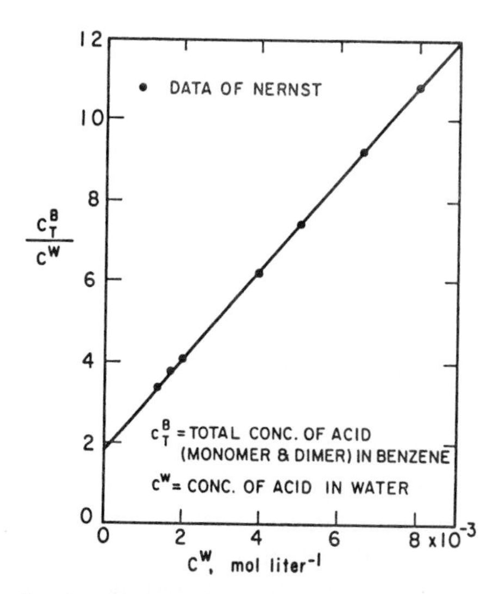

Figure 7-27 Linearization of distribution data with chemical theory: the system benzoic acid/water/benzene at 20°C.

are strongly dimerized in one of the solvents. In the benzoic acid/water/benzene system, dimerization of benzoic acid is moderately strong but the inequality given in Eq. (7-207) is not valid until c^W is at least 10^{-2} mol liter^{-1}.

An example of the applicability of Eq. (7-208) is given in Fig. 7-28. In both alkanoic acid/water/n-dodecane systems (Aveyard and Mitchell, 1970), nearly all acid molecules are dimers in n-dodecane.

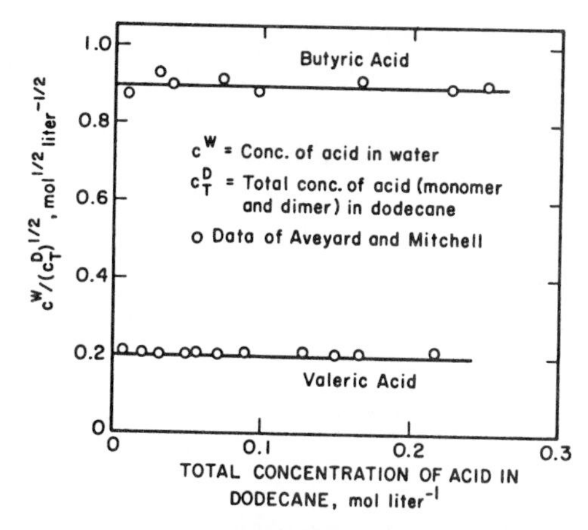

Figure 7-28 Linearization of distribution data for two alkanoic acids distributed between water and n-dodecane at 20°C. Nearly all acid molecules are dimers in the organic phase (Aveyard and Mitchell, 1970).

Numerous other systems have been investigated where deviations from Nernst's distribution law can be reasonably explained by a coupling of chemical and phase-distribution equilibria. For example, data for the distribution of picric acid between water and a nonpolar solvent can be interpreted by taking into account the ionization of picric acid in the aqueous phase. In a similar manner, the distribution (i.e., the solubility) of sulfur dioxide between the gas phase and water can be described quantitatively, as shown in Chap. 10.

7.15 The Generalized van der Waals Partition Function

To calculate fluid-phase equilibria, the main requirement is a good quantitative model for representing thermodynamic properties over a wide range of densities, from zero density (ideal gas) to close-packed density, corresponding to the compressed liquid. No currently available model is entirely satisfactory. Many of the best approximate models and theories for practical purposes are derived starting with an expression for the *canonical partition function*. In many cases, the simplifying concepts utilized are similar to those leading to the *perturbed-hard-sphere* model, first introduced by van der Waals in 1873 and cast into modern form by Zwanzig in 1954. Because so many statistical-thermodynamic models are based on the same fundamental ideas manifested in the *generalized van der Waals partition function*, we discuss it here in some detail.

For a binary mixture, the *partition function* Q depends on temperature T, total volume V, and number of molecules N_1 and N_2. The equation of state is related to Q through

$$P = kT\left(\frac{\partial \ln Q}{\partial V}\right)_{T,N_1,N_2} \tag{7-209}$$

and the chemical potentials are related to Q through

$$\mu_1 = kT\left(\frac{\partial \ln Q}{\partial N_1}\right)_{T,V,N_2} \tag{7-210}$$

$$\mu_2 = kT\left(\frac{\partial \ln Q}{\partial N_2}\right)_{T,V,N_1} \tag{7-211}$$

where k is Boltzmann's constant. If Q is available as a function of T, V, N_1, and N_2, the phase-equilibrium problem is solved, at least in principle.

The *partition function* for a simple pure fluid containing N molecules in total volume V is (Vera and Prausnitz, 1972; Hill, 1986):

$$Q(T,V,N) = \frac{1}{N!}\left(\frac{V}{\Lambda^3}\right)^N (q_{\text{rep}})^N (q_{\text{att}})^N (q_{\text{r,v}})^N \qquad (7\text{-}212)^{31}$$

where Λ is the de Broglie wavelength[32] that depends only on temperature and molecular mass; q_{rep} and q_{att}, respectively, are the contributions from repulsive and attractive intermolecular forces experienced by each molecule due to the presence of all the other molecules; and $q_{\text{r,v}}$ is the contribution per molecule from rotational and vibrational degrees of freedom. For a monatomic ideal gas, the last three terms are unity. For real pure fluids, each of the last three terms depends on temperature and density.

Although van der Waals did not use partition functions, in effect, he suggested that

$$q_{\text{rep}} = \frac{V_f}{V} \qquad (7\text{-}213)$$

and

$$q_{\text{att}} = \exp\left(\frac{-E_o}{2kT}\right) \qquad (7\text{-}214)$$

where the free volume V_f is the volume available to the center of mass of a molecule as it moves about the system, holding the positions of all other molecules fixed; E_o is the intermolecular potential energy experienced by one molecule due to the attractive forces from all other molecules.

Equations (7-213) and (7-214), substituted in Eq. (7-212), lead to the *generalized van der Waals partition function* for a simple pure fluid:

$$\boxed{Q(T,V,N) = \frac{1}{N!}\left(\frac{V}{\Lambda^3}\right)^N \left(\frac{V_f}{V}\right)^N \left[\exp\left(\frac{-E_o}{2kT}\right)\right]^N (q_{\text{r,v}})^N} \qquad (7\text{-}215)$$

The repulsive term is determined by the free volume, V_f, and the attractive term is a Boltzmann factor whose argument is the ratio of the attractive potential energy to the kinetic energy. For simple, small (argon-like) molecules, the final term, $q_{\text{r,v}}$, accounting for molecular rotations and vibrations is only a function of temperature but, as discussed later, for large, polyatomic molecules, $q_{\text{r,v}}$ also depends on density, especially when the molecules deviate significantly from spherical shape.

---

[31] Equation (7-212) is not truly rigorous because it assumes that contributions from repulsive forces, from attractive forces, and those from rotation and vibration can be factored into three separate functions of temperature and density.

[32] $\Lambda = h(2\pi mkT)^{-1/2}$, where h is Planck's constant and m is molecular mass.

To a rough first approximation, the free volume and the potential energy depend only on density; this approximation was used by van der Waals. For hard (impenetrable), spherical molecules, he assumed for V_f the simple relation (strictly valid only at low densities)

$$V_f = V - \frac{N}{N_A} b \tag{7-216}$$

where V is the total volume and $b = (2/3)\pi N_A \sigma^3$, with molecular diameter σ and Avogadro's constant N_A. For E_0 van der Waals assumed

$$E_0 = -\frac{2aN}{VN_A^2} \tag{7-217}$$

where a is a constant representing the strength of the intermolecular forces of attraction. Substituting Eqs. (7-216) and (7-217) into Eq. (7-215), assuming that $q_{r,v}$ is only a function of temperature, and using Eq. (7-209) we obtain the well-known van der Waals equation of state:

$$\frac{Pv}{RT} = \frac{v}{v-b} - \frac{a}{RTv} \tag{7-218}$$

For large, polyatomic molecules, rotational and vibrational degrees of freedom depend on temperature and density. A small molecule (e.g., methane) can rotate and vibrate in a manner essentially independent of its environment but a large molecule (e.g., heptane) can rotate and vibrate easily only when it is remote from nearest neighbors (low density); when neighbors are close (high density), they necessarily interfere with the large molecule's rotational and vibrational freedom. The molecule's ability to rotate and vibrate depends on the "mean-field" exerted on it by the surrounding fluid as well as on internal modes unaffected by the surrounding fluid. For large molecules, therefore, it is not correct to assume that $q_{r,v}$ [Eq. (7-215)] is only a function of temperature and independent of total volume V when the number of molecules N is fixed.

There is no accessible rigorous theory to indicate how $q_{r,v}$ depends on particle density N/V and temperature. Assuming independence of external and internal contributions, $q_{r,v}$ can be factored,

$$q_{r,v} = q_{ext}(V) q_{int}(T) \tag{7-219}$$

where q_{ext} represents contributions from external (density-dependent) rotations and vibrations while q_{int} represents contributions from internal rotations and vibrations; the latter contributions depend only on temperature. The external contributions may also depend on temperature but such dependence is likely to be secondary; it is the density dependence that is of major interest.

To describe the effect of density on rotational and vibrational contributions to the partition function, i.e. to obtain an expression for q_{ext}, Prigogine (1957) introduced the concept of *equivalent translational degrees of freedom*. The total number of *external*[33] degrees of freedom is the sum of the three translational degrees of freedom (possessed by every molecule, regardless of size and shape) and the equivalent translational degrees of freedom (due to rotation and vibration) that become increasingly significant as the size (and nonsphericity) of the molecule rise. Therefore, to obtain a useful partition function, consideration must be given to the external degrees of freedom of large, polyatomic molecules.

First, consider a large rigid molecule with r segments; all bond lengths, bond angles and torsional angles are fixed. This rigid molecule has 3 translational degrees of freedom (the same as for a spherical molecule, one for each translational coordinate) and 2 (if linear) or 3 (if not linear) rotational degrees of freedom, giving a total of 5 or 6 degrees of freedom . Second, consider a large completely flexible molecule with r segments. Here, complete flexibility means that the molecule does not have any restriction on bond length, bond angle and torsional angle. This molecule has a total maximum of $3r$ external degrees of freedom because in this (hypothetical) floppy molecule each segment has 3 degrees of freedom. The total number of degrees of freedom for a real large molecule lies somewhere between the limits set by a completely rigid molecule and by a completely flexible molecule. To approximate the total number of external degrees of freedom while leaving them unspecified, a parameter c is introduced. The total number of "effective" external degrees of freedom per molecule is $3c$, such that $1 < c < r$. For a small spherical molecule (e.g. argon or methane), $r = c = 1$; for more complex molecules, $c > 1$ (e.g. for n-decane, a possible value for c might be 2.7. For isomers of decane, c would be less than 2.7 because a branched paraffin is less flexible than a normal paraffin).

When coupled with Prigogine's approximation that the density-dependent external rotational and vibrational degrees of freedom can be considered as equivalent translational degree of freedom, we obtain

$$q_{ext}(V) = \left(\frac{V_f}{\Lambda^3}\right)^{c-1} \tag{7-220}$$

where $3(c\text{-}1)$ reflects the number of external rotational and vibrational motions, i.e., those rotational and vibrational motions that are affected by the presence of neighboring molecules.

[33] Here, "external" means influenced by density. Prigogine draws a (somewhat arbitrary) line between those degrees of freedom that are affected by density and those that are not. The latter are called internal; they depend only on temperature. External motions have relatively high amplitudes and low frequencies. Internal motions have relatively small amplitudes and high frequencies.

Unfortunately, this expression proposed by Prigogine for q_{ext} does not satisfy some important boundary conditions. First, the ideal gas-limit should be obeyed, i.e. for $V \rightarrow \infty$, $(V_f/\Lambda^3)q_{\text{ext}} = V/\Lambda^3$. Second, when free volume disappears as the system approaches closest–packing volume V_o, the molecules have no external degrees of freedom, i.e. for $V \rightarrow V_o$, $(V_f/\Lambda^3)q_{\text{ext}} = 0$.[34] Therefore, Beret (1975) proposed the function

$$q_{\text{ext}}(V) = \left(\frac{V_f}{V}\right)^{c-1} \tag{7-221}$$

Although this expression meets all necessary boundary conditions, it was modified by Donohue (1978) to account also for the temperature dependence of q_{ext}:

$$q_{\text{ext}}(V,T) = \left[\frac{V_f}{V}\exp\left(\frac{-E_o}{2kT}\right)\right]^{c-1} \tag{7-222}$$

Substituting either Eq. (7-220) or Eq. (7-221) or Eq. (7-222) for q_{ext} into Eq. (7-215) gives a *generalized van der Waals partition function* for polyatomic pure fluids.

Extension of the partition function to mixtures is possible using, e.g., the fundamental idea that a mixture can be considered to be a hypothetical fluid whose characteristic properties are composition averages of the corresponding properties of the mixture's components (*one-fluid theory*). Assuming that the configurational properties of a mixture are given by the one-fluid approximation, the *generalized van der Waals partition function* $Q(T,V,N_i)$ for a mixture containing N_1, N_2, ..., N_m molecules of components 1, 2, ..., m is

$$Q(T,V,N_i) = \frac{1}{\prod_{i=1}^{m}(N_i!)}\left[\prod_{i=1}^{m}\left(\frac{V}{\Lambda_i^3}\right)^{N_i}\right]\left(\frac{\overline{V}_f}{V}\right)^{N}\left[\exp\left(\frac{-\overline{E}_o}{2kT}\right)\right]^{N}\prod_{i=1}^{m}[(\overline{q}_{\text{r,v}})_i]^{N_i} \tag{7-215a}$$

where $N = N_1 + N_2 + \ldots + N_m$ is the total number of molecules in the mixture. Functions Λ_i depend only on temperature and mass m_i. However, $\overline{V}_f$, $\overline{E}_o$, and $(\overline{q}_{\text{r,v}})_i$ are properties of the fluid mixture. $\overline{V}_f$ and $\overline{E}_o$ are composition averages of the corresponding pure-component properties calculated by mixing rules. For mixtures that may contain large polyatomic molecules, the procedure to account for the effect of density on rotational and vibrational contributions to the partition function, is the same as that

[34] Prigogine's original theory is restricted to liquids and liquid mixtures, including polymers, at low or moderate pressures.

for pure components. For component i in the mixture, $(\bar{q}_{r,v})_i$ is given by Eqs. (7-219) and (7-220) [or (7-221) or (7-222)], where V_f is now replaced by $\bar{V}_f$, E_0 by $\bar{E}_0$ and c by c_i. With these changes, substitution of either Eq. (7-220) or Eq. (7-221) or Eq. (7-222) for q_{ext} into Eq. (7-215a) gives a *generalized van der Waals partition function* for mixtures containing simple and/or polyatomic fluids.

To reduce the partition function to practice, we require expressions for free volume V_f and potential energy E_0. For mixtures we need, in addition, mixing rules (e.g. those based on the one-fluid theory) for calculating the composition-dependent properties $\bar{V}_f$ and $\bar{E}_0$ of the mixture. Depending on the particular expression used for q_{ext} and those for V_f and E_0, the generalized van der Waals partition function leads to a variety of models, e.g., the *perturbed-hard-chain* theory discussed in the next section and the *Prigogine-Flory-Patterson* theory discussed in Sec. 8.2. When $c = 1$, the partition function reduces to the classical van der Waals-type *perturbed-hard-sphere* theory.

7.16 Perturbed-Hard-Chain Theory

The serious limitations of Eq. (7-216) were known already in van der Waals' time; attempts to improve it date back to about 1900. However, it was not until the work of Percus and Yevick (1958) and the development of molecular simulation (use of computers to simulate behavior of an assembly of molecules), that a reliable expression for V_f became available, valid at low and at high densities. This expression is (Carnahan and Starling, 1969, 1972):

$$V_f = V \exp\left[\frac{\eta(3\eta - 4)}{(1-\eta)^2}\right] \tag{7-223}$$

where the reduced density $\eta = \eta_{cp}(v_0/v)$ and $v_0 = (\sigma^3/\sqrt{2})N_A$.[35] The significance of v_0 is that Nv_0/N_A is the smallest possible (close-packed) volume that can be occupied by N hard spheres of diameter σ.

In their development of the *perturbed-hard-chain* (PHC) theory, Beret (1975) and Donohue (1978) used this expression to account for molecular repulsion. For q_{ext} they used the expression given by Eq. (7-216). For the potential energy E_0, accounting for molecular attraction, they used an analytical expression obtained from molecular-simulation studies of Alder (Alder *et al.*, 1972) for molecules whose intermolecular forces are represented by the square-well potential:

[35] The reduced density may also be written $\eta = b/4v$, the ratio (volume of molecules)/(volume). The upper limit of v_0/v is unity. The upper limit of η is $\eta_{cp} = \pi\sqrt{2}/6 = 0.7405$, the packing factor for hexagonal closest packing. Here, subscript cp stands for closest packing.

$$\frac{E_o}{2kT} = \sum_{n=1}^{4} \sum_{m=1}^{M} \frac{mA_{nm}}{\tilde{T}^n \tilde{v}^m} \tag{7-224}$$

Unlike the simple expression used by van der Waals [Eq. (7-217)], potential energy E_o given by Eq. (7-224) depends not only on volume but also on temperature. The 24 constants A_{nm} are known numbers obtained from computer-simulation data (upper limit M in the summation term depends on summation index n: $n = 1$, $M = 6$; $n = 2$, $M = 9$; $n = 3$, $M = 5$; $n = 4$, $M = 4$). Here, reduced temperature $\tilde{T} = T/T^* = ckT/\varepsilon q$ where q is the nondimensional external area of the molecule[36] ($q = 1$ for a single arbitrarily chosen reference segment), and ε is the characteristic segment-segment potential energy. Reduced volume $\tilde{v} = v/rv^* = v\sqrt{2}/N_A r\sigma^3$ where v^* is the characteristic hard-core volume per segment[37] and r is the number of segments per molecule.

The resulting equation of state is

$$\frac{\tilde{P}\tilde{v}}{\tilde{T}} = \frac{1}{c} + \frac{4\eta - 2\eta^2}{(1-\eta)^3} + \frac{1}{\tilde{T}\tilde{v}} \sum_{n=1}^{4} \sum_{m=1}^{M} \left(\frac{mA_{nm}}{\tilde{v}^{m-1}}\right)\left(\frac{1}{\tilde{T}^{n-1}}\right) \tag{7-225}$$

where reduced pressure $\tilde{P} = P/P^* = P(rv^*)/\varepsilon q$. The relationship between c and the characteristic quantities P^*, T^* and rv^* is $c = P^*(rv^*)/kT^*$. For each pure fluid, the PHC equation of state contains three molecular parameters εq, rv^* and c obtained from fitting experimental data, usually volumetric and vapor-pressure data.

As shown by Donohue (1978), Kaul *et al.* (1980) and Cotterman *et al.* (1986), PHC theory can be extended to mixtures using mixing rules (e.g. those based on the one-fluid theory) for calculating the composition average of the characteristic parameters $\overline{P}^*$, $\overline{T}^*$ and $\overline{rv}^*$, of the mixture. Liu (1980) showed that it can be used for mixtures of fluids with large differences in molecular size and shape (e.g., ethylene and polyethylene), and also for mixtures whose components differ appreciably in the magnitude and nature of the intermolecular forces.

To illustrate the wide range of applicability of the PHC theory, Fig. 7-29 shows a qualitative plot of molecular complexity versus density from zero density (ideal gas) to liquid-like density. The complexity scale starts with a simple molecule (argon or methane) and rises to a complex molecule, e.g., polyethylene.

PHC theory interpolates between existing knowledge along three edges of the diagram. At low densities, PHC gives a reasonable second virial coefficient and at very low density, it reduces to the ideal-gas law. For negligible molecular complexity ($c = 1$),

---

[36] Parameter q is not to be confused with contribution q to the partition function.

[37] In the PHC equations, $v = V/N$ is the volume *per molecule*. Data reduction for pure fluids gives T^*, P^* and rv^*; r and v^* always appear as a product.

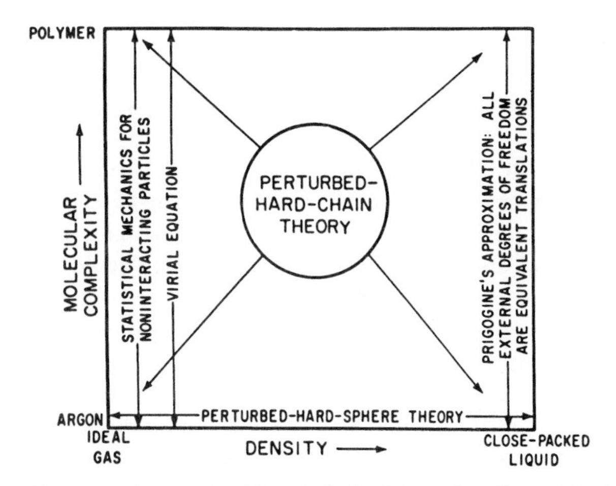

Figure 7-29 Domain of perturbed-hard-chain theory. In effect, this theory interpolates with respect to molecular complexity and fluid density using known results as boundary conditions: ideal-gas limit, perturbed-hard-sphere theory and Prigogine's theory for liquid polymers.

it reduces to perturbed-hard-sphere theory. For high densities, PHC is essentially identical to the Prigogine-Flory-Patterson theory of polymer liquids.

PHC theory provides a useful tool for calculating thermodynamic properties of different types of systems because it is applicable to mixtures of fluids with large differences in molecular size and shape (see Figs. 8-20 and 10-15), and also for mixtures whose components differ appreciably in the magnitude and nature of their intermolecular forces. However, because the attractive term [Eq. (7-224)] uses a fourth-order perturbation expansion, practical calculations require time-consuming computer iterations.

To simplify the equation of state and make it more useful for engineering work, several modifications have been proposed (e.g., Gmehling *et al.*, 1979; Donohue and Vimalchand, 1988; Elliot *et al.*, 1990; van Pelt *et al.*, 1993). Some of these modifications are briefly discussed in Sec. 12.9 where they are used to calculate fluid-phase equilibria at high pressures.

7.17 Hard-Sphere-Chain Models

Since the early 1980's, there has been increased interest in developing an equation of state for pure fluids and mixtures of large polyatomic molecules that does not rely on a lattice description of molecular configurations. A rigorous statistical-mechanical theory for large polyatomic molecules in continuous space is difficult because of their asymmetric structure, large number of internal degrees of freedom, and strong coupling between intra- and intermolecular interactions. Nevertheless, a relatively simple model represents chain-like molecules as freely-jointed tangent hard spheres (Dickman and Hall, 1986; Wertheim, 1987; Chapman *et al.*, 1988; Honnell and Hall, 1989; Chiew, 1990; Mitlin and Sanchez, 1993; Thomas and Donohue, 1993; Song *et al.*, 1994).

A hard-sphere-chain (HSC) equation of state can be used as the reference system in place of the hard-sphere reference used in most existing equations of state for simple fluids. Despite their simplicity, hard-sphere-chain models take into account some significant features of real fluids containing chain-like molecules including excluded-volume effects and chain connectivity.[38]

To describe the properties of fluids consisting of large polyatomic molecules, it is necessary to introduce attractive forces by adding a perturbation to a HSC equation of state. Because the influence of attractive forces on fluid structure is weak, a van der Waals-type or other mean-field term (e.g. square-well fluids) is usually used to add attractive forces to the reference hard-sphere-chain equation of state. Although numerous details are different, most hard-sphere-chain-based equations of state follow from statistical-mechanical perturbation theory; therefore, the equation of state can be written as

$$z = \frac{P}{\rho RT} = z_{\text{ref}} + z_{\text{pert}} = \left(\frac{P}{\rho RT}\right)_{\text{ref}} + \left(\frac{P}{\rho RT}\right)_{\text{pert}} \tag{7-226}$$

where P is pressure, $\rho = N/(VN_A)$ is the molar density, N is the number of molecules, V is the volume of the system, and T is temperature. In Eq. (7-226), the first term represents the reference equation of state, here taken as a fluid of hard-sphere chains $z_{\text{ref}} = z_{\text{HSC}}$, and the second term is the perturbation to account for attractive forces.

A fortunate feature of some hard-sphere-chain-based theories is that the reference equation of state can be extended to hard-sphere-chain mixtures without using mixing rules. Only attractive terms require mixing rules.

We now summarize the reference and perturbation terms for two HSC-based equations of state.

Statistical Associated-Fluid Theory

The *statistical associated-fluid theory* (SAFT) (Chapman *et al.*, 1989, 1990) is based on the first-order perturbation theory of Wertheim (1987). The essence of this theory is that the residual Helmholtz energy[39] is given by a sum of expressions to account not

[38] In a hard-sphere reference system, each sphere is free to move independently subject only to the restrictions of free volume. In a hard-sphere-chain reference system, each sphere (segment) is connected to at least one other sphere; therefore, the spheres (segments) cannot move independently. This lack of independence is denoted by *chain connectivity*.

[39] The definition of residual Helmholtz energy is: $a^R(T,V,N) = a(T,V,N) - a^{\text{id}}(T,V,N)$, where superscript id refers to an ideal gas. See App. B.

only for the effects of short-range repulsions and long-range dispersion forces but also for two other effects: chemically bonded aggregation (e.g. formation of chemically stable chains) and association and/or solvation (e.g. hydrogen bonding) between different molecules (or chains).[40]

For a pure component, Fig. 7-30 shows a three-step process for formation of stable aggregates (e.g. chains) and subsequent association of these aggregates. Initially, a fluid consists of equal-sized, single hard spheres. In the first step, intermolecular attractive forces are added, described by an appropriate potential, such as the square-well potential (Sec. 5.5). Next, each sphere is given one, two or more "sticky spots", such that the spheres can stick together (covalent bonding) to form dimers, trimers and higher stable aggregates such as chains. Finally, specific interaction sites are introduced at some position in the chain such that two chains can associate through some attractive interaction (e.g. hydrogen bonding.) Each step provides a contribution to the Helmholtz energy.

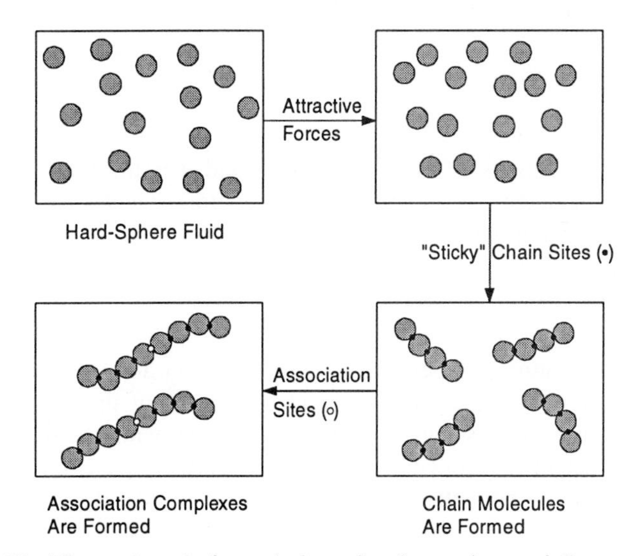

Figure 7-30 Three steps to form chain molecules and association complexes from hard spheres in the SAFT model.

In SAFT, the residual molar Helmholtz energy a^R has contributions from formation of hard spheres and chains, from dispersion (attraction), and from association:

$$a^R = a_{\text{hs}} + a_{\text{chain}} + a_{\text{disp}} + a_{\text{assoc}} \tag{7-227}$$ [41]

---

[40] The literature on SAFT is complex and confusing. The original article by Wertheim, while brilliant, is essentially incomprehensible. Much patience is required to understand what SAFT is, what it can and what it cannot do.

[41] A contribution a_{hs} is included here because of the definition of a^R given in Footnote 39.

Here the sum of the first two terms is the hard-sphere-chain reference system accounting for molecular repulsion and chain connectivity (chemical bonding); the sum of the last two terms is the perturbation accounting for molecular attraction and for association due to specific interactions like hydrogen bonding.

For a pure fluid, formation of one mole of chain molecules, each consisting of r segments, requires r moles of hard spheres. Therefore, prior to chain formation, the *hard-sphere* Helmholtz energy a_{hs}, is (Huang and Radosz, 1990)

$$\frac{a_{hs}}{RT} = r\frac{a'_{hs}}{RT} = r\frac{4\eta - 3\eta^2}{(1-\eta)^2} \tag{7-228}$$

where a'_{hs} is the Helmholtz energy of one mole of the hard-sphere fluid without chain connectivity, derived by Carnahan and Starling (1969). In Eq. (7-228), r is an adjustable pure-component parameter and η, the reduced density (segment packing fraction), is

$$\eta = \frac{\pi N_A}{6}\rho r d^3 \qquad \text{(7-229)}[42]$$

where ρ is the molar density of chain molecules and d is the temperature-dependent effective segment diameter.[43] Based on the work of Barker and Henderson (1967), using a square-well potential, Chen and Kreglewski (1977) obtained

$$d = \sigma[1 - C\exp(-3\varepsilon / kT)] \tag{7-230}$$

where σ is the temperature-independent diameter of a hard-sphere segment (it is d at $T = 0$ K) and ε is the temperature-independent well depth of the square potential. Because Chen and Kreglewski assumed that the ratio of the width of the square-well potential to σ is a constant, σ and ε are the only adjustable pure-component parameters with $C = 0.12$ for all components (Huang and Radosz, 1990). Because a characteristic volume, rather than diameter, is usually selected as a pure-component parameter (e.g., b in van der Waals-type equations of state, v^* in perturbed-hard-chain theory), a temperature-independent segment molar volume v_0 is introduced and used as adjustable parameter:

$$v_0 = \frac{\pi N_A}{6\eta_{cp}}\sigma^3 \tag{7-231}$$

[42] Equation (7-229) gives a definition of η similar but not identical to that used in Eq. (7-223).

[43] A temperature-dependent hard-sphere diameter is to allow some overlap of spheres at high temperatures, that is, to convert hard spheres to soft spheres, for example, in the Lennard-Jones potential.

where η_{cp} is the upper limit (closest packing) of the reduced density. v_0 is the smallest possible (closest packed) volume that can be occupied by N_A hard-sphere segments of diameter σ.

The contribution from *chain formation* a_{chain} is (Huang and Radosz, 1990)

$$\frac{a_{chain}}{RT} = (1-r)\ln\frac{2-\eta}{2(1-\eta)^3} \tag{7-232}$$

where η is the reduced density given by Eq. (7-229). Equation (7-232) follows from Wertheim's association theory (Chapman *et al.*, 1990), where the association bonds are replaced by covalent, chain-forming bonds.[44] It is striking that a_{chain} is calculated from Eq. (7-232) using three pure-component parameters r, ε and v_0 identical to those used for calculating the hard-sphere contribution z_{hs} from Eq. (7-228). No additional parameter is necessary to account for chain connectivity.

For the *dispersion term* a_{disp}, Huang and Radosz (1990) used an analytical expression initially obtained by Alder *et al.* (1972) from fitting molecular-simulation data for a square-well fluid. Alder's expression, also the basis for the perturbed-hard-chain theory discussed in Sec. 7.16, is (per mole of chain molecules each consisting of r segments)

$$\frac{a_{disp}}{RT} = r\sum_n\sum_m D_{nm}\left(\frac{u}{kT}\right)^n\left(\frac{\eta}{\eta_{cp}}\right)^m \tag{7-233}$$

where η is the reduced density given by Eq. (7-229) with upper limit $\eta_{cp} = \pi\sqrt{2}/6$ at closest packing and u is the temperature-dependent depth of the square-well potential describing the nonspecific segment-segment interactions. As given by Chen and Kreglewski (1977), the temperature dependence of u is

$$u = \varepsilon(1 + e/kT) \tag{7-234}$$

where ε is the temperature-independent depth of the square-well potential. Constant e/k in Eq. (7-234) has been related to Pitzer's acentric factor and to critical temperature (Kreglewski, 1984) for various molecules.[45] In SAFT, the energy parameter u is for segments, not for molecules (Huang and Radosz, 1990). In Eq. (7-233), D_{nm} are universal constants fitted to accurate P-V-T, internal energy, and second-virial-coefficient data for argon (Chen and Kreglewski, 1977).

---

[44] Chain formation is athermal; the spheres join together to form chains with no release of energy.

[45] For small molecules (e.g. argon or methane), e/k is set equal to zero. Then u is independent of temperature.

The *association contribution* a_{assoc} in Eq. (7-227) is also obtained from Wertheim's association theory.[46] The number of association sites on a single molecule is unlimited but it must be specified. The location of the association sites, however, is not specified. The various types of sites S are labeled A, B, C, ..., to keep track of the specific site-site interactions. As an example, Fig. 7-31 shows a monomer and a chain molecule, each with two different association sites, A and B. Each association site is assumed to have different interaction with the various sites on another molecule. For example, suppose A is an electron donor (Lewis base) and B is an electron acceptor (Lewis acid). In that event, we expect association whenever an A site on one molecule interacts (attraction) with a B site on another molecule. Geometric factors may hinder site-site interactions. Cluster-structure limitations, steric-hindrance approximations, and size distribution are discussed by Chapman *et al.* (1990).

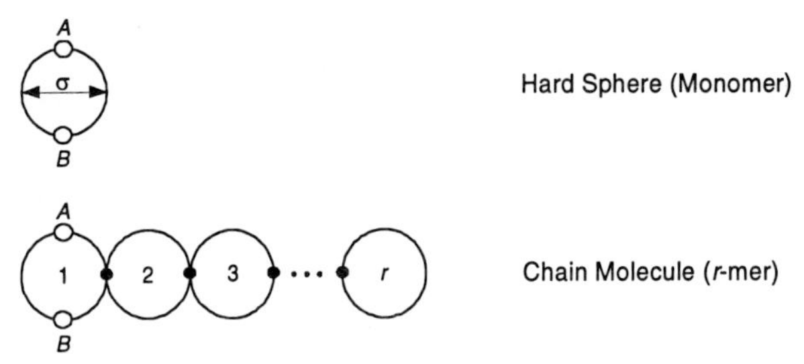

Figure 7-31 Models for a hard-sphere (monomer) and a hard-chain molecule (*r*-mer; • are chemical bonds), with two association sites *A* and *B*. The chain model can represent nonspherical molecules. For example, if sites A and B attract each other, various associated complexes can be formed. This picture is not altered by converting hard spheres to soft spheres according to Eq. (7-230).

The expression for a_{assoc} is

$$\frac{a_{\text{assoc}}}{RT} = \sum_S \left(\ln X^S - \frac{X^S}{2} \right) + \frac{M}{2} \qquad \text{(7-235)}^{47}$$

where M is the number of association sites on each molecule, X^S is the mole fraction of molecules *not* bonded at specific interaction (association) site S, and the summation is over all association sites S on a molecule. X^S is given by

$$X^S = \left\{ 1 + N_A \sum_Y \rho X^Y \frac{2-\eta}{2(1-\eta)^3} (\sigma^3 \kappa^{SY}) \left[\exp\left(\frac{\varepsilon^{SY}}{kT} \right) - 1 \right] \right\}^{-1} \qquad \text{(7-236)}$$

[46] In contrast to chain formation, association of molecules (e.g. through hydrogen bonding) is *not* athermal. When molecules associate, energy is released.

[47] In Eq. (7-235), M is not to be confused with M in Eq. (7-224).

where the summation is over all different types of sites $Y = A, B, C, \ldots$; N_A is Avogadro's constant and ρ, η and σ have the same meanings as before. The two association parameters ε^{SY} and κ^{SY} characterize, respectively, the association energy and the association volume for the specific interaction between association sites S and Y. The temperature-independent segment diameter σ, used in Eq. (7-236) is related to v_0 through Eq. (7-231).

Substitution of the expressions for a_{hs}, a_{chain}, a_{disp}, and a_{assoc} in Eq. (7-227) gives the residual molar Helmholtz energy for a pure chain fluid. Application of Eq. (7-60) gives the SAFT equation of state for pure fluids. We can then write for the compressibility factor of a real fluid

$$z = \frac{P}{\rho RT} = z^{\mathrm{id}} + z_{\mathrm{hs}} + z_{\mathrm{chain}} + z_{\mathrm{disp}} + z_{\mathrm{assoc}} \tag{7-237}$$

with $z^{\mathrm{id}} = 1$ and

$$z_{\mathrm{hs}} = r z'_{\mathrm{hs}} = r\frac{4\eta - 2\eta^2}{(1-\eta)^3} \tag{7-238}$$[48]

$$z_{\mathrm{chain}} = (1-r)\frac{5\eta - 2\eta^2}{(1-\eta)(2-\eta)} \tag{7-239}$$

$$z_{\mathrm{disp}} = r\sum_n \sum_m m D_{nm}\left(\frac{u}{kT}\right)^n \left(\frac{\eta}{\eta_{\mathrm{cp}}}\right)^m \tag{7-240}$$

$$z_{\mathrm{assoc}} = \rho \sum_S \left(\frac{1}{X^S} - \frac{1}{2}\right)\frac{\partial X^S}{\partial \rho} \tag{7-241}$$

The sum of the first three terms on the right side in Eq. (7-237) is the compressibility factor z_{HSC} of the hard-chain reference system accounting for molecular repulsion and chain connectivity; the sum of the last two terms is the perturbation contribution z_{pert} that accounts for molecular attraction and for association of different molecules (chains) due to specific interactions.

In the SAFT equation of state for pure chain fluids there are three essential adjustable parameters for nonassociating components: r, the number of segments per molecule; v_0, the segment molar volume at closest packing; and ε, the temperature-

[48] The hard-sphere contribution z'_{hs} for a hard-sphere fluid without chain connectivity can also be obtained by substituting Eq. (7-223) into Eq. (7-215) with $E_0 = 0$ (no attraction) and $q_{r,v} = 1$ (no rotational and vibrational degrees of freedom) using Eq. (7-209).

independent depth of the square-well potential characteristic of segment-segment interactions.[49] For associating components, there are two additional parameters: the association energy ε^{SY} and the association volume κ^{SY}, characterizing the specific interaction between association sites S and Y.

For mixtures, the general form of the SAFT equation of state is the same as that for pure fluids [Eq. (7-227)]. The hard-sphere contribution for mixtures is based on the theoretical result of Mansoori *et al.* (1971) for the equation of state for hard-sphere mixtures:

$$z_{\text{hs}} = \frac{6}{\pi N_A \rho}\left[\frac{\xi_0 \xi_3}{1-\xi_3} + \frac{3\xi_1 \xi_2}{(1-\xi_3)^2} + \frac{(3-\xi_3)\xi_2^3}{(1-\xi_3)^3}\right] \tag{7-242}$$

with

$$\xi_k = \frac{\pi N_A \rho}{6}\sum_{i=1}^{m} x_i r_i (d_i)^k \qquad k = 0, 1, 2, 3 \tag{7-243}$$

where ρ is the total molar density, x_i is the mole fraction of component i, r_i is the number of segments per molecule i, and d_i is the temperature-dependent segment diameter.

The contribution accounting for the formation of chain molecules of the various components in the mixture is

$$z_{\text{chain}} = \sum_{i=1}^{m} x_i (1-r_i) L(d_i) \tag{7-244}$$

with

$$L(d_i) = \frac{2\xi_3 + 3d_i\xi_2 - 4\xi_3^2 + 2d_i^2\xi_2^2 + 2\xi_3^3 + d_i^2\xi_2^2\xi_3 - 3d_i\xi_2\xi_3^2}{(1-\xi_3)(2-4\xi_3 + 3d_i\xi_2 + 2\xi_3^2 + d_i^2\xi_2^2 - 3d_i\xi_2\xi_3}$$

where ξ_k $(k = 2,3)$ is given by Eq. (7-243). For a pure component, $d_i\xi_2 = \xi_3 = \eta$ given by Eq. (7-229); Eq. (7-244) then reduces to the pure-component expression given by Eq. (7-239). It is remarkable that *no* mixing rules are necessary in Eqs. (7-242) and (7-244). For a mixture of chain fluids, the compressibility factor of the hard-sphere-chain reference system, $z_{\text{HSC}} = 1 + z_{\text{hs}} + z_{\text{chain}}$ can be calculated from the pure-component parameters r_i, ε_i and v_{oi}.[50] Similarly, z_{assoc} can be derived rigorously from

[49] If the hard spheres are converted to soft spheres, there is an additional (softness) parameter C as in Eq. (7-230). If the well depth ε is temperature dependent [as suggested in Eq. (7-240)], there is an additional parameter e/k as in Eq. (7-234).

[50] v_{oi} gives σ_i from Eq. (7-231) and then, together with ε_i and parameter C, Eq. (7-230) gives d_i. If the well depth ε_i is temperature dependent, according to Eq. (7-234) there is an additional parameter e_i/k.

statistical mechanics (Chapman *et al.*, 1990). The relation obtained is a mole-fraction average of the corresponding pure-component equations:

$$z_{\text{assoc}} = \rho \sum_{i=1}^{m} x_i \left[\sum_{S_i} \left(\frac{1}{X^{S_i}} - \frac{1}{2} \right) \frac{\partial X^{S_i}}{\partial \rho} \right] \tag{7-245}$$

where X^{S_i}, the mole fraction of molecules i in the mixture *not* bonded with other components at site S, is given by

$$X^{S_i} = \left(1 + N_{\text{Av}} \sum_{j=1}^{m} \sum_{Y_j} x_j \rho X^{Y_j} W_{ij} \right)^{-1} \tag{7-246}$$

with

$$W_{ij} = \left[\frac{1}{1-\xi_3} + \frac{3d_i d_j}{d_i + d_j} \frac{\xi_2}{(1-\xi_3)^2} + 2\left(\frac{d_i d_j}{d_i + d_j} \right)^2 \frac{\xi_2^2}{(1-\xi_3)^3} \right] \left(\sigma_{ij} \kappa^{S_i Y_j} \right) \left[\exp\left(\frac{\varepsilon^{S_i Y_j}}{kT} \right) - 1 \right]$$

where, as before, ξ_k $(k = 2,3)$ is given by Eq. (7-243) and $\sigma_{ij} = (\sigma_i + \sigma_j)/2$ because hard-sphere diameters are additive. In Eq. (7-246), summation Σ_{Y_j} is over all specific interaction sites on molecule j and summation Σ_j is over all m components.

The association/solvation parameter $\varepsilon^{S_i Y_j}$ and the dimensionless parameter $\kappa^{S_i Y_j}$ characterize, respectively, the association ($i = j$) and solvation ($i \neq j$) energy and volume for the specific interaction between sites S and Y. These parameters are adjustable. Equation (7-246) requires no mixing rules. However, mixing rules are needed to extend Eq. (7-240) to mixtures. For mixtures, z_{disp} depends on the molecular size parameter r and on the segment interaction energy parameter u/kT for the mixture. Huang and Radosz (1991) proposed mixing rules for calculating the composition dependence of these parameters from the corresponding pure-component parameters. The mixing rules of Huang and Radosz are based on the van der Waals one-fluid approximation. However, for the molecular-energy parameter u, they have also proposed a volume-fraction approximation. In either case, only one adjustable binary parameter is used for the energy parameter $u_{ij} = (u_{ii} u_{jj})^{1/2} (1 - k_{ij})$.

The SAFT equation of state has been applied successfully to describe thermodynamic properties and phase behavior of pure fluids and fluid mixtures containing small, large, polydisperse, nonassociating and associating molecules, including supercritical and near-critical solutions of polymers (Huang and Radosz, 1990, 1991; Chen and Radosz, 1992; Gregg *et al.*, 1993; Wu and Chen, 1994; Chen *et al.*, 1992, 1993, 1994, 1995).

The computational requirements for SAFT applications are similar to those for other noncubic equations of state such as the perturbed-hard-chain theory. Pure component parameters for a large variety of nonassociating and associating real fluids have been tabulated (Huang and Radosz, 1990). The two different sets of mixing rules used by Huang and Radosz have been tested to correlate vapor-liquid equilibria (VLE) of fluid mixtures (Huang and Radosz, 1991). For low-pressure VLE of nonassociating and associating binary mixtures, both sets of mixing rules correlated the experimental data well. To illustrate, Fig. 7-32 shows VLE at 323 K for the mixture propanol/*n*-heptane. There is good agreement between experiment and SAFT calculations including the azeotrope.

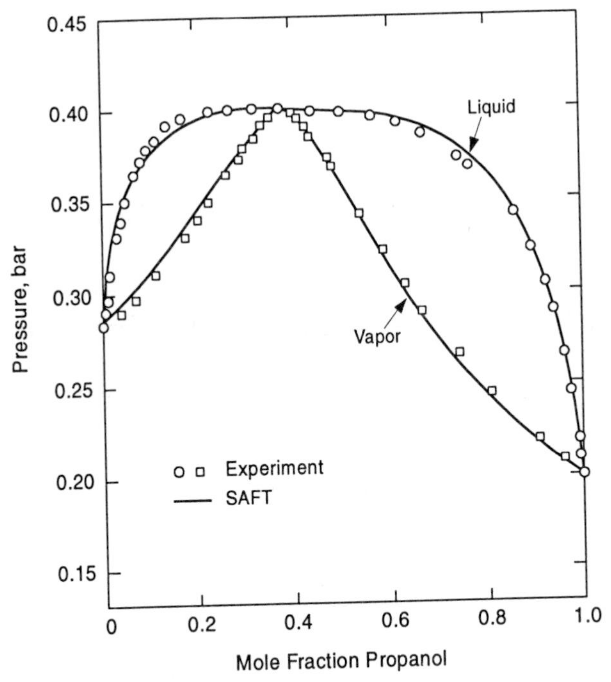

Figure 7-32 Vapor-liquid equilibria at 323 K for the mixture propanol/*n*-heptane (Fu and Sandler, 1995).

High-pressure VLE can also be correlated well with SAFT, except near the critical region. Although the volume-fraction mixing rule represents experiment near critical conditions better than the van der Waals one-fluid mixing rules, the critical region is overpredicted at high pressures and high temperatures as illustrated in Fig. 7-33 for the system CO_2/2-propanol. At 298 K, SAFT prediction agrees well with experiment including the critical region. However, at 394 K agreement is satisfactory only at low pressures.

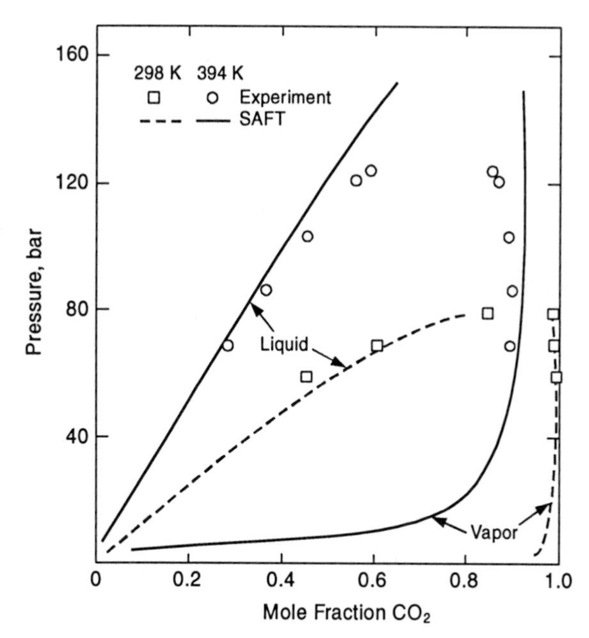

Figure 7-33 Vapor-liquid equilibria at 298 K and 394 K for the system CO_2/2-propanol. SAFT calculations were performed with the volume-fraction mixing rule (Huang and Radosz, 1991).

In general, SAFT correlates VLE experimental data with good accuracy, but significant deviations occur at low pressures and high temperatures, or at high pressures and low temperatures. Such deviations are not surprising because fundamentally, SAFT is a van der Waals-type theory; it has been known for many years that equations of state (or partition functions) of the van der Waals form are based on physical assumptions that are valid only at (relatively) high temperatures and high densities (Kipnis *et al.*, 1996).

Numerous modifications have been proposed either to simplify the SAFT equation of state (e.g., Fu and Sandler, 1995; Blas and Vega, 1998) or to improve it further, e.g., for mixtures containing water (Kraska and Gubbins, 1996; Economou and Tsonopoulos, 1997). A simplified version has been developed by Galindo *et al.* (1996, 1997), suitable for correlating high-pressure phase equilibria of aqueous mixtures. Wu (1998) has presented a SAFT-based extension of the widely used Peng-Robinson equation of state, suitable for calculating vapor-liquid equilibria of systems containing hydrocarbons, water and salt. However, the SAFT equation cannot be used for liquid-liquid equilibria in aqueous systems at normal temperatures because it cannot account for the hydrophobic effect where the dissolution of a nonpolar solute (e.g. methane or ethane) causes a significant change in the molecular structure of water (see Sec. 4.10). The hydrophobic effect tends to disappear at high temperatures say, above 200°C.

The SAFT equation of state and its modifications have been successfully used to model thermodynamic properties and phase behavior for a large variety of simple and complex fluids and fluid mixtures. In Sec. 8.2 we give some examples of the application of SAFT to polymer solutions.

Perturbed Hard-Sphere-Chain Theory

The reference part of the *perturbed hard-sphere-chain* (PHSC) theory equation of state is based on a generalization (Song *et al.*, 1994) of the Percus-Yevick integral-equation theory for hard-sphere chains as obtained by Chiew (1990). A simple van der Waals-type term is used for the perturbation. The PHSC equation of state for pure fluids is (Song *et al.*, 1994)

$$\begin{aligned} z &= \frac{P}{\rho RT} = 1 + z_{hs} + z_{chain} + z_{pert} \\ &= 1 + r\frac{4\eta - 2\eta^2}{(1-\eta)^3} + (1-r)\left[\frac{2-\eta}{2(1-\eta)^3} - 1\right] - \frac{r^2 a\rho}{RT} \end{aligned} \tag{7-247}$$

where the first three terms represent the reference equation of state for hard-sphere chains, while the last term is the van der Waals perturbation for attractive forces. Here, ρ is molar density. The PHSC theory does not include the effect of association.

Because parameter r and reduced density η are the same as those in the SAFT reference equation of state, contribution z_{hs} is also the same; however, contribution z_{chain} is not. In the PHSC theory, parameter $b = (2\pi/3)d^3$ is used instead of segment diameter d to calculate reduced density η defined by Eq. (7-229); η is now given by

$$\eta = \frac{\rho N_A}{4} rb \tag{7-248}$$

In Eq. (7-247), parameter a represents the strength of attractive forces between two non-bonded segments. In PHSC theory, a and b are temperature dependent; they are given by (Song *et al.*, 1994)

$$a(T) = \frac{2}{3}\pi\sigma^3 \varepsilon F_a(kT/\varepsilon) \tag{7-249}$$

$$b(T) = \frac{2}{3}\pi d^3(T) = \frac{2}{3}\pi\sigma^3 F_b(kT/\varepsilon) \tag{7-250}$$

Equations (7-249) and (7-250) follow from the method of Song and Mason (1989) that scales b and a in terms of two potential parameters ε and σ; ε is the well depth of the non-bonded segment-segment pair potential and σ is the distance of separation at the minimum of the potential. In Eqs. (7-249) and (7-250), F_a and F_b are two universal functions of a scaled temperature kT/ε. Because accurate experimental values of ε and σ exist for methane and argon, the experimental thermodynamic properties of these two fluids are used to determine the single-sphere universal functions F_a and F_b; they can be accurately represented by the following empirical formulas:[51]

$$F_a(kT/\varepsilon) = 1.8681\exp[-0.0619(kT/\varepsilon)] + 0.6715\exp[-1.7317(kT/\varepsilon)^{3/2}] \quad (7\text{-}251)$$

$$F_b(kT/\varepsilon) = 0.7303\exp[-0.1649(kT/\varepsilon)^{1/2}] + 0.2697\exp[-2.3973(kT/\varepsilon)^{3/2}] \quad (7\text{-}252)$$

In the PHSC equation of state for pure fluids, there are three adjustable parameters: r, σ and ε. Song *et al.* (1994a) have applied the PHSC equation of state to describe thermodynamic properties of normal fluids and polymers and made parameters available for a wide variety of polymers and normal fluids.

The PHSC equation of state is extended to mixtures according to

$$z = 1 + z_{\text{hs}} + z_{\text{chain}} - \frac{\rho}{RT}\sum_{i,j}^{m} x_i x_j r_i r_j a_{ij} \quad (7\text{-}253)$$

where z_{hs} of the hard-sphere mixture is the same as that in the SAFT equation of state, given by Eqs. (7-242) and (7-243) with $d_i^3 = (3/2\pi)b_i$. In Eq. (7-253), z_{chain}, the contribution accounting for the formation of chain molecules, is

$$z_{\text{chain}} = \sum_{i=1}^{m} x_i(1-r_i)\left[\frac{1}{1-\eta} + \frac{3}{2}\frac{\xi_i}{(1-\eta)^2} + \frac{1}{2}\frac{\xi_i^2}{(1-\eta)^3} - 1\right] \quad (7\text{-}254)$$

where the reduced density of hard-sphere mixtures is given by

$$\eta = \frac{\rho N_A}{4}\sum_{i=1}^{m} x_i r_i b_i \quad (7\text{-}255)$$

and

[51] A simplified version of the PHSC equation is presented here. In previous publications (e.g., Song *et al.*, 1994), the temperature dependencies of the universal functions were scaled by a parameter s, a function of chain length r only. In the present model, this scale factor is removed and the universal functions [Eqs. (7-251) and (7-252)] were determined from the thermodynamic properties of argon and methane over a wide temperature range. Removal of $s(r)$ from the universal functions allows use of simpler combining rules. The simpler rules do not sacrifice accuracy in fitting thermodynamic properties of pure (non-associating) fluids to obtain equation-of-state parameters.

$$\xi_i = \frac{\rho N_A}{4} b_i^{1/3} \sum_{j=1}^{m} x_j r_j b_j^{2/3} \tag{7-256}$$

For one-component systems and for mixtures of equal-sized spheres, $\xi_i = \eta$.

As in Eq. (7-247), the first three terms in Eq. (7-253) represent the reference equation of state for hard-sphere-chain mixtures and the last term is a van der Waals-type perturbation for attractive forces. Like in the SAFT equation of state, for a mixture of chain fluids, the compressibility factor of the hard-sphere-chain reference system, $z_{\text{HSC}} = 1 + z_{\text{hs}} + z_{\text{chain}}$, can be calculated from pure-component parameters r_i, ε_i and b_i only.

However, for each unlike pair of components ($i \neq j$), an additional parameter a_{ij} is needed for the mixture. Its physical meanings is similar to that for pure fluids: a_{ij} reflects the strength of attractive forces between two nonbonded segments i and j. An expression for a_{ij} is obtained by extending Eq. (7-249) to mixtures:

$$a_{ij}(T) = \frac{2}{3}\pi\sigma_{ij}^3\varepsilon_{ij}F_a(kT/\varepsilon_{ij}) \tag{7-257}$$

where σ_{ij} is given by $\sigma_{ij} = (\sigma_i + \sigma_j)/2$ and ε_{ij} is related to the pair potential between unlike segments, given by

$$\varepsilon_{ij} = (\varepsilon_{ii}\varepsilon_{jj})^{1/2}(1 - k_{ij}) \tag{7-258}$$

Therefore, only one binary adjustable parameter k_{ij} is required.

Although the PHSC equation of state can be used to model thermodynamic properties and phase behavior of normal fluids, polymers and their mixtures, it has been primarily applied to polymer mixtures. Song *et al.* (1994a) found that, for nonassociated mixtures, the PHSC equation of state performs as well as the SAFT equation of state, as expected, because these two models share many basic ideas. Section 8.3 presents examples that illustrate the application of the PHSC equation of state to polymer mixtures.

A second-order perturbation theory based on the hard-sphere-chain reference system is given by Hino (1997). Because of the second-order correction, it is better than Song's theory at low temperatures. Also, because Hino's results, based on the work of Chang and Sandler (1994), use a square-well potential of variable width, Hino's work is applicable to a wider variety of mixtures.

7.18 Summary

The theory of solutions is an old subject. Many of the articles in the first volumes of the *Zeitschrift fur Physikalische Chemie* (around 1890) are concerned with the properties of liquid mixtures; since the early days of physical chemistry, thousands of articles have been written in an effort to understand the behavior of mixed fluids. While much progress has been made, we are still far from an adequate theory of liquid mixtures. In this chapter we have indicated a few of the theoretical ideas that have been proposed and it is evident that none of them is sufficiently broad to apply to the general problem; rather, each idea and each model appears to be limited to a particular class of solutions. As a result, while we do not have a general theory of liquid mixtures, we have instead, a variety of restricted theories and models, each useful for a particular type of mixture.

To construct a *theory of liquid mixtures*, we require essentially two kinds of information: We need to know something about the structure of liquids (i.e., the way the molecules in a liquid are arranged in space), and we need to know something about intermolecular forces between like and unlike molecules. Unfortunately, information of either kind is inadequate and, as a result, all of our theories must make simplifying assumptions to overcome this disadvantage. Because simplifying assumptions must be made, it follows that we cannot at this time construct a *general* theory; simplifying assumptions that are reasonable for one type of mixture (e.g., mixtures of hydrocarbons) may be most unreasonable for another (e.g., aqueous mixtures of organic acids) and, because the simplifying assumptions must vary from one type of mixture to another, we inevitably have different theories and models for different applications: The punishment must fit the crime.

Most theoretical work has been concerned with mixtures of liquids whose molecules are nonpolar and spherical. Some of this work has been indicated here: Regular-solution theory, lattice theory and perturbation theory, all start out with simple molecules and are then extended, often semiempirically, to more complicated molecules. Recent theoretical work on mixtures of simple molecules (theories based on the radial distribution function) has not been discussed in detail because an adequate treatment requires more familiarity with statistical mechanics than we can give here. Some of these theories promise to contribute to our understanding of liquid structure, but they do not tell us anything about intermolecular forces between dissimilar molecules. To use these theories, we must include independent information on intermolecular forces.

For mixtures of nonpolar liquids, the *regular-solution theory* of Scatchard and Hildebrand frequently provides a good approximation for the excess Gibbs energy. The most serious simplifying assumption in this theory is the geometric-mean rule for the cohesive energy density of the unlike (1-2) interaction. For application, whenever possible, this geometric-mean rule should be modified empirically by utilizing whatever limited experimental binary data may be available. Two important advantages of the

Scatchard-Hildebrand regular-solution theory are its simplicity and its ease of extension to systems containing more than two components.

Regular-solution theory, like most theories of solution, is more reliable for excess Gibbs energy than for excess enthalpy and excess entropy. All simple theories of solution neglect changes in molecular vibration and rotation that result from the change of molecular environment that is inevitably produced by mixing; these changes, in some cases, affect the excess enthalpy and excess entropy in such a way that they tend to cancel in the excess Gibbs energy, at least to a first approximation.

The *lattice theory* of solutions, although first developed for monatomic molecules, can be extended to molecules of more complex structure using well-defined assumptions, as shown by Guggenheim, Flory, and others. This extension makes it particularly useful for solutions of molecules that differ appreciably in size, such as polymer solutions, as discussed in Chap. 8. However, the concept of a lattice for liquid structure is a vast oversimplification; and as a result, lattice theory becomes increasingly inappropriate as attention is focused on temperatures remote from the melting point. Also, for each binary system, lattice theory requires as an input parameter the interchange energy w (or its equivalent, the Flory parameter χ), that is difficult to predict and that, unfortunately, is temperature-dependent. Because the lattice concept is not truly appropriate for liquids, reduction of experimental data for real mixtures often gives an exchange energy that depends also on composition. In that event, lattice theory requires extensive empirical modifications.

Most simple theories of mixtures assume random mixing of molecules; for strongly nonideal mixtures, this assumption can lead to serious error. Although we do not have a rigorous theory of *nonrandom mixtures*, a fair description is sometimes obtained by using the *quasi-chemical approximation*. A semiempirical generalization of that approximation is provided by the concept of local concentration leading to the equations of Wilson, NRTL, and UNIQUAC. These equations do not have a precise theoretical basis but appear to be of a form that is particularly useful for solutions containing one or more polar components.

For those who favor a philosophy of idealism, it is attractive to do away with nonideality in solutions by claiming that our observations of nonideality are merely apparent, that all solutions are, in fact, ideal if only we use in our calculations the true, rather than the apparent, molecular concentrations. This idealistic view attributes all observed nonideality to formation of new chemical species in solution; by postulating association or solvation equilibria (or both) and then letting equilibrium constants be adjustable parameters, one can indeed fit experimental data for many liquid mixtures. The *chemical theory* of solutions permits us to fit experimental data for any liquid mixture, regardless of complexity, provided that we use a sufficient number of adjustable equilibrium and stoichiometric constants.

The chemical theory of solutions provides a sensible approximation whenever there is independent evidence that strong chemical forces operate in the liquid mixture; for example, whenever there is appreciable hydrogen bonding between like or unlike components (or both), it is reasonable to assume that the formation (or dissociation) of hydrogen-bonded molecules in solution provides the dominant contribution to the solution's thermodynamic properties. If chemical forces are strong, then physical (van der Waals) forces may often be neglected, at least for a first approximation, but careful study has shown that for accurate work both physical and chemical forces must be taken into account. However, the dividing line between physical and chemical forces cannot easily be determined with rigor and as a result, it is often necessary to make an essentially arbitrary decision on where that line is drawn.

The vagueness of the chemical theory of solutions provides a wide range of possible applications. However, we must beware of the strong temptation to use it where, because of other physical evidence, it is inapplicable. Any theory of solutions with a sufficient number of adjustable parameters must always be viewed with suspicion unless supported by independent physicochemical measurements. Nevertheless, when used judiciously, the chemical theory of solutions provides a useful framework for correlating and extending thermodynamic data for strongly nonideal solutions where currently available physical theories are inappropriate.[52]

Reasonable (but nevertheless approximate) theories are now available for mixtures that contain chain-like molecules in addition to "normal", essentially spherical (or globular) molecules. These theories (*perturbed hard chain, statistical associated fluid, perturbed hard chain of spheres*) have a wider range of applicability than those based on a (hole-free) lattice because they are based on equations of state that (unlike a hole-free lattice) give the segment density as a function of temperature, pressure and composition. Further, these equation-of-state theories can incorporate association between like molecules and solvation between unlike molecules. Regrettably, even for nonpolar fluids, these equations-of-state theories require several (typically 3 or 4) pure-component molecular parameters; if the molecules associate, additional pure-component parameters are needed. For simple cases (e.g. mixtures of hydrocarbons), one binary parameter may be sufficient but often two are needed; for solutions where solvation is important, additional binary parameters must be specified. The need for so many parameters follows from our inadequate understanding of intermolecular forces.

Ever-increasing impressive advances in molecular simulation are likely to exert a dramatic influence on our future ability to describe the properties of liquid mixtures. Such simulations not only serve to test the physical significance of analytic models but also to suggest new and improved models. Some day, when computers become even more powerful than they are now, it may be possible to calculate thermodynamic properties of liquid mixtures on a routine basis without any need for equations of state or other analytical models. But as of now (1998), that day is still far in the future.

In general, we may say that theories of solutions are mental crutches that enable us to order, interpret and in a vague sense "understand" thermodynamic data for mixtures. These theories provide a framework that enables us to correlate data in a sensible manner; they tell us what to plot against what, the coordinates we must use to obtain a

[52] A major deficiency of the Dolezalek chemical theory follows from its inability to account for phase separation (demixing). This deficiency, however, is removed by the extensions given in Secs. 7.11 and 7.12.

smooth (and perhaps even straight) line. For engineering work, such a framework is extremely useful because it enables us to interpolate and extrapolate limited experimental results and to make reasonable predictions for systems not previously studied, especially for those systems containing more than two components. Finally, however, it is important to remember, as Scatchard has pointed out,[53] that theories of solution are, essentially, working tools; we must not take any theory too seriously because real liquid mixtures are much more complicated than our oversimplified models.

To make progress, we must keep in mind the simplifying assumptions on which our theories are based, for, as Francis Bacon said many years ago, "Truth is more likely to emerge from error than from confusion."

References

Abrams, D. S. and J. M. Prausnitz, 1975, *AIChE J.*, 21: 16 .

Alder, B. J., D. A. Young, and M. A. Mark, 1972, *J. Chem. Phys.*, 56: 3013.

Aveyard, R. and R. W. Mitchell, 1970, *Trans. Faraday Soc.*, 66: 37.

Barker, J. A., 1963, *Lattice Theories of the Liquid State*. Oxford: Pergamon Press.

Barker, J. A. and D. Henderson, 1967, *J. Chem. Phys.*, 47: 4714.

Barton, A. F. M., 1991, *Handbook of Solubility Parameters*. Boca Raton: CRC Press.

Bazúa, E. R. and J. M. Prausnitz, 1971, *Cryogenics*, 11: 114.

Beret, S. and J. M. Prausnitz, 1975, *AIChE J.*, 21: 1123.

Blanks, R. F. and J. M. Prausnitz, 1964, *Ind. Eng. Chem. Fundam.*, 3: 1.

Blas, F. J. and L. F. Vega, 1998, *Ind. Eng. Chem. Res.*, 37: 660.

Burrell, H., 1955, *Interchem. Rev.*, 14: 3, 31.

Burrell, H., 1968, *J. Paint Technol.*, 40: 197.

Burrell, H., 1975, *Solubility Parameter Values*. In *Polymer Handbook*, 2nd Ed. (J. Brandrup, and E. H. Immergut, Eds.). New York: John Wiley & Sons.

Carnahan, N. F. and G. J. Yevick, 1958, *Phys. Rev.*, 110: 1.

Carnahan, N. F. and K. E. Starling, 1972, *AIChE J.*, 18: 1184.

Chang, J. and S. I. Sandler, 1994, *Mol. Phys.*, 81: 735, 745.

Chao, K. C. and G. D. Seader, 1961, *AIChE J.*, 7: 598.

Chapman, W. G., G. Jackson, and K. E. Gubbins, 1988, *Molec. Phys.*, 65: 1057.

Chapman, W. G., K. E. Gubbins, G. Jackson, and M. Radosz, 1989, *Fluid Phase Equilibria*, 52: 31.

Chapman, W. G., K. E. Gubbins, G. Jackson, and M. Radosz, 1990, *Ind. Eng. Chem. Res.*, 29: 1709.

[53] "The best advice that comes from years of study of liquid mixtures is to use any model insofar as it helps, but not to believe that any moderately simple model corresponds very closely to any real mixture" (1949, *Chem. Rev.*, 44: 7). This advice is a specific example of what A. N. Whitehead has called "the fallacy of misplaced concreteness," i.e., the tempting but erroneous habit of theorists to regard theoretical models as reality, rather than idealized representations of reality.

Chen, S. S. and A. Kreglewski, 1977, *Ber. Bunsenges. Phys. Chem.*, 81: 1048.

Chen, S.-J. and M. Radosz, 1992, *Macromolecules*, 25: 3089.

Chen, S.-J., I. G. Economou, and M. Radosz, 1992, *Macromolecules*, 25: 4987.

Chen, S.-J., I. G. Economou, and M. Radosz, 1993, *Fluid Phase Equilibria*, 83: 391.

Chen, S.-J., Y. C. Chiew, J. A. Gardecki, S. Nilsen, and M. Radosz, 1994, *J. Polym. Sci., Part B: Polym. Phys.*, 32: 1791.

Chen, S.-J., M. Banaszak, and M. Radosz, 1995, *Macromolecules*, 28: 1812.

Cheung, H. and E. H. Zander, 1968, *Chem. Eng. Prog. Symp. Ser.*, 64: 34.

Chiew, Y. C., 1990, *Molec. Phys.*, 70: 129.

Cotterman, R. L., B. J. Schwarz, and J. M. Prausnitz, 1986, *AIChE J.*, 32: 1787.

Croll, I. M. and R. L. Scott, 1958, *J. Phys. Chem.*, 62: 954.

Derr, E. L. and C. H. Deal, 1969, *Inst. Chem. Eng. Symp. Ser., (London)*, 3: 40.

Derr, E. L. and C. H. Deal, 1973, *Adv. Chem. Ser.*, 124: 11.

Dickman, R. and C. K. Hall, *J. Chem. Phys.*, 1986, 85: 4108.

Dolezalek, F., 1908, *Z. Phys. Chem.*, 64: 727.

Donohue, M. D. and J. M. Prausnitz, 1975, *Can. J. Chem.*, 53: 1586.

Donohue, M. D. and J. M Prausnitz, 1978, *AIChE J.*, 24: 849.

Eckert, C. A., 1964, *Dissertation*, University of California, Berkeley.

Economou, I. G. and C. Tsonopoulos, 1997, *Chem. Eng. Sci.*, 52: 511.

Ellis, J. A. and K.-C. Chao, 1972, *AIChE J.*, 18: 70.

Fast, J. D., 1962, *Entropy*, Philips Technical Library. Eindhoven, The Netherlands: Centrex Publishing Co.

Fisher, G. D. and T. W. Leland, 1970, *Ind. Eng. Chem. Fundam.*, 9: 537.

Flory, P. J., 1941, *J. Chem. Phys.*, 9: 660.

Flory, P. J., 1942, *J. Chem. Phys.*, 10: 51.

Flory, P. J., 1944, *J. Chem. Phys.*, 12: 425.

Flory, P. J., 1953, *Principles of Polymer Chemistry*. Ithaca: Cornell University Press.

Fredenslund, Aa., R. L. Jones, and J. M. Prausnitz, 1975, *AIChE* J., 2: 1086.

Fredenslund, Aa., J. Gmehling, M. L. Michelsen, P. Rasmussen, and J. M. Prausnitz, 1977, *Ind. Eng. Chem. Process Des. Dev.*, 16: 450.

Fredenslund, Aa., G. Gmehling, and P. Rasmussen, 1977a, *Vapor-Liquid Equilibria Using UNIFAC*. Amsterdam: Elsevier.[54]

Fredenslund, A. and P. Rasmussen, 1985, *Fluid Phase Equilibria*, 24: 115.

Fredenslund, Aa. and J. M. Sørensen, 1994, *Group-Contribution Estimation Methods*. In *Models for Thermodynamic and Phase-Equilibria Calculations*, (S. I. Sandler, Ed.). New York: Marcel Dekker.

Fu, Y. H. and S. I. Sandler, 1995, *Ind. Eng. Chem. Res.*, 34: 1897.

Funk, E. W. and J. M. Prausnitz, 1970, *Ind. Eng. Chem.*, 62: 8.

[54] The latest compilation of UNIFAC parameters (with pertinent computer programs) is available from J. M. Prausnitz, Department of Chemical Engineering, University of California, Berkeley, CA 94720, USA.

Galindo, A., P. J. Whitehead, G. Jackson, and A. N. Burgess, 1996, *J. Phys. Chem.*, 100: 6781.

Galindo, A., P. J. Whitehead, G. Jackson, and A. N. Burgess, 1997, *J. Phys. Chem. B*, 101: 2086.

Gamson, B. W. and K. M. Watson, 1944, *Natl. Petrol. News,* R623 (August 2; Sept. 6).

Gardon, J. L., 1966, *J. Paint Technol.,* 38: 43.

Gmehling, J., D. D. Liu, and J. M. Prausnitz, 1979, *Chem. Eng. Sci.,* 34: 951.

Gmehling, J., 1986, *Fluid Phase Equilibria*, 30: 119.

Gmehling, J., J. Li, and M. Schiller, 1993, *Ind. Eng. Chem. Res.,* 32: 178.

Gonsalves, J. B. and T. W. Leland, 1978, *AIChE J.,* 24: 279.

González, J. A., I. García de la Fuente, and J. C. Cobos, 1997, *Fluid Phase Equilibria*, 135: 1.

González, J. A., I. García de la Fuente, and J. C. Cobos, 1997a, *J. Chem. Soc. Faraday Trans.*, 93: 3773.

Gregg, C. J., S.-J. Chen, F. P. Stein, and M. Radosz, 1993, *Fluid Phase Equilibria*, 83: 375.

Guggenheim, E. A., 1952, *Mixtures.* Oxford: Oxford University Press.

Guggenheim, E. A., 1966, *Applications of Statistical Mechanics.* Oxford: Oxford University Press.

Gunning, A. J. and J. S. Rowlinson, 1973, *Chem. Eng. Sci.,* 28: 521.

Gupte, P. A. and R. P. Danner, 1987, *Ind. Eng. Chem. Res.,* 26: 2036.

Hansen, C. M., 1967, *J. Paint Technol.,* 39: 104, 505.

Hansen, C. M. and K. Skaarup, 1967, *J. Paint Technol.,* 39: 511.

Hansen, C. M. and A. Beerbower, 1971, *Solubility Parameters.* In *Encyclopedia of Chemical Technology* (Kirk-Othmer), 2nd Ed., Suppl. Vol. (H. F. Mark, J. J. McKetta, and D. F. Othmer Eds.). New York: Wiley-Interscience. See also Barton, 1991.

Hansen, H. K., P. Rasmussen, Aa. Fredenslund, M. Schiller, and J. Gmehling, 1991, *Ind. Eng. Chem. Res.,* 30: 2352.

Harris, H. G. and J. M. Prausnitz, 1969, *Ind. Eng. Chem. Fundam.,* Wilhelm Memorial Issue (May).

Heil, J. F. and J. M. Prausnitz, 1966, *AIChE J.,* 12: 678.

Heintz, A., 1985, *Ber. Bunsenges. Phys. Chem.,* 89: 172.

Heintz, A. and D. Papaioannou, 1998, *Thermochimica Acta,* 310: 69.

Henderson, D. and P. J. Leonard, 1971. In *Physical Chemistry,* Vol. VIIIB: *Liquid State* (D. Henderson, Ed.), Chap. 7. New York: Academic Press.

Hicks, C. P., 1972, *J. Chem. Soc. Faraday Trans II,* 72: 423.

Hildebrand, 1929, J. H., *J. Am. Chem. Soc.,* 51: 66.

Hildebrand, J. H., 1947, *J. Chem. Phys.,* 15: 225.

Hildebrand, J. H. and R. L. Scott, 1950, *The Solubility of Nonelectrolytes,* 3rd Ed. New York: Reinhold.

Hill, T. L., 1986, *An Introduction to Statistical Thermodynamics*. Reading: Addition-Wesley.

Hino, T. and J. M. Prausnitz, 1997, *Fluid Phase Equilibria*, 138: 105.

Honnell, K. G. and C. K. Hall, 1989, *J. Chem. Phys.*, 90: 1841.

Hooper, H. H., S. Michel, and J. M. Prausnitz, 1988, *Ind. Eng. Chem. Res.*, 27: 2182.

Hovorka, F. and D. Dreisbach, 1934, *J. Am. Chem. Soc.,* 56: 1664.

Huang, S. H. and M. Radosz, 1990, *Ind. Eng. Chem. Res.*, 29: 2284.

Huang, S. H. and M. Radosz, 1991, *Ind. Eng. Chem. Res.*, 30: 1994.

Huggins, M. L., 1941, *J. Phys. Chem.*, 9: 440.

Huggins, M. L., 1942, *Ann. N. Y. Acad. Sci.*, 43: 1.

Joffe, J., 1948, *Ind. Eng. Chem.*, 40: 1738.

Johnson, A. I., W. F. Furter, and T. W. Barry, 1954, *Can. J. Technol.*, 32: 179.

Kammerer, K., S. Schnabel, D. Silkenbäumer, and R. N. Lichtenthaler, 1998, *Fluid Phase Equilibria*, in press.

Kaul, B. K., M. D. Donohue, and J. M. Prausnitz, 1980, *Fluid Phase Equilibria*, 4: 171.

Kay, W. B., 1936, *Ind. Eng. Chem.*, 28: 1014.

Kehiaian, H. V., 1983, *Fluid Phase Equilibria*, 13: 243.

Kehiaian, H. V., 1985, *Pure & Appl. Chem.*, 57: 15.

Kipnis, A. Ya., B. E. Yavelov, and J. S. Rowlinson, 1996, *Van der Waals and Molecular Science*. Oxford: Clarendon Press.

Kohler, F., 1957, *Monatsh. Chem.*, 88: 857.

Kortüm, G. and H. Buchholz-Meisenheimer, 1952, *Die Theorie der Destillation und Extraktion von Flüssigkeiten*, Chap. 3. Berlin: Springer. See also the reference in Footnote 22.

Krasha, T. and K. E. Gubbins, 1996, *Ind. Eng. Chem. Res.*, 35: 4727, 4738.

Kreglewski, A., 1984, *Equilibrium Properties of Fluids and Fluid Mixtures*. College Station: Texas A&M University Press.

Kretschmer, C. B. and R. Wiebe, 1954, *J. Chem. Phys.*, 22: 1697.

Larsen, B. L., P. Rasmussen, and A. Fredenslund, 1987, *Ind. Eng. Chem. Res.*, 26: 2774.

Leach, J. W., P. S. Chappelear, and T. W. Leland, 1968, *AIChE J.*, 14: 568.

Lee, B. I. and M. G. Kesler, 1975, *AIChE J.*, 21: 510.

Leland, T. W., P. S. Chappelear, and B. W. Gamson, 1962, *AIChE J.*, 8: 482.

Leland, T. W., J. S. Rowlinson, and G. A. Sather, 1968, *Trans. Faraday Soc.*, 64: 1447.

Leland, T. W., J. S. Rowlinson, G. A. Sather, and I. D. Watson, 1969, *Trans. Faraday Soc.*, 65: 2034.

Lichtenthaler, R. N., D. S. Abrams, and J. M. Prausnitz, 1973, *Can. J. Chem.*, 51: 3071.

Lichtenthaler, R. N., D. D. Liu, and J. M. Prausnitz, 1974, *Ber. Bunsenges. Phys. Chem.*, 78: 470.

Liu, D. D. and J. M. Prausnitz, 1980, *Ind. Eng. Chem. Process Des. Dev.*, 19: 205.

Macedo, E. A., U. Weidlich, J. Gmehling and P. Rasmussen, 1983, *Ind. Eng. Chem. Process Des. Dev.*, 22: 676.

Magnussen, T., P. Rasmussen, and A. Fredenslund, 1981, *Ind. Eng. Chem. Res.*, 20: 331.

Mansoori, G. A., N. F. Carnahan, K. E. Starling, and T. W. Leland, 1971, *J. Chem. Phys.*, 54: 1523.

Mark, H. F., J. J. McKetta, and D. F. Othmer, 1969, (Eds.), *Encyclopedia of Chemical Technology* (Kirk-Othmer), 2nd Ed., Vol. 18, pp. 564-588. New York: Wiley-Interscience.

Marsh, K. N., 1980, *Ann. Rep. Prog. Chemistry, Sec. C, R. Soc. Chem. (Lond.)*, 77: 101.

Martin, R. A. and K. L. Hoy, 1975, *Tables of Solubility Parameters*. Tarrytown: Union Carbide Corp., Chemicals and Plastics, Research and Development Dept. (See also Moelwyn-Hughes, 1940).

Maurer, G. and J. M. Prausnitz, 1978, *Fluid Phase Equilibria*, 2: 91.

McGlashan, M. L. and R. P. Rastogi, 1958, *Trans. Faraday Soc.*, 54: 496.

McGlashan, M. L., 1962, *Chem. Soc. (Lond.), Annu. Rep.*, 59: 73.

Mecke, R., 1948, *Z. Elektrochem.*, 52: 274.

Mitlin, V. S. and I. C. Sanchez, 1993, *J. Chem. Phys.*, 99: 533.

Moelwyn-Hughes, E. A., 1940, *J. Chem. Soc.*, 850.

Nagata, I., 1977, *Fluid Phase Equilibria*, 1: 93.

Nagata, I., 1978, *Z. Phys. Chem. Leipzig*, 259: 1109, 1151.

Nagata, I., K. Tamura, N. Kishi, and K. Tada, 1997, *Fluid Phase Equilibria*, 135: 227.

Nelson, R. C., R. W. Hemwall, and G. D. Edwards, 1970, *J. Paint Technol.*, 42: 636.

Nernst, W., 1891, *Z. Phys. Chem.*, 8: 110.

Null, H. R. and D. A. Palmer, 1969, *Chem. Eng. Prog.*, 65: 47.

Null, H. R., 1970, *Phase Equilibrium in Process Design*. New York: John Wiley & Sons.

Ormanoudis, C. and C. Panayiotou, 1993, *Fluid Phase Equilibria*, 89: 217.

Orye, R. V. and J. M. Prausnitz, 1965, *Ind. Eng. Chem.*, 57: 18.

Panayiotou, C., 1988, *J. Phys. Chem.*, 92: 2960.

Panayiotou, C. and I. C. Sanchez, 1991, *J. Phys. Chem.*, 95: 10090.

Panayiotou, C., 1997, *Fluid Phase Equilibria*, 131: 21.

Pimentel, G. and A. L. McClellan, 1960, *The Hydrogen Bond*. San Francisco: Freeman.

Plöcker, U., H. Knapp, and J. M. Prausnitz, 1978, *Ind. Eng. Chem. Process Des. Dev.*, 17: 324.

Preston, G. T. and J. M. Prausnitz, 1970, *Ind. Eng. Chem. Process Des. Dev.*, 9: 264.

Prigogine, I., V. Mathot, and A. Desmyter, 1949, *Bull. Soc. Chim. Belg.*, 58: 547.

Prigogine, I. and A. Desmyter, 1951, *Trans. Faraday Soc.*, 47: 1137.

Prigogine, I. and R. Defay, 1954, *Chemical Thermodynamics*, Chap. 26. London: Longmans & Green.

Prigogine, I., 1957, *The Molecular Theory of Solutions*. Amsterdam: North-Holland.

Redlich, O. and A. T. Kister, 1947, *J. Chem. Phys.*, 15: 849.

Renon, H. and J. M. Prausnitz, 1967, *Chem. Eng. Sci.*, 22: 299. Errata, 1891.

Renon, H. and J. M. Prausnitz, 1969, *AIChE J.*, 15: 785.

Scatchard, G., 1949, *Chem. Rev.*, 44: 7.

Scott, R. L., 1956, *Annu. Rev. Phys. Chem.*, 7: 43.

Scott, R. L., 1956a, *J. Chem. Phys.*, 25: 193.

Scott, R. L., 1958, *J. Phys. Chem.*, 62: 136.

Skjold-Jørgensen, S., B. Kolbe, J. Gmehling, and P. Rasmussen, 1979, *Ind. Eng. Chem. Process Des. Dev.*, 18: 714.

Smith, F. and I. Brown, 1973, *Aust. J. Chem.*, 26: 691, 705.

Smith, W. R., 1972, *Can. J. Chem. Eng.*, 50: 271.

Song, Y. and E. A. Mason, 1989, *J. Chem. Phys.*, 91: 7840.

Song, Y., S. M. Lambert, and J. M. Prausnitz, 1994, *Macromolecules*, 27: 441.

Song, Y., S. M. Lambert, and J. M. Prausnitz, 1994a, *Ind. Eng. Chem. Res.*, 33: 1047.

Staverman, A. J., 1950, *Recl. Trav. Chim. Pays-Bas*, 69: 163.

Thomas, A. and M. D. Donohue, 1993, *Ind. Eng. Chem. Res.*, 32: 2093.

Thorp, N. and R. L. Scott, 1956, *J. Phys. Chem.*, 60: 670, 1441.

Tompa, H., 1952, *Trans. Farad. Soc.*, 48: 363.

Tucker, E. E. and E. Lippert, 1976. In *The Hydrogen Bond: Recent Developments in Theory and Experiment* (P. Schuster, G. Zundel, and C. Sandorfy, Eds.), Chap. 17. Amsterdam: North-Holland.

Tucker, E. E. and S. D. Christian, 1978, *J. Phys. Chem.*, 82: 1897.

Van Laar, J. J., 1910, *Z. Phys. Chem.*, 72: 723.

Van Ness, H. C., J. van Winkle, H. H. Richtol, and H. B. Hollinger, 1967, *J. Phys. Chem.*, 71: 1483.

Van Pelt, A., C. J. Peters, and J. de Swaan Arons, 1993, *Fluid Phase Equilibria*, 84: 23.

Vera, J. H. and J. M. Prausnitz, 1972, *Chem. Eng. J.*, 3: 1.

Walsh, J. M. and K. E. Gubbins, 1990, *J. Chem. Phys.*, 94: 5115.

Watson, I. D. and J. S. Rowlinson, 1969, *Chem. Eng. Sci.*, 24: 1575.

Weimer, R. F. and J. M. Prausnitz, 1965, *Hydrocarbon Process. Petrol. Refiner.*, 44: 237.

Wertheim, M. S., *J. Chem. Phys.*, 1987, 87: 7323.

Wilson, G. M., 1964, *J. Am. Chem. Soc.*, 86: 127.

Wu, C.-S. and Y.-P. Chen, 1994, *Fluid Phase Equilibria*, 100: 103.

Wu, J. and J. M. Prausnitz, 1998, *Ind. Eng. Chem. Res.*, 37: 1634.

Problems

1. A liquid hydrocarbon A has a saturation pressure of 13.3 kPa at 10°C. Its density at 25°C is 0.80 g cm^{-3} and its molecular weight is 160. This is all the information available on pure liquid A. An equimolar mixture of A in carbon disulfide at 10°C gives an equilibrium partial pressure of A equal to 8 kPa. Estimate the composition of the vapor that at 10 °C is in equilibrium with an equimolar liquid solution of A in toluene, using the following data:

	Carbon disulfide	Toluene
Solubility parameters at 25°C $(J\ cm^{-3})^{1/2}$	20.5	18.2
Liquid molar volumes at 25°C ($cm^3\ mol^{-1}$)	61	107
Saturation pressures at 10°C (kPa)	1.73	25.5

2. Consider a dilute isothermal solution of acetic acid in benzene. For the dilute region (say up to 5 mol % acid), draw schematically curves for $\bar{s}_1^E$ versus x_1 and $\bar{h}_1^E$ versus x_1 where subscript 1 refers to the acid. Briefly justify your schematic graphs with suitable explanations.

3. For distillation-column design, we need K factors ($K_i = y_i/x_i$). A liquid mixture at 50°C contains 30 mol % n-hexane and 70 mol % benzene. Calculate the K factors of n-hexane and benzene in this mixture. Assume that the pressure is sufficiently low to neglect gas-phase corrections and Poynting factors. At 50°C, pure-component vapor pressures are 0.533 bar for n-hexane and 0.380 bar for benzene.
At 25°C, the molar volumes and solubility parameters are:

	v^L (cm^3 mol^{-1})	δ (J cm^{-3})
n-Hexane	132	14.9
Benzene	89	18.8

4. Consider a solution of diethyl ether and pentachloroethane. Draw (schematically) a plot of g^E versus x at constant temperature. Briefly justify your schematic graph with suitable explanations.

5. Liquids A and B when mixed form an azeotrope at 300 K and at a mole fraction $x_A = 0.5$. It is desired to separate a mixture of A and B by distillation, and in order to break the azeotrope it is proposed to add a third liquid C into the mixture. Compute the relative volatility of A to B at 300 K when the ternary mixture contains 60 mol % C and equal molar amounts of A and B. Assume ideal gas behavior and assume that A, B, and C are nonreactive nonpolar substances. The data given below are all at 300 K.

	A	B	C
Liquid molar volume (cm^3 mol^{-1})	100	100	100
Solubility parameter (J cm^{-3})$^{1/2}$	14.3	16.4	18.4

6. A binary liquid mixture contains nonpolar components 1 and 2. The mixture is to be separated by ordinary distillation. To determine if this is feasible, it is necessary to know whether the mixture has an azeotrope. At 300 K the pure-component vapor pressures are $P_1^s = 53.3$ and $P_2^s = 80$ kPa. The pure-component molar volumes are both 160 cm^3 mol^{-1} and the solubility parameters are $\delta_1 = 14.3$ and $\delta_2 = 17.4$ (J cm^{-3})$^{1/2}$. At 300 K, does this mixture have an azeotrope? If so, what is its composition? Assume the vapor phase is ideal.

7. At 380 K, an equimolar liquid mixture of A and B has a total pressure of 0.667 bar. Fluids A and B are simple nonpolar liquids having similar molar volumes. Pure-component vapor pressures (bar) are $P_A^s = 0.427$ and $P_B^s = 0.493$.
If the equimolar mixture is cooled, partial miscibility (two liquid phases) results. Give an estimate of the (upper) critical solution temperature where partial miscibility begins. Explain and justify your method of calculation. Is your estimate likely to be high or low? Give an upper and lower bound of the expected (upper) critical solution temperature.

8. An equimolar liquid mixture of benzene and *n*-butane is fed to an isothermal flash tank operating at 50°C and 1 bar.

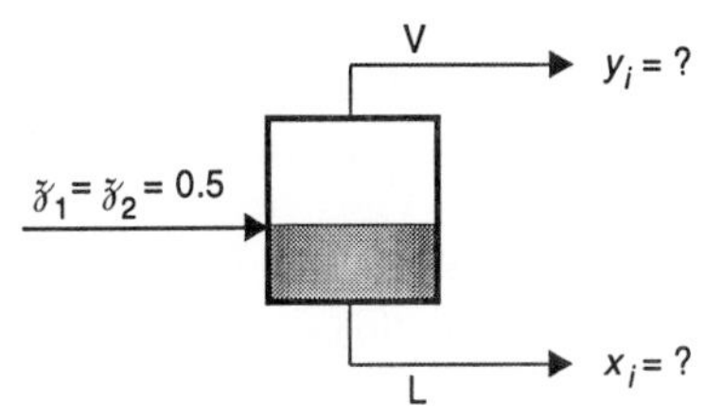

Find the compositions of the two streams leaving the flash tank. Assume that at 1 bar the gas is ideal. Assume also that for the liquid phase, the Scatchard-Hildebrand (regular-solution) equations are valid.
Data (all at 50°C) are as follows:

	Liquid density (g cm^{-3})	Solubility parameter (J cm^{-3})$^{1/2}$	Vapor pressure (torr)
Benzene	0.845	18.8	280
n-Butane	0.548	13.9	3620

9. Estimate the upper critical solution temperature for a binary system containing nonpolar liquids A and B.
Data at 25°C:

Liquid	Liquid molar volume (cm^3 mol^{-1})	Solubility parameter (J cm^{-3})$^{1/2}$
A	180	18
B	180	12

10. At 25°C, carbon disulfide (1) and perfluoro-*n*-heptane (2) are essentially totally immiscible. A small amount of cyclohexane (3) is added to this two-phase mixture. Estimate the distribution coefficient K for cyclohexane [$K = x_3$ (in 1)/x_3 (in 2)].
Pure component data:

Liquid	Liquid molar volume (cm^3 mol^{-1})	Solubility parameter (J cm^{-3})$^{1/2}$
Carbon disulfide	61	20.5
Perfluoro-*n*-heptane	226	12.3

11. Chemical engineers are fond of generalized plots. Show how you would prepare a generalized solubility parameter plot for nonpolar liquids based on Pitzer's three-parameter theory of corresponding states.

12. At room temperature and atmospheric pressure:

(a) Give an order-of-magnitude estimate of $\bar{h}^E$ for methanol dissolved in a large excess of isooctane.

(b) Give an order-of-magnitude estimate of the change in temperature when equal parts of cyclohexane and carbon disulfide are mixed adiabatically. Is ΔT positive or negative?

(c) Name two polar solvents that are likely to be very good and two others that are likely to be very poor for an extraction separation of hexane and hexene. Explain.

13. A dilute solution of picric acid in water is contacted with *n*-hexane. Consider the distribution of picric acid between the two solvents; assume that the acid exists as a monomer in both phases but that it ionizes partially in the aqueous phase. Show that the distribution of the acid should be described by an equation of the form

$$\frac{\sqrt{c_H}}{c_W} = a\left(1 - b\frac{c_H}{c_W}\right)$$

where c_H is the concentration of picric acid in hexane, c_W is the concentration of picric acid in water, and a and b are constants depending only on temperature.

14. Acetaldehyde forms a trimer (paraldehyde) in benzene solution; in excess water, acetaldehyde is completely solvated through hydrogen bonding. Experimental data are available on the distribution of acetaldehyde between benzene and water for small acetaldehyde concentrations. Show how these data should be plotted to yield a straight line, convenient for interpolation and (slight) extrapolation. (Use C for concentration, subscript A for acetaldehyde, and superscripts B and W for benzene phase and water phase, respectively.)

15. (a) Dichloromethane, acetone, and methanol are strongly polar fluids; their molecules have appreciable dipole moments. At 25°C, the following activity coefficients (at infinite dilution) were measured by Smith *et al.* (1983, *J. Chem. Eng. Data,* 28: 412) in binary solutions:

	Dichloromethane	Acetone	Dichloromethane	Methanol
γ^∞	0.59	0.53	2.94	10.3

These data show that binary mixtures of dichloromethane/acetone exhibit appreciable negative deviations from Raoult's law. However, the data for binary mixtures of dichloromethane/methanol exhibit large positive deviations. Why is there such a striking difference between these two binary systems?

(b) At 50°C, we have data for the activity coefficient of nitroethane in benzene and in hexane. When the mole fraction of nitroethane is 0.05, is the activity coefficient of nitroethane larger in benzene or in hexane? Explain.

(c) Near room temperature, we want to dissolve a heavy, aromatic, coal-derived liquid in a volatile solvent. Two solvents are considered: methanol and chloroform. Which solvent is better; that is, in which solvent is the solute likely to be more soluble? Why?

CHAPTER 8

Polymers: Solutions, Blends, Membranes, and Gels

Polymer solutions are liquid mixtures wherein the molecules of at least one component are very much larger than those of the other components. For many linear and branched polymers, liquid solvents are available that dissolve the polymer completely to form a homogeneous solution. However, cross-linked polymers (i.e. networks) only swell when in contact with a compatible liquid solvent. Swelling also occurs when polymeric materials are exposed to solvent vapors or gases that can be absorbed by the polymer.

A polymer blend is a mixture containing two or more polymers and, perhaps, an additional component to enhance polymer compatibility.

The phase behavior of polymer solutions and the swelling of polymers play important roles in polymer processing and in at least some application, primarily because many polymers are produced in solution and therefore the final polymer product may contain some residual solvent. The physical properties of polymers are affected by the amount and type of the low-molecular-weight components they contain. A frequent technical problem is to remove essentially all the low-molecular-weight components; a common procedure is to volatilize them and this removal process is often called *polymer devolatilization*. Total removal of solvent is particularly important for polymeric films used in packaging foods or pharmaceuticals. In other cases, the important technical issue is how much and how fast liquid solvent or vapor or gas is absorbed by a polymer. Processes where sorption behavior is important are, for example, the separation of gaseous and liquid mixtures using nonporous polymeric membranes or the use of supercritical fluids as swelling agents for impregnating polymers with chemical additives (e.g. pigments for color) and, conversely, for extracting low-molecular-weight components from polymeric materials. Qualitative and quantitative description of these processes requires first, knowledge of phase behavior and solubility (equilibrium properties)[1] and second, of diffusivity (transport property). Equilibrium properties must be known to provide a meaningful description of the driving force for a diffusion process.

This chapter presents an introduction to the phase behavior of polymer-solvent solutions and polymer blends. We also briefly address the solubility and diffusivity of low-molecular-weight components in polymeric materials (e.g. membranes) and the unique phase behavior of polymeric gels. However, before these particular topics are addressed, we summarize some special properties of polymers.

8.1 Properties of Polymers

Polymers are large, chain-like molecules composed of many (Greek: *poly-*) structural repeating units, or "mers" (Greek: *meros* meaning part), connected by chemical bonds. These units may be arranged in a variety of ways resulting in various types of molecules with chain-like structure. The simplest is a *linear polymer* where the units are connected to each other in a linear sequence forming a long chain. An alternative to a linear polymer is a *branched polymer*. The branches can be long or short. When branches of different polymers become interconnected *cross-linked* structures (*networks, gels*) are formed. *Homopolymers* contain one type of structural repeating unit and *copolymers* at least two or more. Dendrimers are hyperbranched (tree-like) polymers where branches have branches, etc.

The enormous and intriguing range of physicochemical properties of polymers depend on the arrangements and nature of the repeating units and on the types of intramolecular bonds and intermolecular forces. Due to the large size of polymer molecules, the intermolecular forces (typically, dispersion forces and hydrogen bonding) assume a much greater role in influencing physicochemical properties than they do for substances with small organic molecules. Some of the properties are unique

[1] Methods for calculating equilibrium properties of polymer solutions and a database are presented by R. P. Danner and M. S. High, 1993, *Handbook of Polymer Solution Thermodynamics*, New York: A.I.Ch.E.

to polymers (e.g. *rubber elasticity*) and are simply a consequence of the size, shape and chain-like structure of these large molecules.

Many polymers show little tendency to crystallize or to align the chains in some form of order. Polymers only crystallize if the molecules have regular structure and even then only do so to a limited extent. The remaining material is randomly disordered (*amorphous*); the polymer chains are randomly coiled, as is the case for all molten polymers. For polymers that crystallize at all, the extent of crystallinity may be in the range 30-80%, depending on crystallization conditions and it decreases with increasing structural irregularity. Unlike well-defined low-molecular-weight crystals, polymers do not melt at a precise temperature, but rather over a range of temperature, typically 10-20°C. Nevertheless, the literature often gives a precise melting temperature T_m that usually refers to the highest temperature of the melting range as shown in Fig. 8-1. Melting is a first-order transition and occurs with an abrupt increase in volume, entropy and enthalpy. At temperatures well below the melting range, semi-crystalline polymers are hard and stiff materials. Due to structural irregularity, many polymers remain completely amorphous upon cooling and in this form, a solid polymer resembles glass. When the melt of a non-crystallizable polymer cools, the mobility of the polymer molecules decreases. The lower the temperature, the stiffer the polymers become until the *glass transition* is reached. The temperature where this second-order transition[2] occurs is the *glass-transition temperature* T_g.[3]

Figure 8-1 shows the variation of the specific volume with temperature typically observed for polymers. The melt region corresponds to temperatures above T_m for semi-crystalline polymers and to temperatures above T_g for an amorphous polymer.

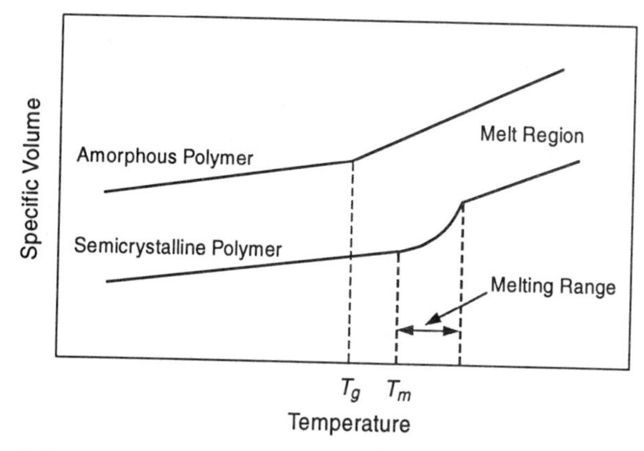

Figure 8-1 Schematic illustration of the variation of the specific volume of polymers with temperature.

[2] In a second-order transition the rate of change of thermodynamic properties (e.g. specific volume and heat capacity) depends upon the rate of temperature change.

[3] Glass transition may occur over a range of a few degrees.

At temperatures well below T_g, amorphous polymers are hard, stiff, *glassy* materials resisting deformation, although they may not necessarily be brittle. At temperatures well above T_g, polymers are in a rubbery or plastic state; large elastic deformations are possible and the polymer is tougher and more pliable. There are, therefore, major changes in the mechanical behavior as the glass transition is traversed. Glass transition also occurs in the amorphous regions of semi-crystalline polymers, always at temperatures lower than their melting temperatures ($T_g < T_m$). Although the importance of glass transition depends on the degree of crystallinity and, although the changes in mechanical properties on traversing T_g are usually not as distinct as observed in fully amorphous polymers, glass transition nevertheless contributes importantly to the overall softening of semi-crystalline polymers.

Polymers are formed by linking together monomer molecules through chemical reactions. Polymers produced synthetically are called synthetic polymers, whereas those produced biologically in nature are biopolymers (or biomacromolecules). In contrast to that what happens in nature, formation of synthetic polymers is governed by random events. As a result, the chains obtained vary in length and, therefore, synthetic polymeric materials consist of a mixture of homologous molecules of different molecular weight, i.e. they have a *molecular weight distribution*. They are *polydisperse* and cannot be characterized by a single molecular weight but must be represented by a statistical average (Cowie, 1991; Rave, 1995). This average can be expressed in several ways. The *number average* is the sum of all the molecular weights of the individual molecules present in a sample divided by their total number. Each molecule contributes equally to the average. If N_i is the number of molecules with molecular weight M_i, the total number of molecules is $\sum N_i$, the total weight of the sample is $\sum N_i M_i$, and the number average molecular weight is

$$\overline{M}_n = \frac{\sum N_i M_i}{\sum N_i} \tag{8-1}$$

Another way to express the molecular weight average is as a *weight average* where each molecule contributes according to the ratio of its particular weight to the total weight:

$$\overline{M}_w = \frac{\sum N_i M_i^2}{\sum N_i M_i} \tag{8-2}$$

$\overline{M}_w$ is more sensitive to high-molecular-weight species than $\overline{M}_n$. Therefore, $\overline{M}_w$ is always larger than $\overline{M}_n$ for a polydisperse polymer. Consequently, the ratio $\overline{M}_w / \overline{M}_n$, greater than unity, is known as the *polydispersity* or *heterogeneity index*.[4] Its value is

[4] Instead of $\overline{M}_w$ the *viscosity-average* molecular weight $\overline{M}_\eta$ is often used because it is easily obtained from viscosity measurements of dilute polymer solutions. $\overline{M}_\eta$ is defined as (Young and Lovell, 1991):

often used as a measure of the width of the molecular-weight distribution; because the wider the distribution of molecular sizes, the greater the disparity between the averages. A perfectly monodisperse polymer would have $\overline{M}_w / \overline{M}_n = 1.00$. For many polydisperse polymers $\overline{M}_w / \overline{M}_n$ is in the range 1.5-2.0. Characterizing a polymer sample with its average molecular weight plus its polydispersity index is better than using only an average molecular weight because the molecular weight distribution is then to some extent taken into account. Two samples of the same polymer, e.g., equal in weight average molecular weight may, exhibit different physicochemical properties if they differ in their molecular-weight distribution.

In general, physicochemical properties of polymers depend on the sizes and shapes of the molecules. They are influenced by the nature of intra- and intermolecular forces, by the degree of symmetry and uniformity in molecular structures, and by the arrangements of the large molecules into amorphous and crystalline regions. All this affects, e.g., melting and glass-transition temperatures, tensile strengths, flexibility, melt and solution viscosities, miscibility with other polymers (*blending*) and solubility in and sorption of low-molecular-weight solvents. These effects and their consequences have to be kept in mind when considering the specific topics in this chapter; these topics are restricted to molten polymers, mixtures of polymers with solvents or other molten polymers (*blends*), and polymeric membranes.

8.2 Lattice Models: The Flory-Huggins Theory

The lattice model, discussed in Sec. 7.4, is particularly useful for describing solutions of polymers in liquid solvents. The *Flory-Huggins theory* is based on this model. That theory is a cornerstone of polymer-solution thermodynamics.

The Gibbs energy of mixing consists of an enthalpy term and an entropy term. The theory of regular solutions for molecules of similar size assumes that the entropy term corresponds to that for an ideal solution and attention is focused on the enthalpy of mixing; however, when considering solutions of molecules of very different size, it is advantageous to assume, at least at first, that the enthalpy of mixing is zero and to concentrate on the entropy of mixing. Solutions with zero enthalpy of mixing are called *athermal solutions* because, when mixed at constant temperature and pressure, there is no liberation or absorption of heat. Athermal behavior is never observed exactly but it is approximated by mixtures of components that are similar in their chemical characteristics even if their sizes are different. Examples of nearly athermal solutions are mixtures of polystyrene with toluene or ethylbenzene and mixtures of polydimethylsiloxane with hexamethyldisiloxane.

$$\overline{M}_\eta = [(\sum_i N_i M_i^{1+\alpha}) / (\sum_i N_i M_i)]^{1/\alpha}$$

where α is a constant. When $\alpha = 1$, then $\overline{M}_\eta = \overline{M}_w$. Typically, $\overline{M}_\eta$ is within 20% of $\overline{M}_w$.

It is convenient to write the thermodynamic mixing properties as the sum of two parts: (1) a *combinatorial contribution* that appears in the entropy (and therefore in the Gibbs energy and in the Helmholtz energy) but not in the enthalpy or in the volume of mixing; and (2) a *residual contribution*,[5] determined by differences in intermolecular forces and in free volumes[6] between the components. For the entropy of mixing, for example, we write[7]

$$\Delta_{\text{mix}} S = \Delta S^C + S^R \tag{8-3}$$

where superscript C stands for combinatorial and superscript R stands for residual.

Consider a mixing process where the molecules of fluids 1 and 2 have no difference in molecular interactions and no difference in free volume. For this case, isothermal, isobaric mixing occurs also at constant volume; the residual mixing properties are zero and we are concerned only with combinatorial mixing properties.

Using the concept of a quasicrystalline lattice as a model for a liquid, an expression for the combinatorial entropy of mixing was derived independently by Flory (1941, 1942) and by Huggins (1942) for flexible chain molecules that differ significantly in size. The derivation, based on statistical arguments and several well-defined assumptions, is not reproduced here. It is presented in several references (Fast, 1962; Flory, 1953); we give here only a brief discussion along with the result.

We consider a mixture of two liquids 1 and 2. Molecules of type 1 (solvent) are single spheres. Molecules of type 2 (polymer) are assumed to behave like flexible chains, i.e., as if they consist of a large number of mobile segments, each having the same size as that of a solvent molecule. Further, it is assumed that each site of the quasilattice is occupied by either a solvent molecule or a polymer segment and that adjacent segments occupy adjacent sites. Let there be N_1 molecules of solvent and N_2 molecules of polymer and let there be r segments in a polymer molecule. The total number of lattice sites is $(N_1 + rN_2)$. Fractions Φ_1^* and Φ_2^* of sites occupied by the solvent and by the polymer are given by

$$\Phi_1^* = \frac{N_1}{N_1 + rN_2} \quad \text{and} \quad \Phi_2^* = \frac{N_2}{N_1 + rN_2} \tag{8-4}$$

[5] The *residual* contribution to a mixing property is defined as the observed change in that property upon mixing (at constant T and P) minus the calculated change in that property upon mixing (at the same T, P, and composition), where the calculation is based on a model that serves as a reference. Two common reference models are the ideal solution and the athermal solution. Residual mixing properties are different from residual properties discussed in App. B.

[6] In general, two pure liquids have different free volumes due to different coefficients of thermal expansion. According to the Prigogine-Flory-Patterson theory, discussed in Sec. 8.2, two liquids with different free volumes experience a net contraction upon mixing and, therefore, negative contributions appear in both $\Delta_{\text{mix}}H$ and $\Delta_{\text{mix}}S$. Contributions from free-volume differences and from differences in contact energy are included in the residual part of a mixing property.

[7] In most, but not all cases, ΔS^C is the dominant term.

Flory and Huggins have shown that if the amorphous (i.e., noncrystalline) polymer and the solvent mix without any energetic effects (i.e., athermal behavior), the change in Gibbs energy and entropy of mixing are given by the remarkably simple expression:

$$-\frac{\Delta G^C}{RT} = \frac{\Delta S^C}{R} = -(N_1 \ln \Phi_1^* + N_2 \ln \Phi_2^*) \tag{8-5}$$

The entropy change in Eq. (8-5) is similar in form to that of Eq. (7-80) for a regular solution except that segment fractions are used rather than mole fractions. For the special case $r = 1$, the change in entropy given by Eq. (8-5) reduces to that of Eq. (7-80), as expected. However, when $r > 1$, Eq. (8-5) always gives a combinatorial entropy larger than that given by Eq. (7-80) for the same N_1 and N_2. Much discussion of these equations has led Hildebrand (1947) to the conclusion that for nonpolar systems, Eq. (7-80) gives a lower limit to the combinatorial entropy of mixing and Eq. (8-5) gives an upper limit; the "true" combinatorial entropy probably lies in between, depending on the size and shape of the molecules.

Modifications of Eq. (7-113) have been presented by several authors, including Huggins (1941, 1942), Guggenheim (1944, 1952), Staverman (1950), Tompa (1952), and Lichtenthaler (1973, 1974).[8]

The modifications introduced by Lichtenthaler, similar to those of Tompa, provide a reasonable method for calculating ΔS^C for mixtures of molecules differing in shape as well as size. The model of Lichtenthaler assumes that the ratio of the molecular (van der Waals) volumes of the polymer and the solvent (regarded as monomer) gives r, the number of segments of the polymer molecule. Similarly, the ratio of the surface areas of the polymer and the solvent gives q, the external surface area of a polymer molecule. The ratio q/r is a measure of the shape of the polymer molecule; for a monomer $q/r = 1$. As r becomes very large, for a linear chain $q/r \rightarrow 2/3$ and for a sphere (or cube) $q/r \rightarrow 0$. For globular molecules, the ratio q/r lies between zero and unity.

Combinatorial entropies of mixing calculated with Lichtenthaler's expression lie between those found from Eqs. (7-80) and (8-5), depending on ratio q/r. If molecules 1 and 2 are identical in size and shape, $q = r = 1$ and the expression of Lichtenthaler reduces to Eq. (7-80). If the coordination number[9] becomes very large, $q/r \rightarrow 1$ and the expression of Lichtenthaler becomes identical to Eq. (8-5), regardless of molecular shape.

[8] Various models are compared in a review by S. G. Sayegh and J. H. Vera, 1980, *Chem. Eng. J.*, 19: 1.

[9] Coordination number is the number of nearest neighbors around a solvent molecule or segment. See Sec. 7.4.

To illustrate, Fig. 8-2 shows excess combinatorial entropies per mole of sites[10] for mixtures of benzene and various forms of polyethylene. This figure shows that the combinatorial entropy is strongly affected by the bulkiness of the large molecule that increases with decreasing q/r. The Flory-Huggins expression [Eq. (8-5)] does not distinguish between the six cases shown in Fig. 8-2.

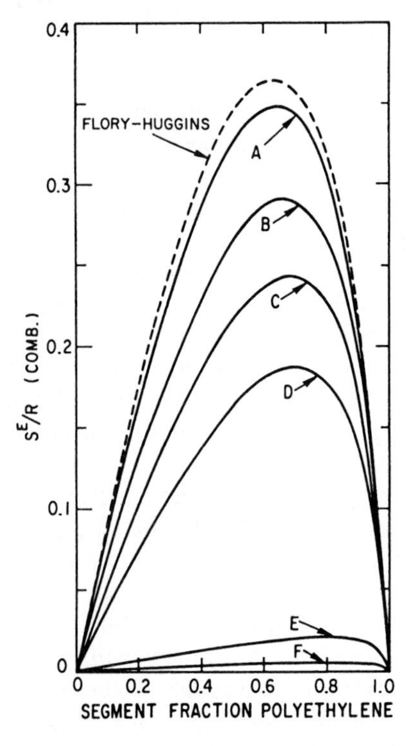

Figure 8-2 Excess combinatorial entropy per mole of sites for benzene/polyethylene. Curves A to F refer to different shapes assumed for polyethylene ($r = 1695$):
A = straight-chain, $(q/r) = 0.788$;
B = double-strand flat ribbon, $(q/r) = 0.591$;
C = quadruple-strand flat ribbon, $(q/r) = 0.493$;
D = rod-like shape, $(q/r) = 0.394$;
E = rod-like shape, but shorter axis and larger cross-section as in D, $(q/r) = 0.127$;
F = cube, $(q/r) = 0.076$.

---

[10] The excess combinatorial entropy is defined, as usual, as the entropy of mixing in excess of that for an ideal system. While entropies usually are calculated per mole of mixture, we can convert to entropies per mole of sites by writing

$$\frac{s^E \text{(per mole of sites)}}{s^E \text{(per mole of mixture)}} = 1 - (1 - \frac{1}{r})\Phi_2^*$$

where Φ_2^* is given by Eq. (8-4).

Since the Flory-Huggins formula depends only on the size ratio r, the same in all six cases, it always gives the same result (dashed curve in Fig. 8-2), independent of molecular shape. Since Flory-Huggins assumes that $q/r = 1$, it gives an upper limit for the combinatorial entropy of mixing; therefore, the results are close to those shown by curve A with the highest value of q/r. For mixtures of bulky molecules, even if they differ significantly in size, q/r is much smaller than 1 (case F) and then the ideal entropy of mixing [Eq. (7-80)] is a much better approximation than Eq. (8-5).

Donohue (1975) has presented a discussion of Lichtenthaler's model for the combinatorial entropy of mixing and has shown how it can be quantitatively transformed into a generalized Flory-Huggins expression.

Although the simple expression of Flory and Huggins does not always give the (presumably) correct, quantitative combinatorial entropy of mixing, it qualitatively describes many features of athermal polymer solutions. Therefore, for simplicity, we use it in our further discussion of polymer solutions in this section.

The expression of Flory and Huggins immediately leads to an equation for the excess entropy that is, per mole of mixture,

$$\frac{s^E}{R} = -x_1 \ln\left[1 - \Phi_2^*\left(1 - \frac{1}{r}\right)\right] - x_2 \ln\left[r - \Phi_2^*(r-1)\right] \tag{8-6}$$

By algebraic rearrangement of Eq. (8-6) and expansion of the resulting logarithmic terms, it can be shown that for all $r > 1$, s^E is positive. Therefore, for an athermal solution of components whose molecules differ in size, the Flory-Huggins theory predicts negative deviations from Raoult's law:

$$\frac{g^E}{RT} = \frac{h^E}{RT} - \frac{s^E}{R} = 0 - \frac{s^E}{R} < 0 \tag{8-7}$$

For an athermal solution, the activity of the solvent from Eq. (8-6) is

$$\boxed{\ln a_1 = \ln(1 - \Phi_2^*) + \left(1 - \frac{1}{r}\right)\Phi_2^*} \tag{8-8}$$

and the corresponding activity coefficient (based on mole fraction) is[11]

[11] In the limit $\Phi_2^* \to 1$, the activity coefficient for a solvent in a polymer solution based on mole fraction is awkward, as in this limit $\ln \gamma_1 \to -\infty$ for large values of r. For polymer solutions, the activity coefficient of the solvent is more conveniently defined on a weight-fraction or volume-fraction basis, as pointed out by D. Patterson, Y. B. Tewari, H. P. Schreiber, and J. E. Guillet, 1971, *Macromolecules,* 4: 356.

$$\ln \gamma_1 = \ln\left[1-\left(1-\frac{1}{r}\right)\Phi_2^*\right]+\left(1-\frac{1}{r}\right)\Phi_2^* \tag{8-9}$$

Figure 8-3 shows activity coefficients for the solvent according to Eq. (8-9) for several values of parameter r that provides a measure of the disparity in molecular size between the two components. The activity coefficient is a strong function of r for small values of that parameter, but for large values ($r \geq 100$) the activity coefficient is essentially independent of r. For solutions of polymers in common solvents, r is a very large number and we can see in Fig. 8-3 that large deviations from ideal-solution behavior result merely as a consequence of differences in molecular sizes even in the absence of any energetic (enthalpy of mixing) effects.

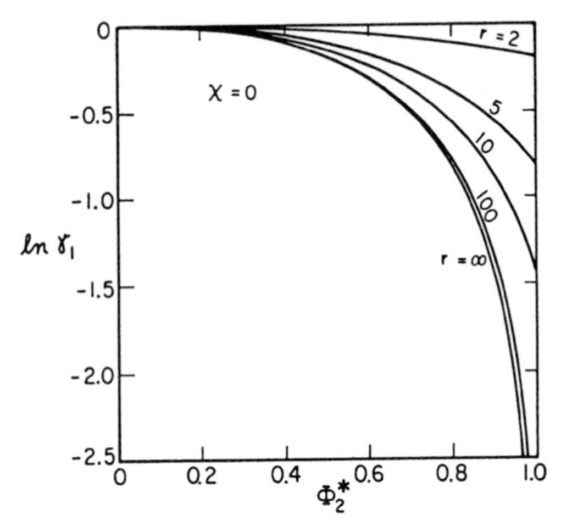

Figure 8-3 Solvent activity coefficient in an athermal polymer solution according to the equation of Flory and Huggins. Parameter r gives the number of segments in the polymer molecule.

To apply the theoretical result of Flory and Huggins to real polymer solutions, i.e., to solutions that are not athermal, it has become common practice to add to the combinatorial part of the Gibbs energy, given by Eq. (8-5), a semiempirical part for the residual contribution. In other words, we add a term that, if there is no difference in free volumes, is given by the enthalpy of mixing. The form of this term is the same as that used in the van Laar-Scatchard-Hildebrand theory of solutions (see Secs. 7.1 and 7.2); the excess enthalpy is set proportional to the volume of the solution and to the product of the volume fractions. The Flory-Huggins equation for real polymer solutions then becomes

$$\frac{\Delta_{\text{mix}}G}{RT}=\frac{\Delta G^C}{RT}+\frac{G^R}{RT}=N_1\ln\Phi_1^*+N_2\ln\Phi_2^*+\chi\Phi_1^*\Phi_2^*(N_1+rN_2) \tag{8-10}$$

The activity of the solvent is given by

$$\ln a_1 = \ln(1-\Phi_2^*) + \left(1-\frac{1}{r}\right)\Phi_2^* + \chi\Phi_2^{*2} \tag{8-11}$$

and the corresponding equation for the activity coefficient of the solvent (based on mole fraction) is

$$\ln\gamma_1 = \ln\left[1-\left(1-\frac{1}{r}\right)\Phi_2^*\right] + \left(1-\frac{1}{r}\right)\Phi_2^* + \chi\Phi_2^{*2} \tag{8-12}$$

where χ, the *Flory-Huggins interaction parameter*, is determined by intermolecular forces. Figure 8-4 shows activity coefficients for the solvent according to the Flory-Huggins equation for real polymer solutions.

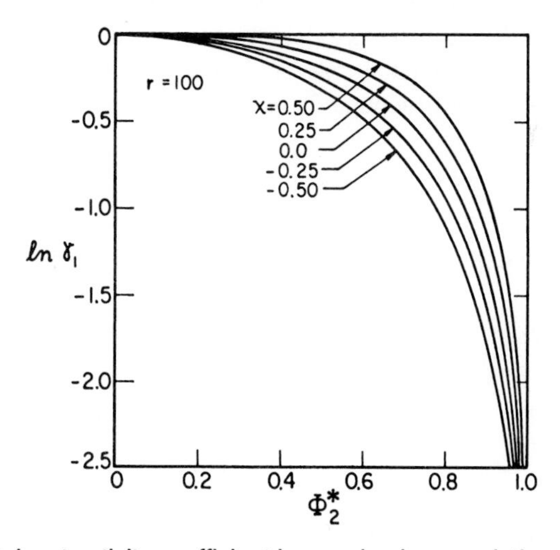

Figure 8-4 Solvent activity coefficient in a real polymer solution according to the equation of Flory and Huggins. Parameter χ depends on the intermolecular forces between polymer and solvent.

Dimensionless parameter χ is assumed to be independent of composition.[12] It is determined by the energies that characterize the interactions between pairs of polymer

[12] However, as found from experiment, and as predicted from more sophisticated theories (compare Sec. 8.3), χ varies with polymer concentration, sometimes appreciably, contrary to the simple Flory-Huggins theory discussed here. The advanced theories are based on an equation of state that, unlike lattice theory, permits components to mix (at constant pressure and temperature) with a change of volume.

segments, between pairs of solvent molecules, and between one polymer segment and one solvent molecule. In terms of the interchange energy [Eq. (7-71)], χ is given by

$$\chi = \frac{w}{kT} \tag{8-13}$$

Assuming that the interchange energy w is independent of temperature, Flory parameter χ is inversely proportional to temperature. In this case, the interchange energy refers not to the exchange of solvent and solute molecules but rather to the exchange of solvent molecules and polymer segments. For athermal solutions, χ is zero, and for mixtures of components that are chemically similar, χ is small compared to unity.

Equations (8-10) to (8-12) have been used widely to describe thermodynamic properties of solutions whose molecules differ greatly in size. For example, Fig. 8-5 shows activity coefficients at infinite dilution for *n*-butane and *n*-octane in *n*-alkane solvents (n-$C_{20}H_{42}$ to n-$C_{36}H_{74}$),[13] measured by gas-liquid chromatography.[14] Figure 8-5 shows that negative deviations from Raoult's law rise with increasing difference in molecular size between solute and solvent. As the components are chemically similar, χ is expected to be small, and therefore deviations from ideal-solution behavior result mainly from differences in molecular size. A similar result is indicated in Fig. 8-6 that shows activity coefficients for *n*-heptane in the *n*-heptane/polyethylene system.

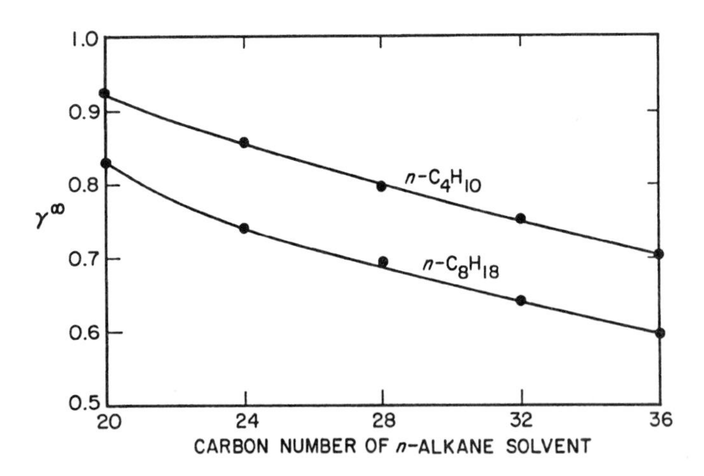

Figure 8-5 Activity coefficients at infinite dilution for *n*-butane and *n*-octane in *n*-alkane solvents near 100°C. Negative deviations from Raoult's law are due to the difference in molecular size.

[13] Data from J. F. Parker *et al.*, 1975, *J. Chem. Eng. Data*, 20: 145.
[14] This experimental method is discussed briefly in App. F.

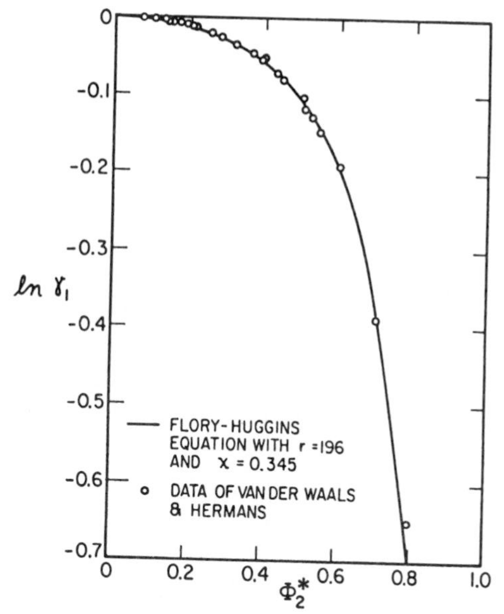

Figure 8-6 Activity coefficients of heptane in the *n*-heptane (1)/polyethylene (2) system at 109°C.

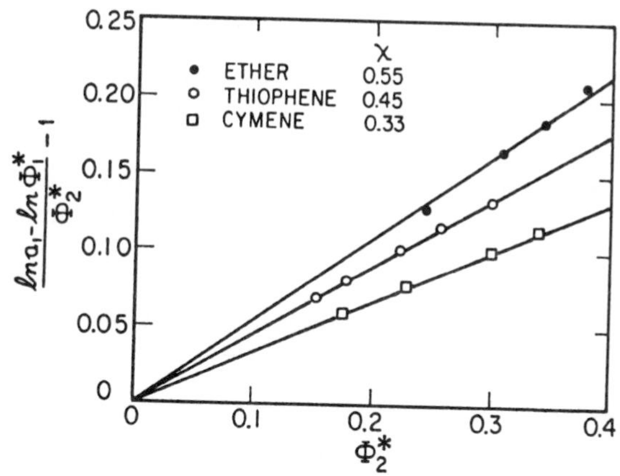

Figure 8-7 Data reduction using the equation of Flory and Huggins. Data are for solutions of rubber near room temperature. Interaction parameters χ are given by the slopes of the lines.

Figure 8-7 shows how Eq. (8-11) may be used to reduce data on rubber solutions to obtain Flory-Huggins parameter χ. In Fig. 8-7, $1/r$ has been set equal to zero. For these systems, Eqs. (8-11) and (8-12) give an excellent representation of the data but for many other systems representation is poor because, contrary to the simple theory, χ varies with polymer concentration.

If we set r equal to the ratio of molar volumes of polymer and solvent, then the segment fraction Φ^* given in Eq. (8-4) is identical to the volume fraction in the Scatchard-Hildebrand theory, Eqs. (7-25) and (7-26). In terms of solubility parameters, it can then be shown that χ is given by

$$\chi = \frac{v_1}{RT}(\delta_1 - \delta_2)^2 \tag{8-14}$$

where v_1 is the molar volume of the solvent and δ_1 and δ_2 are, respectively, the solubility parameters of solvent and polymer. Equation (8-14) is not useful for an accurate quantitative description of polymer solutions but it provides a good guide for a qualitative consideration of polymer solubility. For good solubility, χ should be small or negative, as discussed below. According to the Scatchard-Hildebrand theory for nonpolar components, χ cannot be negative; however, in many polar systems negative values have been observed.

A criterion of a good solvent for a given polymer is

$$\delta_1 \approx \delta_2 \tag{8-15}$$

Equation (8-15) provides a useful practical guide for nonpolar systems (Burrel, 1955; Blanks and Prausnitz, 1964) and for polar systems an approximate generalization of Eq. (8-14) has been suggested by Blanks (1964), Hansen (1967, 1967a, 1971), and Barton (1990).

Table 8-1 gives solubility parameters for some noncrystalline polymers. When these are compared with solubility parameters for common liquids (see Table 7-1), some qualitative statements concerning polymer solubility can easily be made. For example, the solubility parameters show at once that polyisobutylene (δ = 16.4) should be readily soluble in cyclohexane (δ = 16.8) but only sparingly soluble in carbon disulfide (δ = 20.5).

Table 8-1 Solubility parameters for some amorphous polymers near 25°C (Grulke, 1989).

Polymer	δ (J cm^{-3})$^{1/2}$
Teflon	12.7
Poly(dimethyl siloxane)	14.9
Polyethylene	16.2
Polyisobutylene	16.4
Polybutadiene	17.4
Polystyrene	18.6
Poly(methyl methacrylate)	19.4
Poly(vinyl chloride)	19.8
Cellulose diacetate	22.3
Poly(vinylidene chloride)	25.0
Polyacrylonitrile	25.3

It is important to keep in mind that all of the relations given in this section are restricted to amorphous polymers; they are not directly applicable to crystalline, glassy or cross-linked polymers.

Although the Flory-Huggins equation for real polymer solutions does not provide an accurate description of the thermodynamics properties of such solutions, this relatively simple theory contains most of the essential features that distinguish solutions of very large molecules from those containing only molecules of ordinary size.

The addition of a residual term to the theoretical result for athermal mixtures is essentially an empirical modification to obtain a reasonable expression for the Gibbs energy of mixing. According to the theory, χ should be independent of polymer concentration and of polymer molecular weight, but in many, especially polar systems, χ changes considerably with both (Koningsveld *et al.*, 1968, 1970, 1971; Siow *et al.*, 1972; Orwoll, 1977). Further, the theory erroneously assumes that the enthalpy of mixing should be given by the last term in Eq. (8-10); however, calorimetric enthalpy-of-mixing data often give a value of χ significantly different from that obtained when experimental activities are reduced by Eq. (8-11). This follows, in part, because values of χ obtained from Eqs. (8-10) to (8-12) are directly associated with the Flory-Huggins approximation for the combinatorial contribution; a different approximation for the combinatorial contribution necessarily produces a change in χ. Therefore, the value of χ obtained from experimental activities has an entropic as well as enthalpic part. Finally, experimental results by many researchers show clearly that the temperature dependence of χ is not a simple proportionality to inverse temperature.

When applied to liquid-liquid equilibria, the simple Flory-Huggins theory can only explain partial miscibility of polymer/solvent systems at low temperatures. The combinatorial entropy always favors mixing and therefore, if χ is zero or negative, complete miscibility is obtained at any temperature. If $\chi > 0$, however, there exists an upper limit where partial miscibility occurs. When Eq. (8-10) for the Gibbs energy of mixing is combined with the equations for stability (see Sec. 6.12), the condition for complete miscibility of components 1 and 2 is given by

$$\chi \leq \frac{1}{2}\left(1+\frac{1}{\sqrt{r}}\right)^2 \tag{8-16}$$

The equal sign characterizes incipient instability where χ is designated by its critical value χ^c and that occurs at the critical composition

$$\Phi_2^{*c} = \frac{1}{1+\sqrt{r}} \tag{8-17}$$

In polymer solutions, when $r >> 1$, the critical value χ^c is essentially 1/2. The critical composition occurs at a very small polymer concentration, approaching zero as $r \to \infty$. For ordinary solutions, when $r = 1$, the critical values are $\chi^c = 2$ and

$\Phi_2^{*c} = x_2 = 0.5$, in agreement with results discussed in Sec. 7.6. The critical temperature T^c for phase separation is an *upper critical solution temperature* (UCST); i.e. for $T > T^c$ there is only one liquid stable phase (complete miscibility), whereas for $T < T^c$ there are two stable liquid phases. In a polymer/solvent system, the limiting UCST for a polymer of infinite molecular weight is known as the (theta) θ *temperature*. Because χ is inversely proportional to temperature [Eq. (8-13)], it can be expressed as

$$\frac{\chi(T)}{\chi^c(T=\theta)} = 2\chi(T) = \frac{\theta}{T} \qquad \text{(8-18)}^{15}$$

If θ is known for a polymer/solvent system, Eq. (8-18) can be used to calculate $\chi(T)$ if the molecular weight is very high.

Figure 8-8 shows calculated phase diagrams for binary mixtures according to the simple Flory-Huggins theory. From $r = 1$ to $r \to \infty$, the reduced UCST increases by a factor of 4.

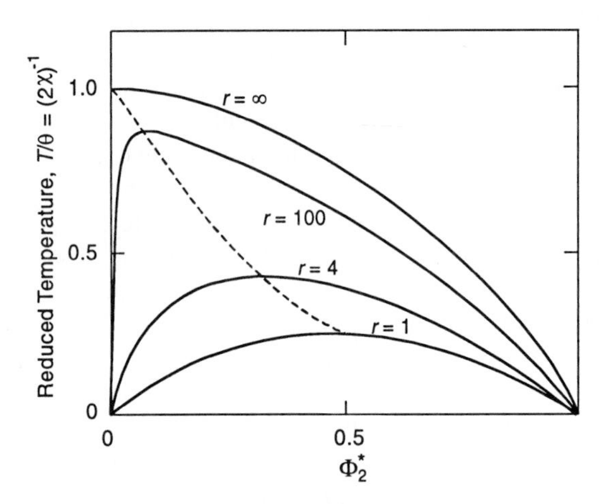

Figure 8-8 Calculated phase diagrams (cloud-point curves) for a binary mixture according to the Flory-Huggins theory. The dotted line shows the locus of upper critical solution temperatures.

Although not predicted by the simple theory, partial miscibility at higher temperatures (LCST) has been observed for many polymer/solvent systems (Siow *et al.*, 1972), as indicated in Fig. 8-9(a). With increasing molecular weight, the difference between LCST and UCST tends to decrease, resulting ultimately in an "hourglass" type of phase diagram, when the two regions of limited miscibility have merged. In that case, complete miscibility is not obtained in the entire temperature range, as shown in

[15] Comparison of Eqs. (8-16) and (8-18) shows that $\chi^c(T = \theta) = 1/2$.

Fig. 8-9(b). However, as discussed in Sec. 6.13, the LCST may occur at temperatures below the UCST, resulting in a closed-loop phase behavior shown in Fig. 8-9(c). A realistic qualitative explanation of this phenomenon was given many years ago by Hirschfelder *et al.* (1937). Closed-loop behavior follows from competition among three contributions to the Helmholtz energy of mixing: dispersion forces, combinatorial entropy of mixing, and highly oriented specific interactions (such as hydrogen bonding). While the dispersion forces energetically favor phase separation, the combinatorial entropy of mixing favors mutual miscibility. The specific interactions are energetically favorable, but entropically unfavorable, because of their highly directional-specific character. Therefore, in the presence of specific interactions between dissimilar components, the mixture could form a single homogeneous phase at low temperatures where the energy of specific interactions compares favorably to thermal energy kT. At moderate temperatures, where neither specific interactions nor combinatorial entropy of mixing dominate, the effect of dispersion forces becomes significant and the mixture exhibits phase separation. At higher temperatures, the combinatorial entropy of mixing becomes dominant and a single homogeneous phase reappears.

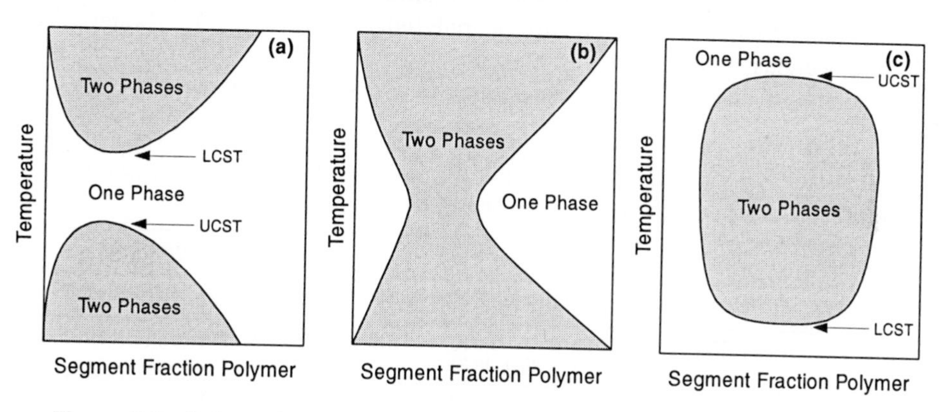

Figure 8-9 Schematic representation of phase stability in three binary polymer solutions. (a) UCST is below LCST; (b) hourglass; (c) closed loop, where LCST is below UCST.

Using essentially empirical arguments, Qian and coworkers (1991; 1991a) introduced into the original Flory-Huggins lattice theory a χ parameter given by the product of two functions, one depending on composition and the other on temperature. This semiempirical model permits fitting most observed types of binary liquid-liquid phase diagrams (UCST, LCST, UCST and LCST, hourglass, and closed loop) by adjusting the coefficients in the functions that give the desired temperature dependence and composition dependence of χ. Extensions and simplifications in Qian's work were reported by Bae and coworkers (1993).

Bae *et al.* (1993) assigned empirical composition and temperature dependencies to Flory parameter χ:

$$\chi(T,\Phi) = D(T)B(\Phi_2) \tag{8-19}$$

with

$$D(T) = d_0 + \frac{d_1}{T} + d_2 \ln T \tag{8-20}$$

and

$$B(\Phi_2) = \frac{1}{1 - b\Phi_2} \tag{8-21}$$

where d_0, d_1, d_2 and b are binary parameters.

Using the adjustable binary parameters given in Table 8-2, Bae *et al.* obtained good fits of LLE data, as Figs. 8-10 and 8-11 show. Figure 8-10 compares calculated and experimental cloud-point curves for the poly(ethylene glycol) (PEG)/water system. Molecular weights of PEG are 3,350, 8,000, and 15,000 g mol^{-1}. Solid lines in the closed-loop phase diagram are calculated; they fit experimental data well.

Table 8-2 Binary parameters* of the empirically extended Flory-Huggins model [Eqs. (8-22) and (8-24)]. PEG = Poly(ethylene glycol); PS = Poly(styrene); PVME = Poly(vinyl methyl ether). The weight-average molecular weight of PVME is $\overline{M}_w$ = 99,000 g mol^{-1}.

	PS/Water				PS/PVME	
	$\overline{M}_w$ of PEG (g mol^{-1})				$\overline{M}_w$ of PS (g mol^{-1})	
	3,350	8,000 (from LLE)	8,000 (from VLE)	15,000	50,000	100,000
b	0.5137	0.4523	0.8750	0.5840	-0.8999	-1.5393
d_0	60.37	46.50	2.3×10^{-7}	25.19	0.0264	0.0443
d_1	-3933.5	-3005.2	-137.25	-1619.6	-9.2490	-16.3640
d_2	-8.3581	-6.4178	0.1161	-3.4401	0	0

* Bae *et al.* (1993).

Figure 8-11 shows the cloud-point curves of two polymer blend systems: poly(styrene) (PS)/poly(vinyl methyl ether) (PVME) for two different molecular weights of PS. This system exhibits LCST that declines with increasing molecular weight of PS. The polydispersity index of PVME ($\overline{M}_w / \overline{M}_n \approx 2.1$) shows that PVME is not a monodisperse polymer, and therefore the equations for a binary system are not truly applicable. Nevertheless, the solid lines calculated using the adjustable parameters in Table 8-2 agree well with experiment.

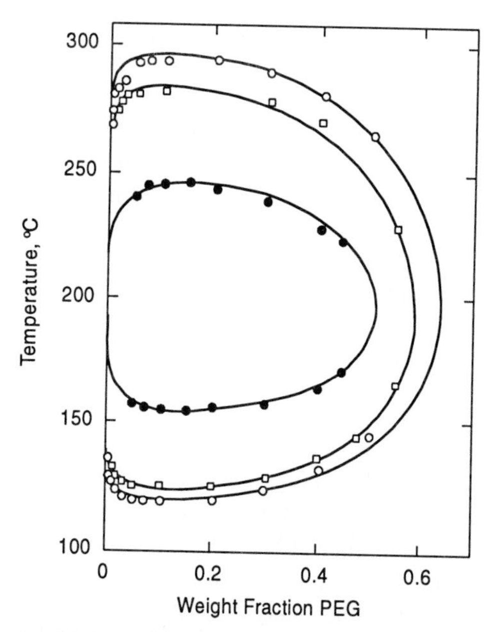

Figure 8-10 Phase diagrams for three PEG/water systems showing cloud-point temperatures as functions of PEG weight fractions. The molecular weight of PEG is 3,350 g mol^{-1} (●); 8,000 g mol^{-1} (□), and 15,000 g mol^{-1} (○). Solid lines are calculated (Bae *et al.*, 1993).

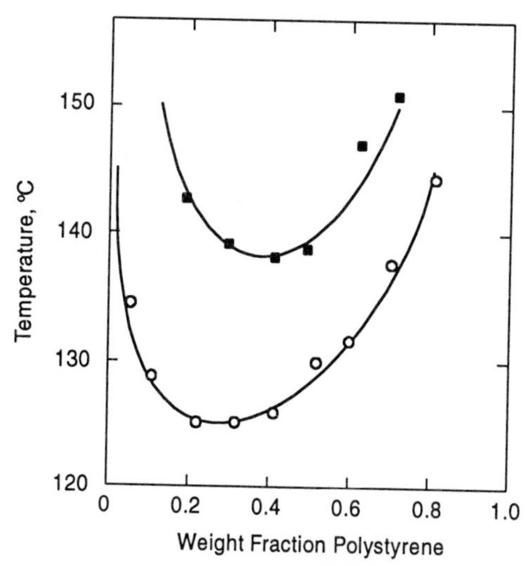

Figure 8-11 Phase diagrams for two PS/PVME systems showing cloud-point temperatures as functions of polystyrene weight fractions. The molecular weights of PS are 50,000 g mol^{-1} (■), and 100,000 g mol^{-1} (○). Solid lines are calculated (Bae *et al.*, 1993).

The empirically extended Flory-Huggins model has essentially no theoretical basis beyond that of the original Flory-Huggins equation. The extended model has a simple algebraic form, uses a few adjustable parameters, and appears to be suitable for representing vapor-liquid and liquid-liquid equilibria, including closed-loop phase diagrams. However, it has no predictive value. Given binary experimental data, the extended Flory-Huggins model is useful only for representing the data in an oversimplified molecular-thermodynamic framework.

A variety of polymer-solution theories has been developed since the original work of Flory and Huggins in the early 1940's. Some proposed theories are extensions of the Flory-Huggins equation, based on a *close-packed lattice*.

For example, Freed and coworkers (Freed, 1985; Bawendi *et al.*, 1987, 1988; Madden *et al.*, 1990, 1990a) developed a *lattice-field theory* for polymer solutions that, in principle, provides an exact mathematical solution of the Flory-Huggins lattice, that is, Freed *et al.* avoid the simplifying assumptions made by Flory and Huggins. In Freed's theory, good agreement was found between calculated values and the computer simulation data of Dickman and Hall (1986). It has also been applied successfully to polymer blends (Dudowicz and Freed, 1991).

Based on Freed's lattice-field theory, Hu and coworkers (1991) reported a double-lattice model for the Helmholtz energy of mixing for binary polymer solutions. This model includes specific interactions such as hydrogen bonding.

A simple molecular-thermodynamic model was developed by Hino *et al.* (1993). Using the incompressible lattice-gas model of ten Brincke and Karasz (1984), Hino *et al.* introduced specific interactions into the expression of the Helmholtz energy of mixing obtained by Lambert *et al.* (1993) by correlating Monte-Carlo-simulation results. Hino's model is conceptionally and mathematically simple. Hino considered a binary mixture of components 1 and 2 that may form specific interactions between similar components and between dissimilar components. Each contact point of a molecule is assumed to interact either in a specific manner with interaction energy $\varepsilon_{ij} + \delta\varepsilon_{ij}$ or in a nonspecific manner with interaction energy ε_{ij}, where i = 1 and j =1 or 2. Both ε_{ij} and $\delta\varepsilon_{ij}$ are negative and independent of temperature. Hino assumed that a fraction, f_{ij}, of the *i-j* interactions is specific and a fraction $1-f_{ij}$ is nonspecific. To obtain a simple expression for the internal energy of mixing, $\Delta_{\text{mix}}U$, Hino also assumed that f_{ii} in the mixture is identical to that in pure i. A similar assumption is made for f_{jj}. These assumptions are consistent with the assumption that f_{ij} depends only on temperature, but is independent of composition, as indicated by Eq. (8-26). With these assumptions, $\Delta_{\text{mix}}U$ is given by

$$\Delta_{\text{mix}}U = \frac{1}{2}N_{12}\omega \tag{8-22}$$

where N_{12} is the total number of 1-2 pairwise contacts and ω is defined by:

$$\omega \equiv \varepsilon + f_{11}(-\delta\varepsilon_{11}) + f_{22}(-\delta\varepsilon_{22}) + f_{12}(2\delta\varepsilon_{12}) \tag{8-23}$$

where ε is an interchange energy:

$$\varepsilon \equiv 2\varepsilon_{12} - \varepsilon_{11} - \varepsilon_{22} \qquad (8\text{-}24)^{16}$$

Further, Hino assumed that f_{ij} is given by the Boltzmann distribution law:

$$\frac{1-f_{ij}}{f_{ij}} = g_{ij} \exp\left[-\frac{\varepsilon_{ij} - (\varepsilon_{ij} + \delta\varepsilon_{ij})}{kT}\right] \qquad (8\text{-}25)$$

where g_{ij} is the ratio of the degeneracy of nonspecific *i-j* interactions; f_{ij} is therefore given by:

$$f_{ij} = \frac{1}{1 + g_{ij} \exp(\delta\varepsilon_{ij} / kT)} \qquad (8\text{-}26)$$

The Helmholtz energy of mixing ($\Delta_{\text{mix}}A$) is obtained by integrating the Gibbs-Helmholtz equation using Guggenheim's athermal entropy of mixing as the boundary condition:

$$\frac{\Delta_{\text{mix}}A}{N_r kT} = \int_0^{1/\tilde{T}} \frac{\Delta_{\text{mix}}U}{N_r \varepsilon} d\left(\frac{1}{\tilde{T}}\right) + \left(\frac{\Delta_{\text{mix}}A}{N_r kT}\right)_{1/\tilde{T}=0} \qquad (8\text{-}27)$$

with

$$\left(\frac{\Delta_{\text{mix}}A}{N_r kT}\right)_{1/\tilde{T}=0} = \frac{\Phi_1^*}{r_1} \ln \Phi_1^* + \frac{\Phi_2^*}{r_2} \ln \Phi_2^* + \frac{z}{2}\left(\Phi_1^* \frac{q_1}{r_1} \ln \frac{\theta_1}{\Phi_1^*} + \Phi_2^* \frac{q_2}{r_2} \ln \frac{\theta_2}{\Phi_2^*}\right) \qquad (8\text{-}28)$$

where N_r is the total number of lattice sites and $\tilde{T}$ is a dimensionless temperature defined as

$$\tilde{T} \equiv \frac{kT}{\varepsilon} \qquad (8\text{-}29)$$

Here, r_i, Φ_i^* and θ_i are, respectively, the number of segments per molecule, volume fraction, and surface fraction of component *i*. Φ_i^* and θ_i are defined by

$$\Phi_i^* \equiv \frac{N_i r_i}{N_1 r_1 + N_2 r_2} \qquad (8\text{-}30)$$

[16] The interchange energy defined here is similar, but not identical, to that given in Eq. (7-71).

$$\theta_i \equiv \frac{N_i q_i}{N_1 q_1 + N_2 q_2} \tag{8-31}$$

where N_i and q_i are, respectively, the number of molecules and the surface area parameter of component i; q_i is related to the number of surface contacts per molecule, zq_i, defined as

$$zq_i \equiv r_i(z-2)+2 \tag{8-32}$$

where z is the lattice coordination number. Hino uses a simple cubic lattice ($z = 6$).

The expression used for N_{12} is based on the expression obtained by Lambert *et al.* (1993) by correlating Monte-Carlo-simulation results for several monomer/r-mer mixtures:

$$N_{12} = N_r \Phi_1^* \Phi_2^* [A + B(\Phi_2^* - \Phi_1^*) + C(\Phi_2^* - \Phi_1^*)^2] \tag{8-33}$$

where

$$A = a_0(r_2) + a_1(r_2)\left[\exp\left(\frac{\omega}{kT}\right) - 1\right] \tag{8-34}$$

$$a_0(r_2) = 6 - \frac{0.9864(r_2 - 1)}{1 + 0.8272(r_2 - 1)} \tag{8-35}$$

$$a_1(r_2) = -1.2374 - \frac{0.09616(r_2 - 1)}{1 + 0.14585(r_2 - 1)} \tag{8-36}$$

$$B(r_2) = \frac{0.8186(r_2 - 1)}{1 + 0.76494(r_2 - 1)} \tag{8-37}$$

$$C = 1.20\left[\exp\left(\frac{\omega}{kT}\right) - 1\right] \tag{8-38}$$

The numerical coefficients in Eqs. (8-35) to (8-38) follow from Monte-Carlo calculations. The temperature dependence of N_{12} is expressed in terms of the dimensionless temperature, now given as kT/ω.

In this model, the number of segments for the smaller molecule, r_1, is always set equal to 1. For mixtures containing low-molecular-weight species, r_2 is determined either from the ratio of UNIQUAC size parameters, proportional to the van der Waals molecular volumes, or r_2 is set equal to the ratio of molar volumes at room temperature. Therefore, size parameter r_2 is not an adjustable parameter, but a preset physical parameter. The adjustable parameters for a binary mixture are energy parameter ε for nonspecific interactions and degeneracy parameter g_{ij} and energy parameter $\delta\varepsilon_{ij}$ for specific interactions, where i =1 or 2 and j = 1 or 2. All these adjustable parameters are determined from experimental data.

For a number of polymer/solvent mixtures, the model of Hino *et al.* (1993) can describe closed-loop phase behavior considering specific interactions between dissimilar components only. Making reasonable assumptions, Hino assigned g_{12} = 5000 and calculated ε from $\tilde{T}^c$ at UCST or LCST and $\delta\varepsilon_{12}$ from the experimental ratio of UCST to LCST. As an example, Fig. 8-12 compares theoretical coexistence curves with experimental data for the system poly(ethylene glycol)/water with two molecular weights of the polymer. Although specific interactions were introduced in a simple way, Hino's model compares favorably with the experimental data.

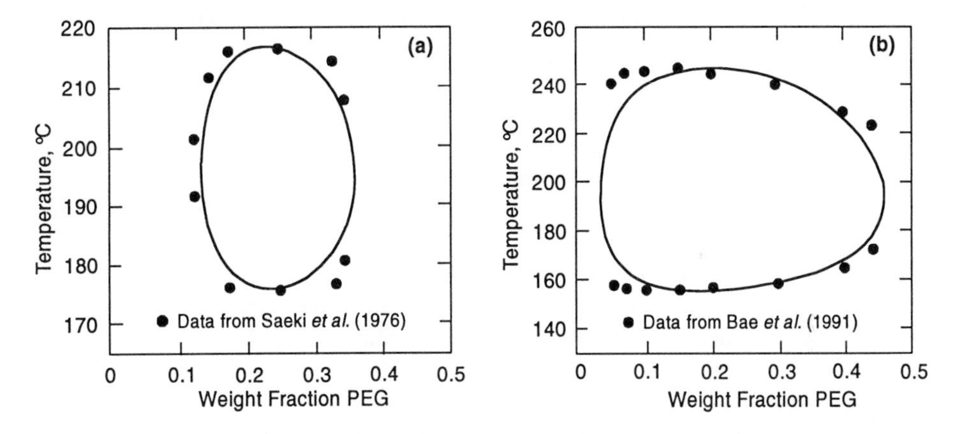

Figure 8-12 Temperature-composition coexistence curves for the system poly(ethylene glycol)/water. (a) $\overline{M}_\eta$ = 2,190 g mol^{-1}; (b) $\overline{M}_n$ = 3,350 g mol^{-1}; —— model of Hino *et al.* (1993).

To account for compressibility and density changes upon isothermal mixing, Sanchez and Lacombe (1976, 1977, 1978), Costas and Sanctuary (1981, 1984), Panayiotou and Vera (1982), and Kleintjens and Koningsveld (1980, 1982) developed new forms of a lattice-fluid model based on Flory-Huggins theory; the central idea here is to use a lattice where some lattice points are not occupied (holes). Sanchez and Balazs (1989) introduced corrections for oriented interactions between dissimilar components. Panayiotou and Sanchez (1991) modified the lattice-fluid theory for polymer solutions to account for strong interactions (hydrogen bonding) between polymer and solvent. Their model is in the form of an equation of state suitable for describing thermodynamic properties of polymer solutions over an extended range of external conditions from the ordinary liquid state to high temperatures and pressures where the solvent may be supercritical, as briefly discussed in the next section.

8.3 Equations of State for Polymer Solutions

Lattice models, discussed in Sec. 8.2, often can describe the main characteristics of liquid mixtures containing nonpolar molecules differing in size and shape. In regular-solution theory, a particular combinatorial entropy of mixing is joined with a simple expression for the energy of mixing obtained from summing the energies of interaction between neighboring sites. However, for quantitative work, this simple approach is inadequate when tested against experimental data. Even for mixtures of *n*-alkanes (Flory *et al.*, 1964, 1965, 1967), the excess thermodynamic properties cannot be described satisfactorily by lattice theory. In particular, changes of volume upon mixing are beyond the scope of such theory. However, experiments show that when liquid mixtures are formed at constant temperature and pressure, small changes of volume upon mixing are the rule rather than the exception, even for mixtures of nonpolar molecules.

Observed excess entropies of mixtures often deviate markedly from calculated combinatorial contributions. That deviation, coupled with observed changes of volume upon mixing, directs attention to the major deficiency of lattice theory: the need to take into account additional properties of the pure components beyond those that reflect molecular size and potential energy. These properties are manifested in *P-V-T* behavior, or in the equation of state.

In general, pure fluids have different free volumes, i.e. different degrees of expansion. When liquids with different free volumes are mixed, that difference contributes to the excess function. Differences in free volumes must be taken into account, especially for mixtures of liquids whose molecules differ greatly in size. For example, in a solution of a polymer in a chemically similar solvent of low molecular weight, there is little dissimilarity in intermolecular interactions but the free volume dissimilarity is significant. The low-molecular-weight solvent may be much more dilated than the liquid polymer; the difference in dilation (or free volume) has an important effect on solution properties. An example is propane and polyethylene at room temperature, where propane, far above its normal boiling point, is much dilated while polyethylene is not.

To develop an equation of state (EOS) for liquids and liquid mixtures, one convenient way is to start with an expression for the canonical partition function utilizing concepts similar to those used by van der Waals. The *Prigogine-Flory-Patterson theory* (Prigogine, 1957; Flory, 1965, 1970; Patterson, 1969, 1970) provides a successful example. Another possibility is to construct a partition function for large molecules to provide an equation of state based on a lattice-with-holes theory, as discussed, for example by Bonner *et al.* (1972) and developed by Eichinger and Flory (1968) and Simha and Somcynsky (1969, 1971); a successful version, developed by Sanchez and Lacombe (1976, 1977), has been extended by Panayiotou and Sanchez (1991) to include associated polymer solutions.

Advances in statistical thermodynamics have brought to the forefront tangent-sphere models of chain-like fluids. These models abandon the lattice picture; they model polymers as freely-jointed tangent-spheres where nonbonded spheres interact through a specified intermolecular potential. These are, for example, the generalized Flory model developed by Hall and coworkers (Dickman and Hall, 1986; Honnell and Hall, 1989; Smith *et al.*, 1995), and, as introduced in Sec. 7.17, the statistical-associated-fluid theory (SAFT) developed by Chapman *et al.* (1990) and the perturbed-hard-sphere-chain (PHSC) theory of Song *et al.* (1994).

Lambert *et al.* (1998) give a comprehensive review of equations of state for molten polymers and for mixtures of polymers with solvents or other polymers. We present here a brief survey of some equations of state and the basic ideas for their derivation.

Prigogine-Flory-Patterson Theory

This statistical-thermodynamic model is based on the fundamental ideas of van der Waals:

- The structure of a fluid is determined primarily by the molecules' repulsive forces.
- The contribution of attractive forces is taken into account by assuming that the molecules are situated in a homogeneous and isotropic field determined by the (attractive) intermolecular potential. This field follows from averaging (smearing) attractive forces. A consequence of this averaging is that, at constant composition, the field is given by simple functions of density and temperature.

Numerous articles have presented van der Waals-type theories for dense fluids containing small, spherical molecules (Henderson, 1974; Swinton, 1976). However, only a few authors (Flory, 1970; Patterson, 1969, 1970) have applied these ideas to fluids containing large molecules, because for such molecules consideration must be given to *external*[17] (rotational, vibrational) degrees of freedom in addition to translational degrees of freedom. An approximation for doing so was suggested by Prigogine (1957) but since this approximation is valid only at high (liquid-like) densities, care must be taken when using Prigogine's suggestion over a wide density range. As pointed out by Scott and van Konynenburg (1970), theories using Prigogine's assumption are qualitatively incorrect at low densities, unless that assumption is modified, as indicated in Sec. 7.15.

[17] Here external refers to those (high-amplitude, low-frequency) rotations and vibrations that depend on density. The division between external and internal is somewhat arbitrary. At very high densities, all degrees of freedom are external.

For liquid mixtures, especially polymer solutions, the Prigogine-Flory-Patterson theory has proved to be useful. We now summarize the essential steps in that theory. First, we discuss pure liquids and then extend the discussion to liquid mixtures.

As shown in Sec. 7.15, the generalized van der Waals partition function for a one-component polyatomic fluid can be written as

$$Q(T,V,N) = \frac{1}{N!}\left(\frac{V_f}{\Lambda^3}\right)^N [q_{\text{ext}}(V)]^N [q_{\text{int}}(T)]^N \left[\exp\left(\frac{-E_0}{2kT}\right)\right]^N \tag{8-39}$$

where N is the number of molecules in total volume V at temperature T, $E_0/2$ is the mean intermolecular potential energy experienced by one molecule due to the attractive forces from all other molecules, and Λ is the de Broglie wavelength that depends only on temperature and molecular mass. V_f is the free volume, i.e., the volume available to the center of mass of one molecule as it moves in volume V. The product $[q_{\text{ext}}(V)\, q_{\text{int}}(T)] = q_{\text{r,v}}$ [see Eq. (7-219)] represents the contribution (per molecule) from rotational and vibrational degrees of freedom, whereas contributions from translational degrees of freedom are given by the first bracketed term of Eq. (8-39).[18]

Following Flory (1964, 1965, 1967, 1970) we subdivide each of the N molecules into r segments. The definition of a segment is essentially arbitrary; e.g., it is appropriately considered to be an isometric portion of a chain molecule.[19]

For large, polyatomic molecules in a condensed phase, the number of external degrees of freedom cannot be estimated from first principles. Instead, we define a parameter c (per segment) such that $3rc$ is the number of *effective* external degrees of freedom per molecule.[20] In this context, "effective" follows from Prigogine's approximation that external rotational and vibrational degrees of freedom can be considered as equivalent translational degrees of freedom. This assumption enables us to postulate a useful expression for $[(V_f\,\Lambda^{-3})q_{\text{ext}}]$:

$$\frac{V_f}{\Lambda^3} q_{\text{ext}} = \left(\frac{V_f}{\Lambda^3}\right)^{rc} \qquad \text{for } rc \geq 1 \tag{8-40}$$

For argon-like molecules ($r = 1$), $rc = 1$, and for all other molecules ($r > 1$), $rc > 1$. Product rc reflects the number of rotational and vibrational motions per molecule that are affected by the presence of neighbors. Indirectly, therefore, rc is often a

[18] The meaning of q used here should not be confused with that used in earlier sections of this chapter.

[19] For $r > 1$, a combinatorial factor Q^C has to be included in Eq. (8-39) to take into account the degeneracy caused by disposition of segments in space. In terms of a lattice model (compare Sec. 8.2), Q^C expresses the number of ways of arranging the segments of N molecules over a spatial array of rN sites. As long as we are concerned only with the equation of state, specification of Q^C is not required; it suffices here to assume that it is independent of volume. In that event, it does not contribute to the equation of state.

[20] Here parameter c is defined *per segment*. In Chaps. 7 and 12, c is defined *per molecule*.

measure of molecular size because a large molecule has more external rotations and vibrations than a small molecule. However, product rc reflects also the looseness (or flexibility) of a molecule. Thus rc for a stiff rod is smaller than that for a soft (rubber-like) rod having the same number of segments.

To reduce the partition function to practice after substituting Eq. (8-40) into Eq. (8-39), we require expressions for the free volume V_f and potential energy E_0.

For the free volume V_f for *one molecule* containing r segments Flory (1970) assumes

$$V_f = \tau r v^* (\tilde{v}^{1/3} - 1)^3 \tag{8-41}$$

where v^* is the characteristic or hard-core volume of a segment, $\tilde{v} = v / v^*$ is the reduced volume, $v = V/(Nr)$ (i.e., the volume available to one segment), and τ is a numerical factor.[21]

In view of the short range of attractive forces operating between uncharged molecules, potential energy E_0 may be considered additive in the molecular surface areas of contact. Therefore Flory proposed for *one molecule*,

$$\frac{E_0}{2} = \frac{-rs\eta}{2v} \tag{8-42}$$

where s is the number of contact sites per segment[22] (proportional to the surface area per segment) and $-\eta/v$ is the intermolecular energy per contact.

Substitution of Eqs. (8-40) to (8-42) into Eq. (8-39) gives a partition function of the form

$$Q = (\text{constant})(\tilde{v}^{1/3} - 1)^{3Nrc} \exp\left(\frac{Nrc}{\tilde{v}\tilde{T}}\right) \tag{8-43}$$

where the reduced temperature is defined by

$$\tilde{T} = \frac{T}{T^*} = \frac{2v^* ckT}{s\eta} \tag{8-44}$$

[21] Specification of τ is not required because it disappears in the process of differentiation for obtaining the equation of state and it cancels in taking differences with respect to the pure components when deriving excess functions for mixtures.

[22] $s \approx q/r$ where q is the number of contact sites per molecule. For a simple long chain, q is related to r by $zq = r(z-2) = 2$, where z is the lattice coordination number; however, for model flexibility, Flory preferred to avoid using an explicit lattice geometry. Therefore, the relation between s and r remained unspecified.

and the "constant" is independent of V.[23] From Eq. (8-43), the equation of state is obtained by differentiation according to Eq. (7-209). Expressed in reduced form, it is[24]

$$\boxed{\frac{\tilde{P}\tilde{v}}{\tilde{T}} = \frac{\tilde{v}^{1/3}}{\tilde{v}^{1/3}-1} - \frac{1}{\tilde{v}\tilde{T}}} \tag{8-45}$$

where the reduced pressure is

$$\tilde{P} = \frac{P}{P^*} = \frac{2Pv^{*2}}{s\eta} \tag{8-46}$$

The characteristic parameters P^*, v^*, and T^* satisfy the equation

$$P^*v^* = ckT^* \tag{8-47}$$

To use the equation of state, we must know the characteristic parameters. For pure fluids, these parameters can be determined from volumetric data in several ways. One method, proposed by Flory (1964, 1965, 1967) is to determine them from data at (essentially) zero pressure for density, thermal expansion coefficient α_p, and thermal pressure coefficient γ. In the limit $P \to 0$ the equation of state takes the simple form

$$\tilde{T} = \frac{\tilde{v}^{1/3}-1}{\tilde{v}^{4/3}} \tag{8-48}$$

For liquids at (essentially) zero pressure, it follows that

$$\tilde{v}^{1/3} - 1 = \frac{\alpha_p T}{3}(1+\alpha_p T)^{-1} \tag{8-49}$$

where

$$\alpha_p = \frac{1}{v}\left(\frac{\partial v}{\partial T}\right)_{P=0} \tag{8-50}$$

If experimental data are available for v as a function of T, these equations suffice to determine $\tilde{v}$ and $\tilde{T}$; then, for a given v and T, we obtain v^* and T^*. Differentiation

[23] While "constant" depends on N and T, this is of no concern in deriving the equation of state.

[24] Because of Eq. (8-40), Eq. (8-45) is restricted to liquids whenever $rc > 1$.

of Eq. (8-45) with respect to temperature, followed by substitutions into Eqs. (8-46) and (8-47), yields

$$P^* = \gamma T \tilde{v}^2 \tag{8-51}$$

where $\gamma \equiv (\partial P/\partial T)_v$, taken in the limit at zero pressure.[25]

The advantage of Flory's method is that volumetric data for liquids are required only at low (e.g., atmospheric) pressure. However, very accurate experimental values of α and γ are necessary and these are often not available. Further, when determined by this method, the parameters are temperature dependent. Table 8-3 gives characteristic parameters of common solvents and Table 8-4 gives parameters for some common polymers at various temperatures.

Another method to determine the characteristic parameters is to fit *P-V-T* data directly to Eq. (8-45) over a wide range of pressures and temperatures. Lichtenthaler *et al.* (1978) used this method to determine characteristic parameters for seven dimethylsiloxane polymers of different molecular weights. Their results are given in Table 8-5, where the last column shows that the Prigogine-Flory equation of state represents all *P-V-T* data with a standard deviation of better than ±0.3%. (Better agreement could be obtained by letting parameters v^*, T^*, and P^* vary slightly with temperature).

To extend Eq. (8-43) to mixtures, we use two assumptions:

1. Hard-core volumes of the components are additive.
2. The intermolecular energy depends in a simple way on the surface areas of contact between solvent molecules and/or segments.

The first assumption is implicit in the partition function. The second assumption was anticipated by Eq. (8-42), expressing the energy as proportional to the surface as measured by the number of contact sites (per segment) s. Equation (8-42) rests on the assumption that intersegmental attractions are short range when compared with segment dimensions.

Mixing is assumed to be random and essentially unaffected by differences in the strength of interaction between neighboring species. For a binary mixture, the segments of all molecules in the mixture are arbitrarily chosen to be of equal core volume; differences in molecular size are then reflected only in parameter r. The partition function for a binary mixture containing N molecules is

$$Q = (\text{constant})\overline{Q}^{C}(\tilde{v}^{1/3} - 1)^{3N\bar{r}\bar{c}} \exp\left(\frac{-\overline{E}_0}{2kT}\right) \tag{8-52}$$

[25] Coefficients α_p and γ are related to isothermal compressibility κ_T through

$$\kappa_T = -\frac{1}{v}\left(\frac{\partial v}{\partial P}\right)_T = \frac{\alpha_p}{\gamma}$$

Table 8-3 Molar volumes and characteristic parameters of some low-molecular-weight liquids.

Liquid	Temp. (°C)	v (cm^3 mol^{-1})	v^* (cm^3 mol^{-1})	T^* (K)	P^* (bar)	Ref.†
CCl_4	25	97.08	75.10	4697	5694	(a)
c-C_6H_{12}	25	108.75	84.27	4719	5317	(a)
c-$CH_3C_6H_{11}$	46	131.57	101.30	4965	4606	(b)
$C(CH_3)_4$	0	118.03	86.24	3762	3945	(a)
n-C_5H_{12}	25	116.10	85.34	4158	4062	(c)
n-C_6H_{14}	25	131.57	99.65	4448	4229	(a)
n-C_7H_{16}	25	147.51	113.70	4652	4292	(c)
n-C_8H_{18}	25	163.57	127.87	4836	4333	(c)
i-C_8H_{18}‡	25	166.09	129.08	4760	3700	(d)
n-$C_{10}H_{22}$	25	195.91	155.75	5093	4459	(c)
n-$C_{12}H_{26}$	25	228.58	184.30	5339	4550	(d)
n-$C_{16}H_{34}$	25	294.19	239.77	5548	4635	(c)
C_6H_6	25	89.40	69.21	4708	6280	(a)
$CH_3(C_6H_5)$	25	106.90	84.58	5049	5610	(e)
$C_2H_5(C_6H_5)$	25	123.07	98.33	5176	5510	(g)
$(C_6H_5)_2$	70	155.28	125.14	6132	6573	(a)
$Cl(C_6H_5)$	20	101.84	82.43	5320	6000	(f)
$SiCl_4$	25	115.47	86.71	4358	4605	(a)
$TiCl_4$	20	109.80	87.88	5118	5485	(a)
$SnCl_4$	20	116.92	91.50	4771	5401	(a)
$CH_3COC_2H_5$	25	90.15	68.94	4557	5820	(f)
n-C_6F_{14}	25	202.23	145.32	3976	3617	(b)
n-C_7F_{16}	50	235.10	168.60	4292	3576	(b)
c-$CF_3C_6F_{11}$	46	202.60	146.40	4299	3890	(b)

† (a) A. Abe and P. J. Flory, 1965, *J. Am. Chem. Soc.*, 87: 1838; (b) A. Abe and P. J. Flory, 1966, *J. Am. Chem. Soc.*, 88: 2887; (c) P. J. Flory, J. L. Ellenson, and B. E. Eichinger, 1968, *Macromolecules*, 1: 279; (d) A. Heintz and R. N. Lichtenthaler, 1980, *Ber. Bunsenges. Phys. Chem.*, 84: 890; (e) R. S. Chahal, W. P. Kao, and D. Patterson, 1973, *J. Chem. Soc. Faraday Trans. 1*, 69: 1834; (f) P. J. Flory and H. Shih, 1972, *Macromolecules*, 5: 761; (g) P. J. Flory and H. Höcker, 1971, *Trans. Faraday Soc.*, 67: 2258, 2270.

‡ 2,2,4-trimethylpentane.

where $\overline{Q}^C$ is the combinatorial factor, $\overline{E}_0/2$ is the mean intermolecular energy for the entire mixture, and

$$N = N_1 + N_2 \tag{8-53}$$

$$\bar{r} = \frac{r_1 N_1 + r_2 N_2}{N} \tag{8-54}$$

$$\bar{c} = \frac{c_1 r_1 N_1 + c_2 r_2 N_2}{\bar{r} N} = c_1 \Phi_1^* + c_2 \Phi_2^* \qquad (8\text{-}55)$$

with $\Phi_2^* = r_2 N_2 / \bar{r} N = 1 - \Phi_1^*$, the segment fraction of component 2, as defined in Eq. (8-4). Equations (8-54) and (8-55) give chain-length parameter $\bar{r}$ and external-degrees-of-freedom parameter $\bar{c}$ as averages of the pure-component parameters.

Table 8-4 Specific volumes and characteristic parameters of common polymers in the amorphous liquid state.

Polymer	Temp. (°C)	v_{sp} (cm^3 g^{-1})	v_{sp}^* (cm^3 g^{-1})	T^* (K)	P^* (bar)	Ref.†
Poly(dimethyl	20	1.0265	0.8381	5494	3430	(a)
siloxane),	25	1.0312	0.8395	5528	3410	(a)
$\overline{M}$ ‡$\approx 10^5$	80	1.0844	0.8563	5907	3120	(a)
	200	1.2140	0.9020	6756	2530	(a)
Polymethylene	25	1.1820	1.0000	6500	4857	(b)
$\overline{M}_n$‡$\approx 10^5$	100	1.2400	1.0130	7010	4731	(b)
	150	1.2830	1.0260	7359	4585	(b)
Polyisobutylene,	0	1.0756	0.9463	7430	4480	(c)
$\overline{M}_\eta \approx 4\times10^4$	25	1.0906	0.9493	7577	4480	(c)
	100	1.1376	0.9597	8029	4312	(c)
	150	1.1706	0.9681	8338	4187	(c)
Poly(*cis*-1,4-isoprene) (natural rubber), $\overline{M}_n \approx 4\times10^4$	25	1.0951	0.9342	6775	5192	(d)
Polystyrene,	0	0.9204	0.8068	7251	5620	(e)
$\overline{M}_n > 5\times10^4$	25	0.9336	0.8098	7420	5470	(e)
	100	0.9746	0.8205	7948	5060	(e)
	200	1.0329	0.8378	8655	–	(e)
Poly(ethylene oxide), $\overline{M}_n \approx 5700$	30	0.9127	0.7532	6469	6720	(f)
Poly(propylene oxide), $\overline{M}_n \approx 5\times10^5$	61	1.0288	0.8424	6330	4986	(f)

† (a) H. Shih and P. J. Flory, 1972, *Macromolecules,* 5: 758; (b) P. J. Flory, B. E. Eichinger, and R. A. Orwoll, 1968, *Macromolecules,* 1: 287; (c) B. E. Eichinger and P. J. Flory, 1968, *Macromolecules,* 1: 285; B. E. Eichinger and P. J. Flory, 1968, *Trans. Faraday Soc.,* 64: 2035; (e) H. Höcker, G. J. Blake, and P. J. Flory, 1971, *Trans. Faraday Soc.,* 67: 2251; (f) C. Booth and C. J. Devoy, 1971, *Polymer,* 12: 309, 320.

‡ For definition of average molecular weight see Sec. 8.1.

Table 8-5 Characteristic parameters for dimethyl siloxanes from *P-V-T* data in the range $298.15 \le T \le 343.15$ K and $1 \le P \le 1000$ bar (Lichtenthaler *et al.*, 1978).

Substance†	$\overline{M}_n$	$\overline{M}_w / \overline{M}_n$‡	v^*_{sp} (cm^3 g^{-1})	T^* (K)	P^* (bar)	$\sigma(v_{sp}) \times 10^4$ § (cm^3 g^{-1})
HMDS	162.38	1.00	0.9995	4468	3253	33
PDMS 3	594	1.12	0.8780	5070	3078	23
PDMS 10	958	1.48	0.8694	5288	3133	20
PDMS 20	1540	1.22	0.8531	5395	3156	18
PDMS 100	4170	1.42	0.8412	5470	3230	19
PDMS 350	6560	1.71	0.8403	5554	3115	18
PDMS 1000	7860	2.17	0.8403	5554	3115	18

† HMDS = hexamethyl disiloxane; PDMS = polydimethyl siloxane.
‡ Polydispersity index (see Sec. 8.1).
§ σ, standard deviation in specific volume v_{sp}.

The definitions used above permit calculation of Φ_i^* in other ways. The molecular characteristic volume of species i is given by $V_i^* = r_i v^*$, and thus $r_2/r_1 = V_2^*/V_1^*$; V_i^* follows from $\tilde{v}_i = V_i / V_i^*$ where $V_i = r_i v$ is the volume per molecule. We may then write for the segment fraction

$$\Phi_2^* = \frac{N_2 V_2^*}{N_1 V_1^* + N_2 V_2^*} = \frac{m_2 v^*_{sp_2}}{m_1 v^*_{sp_1} + m_2 v^*_{sp_2}} = \frac{x_1}{x_1 + (r_2 / r_1) x_2} \tag{8-56}$$

where m_i is the mass of component i in the mixture, $v^*_{sp_i}$ is the characteristic specific volume and $x_i = N_i / N$ is the mole fraction.

Assuming random mixing of surface contacts between molecules, the energy $\overline{E}_o / 2$ of the mixture is given by

$$\frac{\overline{E}_o}{2} = -\frac{(A_{11}\eta_{11} + A_{22}\eta_{22} + A_{12}\eta_{12})}{v} \tag{8-57}$$

where A_{ij} is the number of i-j contacts; each contact is characterized by the energy $-\eta_{ij}/v$. From definitions given above [see Eq. (8-42)], it follows that

$$2A_{11} + A_{12} = s_1 r_1 N_1 \tag{8-58}$$

$$2A_{22} + A_{12} = s_2 r_2 N_2 \tag{8-59}$$

For a random mixture,

$$A_{12} = s_1 r_1 N_1 \theta_2 = s_2 r_2 N_2 \theta_1 \tag{8-60}$$

where the surface fractions θ_1 and θ_2 are defined by

$$\theta_2 = 1-\theta_1 = \frac{s_2 r_2 N_2}{(s_1 r_1 N_1 + s_2 r_2 N_2)} = \frac{(s_2/s_1)\Phi_2^*}{\Phi_1^* + (s_2/s_1)\Phi_2^*} \tag{8-61}$$

Substitution of Eqs. (8-58) to (8-61) into Eq. (8-57) gives

$$\frac{\bar{E}_o}{2} = -(\theta_1\eta_{11} + \theta_2\eta_{22} - \theta_1\theta_2\Delta\eta)\frac{\bar{s}\bar{r}N}{2v} \tag{8-62}$$

where

$$\Delta\eta = \eta_{11} + \eta_{22} - 2\eta_{12} \tag{8-63}$$

and

$$\bar{s} = \frac{s_1 r_1 N_1 + s_2 r_2 N_2}{\bar{r}N} = s_1\Phi_1^* + s_2\Phi_2^* \tag{8-64}$$

By analogy with the energy for a pure component, we define for the mixture

$$\frac{-\bar{E}_o}{2\bar{r}N} = \frac{P^* v^*}{\tilde{v}} \tag{8-65}$$

Comparison of this equation with Eqs. (8-46) and (8-62) gives

$$P^* = P_1^*\Phi_1^* + P_2^*\Phi_2^* - \Phi_1^*\theta_2 X_{12} \tag{8-66}$$

where X_{12} is an *interaction parameter* defined by[26]

$$X_{12} \equiv \frac{s_1\Delta\eta}{2v^{*2}} \tag{8-67}$$

Parameter X_{12} is analogous to interchange energy w of Eq. (7-71), but X_{12} has dimensions of energy density instead of energy.

Since Eq. (8-47) applies also to the mixture, with the aid of Eq. (8-55), we obtain

$$T^* = \frac{P^*}{(P_1^*/T_1^*)\Phi_1^* + (P_2^*/T_2^*)\Phi_2^*} \tag{8-68}$$

[26] Note that $X_{12} \neq X_{21} = \dfrac{s_2\Delta\eta}{2v^{*2}}$.

The partition function for the mixture [Eq. (8-52)] has the same form as that for a pure liquid. We assume that the combinatorial factor $\overline{Q}^C$ is independent of volume and temperature and that the "equation of state" (or residual) part, $(\tilde{v}^{1/3}-1)^{3Nrc}\exp(-E_0/2kT)$, does not depend on the detailed structure of the fluid. The equation of state for the mixture, therefore, is identical to Eq. (8-45). However, reduced variables $\tilde{P}=P/P^*$ and $\tilde{T}=T/T^*$ of the mixture depend on composition, as specified by Eqs. (8-66) and (8-68).

The thermodynamic properties of the mixture are directly related to the partition function[27] and can be calculated using Eq. (8-52). Pure-component properties are obtained in the same way from the partition function, either with $N_1 = 0$ or $N_2 = 0$. Thermodynamic mixing properties can now be calculated in the usual way. For example, the entropy of mixing, $\Delta_{\text{mix}}S$ follows as the sum of two contributions,

$$\Delta_{\text{mix}}S = \Delta S^C + S^R \tag{8-69}$$

where superscripts C and R stand for combinatorial and residual, respectively.

The combinatorial contribution ΔS^C arises from combinatorial factor $\overline{Q}^C$. The residual contribution S^R follows from the "equation-of-state" part of the partition function, determined by differences in intermolecular forces and free volumes. Because $\overline{Q}^C$ is assumed independent of volume and temperature, no combinatorial contribution appears in the enthalpy or in the volume of mixing.

Section 8.2 indicates that it is convenient to split mixing functions into a combinatorial part and a residual part, and allows a choice of several analytical expressions for ΔS^C, the combinatorial contribution to the entropy of mixing. The partition function presented here gives a residual contribution S^R, arising from differences between the equation-of-state parameters for the pure components (Flory, 1970):

$$S^R = -3(N_1V_1^* + N_2V_2^*)\left(\frac{\Phi_1^*P_1^*}{T_1^*}\ln\frac{\tilde{v}_1^{1/3}-1}{\tilde{v}^{1/3}-1} + \frac{\Phi_2^*P_2^*}{T_2^*}\ln\frac{\tilde{v}_2^{1/3}-1}{\tilde{v}^{1/3}-1}\right) \tag{8-70}$$

For the enthalpy of mixing (or excess enthalpy H^E), we obtain

$$\begin{aligned}\Delta_{\text{mix}}H = H^E = H^R &= (N_1V_1^* + N_2V_2^*)[\Phi_1^*P_1^*(\tilde{v}_1^{-1}-\tilde{v}^{-1}) \\ &+ \Phi_2^*P_2^*(\tilde{v}_2^{-1}-\tilde{v}^{-1}) + \frac{\Phi_2^*\theta_2X_{12}}{\tilde{v}}] + P\Delta_{\text{mix}}V\end{aligned} \tag{8-71}$$

where the volume of mixing $\Delta_{\text{mix}}V$ (or excess volume V^E) is given by

[27] See Table B-1 in App. B.

$$\Delta_{\text{mix}} V = V^E = V^R = (N_1 V_1^* + N_2 V_2^*)(\tilde{v} - \Phi_1^* \tilde{v}_1 - \Phi_2^* \tilde{v}_2) \tag{8-72}$$

For a condensed phase at normal (low) pressure, the term $P\Delta_{\text{mix}}V$ is negligible; therefore, at low pressure, we may ignore the distinction between enthalpy and internal energy.

The residual Gibbs energy G^R is obtained by combining Eqs. (8-70) and (8-71):

$$G^R = H^R - TS^R \tag{8-73}$$

From the residual Gibbs energy, we can obtain the activity of component 1. To do so, we redefine N_1 and N_2 to represent numbers of moles. If V_1^* denotes the molar hard-core volume of the solvent (component 1), from Eq. (8-73) the residual part of the activity is

$$\begin{aligned}(\ln a_1)^R = \frac{(\mu_1 - \mu_1^0)^R}{RT} = \frac{\Delta\mu^R}{RT} = \frac{P_1^* V_1^*}{RT}\left[3\tilde{T}_1 \ln \frac{\tilde{v}_1^{1/3} - 1}{\tilde{v}^{1/3} - 1} + (\tilde{v}_1^{-1} - \tilde{v}^{-1})\right] \\ + \frac{V_1^*}{RT}\left(\frac{X_{12}}{\tilde{v}}\right)\theta_2^2 + \frac{P}{RT}\left(\frac{\partial \Delta_{\text{mix}} V}{\partial N_1}\right)_{T,P,N_2}\end{aligned} \tag{8-74}$$

where at normal pressures the last term is negligible.

It is useful to compare $(\ln a_1)^R$, given by Eq. (8-74) and neglecting the last term, with the semiempirical part for $(\ln a_1)^R$ used in the Flory-Huggins theory (Sec. 8.2). In Eq. (8-11) the residual part is given by

$$(\ln a_1)^R = \chi \Phi_2^{*2} \tag{8-75}$$

where χ is the Flory-Huggins interaction parameter. We relate Eq. (8-75) to Eq. (8-74) through identification of χ as the *reduced residual chemical potential* defined by

$$\chi \equiv \frac{\Delta\mu_1^R}{RT\Phi_2^{*2}} \tag{8-76}$$

Contrary to the simple Flory-Huggins theory discussed in Sec. 8.2, χ now varies with composition, as found experimentally.

As formulated above, the Prigogine-Flory-Patterson theory is applicable to solutions of small molecules as well as to polymer solutions. The influence of liquid-state properties on each of the thermodynamic functions H^R, V^R, S^R, and $\Delta\mu_1^R$ is represented by "equation-of-state" terms that depend on the differences of reduced volumes (or their reciprocals) and on characteristic parameters P^* and T^*. These terms depend both on the difference between $\tilde{v}_1$ and $\tilde{v}_2$ and on the residual volume V^R through $\tilde{v}$. In general, they do not vanish for $V^E = 0$. Thus the equation-of-state contributions cannot be interpreted simply in terms of a volume change upon mixing. The equation-of-state terms depend implicitly on X_{12} through $\tilde{v}$. Functions H^R ($\equiv H^E$) and $\Delta\mu_1^R$ include a term that depends explicitly on X_{12}; this term represents an enthalpy contribution that arises from nearest-neighbor interactions even when the mixing process is not accompanied by a volume change.

The Prigogine-Flory-Patterson theory of mixtures requires equation-of-state parameters v^*, T^*, and P^* for the pure components. In addition, two quantities are necessary for the characterization of a binary mixture: Segment surface ratio s_2/s_1 and parameter X_{12} that reflects the energy change upon formation of contacts between unlike molecules (or segments). Both parameters can be chosen to match any two of the several possible experimental thermodynamic properties for the mixture. Because the segments of the two components are chosen to have the same core volume ($v_1^* = v_2^*$), the ratio s_2/s_1 is the ratio of the surfaces per unit core volume that can be estimated from structural data as tabulated, for example, by Bondi (1968). In that case, there remains the single parameter X_{12} to be assigned for a binary mixture. It may be chosen to optimize agreement between calculated and experimental enthalpies of mixing (or dilution) or volumes of mixing, because these properties are independent of the expression used for the combinatorial contribution.

Eichinger and Flory (1968), have investigated the system benzene/polyisobutylene at 25°C. They used experimental data for the volume of mixing, the enthalpy of mixing and solvent activities to test the theory. From structural information they estimated $s_2/s_1 = 0.58$, where subscript 2 refers to the polymer and subscript 1 to the solvent. The enthalpy data and the pure-component parameters from Tables 8-3 and 8-4 yield $X_{12} = 41.8$ J cm^{-3}. With all parameters fixed, values of the reduced residual chemical potential χ were calculated from Eq. (8-74). Figure 8-13 shows the calculated results together with the experimental χ values obtained from solvent activity, using Eq. (8-11). Theory overestimates only slightly the effect of composition on χ.

A modification of the residual chemical potential of the solvent [Eq. (8-74)] is made by appending the term $-(V_1^* / R)Q_{12}\theta_2^2$; Q_{12}, analogous to X_{12}, represents the entropy of interaction between unlike segments and is an entropic contribution to χ, the reduced residual chemical potential [Eq. (8-76)]. The appending term is independent of density and affects only the chemical potential and not the equation of state. If excess-volume data are used to determine X_{12}, the residual chemical potential of the solvent is under-predicted as shown by the dashed line in Fig. 8-14 for natural rubber and benzene (Eichinger and Flory, 1968a). By adjusting Q_{12}, a better representation of χ is obtained without affecting the representation of volumetric properties.

Calculated values of the excess volume, however, do not agree well with experimental data. Theory predicts positive values for V^E but calculated values are too large by a factor of about 2. The excess volume is quantitatively reproduced by theory only with a negative X_{12}. Thus the theory is not able to represent all excess functions with the same binary parameter.

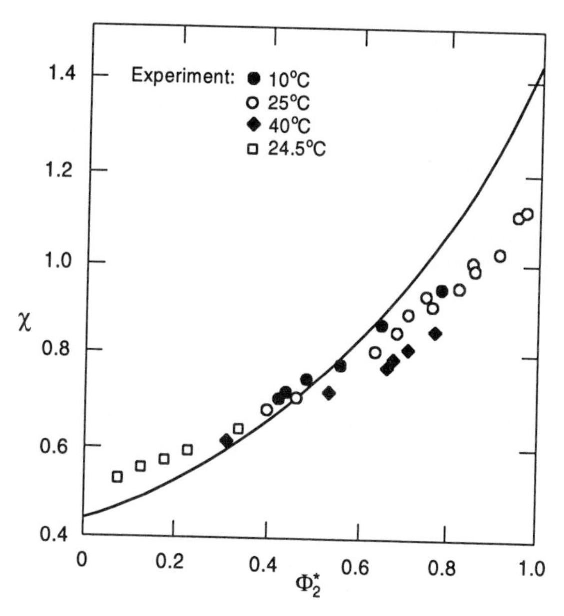

Figure 8-13 Reduced residual chemical potential χ [Eq. (8-76)] of benzene (1) in polyisobutylene (2) at 25°C. Solid curve calculated from Eq. (8-74) with $s_2/s_1 = 0.58$ and $X_{12} = 41.8$ J cm^{-3} (Eichinger and Flory, 1968). The experimental χ was obtained from solvent-activity data using Eq. (8-11).

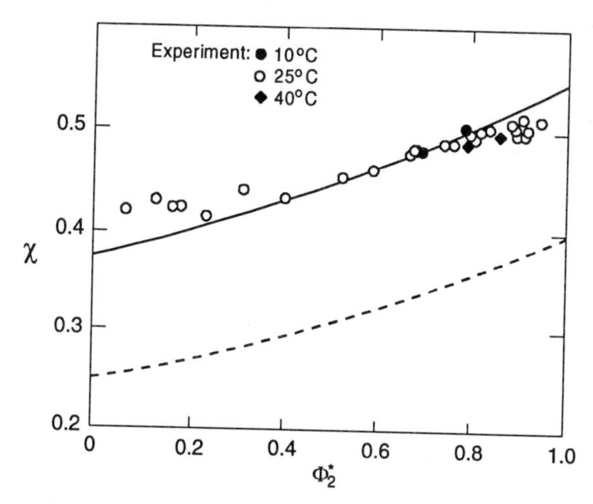

Figure 8-14 Reduced residual chemical potential χ of benzene (1) in natural rubber (2) at 25°C. Dashed curve calculated from Eq. (8-74) with $s_2/s_1 = 0.90$, $X_{12} = 5.86$ J cm^{-3}, and $Q_{12} = 0$. The solid line is calculated with same values of s_2/s_1 and X_{12}, and with $Q_{12} = -0.0184$ J cm^{-3} K^{-1} (Eichinger and Flory, 1968a).

Another deficiency of the theory arises when X_{12} is determined from data at various temperatures. Usually, X_{12} decreases with increasing temperature. For example, Heintz (1977) found such a temperature dependence for cyclohexane/*n*-alkane systems. Heintz (1980) also showed that parameter X_{12} depends on pressure when determined from h^E data at various pressures. The calculated composition dependence of h^E at various pressures is sensitive to variations in X_{12}. To illustrate, Fig. 8-15 shows the difference $\Delta h^E = h^E(P \text{ bar}) - h^E(1 \text{ bar})$ for *n*-dodecane/cyclohexane at 25°C and at 180 and 291 bar. The experimental data show that Δh^E increases with rising pressure. The dashed curves in the upper part of the figure are calculated using Eq. (8-71) with characteristic parameters from Table 8-3, ratio s_2/s_1 = 0.997, and X_{12} adjusted to obtain best agreement. Values for X_{12} are 13.5 and 14.0 J cm^{-3} at 180 and 291 bar, respectively. If calculations at the higher pressures are performed with X_{12} determined from h^E data at ambient pressure (X_{12} = 12.9 J cm^{-3}), the results obtained are shown by the dotted (at 180 bar) and the dashed-dotted (at 291 bar) curves shown in the lower part of Fig. 8-15.

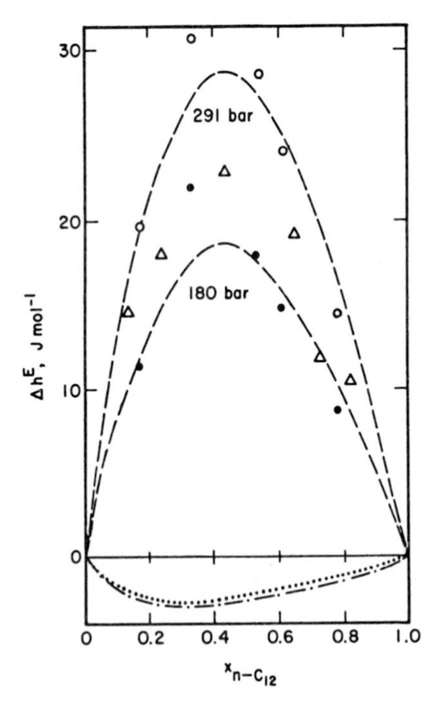

Figure 8-15 Effect of pressure on the excess enthalpy Δh^E [where $\Delta h^E = h^E(P \text{ bar}) - h^E(1 \text{ bar})$] for the system *n*-dodecane/cyclohexane at 25°C. ●, ○ Calorimetric data at 180 and 291 bar, respectively; Δ from volumetric data at 180 bar; – – – from Eq. (8-71) with X_{12} adjusted at 180 bar (X_{12} = 13.5 J cm^{-3}) and at 291 bar (X_{12} = 14.0 J cm^{-3}); ······ (180 bar) and –·–·– (291 bar) from Eq. (8-71) with X_{12} = 12.9 J cm^{-3} obtained at 1 bar (Heintz and Lichtenthaler, 1980).

The quantity Δh^E is predicted poorly by the Prigogine-Flory-Patterson theory when the crucial parameter X_{12} is determined from data obtained at ambient pressures. Heintz attempted to explain the pressure dependence of X_{12} in terms of the orientational order in dense fluids containing long *n*-alkane chains. Using a statistical model for cooperative transitions, he developed an analytical expression for the dependence of X_{12} on temperature and pressure. Incorporation of this model into the Prigogine-Flory-Patterson theory gives better agreement between theory and experiment for all systems investigated by Heintz.

While is evident that Prigogine-Flory-Patterson theory has serious deficiencies, it can explain a phenomenon that has been observed for many polymer/solvent systems (Patterson, 1969, 1970): Partial miscibility at low temperatures and also at high temperatures.[28] The original Flory-Huggins theory can explain only partial miscibility at low temperatures.

We do not here go into details (given by Siow *et al.*, 1972; Zeman and Patterson, 1972) but give only the essential argument. According to traditional Flory-Huggins theory, Flory parameter χ decreases slowly with rising temperature (Patterson, 1969, 1970), as indicated by the enthalpic contribution shown in Fig. 8-16, curve 2. In the traditional theory there is no entropic contribution to χ.

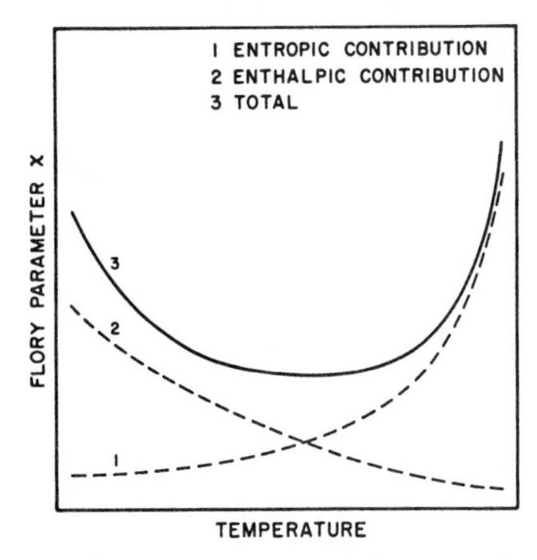

Figure 8-16 Temperature dependence of the χ parameter: curve 1, entropic contribution (due to free volume dissimilarity between polymer and solvent); curve 2, enthalpic contribution (due to contact energy dissimilarity between polymer and solvent); curve 3, total χ.

However, in the Prigogine-Flory-Patterson theory, free-volume (or equation-of-state) effects also contribute to χ. These are entropic contributions; they can be taken into account by parameter Q_{12}, as mentioned above. As temperature rises, the free volume of the solvent increases, especially as the temperature comes close to the critical

[28] For a review, see J. M. Cowie, 1973, *Ann. Rep. Chem. Soc. (Lond.)*, 70A: 173.

of the solvent. However, the free volume of the polymer is nearly constant, increasing only slowly with rising temperature. Because the difference in free volumes increases with temperature, entropic contributions for χ rise with temperature, as indicated by curve 1 in Fig. 8-16. The total value of χ is given by the sum of the enthalpic and entropic contributions, indicated by the top line in Fig. 8-16. At low temperatures and at high temperatures, χ exceeds the limit for complete miscibility; at intermediate temperatures, χ is low enough for the mixture to be miscible in all proportions. The effect of temperature on χ, shown in Fig. 8-16, produces the phase diagram shown in Fig. 8-9(a). Such phase diagrams have been observed for a variety of polymer/solvent systems.

Figures 8-17 and 8-18 show experimental results for the system polystyrene/acetone. The tendency toward limited miscibility rises with the polymer's molecular weight (Siow *et al.*, 1972). Increasing pressure lowers the tendency toward partial miscibility (Zeman and Patterson, 1972), especially at high temperature, because rising pressure decreases the free-volume difference between the two components.

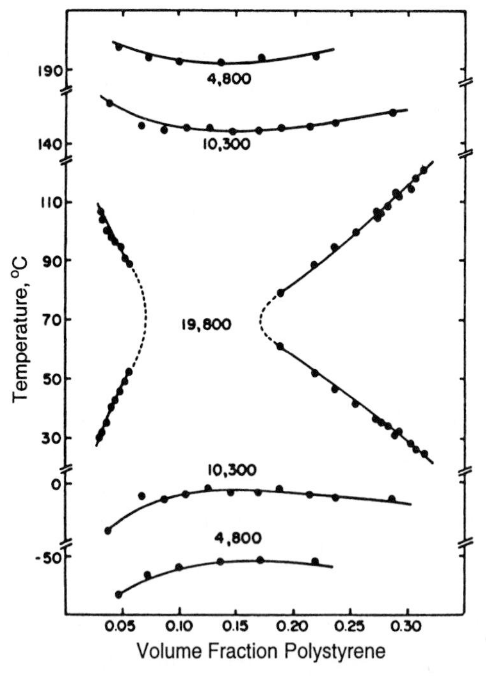

Figure 8-17 Phase diagram (temperature-volume fraction) for the polystyrene/acetone system for indicated polymer weight-average molecular weight (Siow *et al.*, 1972).

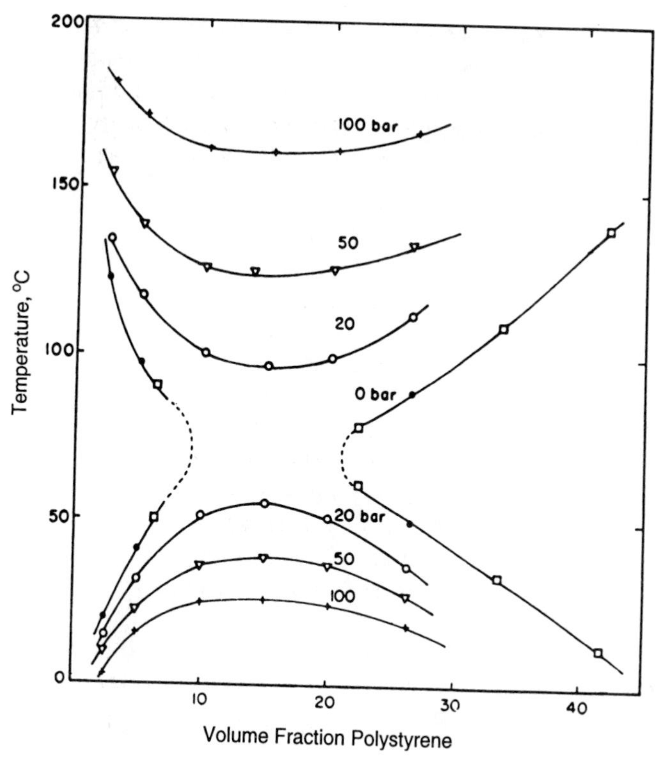

Figure 8-18 Phase diagram (temperature-composition) for a polystyrene fraction of $\overline{M}_w$ = 20,400 g mol^{-1} and for indicated pressures, in acetone (Zeman and Patterson, 1972). □ Results extrapolated from vapor-pressure data to 0 bar (Myrat and Rowlinson, 1965).

The Prigogine-Flory-Patterson theory has also been used to study the effect of compressibility on miscibility in polymer blends. McMaster (1973) used a generalized version of the Prigogine-Flory-Patterson equation of state to show qualitatively how differences in pure-component thermal-expansion coefficients can lead to lower-critical-solution-temperature (LCST) behavior in polymer blends. The effects of molecular weight, pressure and polydispersity were also considered qualitatively. Kammer *et al.* (1989) added a parameter reflecting differences in segment size and illustrated its effect on blend miscibility. In addition to differences in interaction energy and compressibility effects, UCSTs and LCSTs in polymer blends are sensitive to the segment-size parameter. Rostami and Walsh (1985) considered the effect of pressure and molecular weight on the UCST in polymer blends using the Prigogine-Flory-Patterson equation of state. Figure 8-19 illustrates that this equation of state gives reasonable prediction of the increase in UCST with pressure from atmospheric to 1013 bar for polybutadiene ($\overline{M}_n$ = 2350 g mol^{-1}) mixed with polystyrene ($\overline{M}_n$ = 1200 g mol^{-1})

using two binary parameters X_{12} and Q_{12}. X_{12} was determined from fitting experimental enthalpies of mixing and was the same for all molecular weights of either polymer. Q_{12}, determined by fitting the maximum cloud-point temperature at atmospheric pressure, varied with the molecular weights of both polymers.

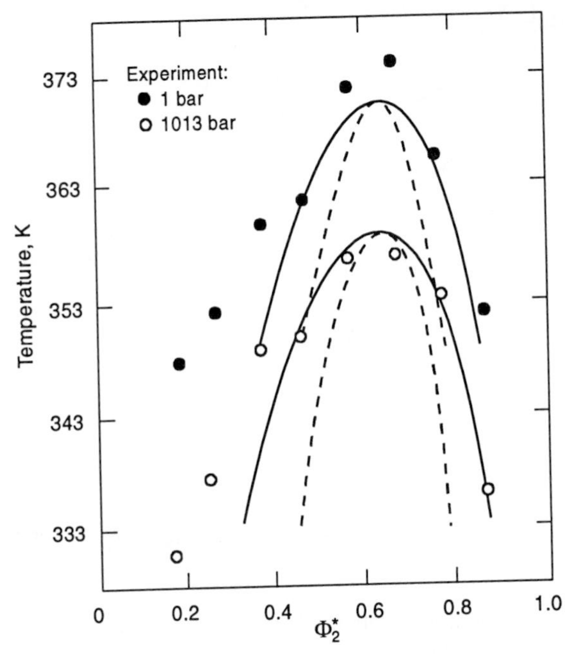

Figure 8-19 Experimental cloud points for polybutadiene (1)/polystyrene (2) blends at 1.013 bar (○) and 1013 bar (●). Binodals (——) and spinodals (– – –) are calculated at each pressure with the Prigogine-Flory-Patterson equation of state with $s_2/s_1 = 0.9$, $X_{12} = 7.0$ J cm^{-3}) and $Q_{12} = 0.0062$ J cm^{-3} K^{-1} (Rostami and Walsh, 1985).

The Prigogine-Flory-Patterson theory corrects two major inadequacies of the conventional Flory-Huggins theory of polymer solutions: First, Flory-Huggins takes into account only the combinatorial entropy and contributions to the enthalpy due to nearest-neighbor interaction in a lattice framework (see Sec. 8.2). The Prigogine-Flory-Patterson theory considers, in addition, contributions to the entropy and enthalpy of mixing [Eq. (8-70)] that follow from differences in liquid-state properties of solute and solvent. This theory can explain negative enthalpies for mixtures of a polymer with a chemically similar solvent. The contribution from nearest-neighbor interactions in nonpolar mixtures is positive, though small; however, there may be large negative equation-of-state contributions. Negative enthalpies of mixing, often observed for polymer solutions, are thus explained (Flory *et al.*, 1968) without invoking an implausible negative interaction parameter χ. Most important, Prigogine-Flory-Patterson theory is able to explain the frequently observed existence of a *lower* critical solution temperature in polymer/solvent systems that lies above the upper critical solution temperature. However, contrary to the original Flory-Huggins theory, the Prigogine-Flory-Patterson theory requires three parameters for each pure component.

These significant improvements are accomplished at considerable cost in algebraic complexity and need for pure-component parameters. Better agreement with experiment doubtless can be achieved with a more refined theory, but the improvement thus gained is likely to require additional binary parameters. An attractive feature of the Prigogine-Flory-Patterson theory is its straightforward extension to multicomponent mixtures (Pouchly and Patterson, 1976) requiring only pure-component and binary parameters.

It is important to remember that the Prigogine-Flory-Patterson theory is incorrect at low densities because, at low densities, Eq. (8-45) does not reduce to the equation of state for an ideal gas. Therefore, this theory is only applicable to fluids and fluid mixtures with liquid-like densities, i.e. liquids or fluids at very high pressures. However, the essential ideas of the Prigogine-Flory-Patterson theory can be generalized to the entire fluid-density range through perturbed hard-chain theory and its variations.

Perturbed-Hard-Chain Theory

The limitations of the Prigogine-Flory-Patterson theory at low densities have been removed by the *perturbed-hard-chain* (PHC) theory. The PHC equation of state and its essential features are presented in Sec. 7.16. Here we only briefly discuss its application to polymer solutions, in particular to the calculation of Henry's constants.

Henry's constant for volatile solute (1) in polymer (2) is defined by

$$H_{1,2} = \lim_{w_1 \to 0} \frac{f_1}{w_1} = \lim_{w_1 \to 0} \frac{RT}{M_1 \upsilon_2} \exp\left(\frac{\mu_1^{HC} + \mu_1^{att} + \mu_1^{SV}}{kT} \right) \tag{8-77}$$

where f_1 is the fugacity and w_1 is the weight fraction of the volatile solute; M_1 is the molar mass of the solute, R is the gas constant, and υ_2 is the specific volume of the polymer. In Eq. (8-77), μ_1^{HC}, μ_1^{att}, and μ_1^{SV} are the chemical potentials from the hard-chain part, attractive part, and second-virial-coefficient part, as calculated from the PCH equation of state for mixtures (Kaul *et al.*, 1980).

Ohzono *et al.* (1984) and Iwai and Arai (1991) applied the PHC equation of state to correlate weight-fraction Henry's constants of hydrocarbon vapors in molten polymers. Figure 8-20 compares calculated and experimental Henry's constants of normal alkanes in polypropylene. With one adjustable binary parameter, Henry's constants are well correlated over a considerable range of temperature. The binary parameter provides a small (but significant) correction to the geometric-mean assumption for calculating the energy of interaction between a solute molecule and a segment of a polymer molecule. Another, similar example is shown in Fig. 10-15.

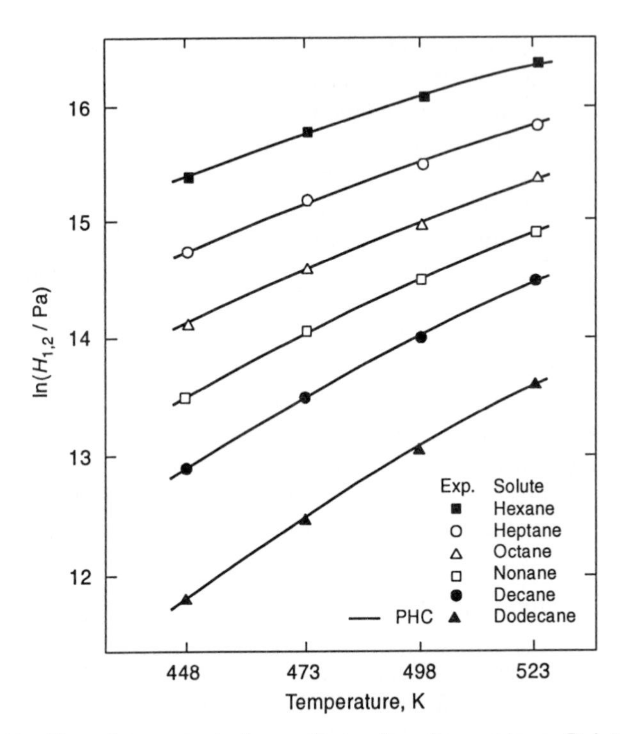

Figure 8-20 Henry's constants for *n*-alkanes in polypropylene. Points are experimental data and curves are calculated using the PHC equation of state (Iwai and Arai, 1991).

Lattice-Fluid Theory

The molecular lattice fluid (LF) theory for liquid and gaseous mixtures developed by Sanchez and Lacombe (1976, 1977, 1987) is formally similar to the Flory-Huggins theory discussed in Sec. 8.2. The essential and important difference is that LF theory includes empty lattice sites (holes). An equation of state characterizes each of the pure components as well as the mixture. Volume changes upon mixing are taken into account, as required for a good description of fluid-phase behavior.

The lattice is occupied by both r-mers and vacant lattice sites. For a pure component, the total number of lattice sites N_r is

$$N_r = N_0 + rN \tag{8-78}$$

where N_0 is the number of vacancies and N is the number of molecules, each with r segments. At closest packing, the volume of a molecule, assumed to be independent of temperature and pressure, is rv^*, where v^* is the volume of a segment, equal to the volume of one lattice site. The total volume of the system is

$$V = (N_0 + rN)v^* \tag{8-79}$$

where V now depends on temperature and pressure through N_0.

A reduced density is defined as the fraction of occupied sites,

$$\tilde{\rho} = \frac{1}{\tilde{v}} = \rho v^* = \frac{v^*}{v} = \frac{rN}{N_0 + rN} \tag{8-80}$$

where $\rho = Nr / V$ is the density of segments, $v = V / Nr$ is the volume per segment, and $\tilde{v}$ is the reduced volume.

The energy of the lattice depends only on nearest-neighbor interactions. For a pure component, the only non-zero interaction energy is the mer-mer pair interaction energy ε. Vacancy-mer and vacancy-vacancy interactions energies are zero. The Sanchez-Lacombe model assumes random mixing of vacancies and mers; therefore, the number of mer-mer nearest-neighbors is proportional to the probability of finding two neighboring mers in the system. The lattice energy is

$$E = -\frac{z}{2} N_r \varepsilon \left(\frac{rN}{N_0 + rN} \right)^2 = -N_r \tilde{\rho}^2 \frac{z\varepsilon}{2} \tag{8-81}$$

where z is the coordination number of the lattice.

The configurational partition function for this system is

$$Q = Q^C \exp(-E / kT) \tag{8-82}$$

The combinatorial factor Q^C is identical to that of the Flory-Huggins incompressible-lattice partition function where the solvent is replaced by a vacancy,

$$Q^C = (\text{constant})^N \frac{(N_0 + rN)!}{N_0 ! N!} \frac{1}{(N_0 + rN)^{N(r-1)}} \tag{8-83}$$

where the constant, r and N are independent of volume. However, N_0 depends on volume [Eq. (8-79)], and therefore Q^C contributes to the equation of state.

Substitution of Eqs. (8-81) and (8-83) into Eq. (8-82) and differentiation according to Eq. (7-209), yields the *Sanchez-Lacombe lattice-fluid equation of state*:

$$\frac{\tilde{P}\tilde{v}}{\tilde{T}} = \frac{1}{r} - \left[1 + \tilde{v}\ln\left(1 - \frac{1}{\tilde{v}}\right)\right] - \frac{1}{\tilde{v}\tilde{T}} \tag{8-84}$$

where the reduced (~) and characteristic (*) temperature and pressure are defined by

$$\tilde{P} = \frac{P}{P^*} = \frac{P}{z\varepsilon / 2v^*} \qquad \tilde{T} = \frac{T}{T^*} = \frac{T}{z\varepsilon / 2k} \tag{8-85}$$

Characteristic parameters P^*, v^*, and T^* obey the equation

$$P^* v^* = kT^* \tag{8-86}$$

Reduced volume $\tilde{v} = 1/\tilde{\rho}$, defined in Eq. (8-80) as the ratio of volume v and close-packed volume v^* per segment, is also given by the ratio of the corresponding specific volumes $v_{sp} = 1/\rho_{sp}$ and $v_{sp}^* = 1/\rho_{sp}^*$, where ρ_{sp} and ρ_{sp}^* are mass densities. Size parameter r in Eq. (8-84) is related to the mass of one molecule, m, and to v_{sp}^* by

$$r = m\frac{v_{sp}^*}{v^*} = m\frac{v_{sp}^* P^*}{kT^*} \tag{8-87}$$

A real fluid is characterized either by the three molecular parameters $z\varepsilon$, v^* and r or by the three equation of state parameters, T^*, P^*, and $v_{sp}^* = 1/\rho_{sp}^*$. In principle, this equation of state is suitable for describing thermodynamic properties of fluids over an extended range of external conditions from the ordinary liquid or gaseous state to high temperatures and pressures where the fluid may be supercritical. To use the equation of state, we must know the characteristic parameters. In principle, any experimental configurational thermodynamic property can be used to determine these parameters. However, vapor-pressure data are particularly useful for solvents because they are readily available for a wide variety of fluids. For polymers, the characteristic parameters can be determined by a nonlinear least-square fit of experimental liquid density data over a range of pressures and temperatures. When only limited *P-V-T* data are available, the parameters can be estimated from experimental values of density, thermal expansion coefficient and compressibility at ambient temperature and pressure. Equation-of-state parameters for many fluids and liquid polymers have been reported by Sanchez and Panayiotou (1994).

Equation (8-84) shows that *P-V-T* data for polymer liquids are relatively insensitive to polymer molar mass. As polymer molecular weight increases, the $1/r$ term becomes insignificant. In the limit of infinite polymer molecular weight, Eq. (8-84) suggests a corresponding-state behavior for polymer liquids, illustrated in Fig. 8-21. The lines for each reduced pressure were calculated using Eq. (8-84) in the limit $r \to \infty$. The points are experimental *P-V-T* data for several polymers reduced by appropriate characteristic parameters. A reduced pressure $\tilde{P} = 0$, is essentially atmospheric pressure; $\tilde{P} = 0.25$ is a pressure of the order 1000 bar.

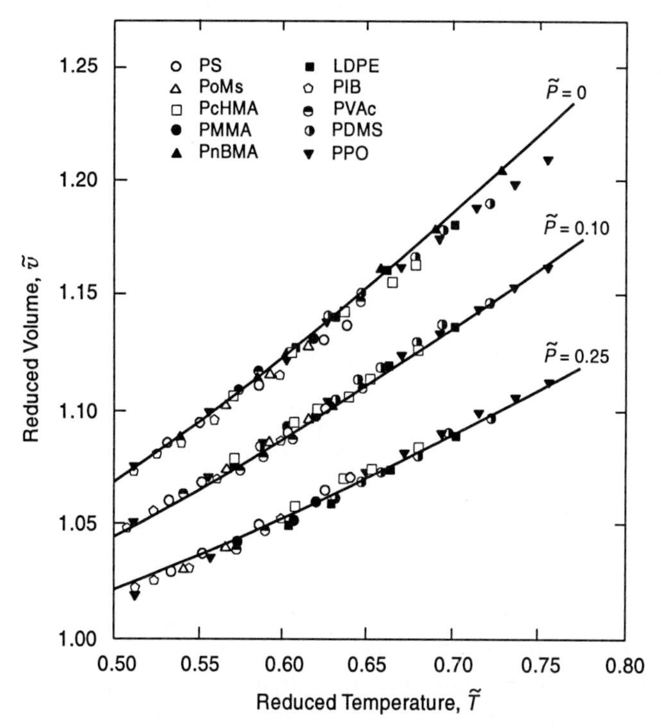

Figure 8-21 Corresponding-states behavior of polymer *P-V-T* data according to the Sanchez-Lacombe equation of state Points are experimental data and lines are calculated from Eq. (8-84) in the limit $r \to \infty$. PS, polystyrene; PoMS, poly(*o*-methyl styrene); PcHMA, poly(cyclohexyl methacrylate); PMMA, poly(methyl methacrylate); PnBMA, poly(*n*-butyl methacrylate); LDPE, low-density polyethylene; PIB, polyisobutylene; PVAc, poly(vinyl acetate); PDMS, poly(dimethyl siloxane); PPO, poly(propylene oxide) (Sanchez and Lacombe, 1978).

To extend the lattice fluid theory to mixtures, we need appropriate mixing rules for the characteristic parameters. There is a fundamental difficulty here because the segment size of any one component is not necessarily equal to that of another component. Each pure component has its own volume v_i^* per segment. In a lattice model, how do we mix segments of different size? Some assumptions are required. The original formulation of the lattice-fluid equation of state for mixtures (Sanchez and Lacombe, 1976, 1978) uses assumptions similar to those used in the Prigogine-Flory-Patterson equation of state. First, the molecular volume of each component is conserved: the molecular volume of a component, determined from its pure-fluid properties $r_i^0 v_i^*$ is equal to its molecular volume in the mixture $r_i v^*$ ($r_i^0 v_i^* = r_i v^*$). Consequently, the chain length of a component in the mixture, r_i, may differ from that for the pure fluid, r_i^0. In addition, the total number of pair interactions in the close-packed mixture is equal to the number of pair interactions of the pure components in their close-packed states. Because the Sanchez-Lacombe equation of state assumes random mixing of segments, these assumptions lead to

$$\frac{1}{v^*} = \sum_i \frac{\Phi_i^*}{v_i^*} \tag{8-88}$$

$$\varepsilon = \sum_i \sum_j \Phi_i^* \Phi_j^* \varepsilon_{ij} = \sum_i \Phi_i^* \varepsilon_{ii} - \left(\frac{2}{z}\right) \sum_{i<j} \sum \Phi_i^* \Phi_j^* \chi_{ij} \tag{8-89}$$

where the segment fraction[29] Φ_i^* is

$$\Phi_i^* = \frac{r_i N_i}{\sum_j r_j N_j} \tag{8-90}$$

and $2\chi_{ij}/z = (\varepsilon_{ii} + \varepsilon_{jj} - 2\varepsilon_{ij})/kT$ is the interaction parameter in terms of the interchange energy [Eq. (7-71)]. ε_{ij} (or equivalently χ_{ij}/z) is an adjustable binary parameter.

With v^* and ε defined by Eqs. (8-88) and (8-89), Eqs. (8-80) and (8-85) define characteristic quantities P^* and T^* for the mixture. The equation of state for the mixture is identical to Eq. (8-84); however, r, explicit in Eq. (8-84), becomes an average value for the mixture, given by

$$\frac{1}{r} = \sum_i \frac{\Phi_i^*}{r_i^0 (v_i^* / v^*)} \tag{8-91}$$

Parameter ε_{ij} characterizes a binary mixture; all other parameters are related to those of the pure components. Alternatively, the characteristic pressure of a binary mixture can be expressed in terms of pure-component characteristic quantities as:

$$P^* = \Phi_1^* P_1^* + \Phi_2^* P_2^* - \Phi_1^* \Phi_2^* \Delta P_{12}^* \tag{8-92}$$

Parameter ΔP_{12}^* is equivalent to parameter χ_{12} in the Prigogine-Flory-Patterson theory [Eq. (8-67)] and has dimensions of energy density. Parameter χ_{12} provides a measure

[29] Sanchez and Lacombe (1978) use close-packed volume fraction, by definition equal to the segment fraction in this chapter.

of the energy change upon formation of contacts between unlike molecules (or segments).

Because the mixing rule for v^* is somewhat arbitrary, a common alternative to Eq. (8-88) is

$$v^* = \sum_i \sum_j \Phi_i^* \Phi_j^* v_{ij}^* \tag{8-93}$$

where $v_{ii}^* = v_i^*$ and v_{ij}^* provides a second binary parameter. Other approximations can be used to count the number of pair interactions. For example, we can assume random mixing of contact sites rather than random mixing of segments (Panayiotou and Vera, 1981, 1982), as well as non-random mixing (Panayiotou and Vera, 1982; Panayiotou 1987), but ε_{ij} typically remains the essential binary parameter.

The lattice-fluid model, summarized above, is applicable to solutions of small molecules as well as to polymer solutions. Like the Prigogine-Flory-Patterson equation of state, the lattice-fluid model and its variations have been used to correlate the composition dependence of the Flory parameter χ (residual chemical potential of solvent) (Panayiotou and Vera, 1982; Panayiotou 1987). These studies show that a binary entropic-interaction parameter (analogous to the Q_{12} parameter in the Prigogine-Flory-Patterson equation of state) is also needed to provide better agreement with measured residual chemical potentials for the solvent.

The essentials of liquid-liquid miscibility are obtained by studying the spinodal condition. According to the lattice-fluid equation of state, immiscibility occurs when the following inequality holds (Sanchez and Lacombe, 1978):

$$\frac{1}{2}\left(\frac{1}{r_1\Phi_1^*} + \frac{1}{r_2\Phi_2^*}\right) - \frac{1}{\tilde{v}}\left(\Delta P_{12}^* v^* / kT + \frac{1}{2}\Psi^2 \tilde{T} P^* \kappa_T\right) < 0 \tag{8-94}$$

where

$$\Psi \equiv \frac{1}{\tilde{v}}\frac{d(1/\tilde{T})}{d\Phi_1^*} - \frac{d(1/r)}{d\Phi_1^*} + \tilde{v}\frac{d(\tilde{P}/\tilde{T})}{d\Phi_1^*} \tag{8-95}$$

and κ_T is the isothermal compressibility of the mixture:

$$\tilde{T}P^*\kappa_T = \tilde{v}[1/(\tilde{v}-1) + 1/r - 2/(\tilde{v}\tilde{T})]^{-1} \tag{8-96}$$

In Eq. (8-94), the first bracketed term is the combinatorial-entropy contribution; $\Delta P_{12}^* v^* / \tilde{v}kT$ is an energetic contribution and $\Psi^2 \tilde{T} P^* \kappa_T / 2\tilde{v}$ is an entropic contribution from the equation of state. Figure 8-22 illustrates the general behavior of these three terms as a function of temperature. The last term makes an unfavorable contribution to the spinodal and favors demixing. Its magnitude increases with increasing temperature and diverges as the vapor-liquid critical temperature T_c is approached, because $\kappa_T \to \infty$ as $T \to T_c$. Hence, according to the lattice-fluid theory, every polymer solution in equilibrium with its vapor should exhibit a LCST prior to reaching its liquid-vapor critical temperature.

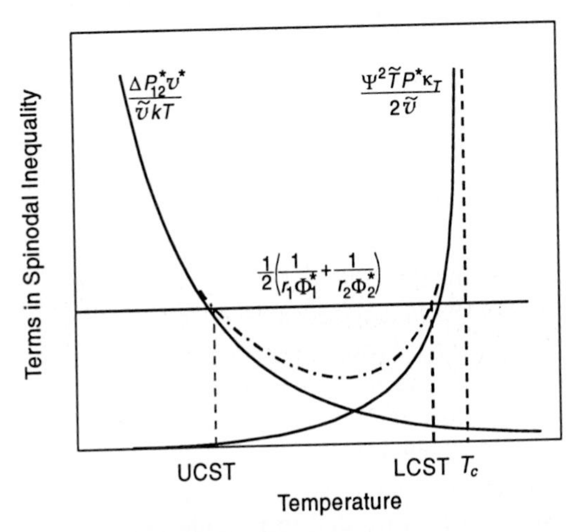

Figure 8-22 Schematic behavior of the three terms in the spinodal inequality [Eq. (8-94)] from the lattice-fluid equation of state as a function of temperature (Sanchez and Lacombe, 1978). The horizontal line represents the combinatorial contribution. The dotted-dashed curve is the sum of the energetic contribution and the entropic contribution. When this sum lies below the combinatorial contribution, the polymer and solvent are miscible. When the sum lies above the combinatorial contribution, the mixture will phase separate. Intersections of the dotted-dashed curve and horizontal line represent upper (UCST) and lower (LCST) critical solution temperatures. T_c is the vapor-liquid critical temperature of the polymer solution.

The lattice-fluid model and its modifications have been successfully used to represent thermodynamic excess properties, vapor-liquid and vapor-liquid-liquid equilibria for a variety of mixtures, as summarized by Sanchez and Panayiotou (1994).

Kim and Paul (1992) investigated the phase behavior of polymer blends containing polycarbonates. Blends of polystyrene (PS) and tetramethyl bisphenol-A polycarbonate (TMPC) show phase separation at temperatures above about 240°C depending on the composition of the blend. The characteristic parameters for PS and TMPC in Table 8-6 were obtained by a non-linear regression fit of Eq. (8-84) to experimental volumetric data in the pressure and temperature ranges 0-500 bar and 220-270°C. The latter includes the range where phase separation of PS/TMPC blends has been observed. Using the pure-component parameters in Table 8-6, an interaction-energy parameter $\Delta P_{12}^* = -0.71$ J cm^{-3} was obtained from the phase-separation data. ΔP_{12}^* was assumed independent of temperature and was found to be independent of blend composition.

Table 8-6 PS and TMPC* parameters for the Sanchez-Lacombe equation of state (Kim and Paul, 1992).

	PS ($\overline{M}_w$ = 330,000 g mol^{-1})	TMPC ($\overline{M}_w$ = 33,000 g mol^{-1})
T^* (K)	810	729
P^* (bar)	3809	4395
$v_{sp}^* = 1/\rho_{sp}^*$ (cm^3 g^{-1})	0.9156	0.8436
r^0	17088	2018

* PS = Polystyrene; TMPC = Tetramethyl bisphenol-A polycarbonate

Figure 8-23 shows the calculated spinodal curve together with experimental phase-separation temperatures obtained by differential scanning calorimetry. The solid line calculated with $\Delta P_{12}^* = -0.71$ J cm^{-3} agrees well with experiment.

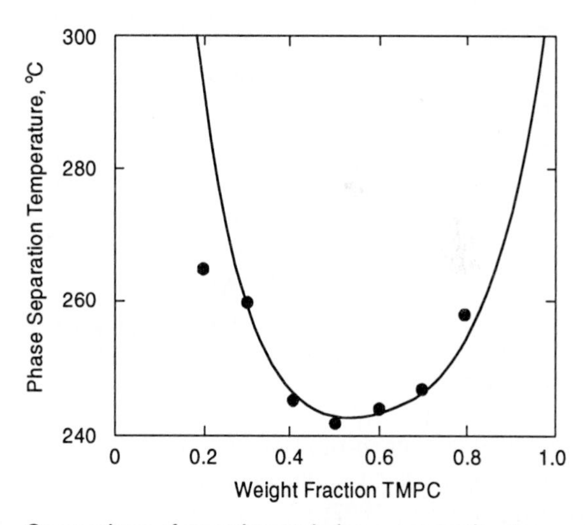

Figure 8-23 Comparison of experimental phase separation temperatures (●) obtained by differential scanning calorimetry with the spinodal curve (——) predicted by the lattice-fluid model (Kim and Paul, 1992).

In addition to mixtures of polymers with normal solvents, the lattice-fluid equation of state has also been applied to polymer-gas systems. Sanchez and Rodgers (1990, 1993) used it to predict gas solubility at infinite dilution. The physical properties of the gas and polymer dominate gas solubility and the gas-polymer interaction plays a secondary role. Using no adjustable parameters, gas solubility was quantitatively predicted for several hydrocarbons and chlorinated hydrocarbons in nonpolar polymers such as polystyrene, poly(1-butene) and atactic polypropylene. For polar polymers, such as poly(vinyl acetate) and poly(methyl methacrylate), the solubilities of hydrocarbons were greatly overestimated, but polar and aromatic gases were correlated reasonably well. Similarly, Pope *et al.* (1991) attempted to predict sorption isotherms of nitrogen, methane, carbon dioxide, and ethylene in silicone rubber. Model parameters were determined only from pure-fluid properties without using any mixture data. Figure 8-24 shows a typical isotherm for ethylene. In general, the experimental isotherm in underpredicted. From the equation of state, expressions for the partial molar volumes and Henry's constant at infinite dilution for the gases can be derived. In general, the Henry's constant is underpredicted, but a good estimate of the partial molar volume at infinite dilution was obtained.

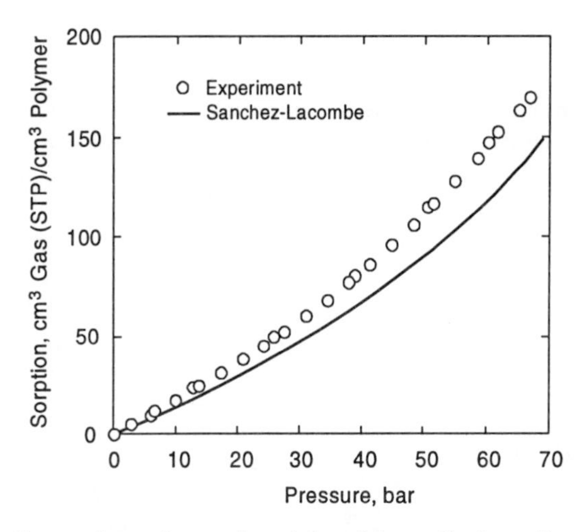

Figure 8-24 Comparison of experimental and theoretical results using Sanchez-Lacombe equation of state (no adjustable parameters) for ethylene sorption in silicone rubber (Pope *et al.*, 1991).

Another application of the lattice-fluid equation of state is provided by describing the effect of a compressed-gas diluent on the behavior of a glassy polymer (Wissinger and Paulaitis, 1991; Condo *et al.*, 1992; Condo and Johnston, 1992; Kalospiros and Paulaitis, 1994). Compressed gases can act as plasticizers when dissolved in a glassy polymer matrix by lowering the polymer's glass-transition temperature (T_g). The glass transition of a polymer or polymer/diluent mixture can be determined by using the Gibbs-Di Marzio criterion (1958, 1963). This criterion states that, at the glass transition, the polymer is essentially "frozen" and has zero configurational entropy. The sorption of the compressed fluid by the polymer and T_g can be calculated simultaneously by solving the equation of state, the condition of equilibrium for partitioning of the diluent between the polymer and the gas phase, and the Gibbs-Di Marzio criterion. Figure 8-25 shows results for poly(methyl methacrylate) (PMMA)/CO_2. Figure 8-25(a) illustrates good agreement between experimental and calculated T_g depression as a function of concentration of dissolved CO_2. However, Fig. 8-25(b), a plot of T_g against

CO_2 pressure shows unexpected phase behavior, first discovered with the model. This unexpected phase behavior is called *retrograde vitrification* (by analogy with retrograde condensation – see Sec. 12.2). For example, at 373 K and a CO_2 pressure of 30 bar, the polymer is a liquid. Decreasing the temperature causes the polymer to undergo a liquid-to-glass transition, as expected; however, a further decrease in temperature causes the glass to become a liquid again. This effect results from the competition between polymer-segment mobility that declines with falling temperature, and diluent solubility that increases with falling temperature.

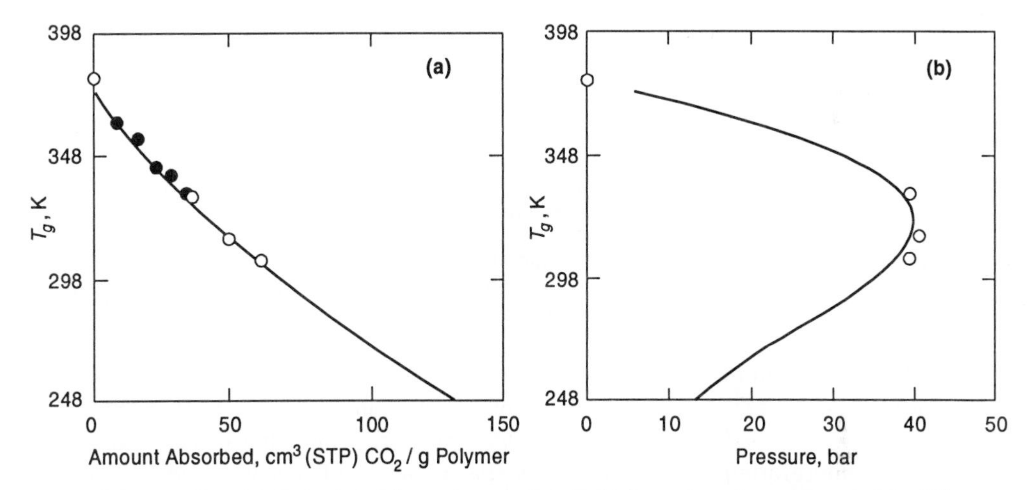

Figure 8-25 (a) Calculated (—— from lattice-fluid theory) and experimental (o, Wissinger and Paulaitis, 1991a; ● Chiou *et al.*, 1985) glass-transition depression of PMMA as a function of CO_2 solubility. (b) Glass-transition temperature depression of PMMA as a function of CO_2 pressure showing retrograde vitrification behavior (Condo *et al.*, 1992). —— Calculated from lattice-fluid theory; o, experiment (Wissinger and Paulaitis, 1991a).

The lattice-fluid equation of state has also been used to describe the phase behavior of supercritical fluid-polymer systems. In these systems, the polymer may completely dissolve in the supercritical-fluid phase, or partition between a polymer-rich phase and a supercritical fluid-rich phase. Figure 8-26 shows an example: poly(ethylene glycol) (PEG, M = 400 g mol^{-1}) mixed with CO_2 to pressures of approximately 260 bar at 50°C (Daneshvar *et al.*, 1990). The upper curve (left-hand axis) gives the weight fraction of CO_2 dissolved in the polymer-rich phase and the lower curve (right-hand axis) gives the weight fraction of polymer dissolved in the supercritical fluid-rich phase. The curves are calculated using the random-mixing version of the lattice-fluid model (Panayiotou, 1987) with one adjustable binary parameter.

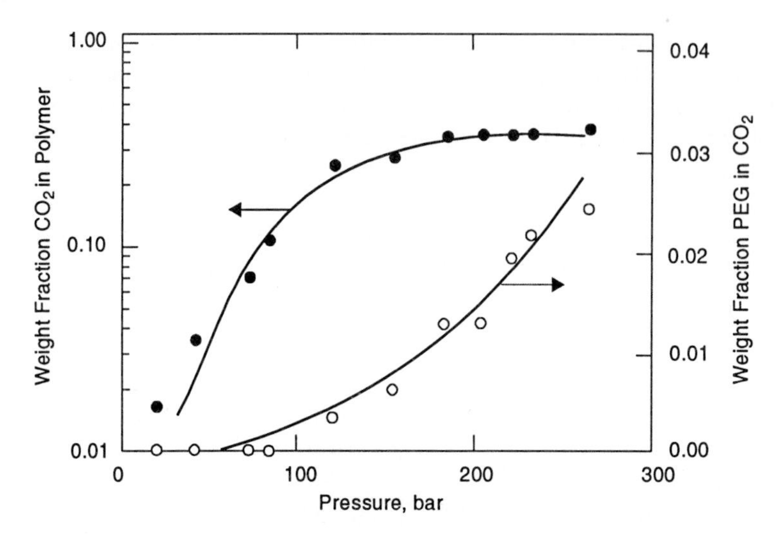

Figure 8-26 Compositions of coexisting phases for PEG/supercritical CO_2 mixtures at 323 K as a function of pressure. —— Calculated from the Panayiotou-Vera version of the lattice-fluid equation of state; ●, ○ experiment (Daneshvar *et al.*, 1990).

The lattice-fluid theory has also been modified to account for strong interactions (hydrogen bonding) between polymer and solvent. This modification has been successfully tested with experimental data for vapor pressures, enthalpies and volumes of mixing for chloroform/polyether solutions (Panayiotou and Sanchez, 1991). The essential features of this modification are discussed by Sanchez and Panayiotou (1994).

Statistical Associated Fluid Theory

The *s*tatistical *a*ssociated-*f*luid *t*heory (SAFT), introduced in Sec. 7.17, has successfully been applied to correlate thermodynamic properties and phase behavior of pure liquid polymers and polymer solutions, including the solubility of gases in polymers, supercritical and near-critical solutions. For example, Wu and Chen (1994) investigated gas solubility in polyethylene. Figure 8-27 shows a comparison of experimental and calculated solubilities of methane and nitrogen in polyethylene using the SAFT equation of state. In their calculations, Wu and Chen used a generalized binary parameter correlation that depends on temperature. Even at very high pressures (600 bar), the agreement between experiment and SAFT is very good.

Numerous applications of the SAFT equation of state to supercritical and near-critical solutions of polymers have been published (Chen and Radosz, 1992; Gregg *et al.*, 1993; Chen *et al.*, 1992, 1993, 1994, 1995; Condo and Radosz, 1996; Banaszak *et al.*, 1996). Figure 8-28 shows experimental and calculated pressure-temperature cloud-point curves for alternating poly(ethylene-propylene) (PEP)/propylene for different PEP

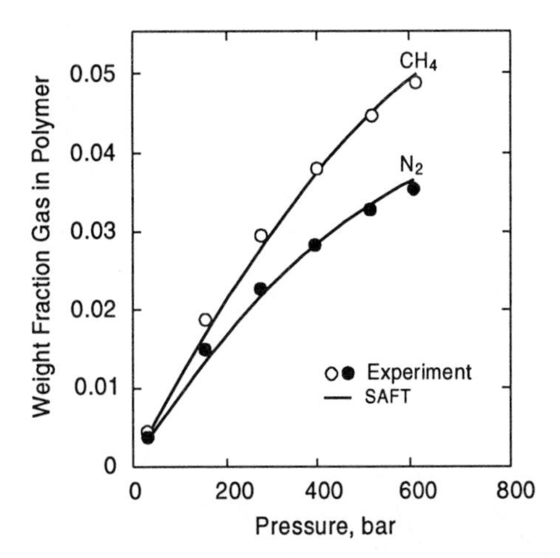

Figure 8-27 Solubilities of methane and nitrogen in polyethylene at 461.4 K. ●, ○ experiment. —— Calculated from SAFT equation of state. The binary parameter is calculated from a temperature-dependent correlation (Wu and Chen, 1994).

molecular weights (Chen and Radosz, 1992). The SAFT equation of state can correlate the experimental data with one binary parameter k_{ij} that depends on, and can be correlated with polymer molecular weight. Figure 8-28 also shows predictions for propylene with PEP ($\overline{M}_w$ = 10,000 and 15,000 g mol^{-1}) using the same empirical correlation for k_{ij}. As shown in the insert of Fig. 8-28, as the polymer molecular weight increases, the LCST curve is shifted to lower temperatures and, simultaneously, the UCST curve is shifted to higher temperatures, i.e., the two curves approach each other and eventually merge into a single curve with a minimum. (This single curve is represented in the insert by a dashed-dotted line.) This behavior is illustrated in the figure with the data for the real system. Calculations from SAFT show that, for propylene/PEP with $\overline{M}_w$ = 10,000 g mol^{-1} (or larger), the LCST and UCST curves have merged, as confirmed by experiment.

Perturbed Hard-Sphere-Chain Theory

The *p*erturbed *h*ard-*s*phere-*c*hain (PHSC) equation of state introduced in Sec. 7.17 can represent all common types of fluid phase diagrams of binary polymer mixtures including vapor-liquid equilibrium[30] (Gupta, 1995, 1996) and, for some cases, liquid-liquid equilibrium (Song *et al.*, 1994a). However, good representation of liquid-liquid

[30] A comprehensive collection of vapor-liquid equilibria data for binary polymer solutions is presented by Ch. Wohlfarth, 1994, *Vapor-liquid Equilibrium Data of Binary Polymer Solutions*, Physical Sciences Data Series No. 44, Amsterdam: Elsevier.

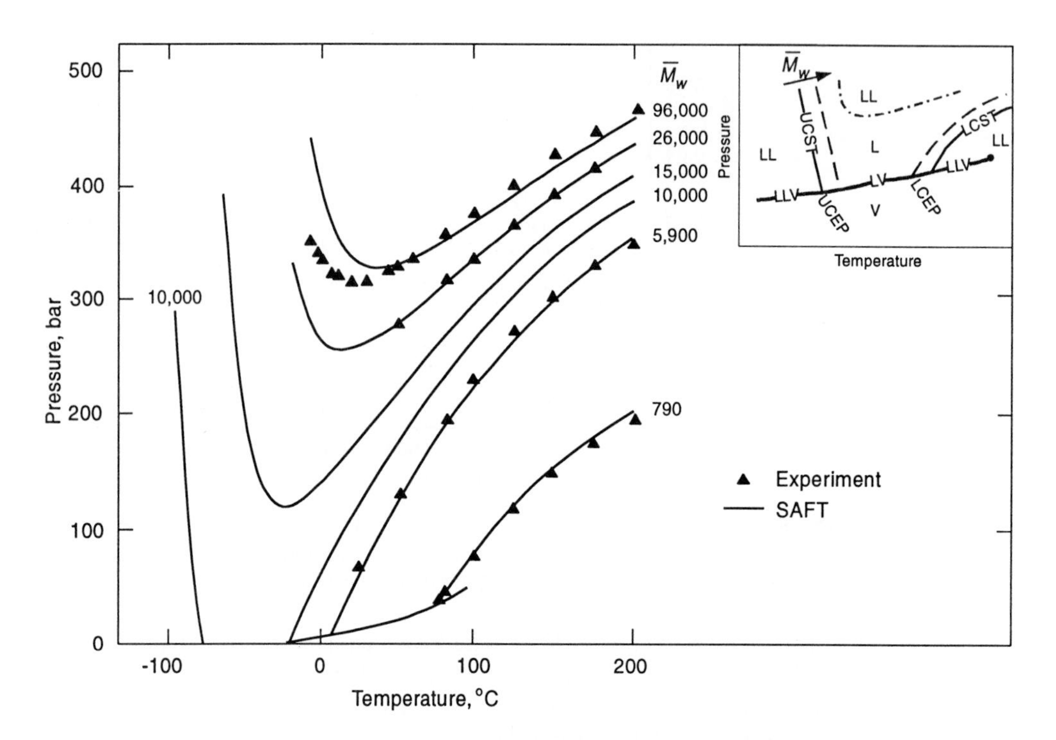

Figure 8-28 Experimental (Chen and Radosz, 1992) and calculated *P-T* phase boundaries for alternating poly(ethylene-propylene) (PEP)/propylene for the indicated PEP molecular weights. Calculations use the binary interaction-parameter correlation developed for this system. For the propylene/PEP ($\overline{M}_w$ = 10,000 g mol^{-1}) system, SAFT predicts a steep upper critical solution temperature (UCST) with an upper critical end point (UCEP) near -80°C. (Chen *et al.*, 1992). ▲ Experiment; —— calculated from SAFT equation of state. The insert shows schematically the influence of increasing polymer molecular weight (represented by the arrow) in the *P-T* diagram. In the insert, the heavy line is the bubble-point curve; in most cases, this curve coincides with the vapor-pressure curve of the solvent.

equilibria is much more difficult than representation of vapor-liquid equilibria; existing theories usually require two or three binary parameters as pointed out by Hino (1996).

Gupta (1996) showed that vapor-liquid equilibrium of polymer solutions can be correlated quantitatively with the PHSC equation of state. However, for accurate results, Gupta relaxed the hard-sphere additivity rule for calculating b_{ij} to introduce a binary size parameter λ_{ij}:

$$b_{ij}(T) = \frac{2\pi}{3}\sigma_{ij}^3 F_b(kT / \varepsilon_{ij}) \tag{8-97}$$

8.4 Nonporous Polymeric Membranes and Polymer Gels

Polymer swelling occurs when polymeric materials are exposed to solvent vapors or gases that can be absorbed by the polymer. Swelling also occurs when cross-linked polymers (polymeric networks) that cannot be dissolved, are in contact with a compatible liquid. The important technical issue often is how fast (determined by diffusion) a low molecular-weight fluid is absorbed by a polymer. Applications where sorption behavior (solubility and diffusivity) is important are, for example, the use of supercritical fluids as swelling agents for impregnating chemical additives (e.g. dyes or pigments for color) into polymers (Berens *et al.,* 1992) and, conversely, for extracting low molecular-weight components from polymeric materials (Cotton *et al.,* 1993) or the separation of gaseous and liquid mixtures using selective nonporous polymeric membranes (Noble and Stern, 1995; Mulder, 1996).

Swelling properties of gels are also of considerable interest for applications in packaging and medicine (e.g. implants in eye surgery). Gels are three-dimensional cross-linked elastic polymers that may either swell or shrink when brought into contact with a liquid. Gels have been proposed as size selective extraction solvents and as environmentally sensitive permeability barriers for a variety of pharmacological agents (DeRossi *et al.*, 1991; Dušek, 1993).

There are numerous aspects of polymer swelling. Here we focus only on the thermodynamics relevant for separation techniques using nonporous membranes and on phase behavior for systems containing gels. The thermodynamics discussed here may be useful for optimum design of membrane separation processes and for design of gels in medicine, pharmaceutics and biotechnology.

Nonporous Membranes

Most nonporous membranes used in industrial applications are composite membranes[31] consisting of a thin, nonporous, polymeric top layer (thickness 0.5-5 μm) covering the surface of a porous support (thickness 100-500 μm). The thin top layer is mainly responsible for the separation. The support has usually no or little influence on the separation characteristics of the composite membranes. Its purpose is to provide mechanical stability in a membrane-separation unit (module).

The separation mechanism of a nonporous membrane is illustrated in Fig. 8-31 for a binary fluid mixture, with different molecules represented by black and white spheres. The *feed* mixture to be separated flows along one side of the membrane while the two feed components are permeating into and through the membrane at different rates. Therefore, the *retentate* leaving the process on the same side of the membrane where the feed enters, is depleted in the component permeating preferentially. Consequently, the *permeate* collected on the other side of the membrane is enriched in the preferentially permeating component.

[31] A review on thin-film composite membranes is given by R. J. Petersen, 1993, *J. Membrane Sci.,* 83: 81.

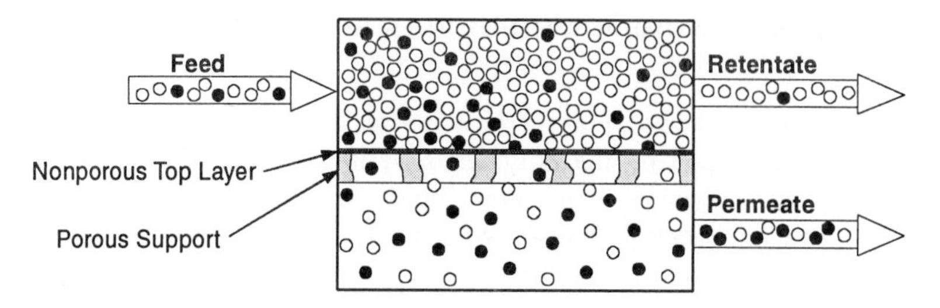

Figure 8-31 Schematic representation of a membrane-separation process using a composite membrane with a thin dense nonporous top layer and a thick porous support layer.

The driving force for any permeating component i is μ_i(feed) − μ_i(permeate) where μ is the chemical potential. This difference in chemical potentials is usually achieved by a permeate pressure much lower than the feed pressure. The resulting driving force causes each component to permeate with an individual velocity from the feed to the permeate. Therefore, at steady state, the compositions of the feed and permeate differ from each other. If the feed is a liquid under its saturation pressure and the permeate is a vapor kept at low pressure (5-20 mbar) by using a vacuum pump combined with a cooling trap for condensing the permeate, the separation process is called *pervaporation* (Feng and Huang, 1997). If both feed and permeate are vapors or gases, the separation process is called *vapor permeation* or *membrane gas separation*.[32] In this case, pressures applied on the feed side are 10-100 bar while the permeate pressure is 1 bar or less. If both feed and permeate are liquid and a pressure in the range 80-100 bar is applied on the feed side while the permeate is kept at ambient pressure, the separation process is called *reverse osmosis* (Amjad, 1993). Reverse osmosis is mainly applied to dilute aqueous solutions to produce purified water or to concentrate aqueous solutions by removing water.

Membrane materials used for the top layer can be divided into two main classes: hydrophilic and organophilic. Hydrophilic membranes are used, for example, in the separation of water from organic mixtures by pervaporation (Huang, 1991), drying of humid gas streams by gas separation or in the production of drinking water from seawater by reverse osmosis. Typical separations using organophilic membranes are removal of organic pollutants from water by pervaporation (Böddeker and Bengtson, 1991) and extraction of aroma compounds from alcoholic beverages by pervaporation (Baudot and Marin, 1997). Some synthetic membrane polymers have particular selectivities for aromatic hydrocarbons to separate them from aliphatic hydrocarbons (Inui *et al.*, 1997; Yoshikawa *et al.*, 1997) or for alcohols to separate them from ethers (Streicher *et al.*, 1995; Nguyen *et al.*, 1997) by pervaporation.

[32] A comprehensive review of membrane-based gas separations is given by W. J. Koros and G. K. Fleming, 1993, *J. Membrane Sci.*, 83: 1.

Other important membrane-separation processes include *dialysis* (particularly hemodialysis) and *micro-*, *ultra-* and *nanofiltration* using porous membranes; *liquid membranes*; and *electrodialysis* where charged membranes are used (Mulder, 1996). We do not discuss these processes here; we restrict our discussion to thermodynamics relevant to pervaporation, gas separation and reverse osmosis.

The driving force for any species permeating is the gradient of the species' chemical potential across the membrane. Therefore, to determine the driving forces for various species we need the chemical potential for each species at state conditions of the feed and the permeate. We consider first a liquid mixture in equilibrium with its vapor. The condition for equilibrium is:

$$\mu_i^G = \mu_i^{0G} + RT\ln(\varphi_i P_i) = \mu_i^L = \mu_i^{0L} + RT\ln(\gamma_i^L x_i^L) \tag{8-99}$$

where μ_i^{0G} is the standard-state (1 bar) chemical potential of the pure gas, φ_i is the fugacity coefficient, P_i is the partial pressure, μ_i^{0L} is the chemical potential of the pure liquid at the same temperature and pressure as those of the mixture, γ_i^L is the activity coefficient, and x_i^L is the mole fraction of i in the liquid mixture.

For the liquid mixture in equilibrium with the swollen polymer, the condition for equilibrium is

$$\mu_i^L = \mu_i^M \tag{8-100}$$

with

$$\mu_i^M = \mu_i^{0M} + RT\ln(\gamma_i^M x_i^M) \tag{8-101}$$

where superscript M indicates the membrane phase. Because the standard state chemical potentials for the liquid phase, μ_i^{0L}, and for the membrane phase, μ_i^{0M}, are identical,

$$\gamma_i^L x_i^L = \gamma_i^M x_i^M \tag{8-102}$$

It is convenient to use molar concentration c_i^M in the membrane phase instead of mole fraction x_i^M; therefore, we rewrite Eq. (8-102)

$$\gamma_i^L x_i^L = \gamma_i^{c,M} c_i^M \tag{8-103}$$

Activity coefficient $\gamma_i^{c,M}$ is based on molar concentration instead of mole fraction. From Eq. (8-103), we obtain

$$\boxed{c_i^M = \frac{\gamma_i^L}{\gamma_i^{c,M}} x_i^L = S_i^L x_i^L} \tag{8-104}$$

where S_i^L is the liquid solubility coefficient of component i.

Phase equilibrium between the gaseous (vapor) phase and the membrane phase requires

$$\mu_i^G = \mu_i^M \tag{8-105}$$

From Eqs. (8-99) and (8-103) we obtain

$$\mu_i^{0G} + RT\ln(\varphi_i P_i) = \mu_i^{0M} + RT\ln(\gamma_i^{c,M} c_i^M) \tag{8-106}$$

or

$$\gamma_i^{c,M} c_i^M = P_i \varphi_i \exp\left(\frac{\mu_i^{0G} - \mu_i^{0M}}{RT}\right) \tag{8-107}$$

or

$$\boxed{c_i^M = S_i^G P_i} \tag{8-108}$$

where the gas solubility coefficient of component i is

$$S_i^G = \frac{\varphi_i}{\gamma_i^{c,M}} \exp\left(\frac{\mu_i^{0G} - \mu_i^{0M}}{RT}\right) \tag{8-109}$$

These equations are useful for the description of transmembrane fluxes within the framework of the *solution-diffusion model*. In this model, the transport of each component i is divided into three steps:

1. The components of the liquid or gaseous feed mixture are absorbed in the membrane. For each component i, there is thermodynamic solubility equilibrium at the phase boundary between the feed mixture and the membrane according to Eqs. (8-104) and (8-108).
2. The absorbed components diffuse across the membrane from the feed side to the permeate side according to Fick's first law of diffusion.
3. The components are desorbed at the phase boundary between the membrane and the liquid or gaseous permeate. Again, it is assumed that for each component i, there is equilibrium at the phase boundary.

We consider first the case of *pervaporation*. Inside the membrane, flux J_i of component i is given by Fick's law:

$$J_i = -D_i \frac{dc_i^M}{d\ell} \tag{8-110}$$

where D_i is the diffusion coefficient (with units m^2 s^{-1} with J_i in mol m^{-2} s^{-1}), and ℓ is the length coordinate perpendicular to the top layer of the membrane with thickness δ_M ($0 \le \ell \le \delta_M$). Integration of Eq. (8-110) (Heintz and Stephan, 1994) gives

$$J_i = \frac{D_i}{\delta_M}(c_{iF}^M - c_{iP}^M) \tag{8-111}$$

where c_{iF}^M and c_{iP}^M are the concentrations (in mol m^{-3}) in the membrane boundaries at the feed side (index F) and the permeate side (index P), respectively. Equation (8-111) is valid only if D_i is independent on concentration; otherwise, D_i is the diffusion coefficient averaged over the concentration across the membrane.

According to the solution-diffusion model, the solubility equilibrium conditions at both phase boundaries justifies substitution of c_{iF}^M from Eq. (8-104) and c_{iP}^M from Eq. (8-108) into Eq. (8-111):

$$\boxed{J_i = \frac{D_i}{\delta_M}(S_i^L x_i^L - S_i^G P_i)} \tag{8-112}$$

At steady state, the mole fraction y_i of the gaseous permeate is obtained from

$$\frac{J_i}{\sum_i J_i} = \frac{P_i}{\sum_i P_i} = y_i \tag{8-113}$$

For a binary mixture, substitution of Eq. (8-112) into (8-113) gives

$$y_1 = \frac{P_1}{P_1 + P_2} = \frac{D_1(S_1^L x_1^L - S_1^G P_1)}{D_1(S_1^L x_1^L - S_1^G P_1) + D_2(S_2^L x_2^L - S_2^G P_2)} \tag{8-114}$$

Because $P_2 = P - P_1$, where P is the total pressure of the permeate, P_1 can be calculated from Eq. (8-114) at given values of $x_2^L = 1 - x_1^L$ and P. Therefore, y_1 is also determined. If the pressure P in the permeate side is kept as low as possible, i.e., $P \approx 0$, then also $P_i \approx 0$. Equation (8-114) then simplifies to

$$y_1 = \frac{D_1 S_1^L x_1^L}{D_1 S_1^L x_1^L + D_2 S_2^L x_2^L} \tag{8-115}$$

A plot of the permeate mole fraction y_1 against liquid-feed mole fraction x_1^L according to Eq. (8-114) or (8-115) is called a *separation diagram.* A measure for the separation effect is the *separation factor* defined in terms of the upstream and downstream mole fractions x_1^L and y_1, respectively:

$$\alpha \equiv \frac{x_1^L / y_1}{x_2^L / y_2} \tag{8-116}$$

If the permeate pressure $P \approx 0$, Eq. (8-115) gives

$$\alpha = \frac{D_2 S_2^L}{D_1 S_1^L} \tag{8-117}$$

where product $D_i S_i^L$ is called the *permeability* of component i. Figure 8-32 shows a separation diagram, $y_1(x_1^L)$, for different values of α. If α is larger than 1, component 2 is enriched in the permeate; if α = 1, no separation effect is observed; if α is smaller than unity, component 1 is enriched in the permeate.

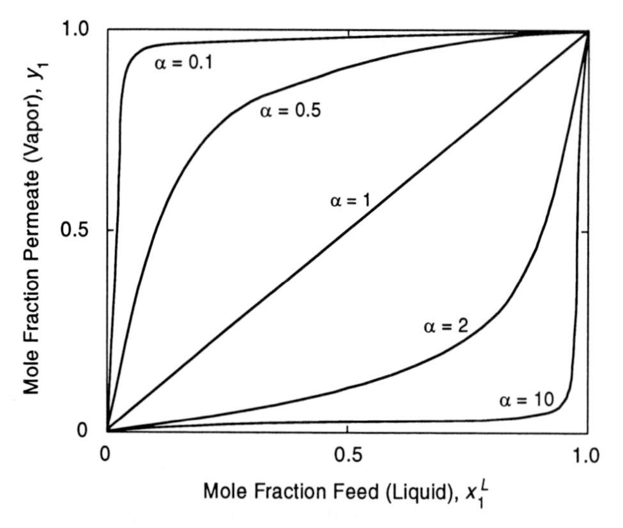

Figure 8-32 Separation diagram in pervaporation calculated from Eq. (8-115) for different values of the separation factor α.

The equations derived above are only valid over the whole range of liquid composition if S_i^L and S_i^G are constants independent of the mixture composition. If S_i^L and S_i^G are functions of x_i^L and y_i, respectively, the shape of the separation diagrams can become more complicated; in extremely nonideal cases, curves with a maximum or a minimum can be obtained. Further complications arise if the diffusive fluxes J_i are no longer independent of each other and diffusion coupling occurs.[33]

In *gas separation*,[34] the flux J_i inside the membrane is also given by Eq. (8-111), with c_{iF}^M and c_{iP}^M from Eq. (8-108):

$$J_i = \frac{D_i}{\delta_M}(S_{iF}^G P_{iF} - S_{iP}^G P_{iP}) \tag{8-118}$$

where P_{iF} and P_{iP} are, respectively, the partial pressure of component i in the pressurized gaseous feed mixture (index F) and in the permeate mixture (index P).

Equation (8-109) shows that S_{iF}^G and S_{iP}^G depend on $\gamma_i^{c,M}$ and φ_i. Both quantities are in principle concentration dependent; in particular, φ_i can change significantly with pressure. Therefore, S_{iF}^G and S_{iP}^G can differ from each other. However, often it is assumed that $S_{iF}^G \approx S_{iP}^G = S_i^G$. In this case, Eq. (8-118) simplifies to

$$\boxed{J_i = \frac{D_i S_i^G}{\delta_M}(P_{iF} - P_{iP})} \tag{8-119}$$

As in pervaporation, we obtain a separation diagram that corresponds to Eq. (8-114), with x_i^L substituted by $y_{iF} = P_{iF} / \sum P_{iF}$ and $y_{iP} = P_{iP} / \sum P_{iP}$. For the simplified case where $P_{iP} << P_{iF}$, we obtain for a binary mixture an equation that is analogous to Eq. (8-115):

$$y_1 = \frac{D_1 S_1^G P_{1F}}{D_1 S_1^G P_{1F} + D_2 S_2^G P_{2F}} \tag{8-120}$$

and the corresponding separation factor is defined by

$$\alpha \equiv \frac{D_2 S_2^G}{D_1 S_1^G} \tag{8-121}$$

As in pervaporation, $D_i S_i^G$ is called *permeability*. In gas separation, special units are used. Flux J_i is expressed[35] in m^3 (STP) m^{-2} s^{-1} instead of mol m^{-2} s^{-1}. Solubility coefficient S_i^G is usually given in m^3 (STP) m^{-3} Pa^{-1} and the diffusion coefficient is

[33] For a discussion and quantitative description of these more complex cases, see Heintz and Stephan (1994).

[34] More detailed information on gas separation is given by Koros and Fleming (1993) and Mulder (1996).

[35] Volume at STP (*s*tandard *t*emperature and *p*ressure) means volume of gas at standard pressure (1 bar) and at 273.15 K.

given in its SI-units, $m^2\ s^{-1}$. The common unit for permeability is the barrer [1 barrer = 10^{-10} cm^3 (STP) cm cm^{-2} s^{-1} $cmHg^{-1}$ = 0.76×10^{-17} m^3 (STP) m m^{-2} s^{-1} Pa^{-1}].

In *reverse osmosis*, high pressures are applied to the liquid-feed side and, therefore, we have to take into account the pressure dependence of the chemical potential,

$$\left(\frac{\partial \mu_i}{\partial P}\right)_{T,x} = \overline{v}_i \tag{8-122}$$

where $\overline{v}_i$ is the partial molar volume of component i in the mixture. Integration of Eq. (8-122) from the standard pressure (P^0 = 1 bar) to P_F and assuming that $\overline{v}_i$ is independent of pressure gives

$$\begin{aligned}\mu_{iF}^L &= \mu_{iF}^L(P^0 = 1\text{ bar}) + \overline{v}_{iF}(P_F - P^0)\\ &= \mu_{iF}^{0L} + RT\ln(\gamma_{iF}^L x_{iF}^L) + \overline{v}_{iF}(P_F - P^0)\end{aligned} \tag{8-123}$$

The chemical potential μ_i^M of component i inside the membrane must also be determined at P_F:

$$\mu_i^M = \mu_i^{0M} + RT\ln(\gamma_i^{c,M} c_i^M) + \overline{v}_i^M(P_F - P^0) \tag{8-124}$$

Assuming that $\overline{v}_{iF} = \overline{v}_i^M = \overline{v}_i$ and taking into account that $\mu_i^{0L} = \mu_i^{0M}$, the equilibrium condition $\mu_i^M = \mu_{iF}^L$ gives the same result as Eq. (8-104) with $S_i^L = S_{iF}^L$.

In reverse osmosis, the pressure drop from the feed side to the permeate side is located at the phase boundary between the membrane and the liquid permeate. Therefore, equality of the chemical potentials requires

$$RT\ln(\gamma_i^{c,M} c_i^M) + \overline{v}_i^M(P_F - P^0) = RT\ln(\gamma_{iP}^L x_{iP}^L) + \overline{v}_i(P_P - P^0) \tag{8-125}$$

or

$$\gamma_{iP}^L x_{iP}^L \exp\left[-\frac{\overline{v}_i(P_F - P_P)}{RT}\right] = \gamma_i^{c,M} c_i^M \tag{8-126}$$

where P_P is the pressure in the liquid permeate (usually 1 bar) and subscript P denotes the permeate. Again, assuming that Fick's law is valid inside the membrane,

$$J_i = \frac{D_{iM}}{\delta_M}(c_{iF}^M - c_{iP}^M) = \frac{D_{iM}}{\delta_M}\{S_{iF}^L x_{iL}^L - S_{iP}^L x_{iP}^L \exp[-\overline{v}_i(P_F - P_P)/RT]\} \tag{8-127}$$

If $S_{iF}^L \approx S_{iP}^L = S_i^L$, an acceptable approximation for many cases,

$$J_i = \frac{D_{iM} S_i^L}{\delta_M} \{x_{iF}^L - x_{iP}^L \exp[-\overline{v}_i (P_F - P_P) / RT]\} \tag{8-128}$$

Equation (8-128) shows that J_i is enhanced as P_F becomes larger than P_P.

Similar to the other separation processes discussed above, at steady state

$$x_{iP}^L = \frac{J_i}{\sum J_i} \tag{8-129}$$

Equation (8-129) says that at given composition x_{iF}^L in the feed mixture, composition x_{iP}^L in the permeate is obtained by substituting Eq. (8-128) into Eq. (8-129).

Reverse osmosis is applied mostly to dilute aqueous solutions. Therefore, for the solvent water, $\overline{v}_w \approx v_w$ is a good approximation and Eq. (8-128) can be rewritten as

$$\boxed{J_w = \frac{D_w}{\delta_M} S_w^L x_{wF}^L \{1 - \exp[-v_w (P_F - P_P - \Delta\pi) / RT]\}} \tag{8-130}$$

where $\Delta\pi$ is given by

$$\Delta\pi = \frac{RT}{v_w} \ln \frac{x_{wP}^L}{x_{wF}^L} \tag{8-131}$$

From Eq. (4-42) that gives the osmotic pressure of a dilute solution, we see that $\Delta\pi$ is the difference in osmotic pressure between feed and permeate.

Equation (8-130) tells us that a positive flow J_w is observed when $(P_F - P_P) > \Delta\pi$; when $(P_F - P_P) = \Delta\pi$, there is osmotic equilibrium with $J_w = 0$; when $(P_F - P_P) < \Delta\pi$, the flux of water is reversed with $J_w < 0$.

When, in the feed and in the permeate, the mole fraction of solute 2 in the binary aqueous solution $x_2^L << 1$, then $\ln(1 - x_2^L) \approx -x_2^L$. Substitution into Eq. (8-131) gives

$$\Delta\pi = RT(c_{2F} - c_{2P}) \tag{8-132}$$

where $c_2 \approx x_2 / v_w$ is the molar concentration of the solute. Equation (8-132) is another form of the van't Hoff equation for the osmotic pressure [Eq. (4-44)], valid only for very dilute solutions.

Reverse osmosis is widely used for the separation of water from aqueous ionic solutions, e.g. desalination of seawater. In these applications, it is important that the transport of salt ions through the membrane is as small as possible. For electrically neutral membranes, the transport of ions is determined by their solubility and diffusivity in the membrane. The driving force for ion transport is proportional to the concentration difference of the ions between feed and permeate.

However, when charged membranes or ion-exchange membranes are used instead of neutral membranes, ion transport is also affected by the presence of the charge and number of ionic groups fixed in the membrane. When an ion-exchange membrane is in contact with an ionic solution, ions with charge of the same sign as that of the fixed ionic groups in the membrane are, in principle, excluded and cannot pass through the membrane. This effect is known as *Donnan exclusion*. Ion-exchange membranes are therefore particularly suitable for separating water from aqueous ionic solutions using reverse osmosis.

Donnan exclusion can be described by equilibrium thermodynamics, as discussed in Sec. 4.11. Let us consider an ion-exchange membrane with fixed negative charges (R^-) and sodium ions Na^+ as counterions in contact with a dilute aqueous sodium chloride (NaCl) solution, as shown schematically in Fig. 8-33. Water, Na^+ and Cl^- can freely diffuse from the solution into the membrane phase, although Na^+ ions can only diffuse in combination with Cl^- ions because electroneutrality must be maintained.

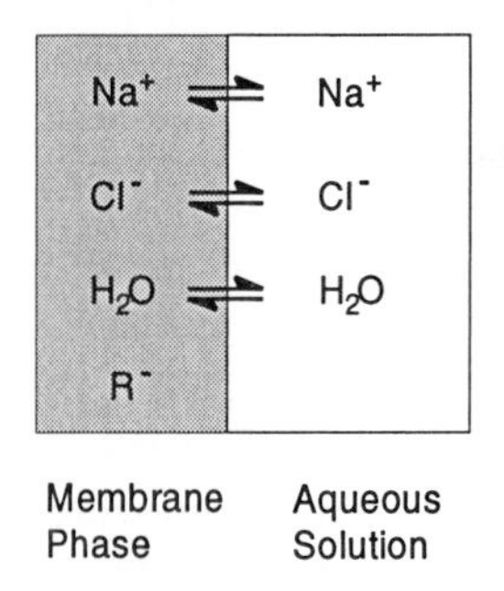

Figure 8-33 Donnan equilibrium when an ionic membrane with fixed negative charges (R^-) is in contact with an aqueous NaCl solution.

At equilibrium, the chemical potential of sodium chloride is the same in both phases. Assuming ideal solution behavior, i.e. solvent activity ($a_s = 1$) and the activity of solute i is equal to its molar concentration ($a_i = c_i$), the equation of phase equilibrium is

$$c^M_{Na^+} c^M_{Cl^-} = c_{Na^+} c_{Cl^-} \qquad \text{(8-133)}^{36}$$

Because electroneutrality must be maintained in the membrane phase,

$$c^M_{Na^+} = c^M_{Cl^-} + c^M_{R^-} \qquad \text{(8-134)}$$

[36] Equation (8-133) assumes complete dissociation of sodium chloride, i.e. $\mu_{NaCl} = \mu_{Na^+} + \mu_{Cl^-}$.

and, similarly, because electroneutrality must be maintained in the aqueous solution,

$$c_{\mathrm{Na}^+} = c_{\mathrm{Cl}^-} \tag{8-135}$$

Substitution of Eqs. (8-134) and (8-135) into Eq. (8-133) gives

$$\frac{c_{\mathrm{Cl}^-}}{c^M_{\mathrm{Cl}^-}} = \left(1 + \frac{c^M_{\mathrm{R}^-}}{c^M_{\mathrm{Cl}^-}}\right)^{1/2} \tag{8-136}$$

For dilute solutions where $c^M_{\mathrm{R}^-} >> c^M_{\mathrm{Cl}^-}$, Eq. (8-136) reduces to

$$\boxed{c^M_{\mathrm{Cl}^-} = \frac{(c_{\mathrm{Cl}^-})^2}{c^M_{\mathrm{R}^-}}} \tag{8-137}$$

Equation (8-137) gives the ionic or *Donnan equilibrium* of anionic solutes in the presence of a charged membrane (or charged macromolecules) with fixed charges R^-. If the concentration in the solution phase is low and the concentration of fixed charges in the membrane is high, Donnan exclusion is very effective. However, with increasing concentration of ions in the solution phase, this exclusion becomes less effective.

The transport equations presented above for pervaporation, gas separation, and reverse osmosis are valid within the frame of the solution-diffusion model in its simplified version, i. e. with constant values of solubility coefficients S_i and neglecting diffusion coupling. An additional problem (not discussed here) is concentration polarization that acts as an additional resistance to the transmembrane fluxes. Concentration polarization is important mainly in reverse osmosis and in special cases of pervaporation (Heintz and Stephan, 1994; Feng and Huang, 1994; Mulder, 1996; Bhattacharya and Hwang, 1997).

The simplified transport equations allow a qualitative (or semiquantitative) description[37] of the separation characteristics of nonporous membranes. To do so, we need the solubility and the diffusivity of each permeating component in the membrane polymer. Solubilities are obtained from gas, vapor or liquid sorption equilibrium data, and diffusion coefficients from the kinetics of the sorption process (Heintz *et al.*, 1991; Neogi, 1996). At low pressures, gas and vapor sorption (swelling behavior) for amorphous rubbery polymers (polymers above their glass transition temperatures) are modeled successfully with a form of Henry's law [Eq. (8-108)]. At higher pressures, where sorption behavior deviates from Henry's law, sorption equilibria can be described, for example, with the Flory-Huggins theory, discussed in Sec. 8.2 (Barbari and Conforti, 1992).

[37] For a good quantitative description, the simplifying assumptions have to be removed.

Vapor sorption and the solubilities of liquids and liquid mixtures can be described well with the UNIQUAC model (Sec. 7.7), as reported by Enneking *et al.* (1993) and Heintz and Stephan (1994). To illustrate, Fig. 8-34 shows a comparison between experimental and calculated solubilities at 333 K for the binary system 2-propanol/water in a thin film of cross-linked poly(vinyl alcohol).[38] In Fig. 8-34, the weight fractions of the two components in the polymer are plotted as functions of the weight fraction of the binary mixture outside the membrane. As Fig. 8-34 shows, the solubility curve for 2-propanol has a maximum, indicating a higher solubility for propanol in the mixture than for pure alcohol. Synergistic solubility effects are evident. Nevertheless, both solubility curves predicted from UNIQUAC are in good agreement with experiment. These calculations were made using only binary interaction parameters between the components in the swollen polymer and between the polymer and each of these components. UNIQUAC parameters were obtained from (membrane-free) binary vapor-liquid equilibrium data for 2-propanol/water at 333 K, and from vapor-sorption isotherms for each of the two pure components in the polymer. Similar results have also been reported for the solubility of other aqueous/organic mixtures in poly(vinyl alcohol) and of multicomponent organic mixtures in hydrophobic polymers (Enneking *et al.*, 1993; Enneking *et al.*, 1996).

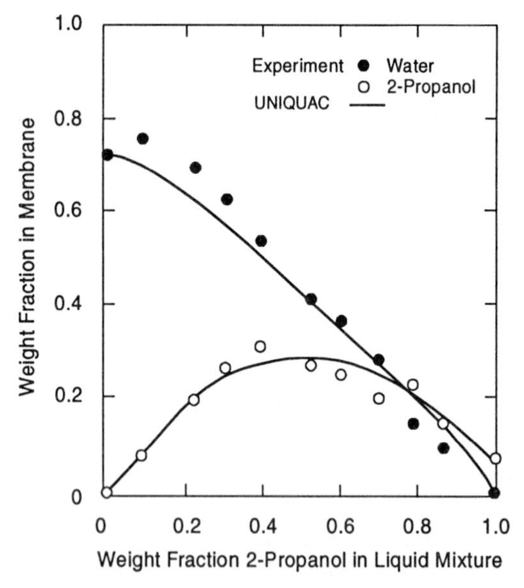

Figure 8-34 Experimental and calculated solubilities for the 2-propanol/water mixture in cross-linked poly(vinyl alcohol) at 333 K. (Heintz and Stephan, 1994). Calculations based on binary data only.

[38] In this cross-linked polymer, about 5% of OH-groups are used for cross-links.

Using solubilities calculated from UNIQUAC, and using the Maxwell-Stefan theory to account for coupling of the diffusion of the two components, Heintz and Stephan (1994) calculated the separation diagram for the mixture 2-propanol/water with the poly(vinyl alcohol) membrane. Figure 8-35 compares the vapor-liquid equilibrium (VLE) curve at 333 K with the pervaporation curves at permeate pressures of 30 and 130 mbar. This comparison shows that water can be removed with high selectivity at feed weight fractions of 0.8-0.9. Within this composition range, the VLE curve shows the azeotropic point, i.e. pervaporation is effective where distillation is not. As Fig. 8-35 shows, lowering the permeate pressure increases the selectivity for water. The selective poly(vinyl alcohol) top layer of the composite membrane has a thickness of 0.13 μm. Transmembrane fluxes of 0.5-1.0 kg m^{-2} h^{-1} are obtained at feed weight fractions 0.8-0.9, providing a yield high enough for possible practical application.

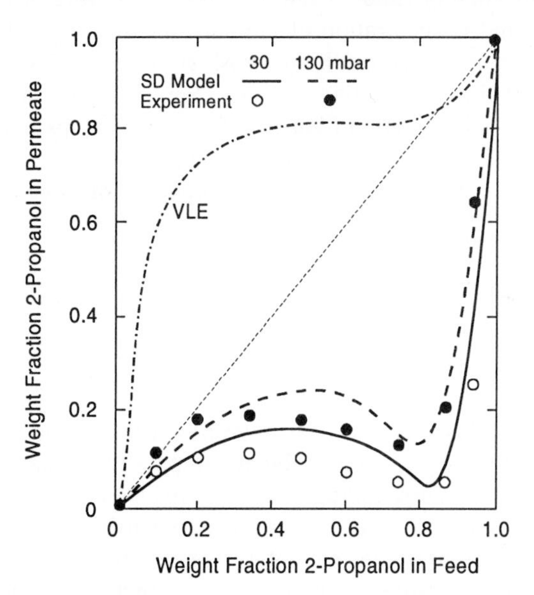

Figure 8-35 Separation diagram for the system 2-propanol/water/poly(vinyl alcohol) at 333 K. ○, ● Pervaporation data (Wesslein *et al.*, 1990) at two permeate pressures; ——, – – – extended solution-diffusion (SD) model (Heintz and Stephan, 1994); – · – · – vapor-liquid equilibrium (VLE) for 2-propanol/water.

This section has presented the essential thermodynamics for nonporous-membrane separation processes. Extension of the simple solution-diffusion model presented above can describe more complex cases with coupling effects and concentration polarization.

Membrane-separation processes are sometimes more economical than conventional methods; the equipment is usually less spacious and in some cases membrane processes are "cleaner" from the environmental point of view. In practical applications, it is sometimes advantageous to use hybrid processes that combine membrane techniques with conventional methods such as distillation (Meckl, 1996; Staudt-Bickel, 1996; Pressly and Ng, 1998). An important task is the synthesis of suitable membranes especially tailored for particular separation problems, having long operation times and high chemical stability in addition to high selectivity and high transmembrane fluxes.

Polymer Gels

Polymer gels are three-dimensional elastic-network materials. In some respects, a gel is similar to a sponge. Brought into contact with a liquid (pure component or solution), a gel may swell or shrink, depending on its initial state and on its interaction with the molecules of the surrounding fluid; if the gel is dry, swelling is likely. Gel swelling depends on the type and concentration of the network-forming polymers (nonionic or ionic homo- and/or copolymers), on the network structure (cross-linking density) and on the composition of the surrounding solution. For ionic systems, swelling also depends on the degree of gel ionization (density of charged groups in the network chains) and on the ionic strength of the surrounding solution.[39]

The heterogeneous system containing a gel consists of three different homogeneous phases:

1. The fluid surrounding the gel.
2. The elastic structure forming the gel.
3. The fluid inside the gel.

The gel-fluid phase is assumed to behave like a bulk fluid phase encaged in an elastic, porous structure, the gel itself. Solvent can pass freely back and forth between the gel-fluid phase and the surrounding solution. However, the porous network structure may allow only selected solute molecules to partition between the two fluid phases, while other solute molecules cannot permeate into the gel (e.g., size exclusion in gel permeation chromatography). A hydrogel is gel that likes water.

Gels may exhibit temperature-induced phase transitions (abrupt changes in volume), as observed for some nonionic hydrogels in aqueous solutions (Ilavsky *et al.*, 1982; Hirokawa and Tanaka, 1984; Marchetti *et al.*, 1990). Figure 8-36 illustrates a hydrogel undergoing such a phase transition. Nonionic poly(*N*-isopropylacrylamide) hydrogel coexists with pure water in a swollen state at temperatures below 35°C, while it collapses at higher temperatures. At about 33°C, there is a phase transition between the shrunken (collapsed) and the swollen state associated with a discontinuous change in volume.

---

[39] The ionic strength of an electrolyte solution is defined in Sec. 9.7 by Eq. (9-45).

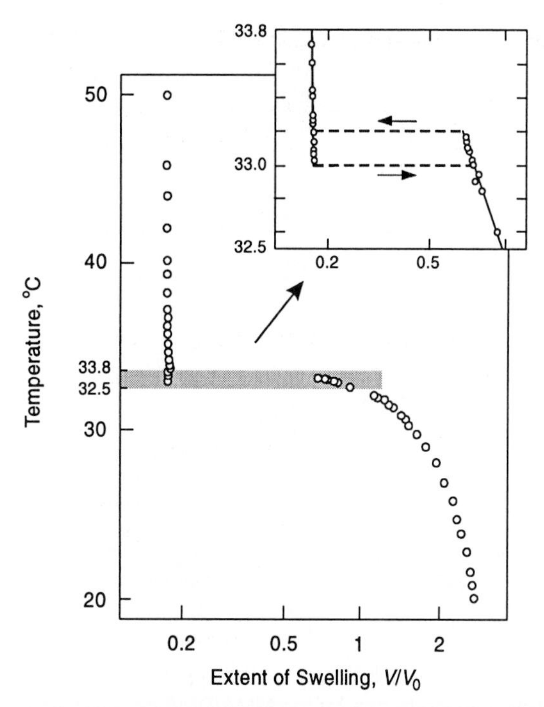

Figure 8-36 Effect of temperature on the volume of a poly(*N*-isopropylacrylamide) hydrogel in pure water. The expanded graph shows that, when the direction of transition is reversed, some hysteresis may be observed (Hirokawa and Tanaka, 1984).

For ionic systems, phase transitions can also be induced by varying the charge density of the gel. If the ionic groups on the polymer correspond to those of a weak acid or base (typically, carboxyl or amino groups), the charge density can be changed by a variation of pH in the surrounding solution (Dušek, 1993).

To illustrate, Fig. 8-37 shows swelling isotherms for four methyl methacrylate (MMA)/dimethylaminoethyl methacrylate (DMA) copolymers of various comonomer composition as a function of pH at 25°C and a total ionic strength of 0.1 M (Siegel and Firestone, 1988). Whereas MMA is insensitive to pH, the amino group in DMA becomes positively charged when the H^+ concentration is large (low pH). The data in Fig. 8-37 show that for pH > 6.6, all gels studied here are collapsed, i.e. compact and hydrophobic, regardless of comonomer composition. These collapsed gels contain 10 weight % water or less at swelling equilibrium. Lowering the pH, a critical value of pH is reached where the equilibrium content of water abruptly increases, giving rise to a highly swollen gel. At still lower pH values, the water content continues to increase but at a more gradual rate. The copolymer composition has a strong impact on the equilibrium swelling behavior of the MMA/DMA gels. Changing the MMA/DMA proportion from 70/30 to 86/14 shifts the transition pH from about 6.5 to 4.8, and the water content in the low-pH range changes from about 90 to 40 weight %. The 93/7 mol/mol gel remains compact at all pH; no volume transition occurs.

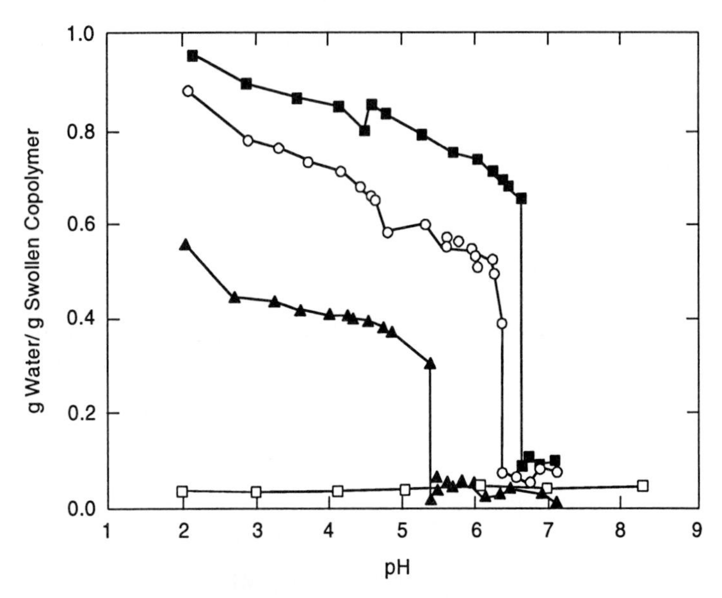

Figure 8-37 Swelling isotherms for the MMA/DMA copolymers of various comonomer composition, determined as a function of pH at 25°C and a total ionic strength of 0.1 M. ■ 70/30 mol/mol, ○ 78/22 mol/mol, ▲ 86/14 mol/mol, □ 93/7 mol/mol (Siegel and Firestone, 1988). MMA = methyl methacrylate (no charge); DMA = dimethylaminoethyl methacrylate (positive charge at low pH).

The abrupt changes in swelling behavior, induced by changes in external conditions, strongly affect the partitioning of solute species between the gel-fluid phase and the surrounding bulk-fluid phase. Therefore, gels may be suitable for chemical-separation systems (e.g. selective extraction) and as environmentally sensitive barriers for pharmaceutical agents (e.g., physiologically controlled drug delivery devices) (DeRossi *et al.*, 1989; Dušek, 1993; Thiel *et al.*, 1995; Dagani, 1997).

When a gel is in equilibrium with its surroundings, we have not only fluid-phase equilibrium but, in addition, mechanical equilibrium between the elastic, porous network structure, the gel fluid and the surrounding fluid. As the network-forming structure is elastic, its energy depends on its expansion as determined by the volume of the gel-fluid phase. The elastic properties of the network cause the pressure in the gel fluid to be larger than that in the surrounding fluid. This pressure difference may be negligible but for strongly swelling gels it cannot be neglected. In a model describing the phase behavior of polymer gels, therefore, polymer-solution thermodynamics must be combined with the theory of elasticity and, for ionic systems, in addition, with the theory of solutions containing electrolytes. Maurer (1996) gives a comprehensive summary of the fundamental relations that govern not only the phase equilibrium of a gel in contact with a fluid phase but also the equilibrium partitioning of solutes between the gel-fluid phase and the surrounding solution.

For most practical applications, attention has been given to hydrogels, that is, to gels made of hydrophilic polymers. Models for describing aqueous polymer gels may require parameters that are different from those for aqueous polymer solutions (Hirotsu *et al.*, 1987; Inomata *et al.*, 1994) or use too many adjustable parameters (Prange *et al.*, 1989). A simple molecular-thermodynamic model, developed by Hino (1998), is applicable to polymer solutions as well as to weakly charged polymer gels that can hydrogen bond with the solvent. Hino's semi-empirical model can be used for predicting the volume-change transition in temperature-sensitive polymer gels, as shown in the following paragraphs.

First we consider a binary polymer solution containing solvent (component 1) and linear polymer (component 2) where solvent molecules are represented by spheres and polymer molecules by freely-jointed chains. We use the extended Flory-Huggins lattice theory discussed in Sec. 8.2 (Qian *et al.*, 1991) to calculate the Helmholtz energy of mixing $\Delta_{\text{mix}} A$:

$$\frac{\Delta_{\text{mix}} A}{kT} = N_1 \ln(1-\Phi_2) + N_2 \ln(\Phi_2) + N_1 \Phi_2 \, g(T, \Phi_2) \tag{8-138}$$

where k is Boltzmann's constant, T is the absolute temperature, N_i is the number of molecules of component i, Φ_2 is the volume fraction of the polymer, and $g(T, \Phi_2)$ is an empirical function that replaces the customary Flory segmental interaction parameter.

We next consider a polymer gel containing solvent (component 1) and a large cross-linked polymer molecule (component 2) having a small number of ionizable segments. The Helmholtz energy of swelling is given by (Flory, 1953; Tanaka, 1980)

$$\Delta A^{\text{swe}} = \Delta_{\text{mix}} A + \Delta A^{\text{elas}} + \Delta A^{\text{ion}} \qquad \text{(8-139)[40]}$$

where $\Delta_{\text{mix}} A$, ΔA^{elas} and ΔA^{ion} represent mixing, elastic, and electrostatic contributions, respectively.

Provided that cross-linking in the gel is modest (i.e. long strands of polymer between cross-link points), the mixing contribution is (Flory, 1953)

$$\frac{\Delta_{\text{mix}} A}{kT} = N_1 \ln(1-\Phi_2) + N_1 \Phi_2 \, g(T, \Phi_2) \tag{8-140}$$

[40] Equation (8-139) assumes that the three contributions are essentially independent of each other. This equation is often referred to as the Flory-Rehner theory.

For the Helmholtz energy change due to elastic deformation, Hino uses an expression given by Birshtein and Pryamitsyn (1991) and Grosberg and Kuznetsov (1992):

$$\frac{\Delta A^{\text{elas}}}{kT} = \frac{3}{2}\nu\left(\alpha^2 + \frac{1}{\alpha^2} - 2\right) + \frac{1}{2}\nu \ln \alpha^3 \tag{8-141}$$

where α is the expansion factor and ν is the total number of chains. The expansion factor is given by (Flory, 1976, 1977; Erman and Flory, 1986; Painter and Shenoy, 1993):

$$\alpha = \left(\frac{\Phi_0}{\Phi_2}\right)^{1/3} \tag{8-142}$$

where Φ_0 is the volume fraction of polymer in the reference state where the conformation of network chains is closest to that of unperturbed Gaussian chains (Khokhlov, 1980); Φ_0 is usually approximated by the volume fraction of polymer at preparation of the gel.

Finally, for a small charge density, Hino expresses the electrostatic effect using the van't Hoff equation (Ricka and Tanaka, 1984; Otake *et al.*, 1989):

$$\frac{\Delta A^{\text{ion}}}{kT} = -m\nu \ln(N_1 + \nu r_n) \tag{8-143}$$

where m is the number of charged segments per network chain between points of cross-linking, and r_n is the number of segments per network chain.

For phase-equilibrium calculations, we introduce the Flory χ parameter, defined by the product of two empirical functions (Qian *et al.*, 1991; Bae *et al.*, 1993):

$$\chi \equiv g - \left(\frac{\partial g}{\partial \Phi_2}\right)_T = D(T)B(\Phi) \tag{8-144}$$

where $B(\Phi)$ is a function of composition and $D(T)$ is a function of temperature.

We also define reduced temperature $\tilde{T}$ and interchange energy ε by

$$\tilde{T} \equiv \frac{kT}{\varepsilon} \tag{8-145}$$

$$\varepsilon \equiv 2\varepsilon_{12} - \varepsilon_{11} - \varepsilon_{22} \tag{8-146}$$

where ε_{ij} $(i, j = 1, 2)$ is the segmental interaction energy for non-specific interactions between components i and j.

For $B(\Phi)$, we use an expression from Bae *et al.* (1993). For $D(T)$, we use an expression based on the work of ten Brinke and Karasz (1984) and Hino *et al.* (1993, 1993a):

$$D(T) = \frac{z}{2}\left[\left(1 + 2\delta\varepsilon_{12}/\varepsilon\right)\left(\frac{1}{\tilde{T}}\right) + 2\ln\left(\frac{1+s_{12}}{1+s_{12}\exp\left(\frac{\delta\varepsilon_{12}}{\varepsilon\tilde{T}}\right)}\right)\right] \tag{8-147}$$

$$B(\Phi) = \frac{1}{1-b\Phi_2} \tag{8-148}$$

where z is the lattice coordination number ($z = 6$ in this work), $\delta\varepsilon_{12}$ is the difference between the segmental interaction energy for specific interactions and that for non-specific interactions, s_{12} is the ratio of degeneracy of non-specific interactions to that of specific interactions, and b is an empirical parameter, all obtained from independent polymer-solution data. Equation (8-147) accounts for specific interactions between unlike molecules. As already discussed in Sec. 8.2, with this expression it is possible to predict lower-critical-solution temperature (LCST) behavior due to specific interactions such as hydrogen bonding.

At equilibrium, μ_1(surrounding pure solvent) is equal to μ_1(in the gel). The chemical potential of the solvent in the gel is found from differentiation of ΔA^{swe} with respect to N_1. We then have

$$\begin{aligned}\Delta\mu_1(\Phi_e) = \left(\frac{\partial \Delta A^{swe}}{\partial N_1}\right)_{T,V,N_2=1} &= \ln(1-\Phi_e) + \Phi_e + \chi\Phi_e^2 \\ &+ \frac{\Phi_0}{r_n}\left[\left(\frac{\Phi_e}{\Phi_0}\right)^{1/3} - \left(\frac{\Phi_e}{\Phi_0}\right)^{5/3} - \left(m - \frac{1}{2}\right)\left(\frac{\Phi_e}{\Phi_0}\right)\right] = 0\end{aligned} \tag{8-149}$$

where $\Delta\mu_1$ is the change in chemical potential of solvent upon mixing, and V is the volume of the gel. For the elastic term in Eq. (8-149), we use Eq. (8-141). For the mixing term, we use the extended Flory-Huggins theory presented above.

Under the specific conditions where two gel phases can coexist at one temperature, a polymer gel exhibits a discontinuous volume change. The conditions for coexistence of the two gel phases are (Marchetti *et al.*, 1990)

$$\mu_1(\Phi_e') = \mu_1(\Phi_e'') \tag{8-150}$$

and

$$\mu_2(\Phi_e') = \mu_2(\Phi_e'') \tag{8-151}$$

where superscripts ' and " denote coexisting phases, i.e. the expanded and the collapsed gel. From the Gibbs-Duhem equation, Eq. (8-151) can be replaced by

$$\int_{\Phi_e'}^{\Phi_e''} \Delta\mu_1 \Phi_2^{-2} d\Phi_2 = 0 \tag{8-152}$$

Hino (1998) applied his model to poly(*N*-isopropylacrylamide) (PNIPAAm) gels in water and to aqueous solutions of non-crosslinked PNIPAAm. The polymer solutions exhibit LCST behavior at about 32°C and from these data the following parameters were obtained: $\varepsilon = 2.92$ kJ mol^{-1}, $\delta\varepsilon_{12}/\varepsilon = -7$, and $b = 0.65$, with s_{12} preset to 5000 as discussed in Sec. 8.2. Using the same parameters for the mixing term, Hino's model can also represent the phase diagram of neutral PNIPAAm gels. As shown in Fig. 8-36, these gels exhibit abrupt volume contractions at about 32.5-33.8°C. Figure 8-38 illustrates that the temperature of this volume transition is represented quantitatively with $m = 0$ (nonionic gel) and with $r_n = 89$ in Eq. (8-149) adjusted such that the swelling ratio is 2.5 at 20°C with $\Phi_0 = 0.07$, as indicated by the PNIPAAm gels studied by Hirotsu *et al.* (1987).

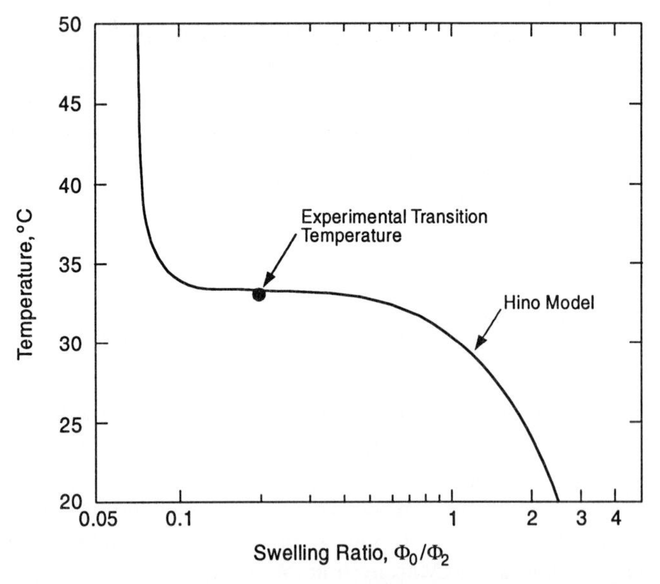

Figure 8-38 Swelling-ratio curve for neutral PNIPAAm gel in water calculated with the model of Hino (1998). Experimental transition temperature range is 32.5-33.8°C (Hirotsu *et al.*, 1987).

PNIPAAm gels copolymerized with ionizable monomers swell more than neutral PNIPAAm gels (Beltran *et al.*, 1990) and exhibit more abrupt volume contractions (Hirotsu *et al.*, 1987; Hino *et al.*, 1993). That observation is predicted by Hino's model with $m > 0$, i.e. for gels with ionizable segments in the network-forming polymer chains.

The experimental phase behavior of neutral and ionic gels has been studied extensively. Applications of gels in chemical engineering, medicine, and biotechnology are, as yet, in an early stage (Dagani, 1997).

8.5 Summary

There has been much progress in the thermodynamics of polymer solutions since the initial path-breaking work of *Flory* and *Huggins*, more than 50 years ago. That initial work was closely tied to the *filled-lattice* concept that limited consideration to those mixtures where there is no volume change upon mixing at constant temperature and pressure. To relieve that severe limitation, much attention has been given toward developing an equation of state suitable for polymers, polymer/solvent mixtures and polymer blends.

While the limitation of the constant-volume assumption of filled-lattice thermodynamics was recognized early, a major stimulus to develop an *equation of state* for polymer-solvent systems was provided over 30 years ago when experimental results indicated the common existence of a lower critical-solution temperature that lies *above* the upper critical-solution temperature. Classical lattice thermodynamics cannot explain such phase behavior. A first necessary test for any proposed equation of state for mixtures is that it must be able to reproduce phase behavior that includes both lower and upper critical-solution temperatures where the lower critical-solution temperature is larger than the upper critical-solution temperature. All proposed equations of state discussed here can meet that test, at least in principle.

Upon introducing *specific attraction* between unlike components (e.g. hydrogen bonding), filled-lattice theory can be modified to give a lower critical-solution temperature that lies *below* the upper critical-solution temperature. Such modification can also be applied to an equation of state with the same result.

As discussed in this chapter, for fluids containing *chain molecules*, equations of state can be derived along several lines: *lattice-fluid* or *hole theories*, where a lattice contains occupied and unoccupied lattice sites (holes); and *free-standing theories*, independent of any geometric construct such as cells or lattices, where the volumetric properties of an assembly of chains are described by a statistical mechanical method (the integral theory of fluids) not discussed here. Analytical results can be obtained from the integral theory of fluids for hard non-attracting chains. For attractive chains, it is necessary to use numerical methods or else to introduce attractive forces as a perturbation about results for non-attractive chains.

All these theories can successfully describe the *P-V-T* properties of pure liquid polymers using (typically) three or four adjustable molecular parameters that, respectively, reflect chain length, segment size, potential energy between two nonbonded segments and (sometimes) chain flexibility or "floppiness" as a measure of external degrees of freedom.

Regrettably, extension to *mixtures* shows serious deficiencies in all of these theories. First, these theories fail to account for long-range correlation; when one segment of a chain moves, that movement influences the movement of other segments in the same chain. Second, these theories are for simple chains and therefore do not account for the effect of chain branching or of rings within a chain. Third, and perhaps most serious, all of these theories are of the mean-field type that fail to account for the contribution of fluctuations in density and composition. It has been known for many years that the mean-field approximation is valid only at high densities and high temperatures. Therefore, when the theories described here are used in the critical region (where fluctuations are large), poor results are often obtained; if model parameters are fixed using critical data (coordinates of upper or lower critical-solution temperatures), the calculated two-liquid region is too narrow; but if parameters are fixed from LLE data remote from critical, erroneous predictions are obtained for critical temperatures and compositions.

For polymer/solvent mixtures, reliable VLE can be calculated easily (except when the solvent is near its critical temperature) because such calculation requires only the chemical potential of the solvent; the chemical potential of the polymer does not enter the calculation because polymers are nonvolatile. However, VLE calculation may depend on the range of composition considered; binary parameters obtained from dilute polymer-solution VLE are often different from those obtained from concentrated polymer-solution VLE.

Accurate calculation of LLE is much more difficult for polymer/solvent mixtures because typically, the polymer is dilute in one liquid phase and concentrated in the other liquid phase. Mean-field approximations are not good for a dilute phase. If binary parameters are found for a solution dilute in polymer, they are likely to be different from those found for a solution where the polymer concentration is large. Mean-field approximations introduced in most equations of state are not valid for dilute solutions. Distinguishing between *intramolecular* and *intermolecular* segment-segment interactions is not as important in concentrated solutions as it is in dilute solutions. Further, unlike VLE, LLE are often sensitive to the polymer's polydispersity.

It is likely that several reasons may be responsible for the failure of EOS theories to represent liquid-liquid equilibrium with good accuracy. One reason may follow from the conventional procedure for finding EOS constants for the pure polymer. That procedure is based on reduction of *P-V-T* data for pure liquid polymers; the characteristic energy parameter obtained from perturbed hard-sphere-chain equations is often too large, perhaps because pure liquid polymers have chains with extensive entanglement. The characteristic energy parameter, therefore, includes the energy of disentanglement. However, the calculated free energy of mixing of a polymer with a solvent makes no provision for disentangling the pure polymer prior to dissolution. Better results for liquid-liquid equilibrium may perhaps, be obtained if the mixture calculations include a contribution that reflects the disentanglement of polymer chains upon mixing. That contribution may be important for polymer-solvent mixtures but probably tends to cancel for polymer-polymer blends.

These problems are not readily observed from model calculations but they become clear when calculated results are compared with experiment. Errors introduced by mean-field approximations are evident when the top of the liquid-liquid coexistence curve is insufficiently flat and whenever calculations of phase equilibria are compared with experiment over a large range of polymer concentration, from very dilute to appreciable polymer concentration. Given a set of molecular parameters, it is possible to fit experimental data for one equilibrium liquid phase but, to fit the other equilibrium liquid phase, a different set of molecular parameters is often required. These problems are much reduced when equations of state are used for polymer blends. For mixtures of polymers, good results can often be obtained because errors introduced by mean-field approximations tend to cancel.

In general, two conceptually different mean-field approximations are invoked during the development of equations of state for chain molecules, regardless of their framework. To calculate the entropy of a collection of polymer molecules, even in the absence of attractive forces, the number of available configurations must be properly calculated without neglect of correlations between segments of a chain that are not nearest neighbors along the chain. This neglect is the essence of the first mean-field approximation. A second mean-field approximation is associated with calculating the potential energy; in that approximation, chain connectivity and the correlation between segments are improperly ignored.

The inadequacy of mean-field approximations was recognized many years ago; it is indicated in the classic text by Flory published in 1953. At present we do not have a useful analytical equation of state for mixtures that overcomes this inadequacy. An alternative to an analytical equation is provided by *molecular simulation* calculations but these are not as yet practical for routine engineering applications. The current literature has reported several simulation calculations for non-attractive mixtures of chains; only very recently have such calculations been made for pure-polymer liquids with attractive forces. It is likely that we will soon see simulation calculations also for mixtures of attractive chains and solvents. Such simulations are likely to provide better agreement with experiment over the entire composition range. As computers become more powerful and as more efficient computing strategies become available, simulation calculations may eventually replace analytical equations of state. However, for the near future, analytical equations of state will remain useful for application provided that their use is restricted to narrow ranges of composition or else, provided that they are modified semi-empirically to overcome limitations imposed by the mean-field approximation.

Finally, there is a geometric consideration that is usually overlooked. In a polymer chain, the distance between bonded segments is probably somewhat smaller than the collision diameter that characterizes the interaction between non-bonded segments. Therefore, the parameter σ used in the reference part of the EOS (i.e. the part for non-attracting chains) is not the same as that used in the perturbation part of the EOS (i.e. the part that corrects for attractive forces). To refrain from using an excessive number of parameters, it is customary to ignore this difference in the two characteristic length parameters. That procedure does not affect the ability of an EOS to fit pure-polymer data but it may have a significant effect on liquid-liquid equilibria for polymer/solvent systems.

While polymers have a multitude of applications in chemical technology, thin films of polymers (*nonporous membranes*) are particularly useful for separation operations. Design of such operations requires transport properties (diffusivities) in addition to thermodynamic properties.

When a model for polymer solutions is combined with a model for polymer network elasticity, it is possible to construct a theory for the properties of a *polymer gel* in contact with a swelling solvent. That theory shows the possibility of a first-order ohase transition in response to a change in temperature. If the polymer gel contains groups that are weakly acid (or basic), a similar phase transition can respond to a change in pH of the solvent. As a result, polymer gels have potential as "smart" materials that can exhibit a strong response to a small change of environment. Thanks to thermodynamic analysis, we now have a good understanding of the fundamental properties of gel-solvent systems but many important details must be clarified before this understanding becomes truly quantitative.

References

Amjad, Z., (Ed.), 1993, *Reverse Osmosis: Membrane Technology, Water Chemistry, and Industrial Applications.* New York: Van Nostrand Reinhold.

Bae, Y. C., J. J. Shim, D. S. Soane, and J. M. Prausnitz, 1993, *J. Appl. Polym. Sci.,* 47: 1193.

Banaszak, M., C. K. Chen, and M. Radosz, 1996, *Macromolecules,* 29: 6481.

Barbari, T. A. and R. M. Conforti, 1992, *J. Polym. Sci., Part B: Polym. Phys.,* 30: 1261.

Barton, A. F. M., 1990, *Handbook of Solubility Parameters.* Boca Raton: CRC Press.

Baudot, A. and M. Marin, 1997, *Trans. I. Chem. Eng.,* 75: 117.

Bawendi, M. G., K. F. Freed, and U. Mohanty, 1987, *J. Chem. Phys.,* 87: 5534.

Bawendi, M. G. and K. F. Freed, 1988, *J. Chem. Phys.,* 88: 2741.

Beltran, S., H. H. Hooper, H. W. Blanch, and J. M. Prausnitz, 1990, *J. Chem. Phys.,* 92: 2061.

Berens, A. R., G. S. Huvard, R. W. Korsmeyer, and F. W. Kunig, 1992, *J. Appl. Polym. Sci.,* 46: 231.

Bhattacharya, S. and S.-T. Hwang, 1997, *J. Membrane Sci.,* 132: 73.

Birshtein, T. M. and V. A. Pryamitsyn, 1991, *Macromolecules,* 24: 1554.

Blanks, R. F. and J. M. Prausnitz, 1964, *Ind. Eng. Chem. Fundam.,* 3: 1.

Böddeker, K.-W. and G. Bengtson, 1991, *Selective Pervaporation of Organics from Water.* In *Pervaporation Membrane Separation Processes,* (R. Y. M. Huang, Ed.). Amsterdam: Elsevier.

Bondi, A., 1968, *Physical Properties of Molecular Crystals, Liquids and Glasses.* New York: John Wiley & Sons.

Bonner, D. C., A. Bellemans, and J. M. Prausnitz, 1972, *J. Polym. Sci., Part C,* 39: 1.

Burrell, H., 1955, *Interchem. Rev.,* 14: 3, 31.

Chapman, W. G., K. E. Gubbins, G. Jackson, and M. Radosz, 1990, *Ind. Eng. Chem. Res.,* 29: 1709.

Chen, S.-J. and M. Radosz, 1992, *Macromolecules,* 25: 3089.

Chen, S.-J., I. G. Economou, and M. Radosz, 1992, *Macromolecules,* 25: 4987.

Chen, S.-J., I. G. Economou, and M. Radosz, 1993, *Fluid Phase Equilibria,* 83: 391.

Chen, S.-J., Y. C. Chiew, J. A. Gardecki, S. Nilsen, and M. Radosz, 1994, *J. Polym. Sci., Part B: Polym. Phys.,* 32: 1791.

Chen, S.-J., M. Banaszak, and M. Radosz, 1995, *Macromolecules,* 28: 1812.

Chiou, J. S., J. W. Barlow, and D. R. Paul, 1985, *J. Appl. Polym. Sci.,* 30: 2633.

Condo, P. D. and K. P. Johnston, 1992, *Macromolecules,* 25: 6730.

Condo, P. D., I. C. Sanchez, C. G. Panayiotou, and K. P. Johnston, 1992, *Macromolecules,* 25: 6119.

Condo, P. D. and M. Radosz, 1996, *Fluid Phase Equilibria,* 117: 1.

Costas, M. and B. C. Sanctuary, 1981, *J. Phys. Chem.,* 85: 3153.

Costas, M. and B. C. Sanctuary, 1984, *Fluid Phase Equilibria,* 18: 47.

Cotton, N. J., K. D. Bartle, A. A. Clifford, and C. J. Dowle, 1993, *J. Appl. Polym. Sci.,* 48: 1607.

Cowie, J. M., 1991, *Polymers: Chemistry and Physics of Modern Materials,* 2nd Ed. London: Chapman & Hall.

Dagani, R., 1997, *Chem. Eng. News,* 75: 26.

Daneshvar, M., S. Kim, and E. Gulari, 1990, *J. Phys. Chem.,* 94: 2124.

DeRossi, D., K. Kajiwara, Y. Osada, and A. Yamauchi, 1991, *Polymer Gels: Fundamentals and Biomedical Applications.* New York: Plenum Press.

Di Marzio, E. A. and J. H. Gibbs, 1963, *J. Polym. Sci., Part A,* 1: 1417.

Dickman, R. and C. K. Hall, 1986, *J. Chem. Phys.,* 85: 4108.

Donohue, M. D. and J. M. Prausnitz, 1975, *Can. J. Chem.,* 53: 1586.

Dudowicz, J. and K. F. Freed, 1991, *Macromolecules,* 24: 5112.

Dušek, K., (Ed.), 1993, *Responsive Gels: Volume Transitions I* and *Volume Transitions II,* Advances in Polymer Science, Vols. 109 and 110. Heidelberg: Springer.

Eichinger, B. E. and P. J. Flory, 1968, *Trans. Faraday Soc.,* 64: 2053.

Eichinger, B. E. and P. J. Flory, 1968a, *Trans. Faraday Soc.,* 64: 2035.

Enneking, L., W. Stephan, and A. Heintz, 1993, *Ber. Bunsenges. Phys. Chem.,* 97: 912.

Enneking, L., A. Heintz, and R. N. Lichtenthaler, 1996, *J. Membrane Sci.,* 115: 161.

Erman, B. and P. J. Flory, 1986, *Macromolecules,* 19: 2342.

Fast, J. D., 1962, *Entropy,* Philips Technical Library. Eindhoven: Centrex Publishing.

Feng, X. and R. Y. M. Huang, 1994, *J. Membrane Sci.,* 92: 201.

Feng, X. and R. Y. M. Huang, 1997, *Ind. Eng. Chem. Res.,* 36: 1048.

Flory, P. J., 1941, *J. Chem. Phys.,* 9: 660.

Flory, P. J., 1942, *J. Chem. Phys.,* 10: 51.

Flory, P. J., 1953, *Principles of Polymer Chemistry.* Ithaca: Cornell University Press.

Flory, P. J., R. A. Orwoll, and A. Vrij, 1964, *J. Am. Chem. Soc.,* 86: 3507, 3515.

Flory, P. J., 1965, *J. Am. Chem. Soc.,* 87: 1833.

Flory, P. J. and R. A. Orwoll, 1967, *J. Am. Chem. Soc.,* 89: 6814, 6822.

Flory, P. J., J. L. Ellenson, and B. E. Eichinger, 1968, *Macromolecules,* 1: 279.

Flory, P. J., 1970, *Discuss. Faraday Soc.,* 49: 7.

Flory, P. J., 1976, *Proc. Roy. Soc. London, Ser. A.,* 351: 351.

Flory, P. J., 1977, *J. Chem. Phys.,* 66: 5720.

Freed, K. F., 1985, *J. Phys. A: Math. Gen.,* 18: 871.

Gibbs, J. H. and E. A. Di Marzio, 1958, *J. Chem. Phys.,* 28: 373.

Gregg, C. J., S.-J. Chen, F. P. Stein, and M. Radosz, 1993, *Fluid Phase Equilibria,* 83: 375.

Grosberg, A. Y. and D. V. Kuzmtsov, 1992, *Macromolecules,* 25: 1970.

Grulke, E. A., 1989, *Solubility Parameter Values.* In *Polymer Handbook,* 3rd Ed., (J. Brandrup and E. H. Immergut, Eds.). New York: John Wiley & Sons.

Guggenheim, E. A., 1944, *Proc. Roy. Soc. A,* 183: 203.

Guggenheim, E. A., 1952, *Mixtures.* Oxford: Clarendon Press.

Gupta, R. B. and J. M. Prausnitz, 1995, *J. Chem. Eng. Data,* 40: 784.

Gupta, R. B. and J. M. Prausnitz, 1996, *Ind. Eng. Chem. Res.,* 35: 1225.

Hansen, C. M., 1967, *J. Paint Technol.,* 39: 104, 505.

Hansen, C. M. and K. Skaarup, 1967a, *J. Paint Technol.,* 39: 511.

Hansen, C. M. and A. Beerbower, 1971, *Solubility Parameters.* In *Encyclopedia of Chemical Technology* (Kirk-Othmer), 2nd Ed., Suppl. Vol. (H. F. Mark, J. J. McKetta, and D. F. Othmer, Eds.). New York: Wiley-Interscience.

Harismiadis, V. I., G. M. Kontogeorgis, Aa. Fredenslund, and D. P. Tassios, 1994, *Fluid Phase Equilibria,* 96: 93.

Heintz, A. and R. N. Lichtenthaler, 1977, *Ber. Bunsenges. Phys. Chem.,* 81: 921.

Heintz, A. and R. N. Lichtenthaler, 1980, *Ber. Bunsenges. Phys. Chem.,* 84: 727.

Heintz, A., H. Funke, and R. N. Lichtenthaler, 1991, *Sorption and Diffusion in Pervaporation Membranes.* In *Pervaporation Membrane Separation Processes,* (R. Y. M. Huang, Ed.). Amsterdam: Elsevier.

Heintz, A. and W. Stephan, 1994, *J. Membrane Sci.,* 89: 143, 153.

Henderson, D., 1974, *Ann. Rev. Phys. Chem.,* 25: 461.

Hildebrand, J. H., 1947, *J. Chem. Phys.,* 15: 225.

Hino, T., S. M. Lambert, D. S. Soane, and J. M. Prausnitz, 1993, *AIChE J.,* 39: 837.

Hino, T., S. M. Lambert, D. S. Soane, and J. M. Prausnitz, 1993a, *Polymer,* 34: 4756.

Hino, T., Y. Song, and J. M. Prausnitz, 1994, *Macromolecules,* 27: 5681.

Hino, T., Y. Song, and J. M. Prausnitz, 1995, *Macromolecules,* 28: 5709, 5717, 5725.

Hino,T. and J. M. Prausnitz, 1998, *Polymer,* 39: 3279.

Hirokawa, Y. and T. Tanaka, 1984, *J. Chem. Phys.*, 81: 6379.

Hirotsu, S., Y. Hirokawa and T. Tanaka, 1987, *J. Chem. Phys.*, 87: 1392.

Hirschfelder, J., D. Stevenson, and H. Eyring, 1937, *J. Chem. Phys.*, 5: 896.

Honnell, K. G. and C. K. Hall, 1989, *J. Chem. Phys.*, 90: 1841.

Hu, Y., S. M. Lambert, D. S. Soane, and J. M. Prausnitz, 1991, *Macromolecules*, 24: 4356.

Huang, R. Y. M., (Ed.), 1991, *Pervaporation Membrane Separation Processes*. Amsterdam: Elsevier.

Huggins, M. L., 1941, *J. Phys. Chem.*, 9: 440.

Huggins, M. L., 1942, *Ann. N. Y. Acad. Sci.*, 43: 1.

Ilavsky, M., J. Hrouz and K. Ulbrich, 1982, *Polym. Bull.*, 7: 107.

Inomata, H., K. Nagahama and S. Saito, 1994, *Macromolecules*, 27: 6459.

Inui, K., H. Okumura, T. Miyata, and T. Uragami, 1997, *J. Membrane Sci.*, 132: 193.

Iwai, Y. and Y. Arai, 1991, *J. Japan Petrol. Inst.*, 34: 416.

Kalospiros, N. S. and M. E. Paulaitis, 1994, *Chem. Eng. Sci.*, 49: 659.

Kammer, H.-W., T. Inoue, and T. Ouginawa, 1989, *Polymer*, 30: 888.

Kaul, B. K., M. D. Donohue, and J. M. Prausnitz, 1980, *Fluid Phase Equilibria*, 4: 171.

Khokhlov, A. R., 1980, *Polymer*, 21: 376.

Kim, C. K. and D. R. Paul, 1992, *Polymer*, 33: 1630, 2089, 4941.

Kleintjens, L. A. and R. Koningsveld, 1980, *Colloid Polym. Sci.*, 258: 711.

Kleintjens, L. A. and R. Koningsveld, 1982, *Sep. Sci. Tech.*, 17: 215.

Koningsveld, R. and A. J. Staverman, 1968, *J. Polym. Sci. Part A-2*, 6: 305, 325, 349.

Koningsveld, R., L. A. Kleintjens, and A. R. Shultz, 1970, *J. Polym. Sci. Part A-2*, 8: 1261.

Koningsveld, R. and L. A. Kleintjens, 1971, *Macromolecules*, 4: 637.

Kontogeorgis, G. M., V. I. Harismiadis, Aa. Fredenslund, and D. P. Tassios, 1994, *Fluid Phase Equilibria*, 96: 65.

Lambert, S. M., D. Soane, and J. M. Prausnitz, 1993, *Fluid Phase Equilibria*, 83: 59.

Lambert, S. M., Y. Song, and J. M. Prausnitz, 1995, *Macromolecules*, 28: 4866.

Lambert, S. M., Y. Song, and J. M. Prausnitz, 1998, *Equations of State for Polymer Systems*. In *Equations of State for Fluids and Fluid Mixtures* (J. V. Sengers, M. B. Ewing, R. F. Kayser, C. J. Peters, and H. J. White, Jr., Eds.). Oxford: Blackwell Scientific.

Lichtenthaler, R. N., D. S. Abrams, and J. M. Prausnitz, 1973, *Can. J. Chem.*, 51: 3071.

Lichtenthaler, R. N., D. D. Liu, and J. M. Prausnitz, 1974, *Ber. Bunsenges. Phys. Chem.*, 78: 470.

Lichtenthaler, R. N., D. D. Liu, and J. M. Prausnitz, 1978, *Macromolecules*, 11: 192 (1978).

Madden, W. G., A. I. Pesci, and K. F. Freed, 1990, *Macromolecules*, 23: 1181.

Madden, W. G., J. Dudowicz, and K. F. Freed, 1990a, *Macromolecules*, 23: 4803.

Marchetti, M., S. Prager, and E. L. Cussler, 1990, *Macromolecules*, 25: 1760, 3445.

Maurer G. and J. M. Prausnitz, 1996, *Fluid Phase Equilibria*, 115:113.

McMaster, L. P., 1973, *Macromolecules*, 6: 760.

Meckl, K. and R. N. Lichtenthaler, 1996, *J. Membrane Sci.*, 113: 81.

Mulder, M., 1996, *Basic Principles of Membrane Technology*, 2nd Ed. Dordrecht: Kluwer Academic Publishers.

Myrat, C. D. and J. S. Rowlinson, 1965, *Polymer,* 6: 645.

Neogi, P., (Ed.), 1996, *Diffusion in Polymers.* New York: Marcel Dekker.

Nguyen, Q. T., C. Léger, P. Billard, and P. Lochon, 1997, *Polymers Adv. Techn.,* 8: 487.

Noble, R. D. and S. A. Stern, (Eds.), 1995, *Membrane Separation Technology: Principles and Applications,* Membrane Science and Technology, Vol. 2. Amsterdam: Elsevier.

Ohzono, M., Y. Iwai, and Y. Arai, 1984, *J. Chem. Eng. Japan,* 17: 550.

Orwoll, R. J., 1977, *Rubber Chem. Technol.,* 50: 452.

Otake, K., H. Inomata, M. Konno, and S. Saito, 1989, *J. Chem. Phys.,* 91: 1345.

Painter, P. C. and S. L. Shenoy, 1993, *J. Chem. Phys.,* 99: 1409.

Panayiotou, C. and J. H. Vera, 1981, *Can. J. Chem. Eng.,* 59: 501.

Panayiotou, C. and J. H. Vera, 1982, *Polym. J.,* 14: 681.

Panayiotou, C., 1987, *Macromolecules,* 20: 861.

Panayiotou, C. and I. C. Sanchez, 1991, *Macromolecules,* 24: 6231.

Park, D.-W. and R.-J. Roe, 1991, *Macromolecules,* 24: 5324.

Patterson, D., 1969, *Macromolecules,* 2: 672.

Patterson, D. and G. Delmas, 1970, *Discuss. Faraday Soc.,* 49: 98.

Pope, D. S., I. C. Sanchez, W. J. Koros, and G. K. Fleming, 1991, *Macromolecules,* 24: 1779.

Pouchly, J. and D. Patterson, 1976, *Macromolecules,* 9: 574.

Prange, M. M., H. H. Hooper and J. M. Prausnitz, 1989, *AIChE J.*, 35: 803.

Pressly, T. G. and K. M. Ng, 1998, *AIChE J.*, 44: 93.

Prigogine, I., 1957, *The Molecular Theory of Solutions.* Amsterdam: North-Holland.

Qian, C., S. J. Mumby, and B. E. Eichinger, 1991, *J. Polym. Sci., Part B: Polymer Physics,* 29: 635.

Qian, C., S. J. Mumby, and B. E. Eichinger, 1991a, *Macromolecules,* 24: 1655.

Rave, A., 1995, *Principles of Polymer Chemistry.* New York: Plenum Press.

Ricka, J. and T. Tanaka, 1984, *Macromolecules,* 17: 2916.

Rodgers, P. A. and I. C. Sanchez, 1993, *J. Polym. Sci. Part B: Polym. Phys.,* 31: 273.

Roe, R.-J. and W.-C. Zin, 1980, *Macromolecules,* 13: 1221.

Rostami, S and D. J. Walsh, 1985, *Macromolecules,* 18: 1228.

Sanchez, I. C. and R. H. Lacombe, 1976, *J. Phys. Chem.,* 80: 2352, 2568.

Sanchez, I. C. and R. H. Lacombe, 1977, *J. Polym. Sci., Polym. Lett. Ed.,* 15: 71.

Sanchez, I. C. and R. H. Lacombe, 1978, *Macromolecules,* 11: 1145.

Sanchez, I. C., 1987, *Encyclopedia of Physical Science and Technology,* Vol. XI. New York: Academic Press.

Sanchez, I. C. and A. C. Balazs, 1989, *Macromolecules,* 22: 2325.

Sanchez, I. C. and P. A. Rodgers, 1990, *Pure & Appl. Chem.,* 62: 2107.

Sanchez, I. C. and C. Panayiotou, 1994, *Equation of State Thermodynamics of Polymer and Related Solutions.* In *Models for Thermodynamic and Phase Equilibria Calculations*, (S. I. Sandler, Ed.). New York: Marcel Dekker.

Scott, R. L. and P. H. van Konynenburg, 1970, *Discuss. Faraday Soc.,* 49: 87.

Siegel, R. A. and B. A. Firestone, 1988, *Macromolecules,* 21: 3254.

Simha, R. and T. Somcynsky, 1969, *Macromolecules,* 2: 342.
Siow, K. S., G. Delmas, and D. Patterson, 1972, *Macromolecules,* 5: 29.
Smith, S. W., C. K. Hall, and B. D. Freeman, 1995, *Phys. Rev. Lett.,* 75: 1316.
Somcynsky, T. and R. Simha, 1971, *J. Appl. Phys.,* 42: 4545.
Song, Y., S. M. Lambert, and J. M. Prausnitz, 1994, *Ind. Eng. Chem. Res.,* 33: 1047.
Song, Y., S. M. Lambert, and J. M. Prausnitz, 1994a, *Chem. Eng. Sci.,* 33: 1047.
Staudt-Bickel, C. and R. N. Lichtenthaler, 1996, *J. Membrane Sci.,* 111: 135.
Staverman, A. J., 1950, *Recl. Trav. Chim. Pays-Bas,* 69: 163.
Streicher, C., L. Asselineau, and A. Forestière, 1995, *Pure & Appl. Chem.,* 67: 985.
Swinton, F. L., 1976, *Ann. Rev. Phys. Chem.,* 27: 153.
Tanaka, T., 1980, *Phys. Rev. Lett.,* 45: 1636.
ten Brinke, G. and F. E. Karasz, 1984, *Macromolecules,* 17: 815.
Thiel, J., G. Maurer, and J. M. Prausnitz, 1995, *Chem.-Ing.-Tech.,* 67: 1567.
Tompa, H., 1952, *Trans. Farad. Soc.,* 48: 363.
Wesslein, M., A. Heintz, and R. N. Lichtenthaler, 1990, *J. Membrane Sci.,* 51: 169.
Wijmans, J. G. and R. W. Baker, 1995, *J. Membrane Sci.,* 107: 1.
Wissinger, R. G. and M. E. Paulaitis, 1991, *Ind. Eng. Chem. Res.,* 30: 842.
Wissinger, R. G. and M. E. Paulaitis, 1991a, *J. Polym. Sci., Part B: Polym. Phys.,* 29: 631.
Wu, C.-S. and Y.-P. Chen, 1994, *Fluid Phase Equilibria,* 100: 103.
Yoshikawa, M., S. Takeuchi, and T. Kitao, 1997, *Angew. Makromol. Chem.,* 245: 193.
Young, R. J. and P. A. Lovell, 1991, *Introduction to Polymers,* 2nd Ed. London: Chapman & Hall.
Zeman, L. and D. Patterson, 1972, *J. Phys. Chem.,* 76: 1214.

Problems

1. In polymer solutions, it is convenient to define an activity coefficient $\Gamma_1 = a_1 / \Phi_1$, where a is the activity, Φ is the volume fraction, and subscript 1 refers to the volatile component. When Φ is very small, $\Gamma_1 \rightarrow \Gamma_1^\infty$; the Flory-Huggins equation gives

$$\ln \Gamma_1^\infty = \left(1 - \frac{1}{r}\right) + \chi$$

where χ is the Flory interaction parameter and $r = v_2/v_1$; v_2 is the molar volume of the polymer and v_1 is that of the solvent. For high-molecular-weight polymers, $r >> 1$.
A film of poly(vinyl acetate) contains traces of isopropyl alcohol. For health reasons, the alcohol content of the film must be reduced to a very low value; government regulations require that $\Phi_1 < 10^{-4}$. To remove the alcohol, it is proposed to evaporate it at 125°C. At this temperature, chromatographic experiments give χ = 0.44 and the vapor pressure of isopropyl alcohol is 4.49 bar. Calculate the low pressure that must be maintained in the evaporator to achieve the required purity of the film. Under the conditions prevailing

here, the activity is given by the ratio of partial pressure to vapor pressure. The polymer is involatile.

2. Estimate the total pressure of a liquid solution containing 50 wt% poly(vinyl acetate) and 50 wt% vinyl acetate at 125°C.
Data (all at 125°C) are as follows:

	Solvent	Polymer
Density ($g\ cm^{-3}$)	0.783	1.11
Vapor pressure (torr)	3340	–

The molecular weight of the polymer is 8.34×10^4. From chromatographic measurements at 125°C, Henry's constant (partial pressure/weight fraction) for vinyl acetate in poly(vinyl acetate) is 18.3 bar. In your calculation, what is the most important simplifying assumption?

3. (a) Derive Flory equation of state [Eq. (8-45)] from the generalized van der Waals partition function given by Eq. (8-39).
(b) Derive Sanchez-Lacombe equation of state [Eq. (8-84)] from the configurational partition function given by Eq. (8-82).

4. In a binary system solvent (1)/polymer (2), the composition at which the UCST occurs depends on polymer molecular weight (see Fig. 8-8). According to the Flory-Huggins theory, the critical segment fraction is $\Phi_2^{*c} = 1/(1+r^{1/2})$. Derive this expression.

5. The characteristic parameters of Flory equation of state for hexamethyl disiloxane (HMDS) and poly(dimethyl siloxanes) (PDMS) of various molecula weights are listed in Table 8-5. For mixtures of HMDS (1)/PDMS (2), Flory parameter $\chi = 0$ is a good approximation. Calculate $\ln \gamma_1$ for the various binary mixtures at $\Phi_2^* = 0.8$ and discuss the dependence on molecular weight.

6. The solubility coefficient (S) and diffusion coefficient (D) of oxygen and nitrogen in silicone rubber at 20°C are:

$$S_{O_2}^G = 1.1\times10^{-6}\ \text{cm}^3\ \text{cm}^{-3}\ \text{Pa}^{-1} \qquad D_{O_2} = 1.6\times10^{-10}\ \text{m}^2\ \text{s}^{-1}$$

$$S_{N_2}^G = 0.7\times10^{-6}\ \text{cm}^3\ \text{cm}^{-3}\ \text{Pa}^{-1} \qquad D_{N_2} = 0.9\times10^{-10}\ \text{m}^2\ \text{s}^{-1}$$

Calculate the separation factor and the fluxes of oxygen and nitrogen through a 5 μm thick membrane of silicone rubber with air at 2 bar as feed and vacuum on the permeate side.

7. A polymeric membrane, mechanically supported by a metal screen, is used to separate a gaseous mixture of carbon dioxide (1) and methane (2) at 300 K, as illustrated below.

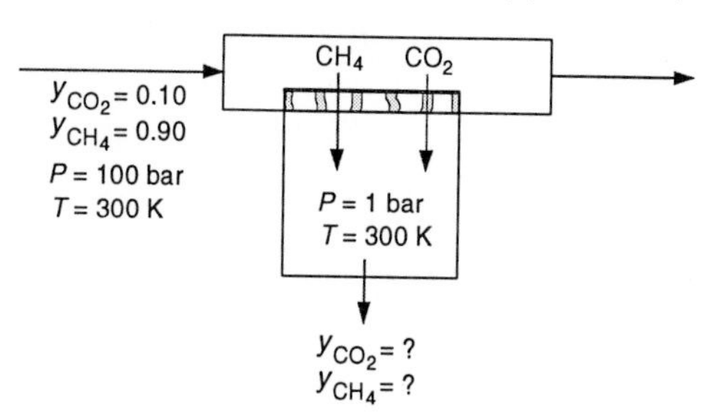

On the upstream side, in contact with the membrane, is a gaseous mixture at 100 bar containing 10 mol % carbon dioxide and 90 mol % methane. The fluxes of carbon dioxide and methane through the membrane follow Fick's law. Diffusion coefficients ($cm^2\ s^{-1}$) are 5×10^{-6} for carbon dioxide and 50×10^{-6} for methane. The membrane's thickness is 1 mm. Downstream preessure is 1 bar. Calculate the composition of the gas downstream from the membrane.

Second virial coefficients ($cm^3\ mol^{-1}$) at 300 K: B_{11} = -121; B_{12} = -55; B_{22} = -41.
At 1 bar and 300 K, the partial molar volumes at infinite dilution and Henry's constants (based on the molar concentration of component i dissolved in the polymeric membrane are:

	$\bar{v}^\infty$ ($cm^3\ mol^{-1}$)	H (bar L mol^{-1})
Carbon dioxide	50	19
Methane	23	50

8. Consider a reverse-osmosis process for producing fresh water from seawater at 25°C. Assume that a perfectly selective membrane is available, i.e., the membrane passes water but no salts.

(a) To achieve a water flux of 7.2×10^{-4} g cm^{-2} s^{-1}, what upstream pressure must be used to operate this process?

(b) How many square feet of membrane area will be required to produce one million gallons of fresh water per day?

The specific permeability of the membrane is 2.6×10^{-7} g cm^{-1} s^{-1}. The thickness of the supported membrane is 10 microns. At 25°C, the vapor pressure of pure water is 0.0312 atm. At the same temperature, the vapor pressure of seawater is 1.84% less than that of pure water.

CHAPTER 9

Electrolyte Solutions

*M*any industrial and natural processes require information on phase equilibria of electrolyte solutions. Some examples are: partitioning processes in biochemical systems; precipitation and crystallization processes in geothermal-energy systems or drilling muds; desalination of water; water-pollution control; salting-in and salting-out effects in extraction and distillation; production of natural gas from high-pressure aquifiers where natural gas is in equilibrium with brines; food processing; and production of fertilizers.

Extensive modifications of traditional activity-coefficient models described in Chap. 6 are required to describe phase equilibria in systems containing electrolytes.

The literature is rich in theoretical and experimental studies of electrolyte solutions. Unfortunately, much of it is confusing, primarily because authors often are not clear in their definitions of thermodynamic functions and because there are no universally accepted standards of notation. One source of confusion concerns standard states but another, often not appreciated, follows from the choice of variables. In mixtures of nonelectrolytes, we use the Lewis-Randall system where the variables are temperature, pressure and concentrations of all components, usually expressed by mole fractions. In electrolyte solutions, especially dilute solutions, the variables may be temperature, concentrations of all solute species, and osmotic pressure. A variety of units is available for solute concentrations; a common one is *molality*, defined as moles of solute per kg of solvent (*not* per kg of solution). This unit is suitable for dilute solutions but not for highly concentrated solutions because it tends to infinity as the solvent-to-solute ratio tends to zero.

Extension of a thermodynamic framework for nonelectrolytes to contain also electrolytes is not a trivial task. It is a common misconception that such extension is only a small detail, a little perturbation, like adding a short tail to a big dog. Not so. Extension to include also electrolytes requires concepts and constraints (e.g. electroneutrality) that can be mastered only with patient and devoted study.

This chapter can only provide an overview. The best thorough discussion of electrolyte-solution thermodynamics is given in the book by Robinson and Stokes (1970). However, this fine book is somewhat out of date. A good supplement, reviewing more recent work, is the monograph edited by Pitzer (1991) and the book by Barthel *et al.* (1998).

The following sections present the thermodynamic basis for describing activities of components in electrolyte solutions, some of the theoretical and semi-empirical models that have been developed for electrolyte solutions and, finally, a few applications of these models to practical phase-equilibrium calculations.

9.1 Activity Coefficient of a Nonvolatile Solute in Solution and Osmotic Coefficient for the Solvent

Before taking into account dissociation of electrolytes into ions, we first discuss the thermodynamics of solutions containing a nonvolatile solute in a volatile solvent. This discussion is relevant because most electrolytes (salts) are essentially nonvolatile at normal temperatures.

As shown in Chap. 6, for a component i at some temperature, pressure, and composition, activity a_i and activity coefficient γ_i are related to chemical potential μ_i by

$$\mu_i = \mu_i^0 + RT \ln a_i = \mu_i^0 + RT \ln(\gamma_i \xi_i) \tag{9-1}$$

where μ_i^0 is the chemical potential of i in some conveniently defined standard state and ξ is some convenient measure of concentration.

In Chap. 6, for liquid mixtures of volatile nonelectrolytes, we define μ_i^0 by choosing as standard state pure liquid i at system temperature and pressure. Mole fractions (or volume fractions) are typically used for ξ.

For a mixture containing a nonvolatile solute dissolved in a solvent, we use Eq. (9-1) for the solvent with the conventional definition of μ_i^0. However, for a nonvolatile solute, pure liquid i at system temperature and pressure is often not a convenient standard state because for most cases, at normal temperatures and pressures, a pure nonvolatile solute cannot exist as a liquid.[1]

For the dissolved solute, the chemical potential is written

$$\mu_i = \mu_i^* + RT \ln a_i = \mu_i^* + RT \ln(\gamma_i \xi_i) \tag{9-2}$$

where μ_i^*, the *standard state* chemical potential of i, is independent of composition but depends on temperature, pressure, and the nature of solute and solvent. A convenient choice for μ_i^* is the chemical potential of i in a *hypothetical ideal solution* of i in the solvent, at system temperature and pressure, and at unit concentration $\xi_i = 1$. In this ideal solution, $\gamma_i = 1$ for all compositions. In the real solution, $\gamma_i \to 1$ as $\xi_i \to 0$.

It is a common misconception to say that the standard state for the solute is the solute at system temperature and pressure and at infinite dilution. That is not correct; at infinite dilution, the chemical potential of the solute is $-\infty$. The standard-state chemical potential for solute i must be at some fixed (non-zero) concentration. That concentration is unit concentration. Unit concentration is used because its logarithm is zero.

For solutions of polymers, the common composition scale is volume fraction but for solutions of other nonvolatile solutes, there are three commonly-used composition scales: *molarity* (moles of solute per liter of solution, c_i); *molality* (moles of solute per kg of solvent, m_i); and *mole fraction*, x_i.

When we set $\xi_i = c_i$, Eq. (9-2) becomes

$$\mu_i = \mu_i^{\Diamond} + RT \ln(\gamma_i^{(c)} c_i) \tag{9-3}$$

where $\gamma_i^{(c)}$ is the is the *molarity* activity coefficient. The standard state is the hypothetical, ideal, 1-molar solution of solute i in solvent j. Because ideal behavior is approached as the real solution becomes infinitely dilute, $\gamma_i^{(c)} \to 1$ as $c_i \to 0$.

It is often more convenient to use the molality scale because it does not require density data. When we set $\xi_i = m_i$, Eq. (9-2) becomes

$$\mu_i = \mu_i^{\otimes} + RT \ln(\gamma_i^{(m)} m_i) \tag{9-4}$$[2]

where $\gamma_i^{(m)}$ is the *molality* activity coefficient. The standard state is the hypothetical, ideal 1-molal solution of solute i in solvent j. In the real solution, $\gamma_i^{(m)} \to 1$ as $m_i \to 0$.[3]

---

[1] However, for some applications it may be useful to define the standard state of the nonvolatile solute as that of the pure liquid at system temperature and pressure, even when that standard state is hypothetical. When the system temperature is not very far below the melting point of the nonvolatile component, it is possible to estimate the chemical potential (or fugacity) of the subcooled liquid by systematic extrapolation as shown in Chap. 11.

[2] For a solution containing a solvent and a solute i, the conversion between molarity (c_i) and molality (m_i) is given by $c_i = d\, m_i/(1 + 0.001\, M_i\, m_i)$, where d is the mass density of the solution and M_i is the molecular mass of solute i. However, in practice this conversion is often made by assuming that $c_i \approx d\, m_i$, a good approximation for dilute solutions.

Molality is an inconvenient scale for concentrated solutions because $m_i \rightarrow \infty$ when we reach the pure solute. For concentrated solutions, the mole fraction is often a more convenient scale. When $\xi_i = x_i$, Eq. (9-2) becomes

$$\mu_i = \mu_i^* + RT\ln(\gamma_i^{(x)} x_i) \tag{9-5}$$

where $\gamma_i^{(x)}$ is the unsymmetric[4] *mole-fraction* activity coefficient. The standard state is the hypothetical, ideal solution where $x_i = 1$. In the real solution, $\gamma_i^{(x)} \rightarrow 1$ as $x_i \rightarrow 0$.

To illustrate Eq. (9-4), Fig. 9-1 shows solute activity plotted against solute molality. The hypothetical ideal solution is shown by a straight line that goes through the coordinates (0,0) and (1,1). The standard-state activity is given by A corresponding to unit molality. If we arbitrarily choose a molality of 1.5, the activity of the solute in the real solution is given by C, while that in the hypothetical ideal solution is given by B. In the hypothetical ideal solution, activity is equal to molality because, in that ideal solution, $\gamma_{(\text{ideal})}^{(m)} = 1$ for all solute concentrations. Therefore, the activity coefficient of the real solution, $\gamma_i^{(m)} = a_i / m_i$, is given by the ratio $\overline{\text{CD}}$ to $\overline{\text{BD}}$.

In Eqs. (9-3), (9-4), and (9-5), $\mu_i^{\Diamond}$, or $\mu_i^{\otimes}$, or μ_i^*, is the chemical potential of the solute in its *thermodynamic standard state*, that is, the state from which we measure or calculate changes in state properties, such as chemical potentials. A standard state need not be physically realizable, but it must be well-defined.[5]

For the solvent, we use the pure liquid at system temperature and pressure as the standard state.

In a given solution, the chemical potential of the solvent and that of the solute are unaffected by the choice of the composition scale but the activity and the activity coefficient are affected not only by the choice of standard state but also by the choice of concentration scale. We can readily derive relationships to convert activity coefficients for one concentration scale to those for another. In a binary mixture of a nondissociating solute 2 in solvent s, these relations are:

$$\gamma_2^{(x)} = \gamma_2^{(m)}(1 + 0.001 M_s m_2) \tag{9-6}$$

$$\gamma_2^{(m)} = \gamma_2^{(c)} c_2 / (m_2 d_s) \tag{9-7}$$

---

[3] Because it is aesthetically preferable to use activity coefficients that are nondimensional, and because m_i has dimensions of mol kg^{-1}, we could write $\mu_i = \mu_i^{\otimes} + RT\ln(\gamma_i^{(m)} m_i / m_0)$ and $\gamma_i^{(m)} = a_i m_0 / m_i$, where m_0 is unit molality, i.e., 1 mol kg^{-1}. For simplicity, we always use Eq. (9-4) that omits m_0. The same applies to Eq. (9-3) where we omit unit molarity c_0.

[4] The convention presented in Sec. 6-4 adds an asterisk (*) to the unsymmetrically normalized activity coefficients. In this chapter, the activity coefficients of the solute are always normalized in this way. However, to clarify notation, instead of γ_i^* we use here $\gamma_i^{(c)}$ or $\gamma_i^{(m)}$ or $\gamma_i^{(x)}$, depending on the variable used for solute concentration.

[5] A mixture at infinite dilution is not appropriate for a standard state because, in the limit of infinite dilution, the chemical potential of the solute approaches $-\infty$.

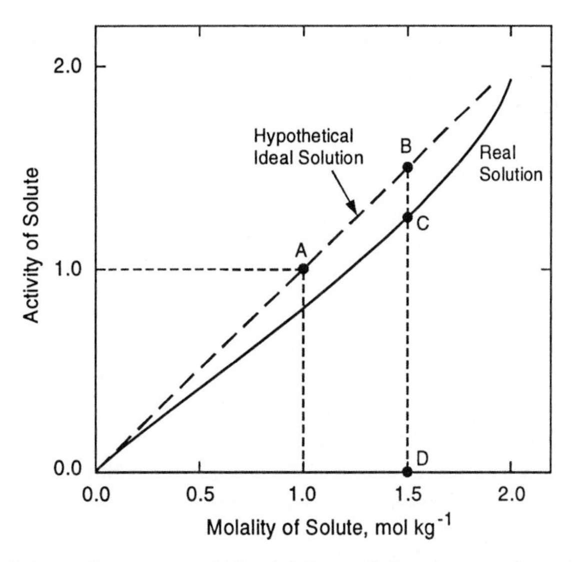

Figure 9-1 Schematic representation of the activity of a nondissociating solute as a function of its molality. Point A shows the standard state. When the solute's molality is 1.5, its activity coefficient is given by $\overline{CD}/\overline{BD}$.

$$\gamma_2^{(x)} = \gamma_2^{(c)}\left[d + 0.001c_2(M_s - M_2)\right]/d_s \tag{9-8}$$

where d is the mass density (g cm^{-3}) of the solution; d_s is the mass density (g cm^{-3}) of the pure solvent; M_2 is the molar mass of the solute and M_s is the molar mass of the solvent.

To obtain the activity of the solvent, we use the Gibbs-Duhem equation that interrelates the activity of the solute and that of the solvent. At constant temperature and pressure, the Gibbs-Duhem equation gives

$$\ln a_s = -\frac{M_s}{1000}m_i - \frac{M_s}{1000}\int_0^{m_i} m_i \left(\frac{\partial \ln \gamma_i^{(m)}}{\partial m_i}\right)_{T,P} dm_i \tag{9-9}$$

The osmotic pressure π of the solution is[6]

$$\pi = -\frac{RT}{v_s}\ln a_s \tag{9-10}$$

where v_s is the molar volume of the solvent.

[6] Osmotic pressure is discussed in Sec. 4.11.

The osmotic coefficient ϕ is the ratio π(real)/π(ideal). To find π(real) and π(ideal), we substitute Eq. (9-9) into Eq. (9-10); for the ideal solution, the second term on the right-hand side of Eq. (9-9) is zero.

The osmotic coefficient is

$$\phi^{(m)} = \frac{\pi(\text{real})}{\pi(\text{ideal})} = \frac{\ln a_s(\text{real})}{\ln a_s^{(m)}(\text{ideal})} = 1 + \frac{1}{m_i}\int_0^{m_i} m_i \left(\frac{\partial \ln \gamma_i^{(m)}}{\partial m_i}\right)_{T,P} dm_i \tag{9-11}$$

Superscript (m) for ϕ and a_s(ideal) indicates that these quantities depend on the choice of concentration scale.[7] However, no superscript is needed for a_s(real) because this quantity is independent of concentration scale.

For representing experimental data of dilute solutions, osmotic coefficients are preferred because they are more sensitive to composition than activity coefficients γ_s. Experimentally, ϕ is often obtained from vapor-pressure measurements where, at modest pressures, the activity is

$$a_s = \frac{P_s}{P_s^{\text{sat}}} \tag{9-12}$$

Here P_s is the partial pressure of the solvent and P_s^{sat} is the vapor pressure of the pure solvent, both at system temperature T. Equations (9-10) and (9-12) give a useful expression to replace the general equation [Eq. (9-11)] for the osmotic coefficient:

$$\phi^{(m)} = -\frac{1000}{M_s m_i} \ln \frac{P_s}{P_s^{\text{sat}}} \tag{9-11a}$$

9.2 Solution of an Electrolyte. Electroneutrality

The equations of Sec. 9.1 define the activity and the activity coefficient of a nondissociating solute. However, in a solution of an electrolyte, the solute dissociates into cations and anions. Cations and anions are not independent components because of electroneutrality. For example, when one mole of a strong electrolyte[8] like NaCl is dissolved in one kilogram of water, we have a one-molal solution of NaCl that is fully dissociated to a one-molal solution of positively-charged sodium ions and a one-molal solution of negatively-charged chloride ions.

[7] If the concentration scale is molality, a_i(ideal) = m_i; if the concentration scale is molarity, a_i(ideal) = c_i.

[8] A strong electrolyte (e.g. NaCl in water) is an electrolyte that in solution is completely dissociated into its constituent ions. By contrast, a weak electrolyte (e.g. acetic acid in water) is only partially dissociated.

Ordinary termodynamic measurements give properties not of individual ionic species, but of the neutral electrolytes formed by cations and anions. In a solution of an electrolyte, *electroneutrality* imposes the condition that the number of moles of the individual ionic species cannot be varied independently. In aqueous NaCl there are three species but only two (not three) components.

When dissolved in a high-dielectric-constant solvent like water, an electrically neutral electrolyte $M_{\nu+}X_{\nu-}$ is dissociated into ν_+ positive ions (cations) each with a charge z_+, and ν_- negative ions (anions) of charge z_-. Charges are given in normalized units where $z_+ = 1$ for a proton. Electrolytic dissociation is represented by

$$M_{\nu+}X_{\nu-} \rightleftharpoons \nu_+M^{z+} + \nu_-X^{z-} \tag{9-13}$$

Electroneutrality requires that

$$\nu_+z_+ + \nu_-z_- = 0 = \nu_+z_+ - \nu_-|z_-| \tag{9-14}$$

For example, in the dissociation

$$H_2SO_4 \rightleftharpoons 2H^+ + SO_4^{2-}$$

we have $\nu_+ = 2$, $z_+ = 1$, $\nu_- = 1$, and $z_- = -2$.

Equation (9-12) expresses a chemical equilibrium. The criterion for chemical equilibrium is

$$\mu_{M_{\nu+}X_{\nu-}} = \nu_+\mu_{M^{z+}} + \nu_-\mu_{X^{z-}} \tag{9-15}$$

Using the molality scale for activity coefficients [Eq. (9-4)], substitution in Eq. (9-14) gives the chemical potential of the electrolyte:

$$\mu_{M_{\nu+}X_{\nu-}} = \mu^{\otimes}_{M_{\nu+}X_{\nu-}} + \nu_+RT\ln(m_+\gamma_+) + \nu_-RT\ln(m_-\gamma_-) \tag{9-16}$$

where we have dropped superscript (m) on activity coefficients γ_+ and γ_- and where

$$\mu^{\otimes}_{M_{\nu+}X_{\nu-}} = \nu_+\mu^{\otimes}_{M^{z+}} + \nu_-\mu^{\otimes}_{X^{z-}} \tag{9-17}$$

In Eq. (9-17), $\mu_i^{\otimes}(M^{z+})$ is the chemical potential of ion M (with charge z_+) in a hypothetical ideal solution where the molality of ion M is unity. A similar definition holds for $\mu_i^{\otimes}(X^{z-})$.

Equation (9-16) can be written in a more efficient form:

$$\mu_{MX} = \mu_{MX}^{\otimes} + RT\ln(m_+^{\nu+} m_-^{\nu-}) + RT\ln(\gamma_+^{\nu+}\gamma_-^{\nu-}) \\ = \mu_{MX}^{\otimes} + RT\ln(a_+^{\nu+} a_-^{\nu-}) \tag{9-16a}$$

where, for convenience, we have dropped the subscripts on MX.

The *mean ionic molality*, $m_{\pm}$, and the *mean ionic activity coefficient*, $\gamma_{\pm}$, are defined by

$$\boxed{m_{\pm} = (m_+^{\nu+} m_-^{\nu-})^{1/\nu}} \tag{9-18}$$

$$\boxed{\gamma_{\pm} = (\gamma_+^{\nu+}\gamma_-^{\nu-})^{1/\nu}} \tag{9-19}$$

where $\nu = \nu_+ + \nu_-$. After substitution, Eq. (9-16a) becomes

$$\mu_{MX} = \mu_{MX}^{\otimes} + \nu RT\ln(m_{\pm}\gamma_{\pm}^{(m)}) \\ = \mu_{MX}^{\otimes} + \nu RT\ln(a_{\pm}^{(m)}) \tag{9-20}$$

with

$$\boxed{a_{\pm} = [(a_+)^{\nu+}(a_-)^{\nu-}]^{1/\nu} = m_{\pm}\gamma_{\pm}^{(m)}} \tag{9-21}$$

where $a_{\pm}$ is the *mean ionic activity*.

For strong electrolytes, where ionization is essentially complete, $m_+ = \nu_+ m_{MX}$ and $m_- = \nu_- m_{MX}$.[9] The mean ionic activity coefficient is

$$\gamma_{\pm}^{(m)} = \frac{a_{\pm}}{m_{MX}(\nu_+^{\nu+}\nu_-^{\nu-})^{1/\nu}} \tag{9-22}$$ [10]

For example, the mean molality and mean ionic activity coefficient for NaCl, a 1-1 electrolyte, are $m_{\pm} = m_{MX}$ and $\gamma_{\pm}^{(m)} = [(\gamma_{Na^+})(\gamma_{Cl^-})]^{1/2}$. Similar relations hold for 2-2 or 3-3 electrolytes. For a 2-1 (or 1-2) electrolyte (e.g. $CaCl_2$), $m_{\pm} = 4^{1/3} m_{MX}$ and $\gamma_{\pm}^{(m)} = [(\gamma_{Ca^{2+}})(\gamma_{Cl^-})]^{1/3}$ where m_{MX} is the molality of the electrolyte. Table 9-1 gives the mean molality for salts with specified stoichiometry.

---

[9] m_{MX} is the molality of the electrolyte as determined from the preparation of the solution, ignoring dissociation.

[10] Similarly, using the molarity concentration scale, this equation becomes $\gamma_{\pm}^{(c)} = a_{\pm} / c_{\pm}$ where $c_{\pm} = [(c_+)^{\nu+}(c_-)^{\nu-}]^{1/\nu} = c_{MX}[(\nu_+)^{\nu+}(\nu_-)^{\nu-}]^{1/\nu}$. Here c_{MX} is the molarity of the salt. For the mole fraction scale, $\gamma_{\pm}^{(x)} = a_{\pm} / x_{\pm}$ where $x_{\pm} = [(x_+)^{\nu+}(x_-)^{\nu-}]^{1/\nu} = x_{MX}[(\nu_+)^{\nu+}(\nu_-)^{\nu-}]^{1/\nu}$. Here x_{MX} is the mole fraction of the electrolyte, ignoring dissociation. However, this definition is arbitrary. Many applications assume complete dissociation of the electrolyte; in these applications, mole fraction may be defined by Eq. (9-26).

Table 9-1 Mean ionic molality ($m_{\pm}$) for several electrolytes.

Type of Solute	Example	$m_{\pm}$
Electrolyte	$M_{\nu+}X_{\nu-}$	m_{MX}
1-1; 2-2; 3-3	NaCl; $ZnSO_4$	m_{MX}
2-1; 1-2	$CaCl_2$	$4^{1/3}m_{MX}$
3-1; 1-3	$AlCl_3$	$27^{1/4}m_{MX}$
4-1; 1-4	$Th(NO_3)_4$	$256^{1/5}m_{MX}$
3-2	$Al_2(SO_4)_3$	$108^{1/5}m_{MX}$

Figure 9-2 shows activity coefficient $\gamma_{\pm}^{(m)}$ as a function of concentration (molality) for a few electrolytes in water at 25°C. By definition, $\gamma_{\pm}^{(m)}$ is unity at zero molality for all electrolytes. In dilute solutions, $\gamma_{\pm}^{(m)}$ decreases rapidly with rising concentration; the steepness of this initial drop varies with the type of electrolyte. However, for a given valence type, Fig. 9-2 shows that for low molalities (say, to about 0.01), $\gamma_{\pm}^{(m)}$ is essentially independent of the chemical nature of the constituent ions. The theoretical basis for this observation is supplied by the *Debye-Hückel theory*, as briefly discussed in Sec. 9.7. For most electrolytes, curves similar to those in Fig. 9-2 show a minimum at intermediate concentrations. At high concentrations, $\gamma_{\pm}^{(m)}$ may be much larger than unity.

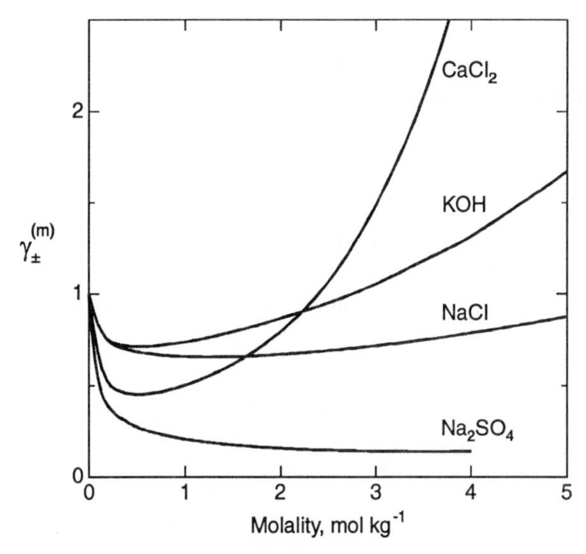

Figure 9-2 Mean ionic activity coefficients (Robinson and Stokes, 1970) for some electrolytes in aqueous solution at 25°C.

Equation (9-19) indicates that the measurable quantity $\gamma_{\pm}^{(m)}$ is determined by individual ion activity coefficients γ_+ and γ_- that, usually are not independently measurable.

Activity data for electrolytes are obtained from measurements of the potentials of electrochemical cells, or from solubility and colligative-property measurements. An important source of experimental data is vapor-pressure measurements.[11] The most accurate relative-vapor-pressure method is the *isopiestic method* discussed elsewhere (Robinson and Stokes, 1970; Rard and Platford, 1991).

For a nondissociating solute, the standard state is at point A in Fig. 9-1. For a dissociating solute (i.e. an electrolyte), the standard state is obtained from a procedure similar to that for a nondissociating solute shown in Fig. 9-1, provided that the appropriate quantities are plotted.

To illustrate, consider a 2-1 electrolyte, such as calcium chloride ($\nu = 3$). Because $m_{\pm} = 4^{1/3} m_{MX}$, the standard state is determined from a plot of $a_{\pm}$ as a function of $4^{1/3} m_{MX}$, as shown in Fig. 9-3 for aqueous $CaCl_2$ solutions at 25°C. As Fig. 9-3 shows, the standard state for a 2-1 electrolyte is a hypothetical ideal dilute solution with unit molality. In that standard state, the mean ionic activity coefficient is 1. For a 2-1 electrolyte, upon selecting the plotted quantities indicated in Fig. 9-3, we assure that $a_{\pm} / m_{\pm} \to 1$ as $m_{\pm} \to 0$. In the hypothetical ideal solution, $a_{\pm} / m_{\pm}$ remains unity for all $m_{\pm}$.

9.3 Osmotic Coefficient in an Electrolyte Solution

An electrolyte MX is dissolved and completely dissociated in solvent s. Using the molality scale, the chemical potential of the solvent μ_s is

$$\mu_s = \mu_s^0(T,P) + RT \ln a_s = \mu_s^0(T,P) + RT\phi^{(m)} \ln a_s^{(m)}(\text{ideal}) \tag{9-23}$$

Here $\mu_s^0(T,P)$ is the chemical potential of pure solvent s at system temperature T and pressure P; $\phi^{(m)}$ is the osmotic coefficient.

If one molecule of salt MX dissociates into ν ions, the ideal-solution activity of the solvent is

$$\ln a_s^{(m)}(\text{ideal}) = -\frac{M_s}{1000} \nu m_{MX} \tag{9-24}$$

[11] Vapor-pressure measurements give the activity of the solvent. To obtain the activity of the solute, we use the Gibbs-Duhem equation that relates the activity of the solvent to that of the solute. See Sec. 9.4.

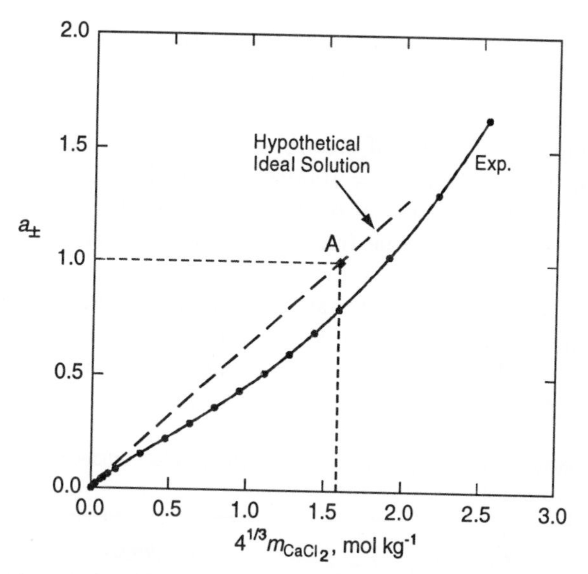

Figure 9-3 Mean ionic activity of aqueous $CaCl_2$ at 25°C as a function of its molality (Robinson and Stokes, 1970). Point A shows the standard state. The (ideal solution) straight line goes through coordinates (0, 0) and (1, $4^{1/3}$). When m_{CaCl_2} = 1 mol kg^{-1}, $4^{1/3}m_{CaCl_2}$ = 1.587 mol kg^{-1}.

For the real solution, substitution of Eq. (9-12) into Eq. (9-23) gives

$$\phi^{(m)} = -\left(\frac{1000}{\nu m_{MX} M_s}\right)\ln a_s \tag{9-25}$$

A superscript (m) is written for ϕ and for a_s(ideal) because these quantities depend on the choice of concentration scale. However, no superscript is needed for a_s because that quantity is independent of concentration scale.

If we use the mole fraction scale, we write for the chemical potential of the solvent, μ_s

$$\mu_s = \mu_s^0(T,P) + RT\ln a_s = \mu_s^0 + RT\ln(\gamma_s x_s) \tag{9-23a}$$

where $\gamma_s \to 1$ as $x_s \to 1$ and the standard state is the pure liquid solvent at the temperature and pressure of the solution.

Assuming complete dissociation of an electrolyte, for aqueous solutions the mole fraction of the electrolyte, x_{MX}, is defined

$$x_{MX} = \frac{n_{MX}}{n_w + \nu n_{MX}} = \frac{m_{MX}}{55.51 + \nu m_{MX}} \qquad (9\text{-}26)^{12}$$

Here n_{MX} is the number of moles of electrolyte per kg of water, n_w is the number of moles of water per kg of water (n_w = 1000/M_w = 55.51 mol, where M_w is the molar mass of water), m_{MX} is the molality of the salt and $\nu = \nu_+ + \nu_-$. According to this definition, $x_w + \nu x_{MX} = 1$; the salt mole fraction can never equal unity because, for pure salt, it becomes $1/\nu$.

In dilute and moderately concentrated electrolyte solutions, the activity coefficient of the solvent, γ_s, is close to unity, as illustrated in Table 9-2 for aqueous sulfuric acid at 25°C. For the calculations in Table 9-2, mole fractions of water, x_w, were obtained from $x_w = n_w/(n_w + \nu n_{MX}) = 55.51/(55.51 + \nu m_{MX})$.[13]

Table 9-2 Activity coefficients (γ), and rational (*g*) and molal osmotic (ϕ) coefficients of aqueous sulfuric acid solutions at 25°C.*

$m_{H_2SO_4}$	$\gamma_\pm^{(m)}$	γ_{H_2O}	g_{H_2O}	ϕ_{H_2O}
0.1	0.266	1.000	0.749	0.680
0.3	0.183	1.005	0.686	0.668
0.5	0.156	1.008	0.708	0.676
1.0	0.132	1.014	0.745	0.721
1.5	0.126	1.015	0.828	0.780

* Robinson and Stokes (1970). The activity coefficient of the solute, $\gamma_\pm^{(m)}$, is based on the molality scale, whereas the activity coefficient of the solvent, γ_{H_2O}, is based on the mole-fraction scale.

Consider a 1-molal solution of sulfuric acid at 25°C. Table 9-2 shows that water has an activity coefficient γ_w = 1.014. This number does not suggest the strong departure from ideal behavior indicated by the mean ionic activity coefficient of the solute, $\gamma_\pm^{(m)}$ = 0.1316. Deviations from ideal behavior are more effectively characterized by the *osmotic coefficient* of the solvent, i.e. be the *rational* osmotic coefficient, *g*, or the more-often-used *practical* (or *molality*) osmotic coefficient, ϕ. The former is related to the chemical potential through

$$\mu_s = \mu_s^0(T,P) + gRT\ln x_s \qquad (9\text{-}27)$$

with $g \to 1$ as $x_s \to 1$. Here x_s is calculated assuming total dissociation of the electrolyte [Eq. (9-26)].

[12] Equation (9-26) is readily extended to the case where the aqueous solution contains two (or more) salts, provided that these salts do not have a common ion. In general, it is better to define mole fractions of ions rather than mole fractions of salts.
[13] An equivalent definition for x_w is $x_w = n_w / (n_w + \sum n_M + \sum n_X)$.

Comparing with Eq. (9-23a), we obtain

$$\ln a_s = \ln(\gamma_s x_s) = g \ln x_s \tag{9-28}$$

Like the molality osmotic coefficient ϕ, the rational osmotic coefficient g is also defined by the ratio π(real)/π(ideal). However, ϕ and g are not identical because π(ideal) using the molality scale is not equal to π(ideal) using the mole-fraction scale.

For the mole fraction scale,

$$\pi(\text{ideal}) = -\left(\frac{RT}{v_s}\right)\ln x_s \tag{9-29}$$

where the definition of mole fraction x_s is given by

$$x_s = \frac{\dfrac{1000}{M_s}}{\dfrac{1000}{M_s} + \nu m_{\text{MX}}} \tag{9-26a}$$

π(real) is found from Eq. (9-10) and experimental data for a_s. The molality osmotic coefficient and the rational osmotic coefficient are related by

$$-g \ln x_s = \frac{\nu m_{\text{MX}} M_s}{1000}\phi \tag{9-30}$$

Compared to γ_s (mole fraction scale), osmotic coefficients are much more sensitive to deviations from ideality, i.e., ϕ differs from unity by a much larger amount than does γ_s, as shown in Table 9-2. For example, for a 1-molal aqueous solution of sulfuric acid at 25°C, while $\gamma_s = 1.014$, $g = 0.745$ and $\phi = 0.721$. The definition of osmotic coefficient assures that it goes to unity at infinite dilution; it recognizes that van't Hoff's equation and Raoult's law for the solvent become exact at infinite dilution.

9.4 Relation of Osmotic Coefficient to Mean Ionic Activity Coefficient

The mean ionic activity coefficient refers to the solute while the osmotic coefficient refers to the solvent. They are inter-related through the Gibbs-Duhem equation.

For a binary system at constant temperature and pressure, the Gibbs-Duhem equation is

$$x_s d\ln a_s + x_{\text{MX}} d\ln a_{\text{MX}} = 0 \tag{9-31}$$

where x_i is the mole fraction of component i, calculated according to Eq. (9-26).

We rewrite Eq. (9-31) in the form

$$d\ln a_s = -(x_{\text{MX}} / x_s) d\ln a_{\text{MX}} = -\frac{m}{\dfrac{1000}{M_s}} d\ln a_{\text{MX}}$$

where, for convenience, we have dropped subscript MX on m.

Substituting Eq. (9-21), we obtain

$$d\ln a_s = -\frac{M_s}{1000} m\, d\ln a_{\text{MX}} = -\frac{M_s}{1000} \nu m\, d\ln a_{\pm} = -\frac{M_s}{1000} \nu m\, d\ln(m_{\pm}\gamma_{\pm}^{(\text{m})})$$

We note that $d\ln m_{\pm} = d\ln m$ because $\ln m_{\pm}$ and $\ln m$ differ only by a constant. From Eq. (9-22), coupled with Eq. (9-25), we obtain

$$d\ln \gamma_{\pm}^{(\text{m})} = d\phi + \frac{1}{m}(\phi - 1) dm \tag{9-32}$$

In the very dilute solution, $\gamma_{\pm}^{(\text{m})} \rightarrow 1$ and $\phi \rightarrow 1$ as $m \rightarrow 0$; therefore, Eq. (9-29) can be integrated to give

$$\ln \gamma_{\pm}^{(\text{m})} = \phi - 1 + \int_0^m \frac{\phi - 1}{m} dm \tag{9-33}$$

For an electrolyte solution, the integrand $(\phi - 1)/m$ diverges to $-\infty$ as $m \rightarrow 0$. This problem is easily solved by changing the independent variable from m to $m^{1/2}$. Then the equation above becomes

$$\ln \gamma_{\pm}^{(\text{m})} = \phi - 1 + 2\int_0^m \frac{\phi - 1}{m^{1/2}} dm^{1/2} \tag{9-34}$$

In Eq. (9-34), the integrand now approaches a finite value $(-A_{\phi})$ as $m \rightarrow 0$ where A_{ϕ} is the Debye-Hückel coefficient for the osmotic coefficient.[14]

From cryoscopic, osmotic-pressure or vapor-pressure measurements, ϕ can be determined as a function of m; Eq. (9-34) can then be used to calculate $\gamma_{\pm}^{(\text{m})}$. However,

---

[14] A_{ϕ} can be calculated from the ionic charges and the solvent's dielectric constant, as shown in Sec. 9.7 [Eq. (9-49a)].

data must be available from (nearly) $m = 0$ and must be spaced closely enough to permit accurate evaluation of the integral in Eq. (9-34).

Alternatively, we may assume a given analytical form to extrapolate the experimental curve $\phi(m)$ to infinite dilution. The limiting law of Debye and Hückel (Sec. 9.7) provides a theoretical basis for this extrapolation. We cannot here go into details. The essential result is that the experimental activity of the *solvent* gives osmotic coefficient ϕ according to Eq. (9-25). The activity coefficient of the *solute* is then calculated from Eq. (9-34). However, for these calculations, it is necessary to determine the activity of the solvent to a very high degree of accuracy, especially in the dilute region.

9.5 Temperature and Pressure Dependence of the Mean Ionic Activity Coefficient

Activity-coefficient data for electrolyte solutions are mostly available at or near 25°C and 1 atmosphere. However, some important applications require data for wide ranges of temperature and pressure. For example, geothermal solutions may exist at temperatures to 600°C and pressures to several kilobars. It is therefore of interest to inquire what thermodynamics can tell us about the temperature and pressure dependence of activity coefficients.

The temperature dependence of the activity coefficient for a nonelectrolyte solute 2 is given by Eq. (6-34).

A similar expression holds for an electrolyte MX:

$$\left(\frac{\partial \ln \gamma_{\pm}^{(m)}}{\partial T}\right)_{P,m} = -\frac{\bar{h}_{MX}^{E*}}{\nu RT^2} \tag{9-35}$$

where ν is the number of ions formed by one molecule of the electrolyte and

$$\bar{h}_{MX}^{E*} = \bar{h}_{MX} - \bar{h}_{MX}^{\infty} \tag{9-35a}$$

Here $\bar{h}_{MX}^{E*}$ is the excess partial molal enthalpy of the dissolved electrolyte, relative to that at infinite dilution. In Eq. (9-35a), $\bar{h}_i^{\infty}$ is the partial molal enthalpy of solute i in an infinitely dilute solution. The partial molal enthalpy in the standard state is the same as that at infinite dilution.

Similar to Eq. (6-36a), at constant temperature and composition, the effect of pressure on the mean ionic activity coefficient for an electrolyte MX is given by

$$\left(\frac{\partial \ln \gamma_{\pm}^{(m)}}{\partial P}\right)_{T,m} = -\frac{\overline{v}_{MX}^{E*}}{\nu RT} \tag{9-36}$$

where

$$\overline{v}_{MX}^{E*} = \overline{v}_{MX} - \overline{v}_{MX}^{\infty} \tag{9-36a}$$

In Eq. (9-36a), $\overline{v}_{MX}$ is the partial molal volume of the electrolyte at the concentration of the solution and $\overline{v}_{MX}^{\infty}$ is the partial molal volume of the electrolyte at infinite dilution, equal to that in the standard state.

Unfortunately, Eqs. (9-35) and (9-36) are of little practical use because experimental partial molal enthalpies and partial molal volumes are rarely available. Further, these quantities depend not only on composition but also on temperature and, to a lesser extent, on pressure.

For some electrolyte solutions, semi-empirical methods have been developed to estimate the effect of temperature and pressure on mean ionic activity coefficients (Millero, 1977).

9.6 Excess Properties of Electrolyte Solutions

As discussed in Sec. 6.2, excess functions are thermodynamic properties of solutions in excess of those of an ideal solution at the same temperature, pressure, and composition. Care must be taken to define *ideal* precisely.

Consider a binary electrolyte solution containing m moles of (totally dissociated) salt MX in 1 kg of solvent s. The number of moles of solvent is $n_s = 1000/M_s$, where M_s is the molar mass of the solvent; if the solvent is water, $n_s = 55.51$ mol. At constant pressure and temperature, the total Gibbs energy G of this solution is

$$G_{\text{solution}} = n_{MX}\mu_{MX} + n_s\mu_s \tag{9-37}$$

where the chemical potential of the salt, μ_{MX}, is given by Eq. (9-20) and the chemical potential of the solvent, μ_s, is given by Eq. (9-23). Substitution of Eqs. (9-21) and (9-22) in the expression for μ_{MX}, and Eq. (9-24) in the expression for μ_s, gives

$$\begin{aligned} G_{\text{solution}} &= n_{MX}\mu_{MX}^{\otimes} + \nu n_{MX}RT\{\ln\gamma_{\pm}^{(m)} + \ln[m(\nu_+^{\nu+}\nu_-^{\nu-})^{1/\nu}]\} \\ &+ n_s\left(\mu_s^{\text{o}} - RT\frac{\nu m M_s}{1000}\phi\right) \end{aligned} \tag{9-37a}$$

where, for convenience, we dropped subscript MX from molality m of the electrolyte.

For an ideal solution, $\gamma_{\pm}^{(m)} = \phi = 1$, and the Gibbs energy is

$$G^*_{\text{ideal solution}} = n_{\text{MX}}\mu^{\otimes}_{\text{MX}} + \nu n_{\text{MX}} RT\{\ln[m(\nu_+^{\nu+}\nu_-^{\nu-})^{1/\nu}]\} + n_s\left(\mu_s^{\text{o}} - RT\frac{\nu m M_s}{1000}\right) \tag{9-38}$$

Therefore, the total *unsymmetric excess Gibbs energy* of the solution containing m moles of salt in 1 kg of solvent is given by

$$G^{E*} = G_{\text{solution}} - G^*_{\text{ideal solution}} = \nu n_{\text{MX}} RT \ln\gamma_{\pm}^{(m)} + n_s RT\nu m\frac{M_s}{1000}(1-\phi) = \nu n_{\text{MX}} RT(\ln\gamma_{\pm}^{(m)} + 1 - \phi) \tag{9-39}$$ [15]

This solution contains n_s moles of solvent and n_{MX} moles of completely dissociated electrolyte. Using Eq. (6.2-9) and Eq. (9-39) gives the partial excess Gibbs energy of the solvent (s) and that of the solute (MX):

$$\left(\frac{\partial G^{E*}}{\partial n_s}\right)_{P,T,n_{\text{MX}}} = \bar{g}_s^{E*} = \left(\frac{\nu m M_s}{1000}\right) RT(1-\phi) \tag{9-40}$$

$$\left(\frac{\partial G^{E*}}{\partial n_{\text{MX}}}\right)_{P,T,n_s} = \bar{g}_{\text{MX}}^{E*} = \nu RT\ln\gamma_{\pm}^{(m)} \tag{9-41}$$

Equations (9-40) and (9-41) can be rewritten in the form

$$\boxed{\phi - 1 = -\frac{1000}{\nu m M_s RT}\left(\frac{\partial G^{E*}}{\partial n_s}\right)_{P,T,n_{\text{MX}}}} \tag{9-42}$$

$$\boxed{\ln\gamma_{\pm}^{(m)} = \frac{1}{\nu RT}\left(\frac{\partial G^{E*}}{\partial n_{\text{MX}}}\right)_{P,T,n_s}} \tag{9-43}$$

where, as before, ν is the total number of ions formed from the dissociation of the salt.

---

[15] The excess Gibbs energy is unsymmetric because the activity coefficients of solute and solvent are not normalized in the same way. See Sec. 6.4. Note that $G^{E*} \to 0$ as $m \to 0$.

For a binary solution containing 1 kg of solvent, the excess enthalpy of an electrolyte solution can be obtained using the Gibbs-Helmholtz relation [Eq. (6-12)],

$$h^{E*} = \nu m R T^2 \left[\left(\frac{\partial \phi}{\partial T} \right)_{P,m} - \left(\frac{\partial \ln \gamma_{\pm}^{(m)}}{\partial T} \right)_{P,m} \right] \tag{9-44}$$

The last term in Eq. (9-44) is related to the partial molal enthalpy of the electrolyte, given by Eq. (9-35).

To determine the excess enthalpy of an electrolyte solution, we need highly accurate data for the osmotic and mean ionic activity coefficients as a function of temperature.[16] More reliable results may be obtained from heat-of-dilution flow calorimetry and heat-capacity measurements (Messikomer and Wood, 1975; Picker *et al.*, 1971).

9.7 Debye-Hückel Limiting Law

At a fixed concentration of ions, electrolytes containing ions with multiple charges have a stronger effect on the activity coefficients of ions than electrolytes containing only singly-charged ions. To express this dependence it is useful to introduce the (molal) *ionic strength* of the solution, I, defined by

$$\boxed{I(\text{mol kg}^{-1}) = \frac{1}{2} \sum_i m_i z_i^2} \tag{9-45}$$

where z_i is the charge on ion i and m_i is its concentration expressed by molality. The summation extends over all ionic species in the solution. For example, for seawater, whose composition is shown in Table 9-3, $I = 0.72$ mol kg^{-1}.

According to its definition, the ionic strength of a 1-1 electrolyte is equal to its molality, m_{MX}; that for a 1-2 electrolyte (e.g. Na_2SO_4) is $3m_{MX}$, and that for a 2-2 electrolyte (e.g. $ZnSO_4$) is $4m_{MX}$.

The activity coefficient of an electrolyte depends strongly on its concentration. As summarized in Chap. 4, charged particles interact with coulombic forces: for two ions with charges z_+ and z_- at a separation r, the force is proportional to $(z_+z_-)/4\pi\varepsilon_0\varepsilon_r r^2$, where ε_r is the dielectric constant of the solvent and ε_0 is the vacuum permittivity. Thus the potential energy of interaction varies inversely with the first power of r; it therefore has a much longer range than other intermolecular forces that depend on higher powers of r^{-1}. Contrary to what is observed with nonelectrolytes

[16] Because the Gibbs-Duhem equation relates ϕ to $\gamma_{\pm}$, we need not measure both. Accurate measurements for either one are sufficient.

Table 9-3 Concentrations of major ions in oceanic seawater (Clegg and Whitfield, 1991).

Component	Molality (mol kg^{-1})
Na^+	0.486
Mg^{2+}	0.055
Ca^{2+}	0.011
K^+	0.010
Cl^-	0.566
SO_4^{2-}	0.029

(where short-range forces are dominant), solutions of electrolytes depend on both long-range electrostatic attractions and repulsions and on short-range interactions between ions and between ions and solvent molecules. Moreover, since $\varepsilon_r > 1$, the potential is reduced from its value in vacuo. For example, for water at 25°C, $\varepsilon_r = 78.41$; at a given separation distance, the Coulomb potential is reduced from its value in vacuo by almost two orders of magnitude. It is therefore not surprising that water is a better solvent for electrolytes than say, benzene. The dielectric constant of water is very much larger than that of benzene (for benzene at 25°C, $\varepsilon_r = 2.27$).

At infinite dilution, the distribution of ions in solution can be considered completely random because the ions are too far apart to exert any significant influence on each other. In this case, the mean ionic activity coefficient of the electrolyte is unity. However, for dilute (not infinitely dilute) solutions, where the ions are no longer "blind" to each other, Coulombic forces become important; in the neighborhood of a negative ion, the local concentration of positive ions is slightly higher than that for the bulk solution.

A slightly positive atmosphere around anion i, and a slightly negative atmosphere around cation j, produce a decrease in attraction between i and j, as illustrated in Fig. 9-4 for an aqueous solution at 25°C of ions Na^+ and Cl^-.[17] This decrease due to preferential spatial distribution of ions gives a shielding effect. To account for shielding, the theory of Debye-Hückel shows that r^{-1} in Coulomb's potential should be multiplied by a "damping factor",

$$r^{-1} \rightarrow (r^{-1})\exp(-r\kappa) \tag{9-46}$$

where κ^{-1} is the *shielding length* commonly called the *Debye length*. The Debye length is a characteristic distance of interaction; it plays an important role in the Debye-Hückel theory. When κ^{-1} is very large, $r\kappa$ is small and the exponential in Eq. (9-46) is close to unity; in that event, we recover the original Coulomb potential that, lest we

[17] As shown in Sec. 4.2, for isolated Cl^- and Na^+ ions in contact (r = 0.276 nm), $\Gamma_{ij} = -8.36 \times 10^{-19}$ J. In water at 25°C, this value is reduced to $\Gamma_{ij} = -0.106 \times 10^{-19}$ J.

forget, is for two *isolated* charges in a continuous medium characterized by ε_r. The substitution indicated by Eq. (9-46) extends Coulomb's potential to the case where charges i and j are no longer the only charged particles in the liquid solution. As Fig. 9-4 shows, when κ^{-1} is small, the shielded potential is weaker than the unshielded potential, even for short distances.

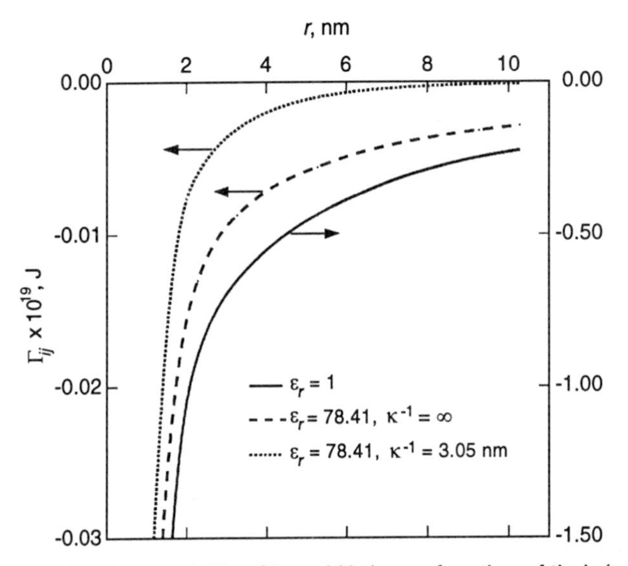

Figure 9-4 Coulomb potential for Cl^- and Na^+ as a function of their (center-to-center) separation distance: in vacuo, $\varepsilon_r = 1$; in an infinitely-dilute aqueous solution at 25°C, $\varepsilon_r = 78.41$, $\kappa^{-1} = \infty$; shielded Coulomb potential in water at 25°C, $\varepsilon_r = 78.41$ and $\kappa^{-1} = 3.05$ nm. Debye length 3.05 nm is that of a 0.01 mol kg^{-1} aqueous NaCl solution at 25°C.

The Debye length is defined as

$$\kappa^{-1} = \left(\frac{\varepsilon_0 \varepsilon_r RT}{2 d_s N_A^2 e^2 I} \right)^{1/2} \tag{9-47}$$

where ε_0 is the vacuum permittivity ($\varepsilon_0 = 8.85419 \times 10^{-12} \mathrm{C^2\ N^{-1}\ m^{-2}}$), ε_r is the relative permittivity or dielectric constant, d_s is the solvent density (g cm^{-3}), N_A is Avogadro's constant, e is the electronic charge ($e = 1.60218 \times 10^{-19}$ C), and I is the ionic strength given by Eq. (9-45).

Equation (9-47) tells us that the Debye length decreases with rising concentration; the higher the concentration of ions (ionic strength), the more effective the shield. For example, for a 1-molal aqueous solution of an 1-1 electrolyte at 25°C, $\kappa^{-1} = 0.30$ nm, giving strong shielding. If the molality is reduced to 0.001 mol kg^{-1}, $\kappa^{-1} = 9.64$ nm, giving almost no shielding.

Because ionic strength depends on the charges of the ions, even a low concentration of highly charged ions may form an effective shield. Also, because Debye length increases with rising dielectric constant, at a fixed ionic strength, shielding in water is less than that in most other solvents whose dielectric constants are below that of water.

Because of long-range Coulombic forces, electrolyte solutions are nonideal even at low electrolyte concentration because electrostatic attractive and repulsive forces between ions are significant even at high distances of separation.

Using well-established concepts from classical electrostatics, Debye and Hückel derived a simple expression (Robinson and Stokes, 1970; Newman, 1991) for the activity coefficient γ_i of an ion with charge z_i in a dilute solution of ionic strength I:

$$\ln \gamma_i^{(c)} = -z_i^2 \frac{e^2 N_A}{8\pi\varepsilon_0\varepsilon_r RT}\kappa \tag{9-48}$$

where κ is the inverse of the Debye length [defined by Eq. (9-47)].

For dilute aqueous solutions near ambient temperature, there is no significant difference between molality and molarity. Switching to molality units for γ and I, substitution of constants in Eq. (9-48) gives

$$\ln \gamma_i^{(m)} = -A_\gamma z_i^2 I^{1/2} \tag{9-48a}$$

where constant A_γ is given by

$$A_\gamma = \left(\frac{e^2}{\varepsilon_0\varepsilon_r RT}\right)^{3/2} \frac{N_A^2}{8\pi}(2d_s)^{1/2} \tag{9-49}$$

Equations (9-48) and (9-48a) give the activity coefficients of ions, not of electrolytes in an electrically-neutral solution. However, the quantity that is usually measured experimentally is the mean ionic activity coefficient, $\gamma_\pm^{(m)}$. For electrolyte $M_{\nu+}X_{\nu-}$, $\gamma_\pm^{(m)}$ is defined by Eq. (9-19). Upon substitution of Eq. (9-48a) in Eq. (9-19) and introduction of the electroneutrality condition expressed by Eq. (9-14), we obtain

$$\ln \gamma_\pm^{(m)} = -A_\gamma |z_+ z_-| I^{1/2} \tag{9-50}$$

where $|z_+ z_-|$ is the absolute value of the product of the charges.

A similar derivation yields for the osmotic coefficient

$$\phi - 1 = -A_\phi |z_+ z_-| I^{1/2} \tag{9-51}$$

where the Debye-Hückel constant A_ϕ is directly related to constant A_γ given by Eq. (9-49):

$$A_\phi = \frac{1}{3} A_\gamma \tag{9-49a}$$

Equation (9-50) is the *Debye-Hückel limiting law*, useful for interpreting the properties of electrolyte solutions. It is an exact limiting law at low concentrations in the same sense that the virial equation of state, truncated after the second virial coefficient, is an exact limiting law for the compressibility factor of a gas at low pressure.

If the solvent is water at 25°C and atmospheric pressure, A_γ = 1.174 mol$^{1/2}$ kg$^{-1/2}$ (ε_r = 78.41;[18] $d_s = 0.997$ g cm^{-3}). Converting to base 10 logarithms, Eq. (9-50) gives

$$\boxed{\log \gamma_{\pm}^{(m)} (25°\text{C}, \text{aqueous}) = -0.510 |z_+ z_-| I^{1/2}} \tag{9-50a}$$

where I is in mol kg^{-1}.

For very dilute solutions, Eq. (9-50a) is in good agreement with experimental data, as shown in Fig. 9-5. This figure also shows that the Debye-Hückel limiting law always predicts negative deviations from ideal-dilute behavior. At low concentrations, $\gamma_{\pm}^{(m)}$ depends on the valence but not on the chemical nature of the electrolyte.

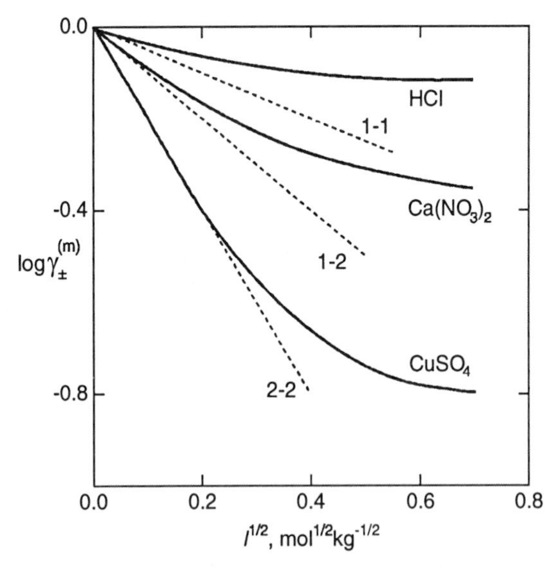

Figure 9-5 Mean ionic activity coefficients as a function of concentration for some aqueous strong electrolytes (Robinson and Stokes, 1970). Full lines show experimental data. Dashed lines are calculated from the Debye-Hückel limiting law [Eq. (9-50a)].

[18] From the 1997 international standard (D. P. Fernández, A. R. H. Goodwin, E. W. Lemmon, J. M. H. Levelt-Sengers, R. C. Williams, 1997, *J. Phys. Chem. Ref. Data,* 26: 1125).

The Debye-Hückel equation is applicable only to solutions at very low concentrations (typically, for ionic strengths to 0.01 mol kg^{-1}). Several factors account for deviations from the Debye-Hückel law at high concentrations; these include ion-ion repulsion due to the finite sizes of the ions and interactions arising from forces other than electrostatic forces (e.g. dispersion forces). An important deviation from the Debye-Hückel law follows from strong and specific ion-solvent solvation forces that invalidate the assumption that the solvent is a dielectric continuum. For realistic applications, it is necessary to derive expressions that apply to more concentrated solutions. For example, the ionic strength of seawater is about 0.7 mol kg^{-1}. For industrial or geothermal applications, the ionic strength may be much larger. For such solutions, the limiting law of Debye-Hückel is insufficient.

For concentrated electrolyte solutions, several semi-empirical corrections to the Debye-Hückel limiting law have been proposed (Zemaitis *et al.*, 1986). A common expression for aqueous solutions with I to 0.1 mol kg^{-1} is

$$\ln \gamma_{\pm}^{(m)} = \frac{-A_{\gamma} |z_+ z_-| I^{1/2}}{1 + I^{1/2}} \qquad (9\text{-}52)$$

For ionic strengths to 1 mol kg^{-1}, a linear term is added, yielding

$$\ln \gamma_{\pm}^{(m)} = \frac{-A_{\gamma} |z_+ z_-| I^{1/2}}{1 + I^{1/2}} + bI \qquad (9\text{-}53)$$ [19]

where b is an adjustable parameter. Table 9-4 compares experimental activity coefficients for aqueous solutions of sodium chloride at 25°C with those calculated from Eqs. (9-50), (9-52), and (9-53).

Table 9-4 Mean-ionic activity coefficients of aqueous sodium chloride at 25°C (Robinson and Stokes, 1970).

m (mol kg^{-1})*	$\gamma_{\pm}^{(m)}$ [Experiment]	$\gamma_{\pm}^{(m)}$ [Eq. (9-50)]	$\gamma_{\pm}^{(m)}$ [Eq. (9-52)]	$\gamma_{\pm}^{(m)}$ [Eq. (9-53) with b = 0.37 kg mol^{-1}]
0.001	0.965	0.964	0.965	0.965
0.005	0.927	0.920	0.925	0.927
0.01	0.902	0.889	0.899	0.902
0.05	0.819	0.769	0.807	0.822
0.1	0.778	0.690	0.755	0.783

* For a 1:1 electrolyte, molality of the electrolyte is equal to ionic strength.

[19] Equation (9-53) [and Eq. (9-52)] may include in the denominator the term $aI^{1/2}$ (instead of $I^{1/2}$), where a is a parameter that reflects the finite size of ions. In practice, however, parameter a is an adjustable parameter or, for simplicity, it is set equal to unity.

Research toward a fundamental theory for concentrated electrolyte solutions is an active topic pursued by physical chemists using modern statistical mechanics. As discussed elsewhere (Mazo and Mou, 1991; Ohtaki and Yamatera, 1992), much progress has been made but, for engineering applications, it is often more useful to utilize a semi-theoretical model. Section 9.10 introduces a few (mostly) empirical models for concentrated electrolyte solutions.

9.8 Weak Electrolytes

The thermodynamic relations presented in the preceding sections are for strong electrolytes, i.e., salts that completely dissociate in the solvent (usually water).

Weak electrolytes are compounds (such as acetic acid) that are only partially dissociated in aqueous solutions. At equilibrium, in addition to the ions, there exists a significant concentration of the molecular (undissociated) electrolyte. The dissociation constant of the weak electrolyte (that depends only on temperature) relates the concentration of the undissociated electrolyte to the concentrations of the ions formed by partial dissociation. This relation, however, also requires activity coefficients for the ions and for the undissociated electrolyte.

Consider electrolyte $M_{\nu_+}X_{\nu_-}$ that dissociates according to

$$M_{\nu+}X_{\nu-} \rightleftharpoons \nu_+M^{z+} + \nu_-X^{z-}$$

The dissociation (or ionization) equilibrium constant is

$$K = \frac{a_+^{\nu+}a_-^{\nu-}}{a_{MX}} = \frac{m_+^{\nu+}m_-^{\nu-}}{m_{MX}}\frac{(\gamma_+^{(m)})^{\nu+}(\gamma_-^{(m)})^{\nu-}}{\gamma_{MX}^{(m)}} = \frac{m_+^{\nu+}m_-^{\nu-}}{m_{MX}}\frac{(\gamma_\pm^{(m)})^{\nu}}{\gamma_{MX}^{(m)}} \tag{9-54}$$

where m_{MX} and $\gamma_{MX}^{(m)}$ are, respectively, the molality and the activity coefficient of the molecular (undissociated) part of the electrolyte. For strong electrolytes, m_{MX} is zero by definition. An equivalent definition of a strong electrolyte is to say it is a solute whose dissociation constant K is infinite.

To illustrate, consider the dissociation constant[20] of acetic acid (HAc) into H^+ and acetate (Ac^-) ions,

$$K = \frac{m_{H^+}\, m_{Ac^-}}{m_{HAc}}\frac{(\gamma_\pm^{(m)})^2}{\gamma_{HAc}^{(m)}}$$

[20] Dissociation (ionization) constants can be obtained from spectroscopic data. Classically, they are obtained from conductivity or electromotive-force measurements. See Robinson and Stokes (1970), Chap. 12.

To calculate vapor-liquid equilibria in aqueous solutions containing a weak electrolyte, it is necessary to know the equilibrium constant in addition to activity coefficients for the aqueous solutes and for water. It is also necessary to know fugacity coefficients for all volatile components that exist in the vapor phase. At normal conditions, we can neglect the concentration of ions in the vapor phase. An application is shown in Sec. 9.17.

9.9 Salting-out and Salting-in of Volatile Solutes

When an appreciable amount of salt dissolves in a liquid, it significantly affects that liquid's vapor pressure. Further, the dissolved salt affects the solubility of a gas (or liquid) in that solvent and finally, if the solvent is a mixture of two (or more) volatile components, the dissolved salt influences the composition of the vapor in equilibrium with the solvent mixture.

The solubility of a gas in a salt solution is usually less than that in salt-free water; this solubility decrease is called *salting-out*. A simplistic but incomplete explanation of salting-out follows from a consideration of hydration forces. Ions (especially cations) like to form complexes with water (hydration), thereby leaving less "free" water available to dissolve the gas. (This explanation, however, is seriously oversimplified because it neglects the subtle effect of ions on water structure). The salting-out influence of an ion usually rises with increasing ionic charge and decreasing ionic radius.[21]

To illustrate, Fig. 9-6 shows experimental results at 80°C for the solubility of carbon dioxide in salt-free water and in aqueous solutions containing sodium sulfate or ammonium sulfate or both. Figure 9-6 shows that when 2 moles of sodium sulfate are added to 1 kg of water at 80°C, the pressure needed to dissolve 0.2 moles of carbon dioxide increases from 17 to 51 bar. Similarly, at the same temperature and at 40 bar, 1 kg of salt-free water dissolves 0.43 moles CO_2 while a 2-molal aqueous sodium sulfate solution containing 1 kg of water dissolves only 0.16 moles CO_2. The same figure also shows that, at the same concentration, sodium ions cause a larger salting-out effect than the larger ammonium ions.

Salting-out effects are often described by an empirical equation proposed many years ago by Setchenov (1889). For a simple derivation of the *Setchenov equation*, consider a three-phase system consisting of a gas phase and two aqueous phases, as schematically shown in Fig. 9-7. One aqueous phase (') contains no salt. The other aqueous phase contains salt with molality m_{MX}. We assume that temperature T is sufficiently low so that water is essentially nonvolatile, i.e. the gas phase contains only solute i.

[21] Salts with large polarizable ions (usually anions) tend to salt in, i.e. to *increase* the solubility of the gas.

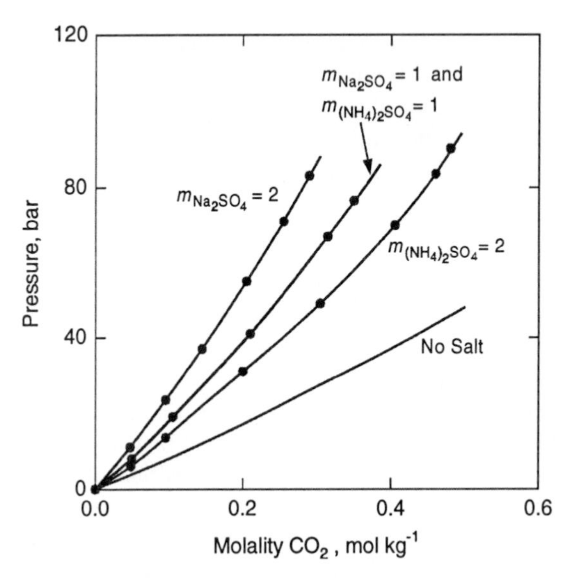

Figure 9-6 Solubility of carbon dioxide in aqueous solutions of sodium sulfate and ammonium sulfate at 80°C (Rumpf and Maurer, 1993). For the salt-containing solutions the ionic strength is 6 mol kg^{-1}.

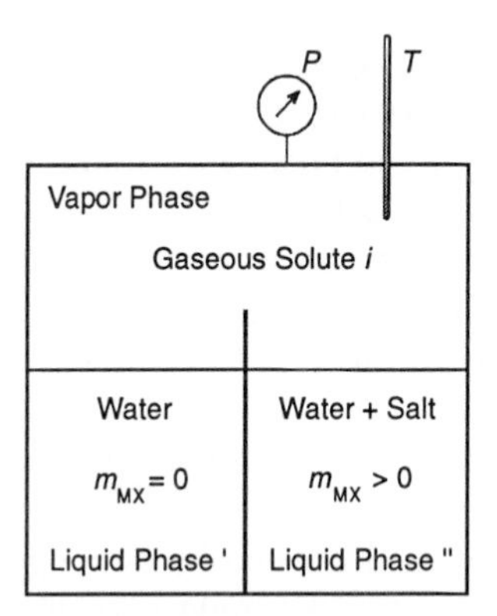

Figure 9-7 Three-phase system consisting of a gaseous solute in equilibrium with two aqueous liquid phases: phase ' contains only water; phase " contains water and a salt with molality m_{MX}. It is assumed that water is nonvolatile.

For solute i, at equilibrium,

$$\mu_i^G = \mu_i' = \mu_i''$$

As discussed earlier, this equation of equilibrium can be written in the form

$$\mu_i^G = \mu_i^{0'} + RT\ln m_i' = \mu_i^{0''} + RT\ln m_i''$$

where m_i' and m_i'' are, respectively, the molalities of solute i in the liquid phase ' (no salt) and in the liquid phase " (with salt).

Now assume that the difference between the standard Gibbs energy change of solution of a solute into a solvent with salt and that into a solution without salt is expressed as a power series of the salt concentration; i.e. assume

$$\mu_i^{0''} - \mu_i^{0'} = RT\ln(k_{MX} m_{MX}) \; + \; \text{higher terms}$$

where k_{MX} is a constant characteristic of the salt.

As an approximation, consider only the first term in the series. For low salt concentrations, the Setchenov equation is then obtained:

$$\boxed{\ln\frac{m_i'}{m_i''} = k_{MX} m_{MX}} \tag{9-55}$$

where m_i' and m_i'' are, respectively, the solubilities (expressed in molality units)[22] of the gas in salt-free solvent (water) and in the aqueous salt solution; k_{MX} is the salting parameter, and m_{MX} is the molality of the salt in the aqueous solution. Strictly, Setchenov's constant, k_{MX}, is independent of m_{MX} only in the limit of infinite dilution of the electrolyte; Eq. (9-55) is therefore a limiting relation. Constant k_{MX} depends on the salt, the solute, and the temperature. Table 9-5 presents Setchenov constants for some common gases in aqueous salt solutions at 25°C.

Table 9-5 Molality Setchenov constants for volatile solutes in aqueous electrolyte solutions at 25°C (Krishnan and Friedman, 1974).

Salt	Gas	k_{MX} (mol kg^{-1})
NaCl	H_2	0.220
	N_2	0.309
	CH_4	0.319
	C_2H_6	0.399
KCl	O_2	0.298
	SO_2	-0.051
$(CH_3)_4NBr$	CH_4	-0.039
	C_2H_6	-0.092
	C_4H_{10}	-0.170

[22] Any consistent set of units may be used for gas solubility. However, constant k_{MX} depends on the units chosen.

Constant k_{MX} can be either positive or negative. If it is positive, the solubility of the gas decreases with rising salt concentration (the gas is salted-out). If it is negative, the solubility of the gas increases with rising salt concentration (the gas is salted-in). Figure 9-6[23] shows that at 80°C, carbon dioxide is *salted-out* significantly by sodium sulfate, less so by ammonium sulfate. Table 9-5 shows that at 25°C, methane, ethane, and butane are *salted-in* by tetramethylammonium bromide.

Several empirical models have been proposed for estimation of Setchenov constants. The model of Schumpe (1993) is the most general because it can also be applied to mixed electrolyte solutions. In this model for gaseous solutes, Setchenov constants are estimated from a set of ion-specific and gas-specific parameters that have been evaluated for 45 ions and 22 gases by non-linear regression of solubility data for salt solutions (Schumpe *et al.*, 1995).

Lang (1996) showed that the Setchenov equation [Eq. (9-55)] and the Schumpe model can also be applied to aqueous solutions containing ionogenic[24] organic compounds, such as amino acids, zwitterion peptides, proteins, and bases. To illustrate, Fig. 9-8 shows that the linear relation [Eq. (9-55)] holds; on a semi-log plot, the relative solubility of oxygen, $(m'/m'')_{O_2}$, is a linear function of the concentration of the organic solute. It is useful to predict the salt effect of an organic solute on the solubility of oxygen, e.g., in aerobic fermentation processes, where the oxygen concentration is a key parameter for optimal control or in medicine, where the solubility of oxygen in blood may change due to dissolved solutes.

A dissolved salt can also have a large effect on the composition of a vapor in equilibrium with an aqueous solution of a volatile liquid (Furter and Cook, 1967; Furter, 1976, 1977). When the dissolved salt solvates preferentially with the molecules of one component (Ohe, 1976, 1991), the salt can have a selective effect on the volatilities of the two liquids, and hence on the composition of the equilibrium vapor, although no salt is present in the vapor phase. For example, a preference for solvation with the less volatile component would result in an increase in the relative volatility of the more volatile component and therefore would enhance separation by distillation. Addition of a soluble salt to a liquid phase of a system may provide a convenient technique for extractive or azeotropic-distillation operations. However, industrial applications of this technique are often hindered by difficulties in salt recovery from the remaining liquid phase and by corrosive properties of salt solutions.

The effect of salt on vapor-liquid equilibria can be described by a Setchenov-type equation proposed by Furter and coworkers (Johnson and Furter, 1960). For a single salt in a binary mixed-solvent at fixed (salt-free) composition, Furter's equation is

$$\ln\frac{\alpha}{\alpha^{o}} = k'_{MX}x_{MX} \tag{9-56}$$

[23] Setchenov's equation cannot be applied to the results shown in Fig. 9-6 because the high salt concentration indicated in Fig. 9-6 requires higher terms in Setchenov's power series in electrolyte concentration.

[24] Organic compounds that are ionogenic show their ionic behavior only if dissolved in aqueous solution.

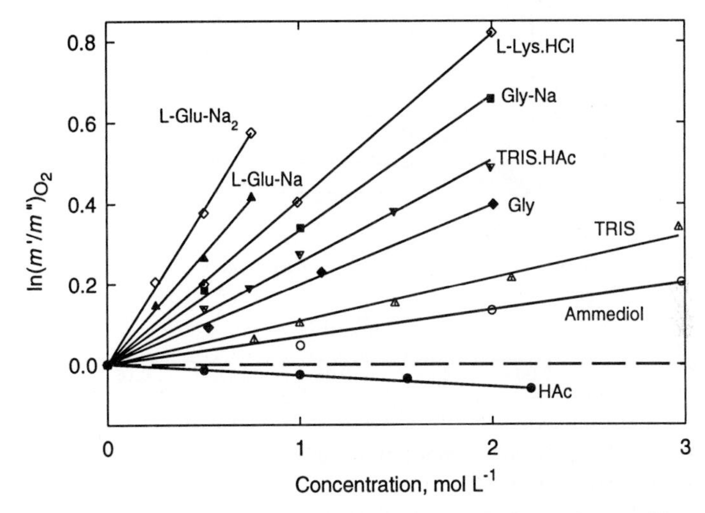

Figure 9-8 Setchenov plot of oxygen in aqueous solutions of some bio-organic compounds at body temperature (37°C).[25] HAc, acetic acid; Ammediol, 2-amino-2-methyl-1,3-propan-diol; TRIS, Tris(hydroxymethyl)amino-methane; Gly, glycine; TRIS.HAc, Tris(hydroxymethyl)amino-methane acetate; Gly-Na, glycine-sodium salt; L-Lys.HCl, L-lysine-monohydrochloride; L-Glu-Na, L-glutamic acid-monosodium salt; L-Glu-Na$_2$, L-glutamic acid-disodium salt. (Lang, 1996).

where α and α^o are the relative volatilities[26] with and without salt, respectively; k'_{MX} is the *salt-effect parameter* (which remains constant for moderate salt concentration), and x_{MX} is the mole fraction of the salt in the liquid phase, given by moles salt/(moles salt + moles solvent).

To illustrate, Fig. 9-9 shows smoothed experimental data (Burns and Furter, 1976) for the effects of KBr and R_4NBr (where R = H, CH_3, C_2H_5, n-C_3H_7, and n-C_4H_9) salts in the ethanol/water system at fixed (salt-free) liquid composition ($x_{ethanol}$ = 0.206). Figure 9-9 shows the large variety of salt effects for vapor-liquid equilibria in the ethanol/water system: they range from a large salting-out effect (KBr) to a large salting-in effect [(n-$C_4H_9)_4NBr$]. Figure 9-9 shows that Eq. (9-56) provides a reasonable method for representing the experimental data.

[25] In Fig. 9-8 the molarity scale is used. To convert from molarity to molality we need the mass density of the solution. See Footnote 2.

[26] In a binary solution, *relative volatility* of i is defined by $\alpha_i = y_i(1-x_i)/x_i(1-y_i)$, where y_i and x_i are, respectively, the mole fraction in the vapor phase and in the liquid phase. Since the salt is nonvolatile, the vapor phase contains only the two volatile species, whereas the liquid phase contains all three components. However, to facilitate direct comparison, the definition of α used here uses liquid compositions on a salt-free basis.

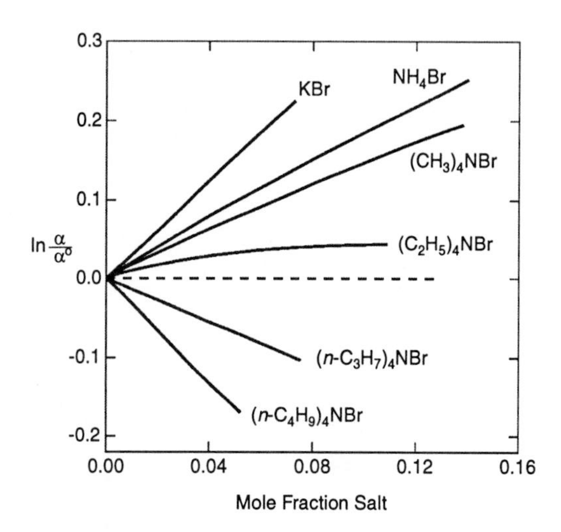

Figure 9-9 Salt effects in ethanol/water system at constant liquid composition $x_{\text{ethanol}} = 0.206$ (salt-free basis).

Another example, shown in Fig. 9-10, concerns vapor-liquid-equilibrium data at 1 bar for acetic acid/water and for acetone/methanol salt-free mixtures; and for saturated mixtures with inorganic salts. Figure 9-10(a) shows that calcium chloride and barium chloride lower the relative volatility of water. In the absence of salt, water (boiling point 100°C) is more volatile than acetic acid (boiling point 118°C). However, in a solution saturated with $CaCl_2$ when $w_{\text{water}} > 0.12$, relative volatility α for acetic acid exceeds unity while α for water is less than unity. Barium chloride has a similar effect but not until $w_{\text{water}} > 0.66$. Here w is weight fraction on a salt-free basis. Figure 9-10(a) indicates that both $CaCl_2$ and $BaCl_2$ induce an azeotrope in this system but at different conditions, respectively, 0.12 weight fraction water and 122.2°C, and 0.66 weight fraction water and 102.3°C. The opposite effect is also possible, as illustrated in Fig. 9-10(b) for acetone/methanol mixtures at 1 bar. The salt-free system has an azeotrope at 55.4°C and a weight fraction for acetone of 0.87. Addition of NaI breaks the azeotrope and produces a salting-out effect on acetone. These examples show how salts may be useful for creating favorable conditions for separation by distillation. However, salts are not often used for that purpose because solids-handling is not convenient and because dissolved salts tend to be corrosive.

A salt dissolved in a mixed solvent may affect the boiling point, the mutual solubilities of the two liquid components, and the equilibrium vapor-phase composition. Generally, the non-dissociated molecules or ions (or both) of dissolved salt tend to attract preferentially one type of solvent molecule, as Fig. 9-10 illustrates. Usually, the molecules of the more polar component are preferentially attracted by the electrostatic field of the ions. In that event, the vapor composition is enriched by the less polar solvent, wherein the salt is less soluble.

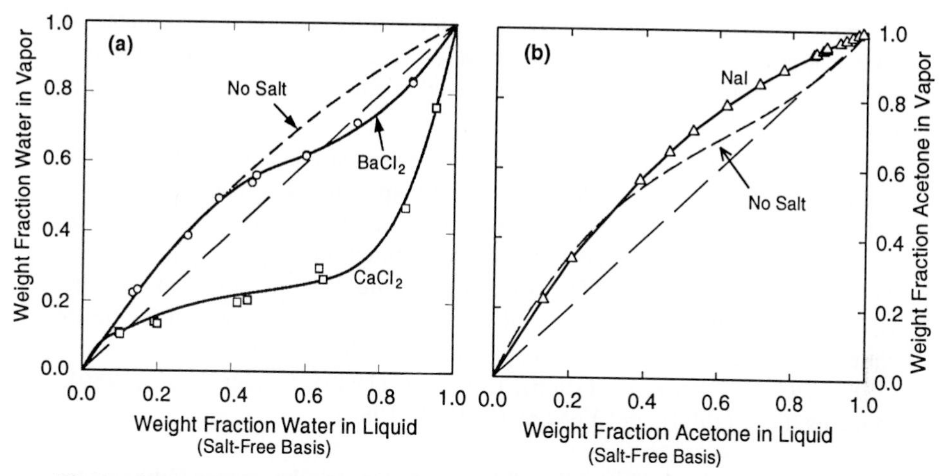

Figure 9-10 A salt may invert relative volatility or eliminate an azeotrope. Vapor-liquid equilibrium at 1 bar for: (a) Acetic acid/water system with no salt (– – –), saturated (about 1.8 mol kg^{-1}) with $BaCl_2$ (○), and saturated (about 7.5 mol kg^{-1}) with $CaCl_2$ (□) (Ramalho *et al.*, 1964).(b) Acetone/methanol system with no salt (– – –), and saturated (about 3.8 mol kg^{-1}) with NaI (△) (Iliuta and Thyrion, 1995).

9.10 Models for Concentrated Ionic Solutions

When ion concentrations are low, the average distance between ions is large; in that case, only long-range electrostatic forces are important. As ion concentration rises, ions begin to interact also with hard-core repulsive forces (leading to excluded-volume effects) and with short-range (van der Waals) attraction forces.

While much effort has been expended by physical chemists toward a fundamental theory for concentrated electrolyte solutions, for most applications it is necessary to resort to semi-empirical models (Rafal *et al.*, 1994). A number of models has been developed based on different assumptions. Typical models are based on perturbation theory (Henderson *et al.*, 1986), or equations of state (Jin and Donohue, 1988, 1988a, 1991), or on solvation concepts leading to chemical theories (Stokes and Robinson, 1973; Schönert, 1986).

To correlate activity coefficients in aqueous electrolyte solutions, semi-empirical models can be divided into three broad categories:

Physical models. Deviations from the Debye-Hückel limiting law are attributed to physical interactions between ions, as, e.g., in the Pitzer model (Pitzer, 1973, Pitzer and Mayorga, 1973, 1974). Here physical interactions refer primarily to excluded-volume and van der Waals-attraction effects.

Chemical models. Nonideal behavior of an electrolyte solution is attributed to chemical reactions that lead to the formation of semi-stable chemical species, particularly solvated ions; the solvation model of Robinson and Stokes (1973) is a typical example.

Local-composition models. The local-composition concept[27] is used to account for nonrandomness. These models are special cases of physical models. The effect of van der Waals interactions between ions is expressed not as a function of bulk composition but as a function of local composition. These models use the NRTL equation, or the Wilson equation, or the UNIQUAC equation for that part of the activity coefficient which is due to short-range forces; examples are provided by the models of Chen *et al.* (1982, 1986), Mock *et al.* (1986), Liu *et al.* (1989), Sander *et al.* (1986), Macedo *et al.*, (1990), and Vera *et al.* (1980, 1988).

In all these models, the key adjustable parameters for the excess Gibbs energy are determined by regression of experimental data for the binary mixture.

We cannot here provide a comprehensive discussion of a very large subject with a huge literature. We provide only a modest introduction.

9.11 Fundamental Models

With few exceptions, theoretical descriptions of electrolyte solutions have been based on the primitive model where the solvent is a dielectric continuum (characterized by its dielectric constant) and the ions are considered to be charged hard spheres. In this model, there are no explicit terms for solvent-solvent and ion-solvent interactions; it is assumed that these interactions are taken into account through the dielectric constant in the ion-ion interaction terms. This crude approximation is satisfactory for dilute solutions or else for solutions of particles (e.g. colloids) that are much larger than the solvent molecules. Nevertheless, it is a bad approximation for concentrated ionic solutions where the size of the solute ion is comparable to that of the solvent molecule.

Henderson *et al.* (1986) developed a non-primitive model where a perturbation expansion is applied to a mixture of dipolar hard spheres (the solvent) and charged hard spheres (the ions) of the same diameter. However, for most practical applications the perturbation expansion converges too slowly.

Using perturbation theory, Jin and Donohue (1988, 1988a, 1991) derived a four-parameter equation of state for aqueous solutions containing strong[28] or volatile weak electrolytes, including multisalt systems. In the model, Jin and Donohue calculate

[27] See Sec. 6.11.

[28] For a review of modeling thermodynamic properties of aqueous strong electrolyte solutions see J. R. Loehe and M. D. Donohue, 1997, *AIChE J.*, 43: 180.

short-range interactions using the *p*erturbed *a*nisotropic-*c*hain-*t*heory (PACT) and long-range coulombic (charge-charge) interactions between ions from a perturbation expansion. The adjustable parameters used are the sizes of the ions.

The *m*ean *s*pherical *a*pproximation (MSA) (Blum, 1980) has also been used to correlate activity coefficients in aqueous electrolyte solutions (see, e.g., Gering *et al.*, 1989). This approximation uses the primitive model of aqueous electrolyte solutions but it takes serious account of the finite sizes of the charged particles (ions). The MSA model reduces to the Debye-Hückel theory for point charges in a very dilute electrolyte solution. Results based on the MSA model are analytical. The MSA yields reasonable activity coefficients of ions to moderate concentrations. It has been applied to describe vapor-liquid equilibria of mixed-solvent electrolyte systems (Wu and Lee, 1992) and combined with an equation of state to gas-electrolyte solution equilibria at high pressures (Harvey and Prausnitz, 1989). Results are very sensitive to ionic diameters. To achieve agreement with experiment, these diameters may depend on electrolyte concentration, reflecting the effect of hydration. That effect, however, is more conveniently taken into account by a chemical theory (Stokes and Robinson, 1973; Schönert, 1986; Zerres and Prausnitz, 1994).

Using powerful computers, molecular simulation provides a promising method for describing the properties of electrolyte solutions without using the primitive-model assumption.

9.12 Semi-Empirical Models

Numerous semi-empirical models have been developed for representing equilibrium properties of electrolyte solutions (for a review see, e.g., Grigera, 1992; Renon, 1986; Maurer, 1983). These models correct the Debye-Hückel theory through additional terms that account for ion-ion interactions and, at high concentrations, for incomplete dissociation that in modern terminology is called *ion pairing*.

In these semi-empirical models, it is customary to assume that the molar excess Gibbs energy of electrolyte solutions is the sum of two contributions, one arising from the *long-range* (LR) coulombic forces (represented by the Debye-Hückel theory or its extension) and the other from *short-range* forces (SR):

$$g^{E*} = g_{\mathrm{LR}}^{E*} + g_{\mathrm{SR}}^{E*} \tag{9-57}$$

Following the relations given in Sec. 9.3, Eq. (9-57) gives two contributions to the unsymmetric mean ionic (mole-fraction-based) activity coefficient:[29]

[29] Mole-fraction based activity coefficients, $\gamma_{\pm}^{(x)}$, can be converted to mean molal activity coefficients, $\gamma_{\pm}^{(m)}$, using an expression similar to the one presented in Sec. 9.1: $\gamma_{\pm}^{(m)} = \gamma_{\pm}^{(x)} / (1 + 0.001 M_s \nu m)$, where $\nu = \nu_+ + \nu_-$. Here, mole fraction x is defined by Eq. (9-26).

$$\ln\gamma_{\pm}^{(x)} = \ln\gamma_{\pm,LR}^{(x)} + \ln\gamma_{\pm,SR}^{(x)}$$

To obtain $\ln\gamma_{\pm}^{(x)}$ from a model of the excess Gibbs energy, we first calculate $\ln\gamma_{+}^{(x)}$ and $\ln\gamma_{-}^{(x)}$, where $\gamma_{+}^{(x)}$ and $\gamma_{-}^{(x)}$ are, respectively, the activity coefficients of the cations and of the anions, and then use Eq. (9-19):

$$\ln\gamma_{\pm}^{(x)} = \frac{1}{\nu}(\nu_{+}\ln\gamma_{+}^{(x)} + \nu_{-}\ln\gamma_{-}^{(x)})$$

Although almost all semi-empirical models use a Debye-Hückel-type term for the long-range contribution, several choices are available to account for the short-range contribution. Possible choices include local-composition expressions (UNIQUAC, NRTL, Wilson) and van Laar or Margules equations. Most models assume complete dissociation of the electrolytes; using at least two adjustable binary parameters, most models are reasonably successful for dilute and semi-concentrated solutions, to about 6 molal.

Long-range forces between ions dominate at dilute electrolyte concentrations, while short-range forces between all species dominate at high electrolyte concentrations. The models of Cruz and Renon (1978), and Chen (Chen *et al.*, 1982, 1986), use an expression for the excess Gibbs energy that includes a contribution derived from the NRTL equation, whereas the model of Liu *et al.* (1989) uses the Wilson equation. In the model of Pitzer (Pitzer, 1973; Pitzer and Mayorga, 1973, 1974) g^{E*} is given by a virial series in molality of the solute.

9.13 Models Based on the Local-Composition Concept

Cruz and Renon and Chen *et al.*, postulate that the local composition of cations around cations is zero and similarly, that the local composition of anions around anions is zero. However, Cruz and Renon also assume that, for a completely dissociated electrolyte, the ions are always completely solvated by solvent molecules. In the more realistic local-composition model of Chen, all ions are surrounded by solvent molecules only in very dilute electrolyte solutions. At higher solute concentrations, the ions are partially surrounded by solvent molecules and partially by other ions of opposite charge. For the long-range contribution, Cruz and Renon use an expression obtained from the Debye-Hückel theory. For the short-range contribution, they use the NRTL model. Cruz and Renon show that their model can represent with very good accuracy the osmotic coefficients of partially or completely dissociated electrolytes using four adjustable binary parameters.

The NRTL model of Chen *et al.* uses Eq. (9-57) with a Debye-Hückel long-range term and a short-range interaction term of the NRTL form.

Chen makes two assumptions to define *local composition*:

Like-ion repulsion assumption. Due to large repulsive forces between ions whose charge is of the same sign, the region immediately surrounding a cation does not contain other cations; similarly, the region surrounding an anion does not contain other anions.

Local electroneutrality assumption. The distribution of cations and anions around a central solvent molecule is such that the net local ionic charge is zero.

With these assumptions, Chen derived an expression for the short-range contribution to the excess Gibbs energy that includes (see Sec. 6.15) two adjustable parameters $\tau_{MX,s}$ and $\tau_{s,MX}$, that are, respectively, the salt-solvent and solvent-salt interaction parameters for a binary pair of a single completely dissociated electrolyte in solution. Chen's model reproduces well the mean ionic activity coefficients of aqueous single electrolytes to a molality of six, as illustrated in Fig. 9-11 for aqueous KOH at 25°C. For multisalt systems, this model requires binary parameters for solvent-salt pairs ($\tau_{s,MX}$), obtained from data correlation (e.g. from osmotic or meanionic activity-coefficient data) of the corresponding binary solvent-salt systems. However, binary salt-salt energy parameters (that have significant effect on the nonideality of the ternary systems) are estimated using binary data for salt solubility in water or ternary solvent/salt(1)/salt(2) activity-coefficient data.

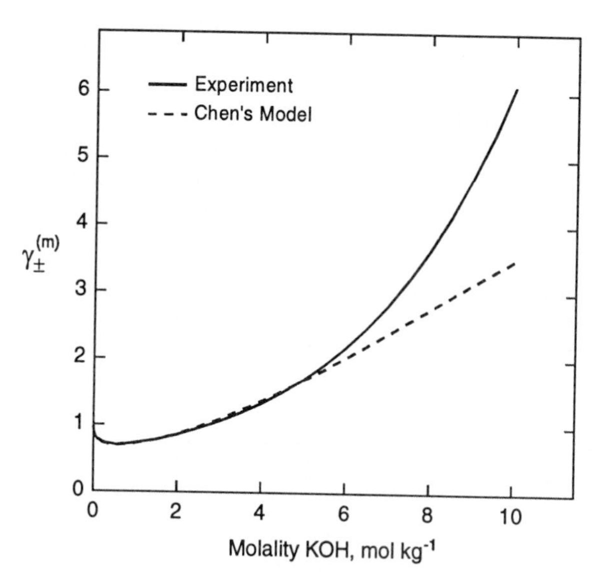

Figure 9-11 Mean ionic activity coefficients for aqueous KOH at 25°C. —— Experiment (Zemaitis *et al.*, 1986); - - - - - Chen's model.

Similar to Chen's model, Haghtalab and Vera (1988) use the NRTL equation [as modified by Panayiotou and Vera (1980)] for short-range interactions, whereas long-range Coulombic interactions are obtained from the Debye-Hückel theory. This model provides good representation of experimental data for the entire range of electrolyte concentration, from the dilute region to saturation, using only two adjustable parameters. However, attempts to use it for multisalt solutions have not been successful.

Liu *et al.* (1989) propose that in Eq. (9-57) g_{LR}^{E*} be obtained from an extended Debye-Hückel theory, and the second term of Eq. (9-57) is obtained from a local-composition expression of the Wilson type. However, contrary to other models, Liu does not assume additivity of long-range and short-range contributions; in Liu's model, the effect of long-range electrostatic forces on local composition is taken into account. The Debye-Hückel term accounts for the contributions due to electrostatic forces between each central ion and all the ions outside the first coordination shell; the local-composition term accounts for the contributions from both the short-range interaction forces of all kinds and the long-range electrostatic forces between each central ion and all ions inside the first coordination shell. While the parameters in Chen's model are salt-specific, in Liu's model they are ion-specific.[30] For a single-electrolyte solution, Liu's model uses only one adjustable energy parameter; other fitted parameters are common to all electrolyte systems containing the same cation and/or the same anion. Extension to multicomponent electrolyte solutions (Liu *et al.*, 1989a) follows without additional assumptions, i.e., the parameters for a multicomponent system are obtained from data correlation of its constituent binary systems; no higher-order parameters are required. Liu's model successfully fits data ($\gamma_{\pm}^{(x)}$) for a variety of concentrated electrolyte solutions. For example, Liu and Grén (1991) use Liu's model to describe vapor-liquid equilibria for the hydrogen-chloride/water system over the temperature range 0-110°C and salt concentrations to 21.55 M. For the calculated results shown in Fig. 9-12, Liu used four ion-specific parameters (the radius of H^+ and the interaction energy parameters g_{H^+,H_2O}, g_{Cl^-,H_2O}, g_{H^+,Cl^-}), given in Table 9-6. The anion radius r_{Cl^-} was 1.81 Å.

Table 9-6 Parameters obtained by Liu and Grén (1991) from fitting $\gamma_{\pm}^{(x)}$ data converted from partial pressure data for the HCl/H_2O system in the temperature range 0-110°C and for HCl concentrations to 21.55 M. Temperature T is in Kelvin.*

$(g_{H^+,H_2O})/RT$	$(g_{H^+,Cl^-})/RT$	$(g_{Cl^-,H_2O})/RT$	r_{H^+}
-1.92597	2.63355	-0.39037	$0.72649 + 0.0019624(T - 273.15)$

* The significance of g here is not to be confused with that used in Eq. (9-27)

[30] Parameters for 9 cations and 8 anions are presented by Y. Liu and U. Grén, 1991, *Chem. Eng. Sci.*, 46: 1815.

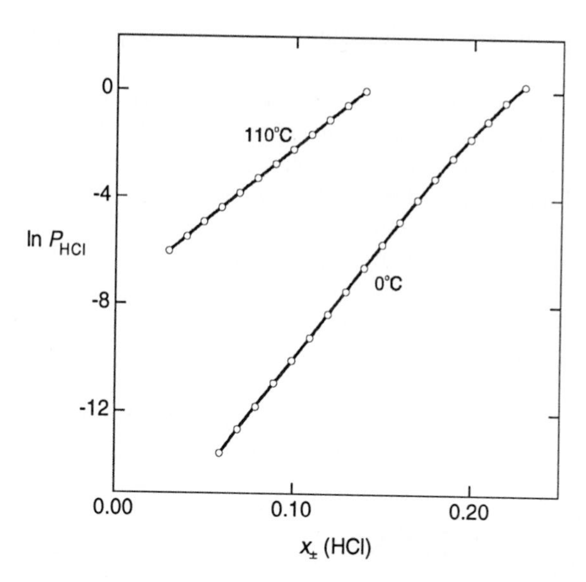

Figure 9-12 Partial pressures of HCl (with pressure in bar) in the HCl/H_2O system as a function of the mean ionic mole fraction $x_\pm$ of HCl [defined by Eq. (9-26)]. o Experiment; —— Liu's model with parameters in Table 9-6.

Sander *et al.* (1986) proposed a method to correlate and predict salt effects in vapor-liquid equilibria for water+cosolvent mixtures. This model combines a term of the Debye-Hückel type with a modified UNIQUAC equation with concentration-dependent parameters. As in the model of Liu, the parameters are ion-specific; no ternary parameters are required. However, a large number of parameters are needed: 7 for a salt/solvent system and 14 for a salt/cosolvent/system. Sander's model, as well as that by Mock *et al.* (1986), suffers from improper combination of Lewis-Randall and McMillan-Mayer formalisms (see Sec. 9.16). This deficiency was corrected in Sander's model by Macedo *et al.* (1990).

9.14 The Model of Pitzer

Since about 1973, the *ion-interaction model* of Pitzer has achieved wide acceptance; it has been applied successfully to several geochemical systems (e.g., equilibria of multicomponent brines with solid phases, solubilities of atmospheric gases in seawater) and to systems of interest in chemical industry. Availability of a substantial model-parameter database and of pertinent computer programs[31] makes Pitzer's model a ver-

[31] See, e.g., *Chemical Modeling of Aqueous Systems*, 1990, Vol. II, Chaps. 8, 9 and 10, (D. C. Melchior and R. L. Bassett, Eds., A.C.S. Symp. Ser. 416, Washington: A.C.S.).

satile and accessible method for describing thermodynamic properties of electrolyte solutions.

The Debye-Hückel theory is valid only at very low electrolyte concentrations. To extend Eq. (9-51) to higher concentrations, terms may be added on an ad-hoc basis to take into account short-range forces and thereby improve agreement with experiment. A common form of an extended Debye-Hückel equation gives the mean ionic activity coefficients, $\gamma_{\pm}^{(m)}$, as a sum of an electrostatic (Debye-Hückel type) term and a virial expansion in electrolyte concentration. Such extended Debye-Hückel equations are useful for representing experimental data.

Pitzer presented a virial expansion that provides very good representation of the properties of electrolyte solutions. Although Pitzer's equation has a theoretical basis, his final equation is at least partly empirical.

For an electrolyte solution containing w_s kilograms of solvent, with molalities m_i, m_j, ..., of solute species i, j, ..., Pitzer assumed that the excess Gibbs energy is given by

$$\frac{G^{E*}}{RTw_s} = f(I) + \sum_i \sum_j m_i m_j \lambda_{ij}(I) + \sum_i \sum_j \sum_k m_i m_j m_k \Lambda_{ijk} + \cdots \qquad (9\text{-}58)^{32}$$

Function $f(I)$ depends on ionic strength I, temperature and solvent properties; it represents long-range electrostatic forces and includes the Debye-Hückel limiting law.

$\lambda_{ij}(I)$ represents the short-range interaction between two solute particles in the solvent; the ionic-strength dependence of λ_{ij} facilitates rapid convergence in the virial expansion.

Λ_{ijk} terms account for three-body ion interactions; they are important only at high salt concentrations. Coefficients $\lambda(I)$ and Λ are analogous to second and third virial coefficients because they represent the effects of short-range forces between, respectively, two and three ions. For highly concentrated solutions, fourth or even higher-order interactions may be required in Eq. (9-58).

Pitzer assumed that the λ and Λ matrices are symmetric, i.e., $\lambda_{ij} = \lambda_{ji}$, and $\Lambda_{ijk} = \Lambda_{ikj} = \Lambda_{jik}$, etc. Functions $f(I)$ and $\lambda(I)$ were established by Pitzer; they are presented elsewhere (Pitzer, 1973, 1973a, 1974; 1991a).

Applying Eq. (9-42) and (9-43) to Eq. (9-58) we obtain expressions for the activity coefficient and for the osmotic coefficient. For a binary (i.e. single electrolyte) solution, they are:

$$\ln \gamma_{\pm}^{(m)} = |z_+ z_-| f^{\gamma} + m\left(\frac{2\nu_+\nu_-}{\nu}\right) B_{\text{MX}}^{\gamma} + m^2 \left[\frac{2(\nu_+\nu_-)^{3/2}}{\nu}\right] C_{\text{MX}}^{\gamma} \qquad (9\text{-}59)$$

[32] The mole-fraction scale could also be used (and it is sometimes used for highly concentrated electrolyte solutions) but molality is the most common concentration scale in the electrolyte literature.

$$\phi - 1 = |z_+ z_-| f^{\phi} + m\left(\frac{2\nu_+\nu_-}{\nu}\right) B^{\phi}_{\text{MX}} + m^2 \left[\frac{2(\nu_+\nu_-)^{3/2}}{\nu}\right] C^{\phi}_{\text{MX}} \tag{9-60}$$

From a systematic analysis using reliable experimental osmotic and activity coefficient data for 1-1, 2-1, 1-2, 3-1, and 4-1 electrolytes, Pitzer found that the best general agreement was obtained when terms f, B_{MX}, and C_{MX} have the form

$$f^{\gamma} = -A_{\phi}\left[\frac{I^{1/2}}{1 + bI^{1/2}} + \frac{2}{b}\ln(1 + bI^{1/2})\right] \tag{9-61}$$

$$B^{\gamma}_{\text{MX}} = 2\beta^{(0)}_{\text{MX}} + \frac{2\beta^{(1)}_{\text{MX}}}{\alpha^2 I}\left[1 - (1 + \alpha I^{1/2} - \frac{\alpha^2 I}{2})\exp(-\alpha I^{1/2})\right] \tag{9-62}$$

$$C^{\gamma}_{\text{MX}} = \frac{3}{2} C^{\phi}_{\text{MX}} \tag{9-63}$$

$$f^{\phi} = -A_{\phi}\frac{I^{1/2}}{1 + bI^{1/2}} \tag{9-64}$$

$$B^{\phi}_{\text{MX}} = \beta^{(0)}_{\text{MX}} + \beta^{(1)}_{\text{MX}}\exp(-\alpha I^{1/2}) \tag{9-65}$$[33]

$$C^{\phi}_{\text{MX}} = \frac{3}{(\nu_+\nu_-)^{1/2}}(\nu_+\Lambda_{\text{MMX}} + \nu_-\Lambda_{\text{MXX}}) \tag{9-66}$$

In Eqs. (9-61) and (9-64), A_{ϕ} is the Debye-Hückel constant for the osmotic coefficient given by Eq. (9-49a) (for water at 25°C, $A_{\phi} = 0.392$ kg$^{1/2}$ mol$^{-1/2}$); b is a universal parameter equal to 1.2 kg$^{1/2}$ mol$^{-1/2}$; and α is another universal parameter equal to 2.0 kg$^{1/2}$ mol$^{-1/2}$ for most electrolytes (2-2 salts are an exception).

Adjustable binary parameters $\beta^{(0)}_{\text{MX}}$, $\beta^{(1)}_{\text{MX}}$, and C^{ϕ}_{MX} are specific for each salt; they have been obtained from least-square fittings to experimental osmotic and activity-coefficient data for aqueous electrolytes at room temperature. A list of these parameters is given elsewhere (Clegg and Whitfield, 1991; Zemaitis *et al.*, 1986; Pitzer, 1991a; Pitzer, 1995). Parameters C^{ϕ}_{MX} depend on triple-ion interactions; they are important only at high concentrations (usually higher than 2 mol kg^{-1}). All parameters (except α and β) are temperature-dependent.

[33] Pitzer gives somewhat different expressions for B^{γ}_{MX} and B^{ϕ}_{MX} for 2-2 electrolytes.

Equations (9-59) and (9-60) give good agreement with experimental data for salt concentrations to about 6 mol kg^{-1}. Therefore, unless higher terms are added, Pitzer's model cannot be applied to highly concentrated electrolyte solutions, e.g. to the limit of saturation for a highly soluble salt. However, using binary parameters regressed over a wide range of electrolyte concentration, significant improvements can be achieved at high concentrations at the cost of less accurate results at low concentrations. To illustrate, Table 9-7 gives standard deviations of $\ln\gamma_{\pm}^{(m)}$ for several electrolytes at 25°C calculated from Pitzer's model with binary parameters obtained from data over different ranges of salt concentration. Parameters from Pitzer and Mayorga (1973, 1974) were obtained from regression of experimental data over a limited range of concentration (up to the lower concentration in the second column of Table 9-7); those of Kim and Frederick (1988, 1988a) were regressed over a range of concentration extended to saturation (higher concentration in the second column of Table 9-7).

Table 9-7 Deviations for estimated versus experimental mean ionic activity coefficients ($\ln\gamma_{\pm}^{(m)}$) at 25°C. Calculations from Pitzer's equation with binary parameters obtained at different conditions: Pitzer and Mayorga from data regression to the lower concentration shown for each electrolyte; Kim and Frederick from data regression to the higher concentration shown for each electrolyte.

Electrolyte	Maximum molality (mol kg^{-1})	Deviation*	
		Pitzer, Mayorga	Kim, Frederick
HCl	16.00	0.22031	0.02854
	6.00	0.00311	0.02956
LiBr	20.00	0.06099	0.07224
	2.50	0.00286	0.06780
$CaBr_2$	7.66	0.46557	0.08760
	2.00	0.00773	0.00732

* Root-mean-square deviation.

Table 9-7 shows that the binary parameters of Kim and Frederick give generally better agreement with experiment over wide ranges of concentration. While the fit of Pitzer and Mayorga gives accurate results at low concentrations, it gives only fair agreement with experiment at high concentrations.

Pitzer's equation has been applied to many aqueous electrolyte solutions, including aqueous mixed-electrolyte solutions. For mixed electrolytes, Pitzer uses additional terms in Eq. (9-58) that require additional interaction parameters, θ_{ij} and ψ_{ijk}, obtained from experimental data for aqueous mixed-electrolyte solutions with a common ion.[34]

[34] Expressions for activity coefficients and osmotic coefficients of multi-electrolyte solutions are presented in App. I together with model parameters and their temperature derivatives for several common aqueous electrolytes.

However, for a multi-electrolyte solution, the principal contributions to G^{E*} usually come from the single-electrolyte parameters; parameters θ_{ij} and ψ_{ijk} have only a small effect, as illustrated in Fig. 9-13. This figure shows contributions to the activity coefficient of NaCl in a multi-electrolyte aqueous mixture containing the same electrolytes as those in seawater (see Table 9-3). Contributions are from the extended Debye-Hückel term [Eq. (9-61)]; from binary (parameters $\beta_{MX}^{(0)}$, $\beta_{MX}^{(1)}$ and C_{MX}^{ϕ}); and from mixed-salt terms (parameters θ_{ij} and ψ_{ijk}) as a function of ionic strength (Clegg and Whitfield, 1991).

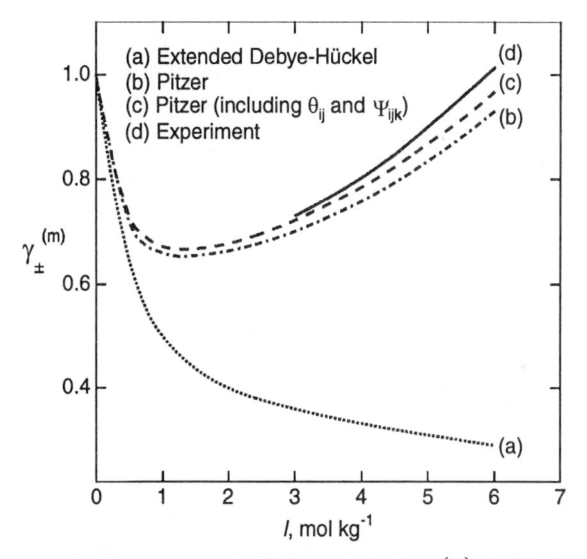

Figure 9-13 Mean ionic activity coefficient of NaCl, $\gamma_{\pm}^{(m)}$, in water at 25°C as a function of ionic strength: (a) Extended Debye-Hückel term [Eq. (9-61)]; (b) Pitzer equation for a multi-electrolyte system with electrolytes as in seawater [Eq. (9-58)], with $\beta_{MX}^{(0)}$, $\beta_{MX}^{(1)}$ and C_{MX}^{ϕ}, but omitting mixed-salt parameters θ_{ij} and ψ_{ijk}; (c) All terms including mixed-salt parameters θ_{ij} and ψ_{ijk}. For comparison, the solid line (d) gives experimental $\gamma_{\pm}^{(m)}$ of NaCl (in the range $3 \le I \le 6$ mol kg^{-1}) in aqueous NaCl at the same ionic strength as that of the multi-electrolyte system. Line (d), based on data for the single-solute solution, closely approximates line (c) for seawater.

As Fig. 9-13 shows, the like-sign interactions have little effect on $\gamma_{\pm}^{(m)}$ of NaCl; even at I = 6 mol kg^{-1}, they account for only 3.5% of the total. This result suggests that, particularly at low concentrations, mixed-salt terms may be omitted without introducing significant error. This omission greatly simplifies Pitzer's model. Because NaCl constitutes about 90 mol % of the salts in seawater, $\gamma_{\pm}^{(m)}$ of NaCl in an aqueous solution of NaCl is similar to $\gamma_{\pm}^{(m)}$ for the multi-electrolyte solution at the same ionic strength.

It is an important characteristic of the Pitzer model that all parameters can be evaluated from measurements for single electrolytes and common-ion two-salt solutions. No new parameters appear for more complex mixtures. Therefore, with Pitzer-model parameters and with solubility products of salts available from the experimental solubility in single-salt solutions, Pitzer's model can be used to predict solid solubilities in mixed-salt systems. For a solid electrolyte $M_{\nu+}X_{\nu-} \cdot nH_2O$ the solubility equilibrium in water is

$$M_{\nu+}X_{\nu-} \cdot nH_2O \rightleftharpoons \nu_+ M^{z+} + \nu_- X^{z-} + nH_2O \tag{9-67}$$

The concentrations of cationic species M^{z+} and anionic species X^{z-} in the liquid phase are determined by the solubility product. If the activity of the pure solid electrolyte is taken as unity, the solubility product is defined as

$$\begin{aligned} K_{sp} &= (a_+)^{\nu+}(a_-)^{\nu-}(a_{H_2O})^n \\ &= (m_+\gamma_+^{(m)})^{\nu+}(m_-\gamma_-^{(m)})^{\nu-}(a_{H_2O})^n \\ &= (m_+)^{\nu+}(m_-)^{\nu-}(\gamma_\pm^{(m)})^{\nu}(a_{H_2O})^n \end{aligned} \tag{9-68}$$

where a_i, m_i, and $\gamma_i^{(m)}$ represent, respectively, the activity, molality and activity coefficient of the aqueous ion i, and $\gamma_\pm^{(m)}$ is the mean ionic activity coefficient defined by Eq. (9-19). The activity of water, a_{H_2O}, is related to the osmotic coefficient ϕ. The Gibbs-Duhem equation relates the composition dependence of ϕ (or a_{H_2O}) to various solute γ_i or $\gamma_\pm$. Because these activity and osmotic coefficients can be calculated with the Pitzer model, it is possible to predict solid-salt solubilities, provided K_{sp} is known.

The solubility product, K_{sp}, can be calculated if the standard-state Gibbs energy of the solid and aqueous species are available at the temperature of interest. The standard state for the aqueous ions and electrolytes is the ideal, molal solution at fixed pressure and temperature. For the solid and solvent, the standard state is the pure phase at the pressure and temperature of interest. At reference temperature $T_r = 298.15$ K and standard pressure, K_{sp} can be calculated from tabulated standard-state values of $\Delta_f g_i^0$, i.e., the Gibbs energies of formation of the various species:

$$\ln K_{sp}(T_r) = -\frac{\Delta g^0(T_r)}{RT_r} \tag{9-69}$$

with

$$\Delta g^0(T_r) = \Delta_f g_M^0 + \Delta_f g_X^0 - \Delta_f g_{MX}^0$$

where, for convenience, we have dropped the superscripts on M and X and the subscripts on MX.[35]

[35] Here, MX stands for a salt that may or may not be hydrated. If the solid salt is not hydrated, $n = 0$ in Eq. (9-67).

Using the Gibbs-Helmholtz equation, the temperature dependence of K_{sp} is

$$\ln K_{sp}(T) = \ln K_{sp}(T_r) - \frac{\Delta h^0(T_r)}{R}\left(\frac{1}{T} - \frac{1}{T_r}\right) + \frac{1}{R}\int_{T_r}^{T}\left[\frac{\int_{T_r}^{T} \Delta c_p^0(T)dT}{T^2}\right]dT \tag{9-70}$$

where $\Delta h^0(T_r) = \Delta_f h_M^0 + \Delta_f h_X^0 - \Delta_f h_{MX}^0$ is obtained from tabulated standard-state values for the enthalpy of formation.

In Eq. (9-70), $\Delta c_p^0(T) = c_{p,M}^0(T) + c_{p,X}^0(T) - c_{p,MX}^0(T)$. If the temperature dependence of $c_{p,i}^0$ (the standard-state heat capacity) is not known or if the difference between T and T_r is not large, Δc_p^0 may be assumed constant. In that case, Eq. (9-70) simplifies to

$$\ln K_{sp}(T) = \ln K_{sp}(T_r) - \frac{\Delta h^0(T_r)}{R}\left(\frac{1}{T} - \frac{1}{T_r}\right) + \frac{\Delta c_p^0(T_r)}{R}\left(\ln\frac{T}{T_r} + \frac{T_r}{T} - 1\right) \tag{9-70a}$$

At typical saturation pressures, the effect of pressure on the solubility of salts is small and can be neglected. At high pressures, however, this effect can be significant and has to be taken into account (Pitzer *et al.*, 1984).

Figures 9-14 and 9-15 present two examples comparing experimental and calculated solubilities (expressed in molalities) of two solid salts in an aqueous ternary mixture.

The system NaCl/KCl in Fig. 9-14 is simple with no intermediate (i.e. hydrated) solid phase but the system $NaCl/Na_2SO_4$ exhibits an intermediate solid phase due to the formation of hydrate $Na_2SO_4 \cdot 10H_2O$. In both examples the two salts have a common ion thereby simplifying the calculations.

Equations (9-68) and (9-70) are applied for each salt. After K_{sp} has been calculated from Eq. (9-70), the molality of one of the non-common ions is fixed and Eq. (9-68) is solved for the molality of the other. An iterative method is required because the three terms on the right side of Eq. (9-68) depend on both molalities. Even when there is no hydrated phase, it is not possible to solve for the molality without iteration. Repetition of this iterative process for various assumed values of one molality gives the curves shown in Figs. 9-14 and 9-15.[36] The intersection of the two curves gives the fixed-point composition where the two solids are in equilibrium with the aqueous solution.

[36] Appendix I gives model parameters and thermodynamic properties of various pertinent species for calculations leading to Figs. 9-14 and 9-15.

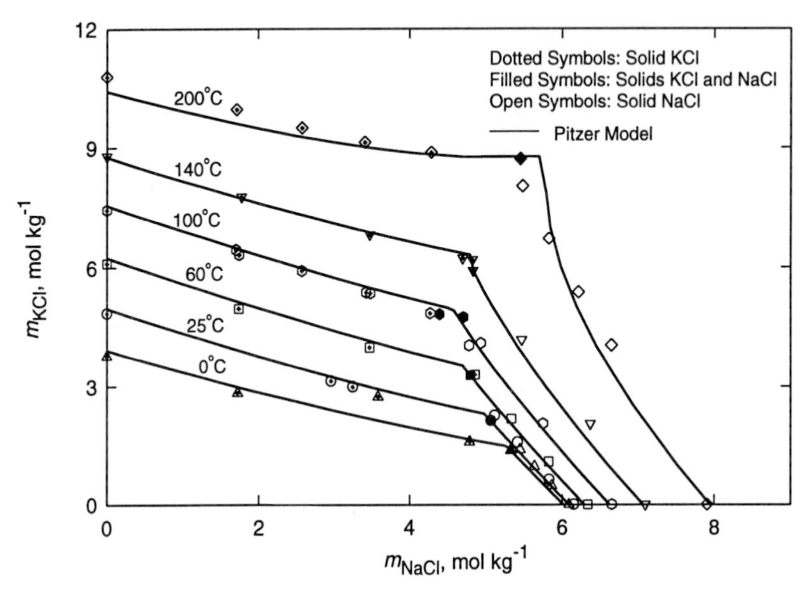

Figure 9-14 Experimental (Linke and Seidell, 1958, 1965) and calculated solubilities in the ternary mixture $NaCl/KCl/H_2O$ at several temperatures. Intersections of isothermal curves represent calculated ternary invariant points, where three phases are in equilibrium: solid NaCl, solid KCl, and aqueous solution.

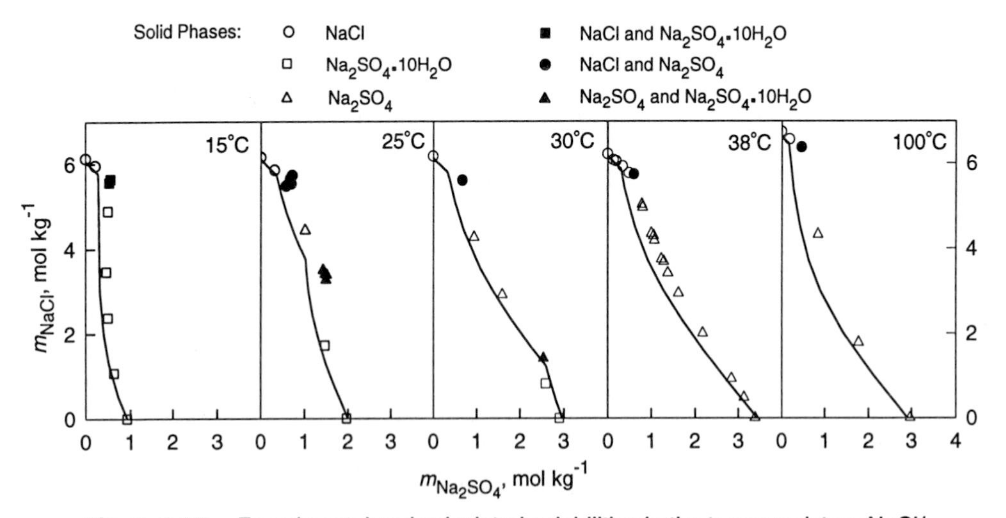

Figure 9-15 Experimental and calculated solubilities in the ternary mixture $NaCl/Na_2SO_4/H_2O$ at several temperatures. Hydrate $Na_2SO_4 \cdot 10H_2O$ is stable only at temperatures below about 38°C. Symbols: experiment (Linke and Seidell, 1958, 1965); —— Pitzer model. Intersections of isothermal curves represent calculated ternary invariant points where three phases are in equilibrium: aqueous solution, solid NaCl, and solid Na_2SO_4 or solid hydrate; or aqueous solution, solid Na_2SO_4 and solid hydrate.

For both examples, agreement with experiment is good, especially considering the large temperature range covered. As expected, the solubilities increase with rising temperature. For the aqueous system $NaCl/Na_2SO_4$ at 15°C the solid phases are NaCl and the hydrate $Na_2SO_4{\bullet}10H_2O$. With rising temperature, the hydrate becomes unstable relative to unhydrated Na_2SO_4 as indicated by the results at 25 and 30°C. At temperatures in excess of 38°C, the hydrate disappears and the only solid phases are NaCl and Na_2SO_4.

If the two salts do not have a common ion or, if there are more than two salts, the calculations become much more complex and require simultaneous solution of several equations.

Another example of application of Pitzer's model to multi-salt mixtures is provided by studies of mineral solubilities in brines by Weare and collaborators (Harvie *et al.*, 1980, 1982, 1987; Weare, 1987). To illustrate Weare's results, Fig. 9-16 compares experimental with calculated solubilities of gypsum ($CaSO_4{\bullet}2H_2O$) in $Na_2SO_4/NaCl$ solutions (Harvie and Weare, 1980, 1987).

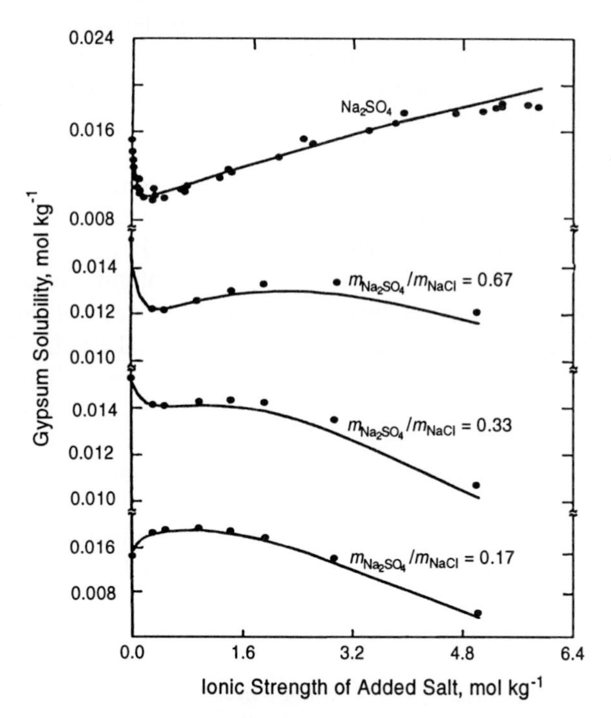

Figure 9-16 Gypsum ($CaSO_4{\bullet}2H_2O$) solubilities in aqueous $Na_2SO_4/NaCl$ solutions at 25°C. ● Experiment; —— Pitzer's model (Harvie *et al.*, 1982).

Because the dissolution of gypsum in brines has a significant effect on ionic strength (and hence on activity coefficients), the solubility must be calculated by an iterative procedure. Weare (Harvie *et al.*, 1982) proposes a convenient method based on a Gibbs-energy-minimization routine.

In Fig. 9-16, the abscissa is the total ionic strength of added salt ($3m_{Na_2SO_4} + m_{NaCl}$). For each curve, the molality ratio of Na_2SO_4 to NaCl is fixed. Figure 9-16 shows how solutions of different compositions affect gypsum solubilities. Calculated gypsum solubilities are in good agreement with experimental data at all ionic strengths. However, for this ternary system the model requires 12 single-electrolyte parameters ($\beta^{(0)}_{MX}$, $\beta^{(1)}_{MX}$ and C^{ϕ}_{MX}) and 5 mixed-electrolyte parameters (θ_{ij} and ψ_{ijk}). The large number of parameters required in Pitzer's model for multi-salt mixtures is its major disadvantage.

9.15 The "Chemical" Hydration Model of Robinson and Stokes

There is much physicochemical evidence to support the idea that when a strong electrolyte dissolves in water, the ions are solvated; water molecules are bound to the ions forming stoichiometric complexes. In some cases the stability of the complex may be sufficiently high to produce a solid hydrated salt. In many aqueous systems, cations are more extensively hydrated than anions.

In a manner similar to that for chemical theories described in Chap. 7, it is possible to relate the activity of the water to the equilibrium constant (or constants) which characterize hydration equilibria. A particularly successful example for a single-solvent solution is provided by the work of Stokes and Robinson (1973) which was later extended (with modifications) to a binary-solvent solution by Zerres (1994). We give here only a summary of the method developed by Stokes and Robinson. We omit the detailed derivation that is clearly presented in the original article.

Consider an aqueous solution of a strong electrolyte, containing n_w moles of free (i.e. not hydrated) water, n_0 moles of anhydrous cations, n_1 moles of singly-hydrated cations, n_i moles of i-hydrated cations and n_A moles of anions, at temperature T and total volume V. We neglect hydration of anions. This solution was prepared by adding c moles of anhydrous salt to water to give a final volume of 1 liter.

We assume stepwise hydration where a hydrate containing $i-1$ molecules of water per cation can add one molecule of water to form a hydrate containing i molecules of water per cation.

The hydration equilibrium is

$$\text{hydrate}\,(i-1) + \text{water} \rightleftharpoons \text{hydrate}\; i$$

and the corresponding equilibrium constant K_i is

$$K_i = \frac{a_i}{a_{i-1}\, a_w} \tag{9-71}$$

and O'Connell (1987), Wu and Lee (1992), Cabezas and O'Connell (1993), Haynes and Newman (1998), and Barthel *et al.* (1998).

Fortunately, for ordinary electrolyte solutions, the inconsistency has no appreciable effect (Cardoso and O'Connell, 1987), on the thermodynamics of single-solvent systems. However, when the empirical models described in previous sections are extended to multi-solvent systems, there is an inconsistency that requires attention, as discussed by Cardoso and O'Connell.

9.17 Phase Equilibria in Aqueous Solutions of Volatile Electrolytes

Waste streams from chemical plants, as well as stack gases from power plants, may contain volatile components (e.g. ammonia, hydrogen sulfide, carbon dioxide, and sulfur dioxide) that ionize, in part, in aqueous solution. Design of operations to remove volatile weak electrolytes from aqueous solutions requires representation of pertinent vapor-liquid equilibria.

In aqueous solution, volatile electrolytes exist in ionic and molecular (undissociated) form, as briefly discussed in Sec. 9.8. At ordinary temperatures and pressures, only the molecular form exists in the vapor. Calculation of vapor-liquid equilibria requires simultaneous solution of phase-equilibrium equations (for the molecular species), chemical-equilibrium equations for the liquid phase, and material balances.

A molecular-thermodynamic framework proposed by Edwards *et al.* (1978) has been successfully used for calculating vapor-liquid equilibria in aqueous solutions containing one or more weak volatile electrolytes for temperatures from 0 to 200°C and for total ionic strengths to more than 6 molal (Bieling *et al.*, 1989; Kurz *et al.*, 1995). Figure 9-19 gives a schematic representation of the system under consideration.

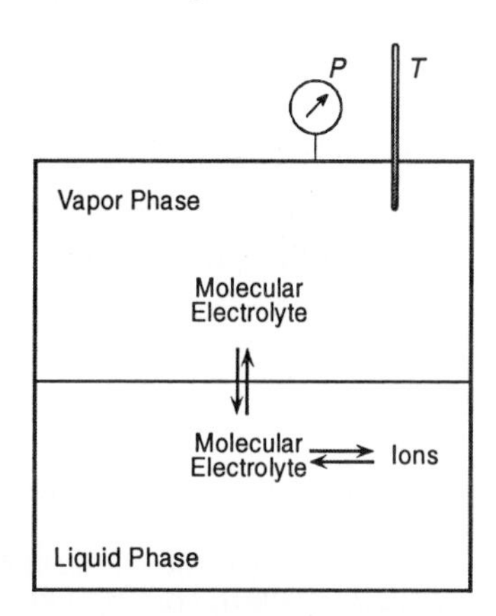

Figure 9-19 Schematic representation of vapor-liquid equilibria for an aqueous solution containing a single volatile weak electrolyte.

At a fixed temperature and pressure, the weak electrolyte is distributed between the vapor phase and the aqueous phase. For some solute i at low concentration of i, the (vertical) phase equilibrium is primarily determined by Henry's constant, H_i, and the molality of molecular (not ionic) solute i. This molality, in turn, is strongly influenced by the (horizontal) chemical dissociation equilibrium, characterized by the dissociation (or ionization) constant, K.[39] At higher concentration of i, Henry's constant H_i must be multiplied by an activity coefficient that depends on the concentrations of *all* solute species and on temperature.

First consider the single-solute case; the multi-solute case is described later. Since ions are not volatile, the (vertical) phase equilibrium is governed only by the molecular (undissociated) concentration of the electrolyte in the liquid phase. Vapor-liquid equilibria for the solvent (water) and for the (molecular electrolyte) solute are described by

$$f_w^V = f_w^L \tag{9-75}$$

$$f_i^V = f_i^L \tag{9-76}$$

where f is fugacity, subscript w refers to water and subscript i to electrolyte.

Equation (9-75) is rewritten (see Sec. 3.3),

$$\varphi_w^V y_w P = a_w P_w^s \varphi_w^s \exp\frac{\upsilon_w (P - P_w^s)}{RT} \tag{9-77}$$

where a_w is the activity of water, P_w^s is the saturation pressure of pure water, and φ_w^s is the fugacity coefficient of pure water at saturation, all at system temperature. Mole fraction y is for the vapor phase.

For the electrolyte, phase equilibrium considers only molecular electrolyte (designated with subscript M); therefore, Eq. (9-76) is rewritten

$$\varphi_{\mathrm{MX,M}}^V y_{\mathrm{MX,M}} P = \gamma_{\mathrm{MX,M}}^{(\mathrm{m})} m_{\mathrm{MX,M}} H_{\mathrm{MX,M}} \tag{9-78}$$

In Eq. (9-78), $y_{\mathrm{MX,M}}P$ is the partial pressure of the molecular weak electrolyte and $\varphi_{\mathrm{MX,M}}^V$ is the vapor-phase fugacity coefficient; $m_{\mathrm{MX,M}}$ is the molality of the electrolyte in molecular form, $\gamma_{\mathrm{MX,M}}^{(\mathrm{m})}$ is the activity coefficient and $H_{\mathrm{MX,M}}$ is Henry's constant for the molecular solute.[40]

---

[39] At moderate pressures, dissociation of the electrolyte in the vapor phase is appreciable only at very high temperatures and therefore is neglected here.

[40] Henry's constant H depends on temperature. When pressure is high, there is also an effect of pressure on H according to the rigorous thermodynamic equation $\partial \ln H_i / \partial p = \overline{\upsilon}_i^\infty / RT$, where i stands for solute and $\overline{\upsilon}_i^\infty$ is the partial molar volume of i in the solvent at infinite dilution. Because solute in Eq. (9-78) is the molecular species (MX,M), here $\overline{\upsilon}_i^\infty$ refers to the partial molar volume of molecular i at infinite dilution.

From a mass balance for the weak electrolyte in the liquid phase, we relate the overall electrolyte concentration (stoichiometric), m_{MX}, to that in molecular form, $m_{MX,M}$ and in chemically reacted form. For example, when NH_3 is dissolved in water,

$$m_{NH_3} = m_{NH_3,M} + m_{NH_4^+} \tag{9-79}$$

In addition, the bulk electroneutrality condition of the liquid phase relates the concentration of cations to those of anions:

$$\sum_i z_i m_i = 0 \tag{9-80}$$

Finally, using Eq. (9-54) the (horizontal) chemical equilibrium relates the molecular concentration of the undissociated electrolyte and the ionic concentrations of the weak electrolyte; for example, for the reaction $MX \rightleftharpoons \nu_+ M^{z+} + \nu_- X^{z-}$,

$$K = \frac{m_+^{\nu+} m_-^{\nu-}}{m_{MX,M}} \frac{(\gamma_{\pm}^{(m)})^{\nu}}{\gamma_{MX,M}^{(m)}} \tag{9-81}$$

where $\gamma_{\pm}^{(m)}$, the mean ionic activity coefficient, is defined by Eq. (9-19).

For very dilute solutions, it is reasonable to set all activity coefficients equal to unity. The important parameters are Henry's constant, H, and dissociation constant, K. Activity coefficients are important only for higher concentrations.

To solve the vapor-liquid equilibrium equations, we need to evaluate $\gamma_{MX,M}^{(m)}$, φ_{MX}^{V}, and $H_{MX,M}$. These quantities are used to solve the phase equilibrium equations given by Eqs. (9-77) and (9-79), coupled with Eq. (9-79), (9-80), and (9-81). However, to find $\gamma_{MX,M}^{(m)}$ from Eq. (9-81), we need information concerning $\gamma_{\pm}^{(m)}$. Further, we need vapor-phase equation-of-state data to find φ_{MX}^{V}.

For aqueous solutions of sulfur dioxide, Fig. 9-20 compares experiment with results calculated using the method of Edwards. Dissociation of sulfur dioxide in the liquid phase was taken into account through the chemical equilibria

$$SO_2 + H_2O \rightleftharpoons HSO_3^- + H^+$$

$$HSO_3^- \rightleftharpoons SO_3^{2-} + H^+$$

$$H_2O \rightleftharpoons H^+ + OH^-$$

Each of these chemical equilibria is characterized by a chemical-equilibrium constant, defined according to Eq. (9-81). Vapor-phase fugacity coefficients were calculated using the method of Nakamura *et al.* (1976). Henry's constants as a function of temperature were obtained from binary-data reduction. Activity coefficients for the electrolyte, $\gamma_{\pm}^{(m)}$, and the activity of water, a_w, were obtained from Edward's extension of Pitzer's model [Eq. (9-58)]. However, because SO_2 is a weak electrolyte with a low dissociation constant, the concentration of ions is so small that Eq. (9-59) reduces to

$$\ln \gamma_{SO_2,M}^{(m)} = 2\beta_{SO_2,SO_2}^{(0)} m_{SO_2,M} \tag{9-82}$$

and for water

$$\ln a_w = -\frac{M_w}{1000}\left(m_{SO_2,M} + \beta_{SO_2,SO_2}^{(0)} m_{SO_2,M}^2\right) \tag{9-83}$$

Combining Eqs. (9-78) and (9-82), the equation for the phase equilibrium of SO_2 is

$$\ln \frac{y_{SO_2}\varphi_{SO_2}P}{m_{SO_2,M}} - \frac{\overline{v}_{SO_2}^{\infty}(P-P_w^s)}{RT} = \ln H_{SO_2,M}^{(P_w^s)} + 2\beta_{SO_2,SO_2}^{(0)} m_{SO_2,M} \tag{9-84}$$

where $H_{SO_2,M}^{(P_w^s)}$ stands for Henry's constant of molecular SO_2 in water at infinite dilution, i.e. when the total pressure is equal to the vapor-pressure of water P_w^s. An independent estimate for $\overline{v}_{SO_2}^{\infty}$ is required.

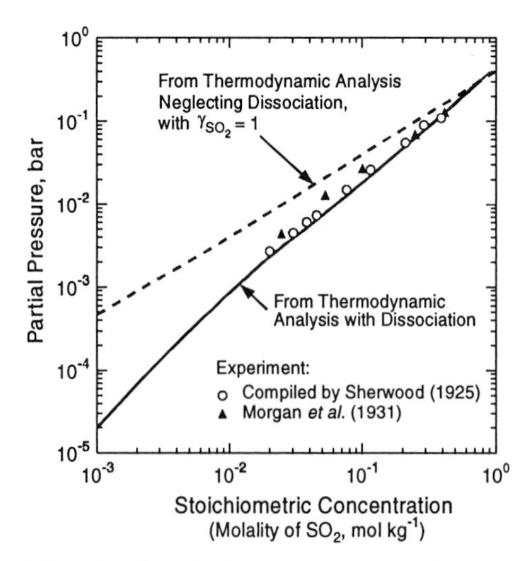

Figure 9-20 Solubility of sulfur dioxide in water at 10°C. Comparison with experiment of the calculated (Edwards *et al.*, 1978) results considering (——) or neglecting (- - -) ionization of SO_2 in water.

Plotting the left-hand side of Eq. (9-84) as a function of the molecular concentration of SO_2, the slope gives the molecule-molecule interaction parameter, $\beta^{(0)}_{SO_2,SO_2}$, and the intercept gives Henry's constant, $H^{(P^s_w)}_{SO_2,M}$.

Figure 9-20 shows that calculations from Edwards' model are in good agreement with experiment. Figure 9-20 also illustrates the importance of considering weak-electrolyte dissociation in vapor-liquid equilibrium calculations. The broken curve presents results calculated with the (erroneous) assumption that sulfur dioxide is a non-dissociating solute in an ideal dilute solution where (in the asymmetric convention) $\gamma_{SO_2} = 1$. The solid curve gives calculated results considering dissociation, as explained above. Large differences are observed at low concentrations where most of the solute is in ionic rather than in neutral molecular form.

The thermodynamic framework described above for a single-solute system can be extended to multisolute systems. The necessary parameters are obtained primarily from binary-data reduction but at high salt concentration, some ternary data are required.

Figure 9-21 compares calculated and experimental results (Rumpf *et al.*, 1993a) for a two-solute system: ammonia and sulfur dioxide in water, from 40 to 100°C at two overall molalities of ammonia (3.2 or 6.1 mol kg^{-1} of water) and at pressures to 22 bar.

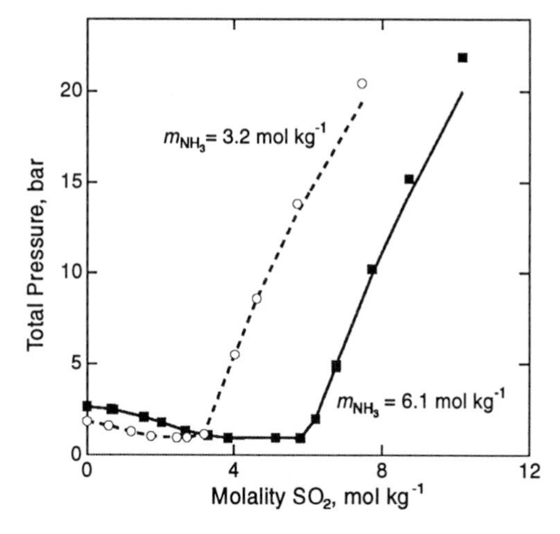

Figure 9-21 Experimental and calculated results for the simultaneous solubilities of ammonia and sulfur dioxide in water at 100°C. Experiment: ○, ■; Calculated from model of Edwards and Maurer: - - - m_{NH_3} = 3.2 mol kg^{-1}; —— m_{NH_3} = 6.1 mol kg^{-1}.

This example is qualitatively different from the previous one because there is strong chemical interaction between the two solutes: acidic sulfur dioxide and basic ammonia. In this example, in addition to water, the liquid phase contains molecular ammonia, molecular sulfur dioxide, ammonium ion, hydrogen ion, hydroxyl ion, sulfite ion, and bisulfite ion, as indicated in Fig. 9-22. As for single-solute systems,

Rumpf *et al.* (1993) wrote chemical-equilibrium expressions for each of the equilibria indicated in Fig. 9-22; applied mass and charge balances, and phase-equilibrium equations similar to Eq. (9-77) for water, and to Eq. (9-78) for each solute (NH_3 and SO_2). The calculations required four temperature-dependent equilibrium constants; activity coefficients of all species present in the liquid phase; Henry's constants for each volatile solute at infinite dilution in water; the vapor pressure, molar volume, dielectric constant of water; partial molar volumes of the dissolved gases; and information on vapor-phase nonideality (obtained from the virial equation). Excepting activity coefficients, all these requirements were obtained from available experimental data. For seven dissolved species (see Fig. 9-22), activity coefficients were calculated from Pitzer's model. If all Pitzer's terms are included, the seven species dissolved in the liquid phase would require 56 binary parameters $\beta_{ij}^{(0)}$ and $\beta_{ij}^{(1)}$, and 84 ternary parameters Λ_{ijk}. Rumpf made many reasonable approximations to reduce the total number of adjustable parameters to 13. Most of these were obtained from reduction of experimental single-solute data; Rumpf required only 5 ternary parameters obtained from reduction of experimental ternary data. The temperature dependence of the binary interaction parameters was taken into account, whereas it was neglected for the ternary parameters.

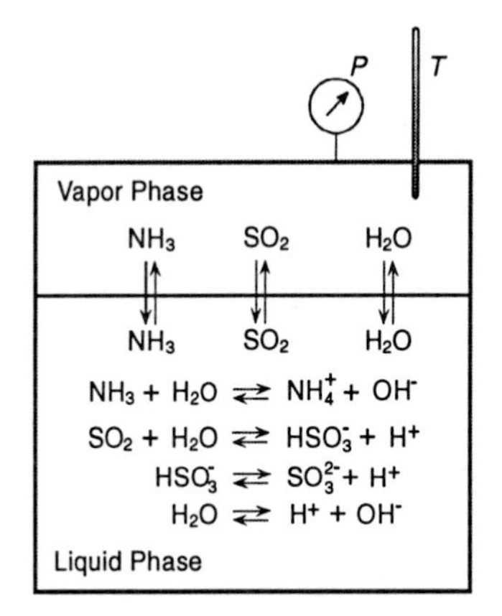

Figure 9-22 Vapor-liquid equilibria and chemical equilibria in the ammonia/sulfur dioxide/water system.

The work of Rumpf and Maurer shows that it is not a simple matter to describe phase equilibria in an aqueous system containing two volatile weak electrolytes over an appreciable range of temperature and solute concentrations, especially when there is strong chemical interaction between the two solutes. The calculations, while not trivial, can be performed with a suitable computer program. However, these calculations require an extensive database that can only be supplied by painstaking experimental measurements.

Rumpf *et al.* (1993, 1994; Bieling *et al.*, 1995) have also measured the solubilities of carbon dioxide and ammonia in aqueous solutions containing salts such as sodium sulfate, ammonium sulfate and ammonium chloride. Experimental results in the temperature range 313 K to 433 K and in the pressure range to 100 bar could again be correlated well with the ion-interaction model of Pitzer.

Coal-gasification and sweetening of natural gases often require removal of acid gases such as carbon dioxide and hydrogen sulfide from gaseous fuels. Such removal is best accomplished by absorption with aqueous alkanolamine solutions. Proper design of absorption equipment requires information on vapor-liquid equilibria, caloric effects and also on the kinetics of mass transfer and of chemical reactions. Because of chemical reactions and strong deviation from ideality in the liquid phase, it is not simple to model the thermodynamic behavior of aqueous mixtures containing alkanolamines and sour gases.

Numerous models (for a review see, e.g., Kohl and Riesenfeld, 1985) have been proposed to describe vapor-liquid equilibria for such systems. Useful models include that of Mather (Deshmukh and Mather, 1981, Xu *et al.*, 1992) and Kuranov *et al.* (1996). The latter, based on Pitzer's model, correlates the solubility of carbon dioxide in aqueous solutions containing *N*-methyldiethanolamine (MDEA).

Silkenbäumer *et al.* (1998) used a similar correlation for the solubility of carbon dioxide in aqueous solutions containing 2-amino-2-methyl-1-propanol (AMP) and the alkanolamines MDEA and AMP. Due to chemical reactions in the liquid phase, carbon dioxide dissolves in both neutral and (non-volatile) ionic forms. Figure 9-23 shows good agreement for calculated and experimental results for the solubility of carbon dioxide in 2.4 molal aqueous AMP solutions at three temperatures. At low loading α (moles of CO_2/moles of AMP), the total pressure of the solution is essentially the vapor pressure of water because all carbon dioxide is absorbed "chemically", i.e., in its non-volatile form. With increasing α, the total pressure rises, in particular when all AMP is "consumed" (at $\alpha = 1$) and more carbon dioxide can only be dissolved physically.

The good agreement between calculation and experiment is achieved only by taking into account all the chemical reactions possible in the liquid phase. In the system CO_2/AMP/H_2O, in addition to the solvent (water), 8 species are present: CO_2, RNH_2, RNH_3^+, $RNHCOO^-$, HCO_3^-, CO_3^{2-}, H^+, and OH^-.[41]

The model calculations require, at any temperature, the equilibrium constants of the pertinent chemical reactions, the activities a_i of all species present in the liquid phase, Henry's constant for the solubility of carbon dioxide in pure water, the vapor pressure, dielectric constant and molar volume of pure water, the partial molar volume of carbon dioxide at infinite dilution in water, and also information on vapor-phase non-ideality.

With equilibrium constants and thermodynamic properties available in the literature, Silkenbäumer calculated activity coefficients of molecular and ionic species

[41] Here R denotes the $HOCH_2C(CH_3)_2$-group in AMP (R-NH_2).

from Pitzer's model.[42] Using phase-equilibrium equations similar to Eq. (9-77) for water and to Eq. (9-78) for the solute carbon dioxide, the calculation procedure is similar to that of Rumpf *et al.* (1993, 1994). Making reasonable approximations, Silkenbäumer reduced the total number of adjustable model parameters to 6 binary and 2 ternary parameters. For the binary parameters, a dependence on temperature was taken into account, but for ternary parameters, it was neglected.

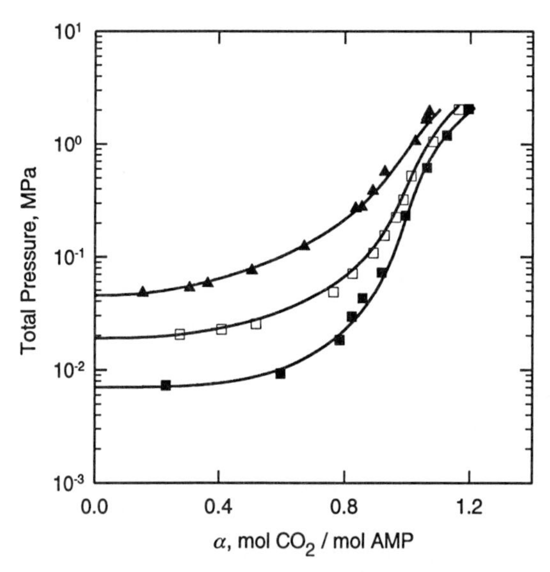

Figure 9-23 Solubility of carbon dioxide in aqueous solutions of 2-amino-2-methyl-1-propanol (AMP). Experiment: ▲ m_{AMP} = 2.44 mol kg^{-1} at 353.15 K; □ m_{AMP} = 2.45 mol kg^{-1} at 333.15 K; ■ m_{AMP} = 2.43 mol kg^{-1} at 313.15 K. —— Pitzer model. (Silkenbäumer *et al.*, 1998).

Figure 9-24 shows predicted molalities of the major molecular species present in the liquid phase as a function of the overall molality of carbon dioxide for a 2.43 molal AMP aqueous solution at 313.15 K. (Results for H^+, OH^- and the carbamate ion $RNHCOO^-$ are not shown because their molalities remain small compared to those of the other species).

As expected, adding carbon dioxide to an AMP-solution reduces the amount of neutral AMP, thereby producing mainly RNH_3^+, HCO_3^-, and CO_3^{2-}. As long as some AMP remains, the molality of molecular carbon dioxide is very small in comparison to the overall amount dissolved. Although it is small, it is very important because it is the concentration of the molecular carbon dioxide, $m_{CO_2,M}$, that mainly determines the total pressure. Only after the molality of neutral AMP approaches zero, does the molality of molecular carbon dioxide increase, along with a strong increase in total pressure.

[42] The equations for multi-electrolyte solutions are given in App. I.

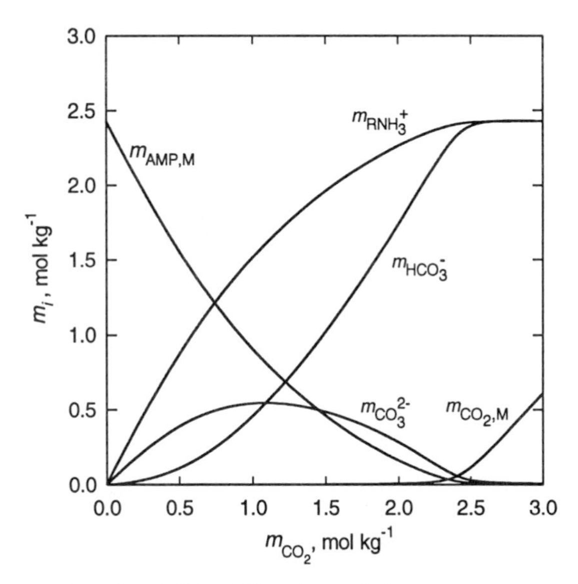

Figure 9-24 Predicted molalities (Pitzer's model) of major species present in aqueous 2.43 molal AMP solutions at 313.15 K as a function of the overall molality of CO_2 (Silkenbäumer *et al.*, 1998).

Silkenbäumer *et al.* (1998) also demonstrated that the solubility of carbon dioxide can be predicted reliably in solutions containing two alkanolamines, MDEA and AMP, using parameters of the Pitzer model obtained from reduction of experimental data for aqueous solutions containing only one of these alkanolamines.

To illustrate, Fig. 9-25 shows the solubility of carbon dioxide in solutions containing either one or both alkanolamines with about the same total molality. Predicted pressures for mixed-amine solutions agree well with experiment. Because AMP is a stronger base than MDEA, at a given total pressure the AMP-solution shows higher carbon dioxide loading than the MDEA-solution. Loadings for the two-amine solution fall in between.

The work described above shows that is possible but not easy to describe phase equilibria of aqueous systems containing weak electrolytes and other solutes that react with those electrolytes. The ion-interaction model of Pitzer is suitable for such calculations; however, the large number of adjustable parameters requires an extensive data base that can be established only by carefully performed experiments.

9.18 Protein Partitioning in Aqueous Two-Phase Systems

Separation of biologically active materials is an important operation in biotechnology. One useful separation process is provided by liquid-liquid extraction using an aqueous

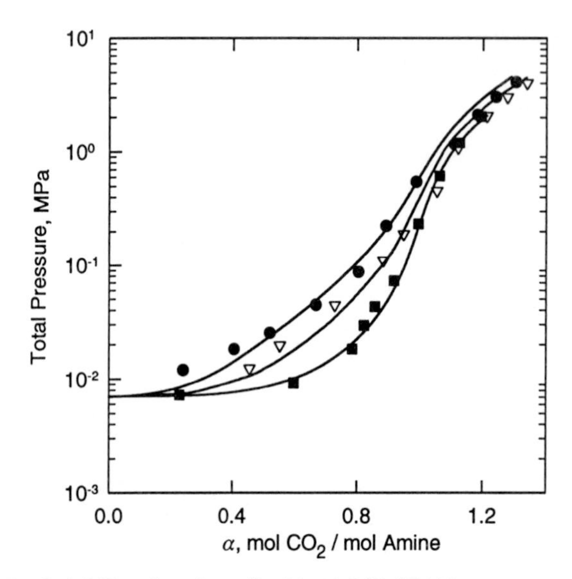

Figure 9-25 Solubility of carbon dioxide at 313.15 K in aqueous alkanolamine solutions. Results for the single-amine and double-amine solutions. Experiment: ● m_{MDEA} = 2.63 mol kg^{-1} (*N*-methyldiethanolamine); ∇ m_{MDEA} = 1.28 mol kg^{-1} + m_{AMP} = 1.27 mol kg^{-1}; ■ m_{AMP} = 2.43 mol kg^{-1} (2-amino-2-methylpropanol). —— Pitzer's model. (Silkenbäumer *et al.*, 1998).

two-phase polymer system formed when two water-soluble polymers (e.g. polyethyleneglycol (PEG) and dextran) are dissolved in excess water. This aqueous two-phase system contains mainly water, with the first polymer predominating in one phase and the second polymer predominating in the other phase as shown schematically in Fig. 9-26.

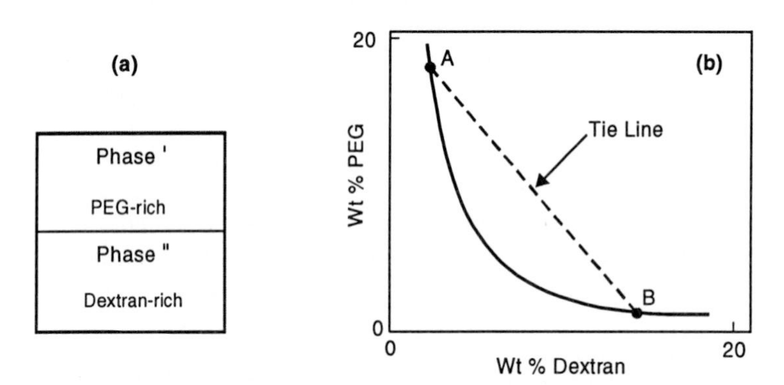

Figure 9-26 Schematics of a two-phase aqueous system H_2O+PEG+dextran for extraction. (a) Phase ': aqueous PEG-rich phase containing only a small amount of dextran; phase ": aqueous dextran-rich phase containing only a small amount of PEG; (b) Calculated phase diagram. Tie-line length is given by $\overline{AB}$, where A and B represent the equilibrium compositions of phases ' and ", respectively.

Due to their high water content, both equilibrium phases provide a suitable environment for biomacromolecules. When, for example, a mixture of proteins is added to a two-phase aqueous system, each type of protein partitions uniquely between the phases. Therefore, separation can be achieved with an extraction process as indicated in Fig. 9-27. To prevent denaturation of the biomacromolecules and to maintain pH control, small amounts of (buffer) salts may be added. A useful feature of such systems is that the partitioning of biomacromolecules between the two phases can be altered by changing the solution pH, ionic strength or the type of salt (electrolyte) added.

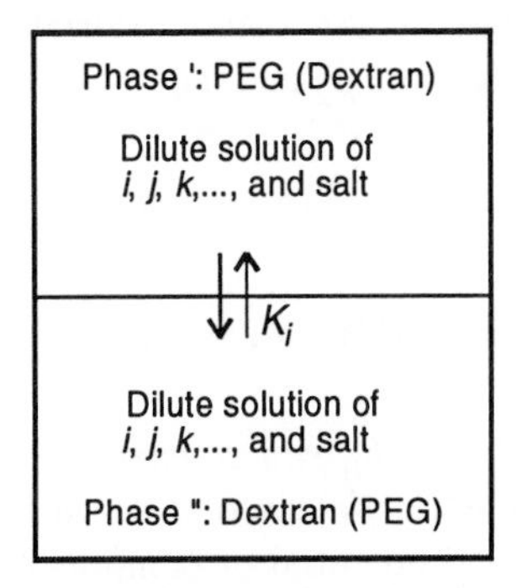

Figure 9-27 Separation of a mixture of proteins $i, j, k, \ldots$ by extraction: partition coefficient K_i = (concentration of protein in phase ')/(concentration of protein in phase ").

Because the compositions of the two aqueous phases are not the same, a salt partitions unequally between the two aqueous polymer phases. The difference in salt concentration establishes an electric-potential difference between the two phases; that difference can greatly affect the partitioning of charged biomacromolecules like proteins. Because the net surface charge of a protein depends on pH, a change in the pH of the solution can result in a significant change in the partitioning behavior of proteins.

While the electric-potential effect is often dominant, partitioning of biomacromolecules also depends on the properties of the phase-forming polymers.

There are many reports on the feasibility of aqueous two-phase systems for extractive biotechnical separations (Kula, 1979; Albertsson, 1986; Fisher and Sutherland, 1989; Prausnitz, 1989). For engineering design, quantitative information is required for thermodynamic properties, especially phase equilibria. Therefore, extensive experimental and theoretical studies have been directed at two-phase equilibria of such systems with and without partitioning biomacromolecules. Some representative articles are those by Curtis *et al.* (1998), Foster (1994), and Rothstein (1994). While details of these studies cannot be presented here, they all have the same objective: to provide a molecular-thermodynamic model for correlating and, perhaps, predicting the partition coefficients of biomacromolecules in aqueous two-phase systems.

The first task is to calculate the liquid-liquid phase diagram formed by water and the two water-soluble polymers in absence of salt or biomacromolecules. This calculation is achieved by expressing the chemical potential of all three components through an osmotic virial expansion[43] in the polymer concentrations; the coefficients in that expansion are best obtained from low-angle laser-light-scattering data (King *et al.*, 1988).

For equilibrium between phase ' and phase '',

$$\mu_1^{'} = \mu_1^{''} \qquad \mu_2^{'} = \mu_2^{''} \qquad \mu_3^{'} = \mu_3^{''} \tag{9-85}$$

where μ is the chemical potential; subscript 1 refers to water and subscripts 2 and 3 refer to the water-soluble polymers. An osmotic virial expansion truncated after the second term gives chemical potentials μ_2 and μ_3:

$$\mu_2 - \mu_2^{\otimes} = RT(\ln m_2 + b_{22}m_2 + b_{23}m_3) \tag{9-86}$$

$$\mu_3 - \mu_3^{\otimes} = RT(\ln m_3 + b_{33}m_3 + b_{23}m_2) \tag{9-87}$$

where m_i is the molality of solute i, b_{ij} is a constant characterizing the interaction between a molecule of polymer i and a molecule of polymer j in the aqueous solvent, and $\mu_i^{\otimes}$ is the standard-state chemical potential of component i (hypothetical state of ideal solution at unit molality). An expression for the chemical potential μ_1 of water is obtained from the Gibbs-Duhem equation,

$$\sum_{i=1}^{3} n_i d\mu_i = 0 \tag{9-88}$$

where n_i is the number of moles of component i. With Eq. (9-86) for μ_2 and Eq. (9-87) for μ_3, the chemical potential μ_1 is

$$\mu_1 - \mu_1^0 = \frac{RTM_1}{1000}\left(m_2 + m_3 + \frac{b_{22}}{2}m_2^2 + \frac{b_{33}}{2}m_3^2 + b_{23}m_2m_3 \right) \tag{9-89}$$

where M_1 is the molar mass of water. In Eq. (9-89) the standard state is pure water at system temperature T.

Interaction parameters b_{22}, b_{33}, and b_{23} are directly related to osmotic second virial coefficients B_{22}^*, B_{33}^*, and B_{23}^* by

$$b_{22} = \frac{2M_2^2 B_{22}^*}{1000^2} \tag{9-90}$$

[43] The osmotic virial expansion is discussed in Sec. 4.11.

$$b_{33} = \frac{2M_3^2 B_{33}^*}{1000^2} \tag{9-91}$$

$$b_{23} = \frac{2M_2 M_3 B_{23}^*}{1000^2} \tag{9-92}$$

where M_i is the molar mass.

With virial coefficients obtained from low-angle laser-light-scattering measurements, it is possible to generate a reliable phase diagram, as shown schematically in Fig. 9-26(b).

Consider now a protein component (subscript 4) distributed between the two aqueous phases. The distribution coefficient K is defined by

$$K_4 = \frac{\text{Concentration of protein in phase '}}{\text{Concentration of protein in phase ''}} \tag{9-93}$$

Depending on pH, the protein may be electrically charged and therefore the presence of ions (salts) must be taken into account. When all proteins in the system are dilute, the distribution coefficient for a particular protein is given by (Haynes *et al.*, 1993),

$$\ln K_4 = \ln \frac{\gamma_4''}{\gamma_4'} + \frac{F z_4 (\phi'' - \phi')}{RT} \tag{9-94}$$

Here F is Faraday's constant, ϕ the electric potential, z the electric charge and γ the chemical activity coefficient, i.e., the coefficient in the absence of electrostatic effects caused by unequal distribution of ions between phase ' and phase ''.

Activity coefficients are found from the osmotic virial expansion with coefficients $B_{22}^*, B_{33}^*, B_{23}^*, B_{24}^*, \text{and } B_{34}^*$; additional terms for protein-salt and polymer-salt interactions are obtained from osmometric data (Haynes *et al.*, 1989).

The electric-potential difference $\Delta\phi = \phi'' - \phi'$ between the two phases arises as the result of the addition of a salt that fully dissociates into ν_+ cations of charge z_+ and ν_- anions of charge z_- but that does not partition equally between the two phases. A direct relation between $\Delta\phi$ and measurable equilibrium properties of the two-phase system can be established through application of quasi-electrostatic potential theory. Applying quasi-electrostatic potential theory gives the relation (Haynes *et al.*, 1991)

$$\Delta\phi = (\phi'' - \phi') = \frac{RT}{(z_+ - z_-)F} \ln \left[\frac{(\gamma_-'' / \gamma_-')}{(\gamma_+'' / \gamma_+')^{z_+ / z_-}} \right] \tag{9-95}$$

where γ_+ and γ_- are the activity coefficients of the cation and anion, respectively. The importance of Eq. (9-95) can be seen more clearly by applying it to the description of a two-phase system at equilibrium containing a 1:1 electrolyte, i.e., $z_+ / z_- = -1$ and $z_+ - z_- = 2$. In this case, Eq. (9-95) reduces to

$$(\phi''-\phi') = \frac{RT}{2F}\ln\left(\frac{\gamma''_-\gamma''_+}{\gamma'_-\gamma'_+}\right) = \frac{RT}{F}\ln\left(\frac{\gamma''_\pm}{\gamma'_\pm}\right) = \frac{RT}{F}\ln(K_s) \tag{9-95a}$$

where K_s is the partition coefficient of the salt and, as usual,

$$\gamma_\pm^{\nu} = \gamma_+^{\nu_+}\gamma_-^{\nu_-} \tag{9-96}$$

Here, $\gamma_\pm$ is the mean ionic activity coefficient for the neutral salt and $\nu = \nu_+ + \nu_-$. Mean ionic activity coefficients are tabulated for most strong electrolytes in water at 25°C; for those salts or temperatures where such data are unavailable, $\gamma_\pm$ can often be estimated using Pitzer's ion-interaction model discussed in Sec. 9.14. The last equality in Eq. (9-95a) holds because, at equilibrium,

$$m'_s\gamma'_\pm = m''_s\gamma''_\pm \tag{9-97}$$

where m_s is the molality of the salt. Equation (9-95a), provides a means for directly calculating the electric-potential difference $\Delta\phi$ from the equilibrium properties of a two-phase system that then can be used for calculating K_4 with Eq. (9-94).

In typical cases, the potential difference is small, perhaps a few millivolts. Nevertheless, that small potential difference can have a large effect. In some cases, the effect of the potential difference on K_4 is dominant, much more important than that of the chemical activity coefficients.

Figure 9-28 compares calculated and experimental partition coefficients for three proteins: albumin, chymotrypsin and lysozyme (Haynes *et al.*, 1991). The horizontal axis of Fig. 9-28 is the tie-line length [see Fig. (9-26)] that provides a measure of how different phase ' is from phase "; when the tie-line length is zero, the two phases are identical.

The molecular-thermodynamic analysis shows that when the protein is charged, the influence of the electric potential is often decisive. This analysis, therefore, suggests that enhanced partition coefficients for a protein could be obtained by raising the asymmetry of partition of the salt. One method for doing so is provided by adding to the two-phase system a very small amount of α-cyclodextrin and a salt whose anion is strongly absorbed by that cyclodextrin. Figure 9-29 shows the dramatic effect on the partition coefficient of chymotrypsin when the salt is KI (Haynes *et al.*, 1991). Because the anion (iodide) is bound by α-cyclodextrin and because α-cyclodextrin is predominantly in the aqueous dextran-rich phase, the salt KI partitions toward that phase. The more asymmetric the partitioning of salt, the larger $\Delta\phi$.

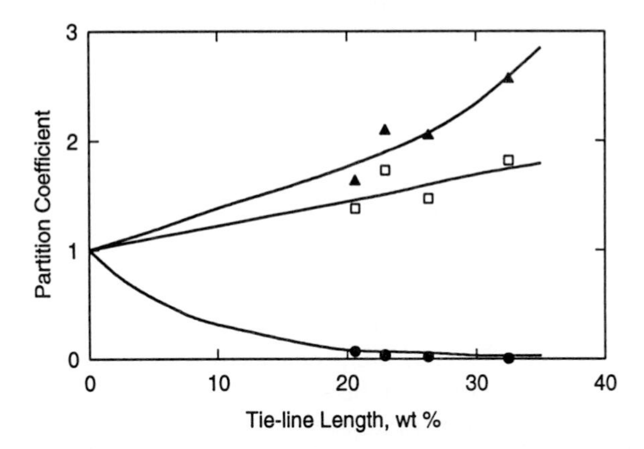

Figure 9-28 Predicted and experimental partition coefficients for a dilute protein mixture in an aqueous two-phase system containing PEG 3350, dextran T-70, and 50 mM KCl (overall) at pH = 7.5 and 25°C. *d*: diameter (Å); *z*: electric charge; ● : albumin (d = 78; z = -8); ■ : chymotrypsin (d = 34; z = 2); ▲ : lysozyme (d = 22; z = 7). Lines are calculated using second osmotic virial coefficients obtained from single-phase light-scattering measurements.

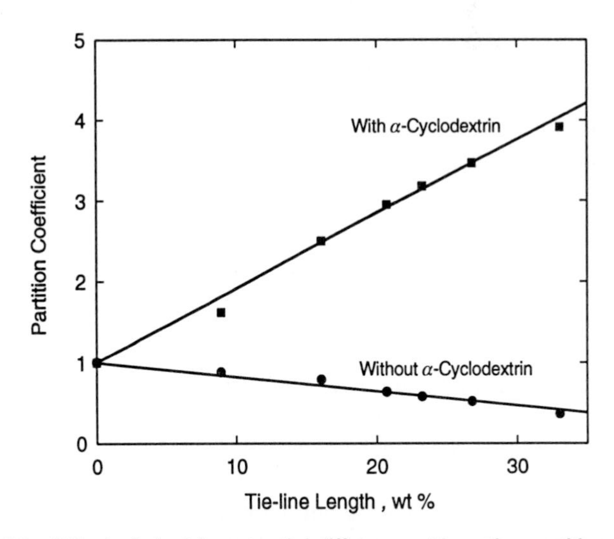

Figure 9-29 Effect of electric-potential difference $\Delta\phi$ on the partition coefficient for chymotrypsin in PEG 3350/dextran T-70/water. z = 5; pH = 7.3; 25°C. ● : 1 mM KI; ■ : 1 mM KI + 1 mM α-cyclodextrin. Tie-line length is not affected by α-cyclodextrin.

We have here a striking example to indicate how molecular thermodynamics can benefit process design. Thermodynamic analysis showed the unexpectedly large influence of salt partitioning on the partition coefficient of a charged protein. Once that influence was identified, the sharpness of separation could be significantly improved by enhancing the electric-potential difference between phase ' and phase " through the addition of a suitably chosen entrainer for "pulling" a salt preferentially into one of the two aqueous phases.

9.19 Summary

The thermodynamics of solutions containing electrolytes is not a simple extension of the thermodynamics of solutions containing nonelectrolytes. Electrolyte solutions require a much more elaborate framework because, in a solvent of high dielectric constant (e.g. water), an electrolyte splits into two or more ions. Therefore, a binary solution, e.g. sodium chloride in water, is in some sense a ternary solution containing water, a cation and an anion. But in another sense, it is not a ternary solution because the concentrations of cation and anion are not two independent variables; if one is fixed, so is the other because of material balance and the requirement of electroneutrality. Because *ionization* and the constraint of *electroneutrality* must be taken into account in the thermodynamics of electrolyte solutions, but not in the thermodynamics of nonelectrolyte solutions, the framework required for electrolyte solutions is necessarily much more elaborate.

For *strong electrolytes* (salts), ionization is usually complete but for *weak electrolytes* (e.g. acetic acid), ionization is only partial. In that event, to obtain a useful thermodynamic description, it is necessary to consider chemical equilibria in addition to one or more material balances and to the constraint of electroneutrality.

To keep this chapter from excessive length, attention is given primarily to solutions containing only one solvent and one solute. Only brief attention is given to multicomponent systems.

Electrolyte-solution thermodynamics often uses *concentration scales* that are different from those used in nonelectrolyte-solution thermodynamics. In the latter, the most common scales are mole fractions or volume fractions that have a desirable characteristic: they vary from zero to unity. For electrolyte solutions, the most popular concentration scale is *molality* (moles of solute per kg of solvent). Molality, however, can go from zero to infinity. Nevertheless, molality is often used because, for most applications, attention is restricted to dilute or moderately concentrated solutions where molality may go as high as 10 or 20 but, with rare exceptions, no higher.

At ordinary temperatures, most strong electrolytes (salts) are solids, not liquids. Therefore, in most cases, activity coefficients in electrolyte solutions are normalized according to the *unsymmetric convention* where the activity coefficients of both solute and solvent go to unity at infinite dilution. This normalization requires that the standard state of the solute be defined not as the pure solute at solution temperature but instead, as an ideal *dilute* solution at *some fixed concentration* at solution temperature. That fixed concentration is conveniently chosen as unit concentration. It is a common error to confuse that standard state with the ideal dilute solution at infinite dilution. The ideal dilute solution at infinite dilution cannot be used as the standard state because the chemical potential of a solute at infinite dilution is minus infinity. If molality is the concentration scale, the standard state for the solute is the ideal dilute solution at system temperature when the solute molality is unity. While the correct standard state for the solute is hypothetical, its properties are experimentally accessible.

In nonelectrolyte solutions of liquid components, no distinction need be made between solute and solvent; they are described with the symmetric convention, as discussed in Chap. 6. For nonelectrolyte solutions of ordinary liquids, the reference system is one that (essentially) obeys Raoult's law and therefore, activity coefficients indicate deviations from ideality as given by Raoult's law. However, for solutions of electrolytes, activity coefficients indicate deviations from ideality as given by Henry's law.

In a solution of electrolytes, it is useful to think of electrolyte activity as analogous to pressure in the virial equation for gases: in both cases, solute molecules are dissolved in a medium. For a gas, the medium is a vacuum. For ions, the medium is a liquid solvent, typically water. In a gas, as in a solution of ions, we are concerned with interactions between solute molecules that "swim" in a medium.

By contrast, when we consider solutions of nonelectrolytes with activity coefficients normalized symmetrically, we are concerned with interactions between solute and solvent molecules. Therefore, the excess Gibbs energy for a solution of nonelectrolyte liquids is necessarily different from that for a solution of electrolyte (ions) in a solvent.

However, the analogy of an electrolyte solution to a gas has a serious deficiency. At ordinary temperatures, a typical gas consists of molecules with no net electric charge. However, ions, by definition, are electrically charged. Intermolecular forces between charged particles are much different from those between uncharged particles, as indicated in Chap. 4. Interactions between uncharged molecules are short-range, whereas those between charged molecules are long-range. Therefore, for nonelectrolyte gases, the ideal-gas law provides a fair approximation even when the concentration (density) reaches say, 5 or 10% of the maximum possible concentration. However, for a solution of ions, deviations from ideality become appreciable at less than 1% of the maximum possible concentration. When two ions are say, 5 diameters apart, there is a strong interaction between them. However, when two uncharged molecules are 5 diameters apart, one molecule hardly knows that the other one is there.

This chapter presents some equations for activity coefficients of ions dissolved in a solvent. However, because of electroneutrality, cations and anions always appear together; while theoretical or semi-empirical equations can be written separately for anions or cations, in a real experiment, it is generally not possible to measure ionic activity coefficients separately, unless we use special techniques, often based on questionable assumptions. In a real situation, we usually measure a *mean ionic activity coefficient*; for a 1-1 electrolyte, this mean activity coefficient is related to the individual ionic activity coefficients by $\gamma_{\pm} = (\gamma_{+}\gamma_{-})^{1/2}$.

In a medium of fixed density, temperature and dielectric constant, at very high dilution, when ions are (on the average) far apart, γ_{+} and γ_{-} are independent of the chemical nature of the ions; they depend only on the ionic valences and on the ionic concentration as expressed by the ionic strength. This result from the *Debye-Hückel theory* is physically reasonable because at large distances of separation, the only important forces between ions are those due to charge-charge interactions; all other intermolecular forces (induction, dispersion, repulsion due to overlap of ionic diameters) are then insignificant. It is therefore not surprising that, for example, in water at 25°C, the mean ionic activity coefficient of 0.01 molar potassium chloride is the same as that of 0.01 molar sodium nitrate.

As the ionic concentration rises, short-range intermolecular forces also become important; indeed, at high ionic concentrations (typically 1 molar and beyond), short-range forces become dominant relative to long-range forces. Several methods have been proposed for taking short-range forces into account; most of these do no more than add terms to the Debye-Hückel expression in such a manner that the added terms tend to disappear at very low ionic concentration where the Debye-Hückel expression must be recovered in the limit. One effective method, initiated many years ago by Guggenheim, uses a power series in electrolyte concentration. This method, systematically developed by Pitzer, gives excellent results and can be extended in a logical manner to multi-salt solutions; however, it requires a large number of parameters that can only be obtained from an extensive data base. Other methods, with fewer parameters, are based on one of the local-concentration models discussed in Chapters 6 and 7. Accuracy is generally not as good as that obtained using *Pitzer's equation* and – depending on model details – extension to multisolute solutions may require some doubtful assumptions.

As in the thermodynamics of nonelectrolytes, it is also possible to describe the effect of short-range interactions through a "chemical" theory that considers ions to be solvated (if water is the solvent, we say hydrated); as ion concentration rises, thereby increasing short-range ion-ion interactions, the extent of solvation declines. The *chemical theory* appears to be particularly useful for describing properties of electrolyte solutions over the entire concentration range when the electrolyte is highly soluble in the solvent.

In addition to the few semi-empirical methods mentioned in this chapter, the scientific literature is rich in theoretical methods based, for example, on the integral-equation theory of fluids, the theory of fluctuations and Monte-Carlo simulations. These theoretical studies are not discussed here in part, because they are beyond the scope of this book and in part, because, as yet, they have only limited utility for typical practical calculations in the applied chemical sciences.

While chemists, chemical engineers and related professionals have given some attention to mixtures containing two (or more) electrolytes in one solvent, little fundamental attention has been given to the thermodynamics of solutions containing one (or more) electrolytes in two (or more) solvents. The literature reports many experimental vapor-liquid-equilibrium (VLE) data for two miscible solvents saturated with an electrolyte but there is little fundamental thermodynamic analysis. More difficult is the case of liquid-liquid equilibria for two partially miscible solvents saturated with an electrolyte because in that case (unlike for VLE) the electrolyte is present in *both* fluid phases.

Finally, there is a fundamental theoretical problem for reconciling a fundamental inconsistency between the thermodynamic framework used for nonelectrolyte solutions (*Lewis-Randall framework*) with that used for electrolyte solutions (*McMillan-Mayer framework*). In principle, this problem has been solved by Friedman but his papers do not make easy reading. Studies by O'Connell and others indicate that, for many cases, the consequences of this theoretical problem are of little importance for single-solvent systems.

This chapter on electrolyte solutions has provided only a short introduction. Electrolyte solutions are of interest in many fields: electrochemistry, geology, biochemistry, metallurgy, electrical engineering, material science, physiology, and more. The technical literature is vast. The main point of this summary is a reminder: the thermodynamics of electrolyte solutions is not a minor extension of the thermodynamics of nonelectrolyte solutions; it is a science of its own. It has much overlap with nonelectrolyte-solution thermodynamics and there are many similarities but the differences are far from trivial.

References

Albertsson, P.-Å, 1986, *Partition of Cell Particles and Macromolecules,* 3rd Ed. New York: Wiley-Interscience.

Barthel, J. M. G., H. Krienke, and W. Kunz, 1998, *Physical Chemistry of Electrolyte Solutions.* Berlin: Springer.

Bieling, V., B. Rumpf, F. Strepp, and G. Maurer, 1989, *Fluid Phase Equilibria,* 53: 251.

Bieling, V., F. Kurz, B. Rumpf, and G. Maurer, 1995, *Ind. Eng. Chem. Res.*, 34: 1449.

Blum, L., 1980. In *Theoretical Chemistry: Advances and Perspectives*, (H. Eyring and D. Henderson, Eds.), Vol. 5, Chap. 1. New York: Academic Press.

Burns, J. A. and W. F. Furter, 1976. In *Thermodynamic Behavior of Electrolytes in Mixed Solvents*, (W. F. Furter, Ed.), Chap. 8, Adv. Chem. Series 155. Washington: American Chemical Society.

Cabezas, H. and J. P. O'Connell, 1993, *Ind. Eng. Chem. Res.*, 32: 2892.

Cardoso, M. J. E. and J. P. O'Connell, 1987, *Fluid Phase Equilibria,* 33: 315.

Chen, C.-C., H. I. Britt, J. F. Boston, and L. B. Evans, 1982, *AIChE J.,* 28: 588.

Chen, C.-C. and L. B. Evans, 1986, *AIChE J.*, 32: 444.

Clegg, S. L. and M. Whitfield, 1991. In *Activity Coefficients in Electrolyte Solutions*, (K. S. Pitzer, Ed.), 2nd Ed., Chap. 6. Boca Raton: CRC Press.

Cruz, J. and H. Renon, 1978, *AIChE J.*, 24: 817.

Curtis, R. A., J. M. Prausnitz, and H. W. Blanch, 1998, *Biotechn. Bioeng.*, 57: 11.

Deshmukh, R. D. and A. E. Mather, 1981, *Chem. Eng. Sci.*, 36: 355.

Edwards, T. J., G. Maurer, J. Newman, and J. M. Prausnitz, 1978, *AIChE J.*, 24: 966.

Fisher, D. and I. A. Sutherland (Eds.), 1989, *Separations Using Aqueous Phase Systems. Application in Cell Biology and Biotechnology.* New York: Plenum Press.

Foster, P. R., 1994. In *Engineering Processes for Bioseparations,* (L. R. Weatherley, Ed.). Oxford: Butterworth-Heinemann.

Friedman, H. L., 1972, *J. Solution Chem., 1,* 387, 413, 419.

Furter, W. F. and R. A. Cook, 1967, *Int. J. Heat Mass Transfer,* 10: 23.

Furter, W. F., 1977, *Can. J. Chem. Eng.*, 55: 229.

Gering, K. L. and L. L. Lee, and L. H. Landis, 1989, *Fluid Phase Equilibria*, 48: 111.

Grigera, J. R., 1992, *Life Sciences,* 50: 1567.

Haghtalab, A. and J. H. Vera, 1988, *AIChE J., 34,* 803.

Hamer, W. J. and Y.-C. Wu, 1972, *J. Phys. Chem. Ref. Data,* 1: 1047.

Harvey, A. H. and J. M. Prausnitz, 1989, *AIChE J.*, 35: 635.

Harvie, C. E. and J. H. Weare, 1980, *Geochim. Cosmochim. Acta,* 44: 981.

Harvie, C. E., H. P. Eugster, and J. H. Weare, 1982, *Geochim. Cosmochim. Acta,* 46: 1603.

Harvie, C. E., J. P. Greenberg, and J. H. Weare, 1987, *Geochim. Cosmochim. Acta,* 51: 1045.

Haynes, C. A., H. W. Blanch, and J. M. Prausnitz, 1989, *Fluid Phase Equilibria,* 53: 463.

Haynes, C. A., J. Carson, H. W. Blanch, and J. M. Prausnitz, 1991, *AIChE J.*, 37: 1401.

Haynes, C. A., F. J. Benitez, H. W. Blanch, and J. M. Prausnitz, 1993, *AIChE J.*, 39: 1539.

Haynes, C. A. and J. Newman, 1998, *Fluid Phase Equilibria,* 145: 255.

Henderson , D., L. Blum, and A. Tani, 1986, *ACS Adv. Chem. Ser.,* 13: 281.

Iliuta, M. C. and F. C. Thyrion, 1995, *Fluid Phase Equilibria,* 103: 257.

Jin, G. and M. D. Donohue, 1988, *Ind. Eng. Chem. Res.*, 27: 1073.

Jin, G. and M. D. Donohue, 1988a, *Ind. Eng. Chem. Res.*, 27: 1737.

Jin, G. and M. D. Donohue, 1991, *Ind. Eng. Chem. Res.*, 30: 240.

Johnson, A. I. and W. F. Furter, 1960, *Can. J. Chem. Eng.*, 38: 78.

Kula, M.-R., 1979, *Appl. Biochem. Bioeng.*, 2: 71.

Kim, H.-T. and W. J. Frederick , Jr., 1988, *J. Chem. Eng. Data,* 33: 177.

Kim, H.-T. and W. J. Frederick , Jr., 1988a, *J. Chem. Eng. Data,* 33: 278.

King, R. S., H. W. Blanch, and J. M. Prausnitz, 1988, *AIChE J.*, 34: 1585.

Kohl, A. L. and F. C. Riesenfeld, 1985, *Gas Purification,* 4th Ed. Houston: Gulf Publ. Co.

Krishnan, C. V. and H. L. Friedman, 1974, *J. Solution Chem.*, 3: 727.

Kuranov, G., B. Rumpf, N. A. Smirnova, and G. Maurer, 1996, *Ind. Eng. Chem. Res.*, 35: 1959.

Kurz, F., B. Rumpf, and G. Maurer, 1995, *Fluid Phase Equilibria,* 104: 261.

Linke, W. F. and A. Seidell, 1958, 1965, *Solubility of Inorganic and Metal-Organic Compounds.* Vol. I (Princeton: D. Van Nostrand). Vol. II (Washington: American Chemical Society).

Liu, Y., A. H. Harvey, and J. M. Prausnitz, 1989, *Chem. Eng. Comm.*, 77: 43.

Liu, Y., M. Wimby, and U. Grén, 1989a, *Computers Chem. Eng.,* 13: 405.

Liu, Y. and U. Grén, 1991, *Fluid Phase Equilibria,* 63: 49.

Lu, X. and G. Maurer, 1993, *AIChE J.,* 39: 1527.

Lu, X., L. Zhang, Y. Wang, J. Shi, and G. Maurer, 1996, *Ind. Eng. Chem. Res.*, 35: 1777.

Macedo, E., P. Skovborg, and P. Rasmussen, 1990, *Chem. Eng. Sci.,* 45: 875.

Maurer, G., 1983, *Fluid Phase Equilibria,* 13: 269.

Mazo, R. M. and C. Y. Mou, 1991. In *Activity Coefficients in Electrolyte Solutions*, (K. S. Pitzer Ed.), 2nd Ed., Chap 2. Boca Raton: CRC Press.

Messikomer, E. E. and R. H. Wood, 1975, *J. Chem. Thermodynamics,* 7: 119.

Millero, F. J., 1977. In *Activity Coefficients in Electrolyte Solutions*, Vol. II, (R. M. Pytkowicz, Ed.), Chap. 2, Boca Raton: CRC Press.

Mock, B., L. B. Evans, and C.-C. Chen, 1986, *AIChE J.,* 32: 1655.

Nakamura, R., G. J. F. Breedveld, and J. M. Prausnitz, 1976, *Ind. Eng. Chem. Proc. Des. Dev.,* 15: 557.

Newman, J. S., 1991, *Electrochemical Systems*, 2nd Ed. Englewood Cliffs: Prentice-Hall.

Ohe, S., 1976. In *Thermodynamic Behavior of Electrolytes in Mixed Solvents*, (W. F. Furter, Ed.), Chap. 5, Adv. Chem. Series 155. Washington: American Chemical Society.

Ohe, S., 1991, *Vapor-Liquid Equilibrium - Salt Effect.* Amsterdam: Elsevier.

Ohtaki, H. and H. Yamatera (Eds.), 1992, *Structure and Dynamic of Solutions*, Chaps. 3 and 4. Amsterdam: Elsevier.

Pailthorpe, B. A., D. J. Mitchell, and B. W. Ninham, 1984, *J. Chem. Soc. Faraday Trans. II,* 80: 115.

Panayiotou, C. and J. H. Vera, 1980, *Fluid Phase Equilibria,* 5: 55.

Picker, C., P.-A. Leduc, P. R. Philip, and J. E. Desnoyers, 1971, *J. Chem. Thermodynamics,* 3: 631.

Pitzer, K. S., 1973, *J. Phys. Chem.,* 77: 268.

Pitzer, K. S. and G. Mayorga, 1973, *J. Phys. Chem.,* 77: 2300.

Pitzer, K. S. and G. Mayorga, 1974, *J. Phys. Chem.,* 3: 539.

Pitzer, K. S., 1980, *J. Am. Chem. Soc.,* 102: 2902.

Pitzer, K. S., J. C. Peiper, and R. H. Busey, 1984, *J. Phys. Chem. Ref. Data,* 13: 1.

Pitzer, K. S. (Ed.), 1991, *Activity Coefficients in Electrolyte Solutions*, 2nd Ed. Boca Raton: CRC Press.

Pitzer, K. S., 1991a. In *Activity Coefficients in Electrolyte Solutions*, (K. S. Pitzer, Ed.), 2nd Ed., Chap. 3. Boca Raton: CRC Press.

Pitzer, K. S., 1995, *Thermodynamics*, 3rd Ed., Apps. 7, 8 and 10. New York: McGraw-Hill.

Prausnitz, J. M., 1989, *Fluid Phase Equilibria,* 53: 439.

Rafal, M., J. W. Berthold, N. C. Scrivner, and S. L. Grise, 1994, *Models for Electrolyte Solutions.* In *Models for Thermodynamic and Phase-Equilibria Calculations*, (S. I. Sandler, Ed.). New York: Marcel Dekker.

Ramalho, R. S., W. James, and J. F. Carnaham, 1964, *J. Chem. Eng. Data,* 9: 215.

Rard, J. A. and R. F. Platford, 1991. In *Activity Coefficients in Electrolyte Solutions*, (K. S. Pitzer, Ed.), 2nd Ed., Chap. 5. Boca Raton: CRC Press.

Renon, H., 1986, *Fluid Phase Equilibria,* 30: 181.

Robinson, R. A. and R. H. Stokes, 1970, *Electrolyte Solutions*, 2nd Ed. London: Butterworths.

Rothstein, F., 1994. In *Protein Purification Process Engineering,* (R. G. Harrison, Ed.). New York: Marcel Dekker.

Rumpf, B. and G. Maurer, 1993, *Ber. Bunsenges. Phys. Chem.,* 97: 85.

Rumpf, B., F. Weyrich, and G. Maurer, 1993a, *Fluid Phase Equilibria,* 83: 253.

Rumpf, B., H. Nicolaisen, and G. Maurer, 1994, *Ber. Bunsenges. Phys. Chem.,* 98: 1077.

Sander, B., A. Fredenslund, and P. Rasmussen, 1986, *Chem. Eng. Sci.,* 41: 1171.

Schönert, H., 1986, *Z. Phys. Chem.,* 150: 163.

Setchenov, J., 1889, *Z. Phys. Chem.,* 4: 117.

Silkenbäumer, D., B. Rumpf and R. N. Lichtenthaler, 1998, *Ind. Eng. Chem. Res.* (in press).

Staples, B. R. and R. L. Nuttall, 1977, *J. Phys. Chem. Ref. Data,* 6: 385.

Stokes, R. H. and R. A. Robinson, 1973, *J. Solution Chem.,* 2: 173.

Waisman, E. and J. L. Lebowitz, 1970, *J. Chem. Phys.,* 52: 4307.

Weare, J. H., 1987, *Rev. Mineral.,* 17: 143.

Wu, R.-S. and L. L. Lee, 1992, *Fluid Phase Equilibria,* 78: 1.

Xu, S., Y.-W. Wang, F. D. Otto, and A. E. Mather, 1992, *Chem. Eng. Proc.,* 31: 7.

Zemaitis, J. F., Jr., D. M. Clark, M. Rafal, and N. C. Scrivner, 1986, *Handbook of Aqueous Electrolyte Thermodynamics.* New York: AIChE.

Zerres, H. and J. M. Prausnitz, 1994, *AIChE J.,* 40: 676.

Problems

1. As determined from emf measurements, the solubility product constant at 25°C of AgCl is $K_{sp} = 1.72\times10^{-10}$ (molal units).

(a) Find the solubility (in mol kg^{-1}) of AgCl in pure water.

(b) If sufficient NaCl is added to the system to form a 0.01 molal solution of NaCl, what is the solubility of NaCl?

(c) What is the solubility of AgCl in a 0.01 molal solution of $NaNO_3$?

2. At 25°C, the solubility of PbI_2 in water is 1.66×10^{-3} mol/kg of water. At the same temperature, what is the solubility of PbI_2 in an aqueous 0.01 molal solution of KI?

3. At 25°C, the solubility of PbI_2 in water is 1.66×10^{-3} mol kg^{-1}, in a 0.01 molal NaCl aqeous solution is 1.86×10^{-3} mol kg^{-1}, and in a 0.01 molal KI aqueous solution is 2.80×10^{-4} mol kg^{-1}. Explain. For these dilute solutions, use the Debye-Hückel limiting law.

4. Acetic acid is a weak electrolyte. Determine the fraction ionized for a 10^{-3} molal aqueous solution at 25°C.

At 25°C, the equilibrium constant is $K = 1.758\times10^{-5}$.

5. Calculate the Debye length of 0.001 M and 0.1 M NaCl solutions at 25°C in:
(a) Water ($\varepsilon_r = 78.4$).
(b) Methanol ($\varepsilon_r = 31.5$)

6. Consider seawater with 3.5 weight % NaCl at 25°C. The density of pure water at 25°C is 0.997 g cm^{-3}.
(a) Calculate the molal osmotic coefficient.
(b) Compute the osmotic coefficient. Compare your result with those listed in Perry (Chemical Engineers Handbook) for osmotic pressures of aqueous sodium chloride solutions at 25°C: 27.12 atm for m_{NaCl} = 0.60 mol kg^{-1} and 0.80 atm for m_{NaCl} = 0.80 mol kg^{-1}.
To obtain $\gamma_\pm$ use Bromley's model:

$$\ln\gamma_\pm = -\frac{A_\gamma I^{1/2}}{1+I^{1/2}} + \frac{(0.138+1.38B)I}{(1+1.5I)^2} + 2.303BI$$

where, for NaCl aqueous solutions at 25°C, A_γ = 1.174 mol$^{1/2}$ kg$^{-1/2}$ and B = 0.0574.

7. For a 0.12 molal K_2SO_4 solution at 25°C, the experimental mean ionic activity coefficient $\gamma_\pm^{(m)}$ is 0.40. Estimate the equilibrium pressure of water above 1.33 molal solution of K_2SO_4 at 25°C.
The vapor pressure of pure water is 0.0317 bar at 25°C.
For a moderately concentrated electrolyte solution, the extended Debye-Hückel equation is:

$$\ln\gamma_\pm^{(m)} = -\frac{A_\gamma |z_+z_-| I^{1/2}}{1+BaI^{1/2}} + bm$$

For water at 25°C, A_γ = 1.174 mol$^{1/2}$ kg$^{-1/2}$ and B = 0.33 mol$^{1/2}$ kg$^{-1/2}$ Å^{-1}. Parameters a and b are specific to K_2SO_4.
When the activity coefficient is given by the above equation, the osmotic coefficient for water, as obtained from the Gibbs-Duhem relation, is:

$$\phi = 1 - \frac{A_\gamma}{3}|z_+z_-| I^{1/2}\sigma(BaI^{1/2}) + \frac{bm}{2}$$

where $\sigma(y)$ is the function:

$$\sigma(y) = \frac{3}{y^3}[1 + y - 2\ln(1+y) - \frac{1}{1+y}]$$

8. An osmometer at 25 °C has two chambers separated by a semi-permeable membrane. One chamber contains 1 M aqueous sodium chloride. The other chamber contains an aqueous solution of bovine serum albumin (BSA) and sodium chloride at pH 7.4; BSA concentra-

tion is 44.6 g L^{-1} and sodium chloride concentration is 1 M. The measured osmotic pressure is 224 mmH_2O.
What is the osmotic second virial coefficient of BSA in this solution? The molar mass of BSA is 66,000 g mol^{-1}. The semi-permeable membrane has a cut-off at molecular weight 10,000. At pH 7.4, the electric charge on BSA is -20. Indicate all simplifying assumptions.

9. Using ion-selective electrodes, Khoshkbarchi and Vera (1996, *AIChE J.*, 42: 249) measured activity coefficients of individual ions in aqueous sodium bromide solutions at 25°C, which then used to obtain mean ionic activity coefficients. These were correlated with a truncated Pitzer equation,

$$\ln \gamma_{\pm} = \frac{-A_x |z_M z_X| \sqrt{I_x} + B_{\pm} I_x^{3/2}}{1 + \rho \sqrt{I_x}}$$

where A_x = 8.766 is the Debye-Hückel constant, ρ = 9 and I_x is the ionic strength expressed in terms of mole fractions. For aqueous solutions of NaBr to 5 molal, they obtained $B_{\pm}$ = 124.598.

10. For NaBr solutions at 25°C with compositions between $m_{NaBr} = 0$ and 5 mol kg^{-1}:
(a) Calculate the activity coefficients of water.
(b) Plot $\gamma_{\pm}$ as a function of m_{NaBr} using Pitzer equation and Debye-Hückel equation.
(c) Obtain the osmotic pressures from van't Hoff equation and determine the range of its validity, i.e., the range of composition where the effect of solution non-ideality can be neglected.
At 25°C, the mass density of pure water is 0.997 g cm^{-3} and those of NaBr aqueous solutions are given by $d(\text{g cm}^{-3}) = 0.997 + 0.0670 m_{NaBr}$.

11. The polar species AB dissociates in water according to the reaction

$$AB \rightleftharpoons A^+ + B^-$$

The equilibrium constant for this reaction at 25°C (with molality as the concentration unit) is

$$K = \frac{(a_{A^+})(a_{B^-})}{a_{AB}} = 5 \times 10^{-3} \text{ mol kg}^{-1}$$

where $a_i = m_i \gamma_i$ is the activity of species *i*. The Henry's constant (based on molality) for molecular AB in water at 25°C is 30 bar. What is the total solubility of AB in water at 25°C and 50 bar? Ignore the vaporization of water, and state clearly any other assumptions you make.
The following additional information is available at 25°C:

Second virial coefficient of AB: B = -200 cm^3 mol^{-1}.
Partial molar volume of AB infinite dilute in water, $\overline{v}_{AB}^{\infty} = 80$ cm^3 mol^{-1}.
Dielectric constant, $\varepsilon_r = 78.41$.
Electron charge, $e = 1.602\times10^{-19}$ C.

12. Consider two reactions in dilute aqueous solution at 25°C:

$$CO[(NH_3)_5Br]^{2+} + OH^- \rightleftharpoons \text{Products} \tag{I}$$

$$[Cr(NH_2CONH_2)_6]^{3+} + H_2O \rightleftharpoons \text{Products} \tag{II}$$

For reaction I and for reaction II, using the absolute (Eyring) theory of reaction rates, calculate the effect on the reaction rate constant k produced by adding an inert salt (e.g., NaC1) to the aqueous solution, that is, by increasing the ionic strength. Assume that the limiting Debye-Hückel relation is valid.
If the molality of NaC1 is 0.01 (and if the molalities of the charged reactants are negligibly small), calculate the change in k. For each reaction, does k increase or decrease upon addition of NaC1?

13. Derive Eqs. (9-59) and (9-60) for a single electrolyte solution from, respectively, Eqs. (I-13) and (I-10).

CHAPTER 10

Solubilities of Gases in Liquids

Numerous examples in nature illustrate the ability of liquids to dissolve gases; indeed, human life would not be possible if blood could not dissolve oxygen, nor could marine life exist if oxygen did not dissolve in water. Gas mixtures can be separated by absorption because different gases dissolve in a liquid in different amounts and therefore, most gaseous mixtures can be separated by contact with a suitable solvent that dissolves one gaseous component more than another. Further, knowledge of gas solubilities in water is important for describing processes that control the environmental distribution and ultimate fate of contaminants such as halogenated hydrocarbons (e.g. freons).

The solubility of a gas in a liquid is determined by the equations of phase equilibrium. If a gaseous phase and a liquid phase are in equilibrium, then for any component i the fugacities in both phases must be the same:

$$f_i^{\text{gas}} = f_i^{\text{liquid}} \tag{10-1}$$

Equation (10-l) is of little use unless something can be said about how the fugacity of component i in each phase is related to the temperature, pressure, and composition of that phase. In Chap. 5 we discussed the fugacity of a component in the gaseous phase. In this chapter we consider the fugacity of a component i, normally a gas at the temperature under consideration, when it is dissolved in a liquid solvent.

10.1 The Ideal Solubility of a Gas

The simplest way to reduce Eq. (10-1) to a more useful form is to rewrite it in a manner suggested by Raoult's law. In doing so, we introduce several drastic but convenient assumptions. Neglecting all gas-phase nonidealities as well as the effect of pressure on the condensed phase (Poynting correction),[1] and also neglecting any nonidealities due to solute-solvent interactions, the equation of equilibrium can be much simplified by writing

$$p_i = x_i P_i^s \tag{10-2}$$

where p_i is the partial pressure of component i in the gas phase,[2] x_i is the solubility (mole fraction) of i in the liquid, and P_i^s is the saturation (vapor) pressure of pure (possibly hypothetical) liquid i at the temperature of the solution. The solubility x_i, as given in Eq. (10-2), is called the *ideal solubility* of the gas.

Aside from the severe simplifying assumptions made in obtaining Eq. (10.2), an obvious difficulty presents itself in finding a value for P_i^s, whenever (as is often the case) the solution temperature is above the critical temperature of pure i. In that case it has been customary to extrapolate the saturation pressure of pure liquid i beyond its critical temperature to the solution temperature; the saturation pressure of the hypothetical liquid is usually found from a straight-line extrapolation on a semilogarithmic plot of saturation pressure versus reciprocal absolute temperature as shown in Fig. 10-1. The use of these particular coordinates does not have any sound theoretical basis but is dictated by convenience.

The ideal solubility, as calculated by Eq. (10-2) and the extrapolation scheme indicated in Fig. 10-1, usually gives correct order-of-magnitude results provided the partial pressure of the gas is not large and provided the solution temperature is well below the critical temperature of the solvent and not excessively above the critical temperature of the gaseous solute. In some cases, where the physical properties of solute and solvent are similar (e.g., chlorine in carbon tetrachloride), the ideal solubility is remarkably close to the experimental value.

---

[1] See Sec. 3.3.

[2] The partial pressure p_i is, by definition, equal to the product of the gas-phase mole fraction and the total pressure: $p_i \equiv y_i P$.

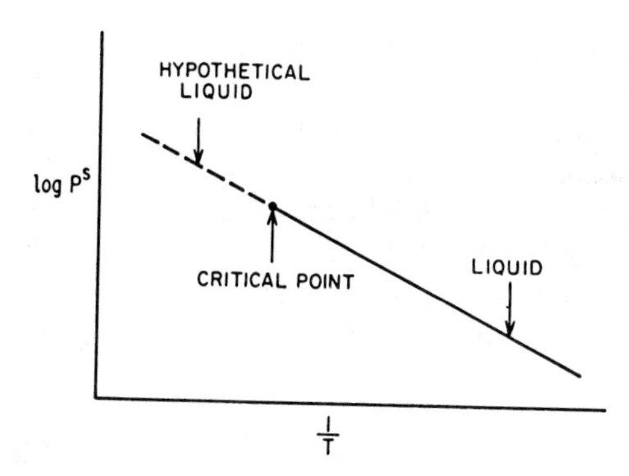

Figure 10-1 Convenient but arbitrary extrapolation of liquid saturation pressure into the hypothetical liquid region.

Table 10-1 compares ideal and observed solubilities of four gases in a number of solvents at 25°C and 1.013 bar partial gas pressure. The ideal solubility is significantly different from observed solubilities, but it is of the right order of magnitude.

Table 10-1 Solubilities (mole fraction x 10^4) of gases in several liquid solvents at 25°C and 1.013 bar partial pressure.

	Ideal	n-C_7F_{16}	n-C_7H_{16}	CCl_4	CS_2	$(CH_3)_2CO$
H_2	8	14.01	6.88	3.19	1.49	2.31
N_2	10	38.7	–	6.29	2.22	5.92
CH_4	35	82.6	–	28.4	13.12	–
CO_2	160	208.8	121	107	32.8	22.3

The ideal solubility given by Eq. (10-2) suffers from two serious defects. First, it is independent of the nature of the solvent; Eq. (10-2) says that a given gas, at a fixed temperature and partial pressure, has the same solubility in *all* solvents. This conclusion is contrary to observation as illustrated by the data in Table 10-1. Second, Eq. (10-2), coupled with the extrapolation scheme shown in Fig. 10-1, predicts that at constant partial pressure, the solubility of a gas always decreases with rising temperature. This prediction is frequently correct but not always; near room temperature the solubilities of the light gases helium, hydrogen, and neon increase with rising temperature in most solvents, and at somewhat higher temperatures the solubilities of gases like nitrogen, oxygen, argon, and methane also increase with rising temperature in many common solvents. Because of these two defects, the ideal-solubility equation has severely limited applicability, it should be used only whenever no more is desired than a rough estimate of gas solubility.

10.2 Henry's Law and Its Thermodynamic Significance

It was observed many years ago that the solubility of a gas in a liquid is often proportional to its partial pressure in the gas phase, provided that the partial pressure is not large. The equation that describes this observation is commonly known as *Henry's law*:

$$p_i = y_i P = kx_i \tag{10-3}$$

where, for any given solute and solvent, k is a constant of proportionality depending only on temperature.[3] Equation (10-3) always provides an excellent approximation when the solubility and the partial pressure of the solute are small and when the temperature is well below the critical of the solvent. Just how small the partial pressure and solubility have to be for Eq. (10-3) to hold, varies from one system to another, and the reasons for this variation will become apparent later. In general, however, as a *rough* rule for many common systems, the partial pressure should not exceed 5 or 10 bar and the solubility should not exceed about 3 mol %; however, in those systems where solute and solvent are chemically highly dissimilar (e.g., systems containing helium or hydrogen) large deviations from Eq. (10-3) are frequently observed at much lower solubilities. On the other hand, in some systems (e.g., carbon dioxide/benzene near room temperature), Eq. (10-3) appears to hold to large partial pressures and solubilities, but such cases are rare; the apparent validity of Eq. (10-3) at large solubilities is usually fortuitous due to cancellation of two (or more) factors that, taken separately, would cause that equation to fail.

The assumptions leading to Eq. (10-3) can readily be recognized by comparing it with Eq. (10-1). A comparison of the left-hand sides shows that in Henry's law the gas phase is assumed to be ideal and thus the fugacity is replaced by the partial pressure; this simplification can be avoided by the methods discussed in Chap. 5. A comparison of the right-hand sides shows that the fugacity in the liquid phase is assumed to be proportional to the mole fraction and that the constant of proportionality is taken as an empirical factor that depends on the natures of solute and solvent and on the temperature. The thermodynamic significance of this constant can be established by comparing the liquid fugacity as given by Henry's law with that obtained in the more conventional manner using the concept of an activity coefficient γ and some standard-state fugacity f^0:

$$f_2^L = kx_2 = H_{2,1}x_2 = \gamma_2 x_2 f_2^0 \tag{10-4}$$

Thus,

$$k = H_{2,1} = \gamma_2 f_2^0 \tag{10-5}$$

[3] For a given binary system, Henry's constant $H_{2,1}$ depends on temperature and, to a lesser degree, on total pressure, as discussed in Sec. 10.3. A precise definition of $H_{2,1}$ is given by Eq. (6-31) or Eq. (10-9).

where 1 stands for solvent and 2 stands for solute.

At a given temperature and pressure, the standard-state fugacity is a constant and does not depend on the solute mole fraction in the liquid phase. Since k does not depend on x_2, it follows from Eq. (10-5) that the activity coefficient γ_2 must also be independent of x_2; it is this feature, the constancy of the activity coefficient, which contains the essential assumption of Henry's law.

The activity coefficient of a solute is nearly independent of the solute's mole fraction provided the latter is sufficiently small. This can be shown from simple mathematical considerations. To fix ideas, take the case where γ_2 is normalized to approach unity as the mole fraction of 2 goes to unity. As shown in Chap. 6, it is convenient to express $\ln\gamma_2$ as a power series in $(1 - x_2)$:

$$RT\ln\gamma_2 = A(1-x_2)^2 + B(1-x_2)^3 + \ldots \tag{10-6}$$

where A, B, ... are constants depending on temperature and on intermolecular forces between solute and solvent. Equation (10-6) shows at once that if $x_2 \ll 1$, then γ_2 is only weakly dependent on x_2 and Henry's law provides a good approximation.

Equation (10-6) gives some insight into the well-known observation that Henry's law is a good approximation to relatively large solubilities for some systems but fails for relatively small solubilities in other systems. Consider the case where only the first term in the series is retained, while higher terms are neglected. Coefficient A is a measure of nonideality; if A is positive, it indicates the "dislike" between solute and solvent, whereas if it is negative, its absolute value may be a measure of the tendency between solute and solvent to form a complex. In any case, it is the absolute value of A/RT that determines the range of validity of Henry's law; in the limit, if $A/RT = 0$ (ideal solution), Henry's law holds for the entire range of composition $0 \leq x_2 \leq 1$. If A/RT is small compared to unity, then activity coefficient γ_2 does not change much even for appreciable x_2, but if it is large, then even a small x_2 can produce a significant change in the activity coefficient with composition. In the limit as x_2 approaches zero, the logarithm of the activity coefficient approaches the constant value A/RT, and therefore Henry's law is valid as a limiting relation.

As indicated by Eq. (10-3), Henry's law assumes that the gas-phase fugacity is equal to the partial pressure. This assumption is not necessary and is easily removed by including the gas-phase fugacity coefficient φ as discussed in detail earlier; more properly, therefore, Henry's law for solute i is

$$\boxed{f_i = \varphi_i y_i P = H_{i,\text{solvent}}\, x_i} \tag{10-7}$$

Table 10-2 presents experimental Henry's constants for four gases in ethylene oxide at three temperatures. These results were calculated from experimental solubility data and from volumetric data shown in Table 10-3. The second virial coefficients are needed to calculate vapor-phase fugacity coefficients and the liquid-phase partial molar volumes of the solutes at infinite dilution are needed to correct for the effect of pressure, as discussed in the next section.

Table 10-2 Henry's constants (bar) for four gases in ethylene oxide.*

Temperature (°C)	N_2	Ar	CH_4	C_2H_6
0	2837	1692	621	85.4
25	2209	1439	622	110
50	1844	1287	603	131

* J. D. Olson, 1977, *J. Chem. Eng. Data,* 22: 326. The estimated experimental uncertainty is about 2%.

Table 10-3 Some volumetric properties of four ethylene oxide (1)/gas (2) systems.*

		cm³ mol⁻¹			
	Temperature (°C)	N_2	Ar	CH_4	C_2H_6
$-B_{22}$	0	10.3	21.5	53.6	223
	25	4.7	15.8	42.8	187
	50	0.3	11.2	34.2	157
$-B_{12}$	0	85.9	119	160	331
	25	69.4	98.6	133	273
	50	56.2	82.3	112	229
$\overline{v}_2^\infty$	0	41.1	35.9	44.4	61.3
	25	43.3	39.5	47.5	64.5
	50	48.5	43.6	52.0	68.6

* J. D. Olson, 1977, *J. Chem. Eng. Data,* 22: 326. Second virial coefficient ($-B_{22}$) from sources quoted in J. H. Dymond and E. B. Smith, 1980, *The Virial Coefficients of Pure Gases and Mixtures* (Oxford: Clarendon Press). Second-virial cross coefficient ($-B_{12}$) are estimated from a correlation by C. Tsonopoulos, 1974, *AIChE J.,* 20: 263. Liquid-phase partial molar volumes at infinite dilution ($\overline{v}_2^\infty$) are estimated from the correlation of E. Lyckman, C. A. Eckert, and J. M. Prausnitz, 1965, *Chem. Eng. Sci.,* 20: 685.

10.3 Effect of Pressure on Gas Solubility

In the preceding section we discussed the essential assumption in Henry's law, *i.e.*, that at constant temperature, the fugacity of solute i is proportional to the mole fraction x_i. The constant of proportionality $H_{i,\text{solvent}}$ is not a function of composition but depends on temperature and, to a lesser degree, pressure. The pressure dependence can be neglected as long as the pressure is not large. At high pressures, however, the effect is not negligible and therefore it is necessary to consider how Henry's constant depends on pressure. This dependence is easily obtained by using the exact equation

$$\left(\frac{\partial \ln f_i^L}{\partial P}\right)_{T,x} = \frac{\overline{v}_i}{RT} \tag{10-8}$$

where $\overline{v}_i$ is the partial molar volume of i in the liquid phase. The thermodynamic definition of Henry's constant is

$$H_{i,\text{solvent}} \equiv \lim_{x_i \to 0} \frac{f_i^L}{x_i} \quad \text{(at constant temperature and pressure)} \tag{10-9}$$

Substitution of Eq. (10-9) into Eq. (10-8) gives

$$\boxed{\left(\frac{\partial \ln H_{i,\text{solvent}}}{\partial P}\right)_T = \frac{\overline{v}_i^\infty}{RT}} \tag{10-10}$$

where $\overline{v}_i^\infty$ is the partial molar volume of solute i in the liquid phase at infinite dilution.[4] Integrating Eq. (10-10) and assuming, as before, that the fugacity of i at constant temperature and pressure is proportional to x_i, we obtain a more general form of Henry's law:

$$\ln \frac{f_i}{x_i} = \ln H_{i,\text{solvent}}^{(P^r)} + \frac{\int_{P^r}^{P} \overline{v}_i^\infty dP}{RT} \tag{10-11}$$

where $H_{i,\text{solvent}}^{(P^r)}$ is Henry's constant evaluated at an arbitrary reference pressure P^r. As $x_i \to 0$, the total pressure is P_1^s, the saturation (vapor) pressure of the solvent; it is often convenient, therefore, to set $P^r = P_1^s$.

If the solution temperature is well below the critical temperature of the solvent, it is reasonable to assume that $\overline{v}_i^\infty$ is independent of pressure. Letting subscript 1 refer to the solvent and subscript 2 to the solute, Eq. (10-11) becomes

$$\boxed{\ln \frac{f_2}{x_2} = \ln H_{2,1}^{(P_1^s)} + \frac{\overline{v}_2^\infty (P - P_1^s)}{RT}} \tag{10-12}$$

[4] See Sec. 12.4 for a discussion of liquid-phase partial molar volumes.

Equation (10-12) is the *Krichevsky-Kasarnovsky equation* (1935), although its first clear derivation was given by Dodge and Newton (1937). This equation is remarkably useful for representing solubilities of sparingly soluble gases to very high pressures. Figures 10-2 and 10-3 show that solubility data for hydrogen and nitrogen in water to 1000 bar are accurately reproduced by the Krichevsky-Kasarnovsky equation; in this case, the vapor pressure of the solvent is negligible in comparison to the total pressure and therefore the abscissa reads P rather than $P - P_1^s$. The intercepts of these plots give $H_{2,1}^{(P_1^s)}$, and the slopes yield the partial molar volumes of the gaseous solutes in the liquid phase. At 25°C, Figs. 10-2 and 10-3 give partial molar volumes; that for hydrogen is 19.5 and that for nitrogen is 32.8 cm^3 mol^{-1}. These results are in fair agreement with partial molar volumes for these gases in water obtained from dilatometric measurements. In his detailed studies of gas solubilities in amines and in alcohols, Brunner (1978, 1979) has shown that Eq. (10-12) gives an excellent representation of the experimental data at high pressures.

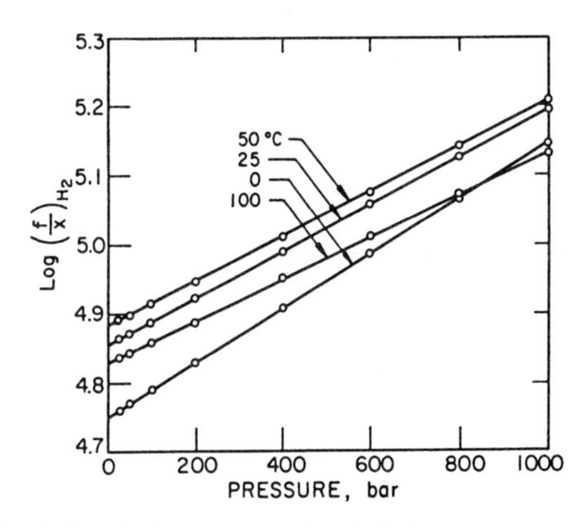

Figure 10-2 Solubility of hydrogen in water at high pressures.

Equation (10-12) can be expected to hold for all those cases that conform to the two assumptions on which the equation rests. One of these is that the activity coefficient of the solute does not change noticeably over the range of x_2 considered; in other words, x_2 must be small, as discussed in the preceding section. The other assumption states that the infinitely dilute liquid solution must be essentially incompressible, very nearly correct at temperatures far removed from the critical temperature of the solution.

To illustrate the use and limitation of the Krichevsky-Kasarnovsky equation, consider the high-pressure solubility data (Wiebe and Gaddy, 1937) for nitrogen in liquid ammonia. These are shown in Fig. 10-4 plotted in the manner indicated by Eq. (10-12). At 0°C, the Krichevsky-Kasarnovsky equation holds to 1000 bar, but at 70°C it breaks down after about 600 bar. This striking difference is readily explained upon considering

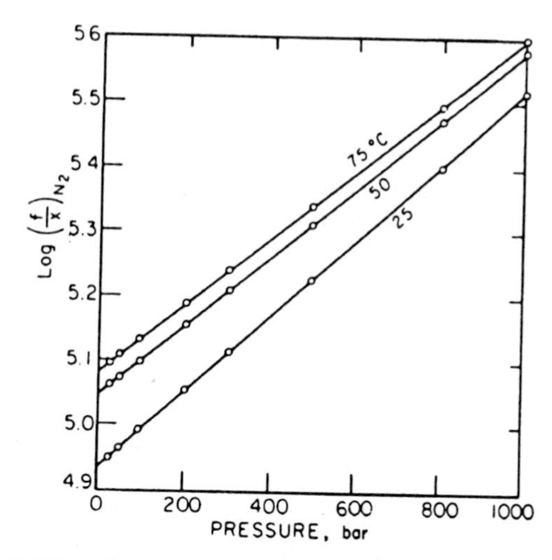

Figure 10-3 Solubility of nitrogen in water at high pressures.

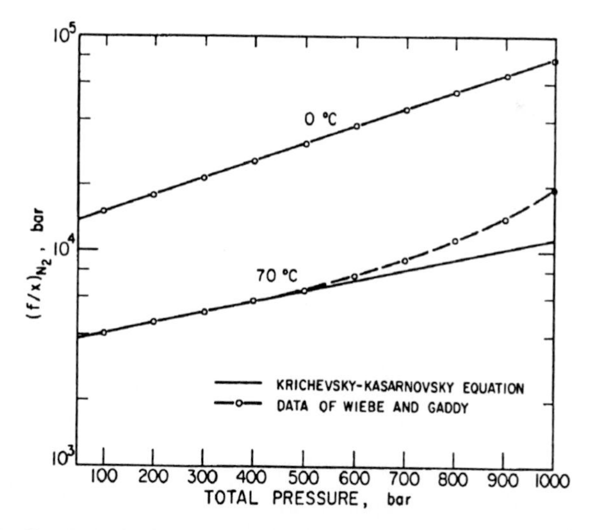

Figure 10-4 Success and failure of the Krichevsky-Kasarnovsky equation. Solubility of nitrogen in liquid ammonia.

the two assumptions just mentioned; at 0°C liquid ammonia is an unexpanded liquid solvent (the critical temperature of ammonia is 132.3°C) and the solubility of nitrogen is small throughout, only 2.2 mol % at 1000 bar. At 0°C, therefore, the assumptions of Eq. (10-12) are reasonably satisfied. However, at 70°C liquid ammonia is already quite expanded (and compressible) and the solubility of nitrogen is no longer small, 12.9 mol % at 1000 bar. Under these conditions, it is not reasonable to expect that the activity coefficient of nitrogen in the liquid phase is independent of composition, nor is it likely that the partial molar volume is constant. As a result, it is not surprising that the Krichevsky-Kasarnovsky equation gives an excellent representation of the data for the entire pressure range at 0°C but fails at higher pressures for the data at 70°C.

Variation of the activity coefficient of the solute with mole fraction can be taken into account by one of the methods discussed in Chap. 6. In the simplest case we may assume that the activity coefficient of the *solvent* is given by a two-suffix Margules equation:

$$\ln \gamma_1 = \frac{A}{RT} x_2^2 \tag{10-13}$$

where A is an empirical constant determined by intermolecular forces in the solution. Typically, A is a weak function of temperature.

The activity coefficient γ_2^* of the solute, normalized according to the unsymmetric convention (see Sec. 6.4), is then found from the Gibbs-Duhem equation; it is given by

$$\ln \gamma_2^* = \frac{A}{RT}(x_1^2 - 1) \tag{10-14}$$

The fugacity of component 2 at pressure P_1^s is

$$f_2 = \gamma_2^* H_{2,1}^{(P_1^s)} x_2 \tag{10-15}$$

and instead of Eq. (10-12) we obtain:[5]

$$\boxed{\ln \frac{f_2}{x_2} = \ln H_{2,1}^{(P_1^s)} + \frac{A}{RT}(x_1^2 - 1) + \frac{\bar{v}_2^\infty (P - P_1^s)}{RT}} \tag{10-16}$$

Equation (10-16) is the *Krichevsky-Ilinskaya equation* (1945). Because of the additional parameter, it has a wider applicability than does Eq. (10-12). It is especially useful for solutions of light gases (such as helium and hydrogen) in liquid solvents where the solubility is appreciable. For example, Orentlicher (1964) found that Eq. (10-16) could be used to correlate solubility data for hydrogen in a variety of solvents at low temperatures and at pressures to about 100 bar. Table 10-4 gives parameters reported by Orentlicher. In the systems studied, the solubility of hydrogen may be as large as 20 mol % and therefore the data could not be adequately represented by the simpler Krichevsky-Kasarnovsky equation.

[5] Equation (10-16) assumes that the partial molar volume of the solute is independent of pressure and composition over the pressure and composition ranges under consideration.

Table 10-4 Thermodynamic parameters for correlating hydrogen solubilities.*

Solvent	T (K)	$H_{2,1}^{(P_1^s)}$ (bar)	A (J mol^{-1})	$\overline{v}_2^\infty$ (cm^3 mol^{-1})
CO	68	648	704±69	31.2
	78	476		32.6
	88	405		34.4
N_2	68	547	704±69	30.4
	79	456		31.5
	95	345		34.4
CH_4	90	1848	1486±297	29.7
	110	1050		31.0
	144	638		36.0
C_2H_6	144	2634	2478±198	37. 9
	200	1672		44.2
	228	1226		54.3
C_3H_8	228	1692	2478±198	50
	255	1317		51
	282	1044		63

* From Orentlicher (1964). Within the accuracy of the data, A is not temperature dependent over the temperature range studied.

When, for a given temperature, gas-solubility data alone are available as a function of pressure, it is difficult to obtain three isothermal parameters ($H_{2,1}^{(P_1^s)}$, $\overline{v}_2^\infty$, and A) from data reduction. When $\ln(f_2/x_2)$ is plotted versus $(P-P_1^s)$, the intercept can give a good value for $H_{2,1}^{(P_1^s)}$ but, since the slope depends on both correction terms in Eq. (10-16), it is often not possible to obtain unique values for $\overline{v}_2^\infty$ and A from gas-solubility data alone. Some other information is needed to establish all three parameters in Eq. (10-16). In the absence of pertinent experimental information, it is possible to use an equation of state as discussed below.

Let $W = \ln(f_2/x_2)$ and consider isothermal changes in Y in the region $P = P_1^s$ and $x_2 = 0$. At a constant temperature T, we write a Taylor series:

$$W(P,x_2) = W(P_1^s,0) + \left(\frac{\partial W}{\partial P}\right)_{P_1^s}(P-P_1^s) + \left(\frac{\partial W}{\partial x_2}\right)_{x_2=0} x_2 \qquad (10\text{-}17)$$

Comparison with Eq. (10-16) shows that

$$W(P_1^s,0) = \ln H_{2,1}^{(P_1^s)} \qquad (10\text{-}18)$$

$$\left(\frac{\partial W}{\partial P}\right)_{P_1^s} = \frac{\overline{v}_2^{\infty}}{RT} \tag{10-19}$$

and

$$\left(\frac{\partial W}{\partial x_2}\right)_{x_2=0} = -\frac{2A}{RT} \tag{10-20}$$

We can calculate parameters $H_{2,1}^{(P_1^s)}$, $\overline{v}_2^{\infty}$, and A from an equation of state. First,

$$H_{2,1}^{(P_1^s)} = P_1^s \varphi_2^{L,\infty} \tag{10-21}$$

where $\varphi_2^{L,\infty}$ is the fugacity coefficient of solute 2 in the liquid phase at temperature T, at infinite dilution ($x_2 = 0$), and at liquid molar volume v_1^s, the saturated molar volume of the solvent at T.

Second, from elementary calculus,

$$\overline{v}_2^{\infty} = -\left[\frac{(\partial P/\partial n_2)_{T,V,n_1}}{(\partial P/\partial V)_{T,n_1,n_2}}\right]_{n_2=0} \tag{10-22}$$

where n is the number of moles and V is the total volume (see Chap. 3). Finally,

$$A = -\frac{RT}{2}\frac{\partial}{\partial x_2}(\ln \varphi_2^L + \ln P)_{T,x_2=0} \tag{10-23}$$

If a reliable equation of state is available for the dilute mixture, the three Krichevsky-Ilinskaya parameters can be calculated. To do so, it is necessary that the equation of state be valid for the entire density range, from zero to $(v_1^s)^{-1}$, because (see Chap. 3) fugacity coefficients depend on an *integral* of the equation of state.

As discussed in other chapters, calculated thermodynamic properties of mixtures often depend strongly on the mixing rules and especially on the cross term for the characteristic energy parameter. For example, in equations of the van der Waals form, the constant a (for a binary mixture) is usually written

$$a = x_1^2 a_1 + x_2^2 a_2 + 2x_1 x_2 (a_1 a_2)^{1/2}(1 - k_{12}) \tag{10-24}$$

where k_{12} is a binary parameter that has a large effect on φ_2^L, especially when x_2 is small.

For some given equation of state, the important binary parameter k_{12} can be found from the experimentally determined Henry's constant or vapor-liquid equilibrium data, using Eq. (10-21); parameter k_{12} is adjusted until the calculated fugacity coefficient $\varphi_2^{L,\infty}$ satisfies Eq. (10-21). Once k_{12} is known, it may also be used to calculate parameters $\bar{v}_2^{\infty}$ and A. Therefore, with the help of an equation of state, gas-solubility data at low pressures ($H_{2,1}^{(P_1^s)}$) may be used to calculate gas solubilities at higher pressures. While such calculations can be done directly (see Chaps. 3 and 12) without an equation such as that of Krichevsky-Ilinskaya, the procedure outlined here shows the direct correspondence between one method, based on an activity coefficient, and another, based on an equation of state. Bender *et al.* (1984) have presented some examples of this correspondence. Using the Redlich-Kwong (or Peng-Robinson) equation of state, Bender *et al.* showed that calculated and observed high-pressure gas solubilities are in good agreement for hydrogen in ethylene diamine and for methane in *n*-hexane.

Another example of this procedure is provided by solubility data for carbon dioxide in phenol (Yau *et al.*, 1992) at 75, 100, 125, and 150°C shown in Fig. 10-5. The equation of state used was that of Redlich-Kwong-Soave (see Sec. 12.8); for each isotherm, interaction parameter k_{12} was obtained from regression of vapor-liquid equilibrium data for the same binary system. Table 10-5 gives the resulting thermodynamic parameters for carbon-dioxide solubility in phenol, $H_{2,1}^{(P_1^s)}$, $\bar{v}_2^{\infty}$, and A, as obtained from Eqs. (10-21), (10-22), and (10-23), respectively. As Fig. 10-5 shows, with the parameters listed in Table 10-5, the Krichevsky-Ilinskaya equation [Eq. 10-16)] represents well the experimental data over the entire temperature and pressure range studied.

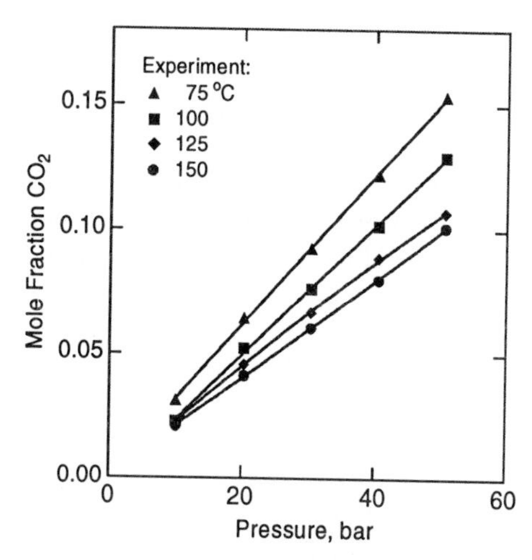

Figure 10-5 Solubility of carbon dioxide in phenol. Solid lines are calculated from the Krichevsky-Ilinskaya equation [Eq. 10-16)] with the parameters listed in Table 10-5 for the Redlich-Kwong-Soave equation of state. Experimental data from Yau *et al.* (1992).

Table 10-5 Thermodynamic parameters for correlating carbon dioxide solubility in phenol. Parameter k_{12} was obtained from regression of vapor-liquid equilibrium data for the carbon dioxide/phenol system at each isotherm.

Temp. (°C)	$k_{12}\times10^2$	$H_{2,1}^{(P_1^s)}$ (bar)	A (J mol^{-1})	$\overline{v}_2^\infty$ (cm^3 mol^{-1})
75	7.53	330	1347	45.3
100	7.72	384	1465	48.6
125	7.19	436	1531	52.1
150	6.31	473	1597	56.3

10.4 Effect of Temperature on Gas Solubility

Many elementary textbooks on chemistry state, without qualification, that as the temperature rises, gas solubility falls. While this statement is often correct, there are many cases where it is not, especially at high temperatures where it is more common for gas solubility to increase with rising temperature.

No simple generalizations can be made concerning the effect of temperature on solubility as indicated by Fig. 10-6 that shows Henry's constants as a function of temperature for nine binary systems. Not only does Henry's constant vary significantly from one system to another but, equally striking, the effect of temperature depends strongly on the properties of the particular system and also on the temperature.

The temperature derivative of the solubility, as calculated from the Gibbs-Helmholtz equation,[6] is directly related to either the partial molar enthalpy or the partial molar entropy of the gaseous solute in the liquid phase. Therefore, if something can be said about the enthalpy or entropy change of solution, insight can be gained on the effect of temperature on solubility. A general derivation of the thermodynamic relations is given elsewhere (Hildebrand and Scott, 1962; Sherwood, 1962); we consider here only the relatively simple case where the solvent is essentially nonvolatile and where the solubility is sufficiently small to make the activity coefficient of the solute independent of the mole fraction. With these restrictions, it can be shown that

$$\left(\frac{\partial \ln x_2}{\partial 1/T}\right)_P = -\frac{\Delta \overline{h}_2}{R} \tag{10-25}$$

and

$$\left(\frac{\partial \ln x_2}{\partial \ln T}\right)_P = -\frac{\Delta \overline{s}_2}{R} \tag{10-26}$$

[6] See Eq. (6-12).

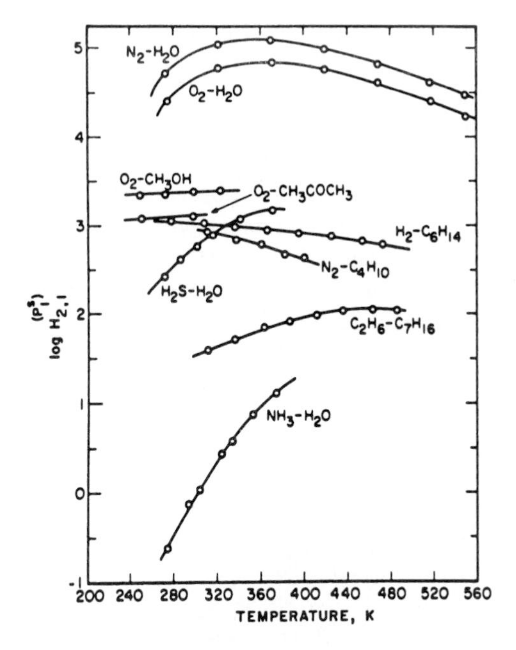

Figure 10-6 Henry's constants (bar) for typical gases range over five orders of magnitude. The effect of temperature differs qualitatively from one system to another.

where x_2 is the mole fraction of gaseous solute at saturation and

$$\Delta\bar{h}_2 \equiv \bar{h}_2^L - h_2^G, \qquad \Delta\bar{s}_2 \equiv \bar{s}_2^L - s_2^G$$

where h_2^G and s_2^G are, respectively, the enthalpy and the entropy of pure 2 gas, at system temperature and pressure.

First, we consider Eq. (10-26); if the partial molar entropy change of the solute is positive, then the solubility increases with rising temperature; otherwise, it falls. To understand the significance of the entropy change, it is convenient to divide it into two parts:

$$\Delta\bar{s}_2 = (s_2^L - s_2^G) + (\bar{s}_2^L - s_2^L) \tag{10-27}$$

where s_2^L is the entropy of the (hypothetical) pure liquid at the temperature of the solution. The first term on the right-hand side of Eq. (10-27) is (essentially) the entropy of condensation of the pure gas and, in general, we expect this term to be negative because the entropy (disorder) of a liquid is lower than that of a saturated gas at the same temperature. The second term is the partial molar entropy of solution of the condensed solute and, assuming ideal entropy of mixing for the two liquids, we can write

$$\bar{s}_2^L - s_2^L = -R \ln x_2 \tag{10-28}$$

Because $x_2 < 1$, the second term in Eq. (10-27) is positive and the smaller the solubility, the larger this term. It therefore follows that $\Delta\bar{s}_2$ should be positive for those gases that have very small solubilities and negative for the others; this result leads to the expectation that sparingly soluble gases (very small x_2) show positive temperature coefficients of solubility, whereas readily soluble gases (relatively large x_2) show negative temperature coefficients. This expectation is observed. This semiquantitative interpretation of the sign of $\Delta\bar{s}_2$ is in good agreement with observed behavior of gas-liquid solutions as shown in Fig. 10-7 where the observed partial molar entropy change [Eq. (10-26)] is related to the ideal partial molar entropy of the condensed solute [Eq. (10-28)]. The plot gives experimental results at 25°C for 14 gases in six solvents at 1.013 bar partial pressure. The figure can be divided into two parts: the upper part corresponds to a positive, and the lower to a negative temperature coefficient of solubility.

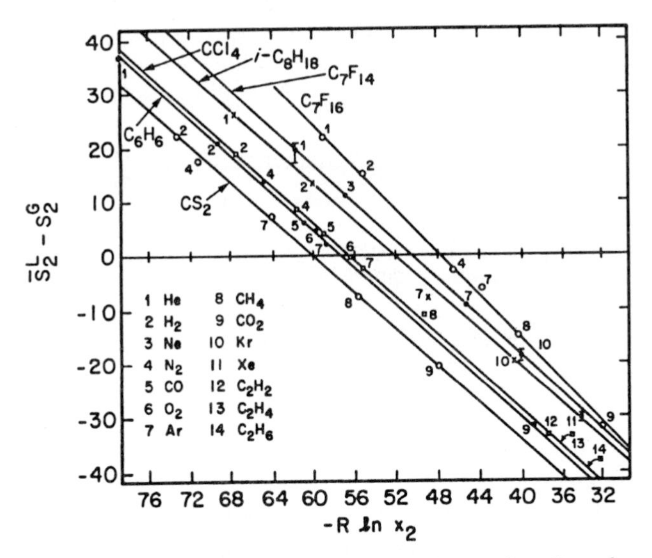

Figure 10-7 Entropy of solution of gases in liquids as a function of gas solubility (mole fraction) x_2 at 25°C and 1.013 bar (Hildebrand and Scott, 1962). Units of entropy and gas constant R are J mol $^{-1}$ K^{-1}.

As suggested by the foregoing discussion, Fig. 10-7 shows that, as a general rule, the solubility of a gas rises with increasing temperature whenever x_2 is small ($-R\ln x_2$ is large), and it falls with increasing temperature whenever x_2 is large ($-R\ln x_2$ is small). This is only a general rule, and there are differences in behavior between different gases and different solvents. For example, consider the solubility that a gas must have to exhibit a zero temperature coefficient of solubility (no change in solubility with temperature). In perfluoroheptane (C_7F_{16}), $\Delta\bar{s}_2$ is zero when $-R\ln x_2 = 47.3$ J mol^{-1} K^{-1}; therefore, in this solvent the temperature coefficient of solubility changes sign when $x_2 = 3.43\times10^{-3}$. In carbon disulfide, however, the corresponding value is

$x_2 = 0.76\times10^{-3}$. Going from left to right, the solvents in Fig. 10-7 are arranged in order of decreasing solubility parameters. At constant solubility (i.e., constant x_2), the temperature coefficient of solubility for a given gas has a tendency to increase (algebraically) as the solubility parameter of the solvent falls.

Some qualitative insight into the effect of temperature on gas solubility can also be obtained from the partial molar enthalpy change [Eq. (10-25)]. It is useful to divide this change into two parts:

$$\Delta\bar{h}_2 = (h_2^L - h_2^G) + (\bar{h}_2^L - h_2^L) \qquad (10\text{-}29)$$

where h_2^L is the enthalpy of the (hypothetical) pure liquid at the temperature of the solution.

The first term in Eq. (10-29) is (essentially) the enthalpy of condensation of pure solute and, because the enthalpy of a liquid is generally lower than that of a gas at the same temperature, we expect this quantity to be negative.[7] The second quantity is the partial enthalpy of mixing for the liquid solute; in the absence of solvation between solute and solvent, this quantity tends to be positive (endothermic) and the theory of regular solutions (Chap. 7) tells us that the larger the difference between the cohesive energy density of the solute and that of the solvent, the larger the enthalpy of mixing. If this difference is very large (e.g., hydrogen and benzene), the second term in Eq. (10-29) dominates; $\Delta\bar{h}_2$ is then positive and the solubility increases with rising temperature. However, if the difference in cohesive energy densities is small (e.g., chlorine in carbon tetrachloride), the first term in Eq. (10-29) dominates; $\Delta\bar{h}_2$ is then negative and the solubility falls as temperature increases.

If there are specific chemical interactions between solute and solvent (e.g., ammonia and water), then both terms in Eq. (10-29) are negative (exothermic) and the solubility decreases rapidly as the temperature rises.

The effect of temperature on solubility is sensitive to the intermolecular forces of the solute-solvent system. When the partial pressure of the solute is small, the solubility typically decreases with temperature, goes through a minimum, and then rises. To illustrate, Fig. 10-8 shows the solubility of methane in *n*-heptane over a wide temperature range. For most common systems the temperature corresponding to minimum solubility lies well above room temperature but for light gases, minimum solubility is often observed at low temperatures.

A thorough study of the solubilities of simple gases in water from 0 to 50°C has been reported by Benson and Krause (1976). Theoretical and empirical evidence show that, although other expressions may provide reasonable approximations, the effect of temperature on Henry's constant over narrow temperature ranges is *best* given by an equation of the form

---

[7] At $T/T_{c2} >> 1$, $(h_2^L - h_2^G)$ may be positive. See, for example, G. J. F. Breedveld and J. M. Prausnitz, 1973, *AIChE J.*, 19: 783, where it is shown that, at very high reduced temperatures, an isothermal increase in the volume of a fluid may lower its enthalpy.

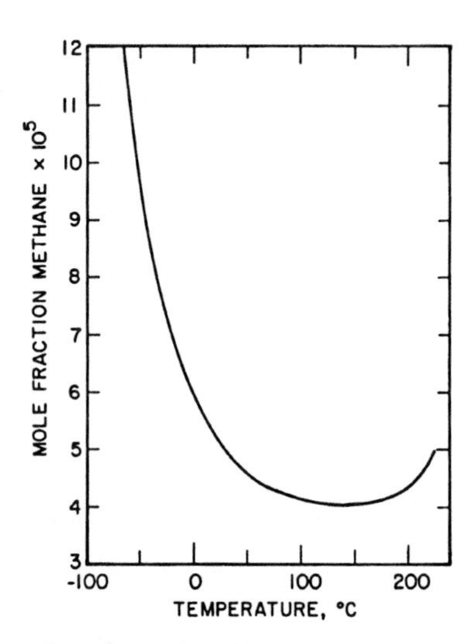

Figure 10-8 Solubility of methane in *n*-heptane when the vapor-phase fugacity of methane is 0.01 bar.

$$\ln H_{2,1} = \alpha_2\left(1-\frac{T_2}{T}\right)-\beta\left(1-\frac{T_2}{T}\right)^2 + \ln 1.01325 \qquad (10\text{-}30)$$

Here, $H_{2,1}$ is Henry's constant (in bar) for solute 2 in solvent 1 (water); α_2 and T_2 are constants specific to the solute. For simple gases, β is a universal constant. Equation (10-30) is based primarily on experimental data in the region 0 to 50°C.

Table 10-6 reports constants α_2 and T_2 for seven solutes; these, in turn, can be related to physical properties of the solute. It is clear from Eq. (10-30) that T_2 is the temperature where $H_{2,1}$ is 1 atm. When $H_{2,1}$ is expressed in atmospheres, T_2 is the normal boiling point of a hypothetical liquid, a condensed solute where each solute molecule is completely surrounded by water. Equation (10-30) gives good results for the partial entropy and enthalpy of solution and, most impressive, for the partial heat capacity of the solute. Therefore, while Eq. (10-30) appears, in essence, to be little more than a concise summary of a large body of high-quality experimental data, it may be useful as a guide toward theoretical understanding of dilute aqueous solutions of nonelectrolytes. To appreciate the possible theoretical significance of this equation, it is necessary to consult the detailed paper by Benson and Krause. It is nevertheless quoted here to point out once again that phase-equilibrium thermodynamics progresses most rapidly when good experimental data are analyzed with attention to primary as well as derivative thermodynamic properties[8] and with ever-open eyes toward establishing connections between theory and experiment.

[8] In this case, the primary property is Henry's constant; the derivative properties are partial molar enthalpy and entropy (first derivative) and partial molar heat capacity (second derivative).

Table 10-6 Parameters for Eq. (10-30) (with $\beta = 36.855$) giving Henry's constants (in bar) for seven gaseous solutes in water in the region 0 to 50°C (Benson and Krause, 1976).

Solute	T_2 (K)	α_2
Helium	131.42	41.824
Neon	142.50	41.667
Nitrogen	162.02	41.712
Oxygen	168.85	40.622
Argon	168.87	40.404
Krypton	179.21	39.781
Xenon	188.78	39.273

Over a wider range of temperatures, simple equations such as Eq. (10-30) are unable to describe Henry's constant. Harvey (1996) developed a semiempirical correlation of Henry's constants over large temperature ranges. To illustrate, Fig. 10-9 shows Henry's constants for six nonpolar gases in water as a function of temperature obtained from Harvey's correlation. In addition to the maximum (corresponding to the minimum solubility), a striking feature is the significant decrease in Henry's constant as the critical temperature of water (647.1 K) is approached.

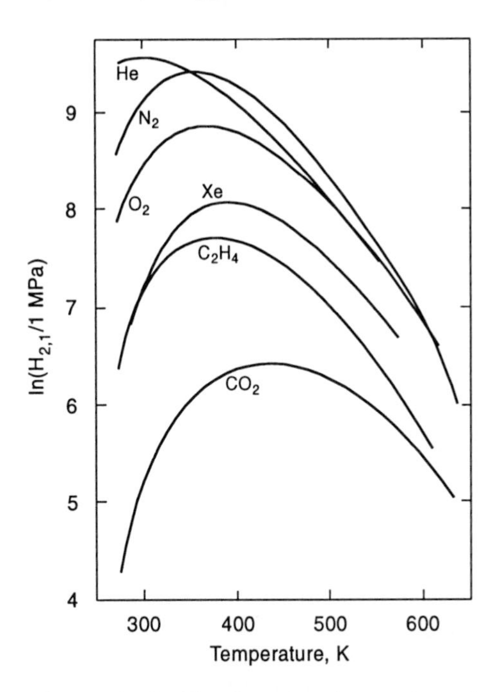

Figure 10-9 Henry's constants $H_{2,1}$ for several gases in water. Curves represent data fitted by the correlation of Harvey (1996).

The behavior of Henry's constant near the solvent critical point has been a source of confusion (for example, erroneous suggestions that $H_{2,1}$ approaches zero or a constant for all solutes in this limit), but it is now understood well. Beutier and Renon (1978) showed that, at the critical point of solvent 1, Henry's constant for solute 2 is given by $H_{2,1} = P_{c_1}\varphi_2^\infty$, where P_{c_1} is the solvent's critical pressure and φ_2^∞ is the solute fugacity coefficient at infinite dilution at the critical temperature and pressure of the solvent. The derivative $dH_{2,1}/dT$ diverges to negative infinity (or positive infinity for some solute/solvent pairs) due to the diverging compressibility of the solvent.

Japas and Levelt Sengers (1989) derived the correct functional form for this divergence. Near the solvent's critical point, a function of Henry's constant is linear in density:

$$T\ln(H_{2,1}/f_1) = A + B(\rho_1 - \rho_{c_1}) \tag{10-31}$$

where f_1 and ρ_1 are, respectively, the fugacity and the density of the pure solvent and ρ_{c_1} is the solvent's critical density. In Eq. (10-31), constant A is related to $H_{2,1}$ at the solvent's critical-point as determined by φ_2^∞; and constant B is related to a thermodynamic derivative called the *Krichevsky parameter* (see, e.g. Levelt Sengers, 1991), the key quantity describing dilute mixtures near the solvent's critical point. Japas and Levelt Sengers have also derived a linear relationship for the *infinite dilution partition coefficient* K^∞,

$$T\ln K^\infty = 2B(\rho_{L_1}^s - \rho_{c_1}) \tag{10-32}$$

where $\rho_{L_1}^s$ is the saturated liquid density of the solvent and K^∞ is defined along the solvent's coexistence curve by

$$K^\infty \equiv \lim_{x_2\to 0}(y_2/x_2) \tag{10-33}$$

While Eqs. (10-31) and (10-32) are only asymptotic results, they describe experimental data over a wide range of conditions. Figure 10-10 shows Henry's-constant data for several solutes in water plotted according to Eq. (10-31). The data display striking linearity (more than one has a right to expect from a result derived only near the critical point) from near-critical temperatures to approximately 100°C.

Although we do not fully understand the reason for such extended linear behavior, it can be used to develop correlations, for example, for Henry's constants (Harvey and Levelt Sengers, 1990; Harvey, 1996) and for infinite-dilution partition coefficients (Alvarez *et al.*, 1994). Because these correlations are "anchored" with the correct near-critical functional form, they can be extrapolated to high temperatures with more confidence than empirical fitting equations. These results demonstrate how we can improve correlations by choosing the proper independent variable (here solvent density) and making reasonable use of theoretical boundary conditions.

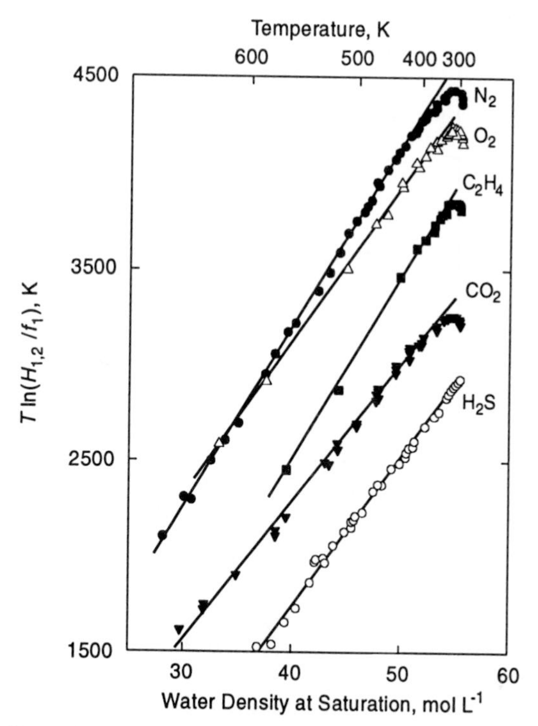

Figure 10-10 Henry's constants for several gases in water plotted according to Eq. (10-31).

10.5 Estimation of Gas Solubility

Reliable data on the solubility of gases in liquids are not plentiful, especially at temperatures well removed from 25°C. Hildebrand and coworkers have obtained a large amount of accurate data in the vicinity of room temperature and therefore we consider first the semiempirical correlations they have established.

Figure 10-11 shows solubilities of 12 gases (at 25°C and 1.013 bar partial pressure) as a function of the solubility parameter of the solvent, and Fig. 10-12 shows solubilities of 13 gases (at the same conditions) in nine solvents as a function of Lennard-Jones energy parameters (ε/k) determined from second-virial-coefficient data for the solutes. These plots, presented by Hildebrand and Scott (1962), indicate that the solubilities of nonpolar gases in nonpolar solvents can be correlated in terms of two parameters: the solubility parameter of the solvent and the Lennard-Jones energy parameter for the solute. Figures 10-11 and 10-12, therefore, may be used with some confidence to predict solubilities in nonpolar systems where no experimental data are available; these predictions are necessarily limited to systems at 25°C, but with the help

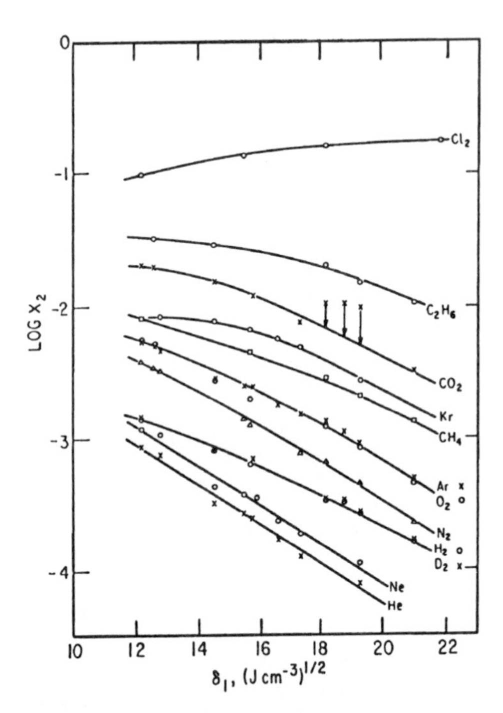

Figure 10-11 Solubilities of gases in liquids at 25°C and at a partial pressure of 1.013 bar as a function of solvent solubility parameter δ_1 (Hildebrand and Scott, 1962).

of Eq. (10-26) and the entropy data shown in Fig. 10-7, it is possible to predict solubilities at other temperatures not far removed from 25°C.

Figures 10-11 and 10-12 are useful, but in addition to their utility we must recognize that even for nonpolar systems at one temperature there are significant deviations from "regular" behavior; the straight lines correlate most of the experimental results but there are notable exceptions. In Fig. 10-11 the solubilities of carbon dioxide in benzene, toluene, and carbon tetrachloride are somewhat larger than those predicted from results obtained in other solvents. The high solubility in aromatic hydrocarbons can probably be attributed to a Lewis acid-base interaction between acidic carbon dioxide and the basic aromatic; apparently there is also some specific chemical interaction between carbon dioxide and carbon tetrachloride that may be related to carbon dioxide's quadrupole moment. In Fig. 10-12 the solubilities of the quantum gases, helium, hydrogen, and neon, are a little higher than expected, the discrepancies becoming larger as the solubility parameter of the solvent increases. These anomalies are not yet fully understood.[9]

[9] Refinements in regular-solution theory for correlating gas solubilities are discussed by J. H. Hildebrand and R. H. Lamoreaux, 1974, *Ind. Eng. Chem. Fundam.*, 13: 110 and by R. G. Linford and D. G. T. Thornhill, 1977, *J. Appl. Chem. Biotechnol.*, 27: 479. Unfortunately, these refinements are confined to temperatures near 25°C.

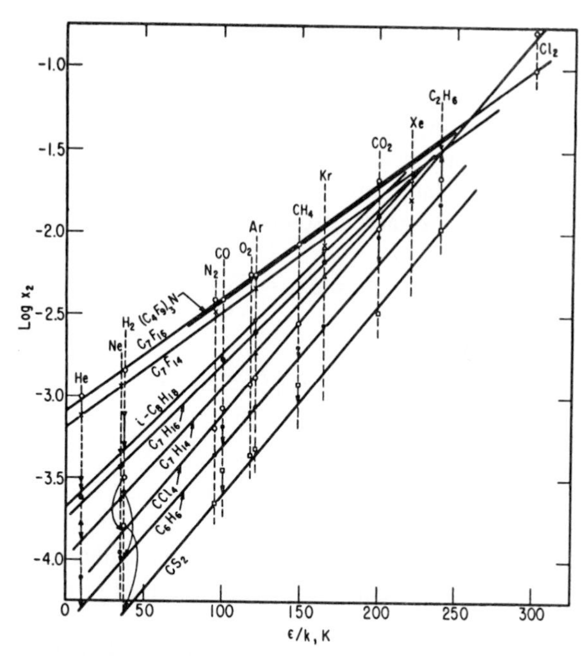

Figure 10-12 Solubilities of gases in liquids at 25°C and at a partial pressure of 1.013 bar as a function of solute characteristic ε/k (Lennard-Jones 6-12-potential) (Hirschfelder *et al.*, 1954).

Hildebrand's correlations provide a good basis for estimating gas solubilities in nonpolar systems at temperatures near 25°C but, since the entropy of solution is temperature-dependent, these correlations are not helpful at temperatures well removed from 25°C. Unfortunately, solubility data at temperatures much larger or smaller than room temperature are scarce and therefore, strictly empirical correlations cannot be used; rather, it is necessary to resort to whatever theoretical methods might be available and appropriate.

A rigorous method for the prediction of gas solubilities requires a valid theory of solutions. While efforts toward such a theory are progressing,[10] they have not yet reached a stage for practical applications. For a semitheoretical description of nonpolar systems, the theory of regular solutions and the theorem of corresponding states can serve as the basis for a correlating scheme (Prausnitz and Shair, 1961) that we now describe.

Consider a gaseous component at fugacity f_2^G dissolved isothermally in a liquid not near its critical temperature. The solution process is accompanied by a change in enthalpy and in entropy, as occurs when two liquids are mixed. However, in addition,

[10] See, for example, L. Lue and D. Blankschtein, 1992, *J. Phys. Chem.*, 96: 8582, who use the integral-equation theory of fluids for describing the solubilities of gases in water.

the solution process for the gas is accompanied by a large decrease in volume because the partial molar volume of the solute in the condensed phase is much smaller than that in the gas phase. This large decrease in volume distinguishes solution of a gas in a liquid from solution of another liquid or of a solid. Therefore, to apply regular solution theory (that assumes no volume change), it is necessary first to "condense" the gas to a volume close to the partial molar volume that it has as a solute in a liquid solvent. The isothermal solution process of the gaseous solute is then considered in two steps, I and II:

$$\Delta g = \Delta g_{\mathrm{I}} + \Delta \bar{g}_{\mathrm{II}} \tag{10-34}$$

$$\Delta g_{\mathrm{I}} = RT \ln \frac{f^L_{\text{pure 2}}}{f_2^G} \tag{10-35}$$

$$\Delta \bar{g}_{\mathrm{II}} = RT \ln \gamma_2 x_2 \tag{10-36}$$

where $f^L_{\text{pure 2}}$ is the fugacity of (hypothetical) pure liquid solute and γ_2 is the symmetrically normalized activity coefficient of the solute referred to the (hypothetical) pure liquid ($\gamma_2 \to 1$ as $x_2 \to 1$).

In the first step, the gas isothermally "condenses" to a hypothetical state having a liquid-like volume. In the second step, the hypothetical, liquid-like fluid dissolves in the solvent. Since the solute in the liquid solution is in equilibrium with the gas that is at fugacity f_2^G, the equation of equilibrium is

$$\Delta g = 0 \tag{10-37}$$

We assume that the regular-solution equation gives the activity coefficient for the gaseous solute:[11]

$$RT \ln \gamma_2 = v_2^L (\delta_1 - \delta_2)^2 \Phi_1^2 \tag{10-38}$$

where δ_1 is the solubility parameter of solvent, δ_2 is the solubility parameter of solute, v_2^L is the molar "liquid" volume of solute, and Φ_1 is the volume fraction of solvent.

Substitution of Eqs. (10-34), (10-35), (10-36), and (10-38) into Eq. (10-37) gives the solubility:

$$\frac{1}{x_2} = \frac{f^L_{\text{pure 2}}}{f_2^G} \exp\left[\frac{v_2^L (\delta_1 - \delta_2)^2 \Phi_1^2}{RT}\right] \tag{10-39}$$

[11] See Sec. 7.2 for a discussion of the regular-solution equation. A list of liquid molar volumes and solubility parameters for some common solvents is given in Table 7-1.

Equation (10-39) forms the basis of the correlating scheme; it requires three parameters for the gaseous component as a hypothetical liquid: the pure liquid fugacity, the liquid volume, and the solubility parameter. These parameters are all temperature dependent; however, the theory of regular solutions assumes that at constant composition

$$\ln \gamma_2 \propto \frac{1}{T} \tag{10-40}$$

and therefore, the quantity $v_2^L(\delta_1 - \delta_2)^2 \Phi_1^2$ is not temperature-dependent. As a result, any convenient temperature may be used for v_2^L and δ_2 provided that the same temperature is also used for δ_1 and v_1^L. (The most convenient temperature, that used here, is 25°C.) The fugacity of the hypothetical liquid, however, must be a function of temperature.

The three correlating parameters for the gaseous solutes were calculated (Prausnitz and Shair, 1961) from experimental solubility data. The molar volume v_2^L and the solubility parameter δ_2, both at 25°C, are given for 11 gases in Table 10-7. However, as explained above, Eq. (10-39) is not restricted to 25°C; in principle, it may be used at any temperature provided that it is well removed from the critical temperature of the solvent. A correlation similar to that of Shair has been given by Yen and McKetta (1962).

Table 10-7 "Liquid" volumes and solubility parameters for gaseous solutes at 25°C.

Gas	v^L (cm^3 mol^{-1})	δ (J cm^{-3})$^{1/2}$
N_2	32.4	5.30
CO	32.1	6.40
O_2	33.0	8.18
Ar	57.1	10.9
CH_4	52	11.6
CO_2	55	12.3
Kr	65	13.1
C_2H_4	65	13.5
C_2H_6	70	13.5
Rn	70	18.1
Cl_2	74	17.8

For nonpolar systems, where the molecular size ratio is far removed from unity, it is necessary to add a Flory-Huggins entropy term to the regular-solution equation for representing gas solubility, as shown by King and coworkers (1977), who correlated solubility data for carbon dioxide, hydrogen sulfide, and propane in a series of alkane solvents (hexane to hexadecane).

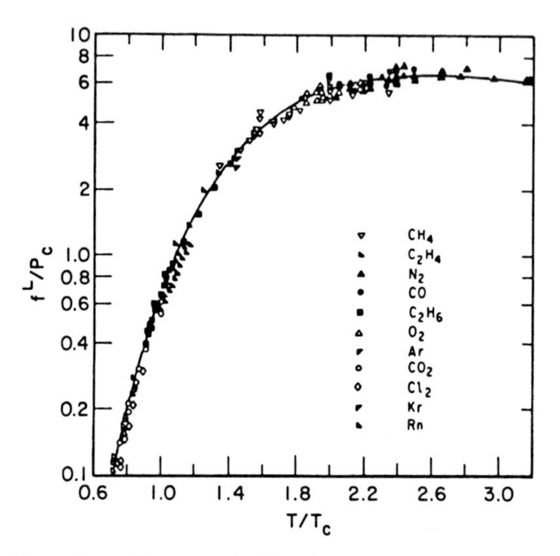

Figure 10-13 Fugacity of hypothetical liquid at 1.013 bar.

The fugacity of the hypothetical pure liquid has been correlated in a corresponding-states plot shown in Fig. 10-13; the fugacity of the solute, divided by its critical pressure, is shown as a function of the ratio of the solution temperature to the solute's critical temperature. The fugacities in Fig. 10-13 are for a total pressure of 1.013 bar. If the solution under consideration is at a considerably higher pressure, the Poynting correction should be applied to the fugacity as read from Fig. 10-13; thus

$$\underset{\text{(at total pressure } P)}{f^L_{\text{pure 2}}} = \underset{\text{(from Fig.10-13)}}{f^L_{\text{pure 2}}} \exp\left[\frac{v_2^L(P-1.013)}{RT}\right] \tag{10-41}$$

where P is in bar.

Equation (10-39) contains Φ_1, the volume fraction of the solvent and, therefore, solving for x_2 requires a trial-and-error calculation; however, the calculation converges rapidly. If the partial pressure of the gas is not large, x_2 is usually very small, and thus setting Φ_1 equal to unity in Eq. (10-39) provides an excellent first approximation.

Shair's technique for correlating gas solubilities with regular-solution theory can readily be extended to mixed solvents. To estimate the solubility of a gas in a mixture of two or more solvents, Eq. (10-39) should be replaced by

$$\frac{1}{x_2} = \frac{f^L_{\text{pure 2}}}{f_2^G} \exp\left[\frac{v_2^L(\delta_2 - \bar{\delta})^2}{RT}\right] \tag{10-42}$$

where $\bar{\delta}$ is an average solubility parameter for the entire solution:

$$\bar{\delta} = \sum_i \Phi_i \delta_i$$

The summation in the definition above is over *all* components, including the solute.

Table 10-7 and Fig. 10-13 do not include information for the light gases, helium, hydrogen, and neon. For these gases, Hildebrand's correlations are useful near room temperature and for low temperatures, a separate correlation for hydrogen is available.[12] Good solubility data for hydrogen, helium, and neon are rare at higher temperatures.

The correlation given by Eq. (10-39), Table 10-7, and Fig. 10-13 gives fair estimates of gas solubilities over a moderate temperature range for nonpolar gases and liquids. It is not to be expected that this simple correlation should give highly accurate results but, in view of the poor accuracy of many of the experimental solubility data reported in the literature, the estimated solubilities may, in some cases, be more reliable than the experimental ones. Whenever really good experimental data are available, they should be given priority, but whenever there is serious disagreement between observed results and those calculated from a pertinent correlation, one must not, without further study, immediately give preference to the experimental value.

Gas solubilities can also be calculated from an equation of state using the methods discussed in Chaps. 3 and 12. The essential requirement is that the equation of state must be valid for the solute-solvent mixture from zero density to the density of the liquid. The equation of state gives fugacity coefficients; Henry's constant for solute 2 in solvent 1 is given by

$$H_{2,1} = \varphi_2^{L,\infty} P \qquad (10\text{-}43)$$

where $\varphi_2^{L,\infty}$ is the fugacity coefficient of the solute in the liquid solvent at infinite dilution. For example, Plöcker *et al.* (1978) calculated Henry's constants for hydrogen in hydrocarbons using Lee and Kesler's form of the Benedict-Webb-Rubin equation (see Sec. 4.14) with results[13] shown in Fig. 10-14. Another example is provided by Liu (1979), who calculated Henry's constants for a variety of solutes in low-density polyethylene using an equation of state based on perturbed-hard-chain theory (see Sec. 12.10); Liu's results are shown in Fig. 10-15. Calculations for Henry's constants based on an equation of state inevitably require at least one adjustable binary parameter.

[12] See Table 10-4

[13] Similar calculations can be made using the corresponding-states tables of Lee and Kesler (see Chap. 4) as shown by Yorizane and Miyano, 1978, *AIChE J.*, 24: 181.

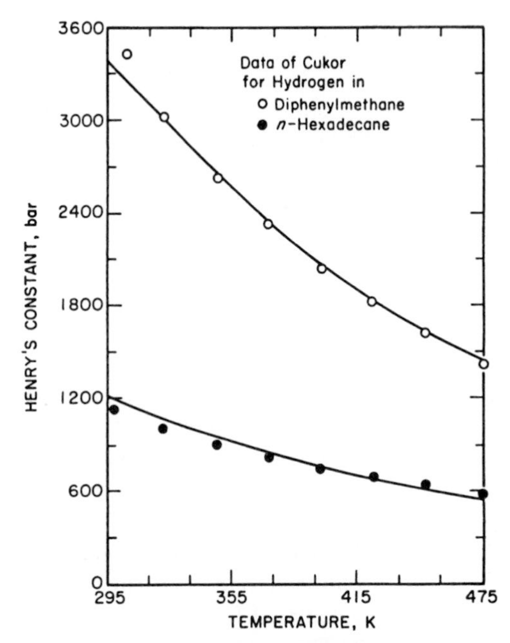

Figure 10-14 Henry's constants for hydrogen calculated from the Lee-Kesler extension of the Benedict-Webb-Rubin equation of state. Each binary system requires one adjustable binary parameter. Experimental data from Cukor (1972).

Gas Solubilities from Scaled-Particle Theory. A statistical-mechanical theory of dense fluids developed by Reiss and coworkers (1960) yields an approximate expression for the reversible work required to introduce a spherical particle of species 2 into a dense fluid containing spherical particles of species 1. This theory, called *scaled particle theory*, serves as a convenient point of departure for correlating gas solubilities, as shown, for example, by Pierotti (1976), Wilhelm and Battino (1971), and Geller *et al.* (1976).

Consider a very dilute solution of nonpolar solute 2 in nonpolar solvent 1 at low pressure and at a temperature well below the critical of the solvent. Henry's constant is given by

$$\ln \frac{H_{2,1}^{(P_1^s)} v_1}{RT} = \frac{\bar{g}_c}{RT} + \frac{\bar{g}_i}{RT} \tag{10-44}$$

where v_1 is the molar volume of the solvent. Equation (10-44) assumes that the dissolution process can be broken into two steps: in the first step, a cavity is made in the solvent to allow introduction of a solute molecule; in the second step, the solute molecule interacts with surrounding solvent. For the solute, the partial molar Gibbs energies $\bar{g}_c$ and $\bar{g}_i$ stand, respectively, for the first step (cavity formation) and for the second step (interaction).

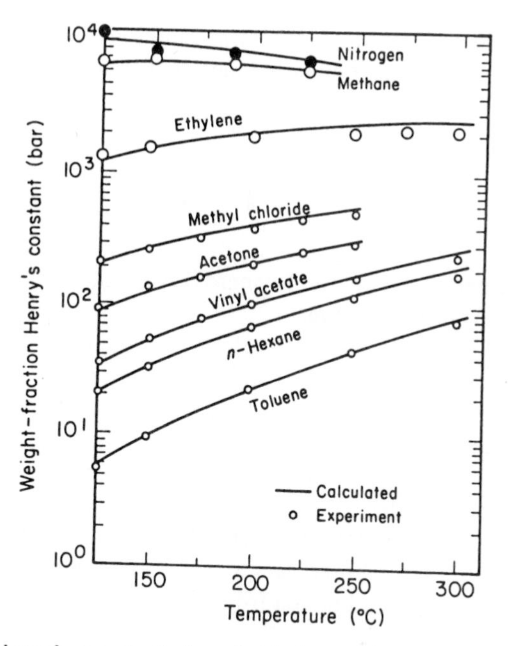

Figure 10-15 Henry's constants in polyethylene calculated from the perturbed-hard-chain equation of state. Each binary system requires one adjustable parameter. Here Henry's constant is defined as the ratio of fugacity to weight fraction of the solute in the limit as the weight fraction approaches zero.

Let a_1 be the hard-sphere diameter for solvent and a_2 that for solute. If the total pressure is low, scaled-particle theory gives for $\bar{g}_c$,

$$\frac{\bar{g}_c}{RT} = \frac{6Y}{1-Y}(2r^2 - r) + \frac{18Y^2}{(1-Y)^2}\left(r - \frac{1}{2}\right)^2 - \ln(1-Y) \tag{10-45}$$

where (N_A is Avogadro's constant)

$$r = \frac{a_1 + a_2}{2a_1}$$

$$Y = \frac{\pi a_1^3 N_A}{6v_1}$$

To obtain an expression for $\bar{g}_i$, we assume first that there are no entropic contributions to the interaction; in other words, we assume that all changes in entropy that result from dissolution of a gas in a solvent are given by the cavity-formation calculation. Second, we assume some potential function for describing solute-solvent intermolecular forces. When the Lennard-Jones-12,6 potential is used, Wilhelm and Battino obtain

$$\bar{g}_i = -\left(\frac{32\pi R}{9v_1}\right)\left(\frac{N_A\sigma_{12}^3\varepsilon_{12}}{k}\right) \tag{10-46}$$

where σ_{12} and ε_{12} are parameters in the Lennard-Jones potential, k is Boltzmann's constant and R is the universal gas constant.

Equations (10-44), (10-45), and (10-46) provide the basis for a correlation of gas solubilities in simple nonpolar systems. Within the approximations used here, it is reasonable to set $\sigma_{12} = 1/2(a_1 + a_2)$. The adjustable parameters, then, are a_1, a_2, and ε_{12}/k. Making the further (drastic) simplification $\varepsilon_{12} = (\varepsilon_1\varepsilon_2)^{1/2}$, Wilhelm and Battino correlate solubilities in terms of pure-component parameters only, as shown in Table 10-8. In this correlation, the parameters for the solutes were determined independently (primarily from second-virial-coefficient data) and those for the solvents were adjusted to give the best fit for the gas-solubility data at 25°C. The adjustable parameters are in reasonable agreement with those obtained from other physical properties.

Table 10-8 Correlating parameters for gas solubilities in nonpolar systems at 25°C using scaled-particle theory (Wilhelm and Battino, 1971).

	a (Å)	ε/k (K)
Solutes		
Helium	2.63	6.03
Neon	2.78	39.9
Argon	3.40	122
Krypton	3.60	158
Xenon	4.06	219
Hydrogen	2.87	29.2
Nitrogen	3.70	95
Oxygen	3.46	118
Carbon dioxide	3.94	195
Methane	3.70	157
Ethane	4.38	236
Carbon tetrafluoride	4.66	134
Solvents		
n-Heptane	6.23	563
n-Octane	6.52	594
Cyclohexane	5.62	540
Benzene	5.25	507
Carbon tetrachloride	5.36	528

It is possible to extend the scaled-particle method to systems containing polar solutes by adding an induction term to Eq. (10-46), as shown by Wilhelm and Battino (1971) and Geller *et al.* (1976). Extension to aqueous systems is discussed by Pierroti (1976).

Scaled-particle theory is a particular form of perturbation theory, i.e., a theory that uses theoretical results for hard-sphere systems (no attractive forces) as a point of departure. Application of perturbation theory to gas solubilities is discussed, for example, by Gubbins and coworkers (Shoor, 1969; Tiepel, 1973), Masterson *et al.* (1969), and Boublik and Lu (1978).

Scaled-particle theory, and most similar perturbation theories, assume that all molecules are rigid, and that the so-called "internal" molecular degrees of freedom (rotation and vibration) are not changed by the dissolution process. This assumption is reasonable only for small spherical molecules. For nonspherical molecules it is likely that, contrary to assumption, solute-solvent interaction forces contribute to the entropy.

When correlating solubility data, failure to take into account changes in the "internal" degrees of freedom is often not important when attention is confined to solubilities at one temperature because the errors are absorbed in the adjustable parameters. However, when solubilities are to be correlated for an appreciable range of temperature, neglecting the entropy of interaction [as was done in deriving Eq. (10-46)] may produce a temperature dependence for adjustable parameters r and ε_{12}. An example of such a correlation is provided by Schulze (1981), who used scaled-particle theory to correlate solubilities of simple gases in water from 0 to (about) 300°C.

10.6 Gas Solubility in Mixed Solvents

Good solubility data for gases in pure liquids are not plentiful, while solubility data in mixed solvents are scarce. With the aid of a simple molecular-thermodynamic model, however, it is often possible to make a fair estimate of the solubility of a gas in a simple solvent mixture, provided that the solubility of the gas is known in each of the pure solvents that comprise the mixture. The procedure for making such an estimate is, essentially, based on Wohl's expansion (Secs. 6.10 and 6.14), as discussed by O'Connell (1964).

Let subscript 2 stand for the gas as before, and let subscripts 1 and 3 stand for the two (miscible) solvents. To simplify matters, we confine attention to low or moderate pressures where the effect of pressure on liquid-phase properties can be neglected. For the ternary liquid phase, we write the simplest (two-suffix Margules) expansion of Wohl for the excess Gibbs energy at constant temperature

$$\frac{g^{E\,\text{(ternary)}}}{RT} = a_{12}x_1x_2 + a_{13}x_1x_3 + a_{23}x_2x_3 \tag{10-47}$$

where each a_{ij} is a constant characteristic of the ij binary pair.

From Eq. (10-47) we can compute the symmetrically normalized activity coefficient γ_2 of the gaseous solute using Eq. (6-25). The unsymmetrically normalized activity coefficient γ_2^* can then be found by

$$\gamma_2^* = \gamma_2 \exp(-a_{12}) \tag{10-48}$$

where the definition of γ_2^* is

$$\gamma_2^* \equiv \frac{f_2}{x_2 H_{2,1}} \tag{10-49}$$

As in previous sections, $H_{2,1}$ stands for Henry's constant of component 2 in solvent 1 at system temperature. O'Connell also has shown that for this simple model, parameters a_{23} and a_{12} are related to the two Henry's constants:

$$a_{23} = a_{12} + \ln\left(\frac{H_{2,3}}{H_{2,1}}\right) \tag{10-50}$$

where $H_{2,3}$ is Henry's constant for the solute in solvent 3 at system temperature.

From Eqs. (10-47) and (10-48), utilizing Eq. (6-25), we obtain

$$\ln \gamma_2^* = a_{12}[x_1(1-x_2)-1] + a_{23}x_3(1-x_2) - a_{13}x_1x_3 \tag{10-51}$$

We now use Eqs. (10-50) and (10-51) to obtain an expression for $H_{2,\text{mixture}}$, Henry's constant for the solute in the mixed solvent. For some fixed ratio of solvents 1 and 3,

$$H_{2,\text{mixture}} \equiv \lim_{x_2 \to 0} \frac{f_2}{x_2} = \lim_{x_2 \to 0} \gamma_2^* H_{2,1} \tag{10-52}$$

From Eq. (10-51),

$$\lim_{x_2 \to 0} \gamma_2^* = (a_{23} - a_{12})x_3 - a_{13}x_1x_3 \tag{10-53}$$

Substitution of Eqs. (10-50) and (10-53) into Eq. (10-52) gives the desired result:

$$\boxed{\ln H_{2,\text{mixture}} = x_1 \ln H_{2,1} + x_3 \ln H_{2,3} - a_{13}x_1x_3} \tag{10-54}$$

Equation (10-54) says that the logarithm of Henry's constant in a binary solvent mixture is a linear function of the solvent composition whenever the two solvents (without solute) form an ideal mixture ($a_{13} = 0$). If the solute-free mixture exhibits positive deviations from Raoult's law ($a_{13} > 0$), Henry's constant in the mixture is smaller (or solubility is larger) than that corresponding to an ideal mixture of the same composition. Similarly, if $a_{13} < 0$, the solubility of the gas is lower than what it would be if the solvents formed an ideal mixture. Constant a_{13} must be estimated from vapor-liquid equilibrium data for the solvent mixture.[14]

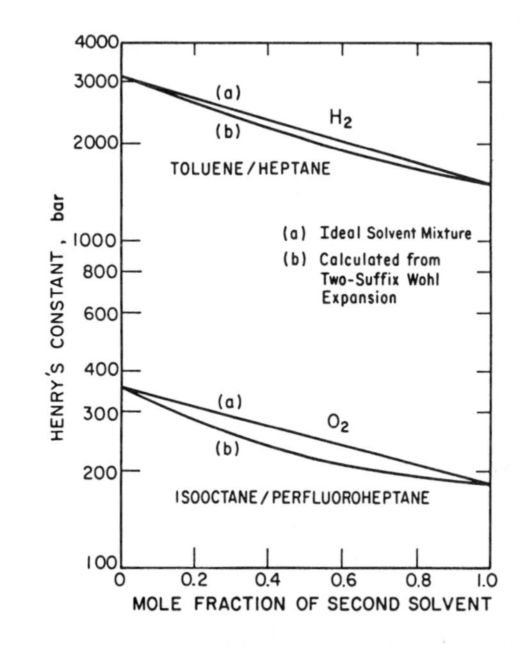

Figure 10-16 Calculated Henry's constants in solvent mixtures at 25°C.

According to Eq. (10-54), the effect of nonideal mixing of the solvents is not large. To illustrate, two calculated examples at 25°C are shown in Fig. 10-16. The first one considers Henry's constants for hydrogen in mixtures of toluene and heptane. Since toluene and heptane show only modest positive deviations from Raoult's law (g^E = 200 J mol^{-1} for the equimolar mixture), a_{13} is small and, as shown in Fig. 10-16, Henry's constant calculated from Eq. (10-54) differs little from that calculated assuming an ideal solvent mixture. The second example considers Henry's constants for oxygen in mixtures of isooctane and perfluoroheptane, a system that exhibits large deviations from Raoult's law; for the equimolar mixture of these solvents g^E = 1380 J mol^{-1}

[14] For nonpolar solvents an estimate can sometimes be made using the theory of regular solutions:

$$a_{13} \approx \frac{(\delta_1 - \delta_3)^2 (v_1^L + v_3^L)}{2RT}$$

where δ is the solubility parameter and v^L the liquid molar volume. See Sec. 7.2.

and at temperatures only slightly below 25°C, these two liquids are no longer completely miscible. In this case there is a more significant difference between Henry's constants calculated with and without nonideality of the solvent mixture. For an equimolar mixture of the two solvents, the calculated solubility of oxygen is enhanced by nearly 20% as a result of the solvent mixture's nonideality.

Equation (10-54) is readily generalized to solvent mixtures containing any desired number of solvents. For an m-component system where the gas is designated by subscript 2, Henry's constant for the gaseous solute is given by

$$\ln H_{2,\text{mixture}} = \sum_{\substack{j=1 \\ j\neq 2}}^{m} x_j \ln H_{2,j} - \sum_{\substack{j=1 \\ j\neq 2}}^{m-1} \sum_{\substack{k>j \\ k\neq 2}}^{m} a_{jk} x_j x_k \tag{10-55}$$

Equation (10-55), like Eq. (10-54), follows directly from the simplest form of Wohl's expansion as given by Eq. (10-47) for a ternary mixture.

Table 10-9 shows results obtained for the solubility of carbon dioxide in an aqueous solution of diglyme (diethylene glycol dimethyl ether with molar mass of 134.2 g mol^{-1}).

Table 10-9 Parameter a_{13} of Eq. (10-54) (with volume fractions instead of mole fractions) for the solubility of carbon dioxide in aqueous diglyme at 25°C (Sciamanna and Lynn, 1988). Henry's constants in bar. Values of $H_{2,\text{mixture}}$ for ideal solutions were obtained from Eq. (10-54) (with mole fractions replaced by volume fractions) and with a_{13} = 0. When the indicated values of a_{13} are used, calculated Henry's constants are identical to those obtained from experiment.

Weight % water in diglyme	$H_{2,\text{mixture}}$ (Experiment)	$H_{2,\text{mixture}}$ (ideal mixture) (Calculated)	a_{13}
0.99	39.74	38.08	-4.4
1.97	41.66	39.70	-2.4
2.97	47.26	41.64	-3.9
3.95	49.34	43.89	-2.6
5.99	57.88	50.24	-1.9

The negative values for a_{13} indicate that the water-solvent mixture exhibits negative deviations from Raoult's law, probably due to hydrogen bonding between water and solvent molecules. However, for the particular case shown in Table 10-9, the ideal-solution mixing rule [Eq. (10-54) with $a_{13} = 0$] is adequate as a first approximation.

In these calculations, the important assumption is that Eq. (10-47) (or a similar relation using volume fractions) gives a valid description of the excess Gibbs energy of the ternary mixture. This assumption provides a reasonable approximation for some solutions of simple fluids, but for mixtures containing polar or hydrogen-bonded liquids a better model is required. For such cases, binary data may not be sufficient; a ternary constant may be necessary.

Nitta *et al.* (1973) have also used Wohl's expansion to obtain an expression for the solubility of a gas in a mixed solvent. As shown below, this expression takes into account the effect of self-association of one of the solvents.

We define a residual quantity $\Re$ by

$$\Re = \ln H_{i,\text{mixture}} - \sum_{j=1}^{m} \Phi_j \ln H_{i,j} \tag{10-56}$$

where $H_{i,\text{mixture}}$ is Henry's constant for solute i in the solvent mixture containing m solvents; $H_{i,j}$ is Henry's constant for solute i in solvent j and Φ_j is the volume fraction of solvent j in the solvent mixture on a solute-free basis.[15]

We can obtain an expression for $\Re$ using Wohl's expansion for the excess Gibbs energy; for all binary pairs i-j, i-k, j-k, ..., we assume the Flory-Huggins equation. Omitting algebraic details, we obtain

$$\Re = (\Re)_{\text{size}} + (\Re)_{\text{physical interaction}} \tag{10-57}$$

where

$$(\Re)_{\text{size}} = \sum_{j=1}^{m} \Phi_j \ln \frac{v_j}{v} \tag{10-58}$$

$$(\Re)_{\text{physical interaction}} = -\frac{v_i}{RT} \sum_{j>k}^{m} \Phi_j \Phi_k \chi_{jk} \tag{10-59}$$

where v, the molar volume of the solute-free solvent mixture, is given by,

[15] It is sometimes customary to define an excess Henry's constant $H^E_{i,\text{mixture}}$

$$\ln H^E_{i,\text{mixture}} = \ln H_{i,\text{mixture}} - \sum_{j=1}^{m} x_j \ln H_{i,j}$$

where x is the mole fraction.

If the solvents form an ideal mixture over the entire range of solvent compositions, $\ln H^E_{i,\text{mixture}} = 0$. In a binary mixture containing solvents a and b, the residual quantity $\Re$ is related to $H^E_{i,\text{mixture}}$ by

$$\ln H^E_{i,\text{mixture}} = \Re + \left(\frac{\Phi_a \Phi_b}{r\Phi_a + \Phi_b}\right)(r-1)(\ln H_{i,b} - \ln H_{i,a})$$

where r is the ratio of liquid molar volumes v_b/v_a. We see that $\ln H^E_{i,\text{mixture}}$ vanishes only when *both* $(\Re)_{\text{size}}$ and $(\Re)_{\text{physical interaction}}$ go to zero.

$$v = \sum_{j=1}^{m} x_j v_j \tag{10-60}$$

with x denoting the mole fraction. In Eq. (10-59), v_i is the molar "liquid" volume of solute i and χ_{jk} is a Flory interaction parameter for solvents j and k, here expressed in units of energy per unit volume.

If one of the solvents (say, solvent k) is an alcohol (or amine) that associates continuously, then an additional term appears in Eq. (10-57):

$$\Re = (\Re)_{\text{size}} + (\Re)_{\text{physical interaction}} + (\Re)_{\text{association}} \tag{10-61}$$

The association term is given by

$$(\Re)_{\text{association}} = -\left(\frac{v_j}{v_k}\right)\Phi_k\left(\frac{2}{1+\sqrt{1+4K_k\Phi_k}} - \frac{2}{1+\sqrt{1+4K_k}}\right) \tag{10-62}$$

where the association constant for solvent k is defined by

$$K_k = \left(\frac{c_n}{v_k}\right)\frac{1}{(c_{n-1}c_1)} \tag{10-63}$$

Here, c_1 stands for the concentration (moles per unit volume) of monomer and c_n for the concentration of polymer of degree n.

To illustrate this discussion of gas solubility in a mixed solvent, Nitta *et al.* (1973) reduced experimental solubility data for nitrogen in the mixed solvent isooctane (j)/n-propanol (k) at 25°C. Parameters are given in Table 10-10 and results are shown in Fig. 10-17.

Table 10-10 Parameters for calculating Henry's constant for nitrogen (i) in the mixed solvent system isooctane (j)/n-propanol (k) at 25°C (Nitta *et al.*, 1973).

Molar liquid volumes ($cm^3\ mol^{-1}$):		
$v_j = 166$	$v_k = 75.2$	v_i (estimated) = 32.4
Association constant:		
$K_k = 110$		
Flory parameter:		
$\chi_{jk} = 4.23\ J\ cm^{-3}$ (from experimental vapor-liquid equilibria for the j-k mixture)		
Henry's constants in the single solvents (bar):		
$H_{i,j} = 662.5$	$H_{i,k} = 2515$	

For positive χ_{jk}, the physical interaction contribution to $\Re$ is negative; the association contribution is also negative. These contributions, therefore, tend to increase gas solubility. However, the size contribution is positive, tending to lower gas solubility.

As indicated in Fig. 10-17, agreement between calculated and observed Henry's constants is remarkable, considering that only binary and single-component data were used in the calculation. In this example, if Henry's constants had been calculated assuming that a plot of $\ln H_{i,\text{mixture}}$ versus x_k is a straight line, the calculated Henry's constants at $x_k = 0.5$ would be about 30% too large. In this case that error is not as large as it might be because of partial cancellation in Eq. (10-61), where the first term on the left is positive while the other two are negative.

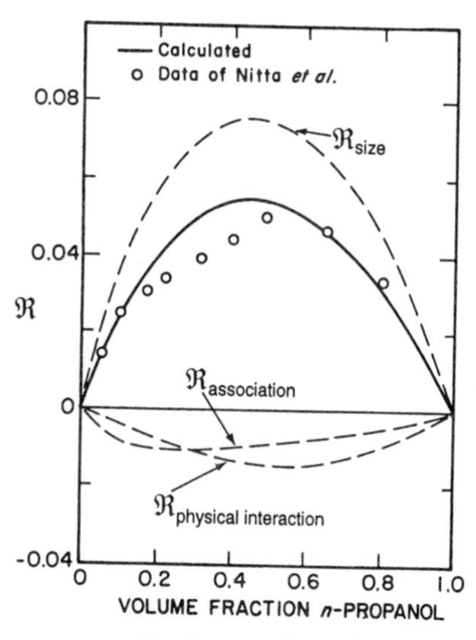

Figure 10-17 Solubility of nitrogen in a mixed solvent containing isooctane and *n*-propanol at 25°C. See Eq. (10-56).

In general it is not possible, without strong simplifying assumptions, to calculate the solubility of a gas in a mixed solvent using only binary data. A more rigorous discussion, based on the direct correlation function of liquids, has been presented by O'Connell (1971) and an extension of perturbation theory to multisolvent systems has been presented by Tiepel and Gubbins (1973).

10.7 Chemical Effects on Gas Solubility

The gas-solubility correlations of Hildebrand and Scott and of Shair, discussed in Sec. 10.5, are based on a consideration of physical forces between solute and solvent; they are not useful for those cases where chemical forces are significant. Sometimes these chemical effects are not large and therefore a physical theory is a good approximation. But frequently chemical forces are dominant; in that case, correlations cannot easily be established because specific chemical forces are not subject to simple generalizations. An extreme example of a chemical effect in a gas-liquid solution is provided by the interaction between sulfur trioxide and water (to form sulfuric acid); somewhat milder examples are the interaction between acetylene and acetone (hydrogen bonding) or that between ethylene and an aqueous solution of silver nitrate (Lewis acid-base complex formation); a still weaker case of the effect of chemical forces was mentioned earlier in connection with Fig. 10-11, where the solubility of (acidic) carbon dioxide in (basic) benzene and toluene is larger than that predicted by a physical correlation. In each of these cases the solubility of the gas is enhanced as a result of a specific affinity between solute and solvent.

Systematic studies of the effect of chemical forces on gas solubility are rare, primarily because it is difficult to characterize chemical forces in a quantitative way; the chemical affinity of a solvent (unlike its solubility parameter) depends on the nature of the solute, and therefore a measure of chemical effects in solution is necessarily relative rather than absolute.[16] A few examples, given below, illustrate how the effect of chemical forces on gas solubility can be studied in at least a semiquantitative way.

The importance of chemical effects is shown by the solubilities of dichlorofluoromethane (Freon-21) at 32.2°C in a variety of solvents. Solubility data in 15 solvents, obtained by Zellhoefer *et al.* (1938) are shown in Fig. 10-18 as a function of pressure. Because the critical temperature of Freon-21 (178.5°C) is well above 32.2°C, it is possible to calculate an ideal solubility [Eq. (10-2)] without extrapolation of the vapor-pressure data; this ideal solubility is shown in Fig. 10-18 by a dashed line. Figure 10-18 indicates that the solvents fall roughly into three groups: In the first group, the solubility is less than ideal (positive deviations from Raoult's law); in the second group, deviations from Raoult's law are very small; and in the third group, the solubility is larger than ideal (negative deviations from Raoult's law). These data can be explained qualitatively by the concept of hydrogen bonding; Freon-21 has an active hydrogen atom and whenever the solvent can act as a proton acceptor, it is a relatively good solvent. The stronger the proton affinity (Brönsted basicity) of the solvent, the better that solvent is, provided that the proton-accepting atom of the solvent molecule is not already "taken" by a proton from a neighboring solvent molecule. In other words, if the solvent molecules are strongly self-associated, they will not be available to form hydrogen bonds with the solute. As a result, strongly associated substances are poor solvents for solutes such as Freon-21 that can form only weak hydrogen bonds and thus cannot compete successfully for proton acceptors. It is for this reason that the glycols are poor solvents for Freon-21; while no data for alcohols are given in Fig. 10-18, it is likely that methanol and ethanol would also be poor solvents for Freon 21.

[16] As shown later, a potentially useful method for characterizing specific solute-solvent interactions is provided by donor-acceptor numbers. See, for example, Viktor Gutmann, 1978, *The Donor-Acceptor Approach to Molecular Interactions,* New York: Plenum Press.

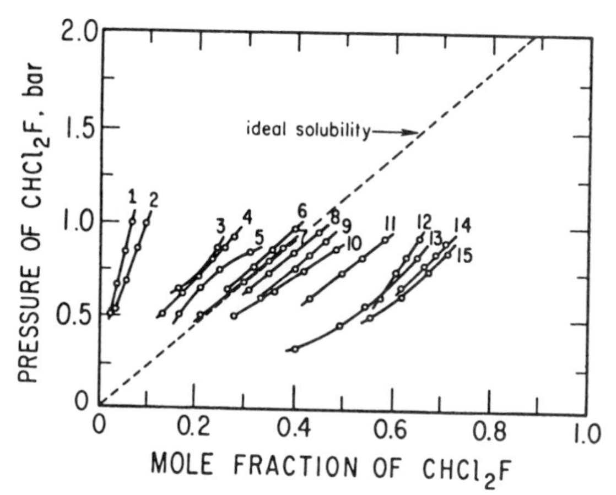

Figure 10-18 Solubility of Freon-21 in liquid solvents at 32.2°C. Solvents are: 1. Ethylene glycol; 2. Trimethylene glycol; 3. Decalin; 4. Aniline; 5. Benzotrifluoride; 6. Nitrobenzene; 7. Tetralin; 8. Bis-β methylthioethyl sulfide; 9. Dimethylaniline; 10. Dioxan; 11. Diethyl oxalate; 12. Diethyl acetate; 13. Tetrahydrofurfuryl laurate; 14. Tetraethyl oxamide; 15. Dimethyl ether of tetraethylene glycol.

Aromatic solvents aniline, benzotrifluoride, and nitrobenzene are weak proton acceptors and it appears that for these solvents, chemical forces (causing negative deviations from Raoult's law) are just strong enough to overcome physical forces (that usually cause positive deviations from Raoult's law), with the result that the observed solubilities are close to ideal. Finally, those solvents that are powerful proton acceptors – and whose molecules are free to accept protons – are excellent solvents for Freon-21. Dimethylaniline, for example, is a much stronger base and hence a better solvent than aniline. The oxygen atoms in ethers and the nitrogen atoms in amides are also good proton acceptors for solute molecules because ethers and amides do not self-associate to an appreciable extent.

Figure 10-19 presents a correlation that provides Henry's constants in terms of "chemical" thermodynamic parameters.[17] This figure shows an empirical correlation of a macroscopic thermodynamic property (infinite-dilution activity coefficients for sulfur dioxide in organic solvents) with specific solvent molecular characteristics (the Gutmann donor[18] number, a basicity scale). As discussed earlier, Henry's constant is directly related to the activity coefficient at infinite dilution; see Eq. (6-43). In this case, Henry's constant is the product of $\gamma^{\infty}_{SO_2}$ and the vapor pressure of pure SO_2 liquid at 25°C.

[17] R. J. Demyanovich and S. Lynn, 1991, *J. Sol. Chem.*, 20: 693.

[18] We follow here the Lewis definition of acids (electron-pair acceptors) and bases (electron-pair donors) to distinguish between acceptors and donors.

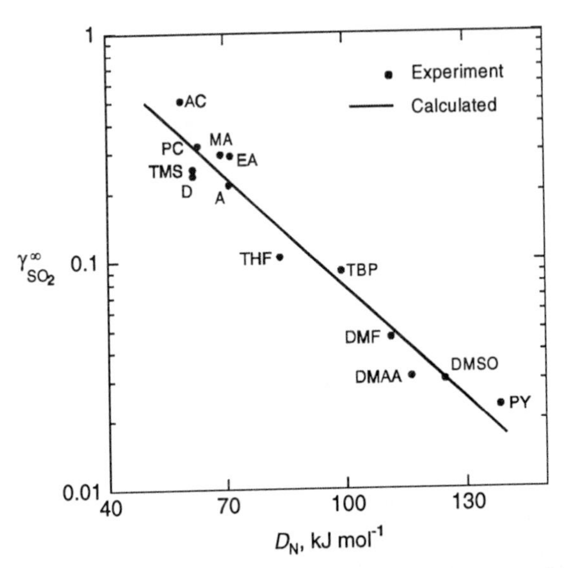

Figure 10-19 Correlation of infinite-dilution activity coefficients of SO_2 with solvent Gutmann donor number at 25°C. The solvents shown are: A - Acetone; AC - Acetonitrile; D - 1,4-Dioxane; DMAA - *N,N*-Dimethylacetamide; DMF - *N,N*-Dimethylformamide; DMSO - Dimethyl sulfoxide; EA - Ethyl acetate; MA - Methyl acetate; PC - Propylene carbonate; PY- Pyridine; TBP - Tributyl phosphate; THF - Tetrahydrofuran; TMS - Tetramethylene sulfone. —— Calculated from Eq. (10-64); ● Experiment.

To characterize the electron-donor ability of a molecule, Gutmann defines the *donor number* D_N (or donicity) as the molar enthalpy value ($-\Delta h$) for the reaction of the donor (D) with $SbCl_5$ as a reference acceptor in a 10^{-3} M solution of $SbCl_5$ in dichloroethane:

$$D + SbCl_5 \rightleftharpoons D{\bullet}SbCl_5 \qquad (-\Delta h_{D\bullet SbCl_5} \equiv D_N)$$

The molar enthalpy of the 1:1 adduct formed is taken as an approximate measure of the energy of the coordinate bond between donor and acceptor. However, the stability of the adduct is a function of Δg^0, the Gibbs energy of adduct formation. If a linear relation exists between Δg^0 and $-\Delta h$, then the enthalpy values (or donor numbers) can be used as a guide to the relative complex stabilities. Gutmann observed a linear relation between $-\Delta h$ and $\ln K$ for the adduct formed by $SbCl_5$ and various donors; here K is the equilibrium constant for adduct formation, related to Δg^0 by $-RT\ln K = \Delta g^0$. Table 10-11 lists the donor numbers for several solvents obtained from calorimetric measurements.

As Fig. 10-19 shows, in the region $60 \leq D_N \leq 150$ kJ mol^{-1}, there is a linear relation between the logarithm of the activity coefficient of SO_2 at infinite dilution and the Gutmann donor number (in kJ mol^{-1}):

Table 10-11 Donor numbers (D_N) for several solvents obtained from calorimetric measurements in 10^{-3} M solutions of $SbCl_5$ in dichloroethane ($SbCl_5$ is the reference acceptor).

Solvent	D_N (kJ mol^{-1})
Benzene	0.4
Nitromethane	11.3
Acetonitrile	60.0
Dioxane	61.9
Acetone	71.1
Water	75.3
Diethyl Ether	80.3
Pyridine	138.5

$$\ln \gamma_{SO_2}^{\infty} = 1.18 - 0.0375 D_N \tag{10-64}$$

Using this linear relation, it is possible to estimate Henry's constants for sulfur dioxide in a variety of "chemical" organic solvents.

While the chemical characteristics of solvents can be used to correlate gas solubilities, it is also possible to use gas-solubility data to characterize solvents as illustrated by solubility data for hydrogen chloride gas in heptane solutions of aromatic hydrocarbons at a low temperature. The solubility data of Brown and Brady (1952) are shown in Fig. 10-20. Because the pressure of hydrogen chloride gas was kept very low, Henry's law holds; Henry's constants (the slopes of the lines in Fig. 10-20) are shown in Table 10-12.

Table 10-12 Henry's constants for solubility of HCl in 5 mol % solutions of aromatics in *n*-heptane, and dissociation constants K for complex formation (all at -78.51°C).

Aromatic solute	H (bar)	K (bar)
None (pure heptane)	6.026	–
Benzotrifluoride	5.626	4.266
Chlorobenzene	5.333	2.000
Benzene	4.666	0.960
Toluene	4.226	0.613
m-Xylene	3.973	0.480
Mesitylene	3.400	0.320

The solubility data show that addition of small amounts of aromatics to heptane increases the solubility of hydrogen chloride. This increase is a result of the electron-donating properties of aromatic molecules that, because of their π-electrons, can act as Lewis bases. The solubility data, therefore, can be explained by postulating the formation of a complex between hydrogen chloride and aromatics; the stability of the complex depends on the ability of the aromatic to donate electrons and this, in turn, is determined by the nature of the substituent on the benzene ring. Halide groups such as CF_3 and Cl withdraw electrons and decrease the electron density on the benzene ring, while methyl groups donate electrons and increase the electron density on the ring. Therefore, relative to benzene, toluene is a stronger Lewis base while chlorobenzene is weaker.

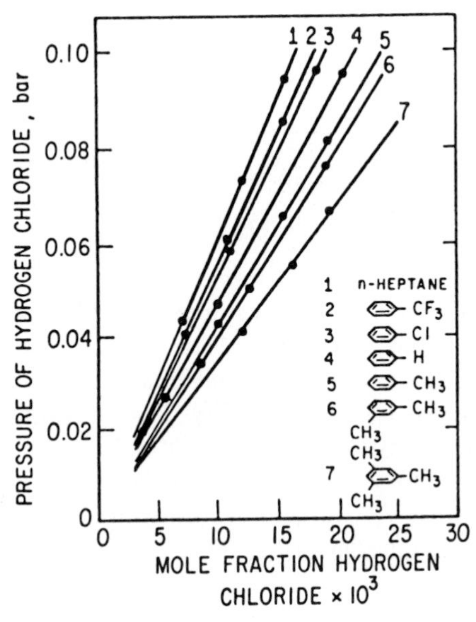

Figure 10-20 Solubility of hydrogen chloride at -78.51°C in *n*-heptane and in 5 mol % solutions of aromatics in *n*-heptane.

The abilities of substituents on benzene to donate or withdraw electrons have also been determined from theoretical calculations, from ionization potentials, and from chemical-kinetic data (Hine, 1962; Hammett, 1970); the results are in excellent agreement with those obtained from solubility studies. Further evidence that hydrogen chloride and aromatics form stable complexes at low temperatures is given by the freezing-point data of Cook *et al.* (1956) for mixtures of hydrogen chloride and mesitylene shown in Fig. 10-21. The maximum at a mole fraction of one-half shows that the stoichiometric ratio of hydrogen chloride to aromatic in the complex is 1:1.

Brown and Brady reduced their solubility data by calculating dissociation equilibrium constants for the complexes; the dissociation constant is defined by

$$K = \frac{x_{\text{free aromatic}}\, p_{\text{HCl}}}{x_{\text{complex}}} \tag{10-65}$$

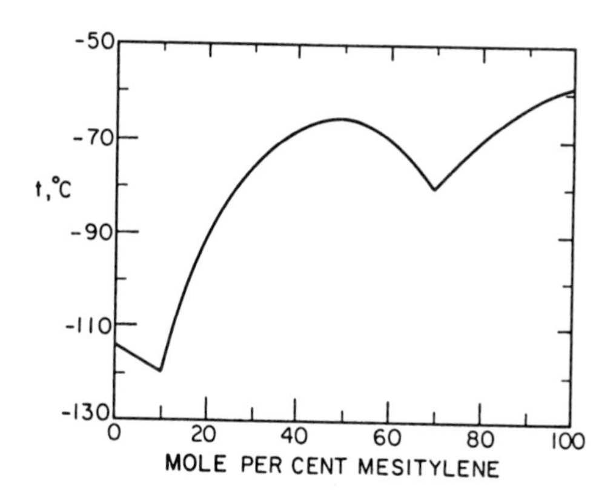

Figure 10-21 Evidence for complex formation: freezing points for hydrochloric acid/mesitylene mixtures.

where x stands for mole fraction and p for partial pressure. According to this definition, the stability of a complex falls as the dissociation constant rises. By simple stoichiometry and by assuming that the complex is nonvolatile, Brown and Brady were able to calculate K from the change in Henry's constant in heptane that results when aromatic molecules are added to the solvent. Their results are shown in Table 10-12. The aromatic components are listed in order of rising basicity. As basicity increases, the solubility also increases and Henry's constant falls. Although the concentration of the aromatic solvent is only 5 mol % in an excess of "inert" heptane, the effect of the aromatic component on the solubility of HCl is pronounced; the presence of 5 mol % mesitylene almost doubles the solubility.

In the preceding paragraphs, we showed the effect of chemical forces on Henry's constant. In the final example, we show how chemical forces can account for deviations from Henry's law; as an illustration we consider the solubility of a gas that dissociates in solution.

Deviations from Henry's law may result from chemical effects even at very low solute concentrations. In Secs. 10.2 and 10.3 we discussed how departures from Henry's law may be due to advanced pressures, to large dissimilarity between solvent and solute, or both. However, whenever the gaseous solute experiences a chemical change such as association or dissociation in the solvent, Henry's law fails because the equilibrium between the vapor phase and the liquid phase is then coupled with an additional (chemical) equilibrium in the liquid phase. We briefly discuss such a case here by analyzing data for the solubility of sulfur dioxide in water at 25°C, shown in Table 10-13.

The first column in Table 10-13 gives the partial pressure of sulfur dioxide that in this case is essentially equal to the fugacity. The second column gives the molality of sulfur dioxide in aqueous solution that at these very low concentrations is proportional to the solute mole fraction. However, when plotted, these data do not yield a straight line; despite the very low concentrations of sulfur dioxide, Henry's law is not valid.

Table 10-13 Solubility of sulfur dioxide in water at 25°C .*

Partial pressure p_{SO_2} (bar)	Molality, m (mol SO_2/1000 g H_2O)	Fraction ionized (α)	Molality of molecular SO_2 (mol SO_2/1000 g H_2O)
0.015	0.0271	0.524	0.0129
0.0456	0.0854	0.363	0.0544
0.0984	0.1663	0.285	0.1189
0.1814	0.2873	0.230	0.2212
0.3374	0.5014	0.184	0.4092
0.5330	0.7643	0.154	0.6470
0.7326	1.0273	0.134	0.8897
0.9312	1.290	0.120	1.134
1.0822	1.496	0.116	1.329

* H. F. Johnstone and P. W. Leppla, 1934, *J. Am. Chem. Soc.*, 56: 2233; W. B. Campbell and O. Maass, 1930, *Can. J. Res.*, 2: 42; O. M. Morgan and O. Maass, 1931, *Can. J. Res.*, 5: 162.

The reason for the failure of Henry's law becomes apparent when we consider that sulfur dioxide plus water produces hydrogen ions and bisulfite ions. When sulfur dioxide gas is in contact with liquid water, we must consider two equilibria:

$$\begin{array}{lll} \text{Gas Phase} & SO_2 & \\ & \updownarrow & \\ \text{Liquid Phase} & SO_2\,(\text{aqueous}) & \rightleftharpoons H^+ + HSO_3^- \end{array}$$

Henry's law governs only the (vertical) equilibrium between the two phases; in this case, Henry's law must be written

$$p = H\,m_M \tag{10-66}$$

where p is the partial pressure of sulfur dioxide, H is a "true" Henry's constant, and m_M is the molality of molecular (nonionized) sulfur dioxide in aqueous solution.

Johnstone and Leppla (1934) have calculated α, the fractional ionization of sulfur dioxide in water, from electrical conductivity measurements; these are given in the third column of Table 10-13. The fourth column gives m_M given by the product of m, the total molality, and $(1 - \alpha)$. When the partial pressure of sulfur dioxide is plotted against m_M (rather than m), a straight line is obtained.

This case is particularly fortunate because independent conductivity measurements are available and thus it was easily possible quantitatively to reconcile the solubility data with Henry's law. In a more typical case, independent data on the liquid solution (other than the solubility data themselves) would not be available; however, even then it is possible to linearize the solubility data by construction of a simple but reasonable model, similar to that given in Sec. 7.14.

For equilibrium between sulfur dioxide in the gas phase and molecular sulfur dioxide in the liquid phase we write

$$p = Hm_{\mathrm{M}} = Hm(1-\alpha) \tag{10-67}$$

For ionization equilibrium in the liquid phase we write

$$K = \frac{m_{\mathrm{H}^+} m_{\mathrm{HSO}_3^-}}{m_{\mathrm{M}}} = \frac{\alpha^2 m^2}{m_{\mathrm{M}}} \tag{10-68}$$

where K is the ionization equilibrium constant. Substituting Eq. (10-67), we have

$$K = \frac{\alpha^2 m^2}{p/H} \tag{10-69}$$

from which we obtain α:

$$\alpha = \frac{\sqrt{p}}{m}\sqrt{\frac{K}{H}} \tag{10-70}$$

Further substitution and rearrangement finally gives

$$\frac{m}{\sqrt{p}} = \frac{\sqrt{p}}{H} + \sqrt{\frac{K}{H}} \tag{10-71}$$

Equation (10-71) shows the effect of ionization on Henry's law; if there were no ionization, $K = 0$ and Henry's law is recovered. The ability of a solute to ionize in solution increases its solubility; however, as the concentration of solute in the solvent rises, the fraction ionized falls. Therefore, the "effective" Henry's constant p/m rises with increasing pressure and a plot of p versus m is not linear but convex toward the horizontal axis.

Figure 10-22 presents solubility data for the sulfur dioxide/water system plotted according to Eq. (10-71). A straight line is obtained. A plot such as this is useful for smoothing, interpolating and cautiously extrapolating limited gas-solubility data for any system where the gaseous solute tends to ionize (or dissociate) in the solvent.

The effect of ionization on solubility is particularly strong when two volatile, ionizing solutes, one basic and one acidic, are dissolved in an ionizing (high-dielectric-constant) solvent such as water (Edwards *et al.*, 1978; Beutier and Renon, 1978a). To illustrate, consider an aqueous solution containing two volatile weak electrolytes: acidic carbon dioxide and basic ammonia. In this ternary mixture, ionization is more drastic than in either binary aqueous mixture because H^+ ions (produced by the reaction

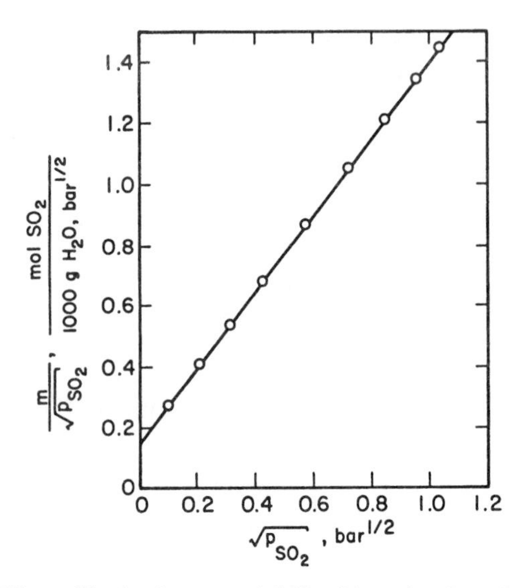

Figure 10-22 Effect of ionization on solubility. Linearization of solubility data for sulfur dioxide in water at 25°C.

of carbon dioxide with water) interact with OH^- ions (produced by the interaction of ammonia with water). Since water has only a small tendency to ionize, the equilibrium

$$H^+ + OH^- \rightleftharpoons H_2O$$

lies far to the right. As a result, the presence of carbon dioxide enhances the solubility of ammonia, and vice versa. This enhancement is shown in Fig. 10-23 that presents calculated partial pressures for the two weak volatile electrolytes in water as a function of carbon dioxide/ammonia ratio at 100°C when the total ammonia concentration is 6.75 molal. The indicated values of K are ionization equilibrium constants for the reactions

$$NH_3 + H_2O \rightleftharpoons NH_4^+ + OH^-$$

and

$$CO_2 + H_2O \rightleftharpoons HCO_3^- + H^+$$

The continuous lines take chemical equilibria into account but the dashed lines do not. Because the ordinate is logarithmic, failure to consider chemical equilibria yields partial pressures that are much too large.

These examples illustrate how chemical effects may have a large influence on solubility behavior. It must be remembered that a solvent is never an inert material that merely acts as a cage for the solute, although it is frequently tempting to think of it that way. Solvent and solute always interact and, historically, when the interaction is strong

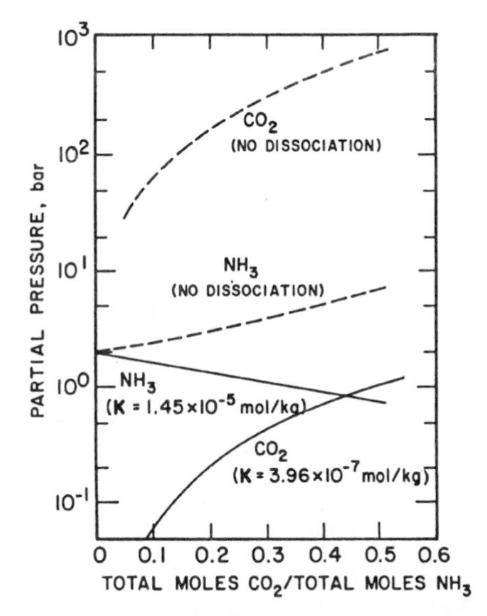

Figure 10-23 Enhancement of solubilities due to chemical interactions. The system carbon dioxide/ammonia/water contains 6.75 molal ammonia at 100°C.

enough to result in a new species, we call it *chemical.* But, as discussed in Secs. 7.9 to 7.14, this is only a convenience, useful for classification. There is no sharp boundary between physical and chemical forces; all molecules have forces acting between them and it is these forces that determine solubility regardless of whether they fall into a physical or chemical classification. For interpreting solubility behavior, however, it is often worthwhile to utilize these classifications by first taking into account the more easily generalized physical forces and then correcting for specific chemical effects. This is essentially what Brown and Brady did since they assumed that the physical forces between hydrogen chloride and hydrocarbons are the same regardless of the nature of the hydrocarbon and that any excess solubility is due to chemical-complex formation. Such a procedure is both simple and practical but its molecular significance must not be taken too seriously unless there is additional, independent evidence for the existence of a complex.

Similarly, the solubility behavior of sulfur dioxide in water can be explained in terms of a physical Henry's law coupled with a chemical equilibrium for a dissociation reaction whose existence is supported by physicochemical data other than the solubility data themselves.

A thorough review of gas solubility is given by Fogg and Gerard (1991). This review clearly shows that despite its obvious importance in biology, medicine, and chemical engineering, our fundamental understanding of gas-liquid equilibria is still in a rudimentary stage. Fogg and Gerrard present a comprehensive description of the experimental methods for measuring gas solubilities. They also discuss factors influencing gas solubility, and present gas-solubility data for various industrial gases.

References

Alvarez, J., H. R. Corti, R. Fernández-Prini, and M. L. Japas, 1994, *Geochim. Cosmochim. Acta*, 58: 2789.

Bender, E., U. Klein, W. Schmitt, and J. M. Prausnitz, 1984, *Fluid Phase Equilibria,* 15: 24.

Benson, B. B. and D. Krause, Jr., 1976, *J. Chem. Phys.*, 64: 689.

Beutier, D. and H. Renon, 1978, *AIChE J.*, 24: 1122.

Beutier, D. and H. Renon, 1978a, *Ind. Eng. Chem. Process Des. Dev.*, 17: 220.

Boublik, T. and B. C.-Y. Lu, 1978, *J. Phys. Chem.*, 82: 2801.

Brown, H. C. and J. O. Brady, 1952, *J. Am. Chem. Soc.*, 74: 3570.

Brunner, E., 1978, *Ber. Bunsenges. Phys. Chem.*, 82: 798.

Brunner, E., 1979, *Ber. Bunsenges. Phys. Chem.*, 83: 715.

Cook, D., Y. Lupien, and W. G. Schneider, 1956, *Can. J. Chem.*, 34: 964.

Cukor, P. M. and J. M. Prausnitz, 1972, *J. Phys. Chem.*, 76: 598.

Dodge, B. F. and R. H. Newton, 1937, *Ind. Eng. Chem.*, 29: 718.

Edwards, T. J., G. Maurer, J. Newman, and J. M. Prausnitz, 1978, *AIChE J.*, 24: 966.

Fogg, G.T. and W. Gerrard, 1991, *Solubility of Gases in Liquids*. Chichester: John Wiley & Sons, 1991. See also Clever and Battino, 1975, *Solutions and Solubilities,* Part 1 (M. R. J. Dack, Ed.), Techniques of Chemistry, Vol.VIII, New York: Wiley-Interscience; W. Gerrard, 1976, *Solubility of Gases and Liquids,* New York: Plenum Press; and W. Gerrard, 1980, *Gas Solubilities,* Oxford: Pergamon Press; IUPAC Solubility Data Series, beginning in 1979, Oxford: Pergamon Press.

Geller, E. B., R. Battino, and E. Wilhelm, 1976, *J. Chem. Thermodynamics,* 8: 197.

Hammett, L. P., 1970, *Physical Organic Chemistry,* 2nd Ed. New York: McGraw-Hill.

Harvey, A. H. and J. M. H. Levelt Sengers, 1990, *AIChE J.*, 36: 539.

Harvey, A. H., 1996, *AIChE J.*, 42: 1491.

Hildebrand, J. H. and R. L. Scott, 1962, *Regular Solutions*. Englewood Cliffs: Prentice-Hall.

Hine, J., 1962, *Physical Organic Chemistry,* 2nd Ed. New York: McGraw-Hill.

Hirschfelder, J. O., C. F. Curtiss, and R. B. Bird, 1954, *Molecular Theory of Gases and Liquids*. New York: John Wiley & Sons.

Japas, M. L. and J. M. H. Levelt Sengers, 1989, *AIChE J.*, 35: 705.

Johnstone, H. F. and P. W. Leppla, 1934, *J. Am. Chem. Soc.*, 56: 2233.

King, M. B., K. Kassim, and H. Al-Najjar, 1977, *Chem. Eng. Sci.*, 32: 1247.

Krichevsky, I. R. and J. S. Kasarnovsky, 1935, *J. Am. Chem. Soc.*, 57: 2168.

Krichevsky, I. R. and A. A. Ilinskaya, 1945, *Zh. Fiz. Khim. USSR,* 19: 621.

Levelt Sengers, J. M. H., 1991, *Supercritical Fluid Technology: Reviews in Modern Theory and Applications*, Chap. 1, (T. J. Bruno and J. F. Ely, Eds.). Boca Raton: CRC Press.

Liu, D. D. and J. M. Prausnitz, 1979, *J. Appl. Poly. Sci.*, 24: 725.

Masterson, W. L., T. P. Lee, and R. L. Boyington, 1969, *J. Phys. Chem.*, 73: 2761.

Nitta, T., A. Tatsuishi, and T. Katayama, 1973, *J. Chem. Eng. Jap.*, 6: 475.

O'Connell, J. P. and J. M. Prausnitz, 1964, *Ind. Eng. Chem. Fundam.*, 3: 347.

O'Connell, J. P., 1971, *AIChE J.*, 17: 658.
Orentlicher, M. and J. M. Prausnitz, 1964, *Chem. Eng. Sci.*, 19: 775.
Pierotti, R. A., 1976, *Chem. Rev.*, 76: 717.
Plöcker, U., H. Knapp, and J. M. Prausnitz, 1978, *Ind. Eng. Chem. Process Des. Dev.*, 17: 324.
Prausnitz, J. M. and F. H. Shair, 1961, *AIChE J.*, 7: 682.
Reiss, H., H. L. Frisch, E. Helfand, and J. L. Lebowitz, 1960, *J. Chem. Phys.*, 32: 119; *ibid.*, 33: 1379.
Schulze, G. and J. M. Prausnitz, 1981, *Ind. Eng, Chem. Fundam.*, 20: 175.
Sciamanna, S.F. and S. Lynn, 1988, *Ind. Eng. Chem. Res.*, 27: 492.
Sherwood, A. E. and J. M. Prausnitz, 1962, *AIChE J.*, 8: 519; Errata, 1963, *ibid.*, 9: 246.
Shoor, S. K. and K. E. Gubbins, 1969, *J. Phys. Chem.*, 73: 408.
Tiepel, E. W. and K. E. Gubbins, 1973, *Ind. Eng. Chem. Fundam.*, 12: 18.
Wiebe, R. and V. L. Gaddy, 1937, *J. Am. Chem. Soc.*, 59: 1984.
Wilhelm, E. and R. Battino, 1971, *J. Chem. Thermodynamics*, 3: 379.
Yau, J.-S., and F.-N. Tsai, 1992, *J. Chem. Eng. Data, 37,* 141; *ibid.*, 1992, *Fluid Phase Equilibria*, 73: 1.
Yen, L. and J. J. McKetta, 1962, *AIChE J.*, 8: 501.
Zellhoefer, G. F., M. J. Copley, and C. S. Marvel, 1938, *J. Am. Chem. Soc.*, 60: 1337.

Problems

1. Compute the K factor for methane at 13.8 bar and 366 K in a hydrocarbon mixture whose composition in mole percent is 20% benzene, 60% *m*-xylene, and 20% *n*-hexane. Use Shair's correlation.

2. At 20°C, the solubility of helium in argyle acetate is $x_2 = 1.00 \times 10^{-4}$ at 25 bar and is $x_2 = 2.86 \times 10^{-4}$ at 75 bar. Estimate the solubility of helium in this solvent at 20°C and 150 bar. State all assumptions made.

3. Estimate the solubility of hydrogen gas (at 1 bar partial pressure) in liquid air at 90 K. The following experimental information is available, all at 90 K: x_{H_2} (1 bar partial pressure) in liquid methane is 0.0549×10^{-2}; x_{H_2} (1 bar partial pressure) in liquid carbon monoxide is 0.263×10^{-2}. Pure-component data (at 90 K):

	Liquid molar volume (cm^3 mol^{-1})	Enthalpy of vaporization (kJ mol^{-1})
CH_4	35.6	8.75
CO	37.0	5.53
N_2	37.5	5.11
O_2	27.9	6.55

4. The Bunsen coefficient is an experimentally obtainable measure of the solubility of a sparingly soluble gas in a liquid solvent. It is defined as the cubic centimeters of gas, measured at 0°C and 1 atm, that can dissolve in 1 cm^3 of liquid, if the partial pressure of the gas above the liquid is 1 atm. Due to its biological importance, the solubility of oxygen in water has been extensively investigated. The following empirical equation gives the Bunsen coefficient (α) of oxygen in water as a function of temperature t (°C):

$$\alpha = 4.9 \times 10^{-2} - 1.335 \times 10^{-3} t + 2.759 \times 10^{-5} t^2 - 3.235 \times 10^{-7} t^3 + 1.614 \times 10^{-9} t^4$$

(a) From the above equation, find Henry's constant for O_2 in an aqueous system at 1 atm and 20°C.

(b) Find the solubility (in mg dm^{-3} or in ppm) of atmospheric oxygen in water at 20°C.

Density of water at 20°C and 1 atm: 0.9982 g cm^{-3}.

Vapor pressure of water at 20°C: 17.5 mmHg.

Dry air can be taken as 20.95% by volume of oxygen.

5. At 25°C a closed vessel of 3 liter volume contains 180 grams of water, 420 grams of cyclohexane, and 28 grams of nitrogen. Estimate the solubility of nitrogen (mole fraction) in the cyclohexane phase and in the water phase. In this calculation assume that nitrogen behaves as an ideal gas.

Data at 25°C:

	Henry's constant for N_2 in solvent (bar)	Density (g cm^{-3})	P^s (bar)
Cyclohexane	1,300	0.774	0.130
Water	86,000	0.997	0.0317

6. Compute the solubility x_2 of hydrogen in nitrogen at 77 K and 100 bar. At these conditions the vapor-phase fugacity coefficient for pure hydrogen is 0.88 and the saturation pressure of pure liquid nitrogen is 1 bar. Use Orentlicher's correlation.

7. Using the gas-solubility correlations of Hildebrand and Scott (not the one of Shair) estimate the solubility of hydrogen in 1-hexene at 0°C when the partial pressure of hydrogen is 2 bar. The solubility parameter of 1-hexene at 25°C is 14.9 (J $cm^{-3})^{1/2}$.

8. Using experimental and calculated results presented by Olson (Tables 10-2 and 10-3), estimate the solubility of methane in ethylene oxide at 10°C and at a total pressure of 25 bar. State all simplifying assumptions on which your estimate is based.

9. Consider a solution of nitrogen (2) dissolved in n-butane (1) at 250 K. Using the Peng-Robinson equation of state (see Sec. 12.7), calculate:

(a) Henry's constant $H_{2,1}$.

(b) The partial molar volume of nitrogen infinitely dilute in n-butane.

(c) Parameter A of the two-suffix Margules equation.

Data (at 250 K) are as follows:

$v_1^s = 93.035\ \text{cm}^3\ \text{mol}^{-1}$

$P_1^s = 0.392$ bar.

Interaction parameter k_{12} obtained from VLE data is 0.0867.

CHAPTER 11

Solubilities of Solids in Liquids

The ability of solids to dissolve in liquids varies enormously; in some cases a solid solute may form a highly concentrated solution in a solvent (e.g., calcium chloride in water) and in other cases the solubility may be barely detectable (e.g., paraffin wax in mercury).

In this chapter we consider some of the thermodynamic principles that govern equilibrium between a solid phase and a liquid phase.

11.1 Thermodynamic Framework

Solubility is a strong function of the intermolecular forces between solute and solvent, and the well-known guide "like dissolves like" is no more than an empirical statement indicating that, in the absence of specific chemical effects, intermolecular forces between chemically similar species lead to a smaller endothermic enthalpy of solution than those between dissimilar species. Since dissolution must be accompanied by a decrease in the Gibbs energy, a low endothermic enthalpy is more favorable than a large one. However, factors other than intermolecular forces between solvent and solute also play a large role in determining the solubility of a solid. To illustrate, consider the solubilities of two isomers, phenanthrene and anthracene, in benzene at 25°C given in Table 11-1. The solubility of phenanthrene is about 25 times larger than that of anthracene even though both solids are chemically similar to each other and to the solvent. The reason for this large difference in solubility follows from something that is all too often overlooked, i.e., that solubility depends not only on the activity coefficient of the solute (that is a function of the intermolecular forces between solute and solvent) but also on the fugacity of the standard state to which that activity coefficient refers and on the fugacity of the pure solid.

Table 11-1 Structures of phenanthrene and anthracene and their solubility in benzene at 25°C.

Solute	Structure	Solubility in benzene (mol %)
Phenanthrene		20.7
Anthracene		0.81

Let the solute be designated by subscript 2. Then the equation of equilibrium is[1]

$$f_{2(\text{pure solid})} = f_{2(\text{solute in liquid solution})} \tag{11-1}$$

or

$$f_{2(\text{pure solid})} = \gamma_2 x_2 f_2^0 \tag{11-2}$$

where x_2 is the solubility (mole fraction) of the solute in the solvent, γ_2 is the liquid-phase activity coefficient, and f_2^0 the standard-state fugacity to which γ_2 refers.

From Eq. (11-2) the solubility is

$$x_2 = \frac{f_{2(\text{pure solid})}}{\gamma_2 f_2^0} \tag{11-3}$$

Thus the solubility depends not only on the activity coefficient but also on the ratio of two fugacities as indicated by Eq. (11-3).

[1] Equation (11-1) is based on the assumption that there is no appreciable solubility of the liquid solvent in the solid phase.

The standard-state fugacity f_2^0 is arbitrary; the only thermodynamic requirement is that it must be at the same temperature as that of the solution. Although other standard states can be used, it is most convenient to define the standard-state fugacity as the fugacity of pure, subcooled liquid at the temperature of the solution and at some specified pressure. This is a hypothetical standard state but it is one whose properties can be calculated with good accuracy provided that the solution temperature is not far removed from the triple point of the solute.

To show the utility of Eq. (11-3), we consider first a very simple case. Assume that the vapor pressures of the pure solid and of the subcooled liquid are not large; in that case we can substitute vapor pressures for fugacities without serious error. This simplifying assumption is an excellent one in the majority of typical cases. Next, let us assume that the chemical natures of the solvent and of the solute (as a subcooled liquid) are similar. In that case we can assume $\gamma_2 = 1$ and Eq. (11-3) becomes

$$x_2 = \frac{P^s_{2(\text{pure solid})}}{P^s_{2(\text{pure, subcooled liquid})}} \tag{11-4}$$

The solubility x_2 given by Eq. (11-4) is called the *ideal solubility*. The significance of Eq. (11-4) can best be seen by referring to a typical pressure-temperature diagram for a pure substance as shown in Fig. 11-1. If the solute is a solid, then the solution temperature is necessarily below the triple-point temperature. The vapor pressure of the solid at T is found from the solid-vapor pressure curve but the vapor pressure of the subcooled liquid must be found by extrapolating the liquid-vapor pressure curve from the triple-point temperature to the solution temperature T. Since the slope of the solid-vapor pressure curve is always larger than that of the extrapolated liquid-vapor pressure curve, it follows from Eq. (11-4) that the solubility of a solid in an ideal solution must always be less than unity, except at the triple-point temperature where it is equal to unity.

Equation (11-4) explains why phenanthrene and anthracene have very different solubilities in benzene: Because of structural differences, the triple-point temperatures of the two solids are significantly different.[2] As a result, the pure-component fugacity ratios at the same temperature T also differ for the two solutes.

The extrapolation indicated in Fig. 11-1 is simple when the solution temperature T is not far removed from the triple-point temperature. However, any essentially arbitrary extrapolation involves uncertainty; when the extrapolation is made over a wide temperature range, the uncertainty may be large. It is therefore important to establish a systematic method for performing the desired extrapolation; this method should be substituted for the arbitrary graphical construction shown in Fig. 11-1. Fortunately, a systematic extrapolation can readily be derived by using a thermodynamic cycle as

[2] The normal melting temperatures for phenanthrene and anthracene are, respectively, 100 and 217°C. These are very close to the respective triple-point temperatures.

indicated in Sec. 11.2. This extrapolation does not require the assumption of low pressures but yields an expression that gives the fugacity, rather than the pressure, of the saturated, subcooled liquid.

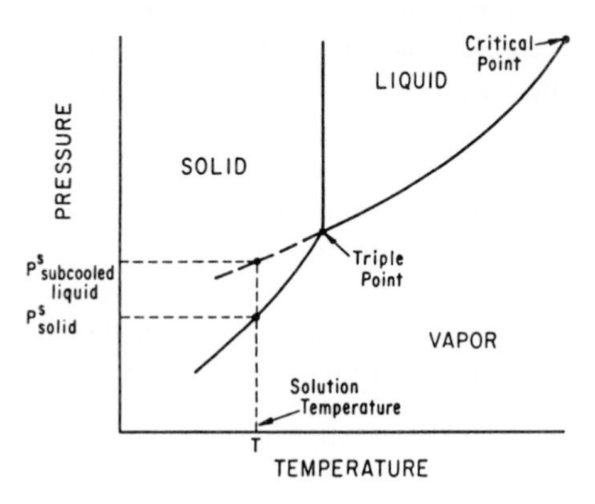

Figure 11-1 Extrapolation (dashed line) of liquid vapor pressure on a pressure-temperature diagram for a pure material (schematic).

11.2 Calculation of the Pure-Solute Fugacity Ratio

For the liquid-phase activity coefficient, we define the standard state as the pure, subcooled liquid at temperature T under its own saturation pressure. Assuming negligible solubility of the solvent in the solid phase, $\mathcal{S}$, the equilibrium equation is

$$x_2 = \frac{f_{2(\text{pure solid})}}{\gamma_2 \, f_{2(\text{pure, subcooled liquid})}} \tag{11-5}$$

To simplify notation, let

$$f_{2(\text{pure solid})} = f_2^{\mathcal{S}}$$

and let

$$f_{2(\text{pure, subcooled liquid})} = f_2^L$$

These two fugacities depend only on the properties of the solute (component 2); they are independent of the nature of the solvent. The ratio of these two fugacities can readily be calculated by the thermodynamic cycle indicated in Fig. 11-2. The molar Gibbs energy change for component 2 in going from a to d is related to the fugacities of solid and subcooled liquid by

$$\underset{a\to d}{\Delta g} = RT\ln\frac{f^L}{f^s} \tag{11-6}$$

where, for simplicity, subscript 2 has been omitted. This Gibbs energy change is also related to the corresponding enthalpy and entropy changes by

$$\underset{a\to d}{\Delta g} = \underset{a\to d}{\Delta h} - T\underset{a\to d}{\Delta s} \tag{11-7}$$

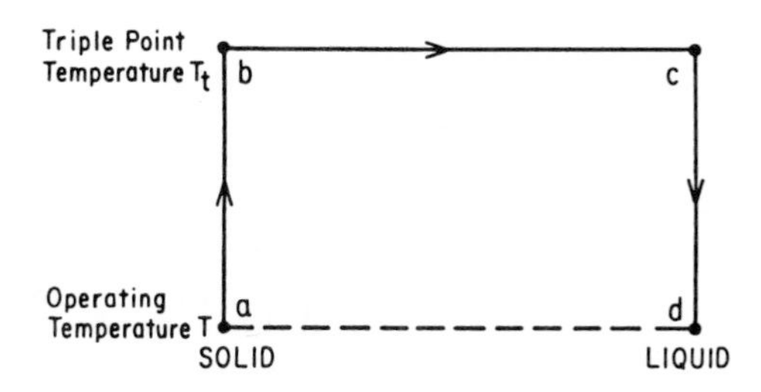

Figure 11-2 Thermodynamic cycle for calculating the fugacity of a pure subcooled liquid.

The thermodynamic cycle in Fig. 11-2 provides a method to evaluate the enthalpy and entropy changes given in Eq. (11-7); because both enthalpy and entropy are state functions independent of the path, it is permissible to substitute for the path $a\to d$ the alternate path $a\to b\to c\to d$. For the enthalpy change from a to d we have

$$\underset{a\to d}{\Delta h} = \underset{a\to b}{\Delta h} + \underset{b\to c}{\Delta h} + \underset{c\to d}{\Delta h} \tag{11-8}$$

Equation (11-8) can be rewritten in terms of heat capacity c_p and enthalpy of fusion $\Delta_{\text{fus}}h$:[3]

$$\underset{a\to d}{\Delta h} = \Delta_{\text{fus}}h_{\text{at } T_t} + \int_{T_t}^{T}\Delta c_p\, dT \tag{11-9}$$

where $\Delta c_p \equiv c_{p(\text{liquid})} - c_{p(\text{solid})}$ and T_t is the triple-point temperature. Similarly, for the entropy change from a to d,

$$\underset{a\to d}{\Delta s} = \underset{a\to b}{\Delta s} + \underset{b\to c}{\Delta s} + \underset{c\to d}{\Delta s} \tag{11-10}$$

[3] Equations (11-9) and (11-11) neglect the effect of pressure on the properties of solid and subcooled liquid. Unless the pressure is large, this effect is negligible.

which becomes[4]

$$\underset{a \to d}{\Delta s} = \Delta_{\text{fus}} s_{\text{at } T_t} + \int_{T_t}^{T} \frac{\Delta c_p}{T} dT \tag{11-11}$$

At the triple point, the entropy of fusion $\Delta_{\text{fus}} s$ is

$$\Delta_{\text{fus}} s = \frac{\Delta_{\text{fus}} h}{T_t} \tag{11-12}$$

Substituting Eqs. (11-7), (11-9), (11-11), and (11-12) into Eq. (11-6), and assuming that Δc_p is constant over the temperature range $T \to T_t$, we obtain[5]

$$\boxed{\ln \frac{f^L}{f^{\Delta}} = \frac{\Delta_{\text{fus}} h}{R T_t}\left(\frac{T_t}{T} - 1\right) - \frac{\Delta c_p}{R}\left(\frac{T_t}{T} - 1\right) + \frac{\Delta c_p}{R} \ln \frac{T_t}{T}} \tag{11-13}$$

Equation (11-13) gives the desired result; it expresses the fugacity of the subcooled liquid at temperature T in terms of measurable thermodynamic properties. To illustrate, Fig. 11-3 shows the fugacity ratio for solid and subcooled liquid carbon dioxide.

Two simplifications in Eq. (11-13) are frequently made; these usually introduce only slight error. First, for most substances there is little difference between the triple-point temperature and the normal melting temperature; also, the difference in the enthalpies of fusion at these two temperatures is often negligible. Therefore, in practice, it is common to substitute the normal melting temperature for T_t and to use for $\Delta_{\text{fus}} h$ the enthalpy of fusion at the melting temperature.[6] Second, the three terms on the right-hand side of Eq. (11-13) are not of equal importance; the first term is the dominant one and the remaining two, of opposite sign, have a tendency approximately to cancel each other, especially if T and T_t are not far apart. Therefore, in many cases it is sufficient to consider only the term that includes $\Delta_{\text{fus}} h$ and to neglect the terms that include Δc_p.

---

[4] See previous footnote.

[5] Equation (11-13) assumes that there are no solid-solid phase transition along path $a \to b$ in Fig. 11-2. When such transitions are present, Eq. (11-13) must be modified, as discussed, for example, by G. T. Preston *et al.*, 1971, *J. Phys. Chem.*, 75: 2345 and by P. B. Choi and E. McLaughlin, 1983, *AIChE J.*, 29: 150.

[6] For cases where no experimental enthalpies (or entropies) of fusion are available, they must be estimated. A group-additivity method for estimating enthalpies (and entropies) of fusion of organic compounds is given by Chickos *et al.*, 1991, *J. Org. Chem.*, 56: 927.

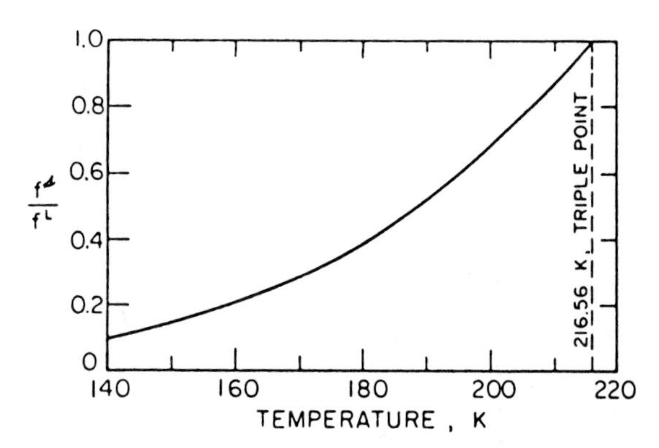

Figure 11-3 Fugacity ratio for solid and subcooled liquid carbon dioxide.

11.3 Ideal Solubility

An expression for the ideal solubility of a solid solute in a liquid solvent has already been given by Eq. (11-4), but no clear-cut method was given for finding the saturation pressure of the subcooled liquid. However, this difficulty can be overcome by substituting Eq. (11-13) into Eq. (11-5); if we assume that the solution is ideal, then $\gamma_2 = 1$ and we obtain for ideal solubility x_2:

$$\ln\frac{1}{x_2} = \frac{\Delta_{\text{fus}}h}{RT_t}\left(\frac{T_t}{T}-1\right) - \frac{\Delta c_p}{R}\left(\frac{T_t}{T}-1\right) + \frac{\Delta c_p}{R}\ln\frac{T_t}{T} \tag{11-14}$$

Equation (11-14) provides a reasonable method for estimating solubilities of solids in liquids where the chemical nature of the solute is similar to that of the solvent. For example, in Sec. 11.1 we quoted the experimental solubilities of phenanthrene and anthracene in benzene at 25°C; they are 20.7 and 0.81 mol %, respectively. The corresponding solubilities calculated from Eq. (11-14) are 22.1 and 1.07 mol %.

Equation (11-14) immediately leads to useful conclusions concerning the solubilities of solids in liquids. Strictly, these conclusions apply only to ideal solutions but they are useful guides for other solutions that do not deviate excessively from ideal behavior:

- For a given solid/solvent system, the solubility increases with rising temperature. The rate of increase is approximately proportional to the enthalpy of fusion and, to a first approximation, does not depend on the melting temperature.

- For a given solvent and a fixed temperature, if two solids have similar enthalpies of fusion, the solid with the lower melting temperature has the higher solubility; similarly, if the two solids have nearly the same melting temperature, the one with the lower enthalpy of fusion has the higher solubility.

A typical application of Eq. (11-14) is provided by the solubility data of McLaughlin and Zainal (1959) for nine aromatic hydrocarbons in benzene in the temperature range 30 to 70°C. With the help of Eq. (11-14) these data can be correlated in a simple manner. To a fair approximation, the terms that include Δc_p in Eq. (11-14) may be neglected; also, it is permissible to substitute melting temperatures for triple-point temperatures. Equation (11-14) may then be rewritten

$$\ln x_2 = -\frac{\Delta_{\text{fus}} s}{R}\left(\frac{T_m}{T} - 1\right) \tag{11-15}$$

where T_m is the normal melting temperature. For the nine solutes considered here, the entropies of fusion do not vary much; an average value is 54.4 J mol^{-1} K^{-1}. Therefore, a semilogarithmic plot of log x_2 versus T_m/T should give a nearly straight line with a slope approximately equal to -(54.4/2.303R) and with intercept log $x_2 = 0$ when $T_m/T = 1$; in this case, R = 8.31451 J mol^{-1} K^{-1}. Such a plot is shown by the dashed line in Fig. 11-4 and it is evident that this line gives a good representation of the experimental data. Thus the ideal-solution assumption is appropriate for these systems. However, for precise work, the assumption of ideality is only an approximation. Solutions of aromatic hydrocarbons in benzene, although the compounds are chemically similar, show slight positive deviations from ideality and therefore the observed solubilities are a little lower than those calculated with Eq. (11-15). The continuous line in Fig. 11-4, determined empirically as the best fit of the data, has a slope equal to -(57.7/2.303R).

Equation (11-14) gives the ideal solubility of solid 2 in solvent 1. By interchanging subscripts, we may use the same equation to calculate the ideal solubility of solid 1 in solvent 2; by repeating such calculations at different temperatures, we can then obtain the freezing diagram of the binary system as a function of composition, as shown in Fig. 11-5 taken from Prigogine and Defay (1954). In these calculations we assume ideal behavior in the liquid phase and complete immiscibility in the solid phase. The left side of the diagram represents equilibrium between the liquid mixture and solid *o*-chloronitrobenzene, while the right side represents equilibrium between the liquid mixture and solid *p*-chloronitrobenzene. At the point of intersection, called the *eutectic point,* three phases are in equilibrium.

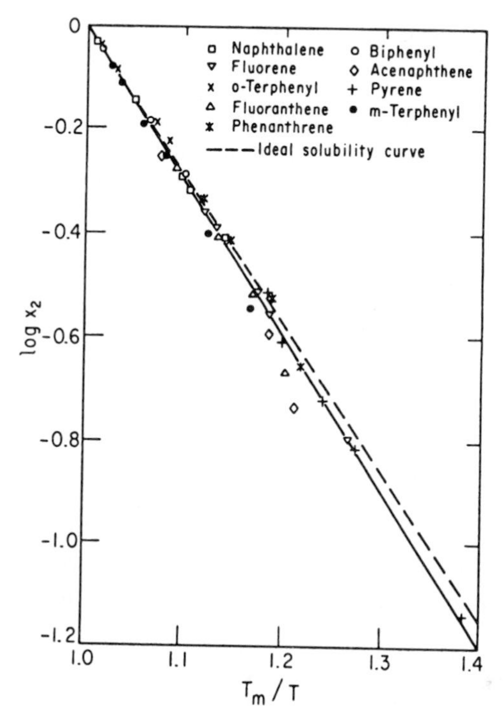

Figure 11-4 Solubility of aromatic solids in benzene.

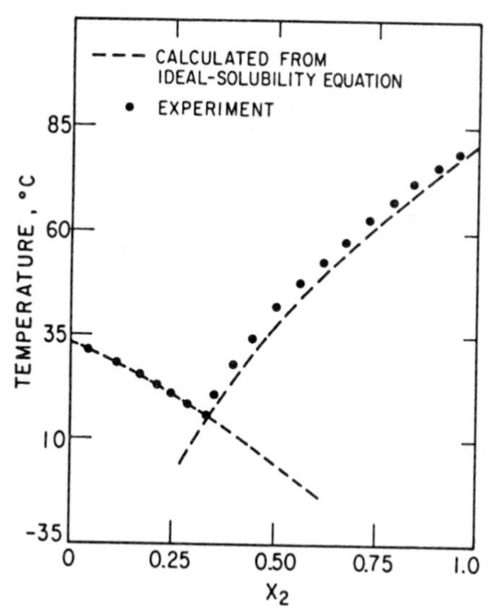

Figure 11-5 Freezing points for the system *o*-chloronitrobenzene (1)/*p*-chloronitrobenzene (2).

11.4 Nonideal Solutions

Equation (11-14) assumes ideal behavior but Eqs. (11-5) and (11-13) are general. Whenever there is a significant difference in the nature and size of the solute and solvent molecules, we may expect that γ_2 is not equal to unity; in nonpolar solutions, where only dispersion forces are important, γ_2 is generally larger than unity (and thus the solubility is less than that corresponding to ideal behavior), but in cases where polar or specific chemical forces are important, the activity coefficients may well be less than unity with correspondingly higher solubilities. Such enhanced solubilities, for example, have been observed for unsaturated hydrocarbons in liquid sulfur dioxide.

An illustration that shows that the ideal-solubility equation [Eq. (11-15)] gives only a crude approximation is provided by the solubility of cholesterol ($C_{27}H_{46}O$, M = 386.67 g mol^{-1}, T_m = 421.7 K, $\Delta_{\text{fus}}h$ = 28.924 kJ mol^{-1}) in different polar solvents, shown in Fig. 11-6. Cholesterol, a lipid sterol, is an important component of cell membranes. From spectroscopic (infrared and NMR) and dielectric-constant measurements, there is evidence that cholesterol shows self-association (Mercier *et al.*, 1983; Góralski, 1993, Jadzyn and Hellemans, 1993). It is likely that hydrogen bonds and van der Waals interactions play an important role in the biological functions of cholesterol.

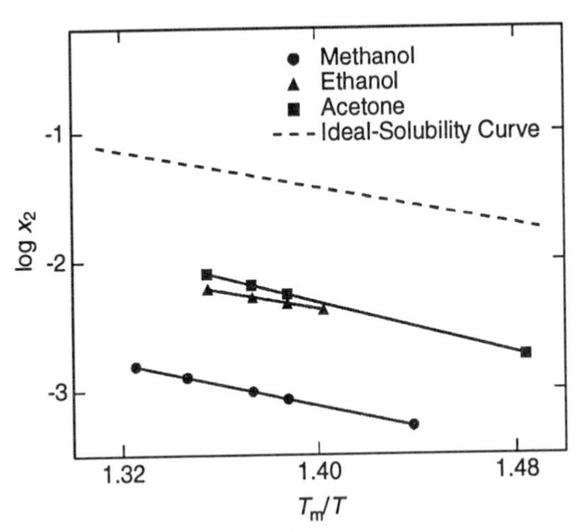

Figure 11-6 Solubility of cholesterol (2) in polar solvents: ● Methanol; ▲ Ethanol; ■ Acetone. Solid line is the best fit through the experimental points; the dashed line is the ideal-solubility curve [Eq. (11-15)].

As Fig. 11-6 shows, the experimental solubility curves (Bar *et al.*, 1984) deviate significantly from the ideal-solubility curve as given by Eq. (11-15). In contrast to the previous example concerned with the solubility of aromatic solids in benzene, for cholesterol/polar-solvent systems, all molecular interactions (solute-solute, solute-solvent, and solvent-solvent) are relatively complex. For these systems, activity coefficients for cholesterol are far in excess of unity.

In Fig. 11-4 we showed the solubilities of nine aromatic hydrocarbons in benzene and it is apparent that the assumption of ideal-solution behavior is approximately valid. By contrast, Fig. 11-7 shows the solubilities of the same nine hydrocarbons in carbon tetrachloride; the data are again from McLaughlin and Zainal (1960). The dashed line shows the ideal solubility curve using, as before, $\Delta_{fus}s = 54.4$ J mol^{-1} K^{-1}; the solid line that best represents the data has a slope $-(66.5/2.303R)$.

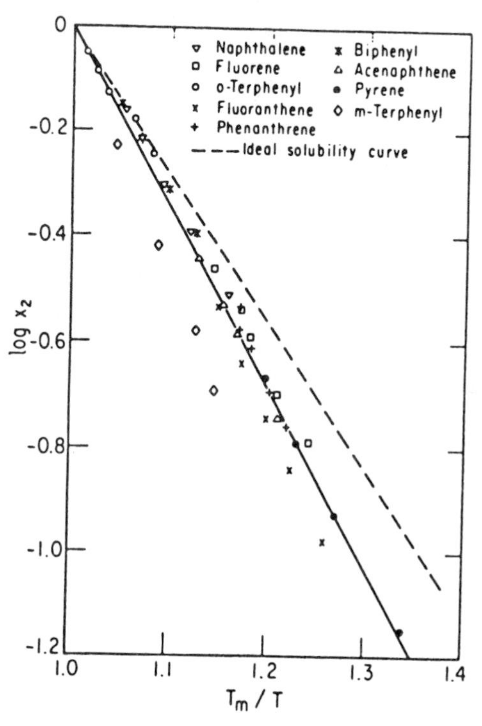

Figure 11-7 Solubility of aromatic solids in carbon tetrachloride.

A comparison of Figs. 11-4 and 11-7 shows that at the same temperature, the solubilities in carbon tetrachloride are lower than those in benzene; in other words, the activity coefficients of the solutes in carbon tetrachloride are larger than those in benzene. If the activity coefficients of the solutes in carbon tetrachloride are represented by the simple empirical relation

$$\ln \gamma_2 = \frac{A}{RT}(1 - x_2)^2 \tag{11-16}$$

then typical values of A are in the range 400 to 1300 J mol^{-1}.

As for solutions of liquid components, there is no general method for predicting activity coefficients of solid solutes in liquid solvents. For nonpolar solutes and solvents, however, a reasonable estimate can frequently be made with the Scatchard-Hildebrand relation

$$\ln \gamma_2 = \frac{v_2^L (\delta_1 - \delta_2)^2 \Phi_1^2}{RT} \tag{11-17}$$

where v_2^L is the molar volume of the subcooled liquid, δ_2 is the solubility parameter of the subcooled liquid, δ_1 is the solubility parameter of the solvent, and

$$\Phi_1 = \frac{x_1 v_1^L}{x_1 v_1^L + x_2 v_2^L}$$

is the volume fraction of the solvent.

The molar liquid volume and solubility parameter of the solvent can be determined from the thermodynamic properties of the solvent, but it is necessary to use a thermodynamic cycle (as illustrated in Fig. 11-2) to calculate these functions for the subcooled liquid solute.

Let $\Delta_{\text{fus}} v$ stand for the volume change of fusion at the triple-point temperature; that is,

$$\Delta_{\text{fus}} v = v_t^L - v_t^{\text{s}} \tag{11-18}$$

where subscript t refers to the triple-point temperature.

Let v^{s} be the molar volume of the solid at temperature T of the solution. The molar volume of the subcooled liquid is then given by

$$v^L = v^{\text{s}} + \Delta_{\text{fus}} v + (v_t^{\text{s}} \alpha^{\text{s}} - v_t^L \alpha^L)(T_t - T) \tag{11-19}$$

where α^{s} and α^L are the volumetric coefficients of expansion of the solid and liquid, respectively.

The energy of vaporization of the subcooled liquid is found in a similar manner. Let $\Delta_{\text{fus}} h$ stand for the enthalpy of fusion of the solid at the triple-point temperature and let $\Delta_{\text{sub}} h$ stand for the enthalpy of sublimation of the solid at temperature T. The energy of vaporization of the subcooled liquid is then

$$\Delta u = \Delta_{\text{sub}} h - \Delta_{\text{fus}} h + \Delta c_p (T_t - T) - P^s (v^G - v^L) \tag{11-20}$$

where P^s is the saturation pressure of the subcooled liquid and v^G is the molar volume of the saturated vapor in equilibrium with the solid, all at temperature T.

In Eqs. (11-19) and (11-20) it is often convenient to replace the triple-point temperature with the melting temperature and at moderate pressures this substitution usually introduces insignificant error; in that case, all subscripts t can be replaced by subscript m. Also, if the temperature T is not far removed from the triple-point (or melting) temperature, the last term in Eq. (11-19) may be neglected, and finally, if the saturation pressure of the subcooled liquid is small, as is usually the case, $v^G >> v^L$, and the last term in Eq. (11-20) may be replaced by RT.

The square of the solubility parameter is defined as the ratio of the energy of complete vaporization to the liquid volume;[7] therefore, if the vapor pressure of the subcooled liquid is large, it is necessary to add a vapor-phase correction to the energy of vaporization given by Eq. (11-20). Such a correction, however, is rarely required and for most cases of interest the solubility parameter of the subcooled liquid is given with sufficient accuracy by

$$\delta_2 = \left(\frac{\Delta_{\text{vap}} u_2}{v_2^L} \right)^{1/2} \tag{11-21}$$

where $\Delta_{\text{vap}} u_2$ is the energy of vaporization, given by the enthalpy of vaporization minus RT.

To illustrate the applicability of Eq. (11-17), we consider the solubility of white phosphorus in n-heptane at 25°C. The melting point of white phosphorus is 44.2°C; the enthalpy of fusion and the heat capacities of the solid and liquid have been measured by Young and Hildebrand (1942). From Eq. (11-14), the ideal solubility at 25°C is $x_2 = 0.942$. A much better approximation can be obtained from regular-solution theory as given by Eq. (11-17). From the extrapolated thermal and volumetric properties, the solubility parameter and molar volume of subcooled liquid phosphorus are, respectively, 27 $(\text{J cm}^{-3})^{1/2}$ and 70.4 $\text{cm}^3\ \text{mol}^{-1}$ at 25°C. The solubility parameter of n-heptane is 15.1 $(\text{J cm}^{-3})^{1/2}$ and therefore one can immediately conclude that subcooled phosphorus and heptane form a highly nonideal liquid solution. When Eqs. (11-17) and (11-13) are substituted into the fundamental Eq. (11-5), the calculated solubility is $x_2 = 0.022$, a result strikingly different from that obtained by assuming liquid-phase ideality. The experimental solubility is $x_2 = 0.0124$.

As discussed in Chap. 7, the regular-solution theory of Scatchard-Hildebrand can be significantly improved when the geometric-mean assumption is not used; in that event, Eq. (11-17) becomes

$$\ln \gamma_2 = \frac{v_2^L [(\delta_1 - \delta_2)^2 + 2 l_{12} \delta_1 \delta_2] \Phi_1^2}{RT} \tag{11-22}$$

[7] See Sec. 7.2.

Preston (1970) has applied Eq. (11-22)[8] to correlate solubilities of nonpolar solids in nonpolar liquids at low temperatures. To do so, it was first necessary to estimate solubility parameters and molar volumes for subcooled liquids; these estimates were obtained by extrapolations as shown in Figs. 11-8 and 11-9. For a given binary, parameter l_{12} must be obtained from some experimental binary datum; it is best to obtain l_{12} from an experimental solubility. Preston gives l_{12} parameters for 25 systems at cryogenic conditions.

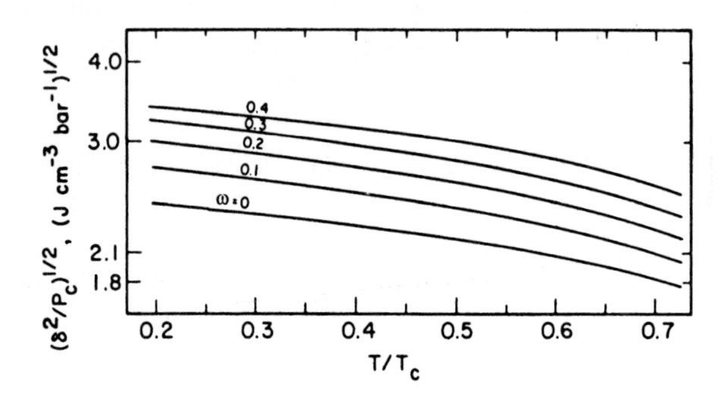

Figure 11-8 Solubility parameters of subcooled liquids.

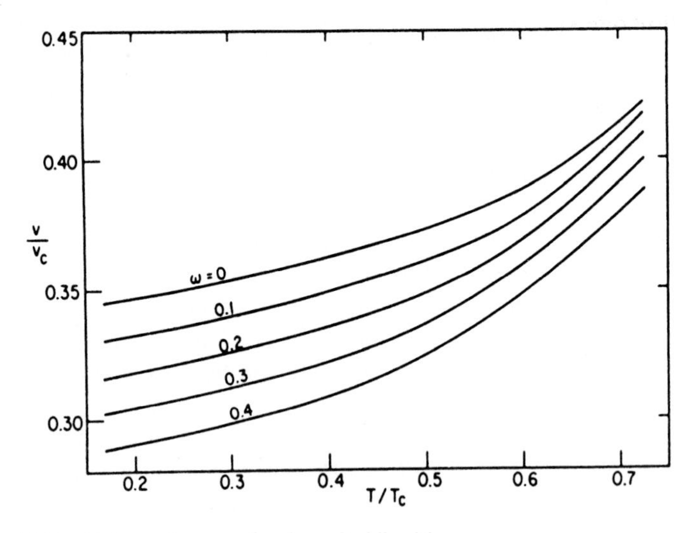

Figure 11-9 Molar volumes of subcooled liquids.

A more fundamental extension of the Scatchard-Hildebrand equation was introduced by Myers (1965) in his study of the solubility of solid carbon dioxide in lique-

[8] This equation was also used by Choi and McLaughlin, 1983, *Ind. Eng. Chem. Fund.*, 22: 46, to correlate solubilities of solid aromatic hydrocarbons in pyridine and in thiophene.

fied light hydrocarbons. Because carbon dioxide has a large quadrupole moment (see Table 4-2), separate consideration was given to the contributions of dispersion forces and quadrupole forces to the cohesive energy density of subcooled liquid carbon dioxide.[9] The energy of vaporization was divided into two parts:

$$\Delta u = \Delta u_{\text{disp}} + \Delta u_{\text{quad}} \tag{11-23}$$

As a result, two cohesive-energy densities can now be computed, corresponding to the two types of intermolecular forces:

$$c_{\text{disp}} \equiv \frac{\Delta u_{\text{disp}}}{v} \tag{11-24}$$

$$c_{\text{quad}} \equiv \frac{\Delta u_{\text{quad}}}{v} \tag{11-25}$$

where superscript L has been omitted. These cohesive-energy densities for carbon dioxide are shown as a function of temperature in Fig. 11-10.

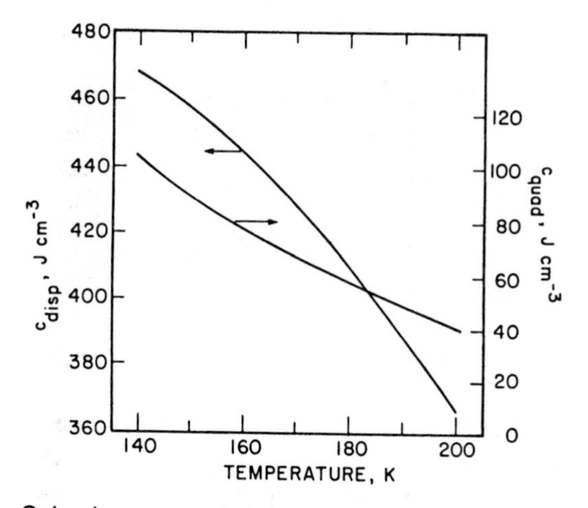

Figure 11-10 Cohesive-energy density due to dispersion forces and due to quadrupole forces for subcooled liquid carbon dioxide.

The activity coefficient of component 2, the solute, dissolved in a nonpolar solvent, is now written

$$RT \ln \gamma_2 = v_2 \Phi_1^2 [c_1 + c_{2\,\text{total}} - 2(c_1 c_{2\,\text{disp}})^{1/2}] \tag{11-26}$$

[9] The solubility parameter is the square root of the cohesive-energy density. See Sec. 7.2.

where $c_{2\,\text{total}} = c_{2\,\text{disp}} + c_{2\,\text{quad}}$. The last term in the brackets follows from the geometric-mean assumption for the attractive dispersion forces between solute and solvent. If $c_{2\,\text{quad}} = 0$, Eq. (11-26) reduces to Eq. (11-17).

Splitting the cohesive-energy density into a dispersion part and a quadrupole part has an important effect on the calculated solubility of solid carbon dioxide. Although the contribution from quadrupole forces is small, appreciable error is introduced by not separately considering this contribution. As an example, consider the solubility of solid carbon dioxide in a nonpolar (and nonquadrupolar) solvent having the (typical) value $c_1 = 251$ J cm^{-3} for its cohesive-energy density at 160 K. If we do not take account of the presence of quadrupolar forces, we calculate, from Eq. (11-17), a solubility $x_2 = 0.067$. However, if we do account for quadrupolar forces [Eq. (11-26)], we obtain the significantly different result $x_2 = 0.016$. In this illustrative calculation we have, for simplicity, assumed that $\Phi_2 = x_2$.

To calculate the cohesive-energy density due to quadrupole forces, Myers derived the relation

$$c_{i\,\text{quad}} = \frac{\beta Q_i^4}{kT\left(\dfrac{v_i}{N_A}\right)^{13/3}} \tag{11-27}$$

where Q_i is the quadrupole moment of species i, v_i is the molar liquid volume, N_A is Avogadro's constant, k is Boltzmann's constant, T is the absolute temperature, and β is a dimensionless constant. If the solvent, component 1, also has a significant quadrupole moment, then an additional term must be added to the bracketed quantity in Eq. (11-26) to account for quadrupole forces between the dissimilar components; further, the geometric-mean term must be modified to include only the dispersion cohesive-energy density of component 1. The bracketed term in Eq. (11-26) then becomes

$$[c_{1\,\text{total}} + c_{2\,\text{total}} - 2(c_{1\,\text{disp}} c_{2\,\text{disp}})^{1/2} - 2c_{12\,\text{quad}}].$$

Using the theory of intermolecular forces, Myers showed that

$$c_{12\,\text{quad}} = \frac{\beta Q_1^2 Q_2^2}{kT\left(\dfrac{v_{12}}{N_A}\right)^{13/3}} \tag{11-28}$$

where β is the same as in Eq. (11-27).

For v_{12} Myers used the combining rule

$$v_{12}^{1/3} = \frac{1}{2}(v_1^{1/3} + v_2^{1/3}) \tag{11-29}$$

The term $c_{12\,\text{quad}}$ is frequently negligible but it is important, for example, in carbon dioxide/acetylene mixtures because in this system both solute and solvent have a significant quadrupole moment (see Table 4-2).

In nonpolar systems, the ideal solubility is generally larger than that observed. This result is correctly predicted by the Scatchard-Hildebrand theory [Eq. (11-17)] because $\gamma_2 \geq 1$; as a consequence, this equation says that the ideal solubility is the maximum possible: the larger γ_2, the smaller x_2. However, whenever there is a tendency for the solvent to solvate with the solute, i.e., whenever there are strong specific forces between the dissimilar molecules, the observed solubility may well be larger than the ideal solubility. Enhanced solubility is observed whenever there are negative deviations from Raoult's law in the liquid solution; such deviations frequently occur in polar systems, especially in systems where hydrogen bonding between solute and solvent is strong. However, even in nonpolar systems, specific solvation forces may, on occasion, be sufficiently strong to result in solubilities above the ideal. For example, Weimer (1965) measured the solubility of hexamethylbenzene (m.p. 165.5°C) in carbon tetrachloride at 25°C and found $x_2 = 0.077$. Contrary to predictions using the Scatchard-Hildebrand theory, this solubility is *larger* than the ideal solubility $x_2 = 0.062$. From spectroscopic and other evidence, we know that carbon tetrachloride forms weak charge-transfer complexes with aromatic hydrocarbons and we also know that the stability of the complex increases with methyl substitution on the benzene ring. For the solutes listed in Fig. 11-6, the complex stability is small, but in the carbon tetrachloride/hexamethylbenzene system the tendency to complex is sufficiently strong to produce an enhanced solubility; in this system, complex formation overshadows the "normal" physical forces that would tend to give a solubility lower than the ideal. Weimer found that in carbon tetrachloride solution the activity coefficient of hexamethylbenzene at saturation (referred to pure subcooled liquid) is $\gamma_2 = 0.79$, indicating negative deviation from Raoult's law.

As briefly indicated in Chap. 6, liquid-phase activity coefficients can sometimes be estimated from a group-contribution method such as UNIFAC. It is therefore possible, in some cases, to use UNIFAC for constructing freezing-point diagrams, as shown by Gmehling *et al.* (1978). To illustrate, Fig. 11-11 shows a diagram for the system benzene/phenol. A simple eutectic diagram is obtained; the lines calculated with UNIFAC are significantly higher than those calculated assuming ideal solubility. Agreement with experiment (Hatcher and Skirrow, 1917; Tsakalotos and Guye, 1910) is good.

In the liquid phase, it is also possible to calculate the fugacity of the dissolved component using an equation of state. For solute 2, the equation of equilibrium is written

$$f_2^{\text{∆}} = \varphi_2 x_2 P \tag{11-30}$$

where φ_2 is obtained from an equation of state, valid for the fluid mixture, over the density range from zero density to liquid density, as discussed in Chaps. 3 and 12.

Equation (11-30) has been used, for example, by Soave (1979) for calculating the solubility of solid carbon dioxide in liquefied natural gas.

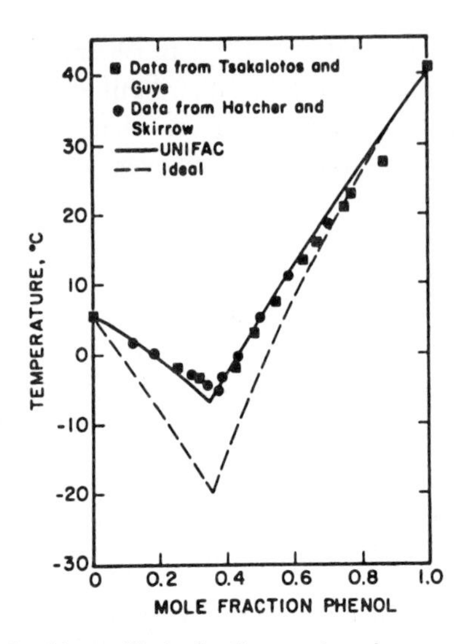

Figure 11-11 Solid-liquid equilibria for the system benzene/phenol.

From experimental solid-liquid equilibrium (SLE) data we can obtain the binary parameters of a particular liquid-phase activity coefficient model, such as Wilson or UNIQUAC. These parameters may give in turn liquid phase activity coefficients at different system conditions that can be used to predict other equilibria, e.g. vapor-liquid equilibria (VLE). Calculation of VLE from experimental SLE is particularly useful for those systems where it is difficult to obtain experimental VLE data, e.g. due to the high-melting points of the system components and to the large difference of vapor pressures, as in systems containing heterocyclic polynuclear aromatics (naphthalene, acenaphthene, fluorene, phenanthrene, etc.) in organic solvents. To illustrate we now consider VLE at 170°C for the systems (Gupta *et al.*, 1991) acenaphthene[10]/tetralin and acenaphthene/cis-decalin, shown in Fig. 11-12. Vapor-liquid equilibria for both systems are predicted using the UNIQUAC equation whose binary interaction parameters a_{12} and a_{21} (see Sec. 6.11) were determined from regression of experimental solid-acenaphthene solubility data in each solvent from room temperature to the melting-point temperature of the solute.

Figure 11-12(a) shows that in the acenaphthene/tetralin system VLE predictions are in good agreement with the experiment. However, this is not always the case; as illustrated in Fig. 11-12(b) for the system acenaphthene/cis-decalin, good agreement with experiment is obtained only if the solid-solubility data and (at least) one vapor-

[10] The melting point of acenaphthene is 93.3°C.

liquid equilibrium data point are combined to obtain an improved set of a_{12} and a_{21} UNIQUAC binary parameters.

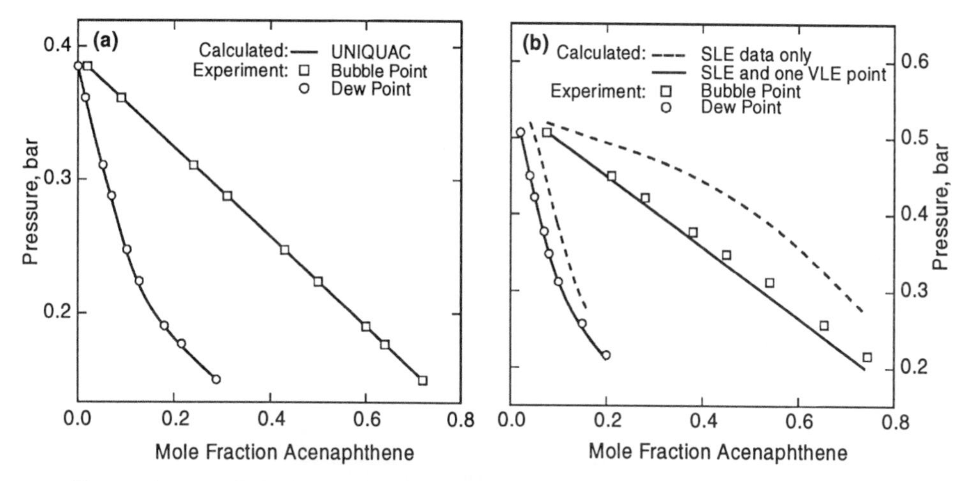

Figure 11-12 Calculated and observed vapor-liquid equilibria at 170°C for: (a) Acenaphthene/tetralin. —— Calculated from UNIQUAC equation and SLE data only. Using also one VLE datum point, the calculated curve is almost indistinguishable from that obtained using SLE data only. (b) Acenaphthene/cis-decalin. Calculated from UNIQUAC equation: - - - - SLE data only; —— SLE and one VLE datum. In this case, use of one VLE datum very much improves agreement with experiment.

11.5 Solubility of a Solid in a Mixed Solvent

Scatchard-Hildebrand theory predicts that the solubility of a solid is a maximum in that solvent whose solubility parameter is the same as that of the (liquid) solute; in that event, the activity coefficient of the solute (referred to the subcooled liquid) is equal to unity. Scatchard-Hildebrand theory suggests, therefore, that when a solid solute is dissolved in a mixture of two carefully selected solvents, a plot of solubility versus (solute-free) solvent composition should go through a maximum.

To test this *maximum-solubility effect*, consider phenanthrene that melts at 100°C; the (liquid) solubility parameter of phenanthrene is 20.3 (J cm^{-3})$^{1/2}$. When phenanthrene is dissolved in a binary solvent mixture, the effective solubility parameter of the mixed solvent varies with the solvent's composition. Let phenanthrene be designated by subscript 1 and the two solvents by subscripts 2 and 3. If $\delta_2 < 20.3$ and $\delta_3 > 20.3$, then there is some mixture of solvents 2 and 3 which is the optimum solvent for phenanthrene.

Gordon and Scott (1952) have measured the solubility of phenanthrene in cyclohexane (2), in methylene iodide (3), and in their mixtures. At 25°C, the solubility parameters are $\delta_2 = 16.8$ and $\delta_3 = 24.4$ (J cm^{-3})$^{1/2}$. Figure 11-13 shows, as expected, that a plot of experimental solubility versus solvent composition (solute-free basis) exhibits a maximum.

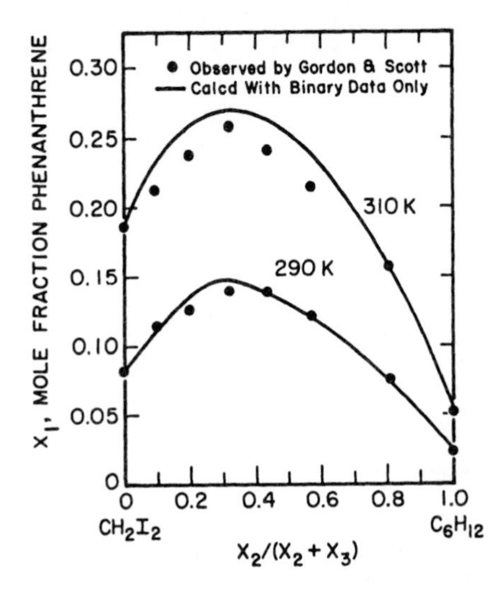

Figure 11-13 Solubility of phenanthrene in a mixed solvent containing cyclohexane and methylene iodide.

Also shown in Fig. 11-13 are calculations based on regular-solution theory without, however, using the geometric-mean assumption. The activity coefficient of phenanthrene (referred to the subcooled liquid) is given by

$$\ln \gamma_1 = \frac{v_1}{RT}[A_{12}\Phi_2^2 + A_{13}\Phi_3^2 + (A_{12} + A_{13} - A_{23})\Phi_2\Phi_3] \tag{11-31}$$

where

$$A_{ij} = (\delta_i - \delta_j)^2 + 2l_{ij}\delta_i\delta_j.$$

Parameters l_{ij}, that give deviations from the geometric mean, are obtained from binary data reported by Gordon and Scott. For the two binaries including phenanthrene, l_{12} and l_{13} are obtained from experimental solubilities of phenanthrene in each of the two solvents at 25°C. For the solvent pair, l_{23} is obtained from binary melting-point data. These parameters are $l_{12} = 0.028$, $l_{23} = 0.010$, and $l_{13} = 0.0028$. For phenanthrene, the enthalpy of fusion is 17991 J mol^{-1} and the (liquid) molar volume at 25°C is estimated to be 150 cm^3 mol^{-1}. Molar volumes for cyclohexane and methylene iodide are 108 and 80 cm^3 mol^{-1}, respectively.

In these calculations, all parameters were taken as independent of temperature. Agreement between calculated and observed solubilities is nearly quantitative; the somewhat larger deviations at 310 K are probably due to uncertainties in l_{23} that was found from data at significantly lower temperatures. It is likely that l_{23} is strongly temperature-dependent near 310 K because cyclohexane and methylene iodide are only partially miscible below the (upper) critical solution temperature near 303 K. The important feature of Fig. 11-13 is that calculations, based on binary data only, correctly predict the maximum solubility that is observed in the ternary system.

A group-contribution method such as UNIFAC may sometimes be used to calculate liquid-phase activity coefficients of solutes in mixed solvents, as discussed by Gmehling *et al.* (1978). To illustrate, consider the solubility of naphthalene in mixed solvents containing water and alcohol. If naphthalene is designated by subscript 2, the equation of equilibrium is given by Eq. (11-5), again assuming that there is negligible solubility of the other components in solid naphthalene. The fugacity ratio for pure naphthalene is given by Eq. (11-13). UNIFAC is used to calculate activity coefficient γ_2.

Table 11-2 gives calculated and experimental solubilities for naphthalene in aqueous solvents containing methanol, ethanol, propanol, or butanol. The ideal solubilities ($\gamma_2 = 1$) are always too large by one order of magnitude. Solubilities calculated with UNIFAC are in semiquantitative agreement with experiment.

Table 11-2 Solubility of naphthalene in alcohol/water systems.

Alcohol	Mole fraction alcohol in solvent mixture†	Temp. (°C)	Solubility (mol %)		
			Ideal	UNIFAC	Exp.*
Methanol	0.922	35.7	39.7	2.8	2.4
	0.922	50.6	55.8	4.8	4.6
Ethanol	0.906	27.5	32.7	2.6	3.4
	0.906	39.5	43.5	3.8	5.5
	0.743	73.0	47.0	2.0	3.8
1-Propanol	0.739	40.9	44.9	3.7	5.6
	0.739	46.7	51.4	4.4	7.4
	0.616	52.1	57.6	3.0	5.7
1-Butanol	0.813	21.8	28.3	3.5	4.3
	0.813	29.6	34.1	4.6	5.8
	0.680	30.7	35.3	2.9	4.7
	0.680	43.5	47.7	4.5	8.0

* O. Mannhardt, R. De Right, W. Martin, C. Burmaster, and W. Wadt, 1943, *J. Phys. Chem.*, 47: 685.

† Solute-free basis.

For a solid dissolved in a binary mixed solvent, *enhanced solubility* (or maximum-solubility effect)[11] has been observed for a large variety of systems. As an example, consider the solubilities of two isostructural aromatic compounds 2-acetyl-1-naphthol (**1**) and 1-acetyl-2-naphthol (**2**), in pure and mixed hydrocarbons and alcohols (Domanska, 1990):

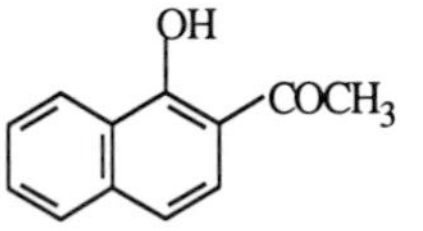

OH COCH$_3$ | $COCH_3$ OH

(1) $t_m = 98.60°C$ $\Delta_{fus}h = 22.52$ kJ mol^{-1}

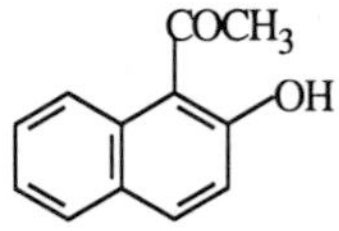

(2) $t_m = 63.75°C$ $\Delta_{fus}h = 21.34$ kJ mol^{-1}

Figure 11-14(a) shows the solubility of (**1**) in cyclohexane, in 1-butanol and in several mixtures of these solvents, at four temperatures. Similarly, Fig. 11-14(b) shows the solubility of (**2**) in cyclohexane, in 1-butanol and in several mixtures of these solvents. Both isomers have enhanced solubility in the binary solvent mixture, and the deviations from additivity[12] rise with increasing temperature. These deviations are probably due to the decrease of self association of 1-butanol molecules with increasing temperature.

The results presented in Fig. 11-14 indicate that solubilities of compound (**2**) in the alcohol are much higher than those in cyclohexane. In contrast, (**1**) exhibits similar solubilities in the alcohol and in the alkane solvents. The large differences observed in the solubilities of these (similar) compounds are attributed to their characteristic behavior in the two solvents. Ultraviolet and infrared absorption spectra of (**1**) and (**2**) reveal that both form intramolecular hydrogen bonds. However, for compound (**1**) this bond is little affected by hydroxylic solvents (such as alcohols), i.e. the long-wavelength bond in its UV spectra is not affected by the solvent, as shown in Fig. 11-15(a). In contrast to (**1**), UV absorption spectra in (**2**) reveal that hydroxylic solvents cause the disruption of intramolecular hydrogen bonds in (**2**), as shown in Fig. 11-15(b).

---

[11] However, in other cases, enhanced solubility is not observed; instead, there may exist *negative synergetic effect* causing a decreased solubility in comparison with those in single solvents. For example, solubility measurements for solid 1-benzoyl-2-naphthol show slight negative deviations from additivity (negative synergetic effects) in the hexane+1-butanol mixed solvent system. Negative synergetic effects on solubility have also been found in solubility measurements for octadecanoic, eicosanoic, and docosanoic acids in mixed solvent systems such as 1,4-dioxane+chloroform and tetrahydrofuran+chloroform.

[12] Additivity [given by the broken lines in Fig. (11-14)] means that the solubility is a linear function of solvent composition.

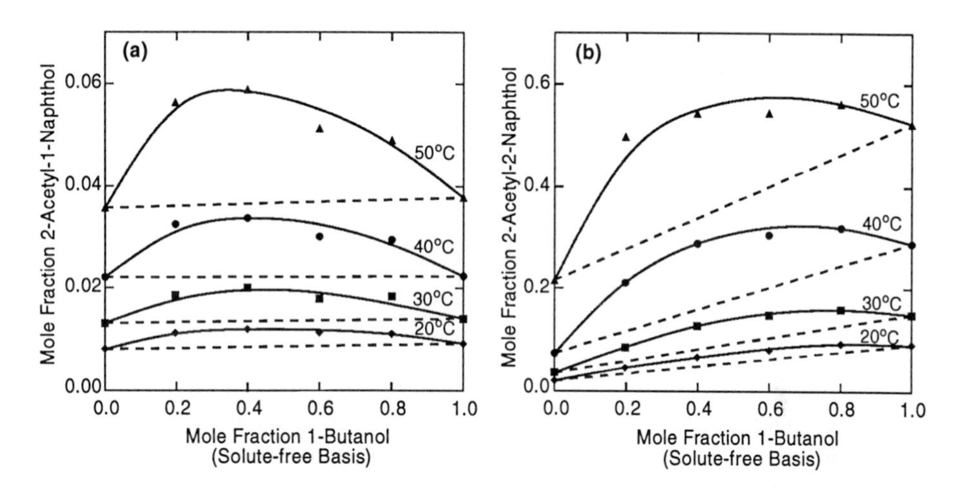

Figure 11-14 Solubility isotherms for (a) 2-acetyl-1-naphthol **(1)** and (b) 1-acetyl-2-naphthol **(2)** in cyclohexane/1-butanol mixed solvent system (Domanska, 1990). —— Smoothed experimental data (symbols); – – – Additivity rule.

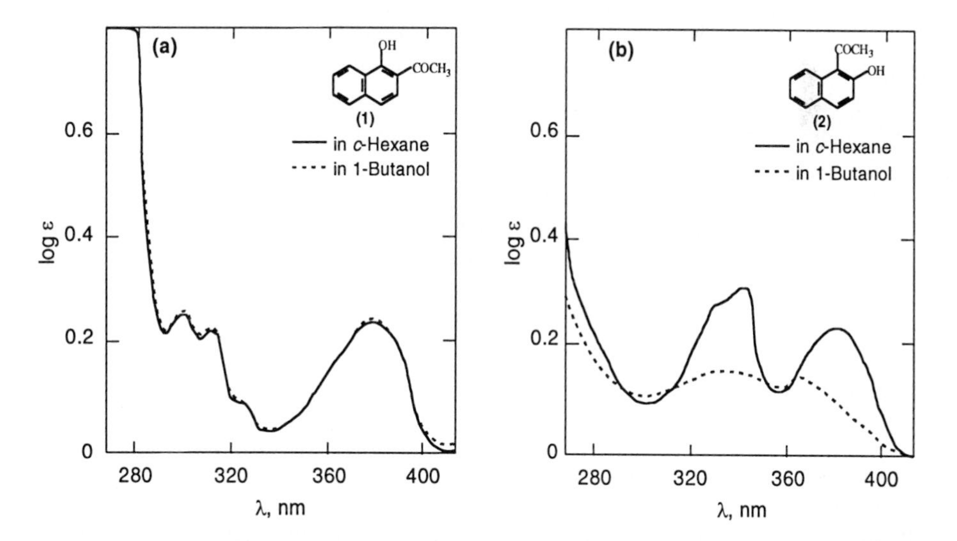

Figure 11-15 Schematics of ultraviolet absorption spectra for (a) 2-acetyl-1-naphthol **(1)** and (b) 1-acetyl-2-naphthol **(2)**, in cyclohexane (——) and in 1-butanol (- - - -). Here ε is the extinction coefficient.

In cyclohexane solutions, molecules of isomeric **(2)** are strongly intramolecular hydrogen bonded and almost planar, but in hydroxylic solvents the intramolecular hydrogen-bond is disrupted due to the formation of intermolecular solute-solvent solvation, as schematically illustrated below:

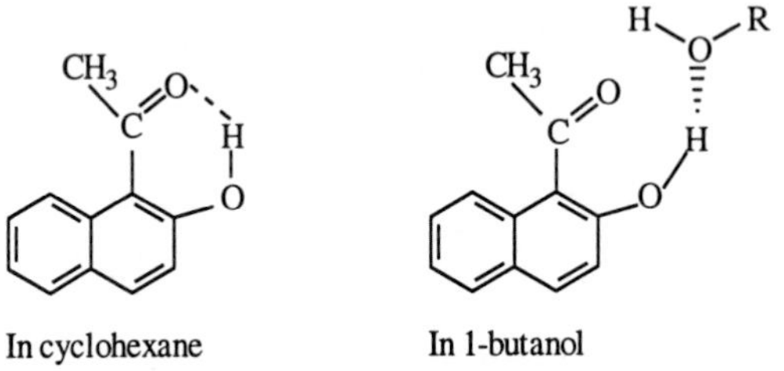

In cyclohexane In 1-butanol

Moreover, dielectric studies also indicate that the intramolecular hydrogen-bond is relatively stable for **(1)** but unstable for **(2)**. Therefore, compound **(1)**, that is able to form strong intramolecular hydrogen bonds, behaves quite differently from its isomer that lacks this ability.

Domanska's studies show once again that understanding of thermodynamic properties (in this case, solubilities) is much enhanced by fundamental physico-chemical measurements that provide information on molecular behavior.

The solubility measurements for solids **(1)** and **(2)** can be described by an equation similar to Eq. (11-14), provided that we do not neglect activity coefficient γ_2. To illustrate, consider again the solubility of **(1)** in the cyclohexane+1-butanol mixed solvent. Domanska (1990) used the Wilson equation to obtain the activity coefficients of each component of the ternary mixture [i.e. Eq.(6-165) with $m = 3$]. Wilson's interaction parameters for the binary cyclohexane+1-butanol solvent mixture were obtained from VLE data, whereas those for solute-solvent were regressed from binary SLE data. Figure 11-16 shows the temperature dependence of the solubility of **(1)** in pure cyclohexane and 1-butanol and also in several of their binary mixtures. As shown, the Wilson equation with temperature-independent parameters can describe well the observed solubilities in the temperature range studied.

11.6 Solid Solutions

In all previous sections we have assumed that whereas the solid has a finite solubility in the liquid solvent, there is no appreciable solubility of the solvent in the solid. As a result, we have been concerned only with the equation of equilibrium for component 2, the solid component. However, there are many situations where components 1 and 2 are miscible not only in the liquid phase but in the solid phase as well; in such cases we must write two equations of equilibrium, one for each component:

$$f_{1\,(\text{solid phase})} = f_{1\,(\text{liquid phase})} \tag{11-32}$$

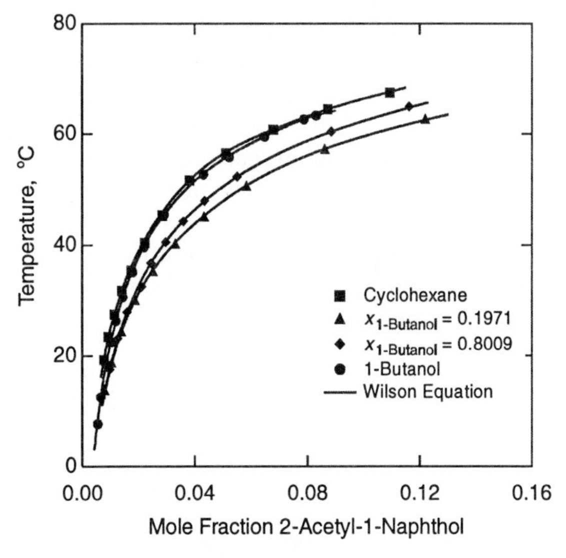

Figure 11-16 Temperature as a function of the solubility of 2-acetyl-1-naphthol **(1)** in: cyclohexane (■); 1-butanol (●); and cyclohexane+1-butanol mixed-solvent system containing $x_{1\text{-butanol}} = 0.1971$ (▲) and $x_{1\text{-butanol}} = 0.8009$ (◆). —— Calculated from Wilson equation.

$$f_{2\,(\text{solid phase})} = f_{2\,(\text{liquid phase})} \tag{11-33}$$

Introducing activity coefficients we can rewrite these equations:

$$\gamma_1^{s} x_1^{s} f_{\text{pure }1}^{s} = \gamma_1^{L} x_1^{L} f_{\text{pure }1}^{L} \tag{11-34}$$

$$\gamma_2^{s} x_2^{s} f_{\text{pure }2}^{s} = \gamma_2^{L} x_2^{L} f_{\text{pure }2}^{L} \tag{11-35}$$

If system temperature T is above the triple-point temperature of component 1 but below that of component 2, then pure solid 1 and pure liquid 2 are both hypothetical. In Sec. 11.2 we showed how to compute the fugacity of a pure (hypothetical) subcooled liquid, and exactly the same principles may be used to calculate the fugacity of a pure superheated solid; the calculation is performed by using a thermodynamic cycle similar to that shown in Fig. 11-2 with temperature T above, rather than below, triple-point temperature T_t. When both components are miscible in both phases, Eqs. (11-34) and (11-35) indicate that information is required for the activity coefficients in both phases. Such information is rarely available. In some systems one may, to a good approximation, assume ideal behavior in both phases ($\gamma_1^{s} = \gamma_2^{s} = \gamma_1^{L} = \gamma_2^{L} = 1$), but in most systems this is a poor assumption.

When Eqs. (11-34) and (11-35) are applied over a range of temperatures, it is possible to calculate the freezing-point diagram for the binary system. If there is miscibility in both phases, such a diagram is qualitatively very different from that shown in Fig. 11-5. For example, Fig. 11-17 gives the experimentally determined phase diagram for the system naphthalene/β-naphthol that exhibits approximately ideal behavior in both phases. On the other hand, the system mercuric bromide/mercuric iodide, shown in Fig. 11-18, exhibits appreciable nonideality resulting in the formation of a hylotrope (no change in composition upon melting). The thermodynamics of solid solutions is of importance in metallurgy and many experimental studies have been made of binary and multicomponent metallic mixtures.

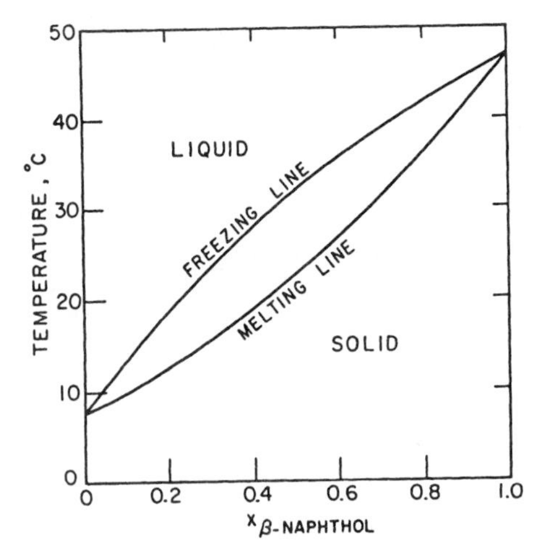

Figure 11-17 Phase diagram for the system naphthalene/β-naphthol.

When calculating the solubility of a solid in a liquid, as discussed in Sec. 11.4, we assume that the solid phase is pure, i.e., $\gamma_1^{s} x_1^{s} = 1$ and $\gamma_2^{s} x_2^{s} = 1$ in Eqs. (11-34) and (11-35). In many cases that is a good assumption, but if it is not, serious error can result. To illustrate, Preston (1970) calculated the solubility of solid argon in liquid nitrogen in the region 64 to 83 K. Using regular-solution theory [Eq. (11-17)] and assuming that solid argon is pure, he obtained the upper line shown in Fig. 11-19. He then repeated the calculation, using Eq. (11-35) coupled with experimental solid-phase composition data, assuming again that the liquid-phase activity coefficient was given by Eq. (11-17) and the solid-phase activity coefficient was unity (ideal solid solution).

Preston then obtained the lower line in Fig. 11-19, in excellent agreement with experimental data (Fedorova, 1938; Long and DiPaolo, 1963; Din *et al.*, 1955). These results show that for the system argon/nitrogen it is necessary to take into account miscibility in the solid phase.

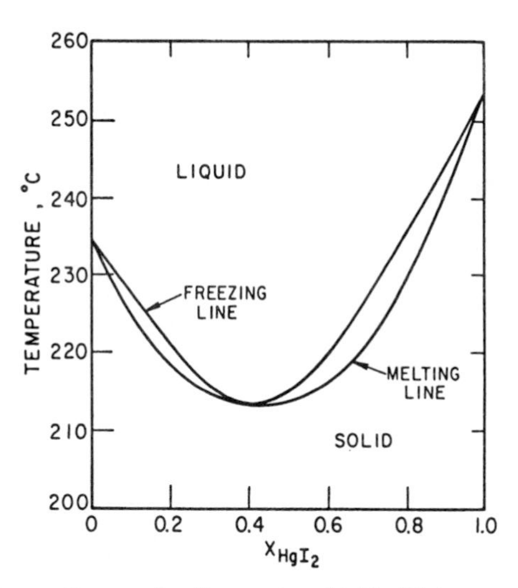

Figure 11-18 Phase diagram for the system $HgBr_2/HgI_2$.

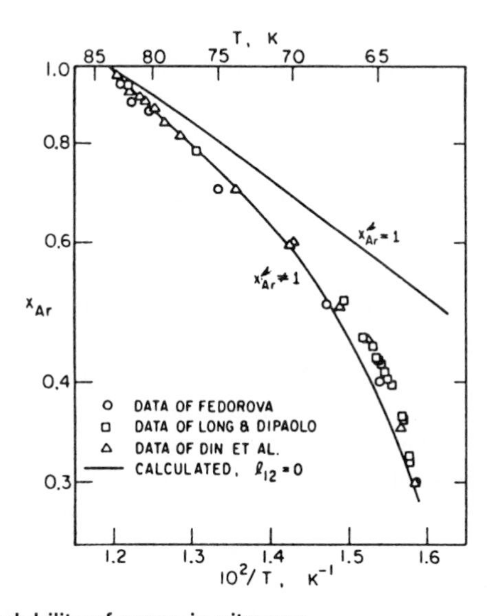

Figure 11-19 Solubility of argon in nitrogen.

It is difficult to say a priori whether or not two components are partially or totally miscible in the solid phase. For nonpolar substances the general rule is that solid-phase miscibility is usually negligible provided that the two components differ appreciably in molecular size and shape. However, other factors than size and shape contribute to determine the formation of more than one solid phase. For example, the fluorene/dibenzofuran system, whose solid-liquid phase diagram is shown in Fig. 11-20

(Sediawan *et al.*, 1989), forms solid solutions at any composition, because, as expected, those two molecules have similar molecular sizes and shapes, as illustrated in Table 11-3.

Table 11-3 Molecular structures and physical properties of pure dibenzofuran, dibenzothiophene, and fluorene.

	Molecular structure	Crystal structure	$\Delta_{fus}h$ (kJ mol^{-1})	t_m (°C)
Dibenzofuran	O	orthorhombic	18.6	82.15
Dibenzothiophene	S	monoclinic	21	98.80
Fluorene	CH_2	orthorhombic	19.6	114.5

However, in spite of the components' similarity in size and shape, in the fluorene/dibenzothiophene system there are two possible mixed-solid phases, as illustrated in Fig. 11-21 (Sediawan *et al.*, 1989). For solid mixtures with fluorene mole fractions smaller than about 0.04, there is one solid phase, α, and for solid mixtures with fluorene mole fractions greater than about 0.18 there is another solid phase, β. In the region with fluorene mole fractions between 0.04 and 0.18, there is a solid immiscibility gap; Table 11-3 indicates that this gap cannot be attributed to differences in size and shape. In this particular case it is likely that the observed solid-phase immiscibility gap is caused by differences in the components' pure crystal structures: solid phase β has a crystal structure similar to that of pure fluorene, while solid phase α has a crystal structure similar to that of dibenzothiophene. Similarly, the binary dibenzothiophene/dibenzofuran shows a small solid-phase immiscibility gap.

Experimental information on binary systems is always necessary to estimate phase equilibria of multicomponent systems. To illustrate, consider the ternary system benzene/fluorene/dibenzothiophene composed of two polynuclear aromatic solutes in an organic solvent. Domanska *et al.* (1993) used binary data to obtain the parameters of the UNIQUAC model. Four parameters per binary are required: two describing the liquid phase, obtained from the liquidus curve, and two describing the solid phase, obtained from the solidus curve.[13] For the ternary system at temperatures in the range 40-60°C, Domanska *et al.* calculate the compositions of the liquid solution in equilibrium with the solid solution, in good agreement with experiment. Their results suggest that it

[13] However, for the two binary eutectic systems (benzene/fluorene and benzene/dibenzothiophene), Domanska and coworkers assumed that solid-phase parameters are zero.

is possible to predict ternary solid-liquid equilibria using UNIQUAC parameters obtained from binary solid-liquid equilibria.

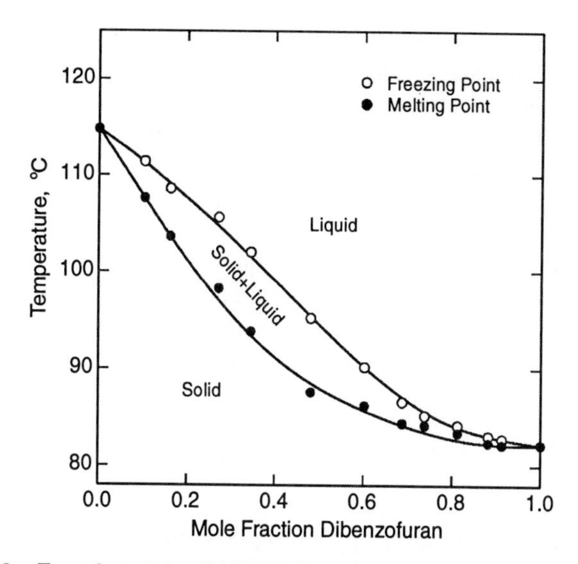

Figure 11-20 Experimental solid-liquid phase diagram for the system fluorene/dibenzofuran (Sediawan *et al.*, 1989). —— Smoothed data.

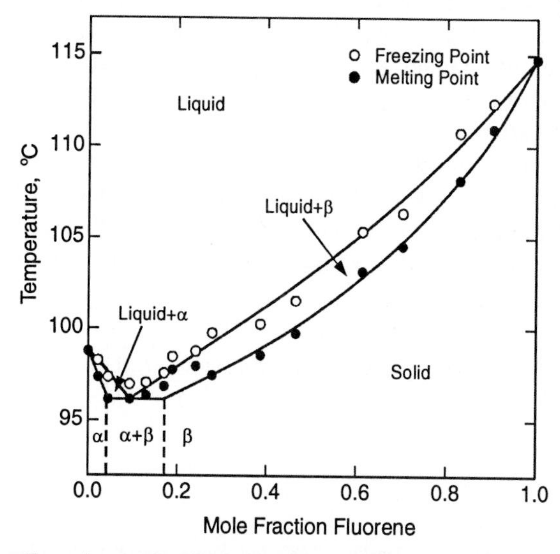

Figure 11-21 Experimental solid-liquid phase diagram for the system fluorene/dibenzothiophene (Sediawan *et al.*, 1989). —— Smoothed data.

Because our knowledge of solid-phase mixtures is so meager, in many typical chemical-engineering calculations for multicomponent systems, it has been customary to assume either that there is complete immiscibility or (more rarely) complete miscibility in the solid phase. (In most cases, it is more realistic to assume no solubility in the solid phase). Unfortunately, calculated phase equilibria are extremely sensitive to this choice of assumption; calculated results depend very strongly on whether we assume complete miscibility or complete immiscibility in the solid phase. An industrially important example is provided by the precipitation of solid waxes from petroleum. For many years it was assumed that waxes (long-chain paraffins) are completely miscible in the solid phase; however, to obtain agreement between calculated and experimental results, elaborate empirical corrections were required. When spectroscopic data showed that long-chain paraffins are, in general, not mutually soluble in the solid phase (unless the chain lengths are similar), subsequent thermodynamic calculations gave much better results, as discussed by Lira-Galeana *et al.* (1996).

11.7 Solubility of Antibiotics in Mixed Nonaqueous Solvents

An industrial application that requires solid-liquid equilibria is provided by separation and recovery processes for amino acids and antibiotics. For some common bioproducts, the cost in bioseparations may reach ninety percent of the total cost of manufacturing. Design of bioseparation units may require phase equilibria for the bioproducts in the solvents used in the production process. Because amino acids and antibiotics have high melting points, the required phase equilibria are mostly solid-liquid equilibria.

A simple solubility model for amino acids and antibiotics was presented by Gupta and Heidemann (1990) using a modified UNIFAC equation.[14] To illustrate, consider the antibiotic carbomycin-A ($C_{42}H_{67}NO_{16}$, M = 841.97 g mol^{-1}, t_m = 214°C). Using available experimental solubilities of the antibiotic in different solvents, and considering the entire antibiotic molecule as a group,[15] Gupta and Heidemann wanted to obtain UNIFAC group-interaction parameters. To do so, they obtained from experimental solubility data the activity coefficient of the solid at saturation, using an equation similar to Eq. (11-15):

$$\gamma_2^{\text{sat}} = \frac{1}{x_2^{\text{sat}}} \exp\left[\frac{-\Delta_{\text{fus}} s}{R}\left(\frac{T_m}{T} - 1\right)\right] \tag{11-36}$$

---

[14] B. L. Larsen, P. Rasmussen, and Aa. Fredenslund, 1987, *Ind. Eng. Chem. Res.*, 26: 2274.

[15] For carbomycin-A, the group volume and the group surface area parameters are, respectively, R = 31.6568 and Q = 5.119, as determined by the van der Waals volumes and surface areas of the constituent conventional UNIFAC groups.

For the entropy of fusion they take 56.51 J mol^{-1} K^{-1}.[16] The optimal UNIFAC parameters, a_{ij}, were then obtained from regression of the objective function $\Sigma_i[\ln(\gamma_i^{sat}) - \ln(\gamma_i^{sat})_{UNIFAC}]^2$, where γ_i^{sat} are obtained from Eq. (11-36) with experimental x_2^{sat}, and where $(\gamma_i^{sat})_{UNIFAC}$ is from the UNIFAC equation. Table 11-4 lists the group-interaction parameters obtained.

Table 11-4 UNIFAC group-interaction parameters a_{IJ} (in Kelvin) between carbomycin-A (2) and the standard alkane (CH_2), alcohol (OH), and aromatic (ACH) groups, obtained from the antibiotic's solubility data in several solvents at 28°C (Gupta and Heidemann, 1990).

a_{2,CH_2}	$a_{CH_2,2}$	$a_{2,OH}$	$a_{OH,2}$	$a_{2,ACH}$	$a_{ACH,2}$
409.2	-1034.4	7181.9	-204.2	451.6	-183.8

Antibiotics are commonly produced from an aqueous fermentation broth as crystalline precipitates by adding solvents wherein the product has a lower solubility. UNIFAC group-interaction parameters can be used to predict solubilities in solvents and in mixed solvents when little or no experimental data are available, thus providing estimates of the required phase equilibria necessary for the design of the separation process. As an illustration, Fig. 11-22 shows calculated solubilities of carbomycin-A in a mixed solvent containing cyclohexane and toluene at 28°C. Calculations were performed with the UNIFAC equation using parameters in Table 11-4. Comparison with experimental data is possible only at the endpoints of the curve, corresponding to the solubility in the pure solvents also at 28°C.

An useful, industrially-oriented way to use calculated solubilities is presented in Fig. 11-23. This figure shows the calculated percent recovery of carbomycin-A that is expected upon diluting a saturated solution in toluene with several solvents. The calculations show that about 90% of the dissolved antibiotic can be recovered as crystalline solid by diluting with *n*-hexadecane on a one-to-one molar basis.

The results presented in Fig. 11-23 are qualitative only; they require experimental confirmation. The model used in the calculations is based on a limited data base and therefore, the results of Fig. 11-23 can only provide guidance toward a possible separation process. Nevertheless, the results indicate how simple thermodynamic calculations, coupled with limited experimental data, may be useful for preliminary chemical-process design.

[16] S. H. Yalkowsky, 1979, *Ind. Eng. Chem. Fundam.*, 18: 109.

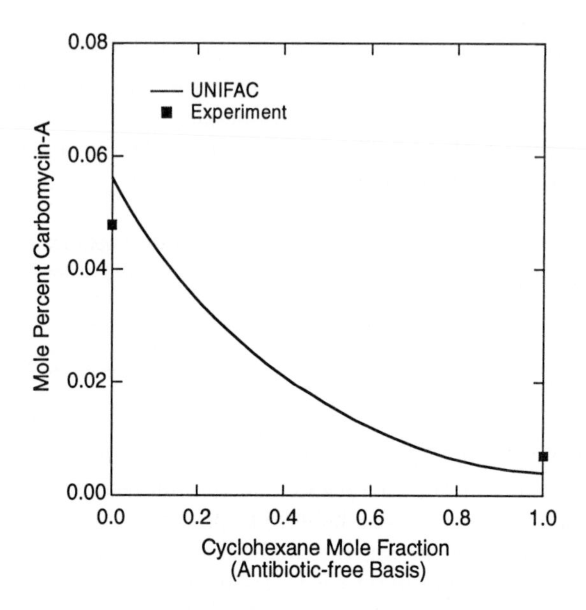

Figure 11-22 Solubility of carbomycin-A in mixtures of toluene and cyclohexane at 28°C. —— UNIFAC; ■ Experiment.

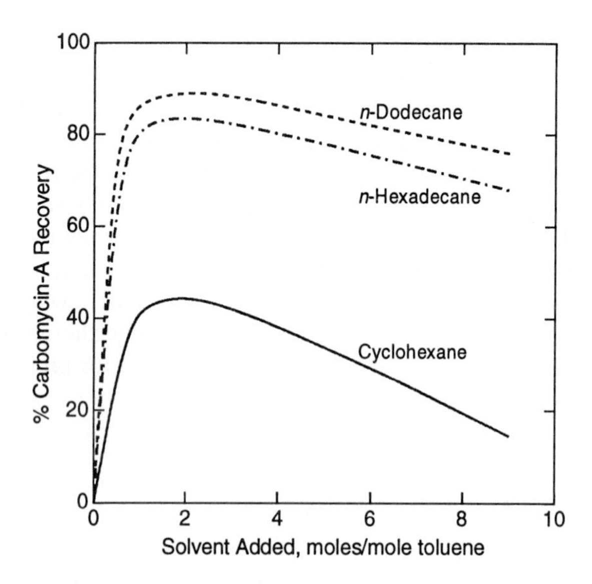

Figure 11-23 Predicted (from UNIFAC) recovery of carbomycin-A from saturated solutions in toluene upon dilution with a selected hydrocarbon.

References

Bar, L. K., N. Garti, S. Sarig, and R. Bar, 1984, *J. Chem. Eng. Data,* 29: 440.

Din, F., K. Goldman, and A. G. Monroe, 1955, *Proc. 9th Int. Conf. Refrig.,* 1: 3.

Domanska, U., 1990, *Fluid Phase Equilibria,* 55: 125.

Domanska, U., F. R. Groves, Jr., and E. McLaughlin, 1993, *J. Chem. Eng. Data,* 38: 88.

Fedorova, M. F., 1938, *Zh. Eksp. Teor. Fiz.,* 8: 425.

Gmehling, J. G., T. F. Anderson, and J. M. Prausnitz, 1978, *Ind. Eng. Chem. Fundam.,* 17: 269.

Góralski, P., 1993, *J. Chem. Thermodynamics,* 25: 367.

Gordon, L. J. and R. L. Scott, 1952, *J. Am. Chem. Soc.,* 74: 4138.

Gupta, A., S. Gupta, F. R. Groves, Jr., and E. McLaughlin, 1991, *Fluid Phase Equilibria,* 64: 201.

Gupta, R. B. and R. A. Heidemann, 1990, *AIChE J.,* 36: 333.

Hatcher, W. and F. Skirrow, 1917, *J. Am. Chem. Soc.,* 39: 1939.

Jadzyn, J. and L. Hellemans, 1993, *Ber. Bunsenges. Phys. Chem.,* 97: 205.

Lira-Galeana, C., A. Firoozabadi, and J. M. Prausnitz, 1996, *AIChE J.,* 42: 239.

Long, H. M. and F. S. DiPaolo, 1963, *Chem. Eng. Prog. Symp. Ser.,* 59: 30.

McLaughlin, E. and H. A. Zainal, 1959, *J. Chem. Soc.,* 863 (March).

McLaughlin, E. and H. A. Zainal, 1960, *J. Chem. Soc.,* 2485 (June).

Mercier, P., C. Sandorfy, and D. Vocelle, 1983, *J. Phys. Chem.,* 87: 3670.

Myers, A. L. and J. M. Prausnitz, 1965, *Ind. Eng. Chem. Fundam.,* 4: 209.

Preston, G. T. and J. M. Prausnitz, 1970, *Ind. Eng. Chem. Process Des. Dev.,* 9: 264

Prigogine, I. and R. Defay, 1954, *Chemical Thermodynamics,* (trans./rev. D. H. Everett). London: Longmans & Green.

Sediawan, W. B., S. Gupta, and E. L. McLaughlin, 1989, *J. Chem. Eng. Data,* 34: 223.

Soave, G., 1979, *Chem. Eng. Sci.,* 34: 225.

Tsakalotos, D. and P. Guye, 1910, *J. Chem. Phys.,* 8: 340.

Weimer, R. F. and J. M. Prausnitz, 1965, *J. Chem. Phys.,* 42: 3643.

Young, F. E. and J. H. Hildebrand, 1942, *J. Am. Chem. Soc.,* 64: 839.

Problems

1. A liquid mixture contains 5 mol % naphthalene and 95 mol % benzene. The mixture is slowly cooled at constant pressure. At what temperature does a solid phase appear? Assume ideal mixing in the liquid phase and total immiscibility in the solid phase. Data are:

	Benzene	Naphthalene
Melting temperature (K)	278.7	353.4
Enthalpy of fusion (J mol^{-1})	9843	19008

2. At 25°C, a solid A is in solution in a liquid solvent B. It is proposed to remove A from the solution by adsorption on a solid adsorbent S that is inert toward B. The adsorption equilibrium constant K is defined by a Langmuir-type expression

$$K = \frac{\theta}{(1-\theta)a_A}$$

where θ is the fraction of surface sites on S that are occupied by A and a_A the activity of A ($a_A = 1$ when $x_A = 1$). At 25°C, $K = 130$.
What fraction of the surface sites on S is covered by A when x_A is one-half of the saturated mole fraction of A in B at 25°C?
The following data are available for pure A:
$\Delta_{fus}h = 19700$ J mol^{-1}
$T_m = 412$ K (melting point)
$c_{p\,(liquid)} = 33.5$ J mol^{-1} K^{-1}
$c_{p\,(solid)} = 26.4$ J mol^{-1} K^{-1}
$P^s_{solid} = 0.28$ kPa
$x_A^s = 0.050$ at 25°C (solubility in B)

3. In a famous paper by Brown and Brady (1952, *J. Am. Chem. Soc.*, 74: 3570) on the basicity of aromatics, the authors comment on Klatt's measurements at 0°C on the solubilities of aromatics in (acidic) liquid hydrogen fluoride. Klatt finds the solubilities are in the order *m*-xylene $<$ toluene $<$ benzene. However, there is ample evidence from other equilibrium measurements, from spectra and from kinetic data that the basicity of the aromatic ring *increases* with methyl substitution. Can you explain the apparent disagreement?

4. Estimate the solubility of naphthalene at 25°C in a mixed solvent consisting of 70 mol % isopentane and 30 mol % carbon tetrachloride.
The following data are available for naphthalene:
Melting point: 80.2°C; enthalpy of fusion: 19008 J mol^{-1}.
$c_{p\,(liquid)} - c_{p\,(solid)}$ is sufficiently small to be negligible for these conditions.
Volume of "liquid" naphthalene at 25°C: 123 cm^3 mol^{-1} (extrapolated).

5. The freezing point of benzene is 278.7 K. At 260 K, an equimolar mixture of *n*-hexane and carbon tetrachloride contains 10 mol % benzene. Calculate the partial pressure of benzene in equilibrium with this mixture.
For benzene: $\Delta_{fus}h = 30.45$ cal g^{-1}; p^s_{solid} (260 K) = 0.0125 bar. At 25°C:

	Molar volume (cm^3 mol^{-1})	Solubility parameter (cal cm^{-3})$^{1/2}$
Benzene	89	9.2
n-Hexane	132	7.3
Carbon tetrachloride	97	8.6

6. A liquid mixture of benzene and n-heptane is to be cooled to as low a temperature as possible without precipitation of a solid phase. If the solution contains 10 mol % benzene, estimate what this lowest temperature is.
The data are as follows:

	Benzene	Heptane
Melting point (K)	278.7	182.6
Enthalpy of fusion (J mol^{-1})	9843	14067
Solubility parameter(J cm^{-3})$^{1/2}$	18.8	15.1
Molar liquid volume (cm^3 mol^{-1})	89	148

7. The Cu_2O/P_2O_5 system has been investigated by X-ray and thermal techniques (M. Ball, 1968, *J. Chem. Soc. (A), Inorg. Phys. Theor.*, 1113). The following table gives liquidus and solidus temperatures for the system, under argon. Melting point of Cu_2O is 1503 K.

	T (K)	
x_{Cu_2O}	Liquidus	Solidus
0.818	1368	1090
0.739	1293	1083
0.666	1200	1080
0.600	1090	1083
0.538	1350	1091
0.482	1490	1203
0.428	1343	1213
0.379	1218	1210
0.333	1363	1211

What physicochemical conclusions can be drawn from these data?

8. Benzene freezes at 5.5°C; its enthalpy of fusion is 9843 J mol^{-1}.
   (a) At -50°C, would you expect benzene to be more or less soluble in CS_2 than in n-octane? Why? (The freezing points of CS_2 and n-octane are -19.5 and -56.5°C, respectively.)
   (b) What composition of a mixed solvent of CS_2 and n-octane is needed such that a solution containing 30 mol % benzene just begins to precipitate at -50°C?

9. The triple point of carbon dioxide is 216.5 K. At 194.3 K, the solubility of carbon dioxide (2) in a solvent (1) is $x_2 = 0.25$, where x_2 is the liquid-phase mole fraction of carbon dioxide. Solvent (1) is insoluble in solid carbon dioxide. Estimate the partial pressure of carbon dioxide in solvent (1) at 194.3 K when $x_2 = 0.05$.
The data (all at 194.3 K) are as follows: Saturation pressure of solid carbon dioxide: 0.99 bar; from enthalpy-of-fusion data, the ratio $(f^{s} / f^{L})_{\text{pure 2}} = 0.56$.

10. At 250 K, a nonpolar solid A has a vapor pressure of 35 torr. The melting temperature of A is 300 K where its enthalpy of fusion is 13 kJ mol^{-1}. Component A is a branched hydrocarbon. At 250 K, a 3 mol % solution of A in carbon tetrachloride has a partial pressure of 5 torr.

What is the partial pressure of A in a 1 mol % solution in *n*-hexane at 250 K? At 25°C, the following data are available for solubility parameters and molar liquid volumes:

	δ $(J\ cm^{-3})^{1/2}$	v $(cm^3\ mol^{-1})$
A	Not known	95
CCl_4	17.6	97
n-Hexane	132	14.9

11. A natural gas, containing mostly methane, contains also small amounts of H_2S. It is proposed to remove the H_2S by an absorption process. Consideration is being given to two liquid solvents A and B whose freezing points are, respectively, 20 and 5°C.

As for most gases, the solubility of gaseous H_2S in either A or B rises with decreasing temperature. To minimize solvent requirements, it is therefore advantageous to operate at a low temperature. However, if A is used, the lowest possible temperature is 20°C and if B is used, the lowest possible temperature is 5°C. To operate at a lower temperature, it is proposed to use for the solvent a mixture of A and B.

To test the feasibility of this proposal, estimate the composition in mole percent of a mixed solvent containing A and B that gives the lowest possible temperature for an absorption process. Also estimate this temperature.

For the preliminary-design purposes considered here, assume that liquids A and B are chemically similar and that they are miscible in all proportions. Also assume that solids A and B are mutually insoluble. The enthalpies of are 8 kJ mol^{-1} for A and 12 kJ mol^{-1} for B.

CHAPTER 12

High-Pressure Phase Equilibria

Numerous chemical processes operate at high pressures and, primarily for economic reasons, many separation operations (distillation, absorption) are conducted at high pressures; further, phase equilibria at high pressures are of interest in geological exploration, such as in drilling for petroleum and natural gas. While the technical importance of high-pressure phase behavior has been recognized for a long time, quantitative application of thermodynamics toward understanding such behavior was not common until about 1940. In the early days of phase-equilibrium thermodynamics, such an attempt could not be made because of computational complexity; realistic thermodynamic calculations for high-pressure equilibria are difficult without computers.

The adjective *high-pressure* is relative; in some areas of technology (e.g., outer-space research) 1 mm of mercury is a high pressure whereas in others (e.g., solid-state research) a pressure of a few hundred bars is considered almost a vacuum. Figure 12-1 (Schneider, 1976) shows a rough comparison between pressures observed in nature and pressures of some common industrial processes. For pure fluids, critical pressures can vary from 2.3 bar (for helium) to 1500 bar (for mercury); the pressure at the bottom of the deepest ocean (about 10 km deep) is about 1150 bar, while at the center of the earth the pressure is estimated to be larger than 4×10^6 bar. On the other hand, for man-made processes, high-pressure techniques can require pressures like 5×10^4 bar (synthesis of diamonds), or even 10^6 bar, as in explosive welding and plating, sophisticated techniques used in the production of tubing for chemical reactors. Pressures between 100 and 1000 bar are used in high-pressure liquid chromatography and in supercritical fluid chromatography, for hydrogenation processes, and for synthesis of products like ammonia, methanol, and acetic acid. Classical production of low-density polyethylene occurs at pressures between 1500 and 3000 bar.

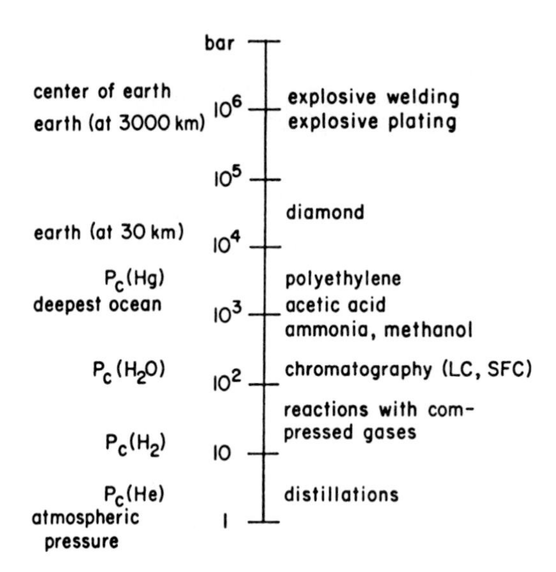

Figure 12-1 Pressure scale for natural (left) and chemical (right) processes (Schneider, 1976).

For the description of these and similar processes, it is necessary to understand the thermodynamic properties of fluids at high pressures, as discussed in this chapter. We designate here as "high-pressure" any pressure sufficiently large to have an appreciable effect on the thermodynamic properties of all the phases under consideration. In vapor-liquid equilibria, a high pressure may be anywhere between (about) 20 to 1000 bar, depending on the system and on the temperature; only in rare cases does the pressure exceed 1000 bar, because in most cases of common interest the vapor-liquid critical condensation pressure of the system is below 1000 bar. In liquid-liquid equilibria or in gas-gas equilibria the pressure may be considerably larger, although experimental studies are rare for fluid mixtures at pressures beyond 1000 bar.

12.1 Fluid Mixtures at High Pressures

Before discussing the thermodynamic relations that govern the behavior of mixtures at high pressures, let us take a brief look at some typical results to obtain a qualitative picture of how mixed fluids behave at pressures well above atmospheric.

For vapor-liquid behavior of a typical simple system, consider mixtures of ethane and *n*-heptane; the critical temperature of ethane is 32.3°C and that of *n*-heptane is 267.0°C. Figure 12-2 shows the relation between pressure and composition at 149°C. The left-hand line gives the saturation pressure (bubble pressure) as a function of liquid composition and the right-hand line gives the saturation pressure (dew pressure) as a function of vapor composition. The two lines meet at the critical point where the two phases become identical. At 149°C the critical composition is 76 mol % ethane and the critical pressure is 88 bar. At this temperature and composition, therefore, only one phase can exist at pressures higher than 88 bar; further, regardless of pressure, it is not possible to have at 149°C a coexisting liquid phase containing more than 76 mol % ethane.

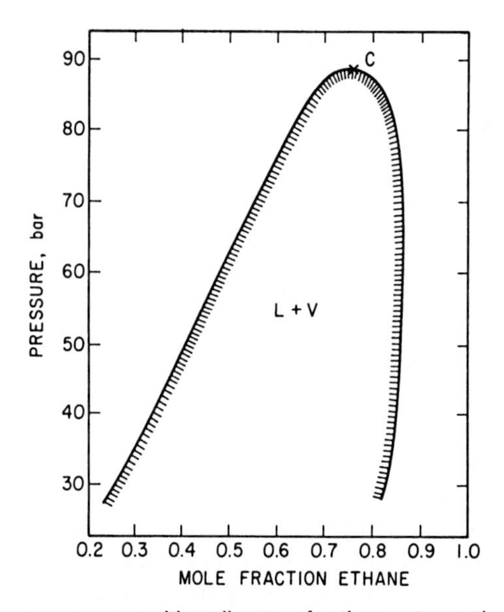

Figure 12-2 Pressure-composition diagram for the system ethane/*n*-heptane at 149°C. The critical point of the mixture is at C (Mehta and Thodos, 1965).

To characterize vapor-liquid equilibria for a binary system, measurements like those shown in Fig. 12-2 must be repeated for other temperatures; for each temperature, there is a critical composition and critical pressure. Figure 12-3 gives experimentally observed critical temperatures and pressures as a function of mole fraction for the ethane/*n*-heptane system. While the critical temperature of this system is a monotonic function of composition, the critical pressure goes through a maximum; many, but by no means all, binary systems behave this way.

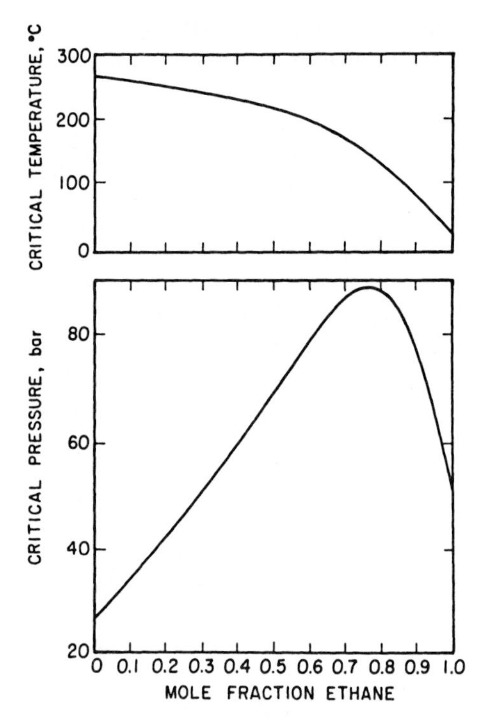

Figure 12-3 Critical temperatures and pressures for the system ethane/*n*-heptane (Mehta and Thodos, 1965).

For typical technical calculations it is convenient to express phase-equilibrium relations in terms of K factors; by definition $K \equiv y/x$, the ratio of the mole fraction in the vapor phase y, to that in the liquid phase x. The definition of K has no thermodynamic significance but K is commonly used in chemical engineering calculations where it is convenient for writing material balances.

Figure 12-4 shows experimental K factors for the two components of the methane/propane system (Sage and Lacey, 1938). The lines for propane start at the saturation pressure for pure propane (where $K = 1$), decrease with rising pressure and, after going through minima, rise to $K = 1$ at the critical point. For methane, K factors decrease in the entire pressure range to $K = 1$ at the critical point.

Because Figs. 12-2 to 12-4 are for simple systems, they do not indicate the variety of phase behavior that is possible in binary systems. That variety increases substantially when we consider also partial immiscibility of liquids and possible occurrence of solid phases, as discussed in the next sections.

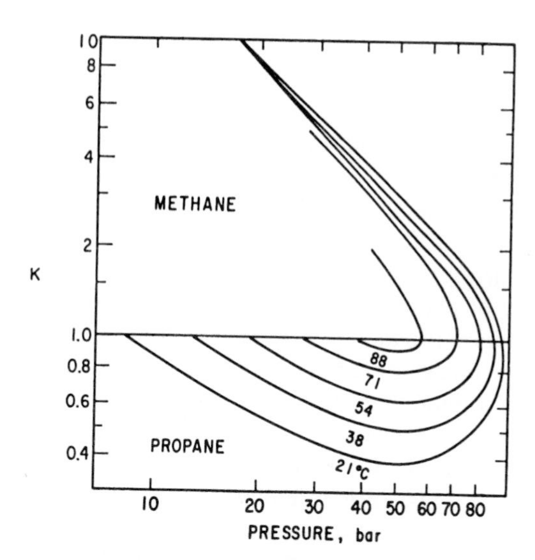

Figure 12-4 *K* factors for the methane/propane system (Sage and Lacey, 1938).

12.2 Phase Behavior at High Pressure

To understand phase behavior at high pressure, we need to know how to calculate and interpret phase diagrams.

Calculation of phase diagrams is based on solving the equation of phase equilibrium; for vapor-liquid equilibrium,

$$f_i^V = f_i^L$$

for every component i in the mixture.

To interpret phase diagrams we need Gibbs' phase rule. We cannot here give a complete discussion; we provide only an introduction. For a review of phase equilibrium diagrams for binary mixtures at high pressures, including a description of experimental and theoretical methods, see, for example, Hicks and Young (1974); Rowlinson and Swinton (1982); Sadus (1992); and Schouten (1992).

Interpretation of Phase Diagrams

To interpret phase diagrams we apply the *Gibbs phase rule* (see Sec. 2.5). For nonreacting systems, that rule is expressed by the simple relation,

$$F = m + 2 - \pi \tag{12-1}$$

where the number of independent variables (or degrees of freedom) F is related to the number of components m and number of phases π. However, at a critical point, the physical properties of the coexisting phases become identical. Criticality therefore imposes an additional $\pi - 1$ constraints that reduce the number of degrees of freedom given by Eq. (12-1). The number of degrees of freedom of a gas-liquid critical point is zero for a one-component system; for a two-component system, the number of degrees of freedom is one (a line), and for a three-component system it is two (a surface).

Pressure and temperature are the most convenient independent variables for the measurement and study of phase equilibria in fluid systems. In fluid systems, changes in pressure and temperature produce large changes in the phase behavior and a three-dimensional diagram in pressure, temperature and a third variable is required for a complete description of a two-component system (see Table 12-1). For qualitative descriptions and comparisons of phase diagrams, the most convenient choice for this third variable is the composition $\mathfrak{z}$, where $\mathfrak{z}$ designates the composition of a phase of any kind (solid, liquid, or gas). Pressure-temperature-composition (P-T-$\mathfrak{z}$) diagrams provide a basis for the design of separation processes such as distillation and extraction. Therefore, three-dimensional drawings are commonly used, together with two-dimensional diagrams cut by planes of constant T or P, or formed by projections of lines and points on the P-T coordinate plane.

Table 12-1 Geometric constraints imposed by the phase rule on phase equilibria.

Number of Components	Number of Phases in Equilibrium			
1	3	2	1	–
2	4	3	2	1
3	5	4	3	2
Degrees of freedom	0	1	2	3
Geometric representation	Point	Line	Surface	Volume

The phase rule is a useful guide for construction and interpretation of phase diagrams; it imposes definite constraints on the geometry of the features that describe the existence and coexistence of a fixed number of phases.[1] For example, Eq. (12-1) tells us that for a two-component system with 3 phases in equilibrium, phase equilibria are represented by lines; phase equilibria for a 1-component system is fully described in a 2-dimensional diagram that contains points, lines, and surfaces, corresponding to the

[1] A useful concept for the analysis of phase-equilibrium diagrams is the distinction between two types of variables: field variables and density variables. *Field variables* are those that take identical values for each phase when two (or more) phases are in equilibrium, such as pressure, temperature, and chemical potential. *Density variables* are those that have different values for each phase, such as density, enthalpy, and composition.

equilibria of, respectively, 3 phases, 2 phases, and 1 phase. Table 12-1 summarizes the results of the phase rule for different types of systems.

Table 12-1 shows that a system of n phases in equilibrium is described by n geometrical features of the appropriate type: a one-component system of three phases is described by three points, a two-component system of three phases is described by three lines, etc. The interpretation of (high pressure) phase diagrams is an important aspect for the proper design and operation of many chemical engineering processes.

Classification of Phase Diagrams for Binary Mixtures

As Table 12-1 shows, complete description of a two-component system requires three-dimensional P-T-z diagrams (a combination of two *field* variables, P and T, and one *density*[2] variable, composition z). In two-component systems, regions of two-, three- and four-phase equilibria are described in P-T-z space by pairs of surfaces, triplets of lines, and quadruplets of points, respectively. However, the geometrical constraints imposed by two field variables require (Rowlinson and Swinton, 1982) that these features have common P-T projections, i.e., two surfaces representing two coexisting phases project as a single surface, three lines representing three coexisting phases project as a single line, etc.

Van Konynenburg and Scott (1980) showed that almost all known types of binary fluid-phase equilibria (vapor-liquid, liquid-liquid, and gas-gas equilibria) can be qualitatively predicted using the van der Waals equation of state[3] and quadratic mixing rules. Their calculations suggested the classification of binary fluid-phase behavior into six types, based on the shape of the mixture critical line and on the absence or presence of three-phase lines. However, due to limitations of the equation of state used, they were able to generate only five of these six types with the van der Waals equation of state. A brief description of each of these types of phase behavior is presented below. More complete descriptions can be found elsewhere (Rowlinson and Swinton, 1982; Schneider, 1978; McHugh and Krukonis, 1994; Sadus, 1994).

Type I-Phase Behavior

To illustrate the variety of phase behavior in binary mixtures, Fig. 12-6 shows a few schematic pressure-temperature diagrams for binary systems. These diagrams are P-T projections of the P-T-z surface, shown schematically in Fig. 12-5 for a typical type I-mixture.

---

[2] See Footnote 1.

[3] Phase diagrams of binary fluid mixtures have been calculated also from other equations of state, namely from those of Redlich-Kwong (Deiters and Pegg, 1989), Carnahan-Starling-Redlich-Kwong (Kraska and Deiters, 1992), and those based on the simplified-perturbed-hard-chain equation (van Pelt *et al.*, 1991).

In the simple case shown in Fig. 12-5, the line ending at C_1 is the vapor-liquid coexistence curve for pure 1 while the line ending at C_2 is the vapor-liquid coexistence curve for pure 2; C_1 and C_2 are the critical points. The dashed line joining these points is the critical locus; each point on that line is the critical point for a mixture of fixed composition. The continuous vapor-liquid critical line and the absence of liquid-liquid immiscibility is often observed for mixtures where the two components are chemically similar and/or their critical properties are comparable. Typical examples are methane/ethane, carbon dioxide/*n*-butane, and benzene/toluene. The critical locus may (but need not) show either a minimum or a maximum. The latter is an indication of large positive deviations from Raoult's law and hence of relatively weak, unlike intermolecular interactions; such behavior is found for binary mixtures of, e.g., a polar with a non-polar fluid, such as methanol/*n*-hexane.

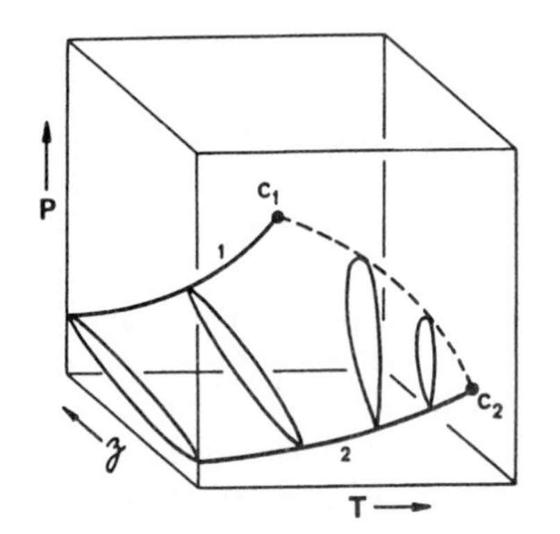

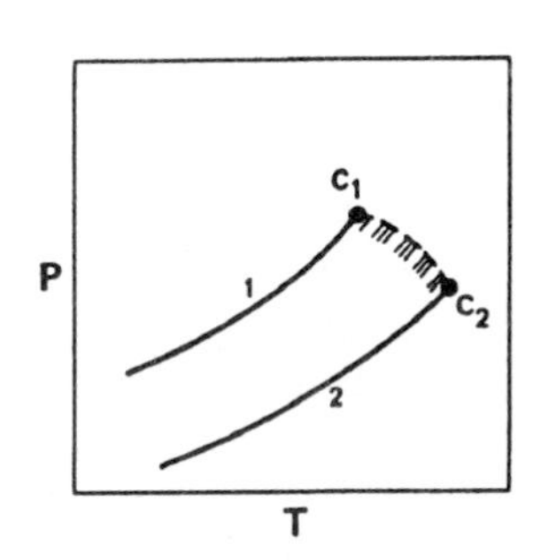

Figure 12-5 *P-T-$\mathfrak{z}$* surface and corresponding *P-T* projection showing vapor-liquid equilibria for a simple binary mixture (type I).

Type II-Phase Behavior

Type II is similar to type I except that, at low temperatures, liquid mixtures of components 1 and 2 are not miscible in all proportions. There is thus one additional critical line. The line labeled LLV gives the locus of a three-phase line where one vapor phase is in equilibrium with two liquid phases. This locus ends at the *upper critical end point*[4] (UCEP) where the two liquid phases merge into one liquid phase; the pressure of the UCEP depends on temperature as shown in Fig. 12-6 by the dashed line curving upward, in this case, with a negative slope. This negative slope indicates that the upper critical solution temperature (see Sec. 6.13) decreases with rising pressure. The phase diagrams of type II-mixtures may be complicated by the presence of an azeotropic line. However, the essential feature of this type (and also of type VI-mixtures) is the con-

[4] Critical end points are limiting end points where two of three coexisting phases become identical.

tinuous vapor-liquid critical line that is distinct from the liquid-liquid critical line. Examples of type II are carbon dioxide/*n*-octane and ammonia/toluene. As shown, the critical line starting at C_2 has an initial negative slope but that is not always so; the slope may initially be positive and become negative at higher pressures. Further, the 3-phase line LLV may in some cases reside above the vapor pressure-curve of component 1, as observed for mixtures of hydrocarbons with the corresponding fully- or near fully-fluorinated fluorocarbons, e.g. methane/trifluoromethane.

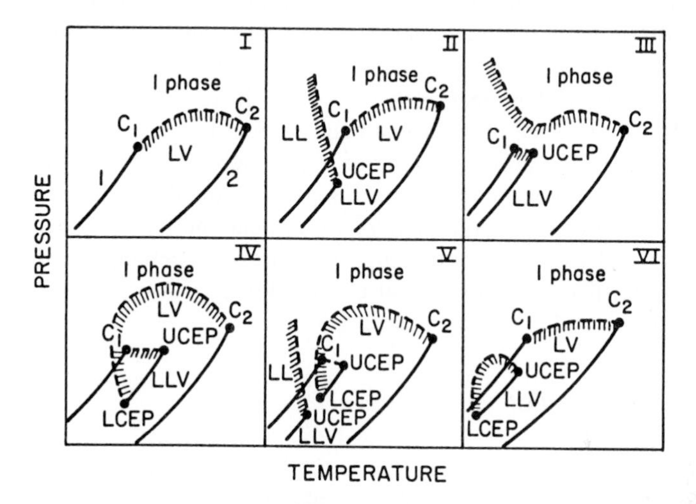

Figure 12-6 Six types of phase behavior in binary fluid systems. C = critical point; L = liquid; V = vapor; UCEP = upper critical end point; LCEP = lower critical end point. Dashed curves are critical lines and hatching marks heterogeneous regions.

Type III-Phase Behavior

For mixtures with large immiscibility such as water/*n*-alkane mixtures, the locus of the liquid-liquid critical lines moves to higher temperatures and may then interfere with the vapor-liquid critical curve. This means that the vapor-liquid critical locus is not necessarily a continuous line connecting C_1 and C_2, as illustrated in type III. In this type, the critical locus has two branches. One branch goes from the vapor-liquid critical point of the more volatile component, C_1, to UCEP, where the gaseous phase and the liquid phase (richer in the more volatile component) have the same composition. The other branch starts at C_2 and then rises with pressure, perhaps with a positive slope (e.g., helium/water) or with a negative slope (e.g., methane/toluene) or with a slope that changes sign (e.g., nitrogen/ammonia and ethane/methanol). The critical line starting from C_2 with a positive slope indicates the existence of what is called *gas-gas equilibria*: two phases at equilibrium at a temperature larger than the critical temperature of either pure component. Gas-gas immiscibility is conveniently classified as either *first kind* (almost exclusively confined to mixtures containing helium as one component, e.g. water/helium and helium/xenon) where the critical curve extends directly from the critical point of the less volatile component with a positive slope; or *second kind*, where the critical line passes first through a temperature minimum and then goes steeply to increasing temperatures and pressures (e.g., methane/ammonia and water/propane).

Experimentally, an unambiguous distinction between, e.g., a type-II and a type-III *P-T* projection is accomplished by direct visual observation to determine the two of the three equilibrium phases that merge at the three-phase LLV critical point. As shown in Fig. 12-6, if the two liquid phases merge in the presence of the equilibrium vapor phase, then the *P-T* projection is of type II; if the vapor phase and one liquid phase merge in the presence of the second equilibrium liquid phase, then the *P-T* projection is of type III. Therefore, the type of *P-T* projection is given by observation of the disappearance of one of the two menisci separating the three equilibrium phases, as well as critical opalescence, at the three-phase critical endpoints. Experimentally, this visualization is usually carried out in a high-pressure view cell equipped with windows (typically sapphire) that allow the visual observation of phase separation, the number of coexisting phases, and critical opalescence.[5]

Type IV-Phase Behavior

Type IV is similar to type V. In both, the vapor-liquid critical line starting at C_2 ends at a LCEP, a terminal point of the LLV three-phase line. However, in type IV the LLV locus has two parts; this means that there is limited liquid-phase miscibility at low temperatures, ending at UCEP. As the temperature and pressure rise, there is a second region of limited miscibility from LCEP to UCEP. Such phase behavior is shown by ethane/1-propanol and carbon dioxide/nitrobenzene.

Type V-Phase Behavior

In type V, the first branch of the critical line goes from C_1 to UCEP as in type III, but the second branch goes from C_2 to the lower critical end point (LCEP). Contrarily to type IV, in type V mixtures, the liquids are completely miscible below LCEP. Figure 12-7 gives a schematic representation of what is visually observed at the LCEP and at the UCEP. Two liquid phases exist only for temperatures between those of the critical endpoints. At the temperature of the LCEP, the meniscus between the two liquids disappears, whereas at that of the UCEP, it is the gas-liquid interface that disappears leaving in both cases one liquid phase in equilibrium with a gas phase.

---

[5] See, e.g., S. P. Christensen and M. E. Paulaitis, 1992, *Fluid Phase Equilibria*, 7: 63; E. Gomes de Azevedo, H. A. Matos, and M. Nunes da Ponte, 1993, *Fluid Phase Equilibria*, 83: 193.

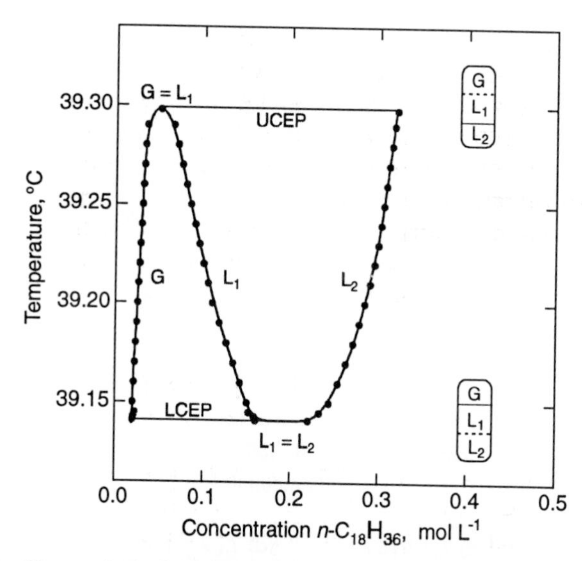

Figure 12-7 Three-phase behavior and schematic representation of critical end points for the binary ethane/*n*-octadecane (Specovius *et al.*, 1985). The dashed line represents a disappearing meniscus, i.e., a critical point: gas-liquid at the UCEP (39.30°C) and liquid-liquid at the LCEP (39.14°C).

Mixtures of *n*-alkanes with large size differences show type-V phase behavior. While the binary methane/*n*-pentane shows type-I phase behavior, type-V behavior occurs in the system methane/*n*-hexane. With ethane as the lighter component, *n*-octadecane is the first long *n*-alkane to exhibit partial miscibility in the liquid state, only 7°C above the critical temperature of ethane (t_c = 32.2°C), as illustrated in Fig. 12-7. The immiscible region for ethane/*n*-octadecane extends over a very narrow temperature range (0.157°C) and all phases are rich in ethane. This temperature range increases to 2.927°C for the binary ethane/*n*-eicosane, ($C_{20}H_{42}$) while complete miscibility is observed for ethane/*n*-heptadecane.

Type V phase behavior is also found in binaries containing alcohols. An example is provided by ethylene/methanol whose phase behavior is shown in Fig. 12-8.

From the *P-T* projection represented in Fig. 12-8(a), several *P-x* diagrams were derived at four characteristic temperatures. At temperature T_i, there is a normal gas-liquid equilibrium [Fig. 12-8(b)]. When the temperature of the LCEP is exceeded, a L_1L_2 region begins to grow out of the bubble curve [Fig. 12-8(c)]. The liquid phase L_1 is rich in methanol and liquid phase L_2 is rich in ethylene. At the critical temperature of pure ethylene, the two-phase L_2-V region separates from the axis corresponding to $x_{C_2H_4} = 1$, and forms an additional critical point [Fig. 12-8(d)]. As the temperature increases further, the L_2-V region becomes smaller until it disappears at the temperature of the UCEP (see Fig. 12-7); regions L_1-V, L_1-L_2, and L_2-V fuse with each other and a normal V-L equilibrium with a supercritical component is now present [Fig. 12-8(e)].

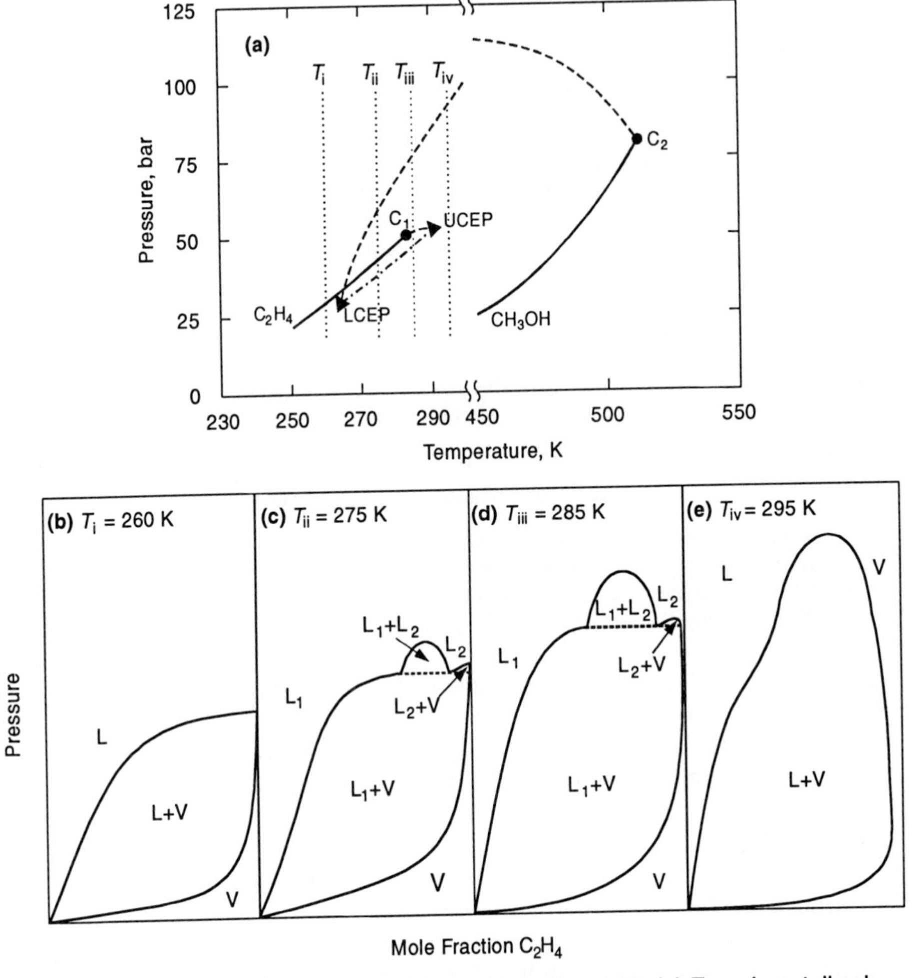

Figure 12-8 Phase-behavior of ethylene (1)/methanol (2): (a) Experimentally obtained *P-T* projection (note the scale change in the temperature axis at about 300 K, required by the large difference between the volatilities of the two components). —— Vapor-pressure curves of the pure components; - - - - critical lines; — - — - LLV three-phase line; (b) to (e) *P-x* isotherms at the temperatures indicated.

By observing the phase changes at the successive boundaries encountered in the *P-T* projection of a binary mixture (illustrated here for the type-V binary ethylene/methanol), we are able to construct the corresponding isothermal *P-x* diagrams (Sadus, 1992). For comparison, Fig. 12-9 shows experimental phase-equilibrium measurements for the same binary at 260, 273, 284.15, and 298.15 K. The 284.15 K isotherm, 2.7 K below the temperature of the UCEP, shows the three two-phase regions L_1-V, L_1-L_2, and L_2-V. Because at this temperature the L_2-V region is very small (extends only over a pressure interval of 0.9 bar and over an ethylene mole fraction interval of 0.04), it is difficult to detect experimentally. The 298.15 K isotherm is 11.3 K above the temperature of the UCEP; it shows only a typical V-L equilibrium with ethylene in the supercritical state. The flat boiling curve observed in the 260 K isotherm suggests two-liquid splitting starting at 263.55 K.

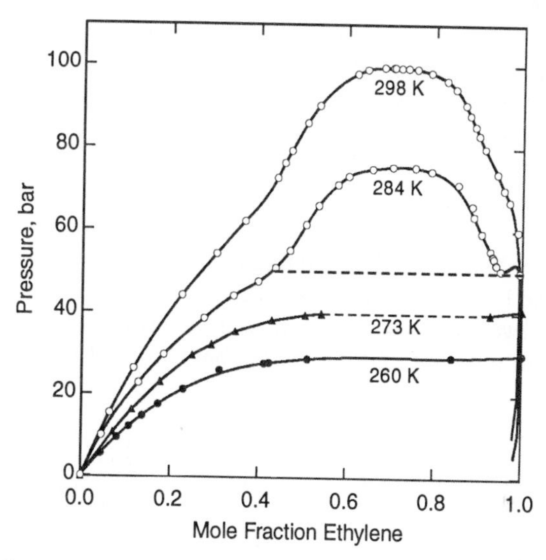

Figure 12-9 Pressure-composition phase-equilibrium data for ethylene/methanol at 260 K and 273 K (Zeck and Knapp, 1986) and at 284 K and 298 K (Brunner, 1985).

Type VI-Phase Behavior

Finally, binary mixtures showing type-VI phase behavior have two critical curves: One connects C_1 and C_2 while another connects UCEP and LCEP. It is the presence of the LCEP that distinguishes type-VI from type-II systems. The closed dome of immiscibility formed between LCEP and UCEP is a consequence of the opposite signs of the slopes $(dT_{\text{LCEP}} / dP)_x$ and $(dT_{\text{UCEP}} / dP)_x$. The two critical curves meet at an upper critical pressure; at higher pressures two liquids are miscible. Examples of this complex behavior are found in mixtures where one (or both) component is self-associated through hydrogen bonding. An example of type VI is water/2-butanol.

Critical Phenomena in Binary Fluid Mixtures

It is a challenge to devise quantitative models that can reproduce fluid-phase behavior as shown in Fig. 12-6 that introduces only some (by no means all) observed phase diagrams. In principle, it should be possible to calculate such diagrams from an equation of state. Van Konynenburg and Scott (1980) have shown that the original van der Waals equation (with conventional mixing rules) can be used to calculate qualitatively nearly all of the observed phase diagrams for binary nonelectrolytes. However, if attention is restricted to conventional equations of state, calculation of such diagrams in quantitative agreement with experiment is at present possible only for type I and type II phase behavior. As better equations of state become available and as computing techniques improve, it is likely that accurate phase diagrams will be calculated not only for all types of binary mixtures but also for multicomponent systems (Sadus, 1992).

While Fig. 12-6 shows only binary fluid phases, some experimental data are also available for binary high-pressure systems where solid phases exist in addition to fluid phases, but we shall not pursue that topic here.

Figure 12-6 gives different types of pressure-temperature diagrams for binary systems with variable composition. When a pressure-temperature diagram is constructed for a simple binary mixture at fixed composition, we have a bubble-point (boiling) curve on the left and a dew-point (condensation) curve on the right, as shown in Fig. 12-10. These two curves come together at the critical point. The dashed lines in Fig. 12-10 are lines of constant *quality*.[6] At the bubble-point line, the entire mixture is liquid and at the dew-point line, the entire mixture is vapor.

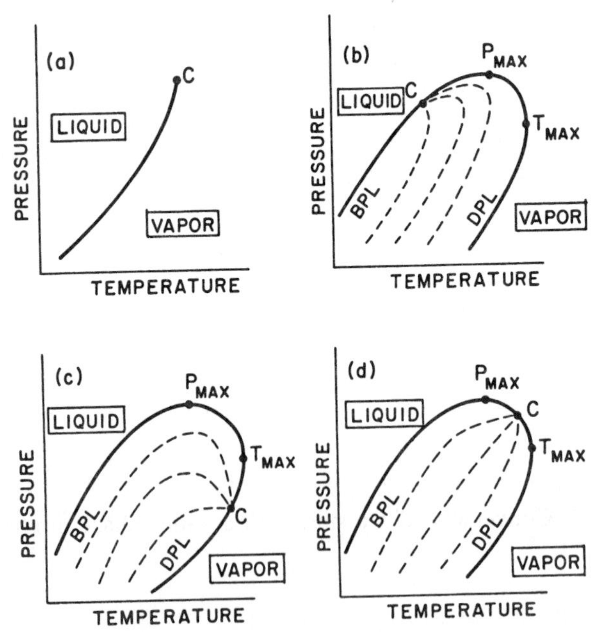

Figure 12-10 Pressure-temperature diagrams for a pure fluid (a) and for fluid mixtures at constant composition (b, c, d). BPL = bubble-point line; DPL = dew-point line; C = critical point. Dashed lines show constant quality.

[6] In a two-phase (vapor-liquid) mixture, quality is defined as the fraction that is in the vapor phase.

For a pure fluid, the bubble-point and dew-point curves collapse into one single line; in that event, the critical point is a maximum in the sense that it is at the highest temperature and pressure where liquid and vapor can coexist at equilibrium. However, in a mixture, the critical point is not necessarily a maximum with respect to either temperature or pressure. In a mixture, liquid and vapor may coexist at equilibrium at temperatures and pressures higher than those corresponding to the critical point. The relative positions of C, T_{max}, and P_{max} (see Fig. 12-10) give rise to a curious but well-known phenomenon: *retrograde condensation.*

Normally, when we compress a vapor mixture at constant temperature, liquid starts to form at the dew point; as compression proceeds, more liquid is formed until (at the bubble point) all vapor has been condensed. However, in the vicinity of the mixture's critical point, isothermal compression may proceed such that an isotherm cuts the dew-point line not once, but twice and does not cut the bubble-point line at all. Such an isotherm is shown in Fig. 12-11.

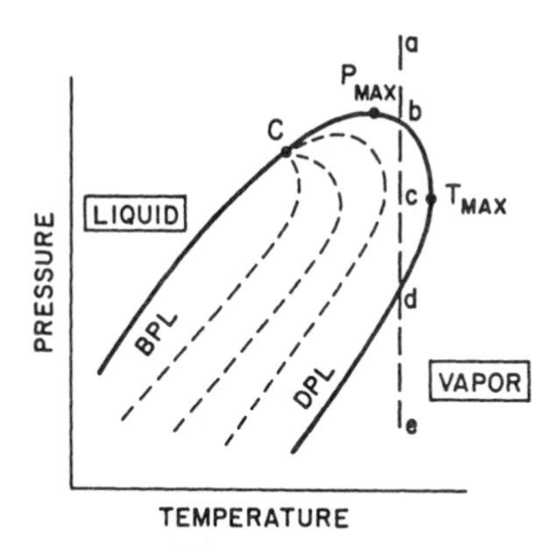

Figure 12-11 Retrograde condensation. The vertical line cuts the dew-point line twice. BPL = bubble-point line; DPL = dew-point line; C = critical point. When gaseous mixture *e* is compressed isothermally, liquid begins to form at *d*; maximum condensation occurs at *c*, and at *b*, all condensate is vaporized. Dashed lines show constant quality.

Suppose a vapor mixture is at *e*. It is isothermally compressed. The first liquid is formed at *d*. More liquid is formed until *c*; thereafter, liquid evaporates until, at *b,* all liquid is evaporated. Conversely, if we start with a vapor-phase mixture at *a* and expand isothermally, the first liquid is obtained at *b;* more liquid is formed until *c* and thereafter, liquid evaporates until all of the mixture is again vapor at *d.*

Similar unexpected effects can be obtained by isobarically cooling or heating a mixture in the critical region. Retrograde phenomena are common in natural-gas reservoirs where proper understanding of retrograde behavior is important for efficient gas production.

Phase behavior of liquid mixtures containing only subcritical components is also affected by pressure, but in this case much higher pressures are usually required to produce significant changes. For example, at ordinary pressures, the system methyl ethyl ketone/water has an upper critical solution temperature and a lower critical solution temperature; in between these temperatures, the two liquids are only partially miscible. As observed by Timmermans (1923), increasing pressure lowers the upper critical solution temperature and raises the lower critical solution temperature as indicated in Fig. 12-12(a).

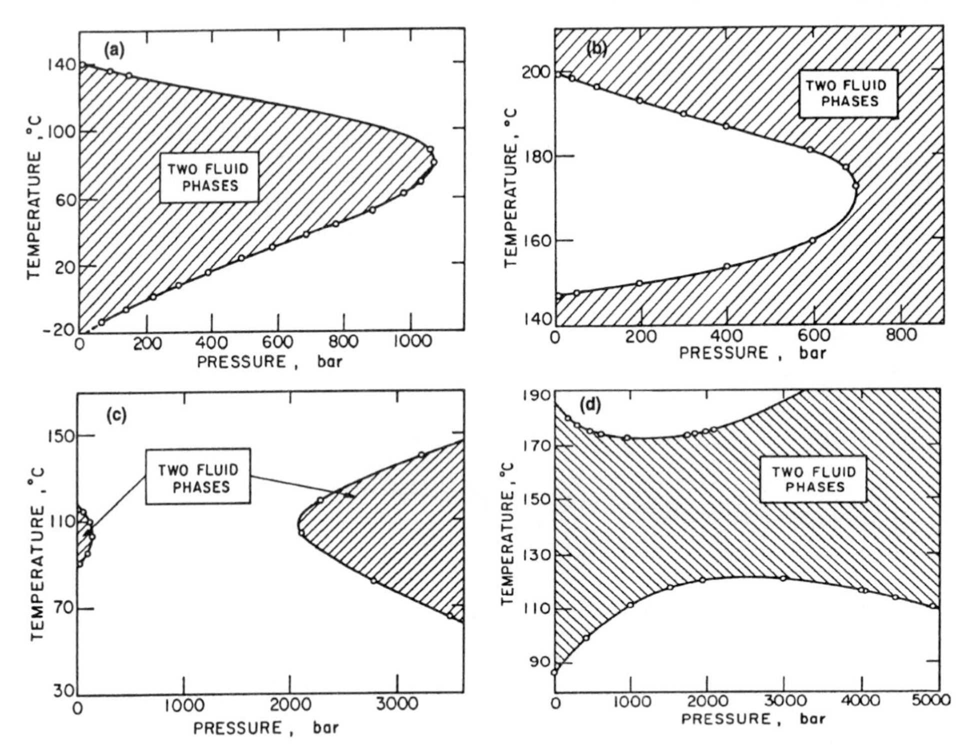

Figure 12-12 Liquid-liquid phase behavior for different binary mixtures (Timmermans, 1923; Schneider, 1966). Lower and upper consolute temperatures as functions of pressure for: (a) methyl ethyl ketone/water; (b) triphenylmethane/sulfur; (c) 2-methylpyridine/deuterium oxide; (d) 4-methylpiperidine/water.

As pressure rises, the region of partial miscibility decreases and at pressures beyond 1100 bar, this region has disappeared. By contrast, the system triphenylmethane/sulfur exhibits the opposite behavior; in this case, increasing pressure reduces the region of complete miscibility as indicated by Fig. 12-12(b). Finally, Figs. 12-12(c) and 12-12(d) illustrate two additional types of phase behavior. Figure 12-12(c) shows that for the system 2-methylpyridine/deuterium oxide, increasing pressure first induces miscibility and then, at still higher pressures, produces again a region of incomplete miscibility. On the other hand, Fig. 12-12(d) shows that for the system 4-methyl-piperidine/water, increasing pressure does not eliminate the region of partial miscibility; very high pressures, after first causing this region to shrink, bring about an increase in the two-phase region. Connolly (1966) has observed behavior similar to that shown in Fig. 12-12(c) for some water/hydrocarbon systems.

Figures 12-2 to 12-12 illustrate only some of the types of phase behavior at high pressures; many other phase phenomena have been observed for binary systems and, when ternary systems are considered, additional phase phenomena become possible, as discussed elsewhere (Rowlinson and Swinton, 1982; Sadus, 1992; Schneider, 1993; de Swaan Arons and de Loos, 1994).

12.3 Liquid-Liquid and Gas-Gas Equilibria

Before examining high-pressure equilibria in systems containing one liquid phase and one vapor phase, we briefly consider the effect of pressure on equilibria between two liquid phases and later, between two gaseous phases.

Liquid-Liquid Equilibria

To illustrate the effect of pressure on liquid-liquid equilibria, Fig. 12-13 shows experimental results for the system carbon tetrafluoride/propane. In this case, rising pressure increases the size of the two-phase region; in other words, rising pressure is unfavorable for miscibility. Thermodynamics can help us to interpret results such as those shown in Fig. 12-13. In particular, we want to examine how pressure may be used to induce miscibility or immiscibility in a binary system.

Two liquids are miscible in all proportions if $\Delta_{\text{mix}}g$, the molar Gibbs energy of mixing at constant temperature and pressure, satisfies the relations

$$\Delta_{\text{mix}}g < 0 \tag{12-2}$$

and

$$\left(\frac{\partial^2 \Delta_{\text{mix}}g}{\partial x^2}\right)_{T,x} > 0 \tag{12-3}$$

for all x. Because $\Delta_{\text{mix}}g$ is a function of pressure, it follows that under certain conditions, a change in pressure may produce immiscibility in a completely miscible system or, conversely, such a change may produce complete miscibility in a partially immiscible system. The effect of pressure on miscibility in binary liquid mixtures is closely connected with the volume change on mixing, as indicated by the exact relation

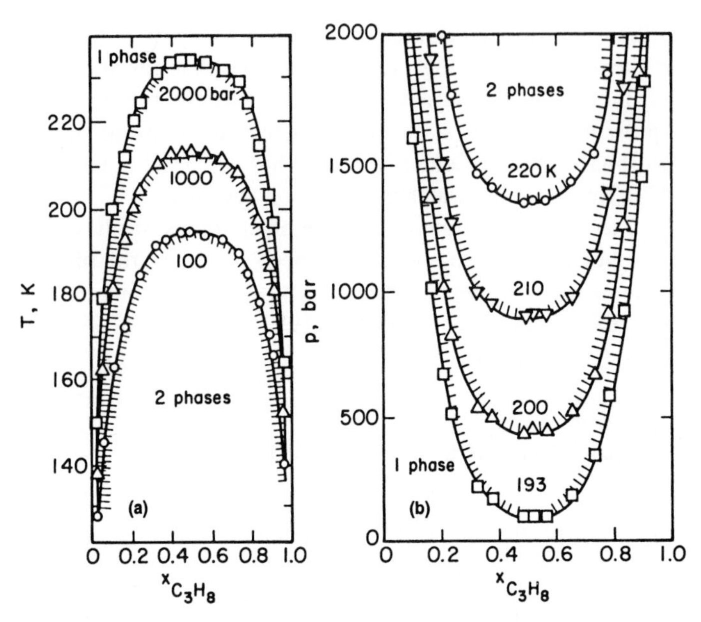

Figure 12-13 Pressure (a) and temperature (b) dependence of liquid-liquid equilibria in the system carbon tetrafluoride/propane (Jeschke and Schneider, 1982).

$$\left(\frac{\partial \Delta_{\text{mix}} g}{\partial P}\right)_{T,x} = \Delta_{\text{mix}} v \tag{12-4}$$

where $\Delta_{\text{mix}} v$ is the volume change on mixing at constant temperature and pressure.

To fix ideas, consider a binary liquid mixture that at normal pressure is completely miscible and whose isothermal Gibbs energy of mixing is given by curve a in Fig. 12-14. Suppose that for this system $\Delta_{\text{mix}} v$ is positive; an increase in pressure raises $\Delta_{\text{mix}} g$, and at some higher pressure the variation of $\Delta_{\text{mix}} g$ with x_1 may be given by curve *b*. As indicated in Fig. 12-14, $\Delta_{\text{mix}} g$ at the high pressure no longer satisfies Eq. (12-3) and the liquid mixture now has a miscibility gap in the composition interval $x_1' < x_1 < x_1''$.

For contrast, consider also a binary liquid mixture that at normal pressures is incompletely miscible as shown by curve *a* in Fig. 12-15. If $\Delta_{\text{mix}} v$ for this system is negative, then an increase in pressure lowers $\Delta_{\text{mix}} g$ and at some high pressure the variation of $\Delta_{\text{mix}} g$ with x_1 may be given by curve *b*, indicating complete miscibility. It follows from these simple considerations that the qualitative effect of pressure on phase stability of binary liquid mixtures depends on the magnitude and sign of the volume change of mixing. To carry out quantitative calculations at some fixed temperature, it is necessary to have information on the variation of $\Delta_{\text{mix}} v$ with x and P in addition to information on the variation of $\Delta_{\text{mix}} g$ with x at a single pressure.

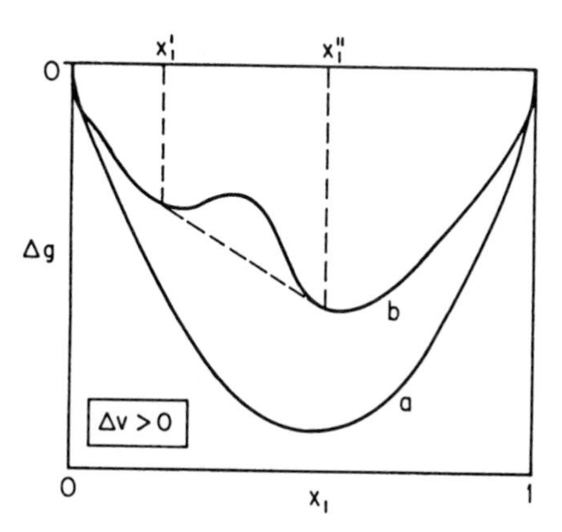

Figure 12-14 Effect of pressure on miscibility: *a* - low pressure, no immiscibility; *b* - high pressure, immiscible for $x_1' < x_1 < x_1''$.

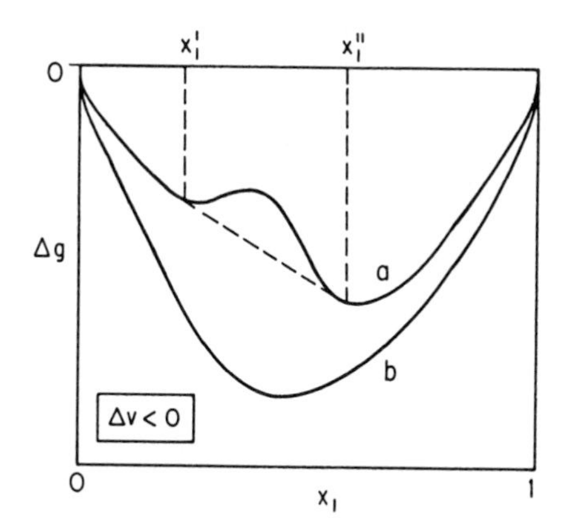

Figure 12-15 Effect of pressure on miscibility: *a* - low pressure, immiscible for $x_1' < x_1 < x_1''$; *b* - high pressure, no immiscibility.

To illustrate, we consider a simple, symmetric binary mixture at some fixed temperature and 1 bar pressure. For this liquid mixture, we assume that

$$\Delta_{\text{mix}} g = RT(x_1 \ln x_1 + x_2 \ln x_2) + A x_1 x_2 \tag{12-5}$$

$$\Delta_{\text{mix}} v = B x_1 x_2 \tag{12-6}$$

where A and B are experimentally determined constants. Further, we assume that the liquid mixture is incompressible for all values of x; i.e., we assume that B is independent of pressure. At any pressure P, we have for $\Delta_{\text{mix}} g$,

$$\Delta_{\text{mix}} g = RT(x_1 \ln x_1 + x_2 \ln x_2) + \left[A + B(P-1)\right] x_1 x_2 \tag{12-7}$$

where P is in bar. Substituting Eq. (12-7) into Eq. (12-3), we find that the mixture is partially immiscible when

$$\frac{A + B(P-1)}{RT} > 2 \tag{12-8}$$

Equation (12-8) tells us that if $A/RT < 2$ (complete miscibility at 1 bar) and if $B > 0$, then there is a certain pressure P (larger than 1 bar) where immiscibility is induced. On the other hand, if $A/RT > 2$ (incomplete miscibility at 1 bar) and if $B < 0$, then there is a certain pressure P (larger than 1 bar) where complete miscibility is attained.

When two liquid phases exist, the compositions of the two phases α and β are governed by equality of fugacities for each component:

$$f_1^\alpha = f_1^\beta \tag{12-9}$$

$$f_2^\alpha = f_2^\beta \tag{12-10}$$

If the same standard-state fugacities are used in both phases, Eqs. (12-9) and (12-10) can be rewritten using activity coefficient γ:

$$(\gamma_1 x_1)^\alpha = (\gamma_1 x_1)^\beta \tag{12-11}$$

$$(\gamma_2 x_2)^\alpha = (\gamma_2 x_2)^\beta \tag{12-12}$$

For the simple mixture described by Eqs. (12-5) and (12-6), we can calculate activity coefficients as discussed in Chapter 6; we substitute into Eqs. (12-11) and (12-12) and we then obtain

$$x_1^\alpha \exp\left[\frac{[A + B(P-1)](1 - x_1^\alpha)^2}{RT}\right] = x_1^\beta \exp\left[\frac{[A + B(P-1)](1 - x_1^\beta)^2}{RT}\right] \tag{12-13}$$

$$x_2^\alpha \exp\left[\frac{[A + B(P-1)](1 - x_2^\alpha)^2}{RT}\right] = x_2^\beta \exp\left[\frac{[A + B(P-1)](1 - x_2^\beta)^2}{RT}\right] \tag{12-14}$$

Simultaneous solution of these equilibrium relations (coupled with the conservation equations $x_1^\alpha + x_2^\alpha = 1$ and $x_1^\beta + x_2^\beta = 1$) gives the coexistence curve for the two-phase system as a function of pressure.[7]

Experimental studies of liquid-liquid equilibria at high pressures were reported many years ago by Roozeboom (1918), Timmermans (1923) and Poppe (1935). Schneider and coworkers (Dahlmann, 1989; Becker, 1993; Ochel *et al.*, 1993; Wallbruch, 1995; Grzanna, 1996), and the thermodynamic group at Delft (Bijl *et al.*, 1983; Gregorowicz *et al.*, 1993; de Loos *et al.*, 1996) have reported similar experimental work.

Winnick and Powers (1963, 1966) made a detailed study of the acetone/carbon disulfide system at 0°C; at normal pressure, this system is completely miscible. For an equimolar mixture the volume increase upon mixing is of the order of 1 cm^3 mol^{-1}, a fractional change of about 1.5%. Winnick and Powers measured the volume change on mixing as a function of both pressure and composition; these measurements, coupled with experimentally determined activity coefficients at low pressure, were then used to calculate the phase diagram at high pressures, using a thermodynamic procedure similar to the one outlined above. The calculations show that incomplete miscibility should be observed at pressures larger than about 3800 bar. Winnick and Powers also made experimental measurements of the phase behavior of this system at high pressures; they found, as shown in the lower part of Fig. 12-16, that their observed results are only in approximate agreement with those calculated from volumetric data. Good quantitative agreement is difficult to achieve because the calculations are extremely sensitive to small changes in $\Delta_{mix}g$, the Gibbs energy of mixing, as shown in the upper part of Fig. 12-16. As indicated by Eqs. (12-3) and (12-4), a small change in the volume change on mixing has a large effect on the second derivative of $\Delta_{mix}g$ with mole fraction.

While gas-gas equilibria are rare, liquid-liquid equilibria are common; as briefly discussed in Sec. 6.13, critical phenomena can exist in liquid-liquid system as well as in liquid-vapor systems. In a binary system, the gas-liquid critical temperature depends on pressure, as illustrated in Fig. 12-3. Similarly, in a binary system with limited liquid-phase miscibility, the liquid-liquid critical temperature also depends on pressure, as shown in Fig. 12-13.

Many binary liquid mixtures have miscibility gaps that depend on temperature; while most of these mixtures have upper critical solution temperatures, some have lower critical solution temperatures and a few have both. When the liquid mixture is subjected to pressure, the critical solution temperature changes;[8] it may increase or decrease. (For systems with upper critical solution temperatures, rising pressure usually increases that temperature.) For typical nonelectrolyte liquid mixtures, the change in critical solution temperature is significant only when the pressure becomes large, at least

---

[7] Because of symmetry, Eq. (12-13) and Eq. (12-14) are always satisfied by the trivial solution $x_1^\alpha = x_2^\alpha = x_1^\beta = x_2^\beta = 1/2$. A useful method for solving Eqs. (12-13) and (12-14) is given by Van Ness and Abbott, 1982, *Classical Thermodynamics of Nonelectrolyte Solutions*, New York: McGraw-Hill.

[8] See the *P-T* projection for type II mixtures shown in Fig. 12-6.

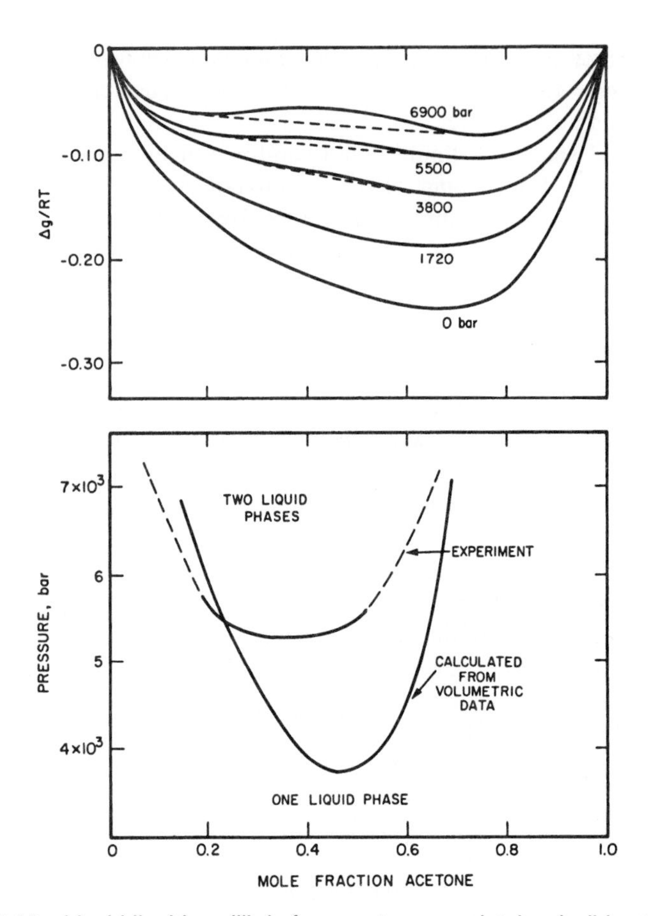

Figure 12-16 Liquid-liquid equilibria for a system completely miscible at normal pressure (Winnick and Powers, 1966). Calculated and observed behavior of the acetone/carbon disulfide system at 0°C.

100 bar. For many typical systems at ordinary temperatures, a rise of 100 bar produces a critical-solution-temperature change in the region 2 to 6°C.

To obtain a useful equation for calculating the effect of pressure on critical solution temperature T^c, we must recall a fundamental feature of classical thermodynamics: the essence of thermodynamics is to relate some macroscopic properties to others, thereby reducing experimental effort; when some properties are measured, others need not be, because they can be calculated. Sometimes, however, the quantity to be calculated requires experimental data that cannot easily be measured. In that event, it is sometimes possible to use instead other measured properties that are more readily available.

The effect of pressure P on critical solution temperature T^c is expressed by dT^c/dP. Thermodynamic analysis gives[9]

$$\frac{dT^c}{dP} = T^c \left[\frac{(\partial^2 v / \partial x^2)_{T,P}}{(\partial^2 h / \partial x^2)_{T,P}} \right]_{x_c} \tag{12-15}$$

where v is the molar volume and h is the molar enthalpy, both at mole fraction x.

Equation (12-15) is not useful because it is difficult to measure the second derivatives of v and h with respect to x. Further, both of these derivatives vanish at the critical point; therefore, Eq. (12-15) must be rewritten

$$\frac{dT^c}{dP} = \lim_{\substack{T \to T_c \\ x \to x_c}} \left[\frac{T(\partial^2 v / \partial x^2)_{T,P}}{(\partial^2 h / \partial x^2)_{T,P}} \right] \tag{12-16}$$

A large experimental effort must be made to evaluate the right side of Eq. (12-16); it would be easier to measure dT^c/dP directly. Nevertheless, Eq. (12-16) suggests an approximation, reasonable for simple mixtures, that allows us to use experimentally available quantities to estimate dT^c/dP. This approximation, suggested by Scott and coworkers (Myers *et al.*, 1966), is a similarity hypothesis: we assume that v and h, plotted against x, show similar shapes. That is, we assume that the molar excess volume v^E and the molar excess enthalpy h^E depend on x and T according to

$$v^E = v_c^E F(x,T) \tag{12-17}$$

$$h^E = h_c^E F(x,T) \tag{12-18}$$

where $F(x,T)$ is an arbitrary function of mole fraction x and temperature T. This assumption, substituted into Eq. (12-16), gives

$$\frac{dT^c}{dP} \approx \frac{T^c v_c^E}{h_c^E} \tag{12-19}$$

where v_c^E and h_c^E are the molar excess volume and the molar excess enthalpy at critical composition x_c and critical solution temperature T^c.

Because v_c^E and h_c^E are not strongly sensitive to T and x (i.e., they do not show anomalous or "catastrophic" behavior at T^c and x_c), they can be estimated from

[9] See page 288 of Prigogine and Defay (1954).

dilatometric and calorimetric data in the critical region without direct measurements at the critical point.

Table 12-2 shows a test of Eq. (12-19). Experimental results for v_c^E and h_c^E are compared with those for dT^c/dP. The test is limited to five systems where the required experimental data are available. Agreement is surprisingly good, even though these systems are far from simple mixtures.

Table 12-2 Predicted and measured effect of pressure on critical solution temperatures in binary systems (Myers *et al.*, 1966; Clerke and Sengers, 1981).

					$10^3 dT^c/dP$ (K bar^{-1})	
Components (1)/(2)	T^c (K)	x_{2c}	v_c^E (cm^3 mol^{-1})	h_c^E (J mol^{-1})	Predicted	Measured
Carbon disulfide/ acetone	222	0.34	0.5	1050	11	10
Triethylamine/water	292	0.90	-1.30	1590	24	21
n-Perfluoroheptane/ isooctane	296	0.62	4.33	2010	65	65
1-Hydro-*n*-perfluoro-heptane/dioxane	308	0.82	1.10	986	35	28
3-Methylpentane/ nitroethane	300	0.50	0.218	1626	4.0	3.67

Equation (12-16), a purely thermodynamic equation, is of no use by itself. However, when combined with physical insight, it produces the much more useful (approximate) Eq. (12-19).

In the preceding paragraphs we indicated how, under certain conditions, pressure may be used to induce immiscibility in liquid binary mixtures that, at normal pressures, are completely miscible. To conclude this section, we consider briefly how the introduction of a third component can bring about immiscibility in a binary mixture that is completely miscible in the absence of the third component. In particular, we are concerned with a binary liquid mixture where the added (third) component is a gas; in this case, elevated pressures are required to dissolve an appreciable amount of the added component in the binary liquid solvent. For the situation to be discussed, phase instability is not a consequence of the effect of pressure on the chemical potentials, as was the case in previous sections, but follows instead from the additional component that affects the chemical potentials of the components to be separated. High pressure enters into our discussion only indirectly, because we want to use a highly volatile substance for the additional component.

The situation under discussion is similar to the familiar *salting-out effect* in liquids where a salt, added to an aqueous solution, serves to precipitate one or more organic solutes. Here, however, instead of a salt, we add a gas; to dissolve an appreciable quantity of gas, the system must be at elevated pressure.

We consider a binary liquid mixture of components 1 and 3; to be consistent with previous notation, we reserve subscript 2 for the gaseous component. Components 1 and 3 are completely miscible at room temperature; the (upper) critical solution temperature T^c is far below room temperature, as indicated by the lower curve in Fig. 12-17. Suppose now that we dissolve a small amount of component 2 in the binary mixture; what happens to the critical solution temperature? This question was considered by Prigogine (1943), who assumed that for any binary pair that can be formed from the three components 1, 2, and 3, the excess Gibbs energy (symmetric convention) is

$$g_{ij}^E = \alpha_{ij} x_i x_j \quad (i,j = 1,2;\ 1,3;\ \text{or } 2,3) \tag{12-20}$$

where α_{ij} is an empirical (Margules) coefficient determined by the properties of the *i-j* binary. Prigogine has shown that, upon adding a small amount of component 2, the rate of change in critical temperature is given by

$$\frac{\partial T^c}{\partial x_2} = -\frac{1}{2R}\frac{(\alpha_{13}-\alpha_{12}+\alpha_{23})(\alpha_{13}-\alpha_{23}+\alpha_{12})}{\alpha_{12}} \qquad (\text{for } x_2 << 1) \tag{12-21}$$

To induce the desired immiscibility in the 1-3 binary at room temperature, we want $\partial T^c / \partial x_2$ to be positive and large.

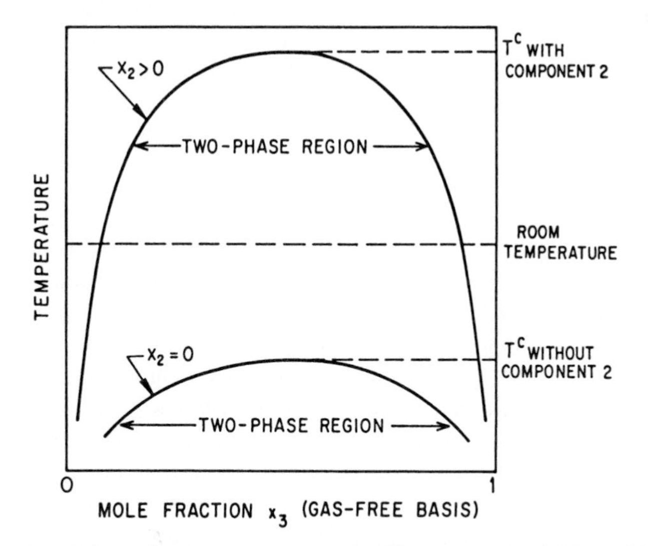

Figure 12-17 Effect of gaseous component (2) on mutual solubility of liquids (1) and (3).

Let us now focus attention on the common case where all three binaries exhibit positive deviations from Raoult's law, i.e., $\alpha_{ij} > 0$ for all *i-j* pairs. If T^c for the 1-3 binary is far below room temperature, then that binary is only moderately nonideal and α_{13} is small. We must now choose a gas that forms a highly nonideal solution with one of the liquid components (say component 3) while it forms with the other component (component l) a solution that is only modestly nonideal. In that event,

$$\alpha_{23} >> \alpha_{12} \tag{12-22}$$

and

$$\alpha_{23} >> \alpha_{13} \tag{12-23}$$

Equations (12-21), (12-22), and (12-23) indicate that under the conditions described, $\partial T^c / \partial x_2$ is both large and positive, as desired; i.e., dissolution of a small amount of component 2 in the 1-3 mixture raises the critical solution temperature as shown in the upper curve of Fig. 12-17. From Prigogine's analysis we conclude that if component 2 is properly chosen, it can induce binary miscible mixtures of components 1 and 3 to split at room temperature into two liquid phases having different compositions.

As shown by Balder (1966) thermodynamic considerations may be used to establish the phase diagram of a ternary system consisting of two miscible liquid components and a supercritical gas at high pressures. Such diagrams have been obtained experimentally (Elgin and Weinstock, 1959; Francis, 1963; Chappelear and Elgin, 1961) for a variety of ternary systems and have led to suggestions for separations using a high-pressure gas as the phase-splitting agent.

Gas-Gas Equilibria

Thermodynamic analysis of incomplete miscibility in liquid mixtures at high pressures [Eqs. (12-2), (12-3), and (12-4)] can also be applied to gaseous mixtures at high pressures. It has been known since the beginning of recorded history that not all liquids are completely miscible with one another, but only in the mid-twentieth century have we learned that gases may also, under suitable conditions, exhibit limited miscibility. The possible existence of two gaseous phases at equilibrium was predicted on theoretical grounds by van der Waals as early as 1894 and again by Onnes and Keesom (1907). Experimental verification, however, was not obtained until about 40 years later.

Immiscibility of gases is observed only at high pressures where gases are at liquid-like densities, as shown in Sec. 12.2. The term *gas-gas equilibria* is therefore misleading because it refers to fluids whose properties are similar to those of liquids, very different from those of gases under normal conditions. For our purposes here, we define two-component, gas-gas equilibria as the existence of two equilibrated, stable, fluid phases at a temperature in excess of the critical of either pure component; both phases are at the same pressure but have different compositions.

Figure 12-18 shows some experimental results for the helium/xenon system. As described in Sec. 12.2, the mixture helium/xenon is of type III and forms gas-gas equilibria of the first kind. At temperatures several degrees above the critical of xenon,[10] the two phase compositions are significantly different, even at pressures as low as 200 bar. However, to obtain the same degree of separation at higher temperatures, much higher pressures are required.

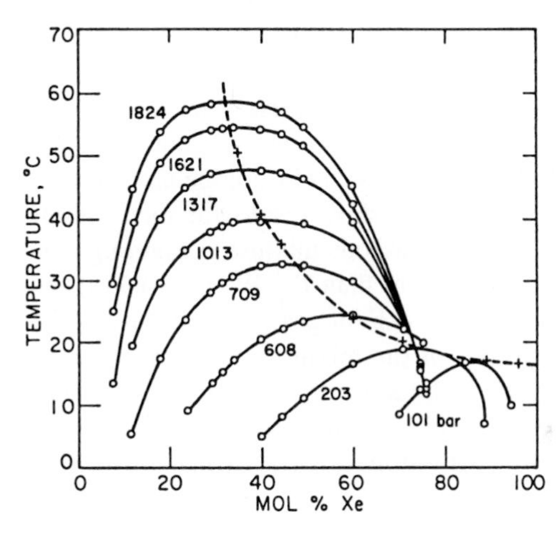

Figure 12-18 Gas-gas equilibria in the helium/xenon system (de Swaan Arons and Diepen, 1966).

A theoretical analysis of the helium/xenon system was reported by Zandbergen *et al.*, (1967) who based their calculations on the Prigogine-Scott theory of corresponding states for mixtures (Prigogine, 1957; Scott, 1956). Zandbergen *et al.* use the *three-liquid* theory to obtain an expression for the volumes of helium/xenon mixtures as a function of temperature, pressure, and composition.

Different models were used by several authors (Kleintjens and Koningsveld, 1980; Trappeniers *et al.*, 1971, 1974) to describe gas-gas immiscibility observed in other systems.

Equations of state provide a logical model for gas-gas equilibrium calculations. If a suitable equation of state is available for a mixture, it is possible to calculate critical lines for that mixture. That is, at least in principle, it is possible to compute phase diagrams such as those shown in Fig. 12-6. Pittion-Rossillon (1980), using an equation of state valid for hard-sphere mixtures combined with the attractive part of the van der Waals equation, predicted immiscibility in binary mixtures at high pressure, in good

[10] The critical temperature of xenon is 16.6°C.

agreement with experimental data. Calculations were reported by Scott and van Konynenburg (1980) with the original van der Waals equation and by Deiters and coworkers with the Redlich-Kwong (Deiters and Pegg, 1989) or with the Carnahan-Starling-Redlich-Kwong (Kraska and Deiters, 1992) equations of state. Using conventional mixing rules, phase diagrams as in Fig. 12-6 can be qualitatively predicted. As expected, however, the original van der Waals equation is not sufficiently accurate to give quantitative agreement with experiment; however, it is sufficiently realistic to reproduce qualitatively most of the binary phase diagrams that have been observed.

Quantitative calculation of critical lines requires not only a good equation of state for mixtures, but also sophisticated computing techniques, as discussed, for example, by Stockfleth and Dohrn (1998), Michelsen (1980, 1981), Heidemann (1980), and Peng and Robinson (1977). The basis for these calculations is the theory of thermodynamic stability for multicomponent systems. A good introduction to that theory is given by Tester and Modell (1997) and by Prigogine and Defay (1954). We do not here discuss stability theory but, instead, show results for some simple cases.

Suppose we have a binary gaseous mixture containing components 1 and 2, such that critical temperature T_{c_2} is larger than T_{c_1}. We now ask: At a temperature T, slightly larger than T_{c_2}, does this mixture exhibit gas-gas equilibria? Following Temkin (1959), we assume that the properties of the mixture are given by the original van der Waals equation with constant b given by a linear function and constant a by a quadratic function of mole fraction. What conditions must be met by pure-component constants a and b and by binary constant a_{12} to produce gas-gas equilibria? Temkin showed that gas-gas equilibria can exist, provided that

$$b_1 \geq 0.42 b_2 \tag{12-24}$$

and that

$$\frac{1}{2}(a_1 + a_2) - a_{12} > \frac{8}{27} a_2 \tag{12-25}$$ [11]

Because component 2 is the heavier component, Eq. (12-24) tells us that gas-gas equilibria near T_{c_2} are unlikely if there is a very large difference in molecular sizes, i.e., b_1 can be smaller than b_2 but not very much smaller. Equation (12-25) tells us that the attractive forces between unlike molecules (given by a_{12}) must be appreciably lower than those calculated by the arithmetic mean of a_1 and a_2. These conclusions are consistent with our intuitive ideas about mixtures. We can imagine that if we shake a box containing oranges and grapefruit [that satisfies Eq. (12-24)], the contents of that box might prefer to segregate into two separate regions, one rich in oranges and the other rich in grapefruit. However, if the box contains grapefruit and walnuts [that does not

[11] Taking into consideration the relations between van der Waals parameters a and b with the critical constants, Eqs. (12-24) and (12-25) can also be written as, $v_{c_1} \geq 0.42\, v_{c_2}$ and $T_{c_1} v_{c_1} < 0.052\, T_{c_2} v_{c_2}$, respectively.

satisfy Eq. (12-24)], segregation is less likely because walnuts, unlike oranges, easily fit into the void regions of an assembly of grapefruit.

Equation (12-25) also is easy to understand. It says that for partial miscibility, the forces of attraction between unlike molecules must be weak; if molecules 1 and 2 do not like each other, they have little tendency to mix, but prefer to stay with neighbors identical to themselves.

If we assume that for nonpolar systems, $a_{12} = (a_1 a_2)^{1/2}$, Eq. (12-25) reduces to

$$a_1 < 0.053 a_2 \tag{12-26}$$

In other words, the intermolecular force parameter a_1 must be less than about 5% of parameter a_2. This is a severe requirement. For example, if component 2 is carbon dioxide, then Eq. (12-26) is satisfied only by helium. For the binary helium/carbon dioxide, Eq. (12-24) is also satisfied and, indeed, gas-gas equilibria have been observed for that system at temperatures slightly above the critical for carbon dioxide.

12.4 Thermodynamic Analysis

Thermodynamic analysis has already been used in previous sections for deriving the equations for liquid-liquid and gas-gas equilibria. We return here to the same subject for a more general discussion.

As discussed in each of the previous chapters, the fundamental equilibrium relation for multicomponent, multiphase systems is most conveniently expressed in terms of fugacities: for any component i, the fugacity of i must be the same in all equilibrated phases. However, this equilibrium relation is of no use until we can relate the fugacity of a component to directly measurable properties. The essential task of phase-equilibrium thermodynamics is to describe the effects of temperature, pressure, and composition on the fugacity of each component in each phase of the system of interest. For any component i in a system containing m components, the total differential of the logarithm of the fugacity f_i is

$$d\ln f_i = \left(\frac{\partial \ln f_i}{\partial T}\right)_{P,x} dT + \left(\frac{\partial \ln f_i}{\partial P}\right)_{T,x} dP + \sum_{j=1}^{m-1} \left(\frac{\partial \ln f_i}{\partial x_j}\right)_{T,P,x_k} dx_j \tag{12-27}$$

$$k = 1,\ldots, j-1, j+1,\ldots, m-1$$

Thermodynamics gives limited information for each of the three coefficients that appear on the right-hand side of Eq. (12-27). The first term can be related to the partial molar enthalpy and the second to the partial molar volume; the third term can be related to the Gibbs energy and that, in turn, can be described by a solution model or an equation of state. For a complete description of phase behavior, we must say something about each of these three coefficients, for each component, in every phase. In high-pressure work, it is important to give particular attention to the second coefficient that tells us how phase behavior is affected by pressure.

When analyzing typical experimental data, it is often difficult to isolate the effect of pressure because, more often than not, a change in the pressure is accompanied by a simultaneous change in either the temperature or the composition, or both. A striking example of such simultaneous changes is given by experimental results in Fig. 12-19 that show the effect of pressure on the melting temperature of solid argon (Mullins and Ziegler, 1964). Line *A* gives the melting line for pure argon as reported by Clusius and Weigand (1940); the melting temperature rises with pressure, as predicted by the Clapeyron equation, because argon expands upon melting. Line *B* gives results of Mullins and Ziegler for the melting temperature of argon in the presence of high-pressure helium gas; these results are similar to those for pure argon, and again the melting temperature rises with increasing pressure. Finally, line *C* gives the melting temperature for argon in the presence of high-pressure hydrogen gas; Mullins and Ziegler also reported these data. We are struck and perhaps puzzled by the completely different behavior of line *C*: The melting temperature now falls with rising pressure.

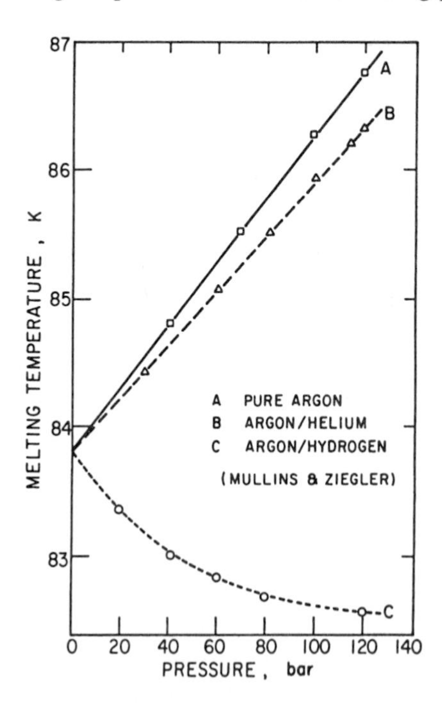

Figure 12-19 Effect of pressure on melting temperature of argon. Qualitative difference between lines *B* and *C* is due to effect of composition on the liquid-phase fugacity of argon.

The qualitative difference between lines B and C can be clarified by thermodynamic analysis when we note that the three-phase equilibrium temperature is determined by two separate effects: first, the effect of pressure on the fugacities of solid and liquid argon is essentially the same for the three cases A, B, and C; and second, there is the effect of liquid composition on the fugacity of liquid argon. It is the effect of composition for case B that is much different from that for case C. At a given pressure, the solubility of hydrogen in liquid argon is much larger than that of helium; because pressure and composition simultaneously influence the fugacity of liquid argon, we find that the presence of hydrogen alters the equilibrium temperature in a manner qualitatively different from that found in the presence of helium. The fugacity of liquid argon is raised by pressure but lowered by solubility of another component. For helium (low solubility), the pressure effect dominates, but for hydrogen (high solubility), the effect of composition is more important. We shall not here go into a more detailed analysis of this particular system. We merely wish to emphasize that the fugacity of a component is determined by the three variables temperature, pressure, and composition; that in a typical experimental situation these influences operate in concert; and that whatever success we may expect in explaining the behavior of a multicomponent, multiphase system is determined directly by the extent of our knowledge of the three coefficients given in Eq. (12-27).

At present we know least about the first coefficient, rigorously related to enthalpy by

$$\left(\frac{\partial \ln f_i}{\partial T}\right)_{P,x} = -\frac{\bar{h}_i - h_i^+}{RT^2} \qquad \text{(12-28)}^{12}$$

where $\bar{h}_i$ is the partial molar enthalpy of i and h_i^+ is the molar enthalpy of i in the ideal-gas state at the same temperature. While we would like to use information on $\bar{h}_i$ to help us toward calculating f_i, we almost never can do so in high-pressure work because so little is known about enthalpies of fluid mixtures at high pressures. (In typical chemical engineering work, it is more common to reverse the procedure, i.e., to differentiate fugacity data with respect to temperature, to estimate enthalpies.) As a result, the best procedure for most cases is to analyze experimental phase-equilibrium data as a function of pressure and composition along an isotherm, and to allow any empirical parameters obtained from such analysis to vary with temperature as dictated by the experimental results.

While formal thermodynamic analysis is equally useful for predicting the effect of pressure, temperature, or composition on phase behavior, it is the effect of pressure that is often simplest to understand because it is directly related to volumetric data through the fundamental relation for molar Gibbs energy g

[12] Equation (12-28) is the same as Eq. (6-3).

$$\left(\frac{\partial g}{\partial P}\right)_{T,x} = v \tag{12-29}$$

that, in turn, leads to

$$\left(\frac{\partial \ln f_i}{\partial P}\right)_{T,x} = \frac{\bar{v}_i}{RT} \tag{12-30}$$

In Sec. 10.3 we briefly discussed the effect of pressure on the solubility of a sparingly soluble gas in a liquid. We here continue that discussion to show how thermodynamic analysis can predict unexpected phase behavior: When the solubility of a sparingly soluble gas is plotted against pressure, the plot may show a maximum.

To illustrate, consider the solubility of nitrogen in water at 18°C (Krichevsky and Kasarnovsky, 1935). Let subscript 2 stand for nitrogen. At equilibrium,

$$f_2^V = f_2^L \tag{12-31}$$

Taking logarithms and total differentials, Eq. (12-31) becomes

$$d\ln f_2^V = d\ln f_2^L \tag{12-32}$$

At constant temperature,

$$d\ln f_2^V = \left(\frac{\partial \ln f_2^V}{\partial P}\right)_{y_2} dP + \left(\frac{\partial \ln f_2^V}{\partial \ln y_2}\right)_P d\ln y_2 \tag{12-33}$$

and

$$d\ln f_2^L = \left(\frac{\partial \ln f_2^L}{\partial P}\right)_{x_2} dP + \left(\frac{\partial \ln f_2^L}{\partial \ln x_2}\right)_P d\ln x_2 \tag{12-34}$$

Because nitrogen is sparingly soluble in water at 18°C, Henry's law holds; therefore, at constant temperature,

$$\left(\frac{\partial \ln f_2^L}{\partial \ln x_2}\right)_P = 1 \tag{12-35}$$

At 18°C, the volatility of water is very low: $y_2 \approx 1$, and therefore the last term in Eq. (12-33) can be neglected.

Substituting Eqs. (12-33), (12-34) and (12-35) into Eq.(12-32), we obtain at constant temperature,

$$\frac{d\ln x_2}{dP} = \left(\frac{d\ln f_2^V}{dP}\right)_{y_2} - \left(\frac{d\ln f_2^L}{dP}\right)_{x_2} = \frac{\bar{v}_2^V - \bar{v}_2^L}{RT} \tag{12-36}$$

Because $y_2 \approx 1$, we can set the vapor-phase partial molar volume of nitrogen equal to the molar volume of pure gaseous nitrogen, $v_{N_2}^G$. Because solubility $x_2 << 1$, we can set

$$\bar{v}_2^L = \bar{v}_2^\infty \tag{12-37}$$

While $v_{N_2}^G$ is a strong function of pressure, $\bar{v}_2^\infty$ is essentially independent of pressure in a solvent (here water) far below its critical temperature. Experimental data (Moors *et al.*, 1985; Angus *et al.*, 1979) for $v_{N_2}^G$ and $\bar{v}_2^\infty$ are shown as a function of pressure in the lower part of Fig. 12-20. As indicated by Eq. (12-36), we expect a maximum in solubility x_2 when $\bar{v}_2^\infty = v_{N_2}^G$.

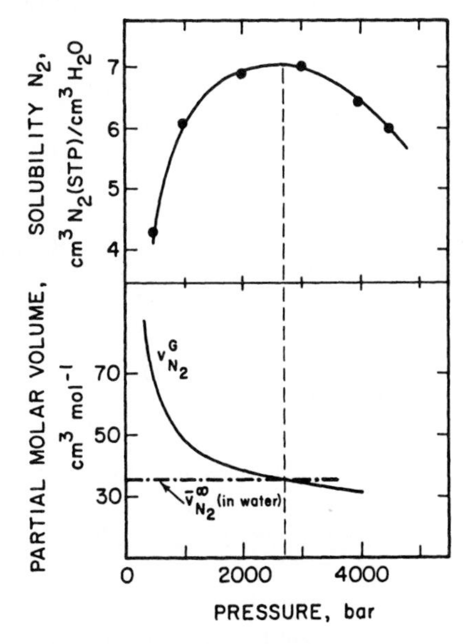

Figure 12-20 Solubility of nitrogen in water at 18°C. Calculation of maximum solubility according to the Krichevsky-Kasarnovsky equation.

The upper part of Fig. 12-20 shows experimental solubility data (Basset and Dode, 1936) for nitrogen in water to high pressures. As predicted by Eq. (12-36) the observed solubility goes through a maximum near 2700 bar, consistent with experimental volumetric data.

Equation (12-30) indicates the important role of the partial molar volume in high-pressure phase equilibria. For mixtures of ordinary liquids remote from critical conditions, the partial molar volume of component i is often very close to the molar volume of pure liquid i at the same temperature; for such cases, in other words, Amagat's law is a good assumption. However, for concentrated solutions of gases in liquids, the assumption is poor; at conditions approaching critical, it is very poor. Partial molar volumes may be positive or negative, and near critical conditions they are usually strong functions of the composition. For example, in the saturated liquid phase of the system carbon dioxide/n-butane at 71.1°C, $\overline{v}$ for butane is 112.4 cm^3 mol^{-1} when the mole fraction of CO_2 is small, and it is -149.8 cm^3 mol^{-1} at critical composition, $x_{CO_2} = 0.71$.

It is difficult to measure partial molar volumes and unfortunately many experimental studies of high-pressure vapor-liquid equilibria report no volumetric data at all; more often than not, experimental measurements are confined to total pressure, temperature, and phase compositions. Even in those rare cases where liquid densities are measured along the saturation curve, there is a fundamental difficulty in calculating partial molar volumes as indicated by the exact relations between partial molar volumes $\overline{v}_1$ and $\overline{v}_2$ and isothermal saturated molar volume v^s in a binary system:[13]

$$\overline{v}_1 = v^s - x_2\left[\left(\frac{\partial v^s}{\partial x_2}\right)_T + v^s\kappa_T\left(\frac{\partial P}{\partial x_2}\right)_T\right] \tag{12-38}$$

$$\overline{v}_2 = v^s + x_1\left[\left(\frac{\partial v^s}{\partial x_2}\right)_T + v^s\kappa_T\left(\frac{\partial P}{\partial x_2}\right)_T\right] \tag{12-39}$$

where compressibility κ_T is defined by

---

[13] To derive Eq. (12-38) we must note that, at constant temperature, v^s is a function of both x_2 and P:

$$dv^s = \left(\frac{\partial v^s}{\partial x_2}\right)_{P,T} dx_2 + \left(\frac{\partial v^s}{\partial P}\right)_{T,x_2} dP \tag{12-38a}$$

Therefore,

$$\left(\frac{\partial v^s}{\partial x_2}\right)_T = \left(\frac{\partial v^s}{\partial x_2}\right)_{T,P} + \left(\frac{\partial v^s}{\partial P}\right)_{T,x_2}\left(\frac{\partial P}{\partial x_2}\right)_T \tag{10-38b}$$

The partial molar volume $\overline{v}_1$ is related to v^s by

$$\overline{v}_1 = v^s - x_2\left(\frac{\partial v^s}{\partial x_2}\right)_{P,T} \tag{12-38c}$$

Simple substitution then gives Eq. (12-38).

$$\kappa_T \equiv -\frac{1}{v^s}\left(\frac{\partial v^s}{\partial P}\right)_{T,x} \tag{12-40}$$

Experimental data for v^s are sometimes available, but experimental compressibilities for mixtures are rare.

For dilute solutions, the literature contains some direct (dilatometric) measurements of $\overline{v}_2$, the partial molar volume of the more volatile component, but the accuracy of these measurements is usually not high. A survey by Lyckman *et al.* (1965) established the rough correlation shown in Fig. 12-21. On the ordinate, the partial molar volume is nondimensionalized with the critical temperature and pressure of component 2; the abscissa is also dimensionless and includes c_{11}, the cohesive-energy density of the solvent, component 1 (see Sec. 7.2). Figure 12-21 is useful for rough approximations in systems remote from critical conditions. For expanded solvents, i.e., for solvents at temperatures approaching T_c, the partial molar volume of the solute tends to be much larger than that suggested by the correlation as indicated in Fig. 12-22. Brelvi and O'Connell (1972) give an alternative correlation for $\overline{v}^\infty$.

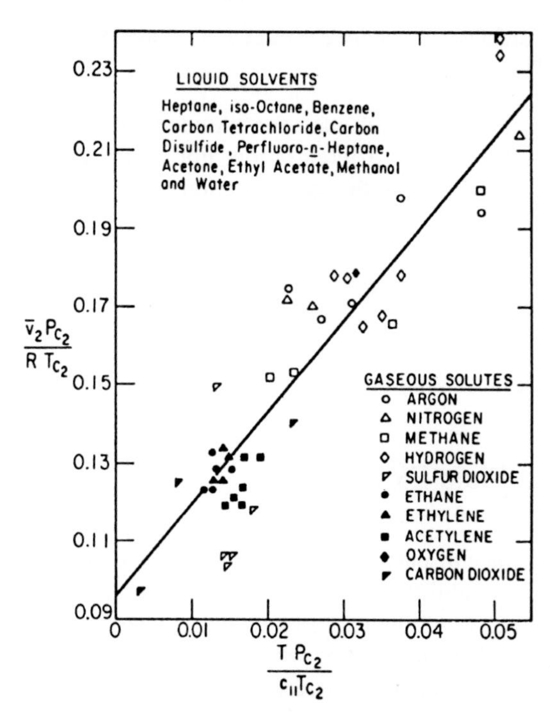

Figure 12-21 Partial molar volumes of gases in dilute liquid solutions.

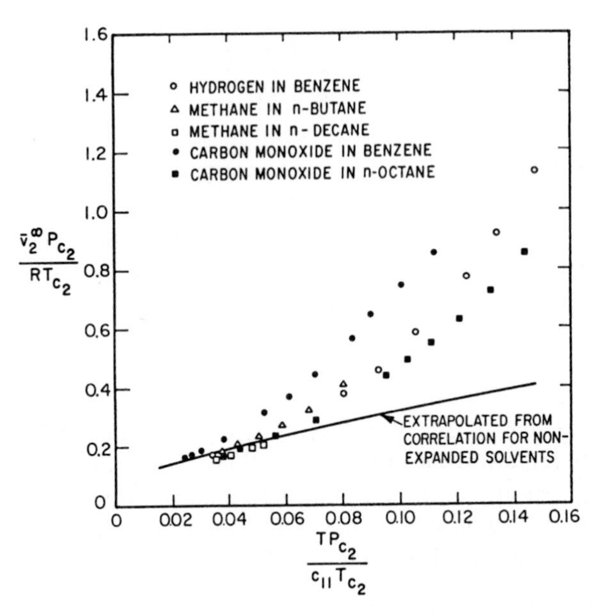

Figure 12-22 Partial molar volumes of gaseous solutes at infinite dilution in expanded solvents.

If we can write an equation of state for liquid mixtures, we can then calculate partial molar volumes directly by differentiation. For a pressure-explicit equation, the most convenient procedure is to use the exact relation

$$\bar{v}_i \equiv -\frac{(\partial P / \partial n_i)_{T,V,n_j}}{(\partial P / \partial V)_{T,\,\text{all}\,n}} \tag{12-41}$$

where V is the total volume of the mixture containing n_1 moles of component 1, etc.

12.5 Supercritical-Fluid Extraction

To illustrate the importance of partial molar volumes, we return to our previous discussion (Sec. 5.14) of the solubility of a solid in a compressed gas. This solubility can become surprisingly large when the compressed gas is near its critical point, leading to a separation operation called *supercritical fluid extraction.* We do not here go into details[14] but show how the essence of supercritical fluid extraction is directly related to

[14] See, e.g., McHugh and Krukonis (1994); E. Kiran and J. F. Brennecke (Eds., 1993), *Supercritical Fluid Engineering Science,* American Chemical Society Symposium Series 514, Washington: ACS; T. J. Bruno and J. F. Ely

the partial molar volume of the solute in the compressed gas. Under certain conditions, this partial molar volume can be negative and extremely large, much larger than the molar volume of the pure solute.

Consider the solubility of a heavy component (2) in a dense gas (1) at temperature T and pressure P. For simplicity, let component 2 be a solid such that the solubility of component 1 in the solid phase is negligible. The equation of equilibrium is

$$f^{s}_{\text{pure }2} = \varphi_2 y_2 P \tag{12-42}$$

where y is the mole fraction in the vapor phase and φ is the vapor-phase fugacity coefficient.

At gas-phase densities approaching the critical, solubility y_2 is sensitive to small changes in T and P because, in the critical region, these have a large effect on the gas-phase density. This sensitivity has applications in separation technology (supercritical extraction)[15], where solvent recovery is an important economic consideration. Figure 12-23 shows a schematic diagram of a particularly simple supercritical-extraction separation process. Suppose that component 2 is in a mixture of mutually insoluble solids; suppose also that dense gas 1 (the solvent) selectively dissolves solid 2 at T and P close to the critical of the solvent. After the solute-saturated solvent is removed, it is slightly expanded (or heated, or both), significantly reducing the solvent's density. That reduction lowers y_2; solid 2 precipitates and the solute-lean solvent is recycled to the extraction unit after recompression (or cooling, or both). The advantage of this separation process is that, under favorable circumstances, energy requirements for solvent recovery may be lower than those required for a conventional extraction process, using liquid solvents.[16]

Extractions with supercritical fluids are potentially useful in many industries, in particular for extraction of bioactive substances from natural products (for a review, see King, 1993 and Rizvi, 1994). In general, an extract obtained by supercritical extraction from natural products is a mixture of various bioactive substances (such as isomers and substituted derivatives). For example, to isolate coumarin (known for its

(Eds., 1991), *Supercritical Fluid Technology: Reviews in Modern Theory and Applications,* Boston: CRC Press; K. P. Johnston and J. M. Penninger (Eds., 1989), *Supercritical Fluid Science and Technology*, American Chemical Society Symposium Series 406, Washington: ACS.

[15] Supercritical fluid solvents also provide unique reaction media because of the strong pressure dependence of their densities and solvent properties (such as viscosity, mass transfer coefficients, etc.). This strong pressure dependence offers the capability of controlling solvent properties, reaction rates, and selectivities through easily conducted pressure regulation.

[16] While the discussion here is limited to a mixture of mutually insoluble solids, supercritical extraction can also be applied to condensed mixtures of soluble (or partially soluble) solids or liquids. The principles are the same, but molecular-thermodynamic analysis is now more complex because, in addition to pure-component, condensed-phase fugacities, we require first, condensed-phase activity coefficients, and second, solubilities for the gaseous solvent if the condensed phase is a liquid. For a review, see M. E. Paulaitis, V. J. Krukonis, and R. C. Reid, 1983, *Rev. Chem. Eng.*, 1: 179.

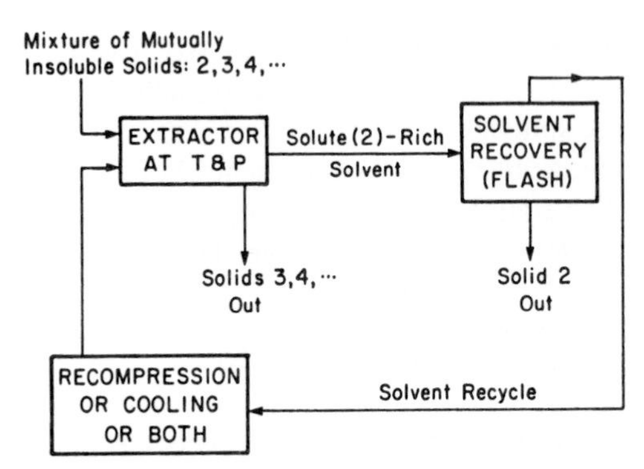

Figure 12-23 Idealized application of supercritical extraction. Temperature T and pressure P are slightly above the solvent's critical. In the recovery step, a small drop in P (or small rise in T) significantly lowers the solubility of solid 2 in the gaseous solvent. (In this idealized example, the solvent is totally selective for solid 2.)

strong antibiotic and anticoagulation properties) for pharmaceutical use by extraction with supercritical CO_2 from a natural plant, the extract also contains appreciable amounts of isomers and numerous coumarin derivatives (Yoo, 1997; Choi, 1998). The design of a supercritical fluid process requires the solubilities of each component in the supercritical fluid. Figure 12-24 compares solubilities at 308.15 K of coumarin and its methoxy-group and hydroxy-group derivatives in supercritical CO_2. As Fig 12-24 shows, coumarin exhibits the highest solubility in CO_2; its solubility is about two orders of magnitude higher than those of its derivatives. This large solubility difference is required for the feasibility of any supercritical-fluid-based extraction process.

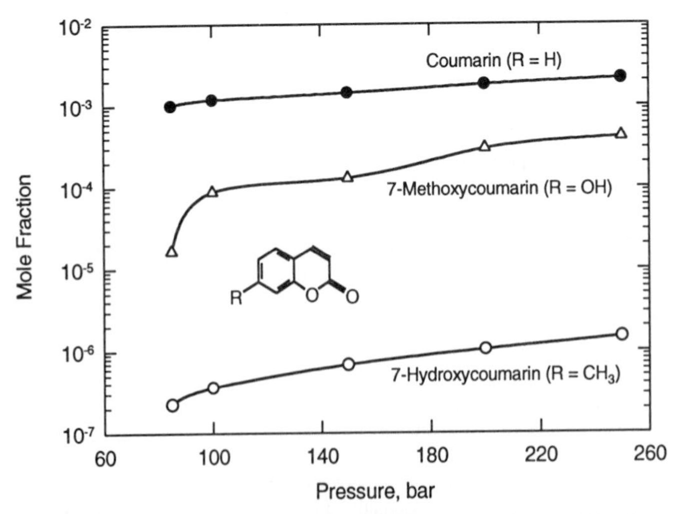

Figure 12-24 Solubility of coumarin, 7-methoxycoumarin and 7-hydroxycoumarin in supercritical CO_2 at 308.15 K (Yoo *et al.*, 1997). Lines are best fit of experimental data.

Figure 12-25 presents an illustration of a practical application of supercritical fluid solvents in the oleochemical industry. This figure shows distribution coefficients (*K* factors) of oleic acid and triolein (a major component of vegetable oils, such as rapeseed and olive oil) in supercritical carbon dioxide at 40°C (Bharath *et al.*, 1992). Carbon dioxide is a convenient solvent for supercritical extraction. Among others, it has the advantage of allowing extractions at temperatures close to room temperature (the CO_2 critical temperature is 31.1°C), particularly useful for thermally labile natural materials such as polyunsaturated fatty acids like oleic acid. A major advantage of CO_2 derives from its environmentally-friendly properties: it is not toxic; it does not burn or explode; it is easily vented into the atmosphere; and it is inexpensive.

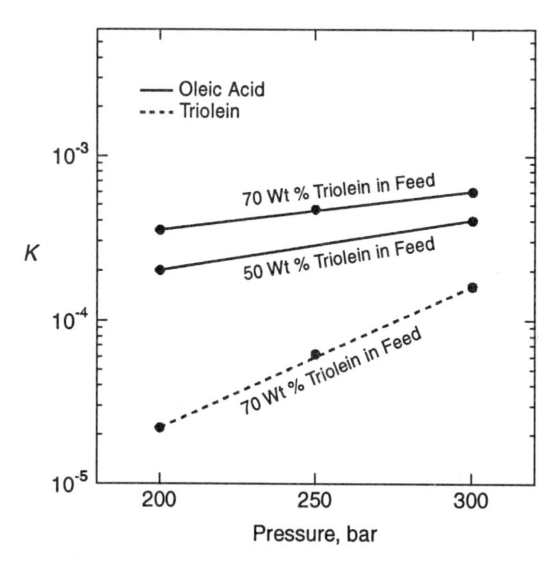

Figure 12-25 *K* factors (Bharath *et al.*, 1992) as a function of pressure (for feeds of 50 weight % and 70 weight % of triolein, CO_2-free basis) for oleic acid (——) and triolein (– – – –) obtained from vapor-liquid-equilibrium data with supercritical CO_2 at 40°C.

Figure 12-25 shows that the distribution coefficients of both oleic acid and triolein vary with pressure and with feed composition. However, *K* for oleic acid is ten times larger than that of triolein, suggesting that extraction with supercritical CO_2 can provide an alternative (Gonçalves *et al.*, 1991) to conventional fractionation processes such as solvent extraction (where, typically, toxic organic solvents are used) and low-pressure distillation (that requires relatively high temperatures).

Another area of rapidly developing possibilities for supercritical fluid technology is in various aspects of environmental control (Abraham and Sunol, 1997) such as removal of pesticides (e.g., DDT and pyrethrin), polyaromatic hydrocarbons, polychlorinated aromatic compounds (PCB's), dioxins, etc. from contaminated soils.

Technologies have been developed for extraction of organic compounds from aqueous and solid environmental matrices. The attractive feature of this process is the ease of separation of the extracted solute from the supercritical fluid that results in the creation of smaller waste volumes of the now concentrated organic, improving the efficiency of subsequent waste-treatment processes such as incineration. Figure 12-26 shows results of supercritical removal of hydrocarbons from soil contaminated with 5 weight % of long-chain alkanes, 4 weight % of monoaromatic, and 2 weight % of polyaromatic hydrocarbons (Firus *et al.*, 1997). Figure 12-26 shows that supercritical water (at 653 K and 25 MPa) removes virtually all the contaminants from the soil. However, supercritical CO_2 (at 353 K and 20 MPa) achieves no more than 21% of extraction (even with large solvent-to-soil ratio), because of the lower operating temperature.

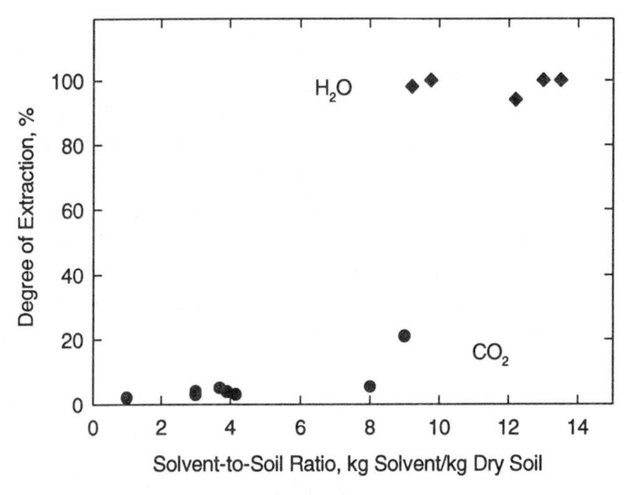

Figure 12-26 Extraction of hydrocarbons from soil with supercritical CO_2 and H_2O as a function of solvent-to-soil ratio. ◆ : H_2O at 653 K and 25 MPa; ●: CO_2 at 353 K and 20 MPa. The critical temperature of CO_2 is 304 K and that of water is 647 K.

In modeling the phase behavior (e.g. an equation of state) of natural products such as those indicated in Figs. 12-24 and 12-25, there is an additional difficulty: due to the complex nature of most natural products, basic physical properties (such as critical data), and constants (such as Antoine vapor-pressure constants), are not available. They are usually estimated from semi-empirical methods such as those described by Reid *et al.* (1987) and Lyman *et al.* (1990).

When component 2 is much heavier than component 1, at temperatures and pressures slightly above the critical of solvent 1, fugacity coefficient φ_2 is almost always very small compared to unity. As indicated by Eq. (12-42), when pressure and temperature are held constant, solubility y_2 is inversely proportional to φ_2. The advantage of supercritical extraction follows from the sensitivity of φ_2 to pressure (and, to a lesser extent, temperature) when the solvent's density is near its critical.

Experimental results have shown that, near the solvent's critical state, $\varphi_2 << 1$, yielding a solubility y_2 much larger than that calculated assuming ideal-gas behavior. Experimental results also show that, near the solvent's critical state, a small decrease in pressure produces a large rise in φ_2; it is this observation that facilitates solvent recovery.

The effect of pressure P on fugacity coefficient φ_2 is directly related to the partial molar volume $\overline{v}_2$:

$$\left(\frac{\partial \ln \varphi_2}{\partial P}\right)_{T,y} = \frac{\overline{v}_2}{RT} - \frac{1}{P} \tag{12-43}$$

When $y_2 << 1$, and when T and P are near the critical of the solvent, $\overline{v}_2$ is large and negative. To illustrate, let component 2 be naphthalene and let component 1 be carbon dioxide. (The critical temperature of carbon dioxide is 31.1°C; the critical pressure is 73.8 bar.) To calculate $\overline{v}_2$ from a pressure-explicit equation of state, we use Eq. (12-41).

Partial molar volumes for naphthalene in carbon dioxide have been calculated using Chueh's modification (Chueh and Prausnitz, 1967) of the Redlich-Kwong equation of state. Results are shown in Fig. 12-27. Because the vapor pressure of solid naphthalene is very small in the range 35 to 50°C, y_2 is also small compared to unity; therefore, the calculated results in Fig. 12-25 are for solutions infinitely dilute in naphthalene ($y_2 = 0$). The calculations were performed using conventional mixing rules (quadratic in mole fraction for a and linear in mole fraction for b) with binary parameter k_{12} = 0.0626 as determined from solubility data. Vapor-phase densities for pure carbon dioxide were obtained from the IUPAC tables (Angus *et al.*, 1976).

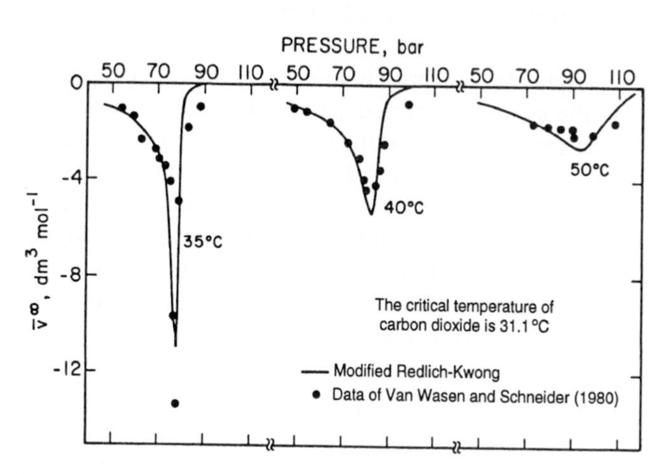

Figure 12-27 Partial molar volumes of naphthalene infinitely dilute in compressed carbon dioxide. Partial molar volumes are large and negative in the critical region.

Figure 12-27 shows that, at constant temperature, $\overline{v}_2^\infty$ is negative and goes through a sharp minimum when plotted against pressure, in agreement with the data of van Wasen (1980). Further, $\overline{v}_2^\infty$ is extremely large, especially at 35°C and (about) 78 bar. At this temperature, the calculated minimum in $\overline{v}_2^\infty$ is -10.8 L mol^{-1}; for comparison, the critical volume of pure carbon dioxide is +0.0944 L mol^{-1} .

Figure 12-27 and Eq. (12-43) indicate that at pressures near 78 bar and at 35°C, a small decrease in pressure very much raises φ_2 thereby very much reducing solubility y_2. The sharp change in φ_2 with pressure is the basis of efficient solute recovery in supercritical fluid extraction.

12.6 Calculation of High-Pressure Vapor-Liquid Equilibria

If a liquid mixture at temperature T is in equilibrium with a vapor at high pressure, it is usually necessary that one of the components must be supercritical; that is, the mixture is likely to contain at least one component i whose critical temperature $T_{c_i} > T$. This follows because the vapor pressures of most liquids are not large (critical pressures of typical pure liquids are in the range 30 to 100 bar) and therefore, with few exceptions, high-pressure vapor-liquid equilibria as found in nature and in the chemical industry apply to mixtures with at least one subcritical component and at least one supercritical component.

The thermodynamics of high-pressure vapor-liquid equilibria is, in principle, similar to that of the solubility of gases in liquids, as discussed in Chap. 10. However, the concepts of Chap. 10 are generally not useful for our purposes here because we are now concerned with a wide range of liquid-phase concentrations, not only with dilute solutions.

It is possible analytically to represent high-pressure vapor-liquid equilibria using, for the liquid phase, the common thermodynamic functions: Henry's constant, activity coefficient and partial molar volume. But experience has shown that in typical cases, these functions are not useful, especially for multicomponent mixtures. A more successful route for quantitatively describing high-pressure vapor-liquid equilibria is provided by the fugacity coefficient, applied to both phases, as indicated in Sec. 3.6.

Consider a binary liquid mixture with mole fractions x_1 and x_2 at temperature T and pressure P; in equilibrium with this liquid mixture is a vapor with mole fractions y_1 and y_2. The equations of equilibrium are

$$f_1^V = f_1^L \quad \text{or} \quad \varphi_1^V y_1 = \varphi_1^L x_1 \tag{12-44}$$

and

$$f_2^V = f_2^L \quad \text{or} \quad \varphi_2^V y_2 = \varphi_2^L x_2 \tag{12-45}$$

where f is the fugacity and φ is the fugacity coefficient. The equilibrium ratios (K factors) are given by

$$K_1 \equiv \frac{y_1}{x_1} = \frac{\varphi_1^L}{\varphi_1^V} \tag{12-46}$$

$$K_1 \equiv \frac{y_2}{x_2} = \frac{\varphi_2^L}{\varphi_2^V} \tag{12-47}$$

12.7 Phase Equilibria from Equations of State

To calculate fugacity coefficients, we can use a partition function (see App. B) or, what is equivalent for our purposes here, an equation of state. Equation-of-state methods provide one of the most useful techniques used in chemical engineering practice for modeling phase equilibria of multicomponent systems.[17]

In the equation-of-state method, a single equation of state is used to represent all fluid phases; for vapor-liquid equilibria, the equation is assumed to be valid for both the vapor-phase mixture and the liquid-phase mixture.

For each component i in the vapor phase,

$$\ln \varphi_i^V = \frac{1}{RT}\int_{V^V}^{\infty}\left[\left(\frac{\partial P}{\partial n_i}\right)_{T,V,n_j} - \frac{RT}{V}\right]dV - \ln\frac{PV^V}{n_T RT} \tag{12-48}$$

as discussed in Chap. 3. Here n_i is the number of moles of component i and n_T is the total number of moles in the vapor phase. Similarly, for each component i in the liquid phase

$$\ln \varphi_i^L = \frac{1}{RT}\int_{V^L}^{\infty}\left[\left(\frac{\partial P}{\partial n_i}\right)_{T,V,n_j} - \frac{RT}{V}\right]dV - \ln\frac{PV^L}{n_T RT} \tag{12-49}$$

where n_i and n_T now refer to the liquid phase.

To use Eqs. (12-48) and (12-49), we require a suitable equation of state that holds for the entire range of possible mole fractions x and y at system temperature T and for the density range between 0 and $(n_T/V)^L$. This last condition is necessary because the integral in Eq. (12-49) goes from the ideal-gas state (infinite volume) to the

[17] For a review see A. Anderko, 1990, *Fluid Phase Equilibria,* 6: 145.

saturated-liquid density. At present, there are no satisfactory equations of state that meet these requirements with generality. However, for many mixtures, we have reasonable, approximate equations of state that provide useful results.

The simplest procedure for using Eqs. (12-48) and (12-49) is to choose an equation of state that holds for pure fluid 1 and for pure fluid 2 and to assume that this same equation of state holds for all mixtures of 1 and 2 by interpolation. To fix ideas, consider, for example, Soave's (1972) modification of the Redlich-Kwong equation:

$$P = \frac{RT}{v-b} - \frac{a(T)}{v(v+b)} \tag{12-50}$$

where $a(T)$ depends on temperature according to

$$a(T) = a(T_c)\,\alpha(T) \tag{12-51}$$

$$a(T_c) = 0.42748 \frac{(RT_c)^2}{P_c} \tag{12-52}$$

where, for normal fluids,

$$\alpha(T) = [1 + (0.480 + 1.574\omega - 0.176\omega^2)(1 - \sqrt{T/T_c})]^2 \tag{12-53}$$[18]

for acentric factors $0 < \omega < 0.5$.

Constant b is given by

$$b = 0.08664 \frac{RT_c}{P_c} \tag{12-54}$$

as in the Redlich-Kwong equation.

Now comes the important step: extension of Eq. (12-50) to the binary mixture. The simplest procedure (but by no means the only possible procedure) is to assume a *one-fluid* theory of mixtures, that is, to assume that the equation of state for the mixture is the same as that for a hypothetical "pure" fluid whose characteristic constants b and $a(T)$ depend on composition.

---

[18] Soave (1993, *Fluid Phase Equilibria,* 84: 339) proposed another expression for the temperature dependence of the attractive parameter of the Soave-Redlich-Kwong equation. This function is a two-parameter extension of the original form and can be applied to polar and nonpolar fluids. The new expression (valid to $\omega = 0.5$) is:

$$\alpha(T) = 1 + m(1 - T/T_c) + n\left(1 - \sqrt{T/T_c}\right)^2$$

For nonpolar or slightly polar fluids, $m = 0.484 + 1.515\omega - 0.044\omega^2$ and $n = 2.756\,m - 0.700$. For strongly polar fluids, m and n must be fitted from experimental data.

One-fluid theory assumes that for the mixture we can generalize Eq. (12-50), i.e. to assume that it holds for the mixtures for both phases. The remaining task is to specify how b and $a(T)$ depend on composition (mixing rules).

The common procedure is to write mixing rules that are quadratic in mole fraction. For a binary mixture,

$$b = z_1^2 b_{11} + 2 z_1 z_2 b_{12} + z_2^2 b_{22} \tag{12-55}$$

$$a(T) = z_1^2 a_{11}(T) + 2 z_1 z_2 a_{12}(T) + z_2^2 a_{22}(T) \tag{12-56}$$

where z is either x or y.

Constants b_{ii} and $a_{ii}(T)$ refer to the pure-component values; however, b_{12} and $a_{12}(T)$ are binary parameters. It is convenient to express these in the form

$$b_{12} = \frac{1}{2}(b_{11} + b_{22})(1 - c_{12}) \qquad (12\text{-}57)^{19}$$

$$a_{12}(T) = [a_{11}(T) a_{22}(T)]^{1/2}(1 - k_{12}) \tag{12-58}$$

where the absolute values of binary parameters $|c_{12}|$ and $|k_{12}|$ are small compared to unity. For simple mixtures of nonpolar components, it is customary to set $c_{12} = 0$ and to fit k_{12} to binary vapor-liquid equilibrium composition data.[20] For simple systems, k_{12} is often nearly independent of temperature. While k_{12} and c_{12} may depend on temperature, they cannot depend on pressure or density without thereby substantially increasing the complexity of the resulting equation for the fugacity coefficient.

For calculating high-pressure vapor-liquid equilibria from an equation of state, the computational procedure is not trivial. In a typical case for a binary mixture, the given quantities may be P and x_1 (and x_2). Equations (12-44), (12-45), (12-48), and (12-49) must then be used to find T, y_1 (and y_2). However, to use Eqs. (12-48) and (12-49), we must know the molar volumes v^V and v^L. To find these, we use the equation of state for the mixture [say, Eq. (12-50)] twice, once for the vapor mixture and once for the liquid mixture.

For the binary case under consideration here, there are four unknowns: y, T, v^V and v^L. To find these, it is necessary to solve simultaneously four independent equations. These are Eqs. (12-44) and (12-45) [with Eqs. (12-48) and (12-49) substituted for the fugacity coefficients] and Eq. (12-50) twice, once for each phase. A similar procedure is used for multicomponent mixtures. Numerical techniques for such

---

[19] If $c_{12} = 0$, Eq. (12-55) reduces to the linear form $b = z_1 b_{11} + z_2 b_{22}$.

[20] For mixtures that are highly asymmetric with respect to size (for example, hydrogen/anthracene), it is sometimes better to adjust c_{12} to experimental data while setting $k_{12} = 0$. See, for example, A. I. ElTwaty and J. M. Prausnitz, 1980, *Chem. Eng. Sci.*, 35: 1765.

calculations are described in numerous references [see, e.g., Anderson (1980, 1980a) and Michelsen (1980); there are many others].

To illustrate this method for calculating high-pressure vapor-liquid equilibria, Fig. 12-28 shows results for the system methane/propane. Calculations are based on Soave's modification of the Redlich-Kwong equation of state with mixing rules given by Eqs. (12-55) and (12-56). In these calculations $c_{12} = 0$. Good agreement with experiment is obtained when $k_{12} = 0.029$. A small change in k_{12} can significantly affect the calculated results.

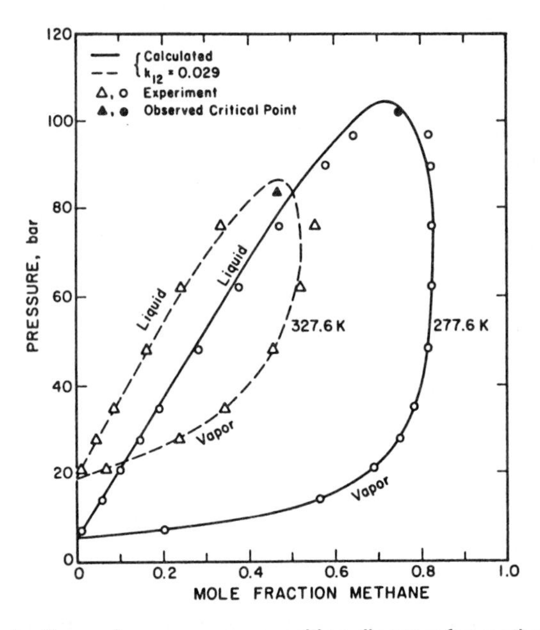

Figure 12-28 Isothermal pressure-composition diagram for methane/propane. Calculations using Redlich-Kwong-Soave equation of state. Experimental data from Reamer *et al.* (1950).

Figure 12-29 shows calculated K factors for nitrogen/propylene, using $c_{12} = 0$ and $k_{12} = 0.0915$. Again, agreement with experiment is good.

Another illustration is given in Fig. 12-30 for the system hydrogen sulfide/water. In this case $c_{12} = 0.08$ and $k_{12} = 0.163$. Pure-component constants were found from Eqs. (12-51) to (12-54), even though Eq. (12-53) is obtained from vapor-pressure data for paraffins. The high value of k_{12} is not surprising because the geometric-mean approximation [Eq. (12-58) is poor for aqueous mixtures. For this system it is likely that the binary constants (especially k_{12}) depend significantly on temperature.

Calculations similar to these can be made with any one of a large number of equations of the van der Waals form. Many such calculations have been reported, notably by Peng and Robinson (1976), Ishikawa *et al.* (1980), and Dohrn and Brunner

(1995). While most work has been limited to binary systems, extension to ternary (or higher) systems is straightforward. If such extension uses mixing rules that are quadratic in mole fraction [Eqs. (12-55) and (12-56)], only pure-component and binary parameters are required.

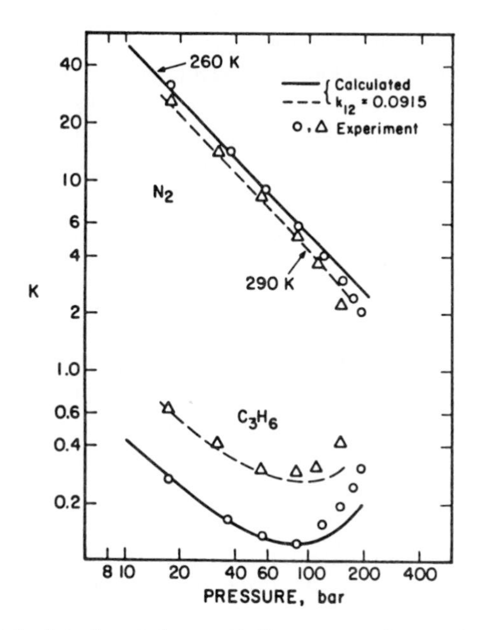

Figure 12-29 Calculated and observed (Grausø *et al.*, 1977) *K* factors for nitrogen/propylene. Calculations using Redlich-Kwong-Soave equation of state.

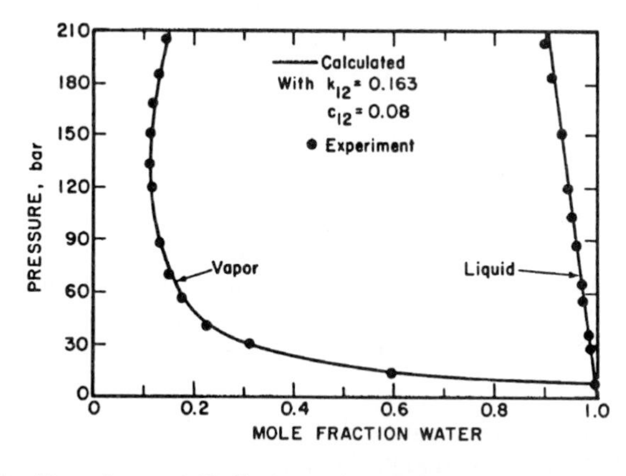

Figure 12-30 Experimental (Selleck *et al.*, 1952) and calculated (Evelein *et al.*, 1976) phase equilibria for the system hydrogen sulfide/water at 171°C.

To calculate vapor-liquid phase equilibria for a binary system, pure-component constants are rarely sufficient; for reliable results it is necessary to use also at least one binary parameter that, with few exceptions, cannot be estimated with sufficient accuracy from unicomponent data alone. However, to calculate vapor-liquid phase equilibria for a ternary (or higher) system, ternary (or higher) parameters are often not required for reliable results. In many cases, ternary (or higher) vapor-liquid equilibria can be predicted using only pure-component and binary parameters.

To illustrate, Fig. 12-31 shows vapor-liquid equilibria obtained by Sandler and coworkers at 410.9 K for the ternary system N_2/CO_2/cyclohexane, as well as for the constituent binaries $N_2/c\text{-}C_6H_{12}$ and $CO_2/c\text{-}C_6H_{12}$. According to the phase rule, a ternary mixture with two phases in equilibrium has 3 degrees of freedom; these are described by volumes (see Table 12-1) in P-T-$\mathfrak{z}$ space. Therefore, a convenient two-dimensional way of representing ternary phase equilibrium data is by using an isothermal pressure-composition[21] (or isobaric temperature-composition) prism with a triangle as a base. Each of the four cross-sectional triangles is an isothermal-isobaric phase diagram, and each side of the prism corresponds to the appropriate binary P-x diagrams, as illustrated in Fig. 12-31. At 410.9 K, both N_2 and CO_2 are supercritical and therefore this binary is not shown in Fig. 12-31.

In addition to experimental data, Fig. 12-31 shows also calculations using the Peng-Robinson equation of state:

$$P = \frac{RT}{v-b} - \frac{a(T)}{v(v+b)+b(v-b)} \tag{12-59}$$

For a pure fluid, constant b is given by

$$b = 0.07780\frac{RT_c}{P_c} \tag{12-60}$$

while $a(T)$, a function of temperature, is given by

$$a(T) = a(T_c)\alpha(T) \tag{12-61}$$

$$a(T_c) = 0.45724\frac{(RT_c)^2}{P_c} \tag{12-62}$$

$$\alpha(T) = \left[1+\beta(1-\sqrt{T/T_c})\right]^2 \tag{12-63}$$

Here, T_c is the critical temperature and β depends on acentric factor ω according to $\beta = 0.37464 + 1.54226\,\omega - 0.26992\,\omega^2$ (for $0 \le \omega \le 0.5$).

[21] Using an isothermal (or isobaric) representation, the degrees of freedom are reduced by one. Using an isothermal and isobaric representation, the degrees of freedom are reduced by two.

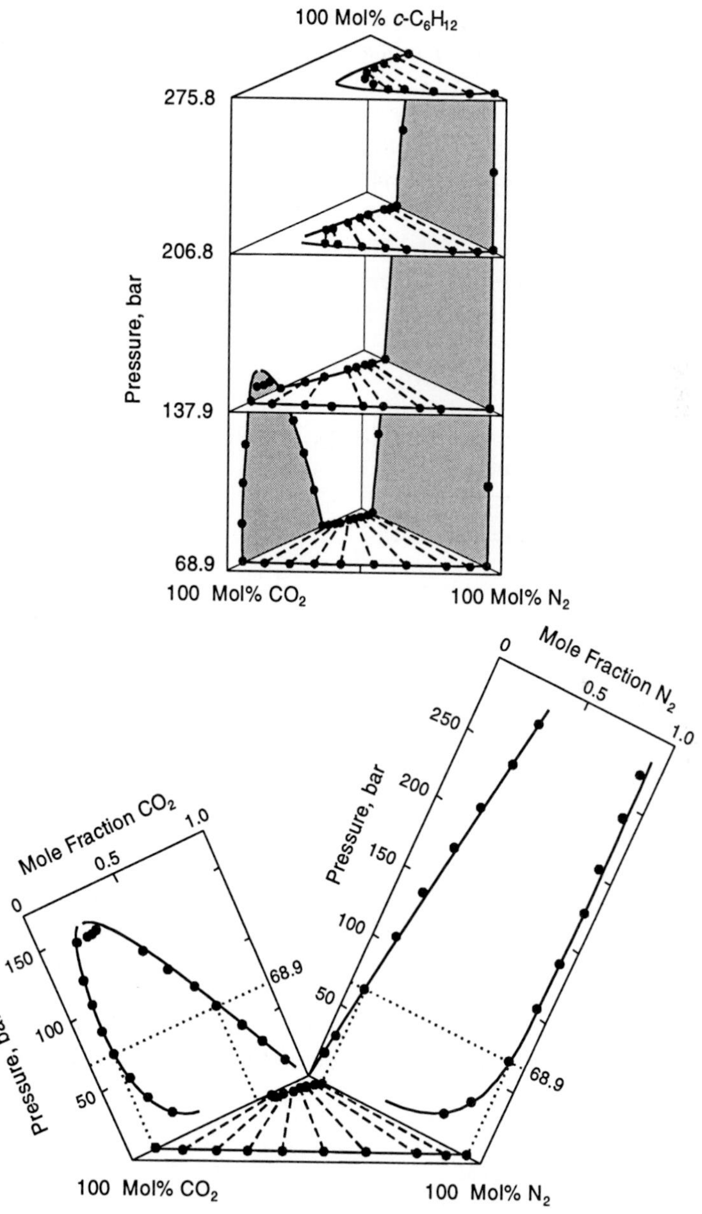

Figure 12-31 Vapor-liquid equilibria for the system $N_2/CO_2/c\text{-}C_6H_{12}$ at 410.9 K. The four triangles shown on top are isothermal-isobaric phase diagrams for the ternary mixture at 410.9 K and at the pressures indicated. At bottom, phase diagrams of the constituent binaries $CO_2/c\text{-}C_6H_{12}$ and $N_2/c\text{-}C_6H_{12}$ (also at 410.9 K) and their relation to the 68.9-bar triangular diagram. ● Experiment; —— Calculated from Peng-Robinson equation with the binary parameters listed in Table 12-3. (Shibata and Sandler, 1989).

Using the conventional mixing rules for a mixture [Eqs. (12-55) and (12-56)], the fugacity coefficient for component k in a mixture is given by

$$\ln\varphi_k = \frac{b_k}{b}\left(\frac{Pv}{RT}-1\right) - \ln\frac{P(v-b)}{RT} - \frac{a}{2\sqrt{2}\,bRT}\left[\frac{2\sum_i z_i a_{ik}}{a} - \frac{b_k}{b}\right]\ln\frac{v+(1+\sqrt{2})b}{v+(1-\sqrt{2})b} \qquad (12\text{-}64)$$

where z_i is the mole fraction of component i.

The phase behavior of the ternary $N_2/CO_2/c\text{-}C_6H_{12}$ system was predicted using the binary parameters k_{ij} given in Table 12-3, determined from binary data only (for the binary N_2/CO_2, k_{12} was set to zero).

Table 12-3 Binary parameters k_{ij} [Eq. (12-58)] in the Peng-Robinson equation as obtained from binary data at 410.9 K.

System	k_{ij}
$CO_2/c\text{-}C_6H_{12}$	0.103
$N_2/c\text{-}C_6H_{12}$	0.076
N_2/CO_2	0

Figure 12-31 shows good agreement between experimental and calculated results obtained for the ternary $N_2/CO_2/c\text{-}C_6H_{12}$ at 410.9 K and at 68.9, 137.9, 206.8, and 275.7 bar, and also for the binaries $N_2/c\text{-}C_6H_{12}$ and $CO_2/c\text{-}C_6H_{12}$ at the same temperature. However, near the critical region agreement between calculated and experimental results is not satisfactory. When calculations are based on an equation of state of the van der Waals form (e.g., Redlich-Kwong-Soave, Peng-Robinson), it is often found that the calculated two-phase region is larger than that given by experiment, as shown in Fig. 12-31. While we now have a good understanding of why equations of state of the van der Waals form fail in the critical region, it is not a simple matter to "fix" these equations for better performance. In the critical region, density fluctuations contribute strongly to thermodynamic properties. These fluctuations, ignored in van der Waals-type equations of state, can be taken into account but the procedure for doing so is complex.[22]

Non-Quadratic Mixing Rules

To extend pure-fluid equations of state to mixtures of nonpolar (or slightly polar) components, it is customary to use classical one-fluid mixing rules, as proposed by van

[22] L. Lue and J. M. Prausnitz, 1998, *AIChE J.*, 44: 1455; Z. Y. Cheng, M. A. Anisimov, J. V. Sengers, 1997, *Fluid Phase Equilibria*, 128: 67.

der Waals. In the one-fluid theory, it is assumed that the properties of a fluid mixture are identical to those of a hypothetical pure fluid, at the same temperature and pressure, whose equation-of-state parameters are functions of mole fraction. In the van der Waals approximation, these functions are quadratic in mole fraction.

For highly nonideal mixtures, however, quadratic mixing rules are inadequate. Several methods have been proposed to modify the original mixing rules. In many cases the modified mixing rules include composition-dependent or density-dependent binary interaction parameters. Density-dependent mixing rules are formulated to obtain a correct representation of mixture properties at both high and low-density regions: in the limit of low density, they reproduce the correct composition dependence of the mixture second virial coefficient (a quadratic function in mole fraction, as shown in Sec. 5.3), whereas in the high-density limit (or infinite pressure), mixing rules are determined to force agreement with some excess-Gibbs-energy model for a liquid mixture (see, for example, Dimitrelis and Prausnitz, 1990). Some authors have suggested mixing rules independent of density, as suggested by excess-Gibbs-energy models. Most of these mixing rules, however, violate the theoretically-based boundary condition that the second virial coefficient must be quadratic in mole fraction.

Promising mixing rules are those proposed by Wong *et al.* (1992, 1992a; Sandler *et al.*, 1994).[23] Their rules produce the correct low-density limit (the mixture second virial coefficient has a quadratic dependence in mole fraction) and a high-density limit consistent with experiment (the equation of state predicts the same excess molar Helmholtz energy at infinite pressure, a_{∞}^{E}, as a function of composition as that obtained from a selected activity-coefficient model). Wong-Sandler mixing rules are not density dependent.

The Wong-Sandler model is applied to a van der Waals-type equation of state with

$$b_{\text{mixt}} = \frac{\sum_i \sum_j x_i x_j \left(b - \frac{a}{RT}\right)_{ij}}{1 + \frac{a_{\infty}^{E}}{CRT} - \sum_i x_i \left(\frac{a_i}{b_i RT}\right)} \tag{12-65}$$

and

$$a_{\text{mixt}} = b_{\text{mixt}} \left[\sum_i x_i \frac{a_i}{b_i} - \frac{a_{\infty}^{E}(x_i)}{C}\right] \tag{12-66}$$

where C is a constant dependent on the equation of state. For the Peng-Robinson equation, $C = [\ln(\sqrt{2} - 1)] / \sqrt{2} = -0.62322$.

[23] For a description of the capabilities and limitations of the Wong-Sandler mixing rule see P. Coutsikos *et al.*, 1995, *Fluid Phase Equilibria*, 108: 59, and M. A. Satyro and M. A. Trebble, 1996, *ibid.*, 115: 135.

The composition-independent cross second virial coefficient $[b - a/(RT)]_{ij}$, obtained from the equation of state, is related to those of the pure components by

$$\left(b-\frac{a}{RT}\right)_{ij}=\frac{1}{2}\left[\left(b_i-\frac{a_i}{RT}\right)+\left(b_j-\frac{a_j}{RT}\right)\right](1-k_{ij}) \qquad (12\text{-}67)$$

where k_{ij} is a binary interaction parameter for the second virial coefficient. This interaction parameter is usually obtained from fitting experimental vapor-liquid data as expressed by the molar excess Gibbs energy g^E as a function of mole fraction x at constant temperature. To fix k_{ij}, it is common to use g^E at $x = 0.5$.

For the high-density limit, Wong and Sandler equate the excess Helmholtz energy at infinite pressure obtained from the equation of state to that obtained from a chosen liquid-phase activity-coefficient model (van Laar or NRTL, etc.), using the approximation

$$g^E(T,P=1\text{ bar},x_i)\approx a^E(T,P=1\text{ bar},x_i)\approx a^E(T,\text{high pressure},x_i)\equiv a_\infty^E(T,x_i)$$

where $a_\infty^E(T,x_i)$ is the molar excess Helmholtz energy for a fixed composition x_i and temperature T in the infinite-pressure limit. In subsequent calculations, it is common practice to neglect the effect of temperature on a_∞^E.

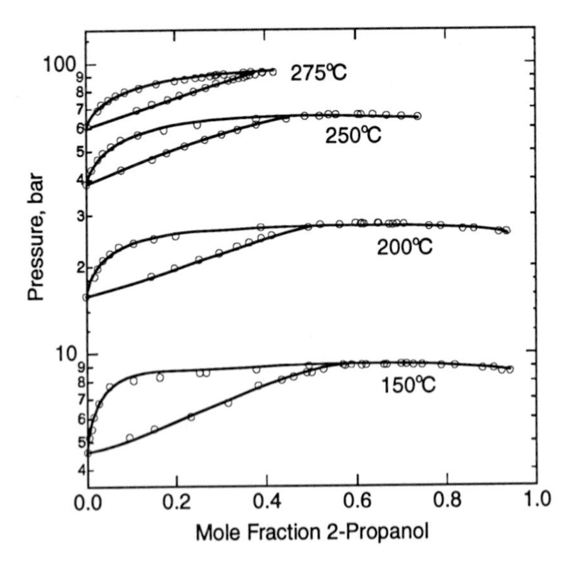

Figure 12-32 High-pressure vapor-liquid equilibria for the binary 2-propanol/water. Calculations from the Peng-Robinson equation, as modified by Stryjek and Vera (1986) using the Wong-Sandler mixing rules (1992) with $k_{12} = 0.326$. —— Calculated; o Experiment (Barr-David and Dodge, 1959).

Figure 12-32 compares measured and calculated vapor-liquid equilibria for the binary 2-propanol/water (Wong *et al.*, 1992a). The calculations were performed with the Peng-Robinson equation of state – with pure component parameters obtained from the correlation of Stryjek and Vera (1986) – using the mixing rules of Wong-Sandler (as described above) with a_∞^E determined from the van Laar equation. The van Laar parameters were obtained by correlating experimental vapor-liquid-equilibrium data at 4.12 bar, reported in the DECHEMA data series (Gmehling and Onken, 1977); binary interaction parameter $k_{12} = 0.326$ was determined by adjusting the excess Gibbs energy calculated from the equation of state (at 4.12 bar) to that calculated from the van Laar model. This parameter was then used to predict the phase behavior at other pressures and temperatures. As Fig. 12-32 shows, with the parameters obtained at conditions where all components are subcritical, this model can predict well vapor-liquid equilibria at conditions where one of the components (2-propanol) becomes supercritical.

The main attraction of the Wong-Sandler mixing method follows from its ability to extrapolate vapor-liquid-equilibrium data to other (usually higher) temperatures and pressures.

12.8 Phase Equilibria from a Corresponding-States Correlation

Corresponding-states correlations can be used to calculate fugacities of components in mixtures of normal (nonpolar or slightly polar) fluids; these fugacities are then used to compute phase equilibria. The general procedure is first, to write a reduced equation of state for compressibility factor z:

$$z = \frac{Pv}{RT} = F_z\left(\frac{T}{T_c}, \frac{P}{P_c}, X\right) \tag{12-68}$$

or a reduced equation of state of the form

$$\frac{P}{P_c} = F_P\left(\frac{T}{T_c}, \frac{v}{v_c}, X\right) \tag{12-69}$$

where T_c, P_c, and v_c are critical temperature, pressure, and volume and where X is some additional characteristic parameter such as the acentric factor.[24] Functions F_z or F_P may be analytic, tabular or graphical; for computer applications, analytical functions are most convenient.[25] From Eq. (12-72) or (12-73), we calculate fugacity coefficients as discussed in Chap. 3 [Eqs. (3-14) and (3-53)].

[24] For a fixed value of acentric factor (see Sec. 4.13), v_c is not an independent parameter when T_c and P_c are specified.

[25] Analytical equations are given by D. R. Schreiber and K. S. Pitzer, 1989, *Fluid Phase Equilibria,* 46: 113.

To extend Eqs. (12-68) or (12-69) to mixtures, we use the pseudocritical hypothesis. The essential step here is to assume some expressions (mixing rules) indicating how the characteristic properties (T_c, X, P_c or v_c) depend on composition. We then find the fugacity coefficient for a component i in the mixture by differentiation; for a binary mixture, fugacity coefficient φ_1 is found from

$$\ln\varphi_1 = \ln\varphi_{\text{mixt}} + (1 - z_1)\left(\frac{\partial \ln\varphi_{\text{mixt}}}{\partial z_1}\right)_{T,P} \tag{12-70}$$

where z is the mole fraction in either the liquid phase or the vapor phase. The condition of vapor-liquid equilibrium is given by Eqs. (12-44) and (12-45).

The literature presents many examples where corresponding-states correlations have been used to calculate vapor-liquid equilibria at high pressures. One of the most successful is the correlation given by Mollerup (1980) for mixtures of natural gases (methane, nitrogen, ethane, propane, and carbon dioxide). Mollerup uses these mixing rules for the critical temperature, critical volume, and acentric factor:

$$T_{c_{\text{mixt}}} = \frac{\sum_i \sum_j z_i z_j v_{c_{ij}} T_{c_{ij}}}{v_{c_{\text{mixt}}}} \tag{12-71}$$

$$v_{c_{\text{mixt}}} = \sum_i \sum_j z_i z_j v_{c_{ij}} \tag{12-72}$$

$$\omega_{\text{mixt}} = \sum_i z_i \omega_i \tag{12-73}$$

with

$$v_{c_{ij}}^{1/3} = \frac{1}{2}(v_{c_i}^{1/3} + v_{c_j}^{1/3}) \tag{12-74}$$

$$T_{c_{ij}} = (T_{c_i} T_{c_j})^{1/2}(1 - k_{ij}) \tag{12-75}$$

where k_{ij} is a binary parameter characteristic of the i-j interaction. Because calculated results are sensitive to k_{ij}, Mollerup has taken much care to find the best values from reliable binary experimental data. Figure 12-33 shows K factors for the methane/propane system calculated by Mollerup. For this simple system, Mollerup obtains excellent agreement with experiment for a wide temperature range.

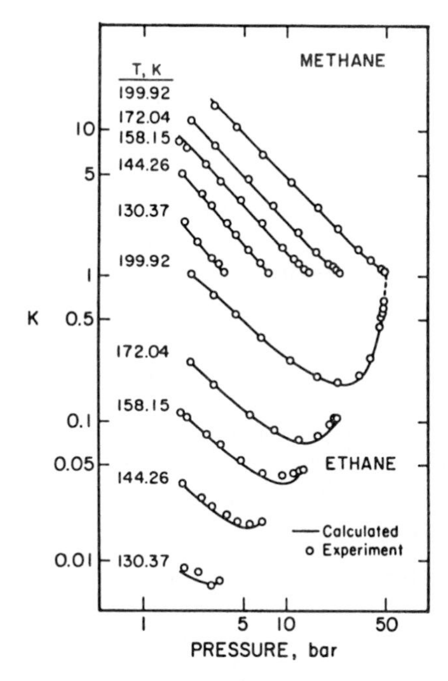

Figure 12-33 *K* factors for the methane/propane system calculated by Mollerup using corresponding-states theory.

Another example of how corresponding states may be used to calculate high-pressure vapor-liquid equilibria is provided by Plöcker's extension (Plöcker *et al.*, 1978) to mixtures of the correlation for pure fluids presented by Lee and Kesler (1975) (see Sec. 4.11). Plöcker's chief interest is in asymmetric mixtures, i.e., in mixtures of normal fluids differing appreciably in molecular size. For such mixtures, Plöcker[26] found good agreement with experiment upon modifying Eq. (12-75) to read

$$T_{c_{\text{mixt}}} = \frac{\sum_i \sum_j z_i z_j v_{c_{ij}}^{1/4} T_{c_{ij}}}{v_{c_{\text{mixt}}}^{1/4}} \tag{12-76}$$

The exponent 1/4 provides an empirical attempt to take into account (crudely) that when a small molecule interacts with a large molecule, the small molecule "sees" only part of the large molecule. Weighting intermolecular attractions as in Eq. (12-71), emphasizes molecular size too strongly; on the other hand, if the exponent 1/4 were replaced by exponent zero, no weight at all would be given to molecular size.

[26] A description similar to that of Plöcker is presented by Plazer and Maurer, 1993, *Fluid Phase Equilibria*, 84: 79, and tested successfully for a large number of systems.

Plöcker also rewrites Eq. (12-75) in the equivalent form

$$T_{c_{ij}} = (T_{c_i} T_{c_j})^{1/2} K_{ij} \tag{12-77}$$

where $K_{ij} = 1 - k_{ij}$.

Figure 12-34 illustrates calculations from Plöcker and coworkers using the correlation of Lee and Kesler. The same authors present a large number of binary parameters K_{ij}. For binaries where limited experimental information is available, they correlated K_{ij} parameters as a function of the dimensionless group $(T_{c_i} v_{c_i}) / (T_{c_j} v_{c_j})$.

Other applications of corresponding states to high-pressure phase equilibria include the shape-factor method of Leland and coworkers (Leach *et al.*, 1968).

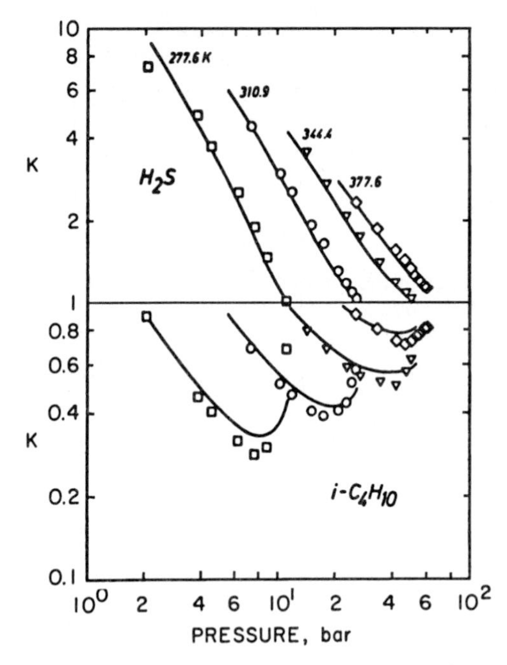

Figure 12-34 Calculated (Plöcker *et al.*, 1978) and experimental (Robinson and Besserer, 1972) *K* factors for the binary system hydrogen sulfide/isobutane.

12.9 Vapor-Liquid Equilibria from the Perturbed-Hard-Chain Theory

The perturbed-hard chain (PHC) theory, discussed in Sec. 7.16, can be used to calculate vapor-liquid equilibria at high pressures. As shown by Liu *et al.* (1980), it is particularly useful for mixtures containing large and small molecules (e.g., ethylene and polyethylene) at high pressures. We describe briefly some simplified models based on PHC theory. Section 12.10 presents a model for fluids with associating molecules.

Kim *et al.* (1982) use the Carnahan-Starling expression for hard-sphere molecular repulsion [Eq. (7-238)] but for the attractive part they use an expression based on the local-composition model of Lee *et al.* (1985); the resulting equation of state has a relatively simple mathematical form. The attractive contribution is

$$z_{att} = \frac{-Z_M v^* \left[\exp\left(\frac{1}{2\tilde{T}}\right) - 1\right]}{v + v^* \left[\exp\left(\frac{1}{2\tilde{T}}\right) - 1\right]} \tag{12-78}$$

where Z_M is the maximum lattice-site coordination number. For the R = 1.5 square-well fluid potential (see Sec. 5.5), Kim *et al.* used the lattice model of Lee with Z_M = 18. For mixtures, Kim used also mixing rules based on van der Waals one-fluid theory. Figures 12-35 and 12-36 show examples of calculations using this simplified form of the PHC theory.

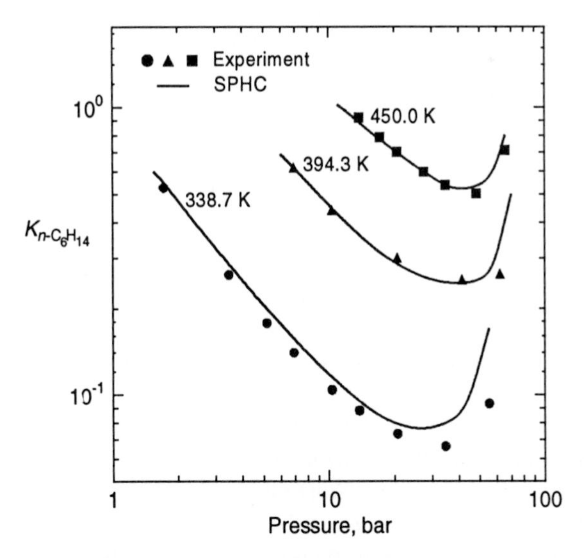

Figure 12-35 Calculated (simplified PHC theory of Kim) and experimental *K* factors for *n*-hexane in the ethane/*n*-hexane binary at 450.0, 394.3, and 338.7 K (Kim *et al.*, 1986).

Figure 12-35 compares calculated with experimental *K* factors for the ethane/*n*-hexane system at three temperatures. Predictions of the simplified PCH theory are in good agreement with experiment. The *K* factors shown in Fig. 12-35 were calculated using only pure-component parameters (T^*, v^*, and c); in this fortuitous case, no adjustable binary interaction parameters are required. However, as Fig. 12-36 shows, to describe vapor-liquid equilibria for mixtures of ethane with a heavy *n*-alkane[27] (eicosane, n-$C_{20}H_{42}$) at 350 K, it is necessary to use a binary parameter k_{12}[28] obtained from data regression of the phase envelopes at 350 K (Peters *et al.*, 1988). Using pure-component parameters only (k_{12} = 0), fair agreement with the experiment (better than 5% in pressure) is obtained. With k_{12} obtained at 350 K, the simplified PHC theory represents well the experimental data at 450 K. For comparison, Fig. 12-36 shows also calculated results from the Redlich-Kwong-Soave equation of state for the same binary at 450 K, but with k_{12} optimized at this temperature.

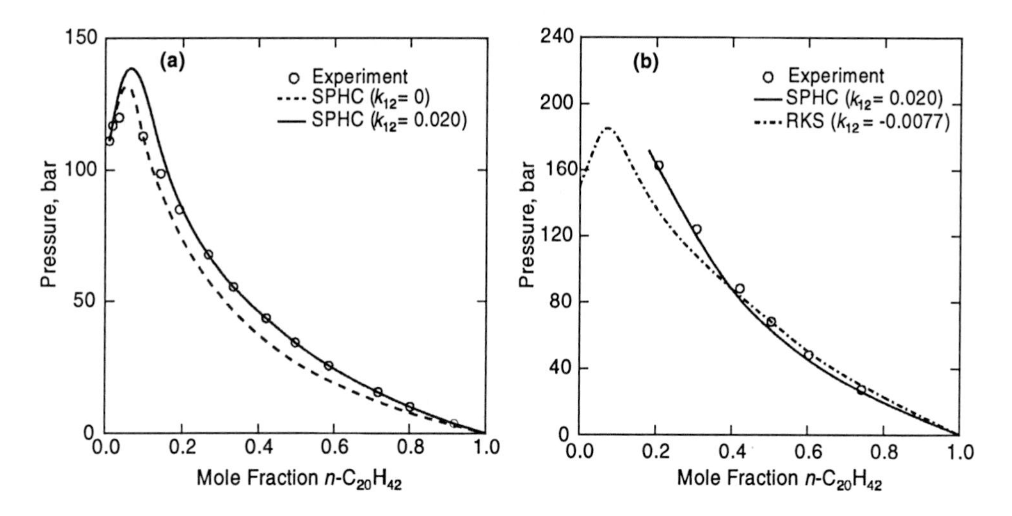

Figure 12-36 Vapor-liquid equilibria for ethane/eicosane at 350.0 K and 450.0 K (Peters *et al.*, 1988). (a) At 350.0 K: o Experiment; - - - - Simplified PHC, k_{12} = 0 ; —— Simplified PHC, k_{12} = 0.020; (b) At 450.0 K: o Experiment; —— Simplified PHC, k_{12} = 0.020; – - – - – Redlich-Kwong-Soave, k_{12} = -0.0077.

In a manner similar to that used by van Konynenburg and Scott with the van der Waals equation of state, and Deiters with the Redlich-Kwong equation (Deiters and Pegg, 1989) and with the Carnahan-Starling-Redlich-Kwong equation (Kraska and Deiters, 1992), van Pelt *et al.* (1991) used the simplified PHC equation to calculate semi-quantitative critical curves and phase behavior of binary fluid mixtures. In addition to the six types described in Fig. 12-6, van Pelt found new types of phase behavior.

[27] In these calculations, pure-component data were obtained by linearly extrapolating plots of Kim's parameters versus carbon number. Parameter Z_M has been set equal to 36.

[28] As usual, binary interaction parameter k_{12} is defined by $\varepsilon_{12} = (\varepsilon_{11}\varepsilon_{22})^{1/2}(1 - k_{12})$ where ε is an energy parameter.

Extensive application of PHC is discussed by Cotterman (1986). To illustrate, Fig. 12-37 shows calculated and observed solubilities (Henry's constants) for five solutes in low-density polyethylene and Fig. 12-38 shows calculated and observed K factors for a 12-component synthetic oil at 322 K with 70 mol % carbon dioxide. The oil contains nitrogen and hydrocarbons from methane to tetradecane.

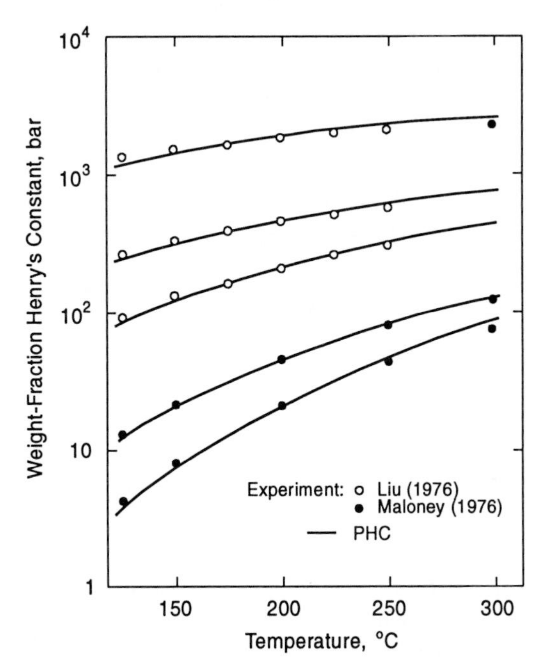

Figure 12-37 Weight-fraction Henry's constants for some solutes in low-density polyethylene. Symbols are experimental data (o Liu, 1976; • Maloney, 1976); —— PHC (Cotterman, 1986) with one binary parameter k_{12} per binary system.

Additional extensions of PHC theory are reported by Donohue and coworkers (Ikonomou and Donohue, 1986, 1988; Donohue and Vimalchand, 1988; Elliot *et al.*, 1990), who give special attention to the effect of polarity and hydrogen bonding.

The literature is rich with variations of PHC theory (or the similar chain-of-rotators equation of Chao and coworkers (Chien *et al.*, 1983). Each one has advantages and disadvantages but, essentially, all give, more or less, similar results. The key to success is to use a good binary parameter that must be fitted to experimental binary data.

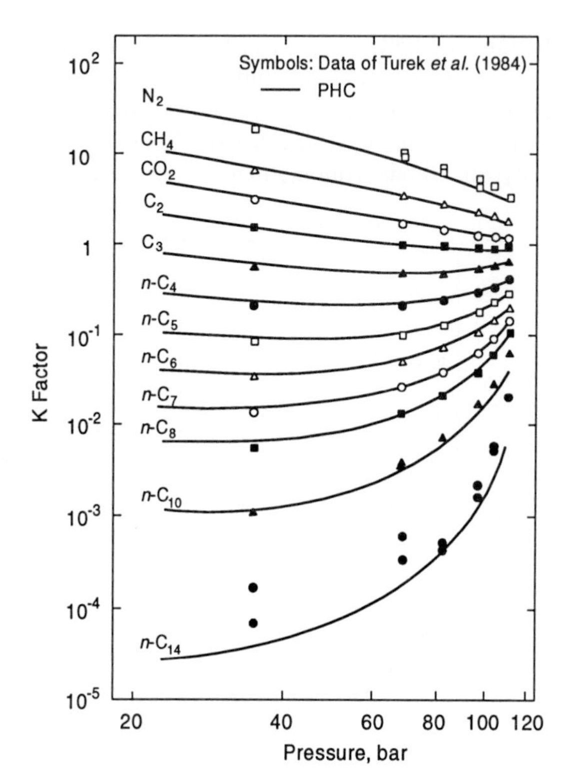

Figure 12-38 *K* factors for 12-component synthetic oil with 70 mol % carbon dioxide at 322 K. Symbols : Experiment (Turek *et al.*, 1984); —— PHC (Cotterman, 1986) with one binary parameter k_{12} per binary system.

12.10 Phase Equilibria Using the Chemical Theory

A classical method to describe strongly nonideal and associating (hydrogen-bonding) mixtures, is provided by coupling an equation of state with a chemical theory of association and solvation, as indicated in Sec. 5.10. An example of this method is the model of Gmehling based on PHC theory.

Gmehling *et al.* (1979)[29] truncated the perturbation expansion at the second-order term, reducing the original 21 universal A_{nm} constants required by Eq. (7-224) to 10 without significant loss of accuracy. They also extended the PHC theory to mixtures containing one (or more) strongly polar component by including chemical dimerization equilibria. The extension takes polar forces into account by assuming that, in addition to the nonideal properties given by the PHC equation of state, strong polar forces can

[29] See also Grenzheuser and Gmehling (1986).

form dimers. Thus, a binary mixture of A and B is considered to be a five-species mixture containing monomers A and B and, in addition, dimers A_2, B_2, and AB. The concentrations of these five species are calculated by material balances and by chemical equilibria:

$$K_{A_2} = \frac{\mathfrak{z}_{A_2}}{\mathfrak{z}_A^2} \frac{\varphi_{A_2}}{\varphi_A^2 P} \tag{12-79}$$

$$K_{B_2} = \frac{\mathfrak{z}_{B_2}}{\mathfrak{z}_B^2} \frac{\varphi_{B_2}}{\varphi_B^2 P} \tag{12-80}$$

$$K_{AB} = \frac{\mathfrak{z}_{AB}}{\mathfrak{z}_A \mathfrak{z}_B} \frac{\varphi_{AB}}{\varphi_A \varphi_B P} \tag{12-81}$$

where $\mathfrak{z}$ represents mole fraction. The equilibrium constants are functions only of temperature. For a binary mixture, fugacity coefficients φ are calculated from the (five-species) partition function. Equilibrium constants K_{A_2} and K_{B_2} can be determined from pure-component properties, but K_{AB} must be obtained from binary data. If only A is strongly polar, $K_B = K_{AB} = 0$.

Figure 12-39 shows calculated and observed equilibria for methanol/water. Gmehling reduced the number of adjustable binary parameters by making some reasonable simplifying assumptions. The results shown in Fig. 12-39 require only two binary parameters for the indicated ranges of temperature, pressure, and composition.

Gmehling's model considers only monomers and dimers; trimers (and higher aggregates) are neglected. The advantage of considering only dimers is that extension to mixtures is straightforward. As shown in the following paragraphs, it is possible to construct a consistent theory that includes trimers, tetramers, etc. but then extension to mixtures is difficult.

In the "chemical" theory of mixtures, molecules may associate or solvate to form new species; chemical equilibria are included in the equation of state. Thermodynamic properties of polar fluids and their mixtures follow from chemical equilibria leading to associated or solvated species and from physical (nonspecific) interactions between all species present in the solution. The *chemical effects*, taken into account by a molecular-based association model, are combined with the *physical effects* into one comprehensive equation of state.

To illustrate, we consider a simple but representative model presented by Heidemann and Prausnitz (1976). This model combines an equation of the van der Waals form with a continuous linear association model. Here a pure fluid is considered to be a mixture of monomers, dimers, trimers, etc. In the definition of reduced density, $\eta = n_T b/V$, the total number of moles n_T is not a constant but depends on the density

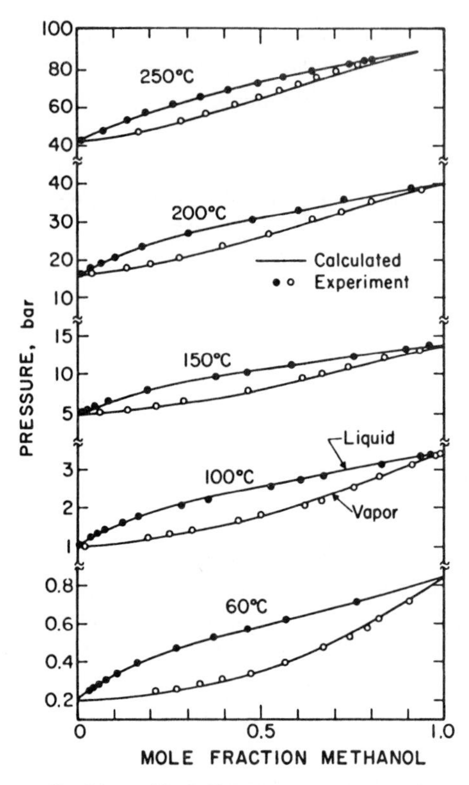

Figure 12-39 Vapor-liquid equilibria for methanol/water (Gmehling *et al.*, 1979).

and temperature. Also, van der Waals parameter b depends on the *true* composition of the fluid; in the low-density limit $b = \pi N_A \sigma^3/6$ (where all molecules are monomers) where N_A is Avogadro's constant and σ is the collision diameter.[30]

The equation of state is separated into its repulsive and attractive contributions,

$$P = P_{\text{rep}} + P_{\text{att}} \tag{12-82}$$

The repulsive term is rewritten in terms of the compressibility factor z_{rep},

$$P_{\text{rep}} = n_T z_{\text{rep}} \frac{RT}{V} \tag{12-83}$$

where V is the total volume.

The attractive-force contribution in the equation of state is rewritten in the form

[30] In the classical van der Waals equation of state, b is four times larger than the b defined in Heidemann's work.

$$P_{att} = -\frac{\hat{a}}{\hat{b}^2}\pi_{att}(\eta) \tag{12-84}$$

where π_{att} is a function only of reduced density η. Constant $\hat{a}$ depends on the composition of the mixture (monomers, dimers, trimers, etc.); in the low-density limit (all molecules are monomers), constant $\hat{a}$ is that of the conventional van der Waals-type equation.

Based on experimental evidence indicating that molecular clusters exist for polar and hydrogen-bonded molecules, Heidemann postulates a linear association model where a monomer molecule X_1 can associate to form dimers, trimers, etc., according to the chemical equilibria

$$\begin{aligned} X_1 + X_1 &\rightleftharpoons X_2 \\ X_1 + X_2 &\rightleftharpoons X_3 \\ \vdots \quad & \quad \vdots \\ X_1 + X_i &\rightleftharpoons X_{i+1} \end{aligned} \tag{12-85}$$

To apply the equation of state to mixtures, the classical mixing rules are

$$\hat{a} = \sum_i \sum_j z_i z_j (a_i a_j)^{1/2} \tag{12-86}$$

$$\hat{b} = \sum_i z_i b_i \tag{12-87}$$

where z_i ($z_i = n_i/n_T$) is the mole fraction of species i (monomer, dimer, etc.) in the mixture, and a and b refer to the monomer.

To simplify the model, Heidemann introduce also the following reasonable assumptions

$$a_i = i^2 a \tag{12-88}$$

$$b_i = ib \tag{12-89}$$

where a and b refer to the monomer.

Conservation of mass relations require that the total number of moles n_T is related to n_0, the number of moles that would exist if there were no association, by

$$n_T \sum_i i z_i = n_0 \tag{12-90}$$

Combination of Eq. (12-90) with Eqs. (12-86) to (12-89) gives

$$\hat{a} = \left(\frac{n_0}{n_T}\right)^2 a \tag{12-91}$$

and

$$\hat{b} = \left(\frac{n_0}{n_T}\right) b \tag{12-92}$$

With the results presented above, we obtain for the equation of state,

$$\boxed{P = \frac{n_T}{n_0}\frac{RT\eta}{b} z_{\text{rep}}(\eta) - \frac{a}{b^2}\pi_{\text{att}}(\eta)} \tag{12-93}$$

In Eq. (12-93), z_{rep} and π_{att} are functions only of reduced density η.

To obtain the ratio n_T/n_0[31] we define the chemical equilibrium constant (a function of temperature only),

$$K_{i+1} = \frac{\gamma_{i+1}}{\gamma_i \gamma_1}\frac{\varphi_{i+1}}{\varphi_i \varphi_1 P} = \exp\left(-\frac{\Delta h^0}{RT} + \frac{\Delta s^0}{R}\right) \tag{12-94}$$

where φ is the fugacity coefficient, obtained from the equation of state [using e.g. Eq. (3-53)], Δh^0 is the standard enthalpy of association, and Δs^0 is the standard entropy of association [see Eq. (5-111)].

Heidemann considers equilibrium constants K independent of the degree of association, i.e., $K_2 = K_3 = \ldots = K$. From material balance equations we find n_T/n_0,

$$\frac{n_T}{n_0} = \frac{2}{1 + \sqrt{1 + \frac{4}{v} RTK \exp(g)}} \tag{12-95}$$

Upon substitution in Eq. (12-93), we obtain the equation of state for associating fluids

[31] The ratio n_T/n_0 is a measure of the extent of association. In the absence of association, n_T/n_0 is unity, whereas for strongly associating species n_T/n_0 is always less than unity except at very low pressures, where this ratio goes to unity.

$$z = \frac{Pv}{RT} = \frac{2z_{\text{rep}}}{1+\sqrt{1+\frac{4}{v}RTK\exp(g)}} - \frac{a\pi_{\text{att}}}{b\eta RT} \tag{12-96}$$

with

$$g = \int_0^{\eta}\left(\frac{z_{\text{rep}}-1}{\eta}\right)d\eta \tag{12-97}$$

In Eq. (12-96), v is the volume per mole of non-associated molecules, that is, $v = V/n_0$. In the limit as $K \to 0$ (no association), Eq. (12-96) reduces to the equation for nonassociating fluids, as expected. Moreover, we can use in Eq. (12-96) any desired explicit form for z_{rep} , e.g. the Carnahan-Starling equation of state gives $g = (4\eta - 3\eta^2)/(1-\eta)^2$. For π_{att} we could use e.g. the Redlich-Kwong equation of state: $\pi_{\text{att}} = \eta^2/(1+4\eta)$.

Heidemann's continuous linear association model briefly described above provides an efficient method for describing phase equilibria of pure highly polar and associating fluids. However, useful extension to mixtures containing two or more associated fluids is not simple without severe simplifying assumptions.

Using the equation of state of Yu and Lu (1987), Heidemann's formalism was extended to mixtures containing one associated and one or more unassociated fluids by Anderko (1989; 1989a). Many other association models have been proposed, notably by Anderko (1990) and by Economou *et al.* (1991, 1992), but all of them are essentially similar to that of Heidemann.

Consider, as an example, the method proposed by Lencka and Anderko (1993) applied to the calculation of vapor-liquid equilibria for the system hydrogen fluoride/$C_2F_3Cl_3$ (Freon-113). This system is of technical interest because HF is a byproduct in the production of a common refrigerant.[32] Anderko assumes that the compressibility factor of an associated mixture is separated into a physical (van der Waals-type) and a chemical (association) contribution,

$$z = z_{\text{phys}} + z_{\text{chem}} - 1 \tag{12-98}$$

In Eq. (12-98), contribution z_{phys} is expressed by a common equation of state for the true monomeric species and z_{chem} is defined as the ratio of the total number of moles of all species to the number that would be present in the absence of association (n_0); that is, $z_{\text{chem}} = n_T/n_0$.

[32] While chlorine-containing freon refrigerants are now in decline because of environmental considerations (decomposition of ozone), production of environmentally acceptable refrigerants may also produce HF as a side product.

For z_{phys} Anderko uses the Peng-Robinson equation of state. To obtain an expression for z_{chem}, Anderko considers that, in the mixture, HF can form linear multimers formed by consecutive self-association reactions,

$$(HF)_j + HF \rightleftharpoons (HF)_{j+1} \quad \text{with} \quad j = 1,2,\ldots,\infty$$

Anderko does not assume that the chemical equilibrium constant is independent of j. Instead, he uses a distribution function $f(j)$:

$$K_{j,j+1} = f(j)K \tag{12-99}$$

where K is a constant. Distribution function $f(j)$ depends on the nature of the associating fluid. Assuming that $f(1) = 1$, then K is the dimerization constant. The equilibrium constant K_i for the i-merization reaction

$$i\text{HF} \rightleftharpoons (HF)_i \qquad i = 2, 3, \ldots, \infty$$

is related to the consecutive association constants [Eq. (12-99)] by

$$K_i = \left[\prod_{j=1}^{i-1} f(j)\right] K^{i-1} \tag{12-100}$$

On the basis of simulation data for hydrogen fluoride, Lencka and Anderko proposed that distribution function $f(j)$ should smoothly cover a small range of j values with a maximum near $j = 6$, as shown in Fig. 12-40.

Anderko obtained his distribution function by fitting the equation of state to isothermal pure HF thermophysical data; the equation above then give z_{chem}. Figure 12-41 shows calculated vapor-liquid-liquid equilibria for the system $HF/C_2F_3Cl_3$ at 383.15 K. Anderko's model represents well the experimental data for this strongly nonideal system. Using no binary parameters, Anderko's model gives a good, nearly quantitative prediction. However, agreement is improved when a small adjustable binary parameter is introduced.

In this binary case, only one of the components (HF) forms dimers, trimers, etc.; the other component ($C_2F_3Cl_3$) does not. Drastic additional simplifying assumptions are required to extend Anderko's model to a binary where both components form dimers, trimers, etc. as well as cross-dimers, cross-trimers, etc.

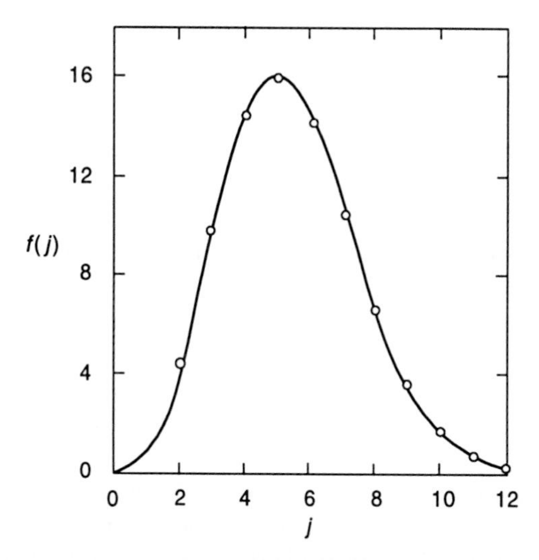

Figure 12-40 Distribution function $f(j)$ proposed by Lencka and Anderko (1993) for hydrogen fluoride. The distribution function shows a maximum near $j = 6$ indicating that HF preferentially forms hexamers.

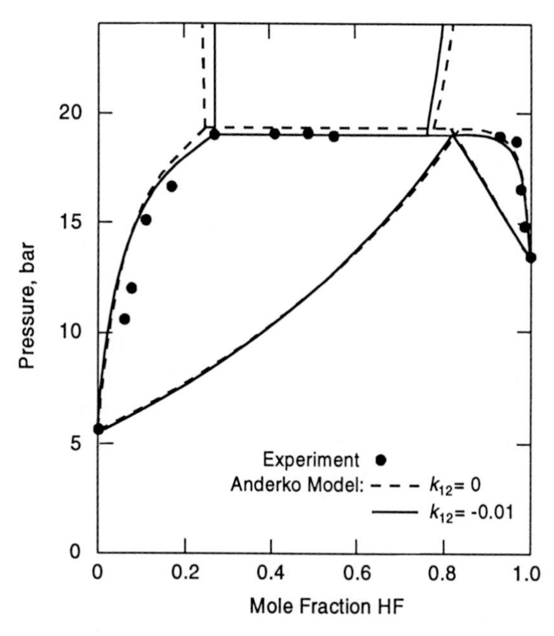

Figure 12-41 Vapor-liquid-liquid equilibria for $HF/C_2F_3Cl_3$ at 383.15 K (Lencka and Anderko, 1993). ● Experiment; —— Association model of Anderko (with $k_{12} = -0.01$); - - - - Association model of Anderko (with $k_{12} = 0$).

12.11 Summary

When binary fluid mixtures are subjected to high pressure, a large variety of phase behavior is possible. Many binary systems have been studied experimentally, giving rise to numerous types of *phase diagrams*; some of these are shown in the early sections of this chapter. When we go to ternary (or higher) fluid mixtures, the possible variety of phase behavior increases further, but for such mixtures experimental data are scarce.

Analysis of high-pressure phase diagrams shows that, at some conditions, *retrograde phenomena* can occur. Such phenomena are commonly found in petroleum and natural-gas reservoirs: at some conditions, contrary to intuition, isothermal compression produces vaporization rather than condensation. For a thermodynamic description of high-pressure phase equilibria, the key thermodynamic equation is the pressure derivative of the *chemical potential* or *fugacity*; that derivative is directly related to the partial molar volume. If the partial molar volume is positive, an increase in pressure raises the fugacity. In an isothermal two-phase system, if we increase the pressure, we affect the distribution of a dilute component i between the two phases. As pressure rises, component i is shifted toward that phase where its partial molar volume is lower. Raising pressure increases the fugacity (escaping tendency) of i; as pressure rises, component i's tendency to escape is larger in the phase where i has the larger partial molar volume.

Interpretation of high-pressure phase-equilibrium data is often not simple because (especially in binary systems) a change in pressure is usually accompanied by a simultaneous change in another intensive property. For example, in isothermal vapor-liquid equilibria for say, methane-toluene, if we raise the pressure, we not only change the densities but also the compositions of both phases. In a two-phase system, it is therefore not possible experimentally to study the effect of pressure alone while keeping the compositions of both phases constant. However, the effect of pressure can be isolated from the effects of other intensive variables through thermodynamic analysis where we consider separately the effects of pressure, composition and temperature on the chemical potential (or fugacity) through the use of partial derivatives.

For a heavy solute i dilute in a light gas near that gas' critical point, the partial molar volume of i is typically large and negative. Under these conditions, it follows that the fugacity coefficient of i is very much smaller than unity and, therefore, the solubility of i in the gas is much larger than that at conditions remote from the gas' critical point. The large negative partial molar volume of the solute provides the thermodynamic basis for *supercritical extraction* processes; such processes enable separation of low-volatile compounds, e.g., biologically active substances in plants or pollutants in soil. Solvent recovery from the extract is simple because the solubility of the solute is highly sensitive to small changes in solvent density; a small decline in pressure precipitates the dissolved solute.

For describing isothermal *liquid-liquid equilibria*, we need expressions for the excess Gibbs energy and for the excess volume, both as functions of composition. Depending on the sign of the excess volume (positive or negative), rising pressure may induce phase splitting in a liquid mixture that is homogeneous at normal pressure or, vice versa, rising pressure may homogenize a liquid mixture that is heterogeneous at normal pressure.

For high-pressure vapor-liquid calculations, we need an equation of state that can describe the properties of both phases. In typical applications, we use an *equation of state* of the van der Waals form coupled with the one-fluid assumption: the configurational properties of a fluid mixture are assumed to be the same as those of a hypothetical pure fluid whose characteristic constants (a, b) are some composition average. The procedure for averaging is given by (mostly) empirical *mixing rules*. Many mixing rules have been proposed but there is no clear conclusion to indicate that any one mixing rule is the best. To obtain good agreement with experiment, details in a reasonable mixing rule are usually less important than the number and reliability of the adjustable binary parameters that appear in the mixing rule.

When strong polarity or *chemical association* are significant (e.g., carboxylic acids, alcohols, hydrogen fluoride), it is necessary to superimpose on the "physical" equation of state the contribution of "chemical" forces. The major contribution of chemical effects on thermodynamic properties follows from a change in the number of particles; for example, for dimerization, two particles join to become one particle giving a decrease in entropy. The extent of association due to chemical equilibria depends not only on temperature and pressure but also on concentration, often strongly so. As a result, when "chemical" forces are significant, complex phase behavior is often observed.

The fundamental thermodynamic relations that govern high-pressure phase equilibria are well understood. But these relations cannot be reduced to practice without good *molecular-thermodynamic models*. Good models are scarce, especially for application to regions both near to and far from critical conditions. Progress in *molecular simulations* is likely to improve our understanding of fluids and fluid mixtures under a variety of conditions, including high pressures. For practical application, the challenge will be to translate that understanding toward establishing reasonably simple, yet sufficiently accurate models for performing reliable, engineering-oriented calculations.

References

Abraham, M. A. and A. K. Sunol, (Eds.), 1997, *Supercritical Fluids: Extraction and Pollution Prevention*. Washington: ACS.

Anderko, A., 1989, *Fluid Phase Equilibria,* 45: 39.

Anderko, A., 1989a, *Chem. Eng. Sci.,* 44: 713.

Anderko, A., 1990, *Fluid Phase Equilibria,* 61: 145.

Anderson, T. F. and J. M. Prausnitz, 1980, *Ind. Eng. Chem. Process Des. Dev.,* 19: 1.

Anderson, T. F. and J. M. Prausnitz, 1980a, *Ind. Eng. Chem. Process Des. Dev.*, 19: 9.

Angus, S., B. Armstrong, and K. M. de Reuck, (Eds.), 1976, *IUPAC - International Thermodynamic Tables of the Fluid State – Carbon Dioxide*. Oxford: Pergamon Press.

Angus, S., B. Armstrong, and K. M. de Reuck, (Eds.), 1979, *IUPAC - International Thermodynamic Tables of the Fluid State – Nitrogen*. Oxford: Pergamon Press.

Balder, J. R. and J. M. Prausnitz, 1966, *Ind. Eng. Chem. Fundam.,* 5: 449.

Barr-David, F. and B. F. Dodge, 1959, *J. Chem. Eng. Data,* 4: 107.

Basset, J. and M. Dode, 1936, C. *R. Acad. Sci. (Paris),* 20: 775.

Becker, P. J. and G. M. Schneider, 1993, *J. Chem. Thermodynamics,* 25: 795.

Bharath, R., H. Inomata, T. Adschiri, and K. Arai, 1992, *Fluid Phase Equilibria,* 81: 307.

Bijl, H., Th. W. de Loos, and R. N. Lichtenthaler, 1983, *Fluid Phase Equilibria,* 14: 157.

Brelvi, S. W. and J. P. O'Connell, 1972, *AIChE J.,* 18: 1239.

Brunner, E., 1985, *J. Chem. Thermodynamics,* 17: 985.

Chappelear, D. C. and J. C. Elgin, 1961, *J. Chem. Eng. Data,* 6: 415.

Chien, C. H., R. A. Greenkorn, K. C. Chao, 1983, *AIChE J.,* 29: 560.

Choi, E. S., M. J. Noh, and K.-P. Yoo, 1998, *J. Chem. Eng. Data,* 43: 6.

Chueh, P. L. and J. M. Prausnitz, 1967, *Ind. Eng. Chem. Fundam.,* 6: 492.

Clerke, E. A. and J. V. Sengers, 1981, *Proc. 8th Symp. Thermophys. Properties Am. Soc. Mech. Eng.*, New York.

Clusius, K. and K. Weigand, 1940, *Z. Phys. Chem.,* B46: 1.

Connolly, J. F., 1966, *J. Chem. Eng. Data,* 11: 13.

Cotterman, R. L. and J. M. Prausnitz, 1986, *AIChE J.,* 32: 1799.

Dahlmann, U. and G. M. Schneider, 1989, *J. Chem. Thermodynamics,* 21: 997.

De Loos, Th. W., L. J. de Graaf, and J. de Swaan Arons, 1996, *Fluid Phase Equilibria,* 117: 40.

De Swaan Arons, J. and T. W. de Loos, 1994, *Phase Behavior: Phenomena. Significance, and Models.* In *Models for Thermodynamic and Phase Equilibria Calculations,* (S. I. Sandler, Ed.). New York: Marcel Dekker.

Deiters, U .K. and I. L Pegg, 1989, *J. Chem. Phys.*, 90: 6632.

Dimitrelis, D. and J. M. Prausnitz, 1990, *Chem. Eng. Sci.,* 45: 1503.

Dohrn, R. and G. Brunner, 1995, *Fluid Phase Equilibria,* 106: 213.

Donohue, M.D. and P. Vimalchand, 1988, *Fluid Phase Equilibria,* 40: 185.

Economou, I. G. and M. D. Donohue, 1991, *AIChE J.,* 37: 1875.

Economou, I. G. and M. D. Donohue, 1992, *Ind. Eng. Chem. Res.,* 31: 1203.

Elgin, J. C. and J. J. Weinstock, 1959, *J. Chem. Eng. Data,* 4: 3.

Elliot, J. R., S. J. Suresh, and M. D. Donohue, 1990, *Ind. Eng. Chem. Res.,* 29: 1476.

Evelein, K. A., R. G. Moore, and R. A. Heidemann, 1976, *Ind. Eng. Chem. Proc. Des. Dev.,* 15: 423.

Francis, A. W., 1963, *Liquid-Liquid Equilibria*. New York: Wiley-Interscience.

Gmehling, J. and U. Onken, 1977, *Vapor-Liquid Equilibrium Data Compilation: DECHEMA Data Series*. Frankfurt: DECHEMA.

Gmehling, J., D. D. Liu and J. M. Prausnitz, 1979, *Chem. Eng. Sci.,* 34: 951.

Gonçalves, M., A. M. P. Vasconcelos, E. Gomes de Azevedo, H. J. Chaves das Neves, and M. Nunes da Ponte, 1991, *J. Am. Oil Chem. Soc.,* 68: 474.

Gonikberg, M. G., 1963, *Chemical Equilibria and Reaction Rates at High Pressures* (trans. from Russian). Washington: Office of Technical Services, U.S. Dept. of Commerce.

Grausø, L., Aa. Fredenslund, and J. Mollerup, 1977, *Fluid Phase Equilibria,* 1: 13.

Gregorowicz, J., Th. W. de Loos, and J. de Swaan Arons, 1993, *Fluid Phase Equilibria,* 84: 225.

Grenzheuser, P. and J. Gmehling, 1986, *Fluid Phase Equilibria,* 25: 1.

Grzanna, R. and G. M. Schneider, 1996, *Z. Phys. Chem.,* 193: 41.

Heidemann, R. A. and J. M. Prausnitz, 1976, *Proc. Nat. Acad. Sci.,* 73: 1773.

Heidemann, R. A. and A. M. Khalil, 1980, *AIChE J.,* 26: 769.

Hicks, C. P. and C. L. Young, 1975, *Chem. Rev.,* 75: 119.

Ikonomou, G. D. and M. D. Donohue, 1986, *AIChE J.,* 32: 1716.

Ikonomou, G. D. and M. D. Donohue, 1988, *Fluid Phase Equilibria,* 39: 129.

Ishikawa, T., W. K. Chung, and B. Lu, 1980, *AIChE J., 26*: 372.

Jeschke, P. and G. M. Schneider, 1982, *J. Chem. Thermodynamics,* 14: 547.

Kim, C.-H., P. Vimalchand, M. D. Donohue, and S. I. Sandler, 1986, *AIChE J.,* 32: 1726.

King., M. B. and T. R. Bott (Eds.), 1993, *Extraction of Natural Products Using Near-Critical Solvents.* London: Chapman & Hall.

Kleintjens, L. A. and R. Koningsveld, 1980, *J. Electrochem. Soc.,* 127: 2352.

Kraska, T. and U. K. Deiters, 1992, *J. Chem. Phys.,* 96: 539.

Krichevsky, I. R. and J. S. Kasarnovsky, 1935, *J. Am. Chem. Soc.,* 57: 2168.

Leach, J. W., P. S. Chappelear, and T. W. Leland, 1968, *AIChE J.,* 14: 568.

Lee, B. I. and M. G. Kesler, 1975, *AIChE J.,* 21: 510.

Lee, K.-H., M. Lombardo, and S. I. Sandler, 1985, *Fluid Phase Equilibria,* 21: 177.

Lencka, M. and A. Anderko, 1993, *AIChE J.,* 39: 533.

Liu, D. D. and J. M. Prausnitz, 1976, *Ind. Eng. Chem. Fundam.,* 15: 330.

Liu, D. D. and J. M. Prausnitz, 1980, *Ind. Eng. Chem. Process Des. Dev.,* 19: 205.

Lyckman, E. W., C. A. Eckert, and J. M. Prausnitz, 1965, *Chem. Eng. Sci.,* 20: 685.

Lyman, W. J., W. F. Reehl, and D. H. Rosenblatt, 1990, *Handbook of Chemical Property Estimation Methods.* Washington: ACS.

Maloney, D. P. and J. M. Prausnitz, 1976, *AIChE J.,* 22: 74.

McHugh, M. A. and V. J. Krukonis, 1994, *Supercritical-Fluid Extraction: Principles and Practice,* 2nd Ed. Boston: Butterworth-Heinemann.

Mehta, V. S. and G. Thodos, 1965, *J. Chem. Eng. Data,* 10: 211.

Michelsen, M. L., 1980, *Fluid Phase Equilibria,* 4: 1.

Michelsen, M. L. and R. A. Heidemann, 1981, *AIChE J.,* 27: 521.

Mollerup, J., 1980, *Fluid Phase Equilibria,* 4: 11.

Moors, J. C., R. Battino, T. R. Rettich, Y. P. Handa, and E. Wilhelm, 1982, *J. Chem. Eng. Data,* 27: 221.

Mullins, J. C. and W. T. Ziegler, 1964, *Int. Ad. Cryog. Eng.,* 10: 171.

Myers, D. B., R. A. Smith, J. Katz, and R. L. Scott, 1966, *J. Phys. Chem.,* 70: 3341.

Ochel, H., H. Becker, K. Maag, and G. M. Schneider, 1993, *J. Chem. Thermodynamics,* 25: 667.

Onnes, H. K. and W. H. Keesom, 1907, *Commun. Phys. Lab. Univ. Leiden,* Suppl. 16.

Peng, D.-Y. and D. B. Robinson, 1976, *Ind. Eng. Chem. Fundam.,* 15: 59.

Peng, D.-Y. and D. B. Robinson, 1977, *AIChE J.,* 23: 137.

Peters, C. J., J. L. de Swaan Arons, J. M. H. Levelt-Sengers, and J. S. Gallagher, 1988, *AIChE J.,* 34: 834.

Pittion-Rossillon, G., 1980, *J. Chem. Phys.,* 73: 3398.

Plöcker, U., H. Knapp, and J. M. Prausnitz, 1978, *Ind. Eng. Chem. Process Des. Dev.,* 17: 324.

Poppe, G., 1935, *Bull. Soc. Chim. Belg.,* 44: 640.

Prigogine, I., 1943, *Bull. Soc. Chim. Belg.,* 52: 115.

Prigogine, I. and R. Defay, 1954, *Chemical Thermodynamics*. London: Longmans & Green.

Prigogine, I., 1957, *The Molecular Theory of Solutions*. Amsterdam: North-Holland.

Reamer, H. H., B. H. Sage, and W. N. Lacey, 1950, *Ind. Eng. Chem.,* 42: 534.

Reid., R. C., J. M. Prausnitz, and B. E. Poling, 1987, *The Properties of Gases and Liquids,* 4th Ed. New York: McGraw-Hill.

Rizvi, S. S. (Ed.), 1994, *Supercritical Fluid Processing of Foods and Biomaterials*. London: Blackie Academic and Professional Co.

Robinson, D. B. and G. J. Besserer, 1972, *Nat. Gas Process. Assoc. Res. Report,* 7.

Roozeboom, H. W., 1918, *Die Heterogenen Gleichgewichte,* Braunschweig.

Rowlinson, J. S. and F. L. Swinton, 1982, *Liquids and Liquid Mixtures,* 3rd Ed. London: Butterworths.

Sadus, R. J., 1992, High Pressure Phase Behaviour of Multicomponent Fluid Mixtures. Amsterdam: Elsevier.

Sadus, R. J., 1994, *AIChE J.,* 40: 1376.

Sage, B. H. and W. N. Lacey, 1938, *Ind. Eng. Chem.,* 30: 1299.

Sandler, S. I., H. Orbey, and B.-I. Lee, 1994. In *Models for Thermodynamic and Phase Equilibria Calculations,* (S. I. Sandler, Ed.), Chap. 2. New York: Marcel Dekker.

Schneider, G. M., 1966, *Ber. Bunsenges. Phys. Chem.,* 70: 497.

Schneider, G. M., 1976, *Pure Appl. Chem.,* 47: 277.

Schneider, G. M., 1978, In *Chemical Thermodynamics, Specialist Periodical Report,* (M. L. McGlashan, Ed.), Vol. 2, page 105. London: Chemical Society.

Schneider, G. M., 1993, *Pure Appl. Chem.,* 65: 173.

Schouten, J. A., 1992, *Physics Reports,* 172: 33.

Scott, R. L., 1956, *J. Chem. Phys.,* 25: 193.

Selleck, F. T., L. T. Carmichael, and B. H. Sage, 1952, *Ind. Eng. Chem.,* 44: 2219.

Shibata, S. and S. I. Sandler, 1989, *J. Chem. Eng. Data,* 34: 419.

Soave, G., 1972, *Chem. Eng. Sci.,* 27: 1197.

Specovius, J., M. A. Leiva, R. L. Scott, and C. M. Knobler, 1981, *J. Phys. Chem.,* 85: 2313.

Stockfleth, R. and R. Dohrn, 1998, *Fluid Phase Equilibria,* 145: 43.

Stryjek, R. and J. H. Vera, 1986, *Can. J. Chem. Eng.*, 64: 323.

Temkin, M. I., 1959, *Russ. J. Phys. Chem.*, 33: 275.

Tester, J. W. and M. Modell, 1997, *Thermodynamics and Its Applications*, 3rd Ed. Englewood Cliffs: Prentice-Hall.

Timmermans, J., 1923, *J. Chim. Phys.*, 20: 491.

Trappeniers, N. J., J. A. Schouten, and C. A. ten Seldam, 1971, *Chem. Phys. Lett.*, 5: 541.

Trappeniers, N. J., J. A. Schouten, and C. A. ten Seldam, 1974, *Physica, 73,* 556.

Turek, E. A., R. S. Metcalfe, L. Yarborough, and R. L. Robinson, Jr., 1984, *Soc. Pet. Eng. J.*, 24: 308.

Van Konynenburg, P. H. and R. L. Scott, 1980, *Philos. Trans. R. Soc.*, 298: 495.

Van Pelt, A., C. J. Peters, and J. de Swaan Arons, 1991, *J. Chem. Phys.*, 95: 7569.

Van Wasen, V. and G. M. Schneider, 1980, *J. Phys. Chem.*, 84: 229.

Wallbruch, A. and G. M. Schneider, 1995, *J. Chem. Thermodynamics*, 27: 377. See also Errata, 1997, *J. Chem. Thermodynamics*, 29: 929.

Winnick, J., 1963, *Dissertation*, University of Oklahoma.

Winnick, J. and J. E. Powers, 1966, *AIChE J.*, 12: 460, 466.

Wong, D. S and S. I. Sandler, 1992, *AIChE J.*, 38: 671.

Wong, D. S, H. Orbey, and S. I. Sandler, 1992a, *Ind. Eng. Chem. Res.*, 31: 2033.

Yoo, K.-P., Shin, H. Y., M. J. Noh, and S. S. You, 1997, *Korean J. Chem. Eng.*, 14: 341.

Yu, J. M. and B. C.-Y. Lu, 1987, *Fluid Phase Equilibria*, 34: 1.

Zandbergen, P., H. F. P. Knaap, and J. J. M. Beenakker, 1967, *Physica*, 33: 379.

Zeck, S. and H. Knapp, 1986, *Fluid Phase Equilibria*, 26: 37.

Problems

1. A typical *P-T* projection for water/*n*-alkane systems is presented below.

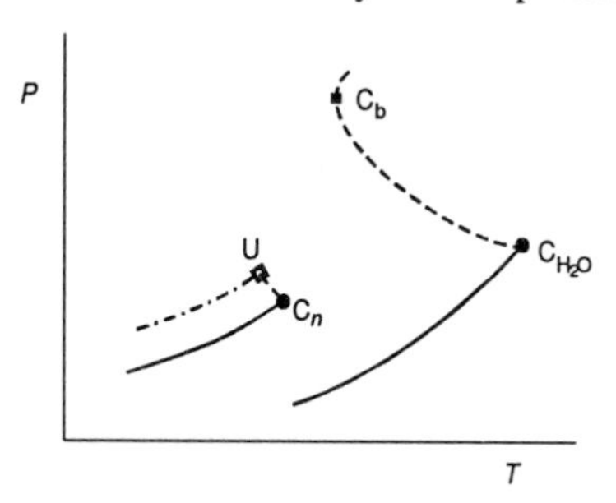

In the above figure, C_n and C_{H_2O} are, respectively, the critical points of the *n*-alkane and of the water; U is the critical-end point of liquid-liquid-gas equilibrium; C_b is the binary homogeneous point (i.e. the point where the two-phase region separates into two parts); solid lines indicate pure-component saturation vapor pressures; dashed lines indicate bi-

nary mixture critical lines; and the dashed-dot line indicates three-phase liquid-liquid-gas equilibrium. Sketch *P-x* diagrams for the temperatures:
(i) $T < T_U$; (ii) $T = T_U$; (iii) $T_U < T < T_{C_n}$; (iv) $T_{C_n} < T < T_{C_b}$; (v) $T = T_{C_b}$; (vi) $T_{C_b} < T < T_{CH_2O}$; (vii) $T > T_{CH_2O}$.

2. Consider the following type II phase diagram:

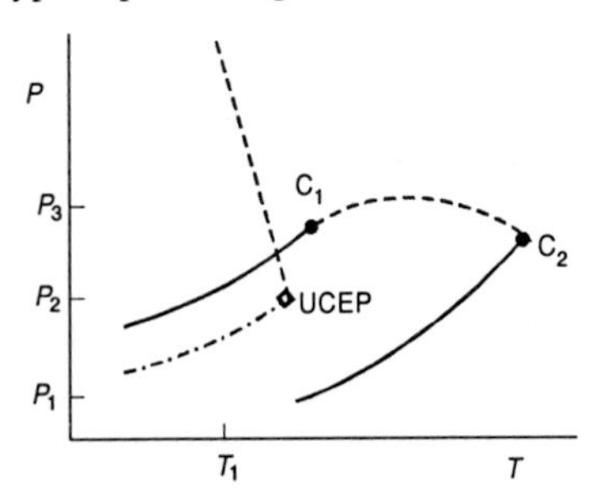

(a) Draw the *P-x* projection for temperature T_1.
(b) Draw *T-x* projections for pressures $P_1 < P_{UCEP}$, $P_2 = P_{UCEP}$, and $P_3 > P_{UCEP}$.

3. An alcohol is distributed between two immiscible fluids, hexane and dimethyl sulfoxide. Calculate the distribution coefficient of the alcohol between the two liquid phases (when the alcohol concentration is very small) at 30°C and 100 bar. The following binary data are available, all at 0°C and 1 bar pressure:

Alcohol/hexane	Alcohol/dimethyl sulfoxide
$g^E = 2400x'_A x'_H$, J mol^{-1}	$g^E = 320x''_A x''_D$ J mol^{-1}
$h^E = 4800x'_A x'_H$, J mol^{-1}	$h^E = 600x''_A x''_D$ J mol^{-1}
$v^E = 16x'_A x'_H$, cm^3 mol^{-1}	$v^E = -10x''_A x''_D$ cm^3 mol^{-1}

In your notation, use ' for the hexane phase and " for the dimethyl sulfoxide phase. State all assumptions made. Distribution coefficient *K* is defined as

$$K = \lim_{\substack{x'_A \to 0 \\ x''_A \to 0}} \frac{x'_A}{x''_A}$$

where subscript A refers to alcohol.

4. At the normal melting point (5.5°C) the density of liquid benzene is 0.891 g cm^{-3} and that of solid benzene is 1.010 g cm^{-3}. Find the melting temperature of benzene under a pressure of 200 bar.
The following vapor-pressure equations are available:

$$\ln P(\text{mm Hg}) = A - \frac{B}{T}$$

	A	B	
Solid	9.846	2310	$T > 243$ K
Liquid	7.9622	1785	$T < 315$ K

5. Consider an equimolar liquid mixture of components A and B at 300 K with a molar volume of 116.6 cm^3 mol^{-1}. Using the Redlich-Kwong equation of state and the data given below, calculate the total pressure.
Henry's constant for A infinitely dilute in B at 300 K is 7.01 bar. The molar volume of pure liquid B at 300 K is 100.3 cm^3 mol^{-1}. Constants for the Redlich-Kwong equation are:

	$a\times10^{-8}$ (bar cm^6 K$^{1/2}$ mol^{-2})	b (cm^3 mol^{-1})
A	4.18	100.2
B	4.53	82.8

6. Using the Redlich-Kwong-Soave equation of state, calculate K factors for ethane and benzene in a binary mixture at 180°C with a liquid composition $x_{C_2H_6} = 0.263$.
Using this equation of state, fugacity coefficients were calculated at 180°C for each component in the vapor phase and in the liquid phase. These fugacity coefficients are expressed in the form $\varphi_i(P, \mathfrak{z}_i) = c^{(0)} + c^{(1)}P + c^{(2)}\mathfrak{z}_i$ (where P is in atm and $\mathfrak{z}_i$ is either x or y). For pressures to 75 atm, the constants are:

	Vapor			Liquid		
	$c^{(0)}$	$c^{(1)}\times10^4$	$c^{(2)}$	$c^{(0)}$	$c^{(1)}\times10^2$	$c^{(2)}$
C_2H_6	1.2545	-2.4580	-0.40910	5.3538	-3.1730	-3.443
C_6H_6	0.74265	-70.069	0.50456	0.26926	-0.25430	0.08750

The experimental values (W. B. Kay and T. D. Nevens, 1952, *Chem. Eng. Prog. Symp. Ser.*, 48: 108) are $K_{C_2H_6} = 2.73$ and $K_{C_6H_6} = 0.41$.

7. For the binary mixture of components 1 and 2, g^E at 0°C and 1 atm is expressed by $g^E/RT = 1.877x_1x_2$. For the same binary, the excess volumes were calculated as a function of pressure and composition; they are given by the equation (with pressure in atm),

$$v^E\ (\text{cm}^3\ \text{mol}^{-1}) = x_1x_2(4.026 - 0.233\ln P) \qquad \text{for } 1 < P < 4000 \text{ atm.}$$

This mixture is totally miscible at 0°C and 1 atm. At 0°C, find the lowest pressure at which phase separation will occur. When the same mixture is under 1500 atm pressure, what are the compositions of the two liquid phases in equilibrium?

8. For the equimolar mixture 2-propanol/*n*-heptane at normal pressure and 333 K, Van Ness *et al.* (1967, *J. Chem. Eng. Data,* 12: 217) report that h^E = 1445 J mol^{-1}. Using the volumetric data from Lichtenthaler *et al.* [Institute of Physical Chemistry, University of Heidelberg, Germany] quoted below, calculate h^E for the equimolar mixture at 333 K and 360 bar. For the equimolar mixture v^E data are given in the form v^E (cm^3 mol^{-1}) = 0.25A, where A depends on temperature and pressure:

	T(K)			
P (bar)	298	323	348	373
1	2.347	3.619	5.483	8.389
100	2 132	3.239	4.779	6.981
250	1.882	2.818	4.057	5.703
500	1.594	2.353	3.316	4.527

9. Consider a binary gaseous mixture of carbon dioxide with a mole fraction of water equal to 0.2 at 150 atm. Find the temperature to which this mixture can be isobarically cooled without condensation. Assume that at this pressure the solubility of CO_2 in liquid water has a negligible effect.
In the calculations, use a modified form of the Redlich-Kwong equation of state where parameter a is given by $a = a^{(0)} + a^{(1)}$ where $a^{(1)}$ depends on temperature. Use $a^{(0)}(H_2O)$ = 35×10^6 and $a^{(0)}(CO_2)$ = 46×10^6 atm (cm^3 mol^{-1}) K$^{1/2}$. Constants for finding $a^{(1)}$ are given in the table below. Use the normal mixing rules, but for the cross term a_{ij} include the equilibrium constant K (for the formation of a water-carbon dioxide complex) as follows: $a_{ij} = [a_i^{(0)}a_i^{(1)}]^{1/2} + 0.5R^2T^{2.5}K$. Constants for the temperature dependence of K (in atm^{-1}) are also given in the table. For water vapor, b = 14.6 cm^3 mol^{-1} and for carbon dioxide, b = 29.7 cm^3 mol^{-1}. Data for pure saturated water can be obtained from the steam tables.
The temperature dependence of function X is given by

$$X = c_0 + c_1/T + c_2/T^2 + c_3/T^3$$

where T is in kelvin.

X	ln K	a_{H_2O}×10$^{-6§}$	a_{CO_2}×10$^{-6§}$
c_0	-11.071	93.415	-7.922
c_1×10^{-3}	5.953	-19.9798	50.4354
c_2×10^{-5}	-27.46	147.3169	-97.7329
c_3×10^{-6}	464.6	7471.12	519.402

§ To find $a^{(1)}$, subtract $a^{(0)}$ from a.

10. For a solute in a supercritical solvent, the solubility minimum and maximum with pressure can be related to the partial molar volume of the solute in the fluid phase.

(a) Show that these solubility extrema can be predicted from

$$\left(\frac{\partial \ln y_2}{\partial P}\right)_T = \frac{\dfrac{v_2^{s} - \bar{v}_2^f}{RT}}{1+\left(\dfrac{\partial \ln \varphi_2}{\partial \ln y_2}\right)_{T,P}}$$

where subscript 2 refers to the solute; v_2^{s} and $\bar{v}_2^f$ are, respectively, the molar volume of the pure solid solute and the partial molar volume of the solute in the supercritical fluid phase; y_2 is the fluid-phase mole fraction; and φ_2 is the fluid-phase fugacity coefficient.

(b) As shown by Kurnik and Reid (1981, *AIChE J.*, 27: 861), for naphthalene dissolved in ethylene, $(\partial \ln \varphi_2 / \partial \ln y_2)_{T,P}$ is always more positive than -0.4 for all pressures to 4 kbar. Using the Redlich-Kwong equation of state and Eq. (12-41), calculate the pressures for solubility minimum and maximum at 318 K. Compare with experimental results.

For naphthalene at 318 K the vapor pressure and the solid density are 0.54 mmHg and 1.144 g cm^{-3}. Constants for Redlich-Kwong equation are: Ω_a = 0.4323, Ω_b = 0.0876 for ethylene, and Ω_a = 0.4497, Ω_b = 0.0915 for naphthalene. The binary parameter k_{12} is equal to -0.0182. Molar volumes of ethylene at 318 K and as a function of pressure are shown in the following table:

P (bar)	$v_{C_2H_4}$ (cm^3 mol^{-1})	P (bar)	$v_{C_2H_4}$ (cm^3 mol^{-1})	P (bar)	$v_{C_2H_4}$ (cm^3 mol^{-1})
10	2520.5	90	137.06	450	63.283
20	1194.3	100	116.35	500	61.994
30	749.27	150	85.441	550	60.878
40	524.10	200	76.817	600	59.897
50	386.41	250	72.135	650	59.022
60	291.65	300	68.992	700	58.234
70	222.56	350	66.654	750	57.518
80	171.72	400	64.805		

APPENDIX A

Uniformity of Intensive Potentials as a Criterion of Phase Equilibrium

As discussed in Sec. 2.3, we use the function U to show that the temperature and pressure and the chemical potential of each species must be uniform throughout a closed, heterogeneous system at equilibrium internally with respect to heat transfer, boundary displacement, and mass transfer across phase boundaries. Since we identify equilibrium processes (variations) with reversible processes, the criterion for equilibrium in a closed system is that U is a minimum, and that any variation in U at constant total entropy and total volume vanishes; i.e.,

$$dU_{S,V} = 0 \tag{A-1}$$

An expression for the total differential dU can be written by summing over all the phases, the extension of Eq. (2-20) to a multiphase system:

$$dU = \sum_\alpha T^{(\alpha)} dS^{(\alpha)} - \sum_\alpha P^{(\alpha)} dV^{(\alpha)} + \sum_\alpha \sum_i \mu_i^{(\alpha)} dn_i^{(\alpha)} \tag{A-2}$$

where α is a phase index, taking values 1 to π, and i is a component index, taking values 1 to m.

On expansion, Eq. (A-2) becomes

$$\begin{aligned} dU = {} & T^{(1)} dS^{(1)} - P^{(1)} dV^{(1)} + \mu_1^{(1)} dn_1^{(1)} + \cdots + \mu_m^{(1)} dn_m^{(1)} \\ & + T^{(2)} dS^{(2)} - P^{(2)} dV^{(2)} + \mu_1^{(2)} dn_1^{(2)} + \cdots + \mu_m^{(2)} dn_m^{(2)} \\ & \quad \vdots \qquad\qquad \vdots \qquad\qquad \vdots \qquad\qquad \vdots \\ & + T^{(n)} dS^{(n)} - P^{(n)} dV^{(n)} + \mu_1^{(n)} dn_1^{(n)} + \cdots + \mu_m^{(n)} dn_m^{(n)} \end{aligned} \tag{A-3}$$

The individual variations $dS^{(1)}$, etc., are subject to the constraints of constant *total* entropy, constant *total* volume, and constant *total* moles of each species (chemical reaction excluded). These may be written as:

$$dS = dS^{(1)} + \cdots + dS^{(n)} = 0 \tag{A-4}$$

$$dV = dV^{(1)} + \cdots + dV^{(n)} = 0 \tag{A-5}$$

$$\sum_\alpha dn_i^{(\alpha)} = dn_i^{(\alpha)} + \cdots + dn_i^{(n)} = 0, \quad i = 1, \ldots, m \tag{A-6}$$

There are thus $\pi(m + 2)$ independent variables in Eq. (A-3), and there are $m + 2$ constraints. The expression for dU may be written in terms of $m + 2$ fewer independent variables by using the constraining equations to eliminate, for example, $dS^{(1)}$, $dV^{(1)}$, and the m $dn_i^{(1)}$. The result is an expression for dU in terms of $(\pi - 1)(m + 2)$ truly independent variables; i.e., all the variations expressed as $dS^{(\alpha)}$, etc., are then truly independent, because the constraints have been used to eliminate certain variables. The resulting expression, if we eliminate $dS^{(1)}$, $dV^{(1)}$, and all $dn_i^{(1)}$, as indicated above, is

$$\begin{aligned} dU = {} & (T^{(2)} - T^{(1)}) dS^{(2)} - (P^{(2)} - P^{(1)}) dV^{(2)} + (\mu_1^{(2)} - \mu_1^{(1)}) dn_1^{(2)} \\ & + \cdots + (\mu_m^{(2)} - \mu_m^{(1)}) dn_m^{(2)} \\ & + (T^{(3)} - T^{(1)}) dS^{(3)} - (P^{(3)} - P^{(1)}) dV^{(3)} + (\mu_1^{(3)} - \mu_1^{(1)}) dn_1^{(3)} \\ & + \cdots + (\mu_m^{(3)} - \mu_m^{(1)}) dn_m^{(3)} \end{aligned} \tag{A-7}$$

All variations $dS^{(2)}$, $dV^{(2)}$, $dn_1^{(2)}$, $dn_2^{(2)}$, etc., are truly independent. Therefore, at equilibrium in the closed system where $dU = 0$, it follows that[1]

$$\frac{\partial U}{\partial S^{(2)}} = 0, \quad \frac{\partial U}{\partial V^{(2)}} = 0, \quad \frac{\partial U}{\partial n_1^{(2)}} = 0, \quad \frac{\partial U}{\partial n_2^{(2)}} = 0, \quad \text{etc.} \tag{A-8}$$

Hence $T^{(2)} - T^{(1)} = 0$, or

$$T^{(2)} = T^{(1)}, \quad T^{(3)} = T^{(1)}, \quad \text{etc.} \tag{A-9}$$

Similarly,

$$P^{(2)} = P^{(1)}, \quad P^{(3)} = P^{(1)}, \quad \text{etc.} \tag{A-10}$$

and

$$\mu_1^{(2)} = \mu_1^{(1)}, \quad \mu_1^{(3)} = \mu_1^{(1)}, \quad \mu_2^{(2)} = \mu_2^{(1)}, \quad \mu_2^{(3)} = \mu_2^{(1)}, \quad \text{etc.} \tag{A-11}$$

Equations (A-9), (A-10), and (A-11) tell us that at internal equilibrium with respect to the three processes (heat transfer, boundary displacement, and mass transfer), temperature, pressure, and chemical potential of each component are uniform throughout the entire heterogeneous, closed system. This uniformity is expressed by Eqs. (2-25), (2-26), and (2-27).

Although chemical reactions have been excluded from consideration in this section, it can be shown that Eq. (2-27) is not altered by the presence of such reactions. For any component i at equilibrium, the chemical potential of i is the same in all phases, regardless of whether or not component i can participate in a chemical reaction in any (or all) of these phases. This is true provided only that all such chemical reactions are also at equilibrium.

However, the existence of chemical reactions does affect the phase rule given by Eq. (2-32). In that equation, m is the number of distinct chemical components only in the absence of chemical reactions. If chemical reactions are considered, then m is the number of *independent* components, i.e., the number of chemically distinct components minus the number of chemical equilibria interrelating these components.

[1] F. B. Hildebrand, 1976, *Methods of Applied Mathematics,* 2nd Ed., (Englewood Cliffs: Prentice-Hall).

APPENDIX B

A Brief Introduction to Statistical Thermodynamics

Statistical mechanics describes the behavior of macroscopic systems in terms of microscopic properties, i.e., those of particles such as atoms, molecules, ions, etc. That part of statistical mechanics that deals with equilibrium states is called *statistical thermodynamics*; this appendix gives a short introduction whose primary purpose is to provide the working equations of statistical thermodynamics. More detailed discussions are available in many textbooks. Some of these are listed at the end of this appendix. The summary given here is similar to the discussions found in Reed and Gubbins (1973) and Everdell (1975).

Thermodynamic States and Quantum States of a System

Thermodynamic specification of a macroscopic system (e.g., the statement that a system has fixed values of energy, volume, and composition) provides only a partial, incomplete description from a molecular point of view. For example, in a pure crystal, at fixed temperature and volume, many distinguishable arrangements of the atoms on the lattice may correspond to the same thermodynamic state.

According to wave mechanics, the most complete description that is possible about a system is a statement of its wave function, the quantity ψ that appears in Schrodinger's equation. When ψ is known as a function of the coordinates of the elementary particles, we have a specification of the *quantum state* of the system. For a macroscopic system ($\approx 10^{24}$ electrons and nuclei), many of these quantum states, indeed astronomically many, may all be compatible with the same total energy, volume, and composition.

When measuring a macroscopic property X (e.g., pressure, density, etc.), the value obtained results from the chaotic motions and collisions of a large number of molecules; when viewed over a short time scale (e.g., 10^{-8} s), property X is a fluctuating quantity. In practice, however, the time required for a macroscopic measurement is usually so much longer than 10^{-8} s that fluctuations are not observed. In other words, the macroscopic properties are *time averages* over the very large number of possible quantum states that a system may assume, even though each quantum state is compatible with the macroscopically observed values.

The object of statistical thermodynamics is to calculate these time averages as a function of molecular properties.

Ensembles and Basic Postulates

To calculate time averages over all possible quantum states, some postulates are necessary. To give an exact formulation of the basic postulates, it is useful to define an *ensemble,* which is a large number of imagined systems. In the ensemble, each system has the same macroscopic properties as those chosen to describe the thermodynamic state of some real system in which we are interested. Although the single systems of the ensemble all have the same macroscopic properties, they may have different quantum states.

If, for example, the total energy of a real system is E, the volume is V, and the number of molecules is N, then every system in the ensemble also has energy E, volume V, and N molecules. Or, for another example, suppose that a real system, having N molecules in volume V with heat-conducting walls, is immersed in a large heat bath. In that event, each system of the ensemble, containing N molecules in volume V with heat-conducting walls, is also immersed in the same large heat bath.

These two examples provide the most frequently encountered types of systems in chemical thermodynamics. The first one is an *isolated system* (constant N, V, and E) and the second is a *closed isothermal system* (constant N, V, and T). The corresponding ensembles are called the *microcanonical* and *canonical* ensemble, respectively.

Having briefly explained what we mean by ensemble, it is now possible to formulate the first postulate of statistical mechanics:

> The time average of a dynamic[1] property of a real system is equal to the ensemble average of that property.

To calculate the ensemble average, it is necessary to know the probabilities of the different quantum states of the systems of the ensemble. These probabilities are given by the second postulate of statistical mechanics:

> All accessible and distinguishable quantum states of a closed system of fixed energy (microcanonical ensemble) are equally probable.

These postulates are expressed by the following equations:

$$X = \sum_i p_i X_i \tag{B-1}$$

and

$$p_1 = p_2 = p_3 = \cdots = p_i = \cdots \tag{B-2}$$

where X is the measured macroscopic dynamic property of the real system and X_i is the value of this property in that system of the ensemble that is in quantum state i. Probability p_i is the probability of quantum state i of the systems of the ensemble, normalized such that $\sum_i p_i = 1$. The notation $\sum_i$ indicates summation over all possible quantum states.

To facilitate understanding of the terms "ensemble" and "ensemble average," consider the following example. In a box (which is the real system), there are six spheres of equal size; one is white, two are red, and three are blue. Without looking into the box, we take out one sphere, note its color, and put it back again. If we do that often enough, e.g., 1000 times, we find that the total numbers of white, red, and blue spheres taken out are in the ratio 1: 2: 3, respectively. Now let us imagine that we have 1000 of these boxes, each containing six spheres as described above; this set of 1000 boxes is the ensemble. We take one sphere out of each box and note the color of the sphere. Again, the ratio of total numbers of white, red, and blue spheres is 1: 2: 3.

[1] A *dynamic* property (e.g. pressure) is one that fluctuates in time, in contrast to a static property (e.g. mass), that is constant in time.

This example illustrates that the result of an experiment often repeated with one box is identical to that obtained with only one experiment with an ensemble of many boxes (systems). In other words, we believe that the time average is equal to the ensemble average. This belief is called the *ergodic hypothesis.*

The Canonical Ensemble

The *canonical ensemble* corresponds to a large number of closed systems, each of fixed volume and fixed number of molecules, which are immersed in a large heat bath. To calculate the ensemble average, we must know the distribution of quantum states, i.e., the probability that any one system of the canonical ensemble is in a particular quantum state. To perform the calculation, we visualize that each of the K systems of the canonical ensemble is a cell of volume V that contains N molecules. All cells are in thermal contact but the ensemble itself is thermally isolated, as illustrated in Fig. B-1.

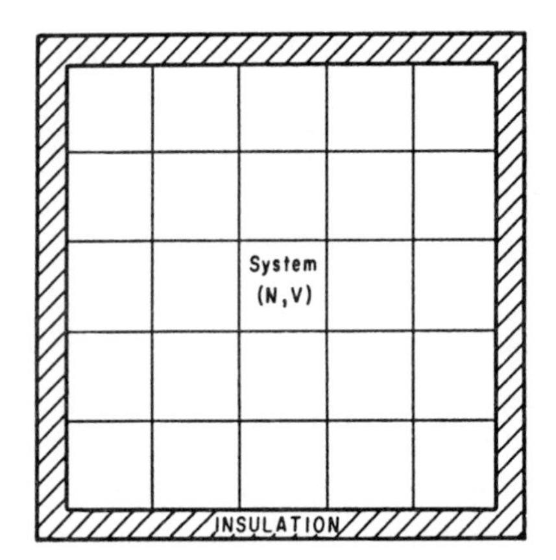

Figure B-1 Canonical ensemble. There are K systems. Each one is in thermal contact with the $K-1$ other systems permitting exchange of energy.

The canonical ensemble is an isolated system of volume KV with KN molecules and a total energy E_t. The K systems may be in different energy states and can exchange energy with each other. Each system is in contact with a large heat bath formed by the $(K-1)$ other systems.

If $n_1, n_2, \ldots, n_i, \ldots$ systems are in quantum states $1, 2, \ldots, i, \ldots$ with energy eigenvalues $E_1, E_2, \ldots, E_i, \ldots$, respectively, then the values $n_1, n_2, \ldots, n_i, \ldots$ determine the distribution of quantum states in the canonical ensemble. There are many distributions that satisfy the relations:

$$\sum_i n_i = K \tag{B-3}$$

$$\sum_i n_i E_i = E_t \tag{B-4}$$

Let n stand for a *particular* distribution of quantum states having n_1 systems in quantum state 1, n_2 systems in quantum state 2, etc. However, a large number of states $\Omega(n)$ of the canonical ensemble is compatible with this particular distribution. $\Omega(n)$ is the number that we have to calculate.

We can reformulate the problem in combinatorial terms; we want to find the number of possible, different arrangements for a total of n elements, from which n_1, n_2, ... elements are not distinguishable. The result is (see App. B-1)

$$\Omega(n) = \frac{(n_1 + n_2 + n_3 + \ldots)!}{n_1!n_2!n_3!\ldots} = \frac{K!}{\prod_i n_i!} \tag{B-5}$$

The *probability* that for a given distribution n, a particular system is in quantum state i, is given by

$$(p_i)_n = \frac{n_i(n)}{K} \tag{B-6}$$

Index n indicates that Eq. (B-6) holds for a given distribution n. However, a large number of distributions is compatible with Eqs. (B-3) and (B-4) and therefore the probability given by Eq. (B-6) is not sufficient. We want a probability averaged over all possible distributions. As the total ensemble is an isolated system of fixed energy, all accessible and distinguishable quantum states of the system are equally probable. It follows that the statistical weight of a particular distribution n *is* proportional to $\Omega(n)$. The *averaged probability* $<p_i>$ *is* then given by

$$< p_i > = \frac{\sum_n (p_i)_n \Omega(n)}{\sum_n \Omega(n)} = \frac{1}{K} \frac{\sum_n n_i(n)\Omega(n)}{\sum_n \Omega(n)} \tag{B-7}$$

where K *is* the total number of systems in the canonical ensemble; $n_i(n)$ is the number of systems of distribution n that are in quantum state i with energy eigenvalue E_i; $\Omega(n)$ is the number of states of the canonical ensemble of distribution n; and Σ_n indicates summation over all distributions that are compatible with the boundary conditions Eqs. (B-3) and (B-4).

In principle, Eq. (B-7) allows calculation of any dynamic property of a closed system with fixed volume V and fixed number of molecules N. For practical purposes, however, we want to avoid carrying out the summations. Fortunately, as the number of systems in the ensemble becomes very large ($K \to \infty$), the maximum-term method (see App. B-2) can be used and the summation can be replaced by the *most probable distribution*, i.e., that distribution in which the number of quantum states in the canonical ensemble has its maximum value. The average probability in Eq. (B-7) can be replaced by its *most probable* value p_i^*:

$$p_i^* = \frac{1}{K} \frac{\Omega(n^*) n_i(n^*)}{\Omega(n^*)} = \frac{n_i(n^*)}{K} \tag{B-8}$$

where $n_i(n^*)$ is the number of systems of distribution n^* that are in quantum state i with energy eigenvalue E_i and where $\Omega(n^*)$ is the number of states of the canonical ensemble in the most probable distribution.

To calculate $\Omega(n^*)$, the maximum value of $\Omega(n)$, Eq. (B-5) is written as

$$\ln \Omega(n) = \ln(K!) - \sum_i \ln(n_i!) \tag{B-9}$$

For the maximum condition [note that ln ($K!$) = constant],

$$\delta \ln \Omega(n^*) = 0 = \sum_i \delta \ln(n_i!) \tag{B-10}$$

Using Stirling's formula (see App. B.3), we obtain

$$\sum_i \delta \ln(n_i!) = \sum_i \ln n_i \delta n_i = 0 \tag{B-11}$$

As K and E_t are constants, Eqs. (B-3) and (B-4) can be written as

$$\delta K = \sum_i \delta n_i = 0 \tag{B-12}$$

$$\delta E_t = \sum_i E_i \delta n_i = 0 \tag{B-13}$$

We now apply Lagrange's[2] method of *undetermined multipliers*. Multiplying Eq. (B-12) by α and Eq. (B-13) by β, and adding both to Eq. (B-11), gives

[2] Lagrange's method is convenient for evaluating the extrema of a function of several variables where additional relations (restraints) are given between these variables.

$$\sum_i \alpha\delta n_i + \sum_i \beta E_i \delta n_i + \sum_i \ln n_i \delta n_i = \sum_i (\alpha + \beta E_i + \ln n_i)\delta n_i = 0 \qquad \text{(B-14)}$$

As the variations of δn_i are independent, Eq. (B-14) is satisfied for any i only when

$$\alpha + \beta E_i + \ln n_i = 0 \qquad \text{(B-15)}$$

or

$$n_i = e^{-\alpha}\, e^{-\beta E_i} \qquad \text{(B-16)}$$

The multiplier α can be eliminated using Eq. (B-3). It then follows that

$$n_i = K \frac{e^{-\beta E_i}}{\sum_i e^{-\beta E_i}} \qquad \text{(B-17)}$$

and, with Eq. (B-8),

$$p_i^* = \frac{e^{-\beta E_i}}{\sum_i e^{-\beta E_i}} \qquad \text{(B-18)}$$

where p_i^* is the probability that a given system of the canonical ensemble is in quantum state i with the energy eigenvalue E_i. Summation $\sum_i$ is over all possible quantum states.

Statistical Analogues of Thermodynamic Properties in the Canonical Ensemble. Using the first postulate of statistical thermodynamics [Eq. (B-1)], any dynamic property of a closed isothermal system can be calculated with Eq. (B-18). For example, the energy E, which in classical thermodynamics is the internal energy U, is given by

$$U = E = \sum_i p_i^* E_i = \frac{\sum_i E_i e^{-\beta E_i}}{\sum_i e^{-\beta E_i}} \qquad \text{(B-19)}$$

The summation term in the denominator is called the *canonical partition function*, Q:

$$\boxed{Q = \sum_i e^{-\beta E_i(V,N)}} \qquad \text{(B-20)}$$

$E_i(V, N)$ is the energy of that system of the ensemble that is in quantum state i; the term in parentheses (V, N) indicates the canonical ensemble. Partition function Q depends on β, V, and N. Partial differentiation of Eq. (B-20) shows that Eq. (B-19) can be written in the form

$$U = E = -\left(\frac{\partial \ln Q}{\partial \beta}\right)_{V,N} \tag{B-21}$$

We now identify the physical significance of coefficient β, introduced formally in Eq. (B-14); β is an intensive property whose value must be the same for all systems of the canonical ensemble because all of these systems are in thermal equilibrium with the same large heat bath. The only property that possesses this characteristic is the temperature. However, the dimensions of β are those of a reciprocal energy, different from those of the thermodynamic temperature. To establish the link between thermodynamic temperature T and statistical-mechanical parameter β we use a simple proportionality

$$T = \frac{1}{k\beta} \tag{B-22}$$

where k is a universal constant known as *Boltzmann's constant*; if T is expressed in kelvins, $k = 1.38066\times10^{-23}$ J K^{-1}.

Other statistical analogues of thermodynamic functions are obtained by mathematical operations. First, differentiation of Eq. (B-19) yields

$$dU = dE = \sum_i E_i dp_i^* + \sum_i p_i^* dE_i \tag{B-23}$$

For a constant number of molecules in the system, dE_i is given by

$$dE_i = \left(\frac{\partial E_i}{\partial V}\right)_N dV \tag{B-24}$$

Further, it can be shown that

$$\sum_i E_i dp_i^* = \beta^{-1} d(-\sum_i p_i^* \ln p_i^*) = Td(-k\sum_i p_i^* \ln p_i^*) \tag{B-25}$$

Substitution of Eqs. (B-24) and (B-25) into Eq. (B-23) gives

$$dU = Td(-k\sum_i p_i^* \ln p_i^*) + \sum_i p_i^* \left(\frac{\partial E_i}{\partial V}\right)_N dV \tag{B-26}$$

From classical thermodynamics,

$$dU = TdS - PdV \tag{B-27}$$

Comparing Eqs. (B-26) and (B-27), the statistical analogue for the entropy S is

$$S = -k\sum_i p_i^* \ln p_i^* \tag{B-28}$$

which can be rewritten as

$$S = k\left[\ln Q + \left(\frac{\partial \ln Q}{\partial \ln T}\right)_{V,N}\right] \tag{B-29}$$

Equation (B-29) is suitable for deriving the remaining analogues upon using the fundamental relations of classical thermodynamics (see Table 2-1). Table B-1 presents several statistical thermodynamic analogues in terms of the canonical partition function Q.

Table B-1 Statistical analogues of thermodynamic functions.

Classical thermo-dynamics	Statistical thermodynamics	
	Canonical ensemble	Grand canonical ensemble
U	$kT\left(\frac{\partial \ln Q}{\partial \ln T}\right)_{V,N}$	$kT\left[\left(\frac{\partial \ln \Xi}{\partial \ln \mu}\right)_{T,V} + \left(\frac{\partial \ln \Xi}{\partial \ln T}\right)_{V,\mu}\right]$
S	$k\left[\ln Q + \left(\frac{\partial \ln Q}{\partial \ln T}\right)_{V,N}\right]$	$k\left[\ln \Xi + \left(\frac{\partial \ln \Xi}{\partial \ln T}\right)_{V,\mu}\right]$
PV	$kT\left(\frac{\partial \ln Q}{\partial \ln V}\right)_{T,N}$	$kT \ln \Xi$
H	$kT\left[\left(\frac{\partial \ln Q}{\partial \ln T}\right)_{V,N} + \left(\frac{\partial \ln Q}{\partial \ln V}\right)_{T,N}\right]$	$kT\left[\ln \Xi + \left(\frac{\partial \ln \Xi}{\partial \ln \mu}\right)_{T,V} + \left(\frac{\partial \ln \Xi}{\partial \ln T}\right)_{V,\mu}\right]$
A	$-kT \ln Q$	$kT\left[\left(\frac{\partial \ln \Xi}{\partial \ln \mu}\right)_{T,V} - \ln \Xi\right]$
G	$-kT\left[\ln Q - \left(\frac{\partial \ln Q}{\partial \ln V}\right)_{T,N}\right]$	$kT\left(\frac{\partial \ln \Xi}{\partial \ln \mu}\right)_{T,V}$

To illustrate the statistical meaning of entropy, we consider the special case of an ensemble where all quantum states are equally probable (microcanonical ensemble):

$$p_1 = p_2 = \cdots = p_i = \cdots = \frac{1}{W} \tag{B-30}$$

where W is the number of possible quantum states of the system. In this case, recalling that $\sum_i p_i = 1$, Eq. (B-28) reduces to the *Boltzmann relation*

$$\boxed{\begin{aligned} S &= -k(-\ln W)\sum_i p_i \\ &= k \ln W \end{aligned}} \tag{B-31}$$

where W is the *thermodynamic probability*. Although Eq. (B-31) is only a special case of the more general Eqs. (B-28) and (B-29), it can be used to draw an important conclusion: If a system is in its only possible quantum state (e.g., a perfect crystal at zero absolute temperature), $W = 1$ and

$$S = k \ln W = 0 \tag{B-32}$$

Equation (B-32) is the statistical form of the *third law* of thermodynamics.

The Grand Canonical Ensemble

For some applications, it is convenient to use an ensemble where the systems can exchange matter as well as energy. Such an ensemble is the *grand canonical ensemble*, that corresponds to a large number of *open* systems, each having fixed volume V, that are in internal equilibrium and that are able to exchange matter (molecules) and energy with their surroundings. To describe the thermodynamic state of such an ensemble, it is convenient to use as independent variables the temperature T, the volume V, and the chemical potentials $\mu_1, \mu_2, \ldots$ of components 1, 2, ... of the systems. For simplicity, we here consider only a system with one component. Generalization to a multicomponent system is straightforward, as discussed in many books on statistical mechanics.

For the calculation of the ensemble average, we must know the quantum-state distribution of the systems. Further, it is necessary to know the probabilities of the quantum states in the systems of the grand canonical ensemble.

The procedure is similar to that used for calculating the properties of the canonical ensemble. We suppose that each of the K systems of the grand canonical ensemble is a cell of volume V. Each cell can exchange matter and energy with the other cells but the total ensemble is completely isolated, as illustrated in Fig. B-2. The dashed lines indicate that, in contrast to Fig. B-1, exchange of matter as well as energy is permitted.

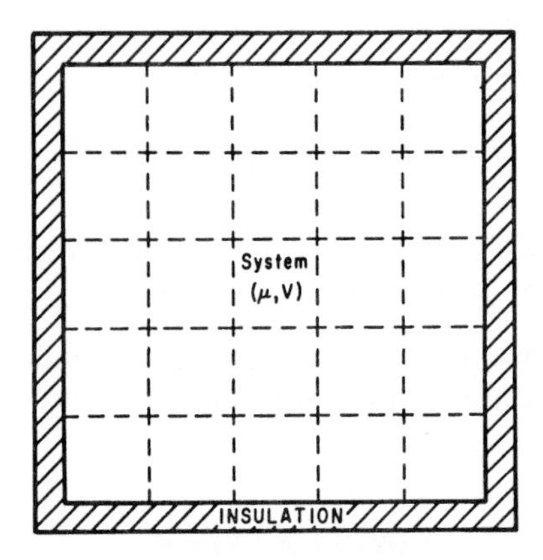

Figure B-2 Grand canonical ensemble. There are K systems. Each one can exchange energy and matter with $K-1$ other systems.

The ensemble is an isolated system of volume KV, total energy E, and total number of molecules N_t. The K systems may be in different energy states, but they can exchange energy and matter with each other because each one is in contact with a large reservoir of heat and matter formed by the $(K-1)$ other systems.

The quantum state of the entire ensemble is determined if the number of molecules N and the energy eigenvalues $E_j(N, V)$ of all its systems are known. If $n_1(N)$, $n_2(N)$, ..., $n_j(N)$, ... systems contain N molecules each with the energy eigenvalues $E_1(N, V)$, $E_2(N, V)$, ..., $E_j(N, V)$, ... respectively, then the values $n_1(N)$, $n_2(N)$, ..., $n_j(N)$, ... determine the distribution of the systems of the grand canonical ensemble with respect to the states of energy for given N and V. In contrast to the canonical ensemble, N is now not fixed. The total number of quantum states $\Omega(n)$ for a given distribution n is

$$\Omega(n)=\frac{\left[\sum_j n_j(N)\right]!}{\prod_j n_j(N)!} \tag{B-33}$$

There are many of these distributions, but all have to fulfill the conditions:

$$\sum_j n_j(N)=K \tag{B-34}$$

$$\sum_j n_j(N)E_j(N)=E_t \tag{B-35}$$

$$\sum_j n_j(N)N=N_t \tag{B-36}$$

The probability that for a given distribution, a particular system contains N molecules and is in quantum state j, is given by

$$p_j(N) = \frac{n_j(N)}{K} \tag{B-37}$$

The probability has to be averaged over all possible distributions. As the number of systems in the ensemble becomes very large ($K \to \infty$), the maximum-term method (see App. B.2) can be used and the average probability can be replaced by the most probable value $p_j^*(N)$:

$$< p_j(N) >= p_j^*(N) = \frac{n_j^*(N)}{K} \tag{B-38}$$

where $n_j^*(N)$ is the number of systems of the grand canonical ensemble that have the most probable distribution. Each of these systems contains N molecules and is in the quantum state with energy eigenvalue $E_j(N)$.

To calculate $\Omega(n*)$, the maximum value of $\Omega(n)$, Eq. (B-33) is written as

$$\ln \Omega(n) = \ln[\sum_j n_j(N)]! - \sum_j \ln n_j(N)! \tag{B-39}$$

The maximum condition is given by

$$\delta \ln \Omega(n^*) = \delta \ln[\sum_j n_j^*(N)]! - \delta \sum_j \ln n_j^*(N)! = 0 \tag{B-40}$$

Using Stirling's formula (see App. B.3) and remembering that K is a constant, Eq. (B-40) becomes

$$\delta \ln \Omega(n^*) = 0 = -\sum_j \ln n_j^*(N) \delta n_j(N) \tag{B-41}$$

Because K, E_t, and N are constants, Eqs. (B-34), (B-35), and (B-36) can be written as

$$\delta K = \sum_j \delta n_j(N) = 0 \tag{B-42}$$

$$\delta E_t = \sum_j E_j(N,V) \delta n_j(N) = 0 \tag{B-43}$$

$$\delta N_t = \sum_j N \delta n_j(N) = 0 \tag{B-44}$$

Using Lagrange's method, i.e., multiplying Eq. (B-42) by α, Eq. (B-43) by β, and Eq. (B-44) by γ, and adding all three to Eq. (B-41) gives

$$\sum_j \alpha\delta n_j(N) + \sum_j \beta E_j(N,V)\delta n_j(N) + \sum_j \gamma N\delta n_j(N) + \sum_j \ln n_j^*(N)\delta n_j(N)$$
$$= \sum_j [\alpha + \beta E_j(N,V) + \gamma N + \ln n_j^*(N)]\delta n_j(N) = 0 \tag{B-45}$$

As variations of $\delta n_j(N)$ are independent, Eq. (B-45) is satisfied for any j when

$$\alpha + \beta E_j(N,V) + \gamma N + \ln n_j^*(N) = 0 \tag{B-46}$$

or

$$n_j^*(N) = e^{-\alpha} e^{-\beta E_j(N,V)} e^{-\gamma N} \tag{B-47}$$

Using Eq. (B-34), the multiplier α can be eliminated and

$$n_j^*(N) = K \frac{e^{-\beta E_j(N,V)} e^{-\gamma N}}{\sum_j e^{-\beta E_j(N,V)} e^{-\gamma N}} \tag{B-48}$$

and, combining with Eq. (B-38),

$$p_j^*(N) = \frac{e^{-\beta E_j(N,V)} e^{-\gamma N}}{\sum_j e^{-\beta E_j(N,V)} e^{-\gamma N}} \tag{B-49}$$

where $p_j^*(N)$ is the probability that a randomly chosen system of the grand canonical ensemble contains N molecules and is in the quantum state with energy eigenvalue $E_j(N, V)$.

Statistical Analogues of Thermodynamic Properties in the Grand Canonical Ensemble. Using Eq. (B-49) with the first postulate, we can calculate any dynamic property of an open isothermal system. For example, the energy E that in classical thermodynamics is the internal energy U, is

$$U = E = \sum_j p_j^*(N) E_j(N,V) = \frac{\sum_j E_j(N,V) e^{-\beta E_j(N,V)} e^{-\gamma N}}{\sum_j e^{-\beta E_j(N,V)} e^{-\gamma N}} \tag{B-50}$$

The summation term in the denominator is called the *grand canonical partition function*, Ξ:

$$\boxed{\Xi = \sum_j e^{-\beta E_j(N,V)} e^{-\gamma N}} \tag{B-51}$$

where Ξ depends on β, γ, V, and N.

We obtain the number of molecules of the open, isothermal system from

$$\overline{N} = \sum_j p_j^*(N) N \tag{B-52}$$

Summation $\sum_i$ refers to all N and to all quantum states at given N. The physical significance of the multiplier β is the same as that given by Eq. (B-22).

Differentiation of Eq. (B-50) yields

$$dU = dE = \sum_j E_j(N,V) dp_j^*(N) + \sum_j p_j^*(N) dE_j(N,V) \tag{B-53}$$

The first term on the right-hand side of Eq. (B-53) can be written as

$$\sum_j E_j(N,V) dp_j^*(N) = \beta^{-1} d[-\sum_j p_j^*(N) \ln p_j^*(N)] - \gamma \beta^{-1} d\overline{N} \tag{B-54}$$

Considering an open, isothermal system that can perform only mechanical (PV) work, the second term on the right-hand side of Eq. (B-53)[3] is given by

$$\sum_j p_j^*(N) dE_j(N,V) = \sum_j p_j^*(N) \frac{\partial E_j(N,V)}{\partial V} dV \tag{B-55}$$

Substitution of Eqs. (B-54) and (B-55) into Eq. (B-53), and using Eq. (B-22), gives

$$dU = Td[-k \sum_j p_j^*(N) \ln p_j^*(N)] + \sum_j p_j^*(N) \frac{\partial E_j(N,V)}{\partial V} dV - \gamma kT d\overline{N} e \tag{B-56}$$

The corresponding relation from classical thermodynamics is

[3] In a group of systems of the ensemble with a given number N, $E_j(N, V)$ can only vary with the volume.

$$dU = TdS - PdV + \mu dN \tag{B-57}$$

where N is again the number of molecules (not moles) and μ is the chemical potential for one molecule. Comparison of Eqs. (B-56) and (B-57) yields the statistical analogues for the entropy and the chemical potential:

$$S = -k\sum_j p_j^*(N)\ln p_j^*(N) \tag{B-58}$$

and

$$\mu = -\gamma\beta^{-1} = -\gamma kT \tag{B-59}$$

Using Eqs. (B-22) and (B-59), the grand canonical partition function [Eq. (B-51)] can be written

$$\Xi = \sum_j e^{-E_j(N,V)/kT} e^{N\mu/kT} \tag{B-60}$$

Partial differentiation of Eq. (B-60) shows that Eq. (B-50) can be rewritten in the form

$$U = kT\left[\left(\frac{\partial \ln \Xi}{\partial \ln \mu}\right)_{T,V} + \left(\frac{\partial \ln \Xi}{\partial \ln T}\right)_{V,\mu}\right] \tag{B-61}$$

and Eq. (B-58) becomes

$$S = k\left[\ln \Xi + \left(\frac{\partial \ln \Xi}{\partial \ln T}\right)_{V,\mu}\right] \tag{B-62}$$

With Eqs. (B-61) and (B-62), the statistical analogues of other thermodynamic properties can be derived in terms of the grand canonical partition function using the relations of classical thermodynamics (see Table 2-1). The results are given in Table B-1.

The Semiclassical Partition Function

When calculating the canonical partition function, it is convenient to examine separately the energy contributions of the various molecular degrees of freedom. The most important factorization separates the *translational* contribution Q_{trans} (due to the positions and motions of the centers of mass of the molecules) from all others, due to other degrees of freedom such as rotation and vibration. These other degrees of freedom are often called *internal* degrees of freedom, although they are internal only for small spherical molecules. In this context, internal means independent of density.

The factored partition function has the form

$$Q = Q_{\text{int}}(N,T)Q_{\text{trans}}(N,T,V) \tag{B-63}$$

where Q_{int} is assumed to be independent of volume, i.e., Q_{int} has the same value for a dense fluid or solid as for an ideal gas. Equation (B-63) is strictly valid only for monatomic fluids (e.g., argon), but also provides a good approximation for molecules of nearly spherical symmetry like CCl_4 and $C(CH_3)_4$. It is, however, not valid for systems containing large asymmetric molecules or molecules interacting through strong dipolar forces or hydrogen bonding as found, for example, in alcohols. Since polar forces have an angular dependence, the rotation of a dipole depends on the positions of the centers of mass.

In the classical approximation, the translational partition function Q_{trans} is split into a product of two factors, one arising from the *kinetic energy* and the other from the *potential energy*. For a one-component system of N molecules, Q_{trans} is given by

$$Q_{\text{trans}} = Q_{\text{kin}}(N,T)Z_N(N,T,V) \qquad \text{(B-64)}^4$$

where

$$Q_{\text{kin}} = \left(\frac{2\pi mkT}{h^2}\right)^{(3/2)N} \frac{1}{N!} \tag{B-65}$$

and

$$\boxed{Z_N = \int_V \cdots \int \exp\left[-\frac{\Gamma_t(\mathbf{r}_1,\ldots,\mathbf{r}_N)}{kT}\right] d\mathbf{r}_1\ldots d\mathbf{r}_N} \tag{B-66}$$

In these equations, m is the molecular mass, k is Boltzmann's constant, and h is Planck's constant; $\Gamma_t(\mathbf{r}_1,\ldots,\mathbf{r}_N)$ is the potential energy of the entire system of N molecules whose positions are described by $\mathbf{r}_1,\ldots,\mathbf{r}_N$. For a given number of molecules, the first factor, Q_{kin}, in Eq. (B-64) depends only on temperature. The second factor, Z_N, the *configurational partition function*, depends on temperature and volume. Hence, the configurational part provides the only contribution that depends on intermolecular forces. However, it should be noted that Z_N is *not* unity for an ideal gas ($\Gamma_t = 0$) because in that event,

$$Z_N^{\text{id}} = V^N \tag{B-67}$$

[4] For a derivation of Eqs. (B-65) and (B-66), see one of the references at the end of this appendix.

The complete canonical partition function may now be written

$$\boxed{Q = Q_{\text{int}}(N,T)Q_{\text{kin}}(N,T)Z_N(N,T,V)} \tag{B-68}$$

Equation (B-68) is the basic relation for the statistical thermodynamics of dense and dilute fluids whose molecules interact with central forces.

Configurational and Residual Properties. From Eq. (B-68), ln Q is the sum of internal, kinetic, and configurational parts,

$$\ln Q = \ln Q_{\text{int}} + \ln Q_{\text{kin}} + \ln Z_N \tag{B-69}$$

When Eq. (B-69) is substituted into any of the equations in Table B-1, we obtain separate contributions to the thermodynamic properties. Thus, for a property X,

$$X = X^{\text{int}}(N,T) + X^{\text{kin}}(N,T) + X^{\text{conf}}(N,T,V) \tag{B-70}$$

where the superscripts indicate the portion of the partition function from which the term is derived. The internal and kinetic parts are identical to those for an ideal gas. The term X^{conf}, called the *configurational property*, arises from $\ln Z_N$, the configurational part of the partition function. Equations for X^{conf} are obtained by replacing Q by Z_N in the equations of Table B-l. Thus, the configurational pressure, internal energy, and Helmholtz energy are

$$P^{\text{conf}} = P = kT\left(\frac{\partial \ln Z_N}{\partial V}\right)_{T,N} \tag{B-71}$$

$$U^{\text{conf}} = kT\left(\frac{\partial \ln Z_N}{\partial \ln T}\right)_{V,N} \tag{B-72}$$

$$A^{\text{conf}} = -kT \ln Z_N \tag{B-73}$$

The configurational properties are the only contributions that depend on intermolecular forces, as indicated in the previous section. It should be noted that these contributions are *not* zero for ideal gases (no intermolecular forces) because then Z_N is given by Eq. (B-67).

The dimensions of Z_N are V^N. Therefore, the value calculated for properties such as A^{conf} or S^{conf}, proportional to ln Z_N, depends on the units chosen for Z_N. These are not arbitrary but must be the same as those chosen for $(Q_{\text{kin}})^{-1}$ because the complete partition function Q is dimensionless. When reporting values of the configurational Helmholtz or Gibbs energy and entropy, it is therefore necessary to state the units of Z_N. This difficulty does not arise for P^{conf}, U^{conf}, and H^{conf} because $d \ln Z_N = d Z_N / Z_N$; in this case the dimensions of Z_N cancel.

It is sometimes useful to introduce an alternative (but equivalent) property called *residual property*.[5] Residual functions are defined such that they give a direct measure of the contribution to the property from intermolecular forces at the given state condition. For any property X of a substance at a particular temperature, volume, and number of molecules, the *residual* X is defined by[6]

$$\boxed{X^R(T,V,N) = X(T,V,N) - X^{\text{id}}(T,V,N)} \tag{B-74}$$

where superscript R means residual. Thus the residual X is obtained by subtracting the ideal-gas property from that for the real fluid, both evaluated at the same T, V, and N. Since X^{id} is the value for the fluid at T, V, and N in the absence of intermolecular forces, X^R represents the contribution to X from "turning on" intermolecular forces. Equations for residual properties can be obtained from those given in Table B-1 together with Eq. (B-67). Thus,

$$\begin{aligned} P^R &= kT\left[\frac{\partial \ln(Q/Q^{\text{id}})}{\partial V}\right]_{T,N} = kT\left[\frac{\partial \ln(Z_N/Z_N^{\text{id}})}{\partial V}\right]_{T,N} \\ &= kT\left(\frac{\partial \ln Z_N}{\partial V}\right)_{T,N} - \frac{NkT}{V} = P^{\text{conf}} - P^{\text{id}} \end{aligned} \tag{B-75}$$

$$U^R = kT\left[\frac{\partial \ln(Q/Q^{\text{id}})}{\partial \ln T}\right]_{V,N} = kT\left(\frac{\partial \ln Z_N}{\partial \ln T}\right)_{V,N} = U^{\text{conf}} \tag{B-76}$$

$$\begin{aligned} A^R &= -kT \ln \frac{Q}{Q^{\text{id}}} = -kT \ln \frac{Z_N}{Z_N^{\text{id}}} \\ &= -kT \ln Z_N + kT \ln V^N = A^{\text{conf}} - A^{\text{conf, id}} \end{aligned} \tag{B-77}$$

---

[5] The residual properties discussed here are *different* from the residual mixing properties discussed in Sec. 8.2.

[6] Some authors define an excess property as the value of X for the real substance at temperature T and pressure P, minus X^{id} for the ideal gas at the same temperature and pressure:

$$X^E(T, P, N) = X(T, P, N) - X^{\text{id}}(T, P, N)$$

This X^E differs from X^R defined in Eq. (B-74).

Residual-property values are independent of the units for Z_N provided that these are consistent with those of V^N.

Appendix B.1: Two Basic Combinatorial Relations

Combinatorics deal with the arrangement of elements according to specified restrictions. An element designates a particle (sphere, molecule, etc.) that is the object of the arrangement.

1. The number of different arrangements $W(n)$ of n distinguishable elements of an assembly is

$$W(n) = n! \tag{B-78}$$

2. The number of different arrangements of an assembly of $n = n_1 + n_2 + \ldots + n_k$ elements where $n_1, n_2, \ldots, n_k$ elements are *not* distinguishable is

$$W(n_1, n_2, \ldots, n_k) = \frac{(n_1 + n_2 + \ldots + n_k)!}{n_1! n_2! \ldots n_k!} = \frac{n!}{\prod_i n_i!} \tag{B-79}$$

Appendix B.2: Maximum-Term Method

In statistical mechanics, the logarithm of a summation often is replaced by the *maximum term* of that summation. This approximation is very good because, for a large number of elements, the most probable distribution outweighs all other distributions to such an extent that it is practically identical with the average distribution.

To fix ideas, consider the following example: An assembly of 2×10^7 elements consists of 2×10^4 groups. We compare the number of possible arrangements for two cases:

(a) Each group contains 1000 elements.

(b) Half of the groups contain 1001 elements and the other half 999 elements per group.

The number of possible arrangements W is [see App. B.1]

$$\text{(a)} \qquad W_a = \frac{(2\times10^7)!}{10^3!10^3!\ldots} = \frac{(2\times10^7)!}{(10^3!)^{20,000}} \tag{B-80}$$

The ratio of the two distributions is

$$\text{(b)} \qquad \frac{W_b}{W_a} = \left[\frac{(1001!)(999!)}{(1000!)(1000!)}\right]^{10,000} = \left(\frac{1001}{1000}\right)^{10,000} \approx 22,000 \qquad \text{(B-81)}$$

Equation (B-81) tells us that distribution (a) is 22,000 times more probable than distribution (b), although the latter is very close to the former.

In this example, we used only 2×10^7 elements and 2×10^4 groups. For systems encountered in statistical mechanics, the number of elements (molecules) and groups (energies) is significantly larger (10^{20} and more). The maximum-term method is therefore justified for realistic problems encountered in practical thermodynamics.

Appendix B.3: Stirling's Formula

Stirling's formula provides an excellent approximation for factorials of large numbers. By definition,

$$n! = 1\times2\times3\times\ldots\times n \qquad \text{(B-82)}$$

Taking logarithms gives

$$\ln n! = \ln 1 + \ln 2 + \ldots + \ln n \qquad \text{(B-83)}$$

For large values of n, the right-hand side of Eq. (B-83) can be replaced by an integral:

$$\ln n! = \int_1^n \ln x\,dx \qquad \text{(B-84)}$$

This integral can be evaluated:

$$\int_1^n \ln x\,dx = [x\ln x]_1^n - \int_1^n x\frac{dx}{x} = n\ln n - n + 1 \qquad \text{(B-85)}$$

As $n >> 1$, Eq. (B-85) can be approximated by

$$\boxed{\ln n! \approx n\ln n - n} \qquad \text{(B-86)}$$

which is *Stirling's formula*. Figure B-3 shows the percentage error of the formula for insufficiently large n. Note that the absolute error $\ln n! - (n\ln n - n)$ increases monotonically with n, although the percent error falls. For $n \geq 100$, the formula provides an excellent approximation (error less than 1%).

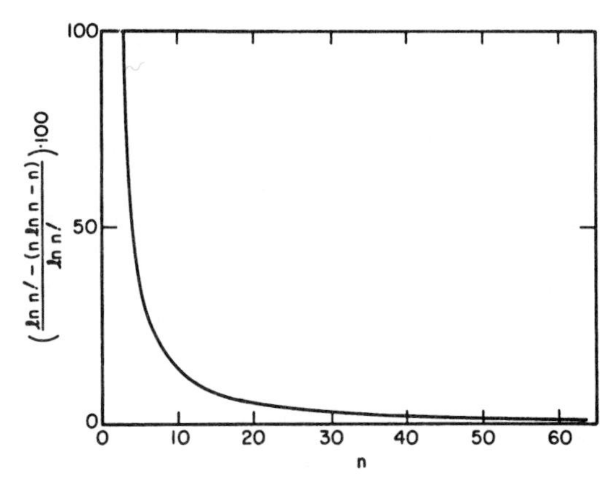

Figure B-3 Percent error in Stirling's formula for low n.

References

Andrews, F. C., 1975, *Equilibrium Statistical Mechanics*, 2nd Ed. New York: John Wiley & Sons.

Chandler, D., 1987, *Introduction to Modern Statistical Mechanics.* New York: Oxford University Press.

Chao, K. C. and R. A. Greenkorn, 1975, *Thermodynamics of Fluids: An Introduction to Equilibrium Theory,* Chap. 2. New York: Marcel Dekker.

Denbigh, K., 1981, *The Principles of Chemical Equilibrium*, 4th Ed., Chaps. 11-14. Cambridge: Cambridge University Press.

Everdell, M. H., 1975, *Statistical Mechanics and Its Chemical Applications.* New York: Academic Press.

Garrod, C., 1995, *Statistical Mechanics and Thermodynamics.* Oxford: Oxford University Press.

Gupta, M. C., 1991, *Statistical Thermodynamics.* New York: John Wiley & Sons

Hill, T. L., 1986, *An Introduction to Statistical Thermodynamics.* Reading: Addison-Wesley.

Maczek., A., 1998, *Statistical Thermodynamics.* Oxford: Oxford University Press.

McQuarrie, D. A., 1976, *Statistical Mechanics.* New York: Harper & Row.

McQuarrie, D. A., 1985, *Statistical Thermodynamics.* Mill Valley: University Science Books.

Reed, T. M. and K. E. Gubbins, 1973, *Applied Statistical Mechanics.* New York: McGraw-Hill (Reprinted by Butterworth-Heinemann, 1991).

Tien, C. L. and J. H. Lienhard, 1988, *Statistical Thermodynamics.* Washington: Hemisphere.

APPENDIX C

Virial Coefficients for Quantum Gases

This appendix presents the relationship between virial coefficients B and C with B' and C'. In addition, it presents second and third virial coefficient data for quantum gases (hydrogen, helium, and neon) as a function of temperature.

Virial Equation as a Power Series in Density or Pressure

The compressibility factor of a gas can be expressed by an expansion using either the density or the pressure as the independent variable:

$$z = \frac{Pv}{RT} = 1 + B\rho + C\rho^2 + D\rho^3 + \ldots \tag{C-1}$$

$$z = \frac{Pv}{RT} = 1 + B'P + C'P^2 + D'P^3 + \ldots \tag{C-2}$$

where, for a pure gas, virial coefficients B, C, D, ... and B', C', D', ... are functions only of temperature. Equation (C-1) is the *Leiden form* and Eq. (C-2) is the *Berlin form* of the virial equation in recognition of early workers in these cities who first used these equations.

We now show how the coefficients of one series are related to those of the other. First, we multiply both equations by $RT\rho$ and obtain

$$P = RT\rho + BRT\rho^2 + CRT\rho^3 + DRT\rho^4 + \ldots \tag{C-1a}$$

and

$$P = RT\rho + B'RT\rho P + C'RT\rho P^2 + D'RT\rho P^3 + \ldots \tag{C-2a}$$

Next, we substitute Eq. (C-1a) into Eq. (C-2a):

$$\begin{aligned} P = {} & RT\rho + B'RT\rho(RT\rho + BRT\rho^2 + CRT\rho^3 + DRT\rho^4 + \ldots) \\ & + C'RT\rho(RT\rho + BRT\rho^2 + CRT\rho^3 + DRT\rho^4 + \ldots)^2 \\ & + D'RT\rho(RT\rho + BRT\rho^2 + CRT\rho^3 + DRT\rho^4 + \ldots)^3 \\ & + \ldots \end{aligned} \tag{C-3}$$

Equations (C-3) and (C-la) are both power series in ρ. If we compare like terms in the two equations, we obtain from terms in ρ^2:

$$BRT\rho^2 = B'(RT)^2\rho^2 \tag{C-4}$$

or

$$\boxed{\frac{B}{RT} = B'} \tag{C-5}$$

From terms in ρ^3:

$$CRT\rho^3 = B'B(RT)^2\rho^3 + C'(RT)^3\rho^3 \tag{C-6}$$

or

$$\boxed{\frac{C - B^2}{(RT)^2} = C'} \tag{C-7}$$

Similarly, from terms in ρ^4 we obtain

$$\boxed{D' = \frac{D - 3BC + 2B^3}{(RT)^3}} \tag{C-8}$$

Equations (C-5), (C-7), and (C-8) are exact if we compare the two *infinite* series given by Eqs. (C-1) and (C-2). In other words, these relations are correct only if we evaluate the coefficients in both series from isothermal experimental compressibility factor (z) data according to

$$B = \lim_{\rho \to 0} \left(\frac{\partial z}{\partial \rho} \right)_T \quad \text{and} \quad B' = \lim_{P \to 0} \left(\frac{\partial z}{\partial P} \right)_T \tag{C-9}$$

$$C = \frac{1}{2!} \lim_{\rho \to 0} \left(\frac{\partial^2 z}{\partial \rho^2} \right)_T \quad \text{and} \quad C' = \frac{1}{2!} \lim_{P \to 0} \left(\frac{\partial^2 z}{\partial P^2} \right)_T \tag{C-10}$$

$$D = \frac{1}{3!} \lim_{\rho \to 0} \left(\frac{\partial^3 z}{\partial \rho^3} \right)_T \quad \text{and} \quad D' = \frac{1}{3!} \lim_{P \to 0} \left(\frac{\partial^3 z}{\partial \rho^3} \right)_T \tag{C-11}$$

In practice, it is not possible to evaluate virial coefficients from experimental data very near zero density or zero pressure because experimental measurements at these conditions are insufficiently accurate, especially for the second and third derivatives. As a result, Eqs. (C-5), (C-7), and (C-8) are only approximations when they are used to convert experimental virial coefficients from one series to those of the other. To illustrate the approximate nature of these equations, we consider two methods of data reduction discussed by Scott and Dunlap (1962).

Suppose we want to evaluate B from low-pressure volumetric measurements of some gas at constant temperature. We can evaluate B either by fitting data to the truncated virial equation

$$\frac{Pv}{RT} = 1 + B\rho \tag{C-12}$$

or else by fitting to the truncated virial equation

$$\frac{Pv}{RT} = 1 + \frac{B}{RT} P \tag{C-13}$$

In either case, we obtain B by fitting the experimental data over a finite region of density or pressure. If we use Eq. (C-12), we are, in effect, assuming that over the density range used, $C = 0$. If we use Eq. (C-13), we are, in effect, assuming that over the pressure range used, $C' = 0$. Since B is generally not zero, we can see from Eq. (C-7) that if one of these assumptions is valid, then the other one is not. Therefore, it is not surprising that the value of B that we obtain by reduction of actual data depends on the method used and we find that Eq. (C-5) cannot, in practice, be satisfied exactly.

Scott and Dunlap made accurate volumetric measurements at low pressures for n-butane at 29.88°C. When they used Eq. (C-12) for data reduction they found

$$B = (-715 \pm 5) \text{ cm}^3 \text{ mol}^{-1}$$

However, when Eq. (C-13) was used, they obtained

$$B = (-745 \pm 6) \text{ cm}^3 \text{ mol}^{-1}$$

We can see from these results that even if experimental uncertainties are considered, Eq. (C-5) is only an approximation.

A preferable procedure for determining B from experimental data is to use for data reduction either one of the virial equations truncated after the third virial coefficient. When this was done, Scott and Dunlap found for the same data, using the density series,

$$B = (-695 \pm 25) \text{ cm}^3 \text{ mol}^{-1}$$

$$C = (-4 \pm 4)\times 10^5 \text{ (cm}^3 \text{ mol}^{-1})^2$$

and, using the pressure series,

$$B = (-691 \pm 26) \text{ cm}^3 \text{ mol}^{-1}$$

$$C = (-10 \pm 5)\times 10^5 \text{ (cm}^3 \text{ mol}^{-1})^2$$

With this method of data reduction, the uncertainties in B are larger than before but agreement between the two methods is now very much improved. The results obtained for C are of little value but for obtaining B, it is preferable to include even a rough estimate for C than to include none at all.

A similar investigation was made by Lichtenthaler and Strein (1971), who concluded that Eq. (C-5) can be used for many typical systems provided that the second virial coefficients have been determined from highly accurate P-V-T data at low pressures, generally well below 1 bar.

Virial Coefficients for Hydrogen, Helium, and Neon

Because of their small masses, the properties of hydrogen, helium, and neon cannot be described by classical statistical mechanics. As discussed elsewhere (Hirschfelder *et al.*, 1954), it is possible to write an expression for the second virial coefficient of light gases based on quantum mechanics:

$$\frac{B}{b_0} = F\left(\frac{kT}{\varepsilon}, \Lambda^*\right) \tag{C-14}$$

where

$$b_0 = \frac{2}{3}\pi N_A \sigma^3$$

and

$$\Lambda^* = \frac{h}{\sigma\sqrt{m\varepsilon}}$$

Λ^* is the *reduced de Broglie wavelength*; it depends on Planck's constant h, molecular mass m, and intermolecular potential parameters ε and σ.

Figure C-1 gives reduced experimental second virial coefficients of hydrogen and helium at very low temperatures. The parameters used for data reduction are:

	ε/k (K)	b_0 (cm^3 mol^{-1})	Λ^*
Hydrogen	37.0	31.67	1.73
Helium	10.22	21.07	2.67

For comparison, Fig. C-1 also shows reduced second virial coefficients calculated from the Lennard-Jones potential using classical statistical mechanics.

The results shown in Fig. C-1 are useful for estimating second virial cross-coefficients B_{12} whenever component 1 or 2 (or both) is a light (quantum) gas. To estimate B_{12} we use the customary combining rules

$$\varepsilon_{12} = (\varepsilon_1\varepsilon_2)^{1/2} \tag{C-15}$$

and

$$\sigma_{12} = \frac{1}{2}(\sigma_1 + \sigma_2) \tag{C-16}$$

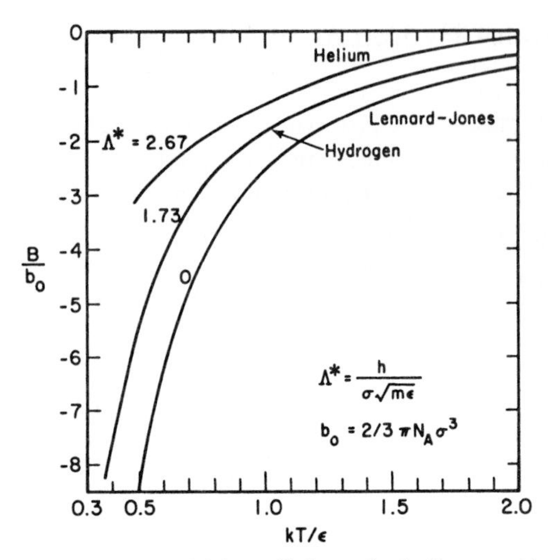

Figure C-1 Reduced second virial coefficients for helium and hydrogen at low temperatures.

and in addition we have

$$\Lambda_{12}^{*} = \frac{h}{\sigma_{12}\sqrt{m_{12}\varepsilon_{12}}} \tag{C-17}$$

where

$$\frac{1}{m_{12}} = \frac{1}{2}\left(\frac{1}{m_1} + \frac{1}{m_2}\right) \tag{C-18}$$

For details, see Hirschfelder *et al.* (1954) and Prausnitz and Myers (1963).

Experimental Results. Dymond and Smith (1980) give a compilation of experimental data for the virial coefficients of pure gases and mixtures. For hydrogen they found that different sets for the second virial coefficient, reported by various authors, all agree within estimated error limits. Table C-1 gives virial coefficients obtained from a smooth curve through all data. This table also gives third virial coefficients for hydrogen reported by Michels *et al.* (1959).

Goodwin and coworkers (1964) established empirical correlations for second and third virial coefficients of hydrogen over a wide temperature range. For the second virial coefficient, their results are given in the form

$$B = \sum_{i=1}^{4} b_i x^{(2i-1)/4} \tag{C-19}$$

Table C-1 Second and third virial coefficients for hydrogen.

T (K)	B (cm^3 mol^{-1})	T (K)	$C^§$ (cm^3 mol^{-1})^2
14	-254 ± 5	98.15	503
15	-230 ± 5	103.15	511
17	-191 ± 5	113.15	506
19	-162 ± 5	123.15	519
22	-132 ± 5	138.15	516
25	-110 ± 3	153.15	480
30	-82 ± 3	173.15	459
40	-52 ± 2	198.15	414
50	-33 ± 2	223.15	406
75	-12 ± 1	248.15	388
100	-1.9 ± 1	273.15	389
150	+7.1 ± 0.5	298.15	356
200	11.3 ± 0.5	323.15	323
300	14.8 ± 0.5	348.15	295
400	15.2 ± 0.5	373.15	290
		398.15	296
		423.15	280

§ The accuracy of these third virial coefficients is low.

where

$$x = \frac{109.83}{T} \quad (T \text{ in kelvin}) \qquad b_1 = +42.464 \qquad b_3 = -2.2982$$
$$b_2 = -37.1172 \qquad b_4 = -3.0484 \quad (\text{cm}^3\text{mol}^{-1})$$

Equation (C-19) covers the temperature range 15 to 423 K and the mean deviation from experimental results is ±0.066 cm^3 mol^{-1}.

The third virial coefficient (in cm^6 mol^{-2}) is given by

$$C = 1310.5x^{1/2}[1 + 2.1486x^3][1 - \exp(1 - x^{-3})] \tag{C-20}$$

Equation (C-20) covers the temperature range 24 to 423 K and the mean deviation from experimental results is ±17.4 cm^6 mol^{-2}.

For helium, agreement between data of various authors for the second virial coefficient is good above 15 K. Below this temperature, there is a significant discrepancy between the results of different sources. Dymond and Smith recommend the virial coefficients given in Table C-2. Third virial coefficients are those reported by Keesom (1942).

Table C-2 Second and third virial coefficients for helium.

T (K)	B (cm^3 mol^{-1})	T (K)	$C^{§}$ (cm^3 mol^{-1})^2
2.0	-174 ± 8	4.0	-302
2.5	-134 ± 5	4.5	+151
2.75	-120 ± 5	5.0	322
3.0	-109 ± 2	6.0	458
3.5	-92.6 ± 2	8.0	513
4.0	-80.2 ± 2	10	518
5.0	-62.7 ± 1	12	508
7.0	-40.9 ± 1	14	498
10.0	-23.1 ± 1	16	488
15.0	-10.8 ± 1	18	478
20.0	-3.4 ± 0.5	20	468
30.0	+2.5 ± 0.5	22	453
50.0	7.4 ± 0.5	30	417
100.0	11.7 ± 0.5	40	382
200.0	12.1 ± 0.5	50	352
400.0	11.2 ± 0.5	60	292
700.0	10.1± 0.5	70	327
		80	307
		90	277
		123.15	241
		173.15	201
		223.15	176
		273.15	156
		323.15	136
		373.15	126
		473.15	101
		573.15	101

§ The accuracy of these third virial coefficients is low.

For high temperatures (300 to 1000 K), virial coefficients of helium have been calculated using results from molecular-beam experiments. Second and third virial coefficients, reported by Harrison (1964) are given by

$$B = 1.3436 \times 10^{-2}(15.8922 - \ln T)^3 - 4.39\exp(-2.4177 \times 10^{-3}T) \tag{C-21}$$

and

$$C = 9.0263 \times 10^{-5}(15.8922 - \ln T)^6 \tag{C-22}$$

where the units are the same as those in Table C-2.

There is good agreement between data of various authors for the second virial coefficient of neon. Dymond and Smith (1980) obtained the virial coefficients given in Table C-3 from a smooth curve drawn through all data. Third virial coefficients are those reported by Crommelin *et al.* (1919) and by Michels *et al.* (1960).

For very low temperatures, estimates of B for neon can be made using Fig. C-1 with parameters ε/k = 34.9 K, b_0 = 27.1 cm^3 mol^{-1}, and Λ^* = 0.593.

Table C-3 Second and third virial coefficients for neon.

T (K)	B (cm^3 mol^{-1})	T (K)	$C^§$ (cm^3 mol^{-1})^2
60	-24.8 ± 1	55.64	900
70	-17.9 ± 1	60.08	483
80	-12.8 ± 1	65.06	534
100	-6.0 ± 1	73.08	523
125	-0.4 ± 1	90.56	510
150	+3.2 ± 1	273.15	246
200	7.6 ± 1	298.15	233
300	11.3 ± 1	323.15	234
400	12.8 ± 1	348.15	236
600	13.8 ± 0.5	373.15	238
		398.15	217
		423.15	208

§ The accuracy of these third virial coefficients is low.

References

Crommelin, C. A., J. P. Martinez, and H. Kammerlingh Onnes, 1919, *Commun. Phys. Lab. Univ. Leiden,* 151a.

Dymond, J. H. and E. B. Smith, 1980, *The Virial Coefficients of Pure Gases and Mixtures.* Oxford: Clarendon Press.

Goodwin, R. D., D. E. Diller, H. M. Roder, and L. A. Weber, 1964, *J. Res. Natl. Bur. Std.,* 68A: 121.

Harrison, E. F., 1964, *AIAA J.,* 2: 1854.

Hirschfelder, J. O., C. F. Curtiss, and R. B. Bird, 1954, *Molecular Theory of Gases and Liquids.* New York: John Wiley & Sons.

Keesom, W. H., 1942, *Helium.* Amsterdam: Elsevier.

Lichtenthaler, R. N. and K. Strein, 1971, *Ber. Bunsenges. Phys. Chem.,* 75: 489.

Michels, A., T. Wassenaar, and P. Louwerse, 1960, *Physica,* 26: 539.

Michels, A., W. de Graaff, T. Wassenaar, J. M. H. Levelt, and P. Louwerse, 1959, *Physica,* 25: 25.

Prausnitz, J. M. and A. L. Myers, 1963, *AIChE J.,* 9: 5.

Scott, R. L. and R. D. Dunlap, 1962, *J. Phys. Chem.,* 66: 639.

APPENDIX D

The Gibbs-Duhem Equation

A brief discussion of the Gibbs-Duhem equation is given in Sec. 2.4 and some applications are given in Chap. 6. In this appendix we give a derivation of the fundamental equation and we present special forms of the equation as applied to activity coefficients.

Let M be an extensive property of a mixture. For a homogeneous phase, M is a function of temperature, pressure, and mole numbers. The total differential of M is given by

$$dM = \left(\frac{\partial M}{\partial T}\right)_{P,\text{all } n} dT + \left(\frac{\partial M}{\partial P}\right)_{T,\text{all } n} dP + \sum_i \overline{m}_i dn_i \tag{D-1}$$

where

$$\overline{m}_i \equiv \left(\frac{\partial M}{\partial n_i}\right)_{T,P,n_j} \tag{D-2}$$

The extensive property M is related to the partial molar properties $\overline{m}_1$, $\overline{m}_2$, ..., by Euler's theorem:

$$M = \sum_i \overline{m}_i n_i \tag{D-3}$$

Differentiation of Eq. (D-3) gives

$$dM = \sum_i \overline{m}_i dn_i + \sum_i n_i d\overline{m}_i \tag{D-4}$$

Equations (D-1) and (D-4) yield the general *Gibbs-Duhem equation*:[1]

$$\boxed{\left(\frac{\partial M}{\partial T}\right)_{P,\text{all } n} dT + \left(\frac{\partial M}{\partial P}\right)_{T,\text{all } n} dP - \sum_i n_i d\overline{m}_i = 0} \tag{D-5}$$

Suppose M is the Gibbs energy G. As indicated in Chap. 2,

$$\left(\frac{\partial G}{\partial T}\right)_{P,\text{all } n} = -S \tag{D-6}$$

$$\left(\frac{\partial G}{\partial P}\right)_{T,\text{all } n} = V \tag{D-7}$$

$$\left(\frac{\partial G}{\partial n_i}\right)_{T,P,n_j} = \mu_i \tag{D-8}$$

where μ is the chemical potential. Equation (D-5) then becomes

$$\boxed{SdT - VdP + \sum_i n_i d\mu_i = 0} \tag{D-9}$$

In terms of excess functions for one mole of mixture, Eq. (D-9) is

$$s^E dT - v^E dP + \sum_i x_i d\mu_i^E = 0 \tag{D-10}$$

[1] Notice that the general Gibbs-Duhem equation applies to any extensive property, not just to the Gibbs energy, as discussed in Sec. 2.4.

The excess chemical potential of component i is related to activity coefficient γ_i by

$$\mu_i^E = RT \ln \gamma_i \tag{D-11}$$

At constant temperature and pressure, we then have

$$\boxed{\sum_i x_i d \ln \gamma_i = 0} \tag{D-12}$$

The phase rule tells us that in a binary, two-phase system it is not possible to change the composition of either phase while holding both temperature and pressure constant. Therefore, in a binary system, experimental data used to compute activity coefficients may be either isothermal or isobaric but not both. Therefore, Eq. (D-12) is not strictly applicable to activity coefficients for a binary system. To obtain an equation similar to, but less restrictive than Eq. (D-12), we consider how Eq. (D-9) can be rewritten in terms of activity coefficients. First, we treat the case of constant pressure and variable temperature and second, the case of constant temperature and variable pressure.

The Isobaric, Nonisothermal Case. Equation (D-9), on a molar basis, now is

$$\sum_i x_i d\mu_i = -s dT \tag{D-13}$$

where s is the entropy per mole of mixture. Introducing the activity coefficient,

$$\frac{\mu_i}{T} - \frac{\mu_i^0}{T} = R \ln \gamma_i x_i \tag{D-14}$$

where superscript 0 stands for the standard state where

$$a_i = \gamma_i x_i = 1$$

Differentiating Eq. (D-14) and rearranging, we have

$$d\left(\frac{\mu_i}{T}\right) = d\left(\frac{\mu_i^0}{T}\right) + R d \ln \gamma_i + R d \ln x_i \tag{D-15}$$

Next, we utilize the Gibbs-Helmholtz equation,

$$d\left(\frac{\mu_i^0}{T}\right) = -\frac{h_i^0}{T^2}dT \tag{D-16}$$

and the mathematical identity

$$d\left(\frac{\mu_i}{T}\right) = \frac{d\mu_i - \frac{\mu_i}{T}dT}{T} \tag{D-17}$$

Substituting Eqs. (D-16) and (D-17) into Eq. (D-15) and solving for $d\mu_i$ yields

$$d\mu_i = -\frac{h_i^0}{T}dT + RTd\ln\gamma_i + RTd\ln x_i + \frac{\mu_i}{T}dT \tag{D-18}$$

To simplify this result we recall that

$$\frac{\mu_i}{T} = \frac{\bar{h}_i}{T} - \bar{s}_i \tag{D-19}$$

Substituting Eqs. (D-18) and (D-19) into Eq. (D-13) and using the relations

$$s = \sum_i x_i \bar{s}_i \qquad \text{and} \qquad \sum_i x_i d\ln x_i = 0$$

we finally obtain

$$\boxed{\sum_i x_i d\ln\gamma_i = -\frac{h^E}{RT^2}dT} \tag{D-20}$$

where $h^E = h - \sum_i x_i h_i^0$ and h is the molar enthalpy of the mixture. Equation (D-20) is the desired result. It shows that the activity coefficients of a multicomponent system at constant pressure are related to one another through a differential equation that includes the enthalpy of mixing.

In many cases the standard state for component i is taken as pure liquid i at the temperature and pressure of the system. In that case, h^E is the enthalpy change that results upon mixing the pure liquids isothermally and isobarically to form the solution. However, in some cases when one of the components is a gaseous (or solid) solute, the standard-state fugacity for the solute is often taken to be Henry's constant evaluated at the system temperature and pressure. In that case, for the solute, $h_i^0 = \bar{h}_i^\infty$, the partial molar enthalpy of i in an infinitely dilute solution at system temperature and pressure.

The Isothermal, Nonisobaric Case. Equation (D-9) on a molar basis now is

$$\sum_i x_i d\mu_i = v dP \tag{D-21}$$

where v is the molar volume of the mixture.

Again, we introduce the activity coefficient

$$\mu_i - \mu_i^0 = RT \ln \gamma_i x_i \tag{D-22}$$

Differentiating Eq. (D-22) at constant temperature,

$$d\mu_i = d\mu_i^0 + RTd\ln\gamma_i + RTd\ln x_i \tag{D-23}$$

To say something about $d\mu_i^0$ we must now distinguish between two cases that we call case A and case B. These cases correspond to our choice of pressure for the standard state.

Case A. Let the standard state for component i be at the system temperature, at a fixed composition, and at some constant pressure that does not vary with composition. Then $d\mu_i^0 = 0$ and Eqs. (D-21) and (D-23) combine to

$$\boxed{\sum_i x_i d\ln\gamma_i = \frac{v dP}{RT}} \quad (P^0 = \text{constant}) \tag{D-24}$$

Case B. Let the standard state for component i be at the system temperature, at a fixed composition, and at the total pressure P of the system, which is not constant but varies with composition. In this case,

$$d\mu_i^0 = v_i^0 dP \tag{D-25}$$

where v_i^0 is the molar volume of component i in its standard state.

Substitution of Eqs. (D-23) and (D-25) into Eq. (D-21) now gives

$$\boxed{\sum_i x_i d\ln\gamma_i = \frac{v^E dP}{RT}} \quad (P^0 = P_{\text{system}}) \tag{D-26}$$

where

$$v^E = v - \sum_i x_i v_i^0$$

Equations (D-24) and (D-26) are the desired result. They show that activity coefficients of a multicomponent system at constant temperature are related to one another through a differential equation that contains volumetric properties of the liquid mixture.

Frequently, the standard state is chosen as the pure liquid at the temperature and pressure of the mixture. In that case, v^E is the change in volume that results when the pure liquids are mixed at constant temperature and constant (system) pressure. An alternative application of Eq. (D-26) might be for a solution of a gas (or solid) in a liquid where the standard-state fugacity of the solute may be set equal to Henry's constant evaluated at system temperature and total pressure. In that case, for the solute, $v_i^0 = \overline{v}_i^\infty$, the partial molar volume of i in an infinitely dilute solution at system temperature and total pressure.

At low or moderate pressures, the right-hand side of Eq. (D-26) is often so small that Eq. (D-12) provides an excellent approximation.

APPENDIX E

Liquid-Liquid Equilibria in Binary and Multicomponent Systems

Many liquids are only partially miscible. In this appendix we briefly discuss the thermodynamics of partially miscible liquid systems with particular emphasis on the relation between excess Gibbs energy and mutual solubilities.[1]

At a fixed temperature and pressure, we consider a binary system containing two liquid phases at equilibrium. Let ' (prime) designate one phase and let " (double prime) designate the other phase. For component 1, the equation of equilibrium is

[1] For a more detailed review, see J. M. Sørensen, T. Magnussen, P. Rasmussen, and Aa. Fredenslund, 1979, *Fluid Phase Equilibria,* 2: 297; *ibid.*, 3: 47; *ibid.*, 1980, 4: 151; and J. P. Novak, J. Matous, and J. Pick, 1987, *Liquid-Liquid Equilibria*, Amsterdam: Elsevier. For a general discussion of thermodynamic stability in multicomponent systems, see J. W. Tester and M. Modell, 1997, *Thermodynamics and Its Applications,* 3rd Ed., Englewood Cliffs: Prentice-Hall.

$$f_1' = f_1'' \tag{E-1}$$

Because both phases are liquids, it is convenient to use the same standard-state fugacity for both phases. Equation (1) can then be rewritten

$$a_1' = a_1'' \tag{E-2}$$

$$\gamma_1' x_1' = \gamma_1'' x_1'' \tag{E-3}$$

where x_1' and x_1'' are equilibrium mole fractions of component 1 in the two phases. Similarly, for component 2,

$$\gamma_2' x_2' = \gamma_2'' x_2'' \tag{E-4}$$

where x_2' and x_2'' are equilibrium mole fractions of component 2 in the two phases.

For a given binary system at a fixed temperature and pressure, we can calculate mutual solubilities if we have information concerning activity coefficients. Suppose we have such information in the form

$$g^E = F(x_1, A, B, \ldots) \tag{E-5}$$

where g^E, the molar excess Gibbs energy, is a function of composition, with parameters A, B, ... depending only on temperature (and to a lesser extent, on pressure). From Eq. (6-25) we can readily obtain activity coefficients of both components. Upon substitution, Eqs. (E-3) and (E-4) are of the form

$$F_1(x_1')x_1' = F_1(x_1'')x_1'' \tag{E-6}$$

$$F_2(x_2')x_2' = F_2(x_2'')x_2'' \tag{E-7}$$

where functions F_1 and F_2 are obtained upon differentiating Eq. (E-5) as indicated by Eq. (6-25). There are two unknowns: x_2' and x_1''.[2] These can be found from the two equations of equilibrium, Eqs. (E-6) and (E-7). For example, suppose we use a two-suffix Margules equation [Eq. (6-46)] for the excess Gibbs energy in Eq. (E-5). We then have for our two equations of equilibrium:

[2] By stoichiometry, $x_1' + x_2' = 1$ and $x_1'' + x_2'' = 1$. If phase ' is rich in component 1 and phase " is rich in component 2, then x_2' and x_1'' are the two mutual solubilities.

$$\left[\exp\left(\frac{A(1-x_1')^2}{RT}\right)\right]x_1' = \left[\exp\left(\frac{A(1-x_1'')^2}{RT}\right)\right]x_1'' \qquad \text{(E-8)}$$

$$\left[\exp\left(\frac{Ax_1'^2}{RT}\right)\right](1-x_1') = \left[\exp\left(\frac{Ax_1''^2}{RT}\right)\right](1-x_1'') \qquad \text{(E-9)}$$

For a given A/RT, Eqs. (E-8) and (E-9) give a solution for x_2' and x_1''. (To be sure that x_2' and x_1'' fall into the interval between zero and one, it is necessary that $A/RT \geq 2$.)

We have just described how mutual solubilities may be found if the excess Gibbs energy is known. Frequently, however, it is desirable to reverse the procedure (i.e., to calculate excess Gibbs energy from known mutual solubilities) because it is often a relatively simple matter to obtain mutual solubilities experimentally.

To calculate excess Gibbs energy from measured x_2' and x_1'', we must first choose a particular function for the excess Gibbs energy [Eq. (E-5)] containing no more than two unknown parameters, A and B. We can then find A and B by simultaneous solution of Eqs. (E-6) and (E-7). Once A and B are known, we can then calculate activity coefficients for both components in the two miscible regions. Therefore, mutual solubility data may be used to calculate vapor-liquid equilibria.[3]

Calculation of excess Gibbs energies from mutual-solubility data is tedious but straightforward. The accuracy of such calculation is sensitive to the accuracy of the mutual solubility data but, even if these are highly accurate, the results obtained are likely to be only approximate because, without additional information, the estimated excess Gibbs energy function may contain no more than two parameters. Further, the results obtained depend strongly on the arbitrary algebraic function chosen to represent the excess Gibbs energy. This sensitivity is illustrated by calculations reported by Brian (1965)[4] for five partially miscible binary systems shown in Table E-1. Using experimental mutual-solubility data, excess Gibbs energy parameters A and B were calculated once using the van Laar equation and once using the three-suffix Margules equation (see Sec. 6.10). With these parameters, Brian calculated activity coefficients at infinite dilution for both components. Although the same experimental data were used, results obtained with the van Laar equation differ markedly from those obtained with the Margules equation. Brian found that the differences are especially large in strongly asymmetric systems, i.e., in systems where the ratio x_2'/x_1'' is far removed from unity. In the last system (propylene oxide/water), where this ratio is approximately 2, results obtained with the van Laar equation are close to those obtained with the Margules equation; however, for the system phenol/water, where the ratio is ap-

---

[3] However, A and B are functions of temperature.

[4] Similar calculations for aqueous systems are presented by D. L. Bergmann and C. E. Eckert, 1991, *Fluid Phase Equilibria*, 63: 141.

proximately 35, the limiting activity coefficient for water obtained with the van Laar equation is several orders of magnitude larger than that obtained with the Margules equation. Brian found that when predicted limiting activity coefficients were compared with experimental results, the Margules equation frequently gave poor results. The van Laar equation gave reasonable results but quantitative agreement with experiment was at best fair.

Table E-1 Limiting activity coefficients as calculated from mutual solubilities in five binary aqueous systems.

		Solubility limits		$\log\gamma_1^\infty$		$\log\gamma_2^\infty$	
Component (1)	Temp. (°C)	x_1''	x_2'	van Laar	Margules	van Laar	Margules
Aniline	100	0.01475	0.372	1.8337	1.5996	0.6076	-0.4514
Isobutyl alcohol	90	0.0213	0.5975	1.6531	0.6193	0.4020	-3.0478
1-Butanol	90	0.0207	0.636	1.6477	0.2446	0.3672	-4.1104
Phenol	43.4	0.02105	0.7325	1.6028	-0.1408	0.2872	-8.2901
Propylene oxide	36.3	0.166	0.375	1.1103	1.0743	0.7763	0.7046

Renon found that the NRTL equation may be useful for calculating excess Gibbs energies from mutual solubility data. The NRTL equation [Eq. (6.11-6)] contains three parameters, but one of them, the nonrandomness parameter α_{12}, can often be estimated for a given binary mixture from experimental results for other mixtures of the same class. Once a value of α_{12} has been chosen,[5] the two remaining parameters can be obtained from simultaneous solution of Eqs. (E-6) and (E-7).

For phase separation, α_{12} may not exceed α_{crit} that depends on mole fraction x_1 where splitting occurs and on β, the ratio of NRTL parameters τ_{21} and τ_{12}. For several values of x_1 and β, α_{crit} has been calculated[6] as shown in Table E-2.

In mixtures with miscibility gaps, it is difficult to find an expression for the molar excess Gibbs energy g^E that yields simultaneously correct activity coefficients in the miscible region and correct activity coefficients at the solubility limits. To illustrate, Table E-3 compares observed solubility limits with those calculated from a two-parameter expression for g^E, where the parameters are evaluated from experimental data for activity coefficients at infinite dilution at the same temperature. Solubilities calculated with the van Laar equation or the UNIQUAC equation are in poor agreement with experiment. Table E-4 shows the reverse calculation: Experimental solubilities are used to determine two parameters in an expression for g^E that then yields lim-

[5] Because these calculations are tedious, Renon (1969) has computer-programmed the equations of equilibrium and has presented the calculated results in graphical form.

[6] Debenedetti and Reid (MIT), personal communication.

iting activity coefficients at the same temperature. Again, agreement with experiment is not good.

Table E-2 Calculated α_{crit} for different values of x_1 and β.

	$\beta = \tau_{12}/\tau_{21}$		
x_1	0.5	1.0	2.0
0.10	0.6109	0.6177	0.6362
0.50	0.3926	0.4270	0.3926
0.90	0.6362	0.6177	0.6109

It appears that, at present, for complex systems it is not possible to describe accurately both vapor-liquid and liquid-liquid equilibria using a simple expression for g^E containing only two adjustable binary parameters. To describe both kinds of equilibria, we probably need an expression for g^E that contains more than two binary parameters and, more important, some experimental data for both kinds of equilibria for evaluating these binary parameters.

Ternary (and higher) liquid-liquid equilibria. If an expression is available for relating molar excess Gibbs energy g^E to composition, the activity coefficient of every component is readily calculated as discussed in Sec. 6.3. For liquid-liquid equilibria in a system containing m components, there are m equations of equilibrium. Assuming that the standard-state fugacity for every component i in phase ' is the same as that in phase ", these equations are of the form indicated by Eq. (E-6), written m times, once for each component.

If an appropriate expression for g^E is available, it is not immediately obvious how to solve these m equations simultaneously when $m > 2$. To fix ideas, consider a ternary system at a fixed temperature and pressure. We want to know coordinates x_1' and x_2' on the binodal curve that are in equilibrium with coordinates x_1'' and x_2'', also on the connodal curve. (The line joining these two sets of coordinates is a *tie line.*) Therefore, we have four unknowns.[7] However, there are only three equations of equilibrium. To find the desired coordinates, therefore, it is not sufficient to consider only the three equations indicated by Eq. (E-6). To obtain the coordinates, we must use a material balance by performing what is commonly known as an *isothermal flash calculation.*[8]

[7] x_3' and x_3'' are not independent unknowns because $\sum_i x_i' = 1$ and $\sum_i x_i'' = 1$.

[8] For a discussion of adiabatic or isothermal flash calculations, see Prausnitz *et al.* (1980).

Table E-3 Solubility limits calculated from activity coefficients at infinite dilution using van Laar or UNIQUAC (Lobien, 1980, 1982).

		Solubility limits*					
		$100x_1'$			$100x_2''$		
Components	Temp. (°C)	van Laar	UNIQUAC	Obs.	van Laar	UNIQUAC	Obs.
n-Heptane (1)/ 2-methoxyethanol (2)	47	27.0	TM	19.2	11.0	TM	26.2
	67	40.0	TM	TM	16.0	TM	TM
	77	50.0	TM	TM	25.0	TM	TM
2,4-Pentanedione (1)/ water (2)	70	4.5	7.5	6.3	30.5	53.0	48.0
	80	5.5	8.5	8.5	39.5	59.0	59.9
	90	6.0	10.0	TM	51.5	67.5	TM
n-Butanol (1) /water (2)	70	1.0	2.0	1.7	45.0	58.0	58.1
	80	2.5	3.0	1.8	49.0	63.0	59.7
	100	5.0	6.0	2.4	58.5	70.5	67.0
Methanol (1)/ /cyclohexane (2)	35	2.0	16.0	16.1	7.0	18.0	22.2
	45.1	3.0	23.0	37.7	8.0	23.0	39.6
	60	5.0	30.0	TM	9.0	26.0	TM

* TM = total miscibility.

Table E-4 Activity coefficients at infinite dilution calculated from solubility data using van Laar or UNIQUAC (Lobien, 1980, 1982).

		γ_1^∞			γ_2^∞		
Components	Temp. (°C)	van Laar	UNIQUAC	Obs.	van Laar	UNIQUAC	Obs.
n-Heptane (1)/ 2-methoxyethanol (2)	47	10.9	17.8	7.32	8.21	25.9	16.1
2,4-Pentanedione (1)/ water (2)	70	23.9	30.6	28.6	3.83	5.63	5.47
	80	20.2	25.6	26.1	3.20	4.44	4.60
n-Butanol (1)/water (2)	70	52.6	76.6	67.8	2.54	3.24	3.27
	80	50.9	74.6	46.5	2.47	3.09	3.12
Methanol (1)/ /cyclohexane (2)	35	12.0	75.8	54.2	8.97	17.8	18.9
	45.1	7.86	44.1	38.0	7.45	15.3	17.4

One mole of a liquid stream with overall composition x_1, x_2 is introduced into a flash chamber where that stream isothermally separates into two liquid phases ' and ". The number of moles of phases ' and " are designated by L' and L''. We can write three independent material balances:

$$L' + L'' = 1 \tag{E-10}$$

$$x_1' L' + x_1'' L'' = x_1 \tag{E-11}$$

$$x_2' L' + x_2'' L'' = x_2 \tag{E-12}$$

We now have six equations that must be solved simultaneously: Three equations of equilibrium and three material balances. We also have six unknowns; they are x_1', x_2', x_1'', x_2'', L', and L''. In principle, therefore, the problem is solved, although the numerical procedure for doing so efficiently is not necessarily easy. This flash calculation for a ternary system is readily generalized to systems containing any number of components. When m components are present, we have a total of $2m$ unknowns: $2(m - 1)$ compositions and two mole numbers, L' and L''. These are found from m independent equations of equilibrium and m independent material balances.

To calculate ternary liquid-liquid equilibria at a fixed temperature, we require an expression for the molar excess Gibbs energy g^E as a function of composition;[9] this expression requires binary parameters characterizing 1-2, 1-3, and 2-3 interactions. Calculated ternary equilibria are strongly sensitive to the choice of these parameters. The success of the calculation depends directly on the care exercised in choosing these binary parameters from data reduction.

Unfortunately, realistic data reduction for a typical binary system cannot yield a *unique* set of binary parameters. To illustrate, consider data reduction for the system acetone (1)/methanol (2) reported by Anderson *et al.* (1978). Isobaric vapor-liquid equilibrium data have been obtained by Othmer at 1.006 bar; the variation in temperature is not large, from about 55 to 65°C. Using a statistical technique (the *principle of maximum likelihood*), Anderson fit Othmer's data to the van Laar equation [Eq. (7-13)] and obtains

$$A' = \frac{A}{RT} = 0.85749 \qquad B' = \frac{B}{RT} = 0.68083$$

In his analysis of the data, Anderson took experimental error into account; for each measured quantity he assigned a variance: 0.003 bar for pressure, 0.2°C for temperature, 0.005 for liquid-phase mole fraction x, and 0.01 for vapor-phase mole fraction y.

[9] At ordinary pressures and at conditions remote from critical, liquid-liquid equilibria are insensitive to the effect of pressure.

In view of the experimental uncertainties and possible unsuitability of the van Laar model for representing the data, how much confidence can we have in the values A' and B'? Are there other sets of parameters that can equally well represent the experimental data? Indeed there are, as was shown by Anderson; his calculations are summarized in Fig. E-1 that shows permissible deviations in A' and B' at two confidence levels. The smaller ellipse (87% confidence limit) gives a region in the $A'B'$ plane wherein any point gives an "acceptable" set of A' and B' parameters; in this context, "acceptable" means that the vapor-liquid equilibria calculated with an acceptable A', B' set reproduce at least 87% of the experimental data to within the estimated experimental uncertainty.

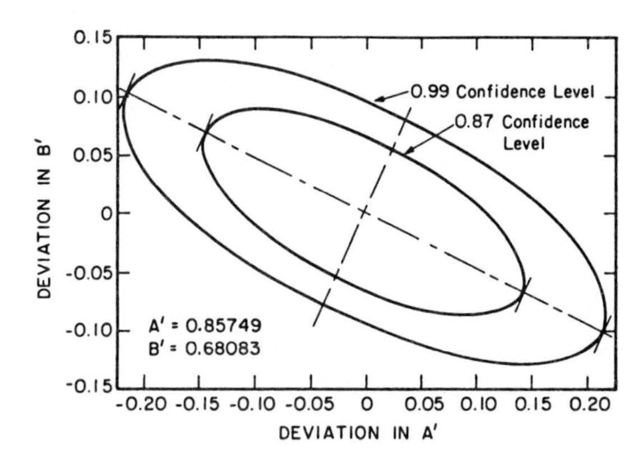

Figure E-1 Data reduction for the system acetone (1)/methanol (2). Confidence ellipses for van Laar constants.

The elliptical form of the deviation plot indicates that the two parameters are correlated; i.e., parameters A' and B' are partially coupled. Having decided on one of the parameters, only a small variation is then permissible in the other.

In this example, the confidence ellipses are large because the data indicate appreciable random experimental error. The ellipses become smaller as the quality and/or quantity of the experimental data rises and as the suitability of the model for g^E increases. However, in practice, regardless of how good and plentiful the data are, and regardless of how suitable the model for g^E may be, there is always some ambiguity in the binary parameters. In practice, no set of binary data is completely free of error and no model is perfectly suitable. Therefore, it is not possible to specify a *unique* set of binary parameters.

To calculate phase equilibria for a ternary mixture, it is necessary to estimate binary parameters for each of the three binaries that constitute the ternary. Even if good experimental data are available for each of the binaries, there is always some uncertainty in the three sets of binary parameters. To obtain reliable calculated ternary liquid-liquid equilibria, the most important task is to choose the best set of binary parameters. This choice can only be made if a few ternary liquid-liquid data are available, as discussed, for example, by Bender and Block (1975), who used the NRTL equation, and by Anderson (1978a), who used the UNIQUAC equation.

To illustrate, consider the ternary system water (1)/TCE (2)/acetone (3)[10] at 25° C. Water and TCE are only partially miscible, but the other two pairs are completely miscible. For the first pair, binary parameters are obtained from mutual solubilities and for the other pairs, from vapor-liquid equilibria. For the two completely miscible pairs, Bender and Block set NRTL parameters $\alpha_{13} = \alpha_{23} = 0.3$

To optimize the fit of the ternary data, Bender and Block first adjusted τ_{13}, τ_{31} and τ_{23}, τ_{32}, the NRTL parameters for the two miscible pairs. The adjustment was based on fitting the distribution coefficient of acetone at infinite dilution. Then Bender and Block adjusted α_{12} to give good ternary tie lines. (Once α_{12} is fixed, τ_{12} and τ_{21} immediately follow from the binary mutual solubility data.) Table E-5 gives NRTL parameters and Fig. E-2 compares calculated and experimental distribution coefficients for acetone.

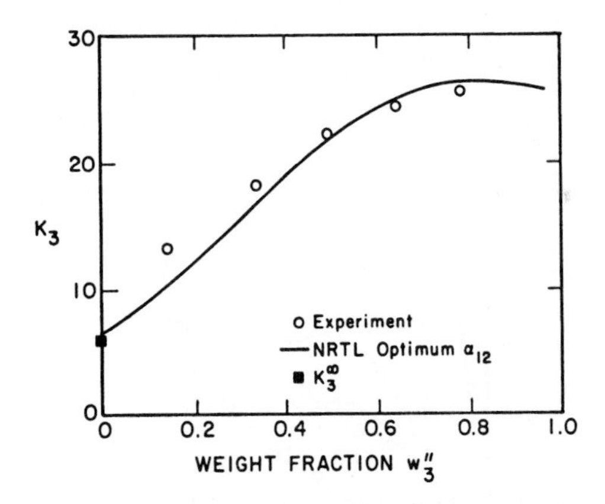

Figure E-2 Calculated and experimental distribution coefficients for acetone in the system water (1)/1,1,2-trichloroethane (2)/acetone (3). The distribution coefficient is defined with weight fractions.

Anderson(1978) used the principle of maximum likelihood to obtain UNIQUAC binary parameters from data reduction; he used mutual-solubility data for partially miscible pairs and vapor-liquid equilibrium data for completely miscible pairs. In addition, he used a few (typically, one or two) ternary tie-line data. Figure E-3 shows calculated and experimental results for four ternary systems; the corresponding binary parameters are shown in Table E-6. No ternary (or higher) parameters are required because binary parameters were chosen with care, that is, with consideration of a few ternary data.

[10] TCE = 1,1,2-trichloroethane.

Table E-5 Binary NRTL parameters used to calculate ternary liquid-liquid equilibria (Bender and Block, 1975).

System: Water (1)/TCE* (2)/Acetone (3) at 25°C				
Component				
i	j	τ_{ij}	τ_{ji}	α_{ij}
1	2	5.98775	3.60977	0.2485
1	3	1.38800	0.75701	0.3
2	3	-0.19920	-0.20102	0.3

* TCE = 1,1,2-trichloroethane.

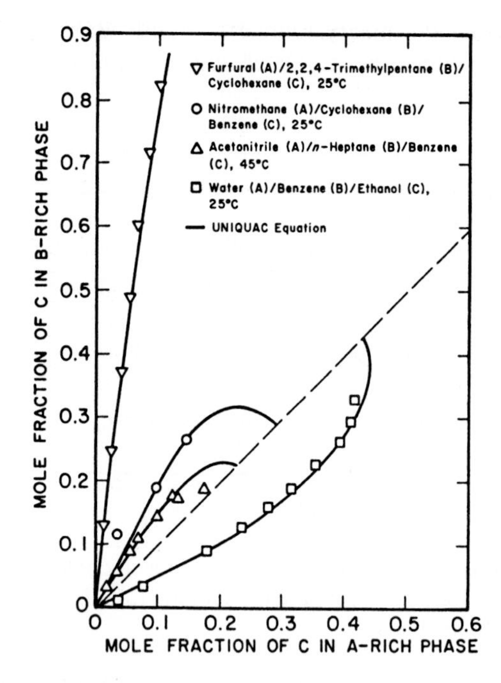

Figure E-3 Distribution of third component in four ternary systems.

Anderson (1978a) also presented a procedure for calculating liquid-liquid equilibria in systems containing four (or more) components. When all binary parameters are chosen with care, it is often possible to calculate multicomponent liquid-liquid equilibria in good agreement with experiment.

An alternate procedure, presented by Cha (1985), includes a ternary correction factor in the expression for g^E, Like Anderson's method, Cha's requires some ternary liquid-liquid equilibrium data but because it includes ternary constants, calculated results are no longer highly sensitive to the choice of binary parameters.

Table E-6 Binary UNIQUAC parameters used to calculate ternary liquid-liquid equilibria (Anderson, 1978a).

	UNIQUAC parameter* (K)					
Components	a_{12}	a_{21}	a_{13}	a_{31}	a_{23}	a_{32}
Benzene (l)/water (2)/ethanol (3)	2057.42	115.13	1131.13	-149.34	573.61	-163.72
Acetonitrile (1)/*n*-heptane (2) /benzene (3)	23.71	545.71	60.28	89.57	245.42	-135 93
Cyclohexane (1)/nitromethane (2) /benzene (3)	517.19	105.01	335.25	-173.60	73.79	82.20
2,2,4-Trimethylpentane (1) /furfural (2) / cyclohexane (3)	410.08	-4.98	141.01	-112.66	41.17	354.83

* See Eqs. (6-120) and (6-121).

References

Anderson, T. F., D. S. Abrams, and E. A. Grens, 1978, *AIChE J.*, 24: 20.

Anderson, T. F. and J. M. Prausnitz, 1978a, *Ind. Eng. Chem. Process Des. Dev.*, 17: 561.

Bender, E. and U. Block, 1975, *Verfahrenstechnik*, 9: 106.

Brian, P. L. T., 1965, *Ind. Eng. Chem. Fundam.*, 4: 101.

Cha, H. and J. M. Prausnitz, 1985, *Ind. Eng. Chem. Process Des. Dev.*, 24: 551.

Lobien, G., 1980, *Dissertation*, University of California.

Lobien, G. and J. M. Prausnitz, 1982, *Ind. Eng. Chem. Fundam.*, 21: 109.

Prausnitz, J. M., T. F. Anderson, E. A. Grens, C. A. Eckert, R. Hsieh, and J. P. O'Connell, 1980, *Computer Calculations for Multicomponent Vapor-Liquid and Liquid-Liquid Equilibria.* Englewood Cliffs: Prentice-Hall.

Renon, H. and J. M. Prausnitz, 1969, *Ind. Eng. Chem. Process Des. Dev.*, 8: 413.

APPENDIX F

Estimation of Activity Coefficients

Chapter 6 presents several equations for representing activity coefficients as a function of composition. These equations contain binary parameters that are obtained from fitting to binary data. While many binary systems have been studied experimentally (compilations have been published, for example, by Wichterle *et al.*, 1973; many volumes by Gmehling and Onken, starting in 1977; Hirata *et al.*, 1975; many volumes by Maczynski *et al.*, starting in 1976; Ohe, 1989, and other titles in Physical Sciences Data Series; continuing compilations by DIPPR-AIChE, New York), there are also many other binaries where only fragmentary data, or no data at all, are available.

When interest is directed at a binary where data are absent or incomplete, one should not immediately dismiss the possibility of obtaining some data oneself. While making reliable experimental measurements is always a challenge, modern instrumentation often simplifies the experimental effort. For a typical binary mixture of nonelectrolytes, only a few experimental data are often sufficient to yield reasonable binary parameters. In any event, there is a vast difference between a few reliable data and none at all.

Estimation from Activity Coefficients at Infinite Dilution

For partially miscible binary mixtures, binary parameters can be obtained from mutual-solubility data, as discussed in App. E, and for binary mixtures that form azeotropes, binary parameters can be calculated from azeotropic data (Gmehling *et al.*, 1994). For all binary mixtures, regardless of the extent of miscibility and regardless of whether or not they form an azeotrope, binary parameters can be calculated from *activity coefficients at infinite dilution*. In many cases, when only limited experimental data are available, activity coefficients at infinite dilution provide the most valuable experimental information (Sandler, 1996). Fortunately, there are convenient experimental methods for measuring infinite-dilution activity coefficients (Dallinga *et al.*, 1993; Eckert and Sherman, 1996; Kojima *et al.*, 1997).

For liquid mixtures where the two components do not differ greatly in volatility, activity coefficients at infinite dilution can be obtained for both components by *differential ebulliometry* (Nicolaidis and Eckert, 1978; Kojima *et al.*, 1997). For liquid mixtures where one component is much more volatile than the other, the infinite-dilution activity coefficient of the more volatile component may be obtained by *gas-liquid chromatography* (Laub and Pecsok, 1978; Kojima *et al.*, 1997). In this technique, the chromatograph is used as the equilibrium cell; the heavy component is the substrate and an inert carrier gas flows over that substrate. A small amount of light component is injected and the important measured quantity is the retention time, i.e., the time that the light component is "retained" as a result of its contact with the heavy component. For example, Comanita *et al.* (1976) measured retention times yielding K factors[1] for propanol (2) at infinite dilution in n-hexadecane using helium as a carrier gas; their results are shown in Fig. F-1. From these data, the activity coefficient γ_2 is found from

$$\gamma_2 = \frac{\varphi_2 P K_2}{f^L_{\text{pure }2}} \tag{F-1}$$

Here P is the total pressure, φ_2 is the fugacity coefficient of propanol in the carrier gas, and $f^L_{\text{pure }2}$ is the fugacity of pure liquid propanol, both at system temperature and pressure.

When component 2 is much more volatile than component 1, the activity coefficient at infinite dilution, γ_2^∞, is found from a plot of γ_2 versus P, according to

$$\gamma_2^\infty = \lim_{P \to P_1^s} \gamma_2 \tag{F-2}$$

From the data shown in Fig. F-1, the activity coefficients for propanol at infinite dilution are 4.32 (100°C), 3.24 (120°C), 2.90 (130°C), and 2.58 (140°C).

[1] $K \equiv y/x$, where y is the vapor-phase mole fraction and x is the liquid-phase mole fraction

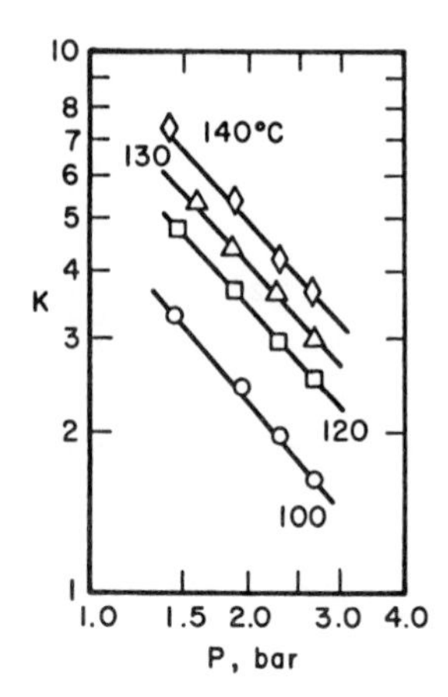

Figure F-1 *K* factors for propanol in *n*-hexadecane from gas-liquid chromatography. (Comanita *et al.*, 1976).

In this example, measurements were made to obtain the infinite-dilution activity coefficient for the more volatile component, propanol; the chromatographic method cannot be used to measure also the infinite-dilution activity coefficient of the heavy component, *n*-hexadecane. Therefore, in this case, the experimental data can be used to obtain only one parameter, e.g., the χ parameter in the Flory-Huggins equation [Eq. (8-12)] or a one-parameter form of the UNIQUAC equation.[2]

Another convenient experimental method for obtaining activity coefficients at infinite dilution is by *head-space analysis* (Hackenberg and Schmidt, 1976; Kieckbusch and King, 1979; Kolb and Ettre, 1997). In this procedure, a chromatograph is used as an analytical tool to determine the vapor-phase concentration of a trace component in equilibrium with a liquid phase of known composition. Figure F-2 shows limiting activity coefficients for some organic solutes in aqueous sucrose solutions obtained by headspace analysis (Chandrasekaran and King, 1971). This experimental method can be used to obtain infinite-dilution activity coefficients for both components in a binary mixture if both pure components have a measurable vapor pressure.

All the methods discussed above for determining infinite-dilution activity coefficients require a solute with appreciable volatility. For slightly volatile components, Trampe and Eckert (1993) have developed a method for measuring infinite dilution activity coefficients based on the accurate determination of the dew point of a mixture, because dew points are very sensitive to small amounts of low-volatile components.

---

[2] In the one-parameter form we assume that $u_{11} = -\Delta_{vap}U_1/q_1$ and $u_{22} = -\Delta_{vap}U_2/q_2$, where $\Delta_{vap}U$ is the energy of complete vaporization and q is a pure-component molecular-structure constant depending on molecular size and external surface area. At conditions remote from critical, $\Delta_{vap}U \approx \Delta_{vap}H - RT$, where $\Delta_{vap}H$ is the enthalpy of vaporization. Further, we assume $u_{12} = u_{21} = -(u_{11}u_{22})^{1/2}(1 - c_{12})$, where c_{12} is the single adjustable binary parameter (Bruin, 1971).

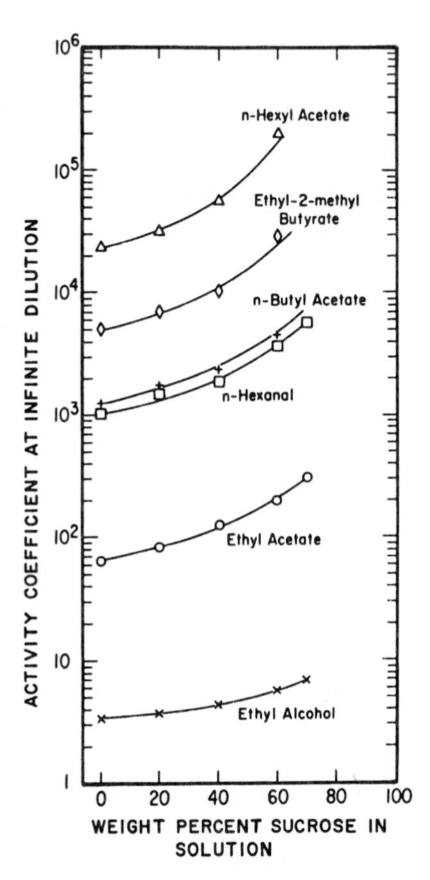

Figure F-2 Activity coefficients at infinite dilution for organic solutes in aqueous sucrose solutions at 25°C obtained by gas-liquid chromatography (Chandrasekaran and King, 1971).

When activity coefficients at infinite dilution are available for both components in a binary mixture, activity coefficients at intermediate compositions can be estimated easily. For a binary mixture, we choose some expression F for the molar excess Gibbs energy g^E containing two binary parameters A and B,

$$g^E = F(x, A, B) \qquad \text{(F-3)}^3$$

where x is the mole fraction and parameters A and B depend on temperature. From Eq. (6-25) we then obtain

[3] See Sec. 6.10.

$$RT \ln \gamma_1 = F_1(x, A, B) \tag{F-4}$$

$$RT \ln \gamma_2 = F_2(x, A, B) \tag{F-5}$$

where functions F_1 and F_2 follow from the choice of function F in Eq. (F-3).

At infinite dilution,

$$\gamma_1 \to \gamma_1^\infty \quad \text{as} \quad x_1 \to 0$$

$$\gamma_2 \to \gamma_2^\infty \quad \text{as} \quad x_2 \to 0$$

Therefore, experimental values for γ_1^∞ and γ_2^∞ are sufficient to yield binary parameters A and B by simultaneous solution of Eqs. (F-4) and (F-5). In some fortunate cases, simultaneous solution is not necessary; e.g., if the van Laar equation[4] is used for Eq. (F-3), we obtain

$$\ln \gamma_1^\infty = \frac{A}{RT} = A' \tag{F-6}$$

$$\ln \gamma_2^\infty = \frac{B}{RT} = B' \tag{F-7}$$

In the general case, however, $\ln \gamma_1^\infty$ and $\ln \gamma_2^\infty$ depend on *both* parameters A and B, requiring simultaneous solution. For example, if we choose Wilson's equation[5]

$$\ln \gamma_1^\infty = -\ln \Lambda_{12} + 1 - \Lambda_{21} \tag{F-8}$$

$$\ln \gamma_2^\infty = -\ln \Lambda_{21} + 1 - \Lambda_{12} \tag{F-9}$$

where binary parameters Λ_{12} and Λ_{21} have replaced A and B. In this case, $\ln \gamma_1^\infty$ and $\ln \gamma_2^\infty$ depend on both binary parameters.

Once the binary parameters are known, activity coefficients can be predicted over the entire range of liquid composition. When experimental values for γ_1^∞ and γ_2^∞ are reliable, such predictions tend to be very good. To illustrate, Figure F-3 indicates that when only γ_1^∞ and γ_2^∞ are used, good agreement is obtained with experiment for the system acetone/benzene at 45°C (Lobien, 1982).

[4] See Eqs. (6-97) and (6-98).
[5] See Eqs. (6-105) and (6-106).

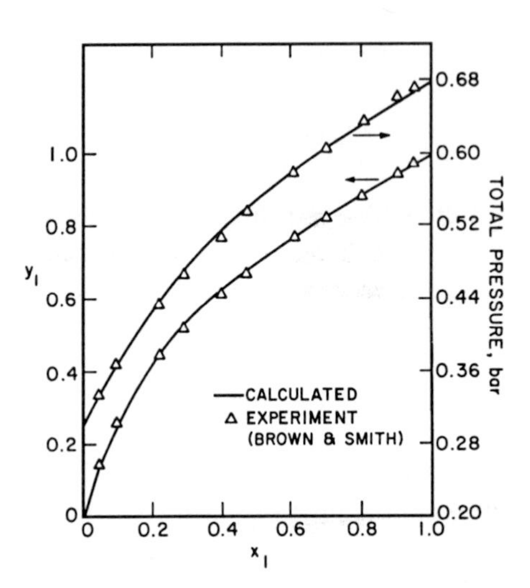

Figure F-3 Vapor-liquid equilibria for the system acetone (1)/benzene (2) at 45°C. Calculations based on Wilson's equation using $\gamma_1^\infty = 1.65$ and $\gamma_2^\infty = 1.57$ (Lobien, 1982).

Estimation from Group-Contribution Methods

An attractive (but not always accurate) method for estimating activity coefficients is provided by equations based on the *group-contribution* concept presented in Sec. 7.8. In a group-contribution method, a molecule is divided (somewhat arbitrarily) into functional groups; molecule-molecule interactions are given by properly weighted sums of group-group interactions. A review of group-contribution methods is given by Fredenslund and Sørensen (1994) and by Gmehling (1998).

From the many different models proposed, ASOG and UNIFAC have the widest practical interest. For both methods, group-group interaction parameter tables are available, e.g. those by Tochigi *et al.* (1990, 1998) for ASOG and those by Hansen *et al.* (1991), Gmehling *et al.* (1993) and Fredenslund and Sørensen (1994) for UNIFAC. UNIFAC is successful for semiquantitative predictions of vapor-liquid equilibria for a wide variety of mixtures, including those containing polymers. It also has been used to predict activity coefficients at infinite dilution (Voutsas and Tassios, 1996; Möllmann and Gmehling, 1997; Zhang *et al.*, 1998). Some extensions of UNIFAC consider the solubilities of gases and light hydrocarbons in water (Li *et al.*, 1997) and aqueous solutions of biochemicals, such as sugars, imino acids and organic and inorganic salts (Kuramochi *et al.*, 1997, 1998).

The most common application of UNIFAC is for prediction of VLE of ordinary solutions. To illustrate, Fig. F-4 compares UNIFAC predictions with experimental vapor-liquid equilibrium for the system methanol/water at 50°C. In the entire composition range agreement is very good. However, such good agreement is not typical.

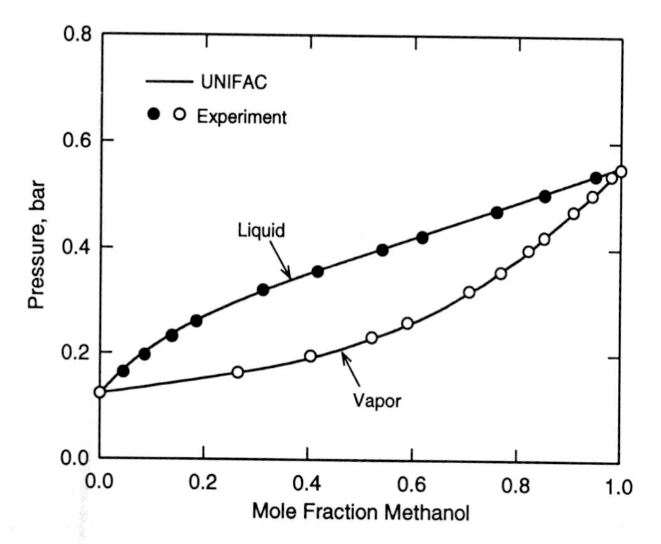

Figure F-4 Vapor-liquid equilibria at 50°C for the system methanol/water. Experimental data from McGlashan and Williamson (1976). In this case agreement is very good but for many other cases it is not.

A second example is shown in Fig. F-5 for the system *n*-hexane/methyl ethyl ketone. Here agreement is not good, indicating that for numerous systems, UNIFAC gives only a semi-quantitative prediction.

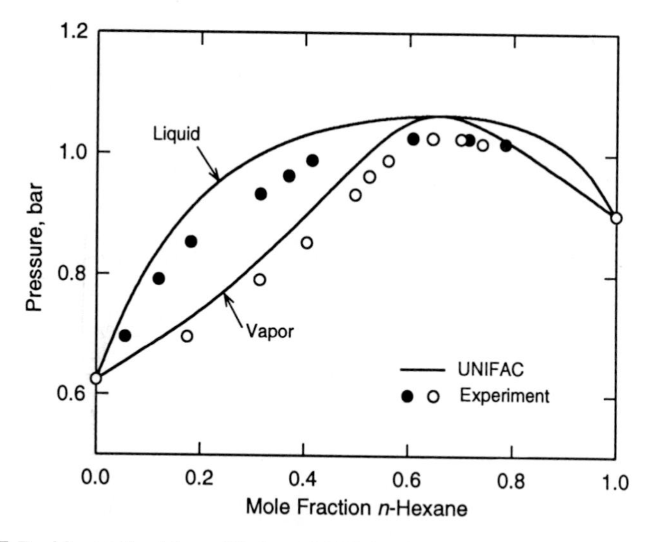

Figure F-5 Vapor-liquid equilibria at 65°C for the system *n*-hexane/methyl ethyl ketone. Experimental data from Maripuri and Ratcliff (1972). Agreement is only semi-quantitative.

UNIFAC with parameters based on vapor-liquid equilibrium data does not yield reliable predictions for liquid-liquid equilibria. Estimates of liquid-liquid equilibria are only possible with UNIFAC parameters determined especially for describing such equilibria (Magnussen *et al.*, 1981; Gupte and Danner, 1987). A correlation restricted to water-hydrocarbon liquid-liquid equilibria was presented by Hooper *et al.* (1988).

UNIFAC has also been used to estimate binary constants in an equation of state. For example, in the quadratic mixing rule for constant a in equations of the van der Waals form, UNIFAC can be used to estimate a_{ij} (where $i \neq j$).

Numerous authors have discussed *group-contribution equations of state*. Illustrative of their work is that of Gmehling and coworkers (Holderbaum and Gmehling, 1991; Fischer and Gmehling, 1996; Gmehling *et al.*, 1997; Li *et al.*, 1998). These authors developed a group-contribution equation of state based on the Soave-Redlich-Kwong equation for the prediction of gas solubilities and vapor-liquid equilibria at low and high pressures.

Oishi (1978) extended UNIFAC to polymer solutions by adding a free-volume contribution, as suggested by Prigogine-Flory-Patterson theory (Sec. 8.3), to account for the free-volume difference between polymer and solvent molecules. While this difference is usually insignificant for liquid mixtures of small molecules, it is important for polymer/solvent systems.

Another extension of UNIFAC to polymer solutions is given by Holten-Andersen *et al.* (1987) who obtain the free-volume contribution from an equation of state similar to that based on perturbed-hard-chain theory (Sec. 7.16). This equation of state has a simplified attraction term that contains a UNIQUAC-like expression for mixtures. Parameters are obtained from a group-contribution correlation. To illustrate, Table F-1 shows experimental and calculated infinite-dilution solvent activity coefficients for a variety of polymer/solvent systems. While there are some exceptions, agreement between experimental and calculated results is within 10%.

Goydan *et al.* (1989) have compared predictions of the UNIFAC free-volume model of Oishi and the Holten-Andersen equation of state for solvent activities in a variety of polymer solutions. They found that the Holten-Andersen correlation is somewhat more accurate, while Oishi's model, also fairly accurate, is more widely applicable. In a similar investigation, High and Danner (1990) also found that both models predict solvent activities reasonably well.

Elbro *et al.* (1990) propose to include free-volume contributions to solvent activities in polymer solutions in a remarkably simple way. Starting with a generalized van der Waals partition function (Sec. 7.15), the combinatorial and free-volume terms are combined to give for an athermal solution:

$$\frac{s^{E,\text{fv}}}{R} = -\frac{g^{E,\text{fv}}}{RT} = -\sum_i x_i \ln \frac{\Phi_i^{\text{fv}}}{x_i} \tag{F-10}$$

where superscript fv denotes free volume and Φ_i^{fv} is the free-volume fraction of component i, given by

Table F-1 Experimental and predicted infinite-dilution activity coefficients ($\gamma_1 = a_1/w_1$) (Holten-Andersen *et al.*, 1987).

Component 1*	Component 2*	T (K)	γ_1^∞ (Exp.)	γ_1^∞ (Calc.)	Ref.
n-C_6H_{14}	n-$C_{16}H_{34}$	293	2.40	2.40	(a)
n-C_7H_{16}	n-$C_{32}H_{66}$	346	3.06	3.10	(a)
n-C_6H_{14}	PIB	298	6.4	5.8	(b)
n-C_7H_{16}	PIB	298	5.8	5.3	(b)
n-C_8H_{18}	PIB	298	5.4	5.1	(b)
Benzene	PEO	361	4.7	4.5	(c)
MEK	PEO	361	4.6	5.2	(c)
Butyl Acetate	PEO	361	5.0	5.7	(c)
Ethanol	PEO	373	7.1	9.1	(d)
1-Propanol	PEO	373	5.5	5.8	(d)
Octane	PEO	373	28.0	44.0	(d)
Toluene	PS	423	5.2	5.1	(e)
Toluene	PVAC	423	6.4	5.8	(f)
MEK	PVAC	423	6.3	6.2	(f)
Butyl acetate	PVAC	423	6.9	7.2	(f)
2-Propanol	PVAC	423	7.5	7.9	(f)
1-Butanol	PVAC	423	6.6	6.3	(f)
Hexane	PBMA	423	10.4	9.8	(f)
Benzene	PBMA	423	4.6	4.6	(f)
MEK	PBMA	423	7.4	6.9	(f)
Butyl acetate	PBMA	423	5.9	5.6	(f)
1- Butanol	PBMA	423	7.8	8.6	(f)

* MEK = methyl ethyl ketone; PIB = polyisobutylene; PEO = poly(ethylene oxide); PS = polystyrene; PVAC = poly(vinyl acetate); PBMA = poly(butyl methacrylate).

(a) P. J. Flory and R. A. Orwoll, 1967, *J. Am. Chem. Soc.* 89: 76, 6822; (b) Y. Leung and B. E. Eichinger, 1974, *Macromolecules* 7: 685; (c) Y. L. Cheng and D. C. Bonner, 1974, *Macromolecules*, 7: 687; (d) M. T. Rätzsch, P. Glindemann, and E. Hamann, 1980, *Acta Polym.*, 31: 377; (e) F. H. Cowitz and J. W. King, 1972, *J. Polym. Sci., Part A-1*, 10: 689; (f) R. D. Newman and J. M. Prausnitz, 1973, *J. Paint Technol.*, 45: 33.

$$\Phi_i^{\text{fv}} = \frac{x_i(v_i - v_i^*)}{\sum_j x_j(v_j - v_j^*)} \tag{F-11}$$

where v_i and v_i^* are, respectively, the pure-component molar volumes and the hard-core volumes; x_i is the mole fraction of component i. In Eq. (F-11), if the volume differences $(v_i - v_i^*)$ are replaced by the hard-core volumes v_i^*, the free-volume fraction Φ_i^{fv} is identical to the segment fraction Φ_i^* given in Eq. (8-4). In that case, Eq. (F-10) gives the combinatorial Flory-Huggins excess entropy, i.e. for a binary mixture Eq. (F-10) is then identical to Eq. (8-6).

For the activity coefficient of an athermal solution, Elbro obtained:

$$\ln\gamma_i^{\text{fv}} = \ln\left(\frac{\Phi_i^{\text{fv}}}{x_i}\right) + 1 - \frac{\Phi_i^{\text{fv}}}{x_i} \tag{F-12}$$

In Eq. (F-12), $\gamma_i^{\text{fv}} = a_i^{\text{fv}} / x_i$, where a_i^{fv} is the activity of component i due to combinatorial and free-volume effects.

Equation (F-12) reflects only the combinatorial and the free-volume contributions and, therefore, for real (non-athermal) polymer solutions a residual term has to be added, as discussed in Sec. 8.2. The total activity coefficient is given by

$$\ln\gamma_i = \ln\gamma_i^{\text{fv}} + \ln\gamma_i^{R} \tag{F-13}$$

Elbro *et al.* (1990) adopted the UNIQUAC residual term that gives the residual activity coefficient[6] γ_i^R for a binary *ij*,

$$\boxed{\ln\gamma_i^R = -q_i' \ln(\theta_i' + \theta_j'\tau_{ji}) + \theta_j' q_i'\left(\frac{\tau_{ji}}{\theta_i' + \theta_j'\tau_{ji}} - \frac{\tau_{ij}}{\theta_j' + \theta_i\tau_{ij}}\right)} \tag{F-14}$$

with

$$\theta_i' = \frac{x_i q_i'}{x_i q_i' + x_j q_j'} = 1 - \theta_j' \tag{F-15}$$

and

$$\tau_{ij} = \exp\left(-\frac{a_{ij}}{T}\right) \quad \text{and} \quad \tau_{ji} = \exp\left(-\frac{a_{ji}}{T}\right) \tag{F-16}$$

where x_i is the mole fraction, parameters q' are pure-component molecular-structure constants depending on molecular size and external surface area, a_{ij} and a_{ji} are characteristic energy parameters dependent on temperature but often that dependence is small.

Parameters a_{ij} and a_{ji} may be estimated from UNIFAC. For polymer/solvent systems they also can be obtained from vapor-liquid-equilibrium data for binary mixtures of the solvent and small-molecule homologues of the polymer. To illustrate, Fig. F-6 compares calculated and experimental solvent activities as a function of solvent weight fraction for the system polystyrene/tetrachloromethane. The curves were calculated

[6] See Eq. (6-122)

with energy parameters obtained from VLE at 293 K for the binary ethylbenzene/tetrachloromethane. Elbro's model gives excellent results; when the simple Flory-Huggins combinatorial term (without free-volume correction) is combined with the UNIQUAC residual contribution, calculated results are poor. Figure F-6 shows the importance of free-volume effects when "scaling-up" data from a low-molecular-weight mixture to predict properties of a polymer solution. In general, Elbro's model compares favorably with the Holten-Andersen equation of state and with Oishi's method.

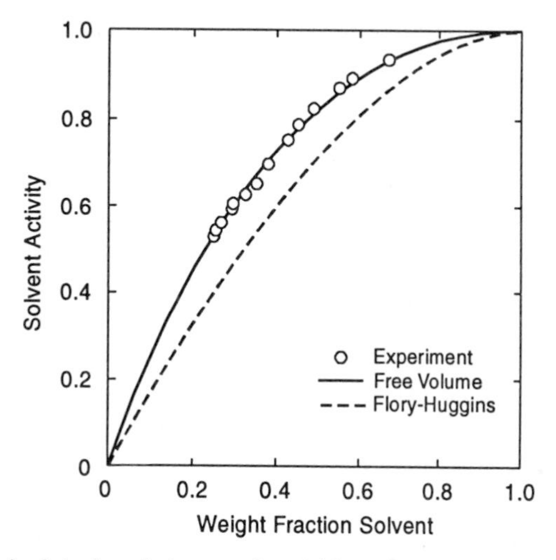

Figure F-6 Calculated and observed activities of tetrachloromethane in polystyrene at 293 K. In both calculations, the residual contribution was calculated with energy parameters obtained from VLE at 293 K for the binary ethylbenzene/ tetrachloromethane (Elbro *et al.*, 1990).

Other group-contribution methods for polymer/solvent systems include those by Chen *et al.* (1990), High and Danner (1990a) and Lee and Danner (1996, 1996a, 1996b, 1997). At present all of the published group-contribution methods provide good estimates of solvent activities for those solvent/polymer systems where the polymer is readily soluble and where both components are nonpolar or weakly polar. Hydrogen bonded systems often require modifications with special parameters. At present, there is no reliable method for estimating liquid-liquid equilibria in polymer/solvent systems.

Group-contribution methods are attractive because they are based on the assumption of group independence; this assumption says, for example, that the properties of say, a carbonyl (C=O) group in a monoketone are the same as those in a diketone, and that the properties of this carbonyl group in a linear ketone (e.g. hexanone) are the same as those in a cyclic ketone (e.g. cyclohexanone). Also, a group-contribution method does not distinguish between hexanone-1 and hexanone-3; the position of the carbonyl group in the molecule is not considered. These assumptions are, of course, not correct. It is possible to relax these assumptions by introducing more groups, e.g. by distinguishing between a carbonyl group in a linear molecule and that in a cyclic molecule. Agreement with experiment can then be improved but at a high cost: more parameters are required; to obtain them we need more experimental data. In the limit, as we define more and more groups, the advantage of the group-contribution method is lost. A useful group-contribution method is a compromise between on the one hand, using many different groups as suggested by our knowledge of molecular structure, and on the other, using only a limited number of groups, as required by a limited bank of experimental data.

An exciting method for improving accuracy and for increasing the range of applicability of group-contribution methods is provided by quantum-chemical calculations to give group-group interaction parameters from first principles. Initial efforts toward that end have been reported by Sandler and coworkers (Wu and Sandler, 1991; Wolbach and Sandler, 1997).

The objective of group-contribution methods is to use existing phase equilibrium data to predict phase equilibria for systems where no data are available. While such predictions may be used for preliminary design purposes, group-contribution methods yield only approximate vapor-liquid equilibria. Whenever reliable experimental data are available, these should be used instead of group-contribution predictions. While Raoult's law provides a zeroth approximation for phase equilibria, group-contribution methods often provide no more than a first approximation.

References

Bruin, S. and J. M. Prausnitz, 1971, *Ind. Eng. Chem. Process Des. Develop.,* 10: 562.

Chandrasekaran, S. K. and C. J. King, 1971, *Chem. Eng. Symp. Ser.,* 67: 122.

Chen, F., Aa. Fredenslund, and P. Rasmussen, 1990, *Ind. Eng. Chem. Res.,* 29: 875.

Comanita, V. J., R. A. Greenkorn, and K. C. Chao, 1976, *J. Chem. Eng. Data,* 21: 491.

Dallinga, L., M. Schiller, and J. Gmehling, 1993, *J. Chem. Eng. Data,* 38: 147.

Eckert, C. A. and S. R. Sherman, 1996, *Fluid Phase Equilibria,* 116: 333.

Elbro, H. S., Aa. Fredenslund, and P. Rasmussen, 1990, *Macromolecules,* 23: 4707.

Fischer, K. and J. Gmehling, 1996, *Fluid Phase Equilibria,* 121: 185.

Fredenslund, Aa. and J. M. Sørensen, 1994, *Group-Contribution Estimation Methods.* In *Models for Thermodynamic and Phase-Equilibria Calculations,* (S. I. Sandler, Ed.). New York: Marcel Dekker.

Gmehling, J. and U. Onken, 1977, *Vapor-Liquid Equilibrium Data Collection* (Chemistry Data Series). Frankfurt: DECHEMA. (Series of volumes starting in 1977).

Gmehling, J., J. Li, and M. Schiller, 1993, *Ind. Eng. Chem. Res.,* 32: 178.

Gmehling, J., J. Menke, K. Fischer, and J. Krafczyk, 1994, *Azeotropic Data.* New York: John Wiley & Sons.

Gmehling, J., J. Li, and K. Fischer, 1997, *Fluid Phase Equilibria,* 141: 113.

Gmehling, J., 1998, *Fluid Phase Equilibria,* 144: 37.

Goydan, R., R. C. Reid, and H.-S. Tseng, 1989, *Ind. Eng. Chem. Res.,* 28: 445.

Gupte, P. A. and R. P. Danner, 1987, *Ind. Eng. Chem. Res.,* 26: 2036.

Hackenberg, H. and A. P. Schmidt, 1976, *Gas Chromatographic Headspace Analysis.* London: Heyden & Son.

Hansen, H. K., P. Rasmussen, Aa. Fredenslund, M. Schiller, and J. Gmehling, 1991, *Ind. Eng. Chem. Res.,* 30: 2352.

High, M. S. and R. P. Danner, 1990, *Fluid Phase Equilibria,* 55: 1.

High, M. S. and R. P. Danner, 1990a, *AIChE J.,* 36: 1625.

Hirata, M., S. Ohe, and K. Nagahama, 1975, *Computer-Aided Data Book of Vapor-Liquid Equilibria.* Amsterdam: Elsevier.

Holderbaum, T. and J. Gmehling, 1991, *Fluid Phase Equilibria,* 70: 251.

Holten-Andersen, J., P. Rasmussen, and Aa. Fredenslund, 1987, *Ind. Eng. Chem. Res.,* 26: 1382.

Hooper, H. H., S. Michel, and J. M. Prausnitz, 1988, *Ind. Eng. Chem. Res.,* 27: 2182.

Kieckbusch, T. G. and C. J. King, 1979, *Chromatogr. Sci.,* 17: 273; *ibid., J. Agric. Food Chem.,* 17: 504.

Kojima, K., S. J. Zhang, and T. Hiaki, 1997, *Fluid Phase Equilibria,* 131: 145.

Kolb, B. and L. S. Ettre, 1997, *Static Headspace-Gas Chromatography: Theory and Practice.* New York: John Wiley & Sons.

Kuramochi, H., H. Noritomi, D. Hoshino, and K. Nagahama, 1997, *Fluid Phase Equilibria,* 130: 117.

Kuramochi, H., H. Noritomi, D. Hoshino, S. Kato, and K. Nagahama, 1998, *Fluid Phase Equilibria,* 144: 87.

Laub, R. J. and R. L. Pecsok, 1978, *Physicochemical Applications of Gas Chromatography.* New York: John Wiley & Sons.

Lee, B.-C. and R. P. Danner, 1996, *AIChE J.,* 42: 837.

Lee, B.-C. and R. P. Danner, 1996a, *Fluid Phase Equilibria,* 117: 33.

Lee, B.-C. and R. P. Danner, 1996b, *AIChE J.,* 42: 3223.

Lee, B.-C. and R. P. Danner, 1997, *Fluid Phase Equilibria,* 128: 97.

Li, J., I. Vanderbeken, S. Ye, H. Carrier, and P. Xans, 1997, *Fluid Phase Equilibria,* 131: 107.

Li, J., K. Fischer, and J. Gmehling, 1998, *Fluid Phase Equilibria,* 143: 71.

Lobien, G. and J. M. Prausnitz, 1982, *Ind. Eng. Chem. Fund.,* 21: 109.

Maczynski, A., *et al.*, 1976, *Verified Vapor-Liquid Equilibrium Data,* Thermodynamic Data for Technology, Series A, PWN. Polish Scientific Publishers. (Series of volumes starting in 1976).

Magnussen, T., P. Rasmussen, and A. Fredenslund, 1981, *Ind. Eng. Chem. Res.,* 20: 331.

Maripuri, V. C. and G. A. Radcliff, 1972, *J. Appl. Chem. Biotechnol.,* 22: 899.

McGlashan, M. L. and A. G. Williamson, 1976, *J. Chem. Eng. Data,* 21: 196.

Möllmann, C. and J. Gmehling, 1997, *J. Chem. Eng. Data,* 42: 35.

Nicolaides, G. L. and C. A. Eckert, 1978, *Ind. Eng. Chem. Fundam.,* 17: 331; see also, C. A. Eckert *et al.*, 1982, *J. Chem. Eng. Data,* 27: 233, 399.

Ohe, S., 1989, *Vapor-Liquid Equilibrium Data,* Physical Sciences Data Series, Vol. 37. Amsterdam: Elsevier.

Oishi, T. and J. M. Prausnitz, 1978, *Ind. Eng. Chem. Process Des. Dev.,* 17: 333.

Panayiotou, C. and J. H. Vera, 1982, *Polymer Journal,* 14: 681.

Sandler, S. I.. 1996, *Fluid Phase Equilibria,* 116: 343.

Tochigi, K., D. Tiegs, J. Gmehling, and K. Kojima, 1990, *J. Chem. Eng. Japan*, 23: 453.

Tochigi, K., 1998, *Fluid Phase Equilibria,* 144: 59.

Trampe, D. B. and C. A. Eckert, 1993, *AIChE J.,* 39: 1045.

Voutsas, E. C. and D. P. Tassios, 1996, *Ind. Eng. Chem. Res.,* 35: 1439; Errata, *ibid.,* 36: 936.

Wichterle, I., J. Linek, and E. Hala, 1973, *Vapor-Liquid Equilibrium Data Bibliography,* (and later supplements). Amsterdam: Elsevier.

Wolbach, J. P. and S. I. Sandler, 1997, *AIChE J.,* 43: 1589, 1597.

Wu, H. S. and S. I. Sandler, 1991, *Ind. Eng. Chem. Res.,* 30: 881, 889.

Zhang, S., T. Hiaki, M. Hongo, and K. Kojima, 1998, *Fluid Phase Equilibria,* 144: 97.

APPENDIX G

A General Theorem for Mixtures with Associating or Solvating Molecules

Many fluid mixtures of practical interest contain molecules that exhibit strong interactions (e.g., hydrogen bonding). The properties of such mixtures can often be interpreted by assuming that the molecules in the mixture are not only monomers – as given by the "apparent" (stoichiometric) composition – but also dimers, trimers, etc. and complexes containing dissimilar components. It is further assumed that all "true" species are in chemical equilibrium.

These attractive assumptions lead to an immediate problem: What is the relation between the chemical potential of the "apparent" monomer and that of the "true" monomer? Standard thermodynamic measurements give us only the composition and chemical potential of the "apparent" monomer. How are these related to the composition and chemical potential of the "true" monomer that we (usually) do not know experimentally? The simple proof below shows that without any additional assumptions, the chemical potential of the "apparent" monomer is always equal to that of the "true" monomer.

Consider a binary mixture containing n_A moles of component A and n_B moles of component B. The chemical potentials are μ_A and μ_B. These quantities (n_A, n_B, μ_A, and μ_B) are obtained in typical thermodynamic measurements. We assume that component A exists not only as monomer A_1 but also, because of association, as dimer A_2, trimer A_3, etc. We make a similar assumption for component B.

In addition, we assume that molecules of components A and B may solvate to form complexes of the type A_iB_j, where i and j are positive integers. For each component there is a material balance

$$n_A = \sum_i i n_{A_i} + \sum_i \sum_j i n_{A_iB_j} \tag{G-1}$$

$$n_B = \sum_i i n_{B_i} + \sum_i \sum_j i n_{A_iB_j} \tag{G-2}$$

By assumption, all "true" species are in chemical equilibrium. That is, every *association reaction*

$$iA_1 \rightleftharpoons A_i \quad \text{and} \quad iB_1 \rightleftharpoons B_i$$

and every solvation reaction

$$iA_1 + jB_1 \rightleftharpoons A_iB_j$$

attains its equilibrium state:

$$\mu_{A_i} = i\mu_{A_1} \tag{G-3}$$

$$\mu_{B_i} = i\mu_{B_1} \tag{G-4}$$

$$\mu_{A_iB_j} = i\mu_{A_1} + j\mu_{B_1} \tag{G-5}$$

To relate μ_A to μ_{A_1} and μ_B to μ_{B_1}, we use the exact differential of the Gibbs energy at constant temperature and pressure. First, we write this differential for the "apparent" (stoichiometric) case, i.e., where we are concerned only with components A and B as such, without consideration of the molecular forms of these components:

$$dG = \mu_A dn_A + \mu_B dn_B \tag{G-6}$$

Next, we write the same differential for the "true" case, i.e., where we postulate the existence of dimers, trimers, etc.:

$$dG = \sum_i \mu_{A_i} dn_{A_i} + \sum_i \mu_{B_i} dn_{B_i} + \sum_i \sum_j \mu_{A_iB_j} dn_{A_iB_j} \tag{G-7}$$

Substituting the chemical equilibria [Eqs. (G-3), (G-4), and (G-5)] into Eq. (G-7), we obtain

$$dG = \mu_{A_1}(\sum_i i dn_{A_i} + \sum_i \sum_j i dn_{A_iB_j}) + \mu_{B_1}(\sum_i i dn_{B_i} + \sum_i \sum_j j dn_{A_iB_j}) \qquad \text{(G-8)}$$

Substituting the material balances [Eqs. (G-1) and (G-2)] into Eq. (G-8), we obtain

$$dG = \mu_{A_1} dn_A + \mu_{B_1} dn_B \qquad \text{(G-9)}$$

We now compare Eq. (G-6) with Eq. (G-9). Because the two equations must be identical for all values of dn_A and dn_B it follows that

$$\boxed{\begin{aligned} \mu_A &= \mu_{A_1} \\ \mu_B &= \mu_{B_1} \end{aligned}} \qquad \text{(G-10)}$$

For convenience, the proof given here is for a binary mixture. However, the same proof can be extended to mixtures containing any number of components, yielding the same results.

Equation (G-10) is important because it relates a readily measurable quantity (left-hand side) to another, not readily measured quantity that is useful for construction of "chemical" models to explain nonideal behavior. Equation (G-10) provides the key for relating common thermodynamic quantities (such as fugacity coefficients φ_A and φ_B or activity coefficients γ_A and γ_B) to mixture models that postulate the existence of associated or solvated molecules.

Equation (G-10) assumes *only* that all postulated monomers, dimers, etc. are in chemical equilibrium. It is independent of any assumption concerning the mode of association (linear or cyclic) or of any assumption concerning physical interactions between the postulated "true" species. In particular, Eq. (G-10) is *not* limited to the so-called "ideal" associated (or solvated) mixture where the "true" species form an ideal mixture, i.e., one where, at constant temperature and pressure, the fugacity of a true species is proportional to its concentration.

Although Eq. (G-10) was derived early in the twentieth century, it was not until 1954 that its importance became well known. In that year, D. H. Everett published his edited translation of *Chemical Thermodynamics* by I. Prigogine and R. Defay (1954),[1] originally published in French. Chapter 26 of that splendid book gives an excellent discussion of the properties of associated and solvated liquid mixtures.

[1] (London: Longmans & Green).

APPENDIX H

Brief Introduction to Perturbation Theory of Dense Fluids

Statistical mechanics provides a powerful tool for representing the equilibrium properties of pure and mixed fluids because it provides the link between microscopic and macroscopic properties; if we can establish some quantitative relations concerning the properties of a small assembly of molecules, statistical mechanics provides a method for "scaling up" those relations to apply to a very large number of molecules. These scale-up relations can then be compared with ordinary thermodynamic properties as measured in typical (macroscopic) experiments.

Because we do not have a satisfactory *physical* understanding of dense fluids, we cannot establish a truly satisfactory theory of dense fluids (except for very simple cases). This difficulty does not follow from inadequate statistical mechanics but from inadequate knowledge of fluid structure and intermolecular forces. To overcome this difficulty, it has become common practice, first, to focus attention on the properties of some idealized dense fluid, and second, to relate the properties of a real dense fluid to those of the idealized fluid. This procedure is the essence of *perturbation theory*.

The philosophical basis of perturbation theory is very old; the fundamental ideas can be found in ancient Greek texts (e.g., Plato) and they have had a profound influence on science: Because the properties of nature are not easily understood, it is useful to postulate an idealized nature (whose properties can be specified) and then to establish "corrections" that take into account differences between real and idealized nature. In this context, the words "correction" and "perturbation" are equivalent. If we choose a good idealization, the "corrections" will be small; in that event, first-order perturbations will be sufficient.

In statistical thermodynamics, the best-known perturbation theory is the virial equation for imperfect gases at modest densities. The reference system (idealized fluid) is the ideal gas; compressibility factor z is written as an expansion in density ρ about z_0, where subscript 0 stands for the reference system.

For a fluid at constant temperature and composition,

$$z = \frac{P}{\rho RT} = z_0 + z_1 + z_2 + \ldots \tag{H-1}$$

where

$$z_0 = z(\text{ideal gas}) \quad (\text{i.e., } z \text{ when } \rho = 0)$$

$$z_1 = \left(\frac{\partial z}{\partial \rho}\right)_{\rho=0} \rho$$

$$z_2 = \frac{1}{2!}\left(\frac{\partial^2 z}{\partial \rho^2}\right)_{\rho=0} \rho^2, \text{ etc.}$$

As discussed in Chap. 5, the first perturbation term (z_1) leads to the second virial coefficient, the second perturbation term leads to the third virial coefficient, and so on. Because the reference system is an ideal gas, Eq. (H-1) is useful only for gases not excessively far from the ideal-gas state. The two most important requirements for a good perturbation theory are that the reference system is as close as possible to the real system and that the properties of the reference system are known accurately.

For dense fluids, it is clear that an ideal gas would be a very bad reference system. To describe dense fluids well, the key problem is to choose a suitable (idealized) reference fluid. Much attention has been given to this problem, as discussed in numerous references. Here we give only a short introduction to indicate some basic ideas.

Consider N molecules of a pure fluid at volume V and temperature T. As discussed in any book on statistical mechanics and as briefly reviewed in App. B, the Helmholtz energy A can be divided into two contributions: The first depends only on temperature, while the second depends also on density. It is only the second part, called the *configurational Helmholtz energy* that is of interest here, because only configurational properties depend on intermolecular forces.

Consistent with the purposes of this introduction, we now limit attention to fluids containing small spherical molecules.[1] The configurational Helmholtz energy of a classical fluid[2] is written[3]

$$A = -kT \ln \int e^{-U/kT} d\mathbf{r}_1 ... d\mathbf{r}_N \tag{H-2}$$

where U is the total potential energy of the fluid and $\mathbf{r}_1$ is the position of molecule 1, etc.

We want to expand A as a power series in some perturbing parameter λ such that at constant temperature and density,

$$A = A_0 + A_1 + A_2 + ... \tag{H-3}$$

where subscript 0 refers to the reference system and where

$$A_1 = \left(\frac{\partial A}{\partial \lambda}\right)_{\lambda=0} \lambda \tag{H-4}$$

$$A_2 = \frac{1}{2!}\left(\frac{\partial^2 A}{\partial \lambda^2}\right)_{\lambda=0} \lambda^2, \text{ etc.} \tag{H-5}$$

The perturbation parameter λ is defined through potential energy U,

$$U_\lambda = U_0 + \lambda U_p \tag{H-6}$$

where subscript p stands for perturbation. Here U_0 is the (total) reference potential and U_p is the total perturbation potential such that when $\lambda = 1$, U_λ is the total potential of the real fluid.

When $\lambda = 1$, we obtain the desired perturbation expansion,

$$A = A_0 + \left(\frac{\partial A}{\partial \lambda}\right)_{\lambda=0} + \frac{1}{2!}\left(\frac{\partial^2 A}{\partial \lambda^2}\right)_{\lambda=0} + ... \tag{H-7}$$

---

[1] In general, perturbation methods are not limited to small spherical molecules; in principle they are applicable to fluids containing molecules of arbitrary complexity.

[2] A classical fluid is one whose properties are described by classical (as opposed to quantum) mechanics.

[3] To simplify notation, we use only a single integral sign to denote multiple integration over all coordinates.

To obtain the first-order perturbation, we first write

$$A = -kT \ln Z_\lambda \tag{H-8}$$

where

$$Z_\lambda \equiv \int e^{-U_\lambda/kT} d\mathbf{r}_1 \ldots d\mathbf{r}_N \tag{H-9}$$

Differentiating with respect to λ yields

$$\frac{\partial A}{\partial \lambda} = -kTZ_\lambda^{-1} \int -(kT)^{-1} \frac{\partial U_\lambda}{\partial \lambda} e^{-U_\lambda/kT} d\mathbf{r}_1 \ldots d\mathbf{r}_N \tag{H-10}$$

Because $\partial U_\lambda / \partial \lambda = U_p$, Eq. (H-10) can be rewritten

$$\frac{\partial A}{\partial \lambda} = <U_p>_\lambda \tag{H-11}$$

where < > designates an ensemble average:

$$<U_p>_\lambda = \int U_p \frac{e^{-U_\lambda/kT}}{Z_\lambda} d\mathbf{r}_1 \ldots d\mathbf{r}_N \tag{H-12}$$

To find A_1, we set $\lambda = 0$ because

$$A_1 = \lim_{\lambda \to 0} \left(\frac{\partial A}{\partial \lambda} \right) = <U_p>_{\lambda=0} = \int U_p \frac{e^{-U_0/kT}}{Z_0} d\mathbf{r}_1 \ldots d\mathbf{r}_N \tag{H-13}$$

Equation (H-13) says that the first-order perturbation term A_1 can be found by calculating an ensemble average of the perturbation energy U_p where the ensemble average is calculated using properties (U_0 and Z_0) of the reference system. To illustrate this important feature of perturbation theory, consider the case where the total perturbation energy is assumed to be the sum of all pair perturbation energies:

$$U_p = \sum_{i<j}\sum \Gamma_{ij_p}(r_{ij}) \tag{H-14}$$

where r_{ij} is the center-to-center distance between any two molecules i and j. The pair perturbation potential, Γ_{ij_p}, is defined in a manner analogous to Eq. (H-6).

For any pair of molecules,

$$\Gamma_\lambda = \Gamma_0 + \lambda \Gamma_p \tag{H-15}$$

When $\lambda = 1$, Γ_λ is the pair potential of the real system. It can then be shown (Smith, 1973) that

$$A_1 = 2\pi N\rho \int_0^\infty \Gamma_p(r) g_0(r) r^2 dr \tag{H-16}$$

where ρ is the number density, $\rho = N/V$, and where r is the center-to-center distance between any arbitrary molecule and some other arbitrary fixed molecule whose center is chosen to be the origin of the coordinate system.

To understand the physical meaning of A_1 let us first ignore $g_0(r)$, the *radial distribution function* of the reference system. Imagine a sphere of radius r whose center is the center of an arbitrarily chosen fixed molecule. The area of that sphere is $4\pi r^2$. Now imagine a thin shell of height dr built on that sphere's surface; the volume of that shell is $4\pi r^2 dr$. The number of molecular centers in that shell is the volume of the shell multiplied by ρ.

Focusing on the molecule at the center, $\Gamma_p(r)$ is the perturbation contribution of any *one* molecule at some distance r. The integral gives the contributions of all molecules, at all distances r. The factor 2 (instead of 4) arises in Eq. (H-16) because $\Gamma_p(r)$ is the perturbation potential for two molecules.

Because molecules have a nonzero size, the number of molecular centers in the shell at distance r is not necessarily $N\rho$. For example, suppose r is very small, smaller than molecular diameter σ. In that event, the number of molecular centers in the shell is not $N\rho$ but zero. For an assembly of molecules whose diameters are σ, where $\sigma > 0$, the number of molecular centers in a shell at distance r must depend on the dimensionless distance r/σ.

Given that a molecular center is at $r = 0$, we must ask: What is the number of molecular centers in a shell of thickness dr at position r? This number is related to the radial distribution function $g(r)$:

Number of molecular centers between r and $r + dr$

$$= 4\pi r^2\, dr\, \rho\, g(r) \quad \text{(given that one molecular center is at } r = 0\text{).}$$

Radial distribution function $g(r)$ gives information on the microscopic structure of an assembly of molecules.[4] For a structureless fluid (e.g., ideal gas), $g(r) = 1$ and for a lattice $g(r)$ is a periodic function. For real fluids ($\sigma > 0$), $g(r)$ is zero at small r, rises to above unity in the vicinity where r is just slightly larger than σ (shell of first neighbors), and then oscillates about unity, asymptotically approaching unity as $r >> \sigma$, as illustrated in Fig. H-1.

[4] More properly, for a pure fluid $g(r)$ it should be written $g(r, \rho, T)$ because it depends not only on r, but also on density ρ and temperature T. For a mixture, it should be written $g(r, \rho, T, x)$ because it also depends on composition x.

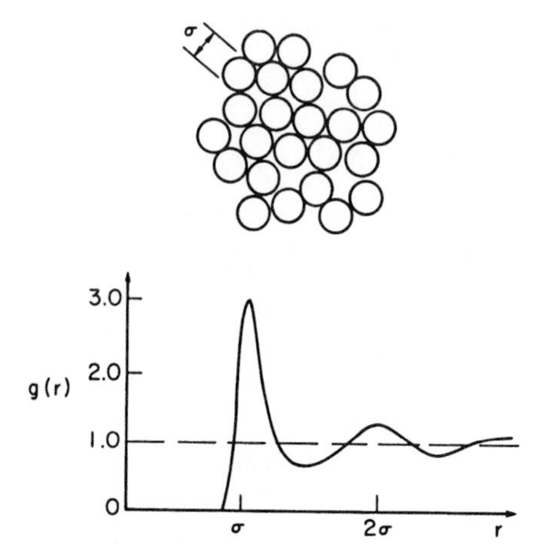

Figure H-1 Radial distribution function for a simple dense fluid.

The important feature of Eq. (H-16) is that the radial distribution function in the integral, $g_0(r)$, is not that of the real system, that is not known, but that of the reference system, that is known. This follows directly from Eq. (H-12), where the averaging procedure to find $<U_p>$ is based on properties of the reference system.

Although this discussion has been limited to pure fluids containing spherical molecules, the same ideas can be applied to pure or mixed fluids containing complex molecules, as discussed elsewhere (Reed and Gubbins, 1973; Gray and Gubbins, 1984; Lucas, 1991). Further, while the discussion here has been limited to first-order perturbation theory, it is possible to construct perturbation theories for second, third, and higher orders. However, when we consider nonspherical molecules and when we consider second-order (or higher) perturbations, not only does mathematical complexity rise but also much more detailed information on fluid structure and intermolecular forces is required.

It is likely that perturbation theories will become increasingly useful in phase-equilibrium thermodynamics. However, when applied to mixtures of real fluids, calculations are not only long but also nonanalytic; numerical integrations are required for every temperature, density, and composition. Further, when applied to real mixtures, considerable ingenuity is required to establish a useful reference system, and intermolecular forces must be characterized through credible potential functions. With increasingly efficient computers and with advances in molecular physics, perturbation theory of fluids is likely to provide an increasingly powerful tool for molecular thermodynamics.

References

Chandler, D., 1987, *Introduction to Modern Statistical Mechanics*. New York: Oxford University Press.

Gray, C. G. and K. E. Gubbins, 1984, *Theory of Molecular Fluids*. Oxford: Clarendon Press.

Lucas, K., 1991, *Applied Statistical Thermodynamics*. Berlin: Springer.

McQuarrie, D. A., 1976, *Statistical Mechanics*. New York: Harper & Row.

McQuarrie, D. A., 1985, *Statistical Thermodynamics*. Mill Valley: University Science Books.

Reed, T. M. and K. E. Gubbins, 1973, *Applied Statistical Mechanics*. New York: McGraw-Hill (Reprinted by Butterworth-Heinemann, 1991).

Smith, W. R., 1973. In *Statistical Mechanics*, (K. Singer, Ed.), Specialist Periodical Report. London: The Royal Society of Chemistry.

APPENDIX I

The Ion-Interaction Model of Pitzer for Multielectrolyte Solutions

Section 9.14 gives a brief discussion of the ion-interaction model of Pitzer and expressions for the activity coefficient and the osmotic coefficient of binary (single-electrolyte) solutions. This Appendix gives the general expressions for these coefficients for multielectrolyte solutions. Also given are tables that list the ion-interaction parameters and their temperature dependence for several common aqueous electrolytes.

For an electrolyte solution containing w_s kg of solvent s and solutes $i, j, \ldots$ with molalities m_i, $m_j, \ldots$, Pitzer (1995) assumed that the excess Gibbs energy is

$$\frac{G^{E*}}{RTw_s} = f(I) + \sum_i \sum_j m_i m_j \lambda_{ij}(I) + \sum_i \sum_j \sum_k m_i m_j m_k \Lambda_{ijk} + \ldots \qquad \text{(I-1)}[1]$$

[1] Equation (I-1) is the same as Eq. (9-58).

Function $f(I)$ depends on ionic strength I, temperature and solvent properties; it is a modified Debye-Hückel term

$$f(I) = -A_\phi \frac{4I}{b} \ln(1 + b\sqrt{I}) \quad \text{(I-2)}$$

with

$$I = \frac{1}{2} \sum_i m_i z_i^2 \quad \text{(I-3)}^2$$

The empirical universal constant $b = 1.2$ kg$^{1/2}$ mol$^{-1/2}$ is used for all electrolytes. A_ϕ is the Debye-Hückel parameter for the osmotic coefficient given by

$$A_\phi = \frac{1}{3}(2\pi N_A d_s)^{1/2} \left(\frac{e^2}{4\pi\varepsilon_0 \varepsilon_r kT} \right)^{3/2} \quad \text{(I-4)}$$

where N_A is Avogadro's constant, d_s is the solvent density in g cm^{-3}, e is the electronic charge ($e = 1.60218\times10^{-19}$ C), ε_0 the permittivity of free space ($\varepsilon_0 = 8.85419\times10^{-12}$ C^2 N^{-1} m^{-2}), ε_r is the dielectric constant or relative permittivity of the solvent, k is Boltzmann's constant and T is absolute temperature.[3]

Interaction parameters $\lambda_{ij}(I)$ and Λ_{ijk} are analogous to second and third virial coefficients; they represent the effects of short-range forces between, respectively, two and three ions. For highly concentrated solutions, fourth or higher-order interactions may be required in Eq. (I-1). The λ_{ij} and Λ_{ijk} matrices are symmetric, with $\lambda_{ij} = \lambda_{ji}$, $\Lambda_{ijk} = \Lambda_{ikj} = \Lambda_{jik}$, etc. Parameters λ_{ij} and Λ_{ijk} are also used for ion-neutral and neutral-neutral interactions in the solvent. For the ion-neutral case, theory suggests an ionic-strength dependence when the neutral species has a significant dipole moment. However, these parameters are usually considered to be constants at a given temperature, independent of ionic strength.

For electrolytes with cations c, c', ... and anions a, a', ... only some combinations of λ's and Λ's are measurable. The following definitions simplify the final equations for activity coefficients:

$$B_{ca} = \lambda_{ca} + \frac{\nu_c}{2\nu_a}\lambda_{cc} + \frac{\nu_a}{2\nu_c}\lambda_{aa} \quad \text{(I-5)}$$

---

[2] Equation (I-3) is the same as Eq. (9-43).

[3] If the solvent is water ($\varepsilon_r = 78.41$), A_ϕ is 0.392 kg$^{1/2}$ mol$^{-1/2}$ at 25°C. A_ϕ increases with temperature to 0.462 at 100°C and to 0.962 at 300°C.

$$\Phi_{cc'} = \lambda_{cc'} - \frac{z_{c'}}{2z_c}\lambda_{cc} - \frac{z_c}{2z_{c'}}\lambda_{c'c'} \tag{I-6}$$

$$C_{ca} = \frac{3}{2}\left(\frac{\Lambda_{cca}}{z_c} + \frac{\Lambda_{caa}}{z_a}\right) \tag{I-7}$$

$$\Psi_{cc'a} = 6\Lambda_{cc'a} - \frac{3z_{c'}}{z_c}\Lambda_{cca} - \frac{3z_c}{z_{c'}}\Lambda_{c'c'a} \tag{I-8}$$

with corresponding expressions for $\Phi_{aa'}$ and $\Psi_{aa'c}$. Because parameters λ_{ij} are functions of ionic strength, $B_{ca'}$, $\Phi_{cc'}$ and $\Phi_{aa'}$ are also functions of I. For brevity, however, the explicit notation $B(I)$, etc. is omitted except for $f(I)$.

The excess Gibbs energy becomes

$$\begin{aligned}\frac{G^{E*}}{RTw_s} &= f(I) + \sum_c\sum_a m_c m_a[B_{ca} + (\sum_c m_c z_c)C_{ca}] \\ &+ \sum\sum_{c<c'} m_c m_{c'}(2\Phi_{cc'} + \sum_a m_a \Psi_{cc'a}) \\ &+ \sum\sum_{a<a'} m_a m_{a'}(2\Phi_{aa'} + \sum_c m_c \Psi_{caa'}) + 2\sum_n\sum_c m_n m_c \lambda_{nc} \\ &+ 2\sum_n\sum_a m_n m_a \lambda_{na} + 2\sum\sum_{n<n'} m_n m_{n'}\lambda_{nn'} + \sum_n m_n^2 \lambda_{nn} + \ldots\end{aligned} \tag{I-9}$$

The double-summation indices, $c < c'$, $a < a'$, and $n < n'$ denote sums over all distinguishable pairs of different cations, anions, and neutral solutes, respectively.

Applying Eqs. (9-42) and (9-43) to Eq. (I-9), gives expressions for the osmotic coefficient and the mean ionic activity coefficients of the various electrolytes in a multielectrolyte mixture. To obtain the mean ionic activity coefficient $\gamma_{\pm,MX}$ for an electrolyte $M_{\nu+}X_{\nu-}$, it is preferable to obtain initially the expressions for the activity coefficients $\gamma_{+,M}$ and $\gamma_{-,X}$ of the individual ions M_M^{z+} and X_X^{z-}; the expression for $\gamma_{\pm,MX}$ is then obtained in a subsequent step. The results are:

$$\begin{aligned}\phi - 1 &= \frac{2}{\sum_i m_i}\Big[\frac{-A_\phi I^{3/2}}{1+bI^{1/2}} + \sum_c\sum_a m_c m_a (B_{ca}^\phi + ZC_{ca}) \\ &+ \sum\sum_{c<c'} m_c m_{c'}(\Phi_{cc'}^\phi + \sum_a m_a \Psi_{cc'a}) + \sum\sum_{a<a'} m_a m_{a'}(\Phi_{aa'}^\phi + \sum_c m_c \Psi_{caa'}) \\ &+ \sum_n\sum_c m_n m_c \lambda_{nc} + \sum_n\sum_a m_n m_a \lambda_{na} + \sum\sum_{n<n'} m_n m_{n'}\lambda_{nn'} + \frac{1}{2}\sum_n m_n^2 \lambda_{nn} + \ldots\Big]\end{aligned} \tag{I-10}$$

$$\ln\gamma_{+,\mathrm{M}} = z_+^2 F + \sum_a m_a(2B_{\mathrm{M}a} + ZC_{\mathrm{M}a}) + \sum_c m_c\left(2\Phi_{\mathrm{M}c} + \sum_a m_a \Psi_{\mathrm{M}ca}\right) + \sum_{a<a'}\sum m_a m_{a'} \Psi_{\mathrm{M}aa'} + z_+ \sum_c \sum_a m_c m_a C_{ca} + 2\sum_n m_n \lambda_{n\mathrm{M}} + \ldots \tag{I-11}$$

$$\ln\gamma_{-,\mathrm{X}} = z_-^2 F + \sum_c m_c(2B_{c\mathrm{X}} + ZC_{c\mathrm{X}}) + \sum_a m_a\left(2\Phi_{\mathrm{X}a} + \sum_c m_c \Psi_{c\mathrm{X}a}\right) + \sum_{c<c'}\sum m_c m_{c'} \Psi_{cc'\mathrm{X}} + |z_-| \sum_c \sum_a m_c m_a C_{ca} + 2\sum_n m_n \lambda_{n\mathrm{X}} + \ldots \tag{I-12}$$

$$\begin{aligned}\ln\gamma_{\pm,\mathrm{MX}} = {} & |z_+ z_-| F + \frac{\nu_+}{\nu} \sum_a m_a \left(2B_{\mathrm{M}a} + ZC_{\mathrm{M}a} + 2\frac{\nu_-}{\nu_+}\Phi_{\mathrm{X}a}\right) \\ & + \frac{\nu_-}{\nu} \sum_c m_c \left(2B_{c\mathrm{X}} + ZC_{c\mathrm{X}} + 2\frac{\nu_+}{\nu_-}\Phi_{\mathrm{M}c}\right) \\ & + \sum_c \sum_a m_c m_a \nu^{-1}(2\nu_+ z_+ C_{ca} + \nu_+ \Psi_{\mathrm{M}ca} + \nu_- \Psi_{ca\mathrm{X}}) \\ & + \sum_{c<c'}\sum m_c m_{c'} \frac{\nu_-}{\nu} \Psi_{cc'\mathrm{X}} + \sum_{a<a'}\sum m_a m_{a'} \frac{\nu_+}{\nu} \Psi_{\mathrm{M}aa'} \\ & + \frac{2}{\nu} \sum_n m_n (\nu_+ \lambda_{n\mathrm{M}} + \nu_- \lambda_{n\mathrm{X}})\end{aligned} \tag{I-13}$$

where second-virial terms for neutral species have been added but third-virial terms for neutrals are omitted. Terms B and C can be evaluated empirically from data for binary (single electrolyte + solvent) systems. Terms Φ and Ψ arise only for multi-electrolyte solutions; they can best be determined from data for common-ion mixtures. Quantity F includes the Debye-Hückel term and other terms as follows:

$$F = -A_\phi \left[\frac{I^{1/2}}{1 + bI^{1/2}} + \frac{2}{b}\ln(1 + bI^{1/2})\right] + \sum_c \sum_a m_c m_a B'_{ca} + \sum_{c<c'}\sum m_c m_{c'} \Phi'_{cc'} + \sum_{a<a'}\sum m_a m_{a'} \Phi'_{aa'} \tag{I-14}$$

Also,

$$Z = \sum_i m_i |z_i| \tag{I-15}$$

$$B^{\phi}_{ca} = B_{ca} + IB'_{ca} \tag{I-16}$$

$$\Phi^{\phi}_{cc'} = \Phi_{cc'} + I\Phi'_{cc'} \tag{I-17}$$

A corresponding expression is used for $\Phi^{\phi}_{aa'}$. B' and Φ' are the ionic-strength derivatives of B and Φ. The sums over i include all solute species; uncharged species do not contribute to I or Z. Parameter C_{MX} is related to the commonly tabulated C^{ϕ}_{MX} by

$$C_{MX} = \frac{C^{\phi}_{MX}}{2|z_+z_-|^{1/2}} \tag{I-18}$$

The ionic-strength dependence of B terms is taken into account by

$$B^{\phi}_{MX} = \beta^{(0)}_{MX} + \beta^{(1)}_{MX}\exp(-\alpha_1 I^{1/2}) + \beta^{(2)}_{MX}\exp(-\alpha_2 I^{1/2}) \tag{I-19}$$

$$B_{MX} = \beta^{(0)}_{MX} + \beta^{(1)}_{MX} g(\alpha_1 I^{1/2}) + \beta^{(2)}_{MX} g(\alpha_2 I^{1/2}) \tag{I-19a}$$

$$B'_{MX} = \frac{\beta^{(1)}_{MX} g'(\alpha_1 I^{1/2}) + \beta^{(2)}_{MX} g'(\alpha_2 I^{1/2})}{I} \tag{I-19b}$$

where functions g and g' are given by

$$g(x) = \frac{2[1-(1+x)\exp(-x)]}{x^2} \tag{I-20}$$

$$g'(x) = \frac{-2[1-(1+x+x^2/2)\exp(-x)}{x^2} \tag{I-20a}$$

$\beta^{(0)}_{MX}$, $\beta^{(1)}_{MX}$ and $\beta^{(2)}_{MX}$ are solute-specific parameters fitted to isothermal or isobaric data for single-electrolyte solutions. $\beta^{(2)}_{MX}$ is important only for 2-2 or higher-valence electrolytes that show a tendency toward electrostatic ion pairing. For solutions of electrolytes containing at least one univalent ion, α_1 is taken to be 2.0 $kg^{1/2}\ mol^{-1/2}$ and $\alpha_2 = 0$. For 2-2 electrolytes at 25°C, the optimized values of α_1 and α_2 are 1.4 and 12 $kg^{1/2}\ mol^{-1/2}$, respectively. For many applications these values can be assumed independent of temperature and pressure, but there are theoretical reasons (Pitzer, 1991) for setting α_2 proportional to the Debye-Hückel parameter A_{ϕ}.

The Φ terms have a strong ionic-strength dependence for unsymmetric cases (e.g. Na^+ with Mg^{2+} or Cl^- with SO_4^{2-}) arising from long-range electrostatic forces. The expressions for Φ_{ij} are

$$\Phi_{ij} = \theta_{ij} + {}^{E}\theta_{ij}(I) \tag{I-21}$$

$$\Phi'_{ij} = {}^{E}\theta'_{ij} \tag{I-21a}$$

$$\Phi^{\phi}_{ij} = \theta_{ij} + {}^{E}\theta_{ij}(I) + I\,{}^{E}\theta'_{ij}(I) \tag{I-21b}$$

where ${}^{E}\theta(I)$ and ${}^{E}\theta'(I)$ account for electrostatic unsymmetric mixing effects and depend only on charges of ions i and j, the total ionic strength, and on the density and dielectric constant of the solvent (i.e. on the temperature and pressure). Equations for calculating these terms have been developed; they are given elsewhere (Pitzer, 1991). The remaining term θ_{ij}, arising from short-range forces, is taken as a constant for any particular c, c' or a, a' at a given temperature and pressure. Terms ${}^{E}\theta(I)$ and ${}^{E}\theta'(I)$ are often omitted for solubility calculations.

Pitzer's ion-interactive model gives expressions for the osmotic and activity coefficients of multi-electrolyte mixtures in terms of six types of empirical parameters, viz. $\beta^{(0)}_{MX}$, $\beta^{(1)}_{MX}$, $\beta^{(2)}_{MX}$, C^{ϕ}_{MX}, θ_{ij}, and ψ_{ijk}. Provided that the temperature and pressure dependencies of these parameters are known, solubilities in binary, ternary and higher mixtures can be calculated.

The excess Gibbs energy given by Eq. (I-1) yields other thermodynamic properties such as excess enthalpies and heat capacities by appropriate differentiation. These other excess properties can be measured directly. In the literature there is a wide array of experimental data for obtaining ion-interaction parameters and their dependence on temperature and pressure. Tables published elsewhere (Pitzer, 1991, 1995; Clegg and Whitfield, 1991; Zemaitis *et al.*, 1986) give ion-interaction parameters for numerous aqueous solutions of electrolytes. Table I-1 gives parameters at 25°C for a few common aqueous electrolytes and Table I-2 gives the temperature dependence of parameters for those electrolytes used in Sec. 9.14 for model calculations. The mixed-electrolyte parameters for variable-temperature calculations are shown in Table I-3.

With model parameters determined as discussed above, and with experimental solubility products, Pitzer's model can be used to predict solid solubilities in aqueous mixed-salt systems. It is relatively easy to calculate solubilities in two-salt solutions with a common ion. If the two salts do not have a common ion or, if there are more than two salts, the calculations become complex; they require solution of numerous simultaneous equations.

Section 9.14 presents two examples to illustrate how Pitzer's equation, coupled with solubility products, can be used to calculate solid-liquid equilibria in aqueous systems containing two salts. Table I-4 gives standard-state chemical potentials, enthalpies of formation, and entropies at 25°C of the ions and solids considered there and for some other species of common interest. It is an important characteristic of the equations of the ion-interaction model of Pitzer that, for calculating solid-liquid equilibria, all parameters can be evaluated from measurements for single-electrolyte solutions and solutions containing two electrolytes with a common-ion. No new parameters are required. Therefore, calculations for multi-salt systems are predictions.

Table I-1 Ion-interaction parameters at 25°C. Parameters only apply to a maximum molality of 4-6 mol kg^{-1} (from Pitzer, 1995).

Electrolyte	$\beta^{(0)}_{MX}$	$\beta^{(1)}_{MX}$	$\beta^{(2)}_{MX}$	$C^{(\phi)}_{MX}$
HCl	0.1775	0.2945	–	0.00080
LiCl	0.1494	0.3074	–	0.00359
NaCl	0.0765	0.2664	–	0.00127
KCl	0.0484	0.2122	–	-0.00084
HBr	0.2085	0.3477	–	0.00152
LiBr	0.1748	0.2547	–	0.00530
NaBr	0.0973	0.2791	–	0.00116
KBr	0.0569	0.2212	–	-0.00180
HI	0.2211	0.4907	–	0.00482
LiI	0.2104	0.3730	–	–
NaI	0.1195	0.3439	–	0.00180
NaOH	0.0864	0.2530	–	0.00440
KOH	0.1298	0.3200	–	0.00410
HNO_3	0.1168	0.3456	–	-0.00539
$LiNO_3$	0.1420	0.2780	–	-0.00551
$NaNO_3$	0.0068	0.1783	–	-0.00072
KNO_3	-0.0816	0.0494	–	0.00660
NH_4NO_3	-0.0154	0.1120	–	-0.00003
Na_2SO_4	0.0187	1.0994	–	0.00555
K_2SO_4	0.4995	0.7793	–	–
$MgCl_2$	0.3509	1.6508	–	0.00651
$FeCl_2$	0.3359	1.5323	–	-0.00861
$NiCl_2$	0.3499	1.5300	–	-0.00471
$CaCl_2$	0.3159	1.6140	–	-0.00034
$Mg(NO_3)_2$	0.3671	1.5848	–	-0.02062
$MgSO_4$	0.2210	3.3430	-37.23	0.02500
$CuSO_4$	0.2340	2.5270	-48.33	0.00440
$NiSO_4$	0.1702	2.9070	-40.06	0.03660
$AlCl_3$	0.6993	5.8447	–	0.00372

Table I-2 Standard-state heat capacity and temperature dependence of ion-interaction parameters for aqueous NaCl, KCl, and Na_2SO_4 (from Pitzer, 1995).

Parameters for NaCl (aq.)

Standard-state heat capacity (fitted to data from 273 to 573 K and 1 bar or saturation pressure):

$$c_p^0 = -1.848175 \times 10^6 + \frac{4.411878 \times 10^7}{T} + 3.390654 \times 10^5 \ln T - 8.893249 \times 10^2 T + 4.00577 \times 10^{-1} T^2 - \frac{7.244279 \times 10^4}{T-227} - \frac{4.098218 \times 10^5}{647-T} \quad \text{(with } c_p^0 \text{ in J K}^{-1}\text{ mol}^{-1} \text{ and } T \text{ in K)}$$

Equation for $\beta_{MX}^{(0)}$, $\beta_{MX}^{(1)}$, and $C_{MX}^{(\phi)}$, valid to 40 bar:

$$f(T) = \frac{a_1}{T} + a_2 + a_3 P + a_4 \ln(T) + (a_5 + a_6 P)T + (a_7 + a_8 P)T^2 + \frac{a_9 + a_{10} P}{T-227} + \frac{a_{11} + a_{12} P}{680-T} \quad \text{(with } P \text{ in bar and } T \text{ in kelvin)}$$

	$\beta_{MX}^{(0)}$	$\beta_{MX}^{(1)}$	$C_{MX}^{(\phi)}$
a_1	-656.81518	119.31966	-6.1084589
a_2	24.8691295	-0.48309327	0.40217793
a_3	$5.381275267 \times 10^{-5}$	0	2.2902837×10^{-5}
a_4	-4.4640952	0	-0.075354649
a_5	0.01110991383	1.4068095×10^{-3}	$1.531767295 \times 10^{-4}$
a_6	$-2.657339906 \times 10^{-7}$	0	$-9.0550901 \times 10^{-8}$
a_7	$-5.307012889 \times 10^{-6}$	0	$-1.53860082 \times 10^{-8}$
a_8	$8.634023325 \times 10^{-10}$	0	8.69266×10^{-11}
a_9	-1.579365943	-4.2345814	0.353104136
a_{10}	$2.202282079 \times 10^{-3}$	0	$-4.3314252 \times 10^{-4}$
a_{11}	9.706578079	0	-0.091871455
a_{12}	-0.02686039622	0	5.1904777×10^{-4}

Parameters for KCl (aq.)

Standard-state heat capacity (fitted to data from 273 to 573 K and 1 bar or saturation pressure):

$$c_p^0 = -991.51 + 5.5645T - 0.00853T^2 - \frac{686}{T-270} \quad \text{(with } c_p^0 \text{ in J K}^{-1}\text{ mol}^{-1} \text{ and } T \text{ in K)}$$

Parameters for KCl (aq.) (continued)

Equation for $\beta^{(0)}_{MX}$, $\beta^{(1)}_{MX}$, and $C^{(\phi)}_{MX}$:

$$f(T) = b_1 + b_2\left(\frac{1}{T}-\frac{1}{T_r}\right) + b_3 \ln\left(\frac{T}{T_r}\right) + b_4(T-T_r) + b_5(T^2-T_r^2)$$

$$+ b_6 \ln(T-260) \quad \text{(with } T \text{ in K and } T_r = 298.15 \text{ K)}$$

	$\beta^{(0)}_{MX}$	$\beta^{(1)}_{MX}$	$C^{(\phi)}_{MX}$
b_1	0.04808	0.0476	-7.88×10^{-4}
b_2	-758.48	303.9	91.27
b_3	-4.7062	1.066	0.58643
b_4	0.010072	0	-0.001298
b_5	-3.7599×10^{-6}	0	4.9567×10^{-7}
b_6	0	0.047	0

Parameters for Na_2SO_4 (aq.)

Standard-state heat capacity (fitted to data from 273 to 573 K and 1 bar or saturation pressure):

$$c_p^0 = -1206.2 + 7.6405T - 1.23672\times10^{-2}T^2 - \frac{6045}{T-263} \quad \text{(with } c_p^0 \text{ in J K}^{-1}\text{ mol}^{-1} \text{ and } T \text{ in K)}$$

Equation for $\beta^{(0)}_{MX}$, $\beta^{(1)}_{MX}$, and $C^{(\phi)}_{MX}$ (note that $\alpha = 1.4$ kg$^{1/2}$ mol$^{-1/2}$ instead of the normal value of 2.0 kg$^{1/2}$ mol$^{-1/2}$):

$$f(T) = c_1 + c_2\left(T_r - \frac{T_r^2}{T}\right) + c_3\left(T^2 + \frac{2T_r^3}{T} - 3T_r^2\right) + c_4\left(T + \frac{T_r^2}{T} - 2T_r\right) + c_5\left[\ln\left(\frac{T}{T_r}\right) + \left(\frac{T}{T_r}\right)^{-1}\right]$$

$$+ c_6\left(\frac{1}{T-263} + \frac{263T - T_r^2}{T(T_r-263)^2}\right) + c_7\left(\frac{1}{680-T} + \frac{T_r^2 - 680T}{T(680-T_r)^2}\right) \quad \text{(with } T \text{ in K and } T_r = 298.15K)$$

	$\beta^{(0)}_{MX}$	$\beta^{(1)}_{MX}$	$C^{(\phi)}_{MX}$
c_1	-1.727×10^{-2}	0.7534	1.1745×10^{-2}
c_2	1.7828×10^{-3}	5.61×10^{-3}	-3.3038×10^{-4}
c_3	9.133×10^{-6}	-5.7513×10^{-4}	1.85794×10^{-5}
c_4	0	1.11068	-3.92×10^{-2}
c_5	-6.552	-378.82	14.213
c_6	0	0	0
c_7	-96.90	1861.3	-24.95

Table I-3 Mixed-electrolyte parameters for variable-temperature calculations (T in kelvin) (Pabalan and Pitzer, 1991).

i	j	k	θ_{ij}	Ψ_{ijk}
Na^+	K^+	Cl^-	-0.012	$-6.809\times10^{-3} + 1.680\times10^{-5}T$
Na^+	Mg^{2+}	Cl^-	0.070	$1.99038\times10^{-2} - 9.51213/T$
K^+	Mg^{2+}	Cl^-	0.000	$2.58557\times10^{-2} - 14.26819/T$
Cl^-	SO_4^{2-}	Na^+	0.030	$-1.6958\times10^{-2} + 3.13544/T + 2.16352\times10^{-5}T - 1.31254\times10^5/(647-T)^4$
Cl^-	SO_4^{2-}	K^+	0.030	5.0×10^{-3}
Cl^-	SO_4^{2-}	Mg^{2-}	0.030	$-1.17457\times10^{-1} + 32.6347/T$
Cl^-	OH^-	Na^+	-0.050	$7.93217\times10^{-2} - 1.89664\times10^1/T - 7.28094\times10^{-5}T$
SO_4^{2-}	OH^-	Na^+	-0.013	$7.94135\times10^{-2} - 19.9387/T - 7.21586\times10^{-5}T - 3.649\times10^5/(647-T)^4$

Table I-4 Standard-state chemical potentials, enthalpies of formation, and entropies of aqueous species and solid salts at 25°C (Pitzer, 1995). Values in parentheses have large uncertainties.

Species	$-\mu^0/RT = -\Delta_f g^0/RT$	$-\Delta_f h^0 / RT$	s^0/R
H_2O	95.6635	115.304	8.409
H^+	0	0	0
Na^+	105.651	96.865	7.096
K^+	113.957	101.81	12.33
Mg^{2+}	183.468	188.329	-16.64
Ca^{2+}	223.30	219.0	-6.4
OH^-	63.435	95.666	-1.29
Cl^-	52.955	67.432	6.778
SO_4^{2-}	300.386	366.800	2.42
K_2SO_4	532.39	580.01	21.12
$MgCl_2\cdot6H_2O$	853.1	1008.11	44.03
$Na_2Mg(SO_4)_2\cdot4H_2O$	1383.6	–	–
$KMgCl_3\cdot6H_2O$	1020.3	1184.85	55.53
$MgSO_4\cdot7H_2O$	1157.83	(1366.3)	44.79
$CaSO_4\cdot2H_2O$	725.67	815.9	23.35
$NaCl$	154.99	165.88	8.676
$MgSO_4\cdot6H_2O$	1061.60	(1244.8)	41.87
$MgSO_4\cdot H_2O$	579.80	649.34	(14.99)
$K_2Mg(SO_4)_2\cdot4H_2O$	1403.97	(1592.4)	–
$Na_2SO_4\cdot10H_2O$	1471.15	1475.75	71.21
KCl	164.84	176.034	9.934
Na_2SO_4	512.35	559.55	17.99

The equations presented in this appendix are complex but they are all analytical. For application, a suitable computer program is required.

The practical applications of Pitzer's method are not due to mathematical complexity but follow from the need for a large experimental-data base to fix parameters.

References

Clegg, S. L. and M. Whitfield, 1991. In *Activity Coefficients of Electrolyte Solutions,* (K. S. Pitzer, Ed.), 2nd Ed., Chap. 6. Boca Raton: CRC Press.

Pabalan, R. T. and K. S. Pitzer, 1991. In *Activity Coefficients of Electrolyte Solutions,* (K. S. Pitzer, Ed.), 2nd Ed., Chap. 7. Boca Raton: CRC Press.

Pitzer, K. S., 1991. In *Activity Coefficients in Electrolyte Solutions*, (K. S. Pitzer, Ed.), 2nd Ed., Chap. 3. Boca Raton: CRC Press.

Pitzer, K. S., 1995, *Thermodynamics*, 3rd Ed. New York: McGraw Hill.

Zemaitis, R. M., Jr., D. M. Clark, M. Rafal, and N. C. Scrivner, 1986, *Handbook of Aqueous Electrolyte Thermodynamics*. New York: A.I.Ch.E.

APPENDIX J

Conversion Factors and Constants

SI Units and Conversion Factors

Table J-1 shows the basic units and some derived units of the International System of Units (SI).[1] The main advantage of this system is internal coherence; i.e., no conversion factors are needed when using basic or derived SI units.

Table J-1 Basic and derived SI units.

Quantity	SI unit	Symbol
Basic		
Length	meter	m
Mass	kilogram	kg
Time	second	s
Electric current	ampere	A
Thermodynamic temperature	kelvin	K
Amount of substance	mole	mol
Luminous intensity	candela	cd
Derived		
Force	newton (kg m/s^2)	N
Energy	joule (N m)	J
Pressure	pascal (N/m^2)	Pa
Power	watt (J/s)	W
Volume	cubic meter	m^3
Density	kilogram/cubic meter	kg/m^3

[1] A guide for using the SI system is given by B. N. Taylor, 1995, NIST Special Publication 811, Washington: U.S. Government Printing Office.

In SI, mass is expressed in kilograms, force in newtons, and pressure in newtons per square meter (pascal). Usually, a prefix is attached to SI units. Common prefixes for powers of ten are listed in Table J-2.

Table J-2 SI prefixes.

Factor	Prefix	Symbol
10^9	giga	G
10^6	mega	M
10^3	kilo	k
10^{-1}	deci	d
10^{-2}	centi	c
10^{-3}	milli	m
10^{-6}	micro	μ
10^{-9}	nano	n

Because other units remain in common usage, Table J-3 gives conversion factors to SI units for selected quantities.

Table J-3 Conversion factors to SI units for selected quantities.

To convert from:	To:	Multiply by:§
ångstrom	meter	1.0000000*E-10
atmosphere (standard)	pascal	1.0132500*E+05
bar	pascal	1.0000000*E+05
barrel (for petroleum, 42 gallons)	meter^3	1.5898729E-01
British thermal unit (Btu_{IT}, International Table)	joule	1.0550559E+03
British thermal unit (Btu_{th}, thermochemical)	joule	1.0543503E+03
Btu_{IT}/(pound-mass•°F)	joule/(kilogram•kelvin)	4.1868000*E+03
Btu_{th}(pound-mass•°F)	joule/(kilogram•kelvin)	4.1840000*E+03
Btu_{IT}/second	watt	1.0550559E+03
calorie (cal_{IT}, International Table)	joule	4.1868000*E+00
calorie (cal_{th}, thermochemical)	joule	4.1840000*E+00
cal_{IT}/(gram•°C)	joule/(kilogram•kelvin)	4.1868000*E+03
cal_{th}/(gram•°C)	joule/(kilogram•kelvin)	4.1840000*E+03
centimeter of water (4°C)	pascal	9.80638E+01
centipoise	pascal•second	1.0000000*E-03
degree Fahrenheit (°F)	kelvin	$T_K = (T_F + 459.67)/1.8$
degree Rankine (°R)	kelvin	1/1.8
dyne	newton	1.0000000*E-05
erg	joule	1.0000000*E-07

To convert from:	To:	Multiply by:[§]
electron volt (eV)	joule	1.602177E-19
fluid ounce (U.S.)	$meter^3$	2.9573530E-05
foot	meter	3.0480000*E-01
foot•pound-force	joule	1.3558179E+00
gallon (U.S. liquid)	$meter^3$	3.7854118E-03
horsepower[#]	watt	7.4569987E+02
inch	meter	2.5400000*E-02
kilogram-force	newton	9.8066500*E+00
kilogram-force/cm^2	pascal	9.8066500*E+04
liter	$meter^3$	1.0000000*E-03
mile (U.S. statute)	meter	1.6093440*E+03
mile/hour	meter/second	4.4704000*E-01
millimeter of mercury (0°C)	pascal	1.3332237E+02
pint (U.S. liquid)	$meter^3$	4.7317647E-04
pound-force•second/$foot^2$	pascal•second	4.7880258E+01
pound-mass (lbm avoirdupois)	kilogram	4.5359237*E-01
pound-mass/$foot^3$	kilogram/$meter^3$	1.6018463E+01
pound-mass/(foot•second)	pascal•second	1.4881639E+00
psia	pascal	6.8947573E+03
quart (U.S. liquid)	$meter^3$	9.4635295E-04
ton (long, 2240 lbm)	kilogram	1.0160469E+03
ton (short, 2000 lbm)	kilogram	9.0718474*E+02
torr (mmHg, 0°C)	pascal	1.3332237E+02
watt•hour	joule	3.6000000*E+03
yard	meter	9.1440000*E-01

[§]An asterisk after the seventh decimal place indicates that the conversion factor is exact and all subsequent digits are zero.

[#]1 horsepower = 550 foot•pound-force/second.

Some Fundamental Constants in Various Units

Acceleration of gravity, g

$m\ s^{-2}$	9.80665

Atomic Mass Unit, *amu*

kg	1.66054×10^{-27}

Avogadro's constant, N_A

mol^{-1}	6.02214×10^{23}

Boltzmann's constant, k	
$J\ K^{-1}$	1.38066×10^{-23}
$erg\ K^{-1}$	1.38066×10^{-16}
$eV\ K^{-1}$	8.61739×10^{-5}
Electron charge, e	
C	1.60218×10^{-19}
Gas constant, R	
$J\ mol^{-1}\ K^{-1}$	8.31451
$Pa\ m^3\ mol^{-1}\ K^{-1}$	8.31451
$atm\ cm^3\ mol^{-1}\ K^{-1}$	82.0578
$atm\ liter\ mol^{-1}\ K^{-1}$	0.0820578
$atm\ ft^3\ lb\text{-}mol^{-1}\ {}^\circ R^{-1}$	0.7302
$bar\ cm^3\ mol^{-1}\ K^{-1}$	83.1451
$Btu_{IT}\ lb\text{-}mol^{-1}\ {}^\circ R^{-1}$	1.98592
$cal_{th}\ mol^{-1}\ K^{-1}$	1.98721
$erg\ mol^{-1}\ K^{-1}$	8.31451×10^{7}
$hp{\cdot}h\ lb\text{-}mol^{-1}\ {}^\circ R^{-1}$	7.805×10^{-4}
$kW{\cdot}h\ lb\text{-}mol^{-1}\ {}^\circ R^{-1}$	5.820×10^{-4}
$mmHg\ liter\ mol^{-1}\ K^{-1}$	62.3640
$psia\ ft^3\ lb\text{-}mol^{-1}\ {}^\circ R^{-1}$	10.73
Permittivity of vacuum, ε_o	
$C^2\ N^{-1}\ m^{-2}$	8.85419×10^{-12}
Planck's constant, h	
J s	6.62608×10^{-34}

Critical Constants and Acentric Factors for Selected Fluids

Table J-4 Critical temperature, pressure and volume and acentric factors for selected fluids.[2]

	T_c (K)	P_c (bar)	v_c ($cm^3\ mol^{-1}$)	ω
Elements				
Argon	150.8	48.7	74.9	0.001
Bromine	588.	103.	127.2	0.108
Chlorine	416.9	79.8	123.8	0.090
Fluorine	144. 3	52.2	66.3	0.054
Helium-4	5.19	2.27	57.4	-0.365

[2] Taken from a more complete list given in R. C. Reid, J. M. Prausnitz, and B. E. Poling, 1987, *The Properties of Gases and Liquids,* 4th Ed. New York: McGraw-Hill.

	T_c (K)	P_c (bar)	v_c (cm^3 mol^{-1})	ω
Hydrogen	33.0	12.9	64.3	-0.216
Iodine	819.	116.5	155	0.229
Krypton	209.4	55.0	91.2	0.005
Neon	44.4	27.6	41.6	-0.029
Nitrogen	126.2	33.9	89.8	0.039
Oxygen	154.6	50.4	73.4	0.025
Xenon	289.7	58.4	118.4	0.008
Hydrocarbons				
Acetylene	308. 3	61.4	112.7	0.190
Benzene	562.1	48.9	259.	0.212
n-Butane	425.2	38.0	255.	0.199
l-Butene	419.6	40.2	240.	0.191
Cyclobutane	460.0	49.9	210.	0.181
Cyclohexane	553.8	40.7	308.	0.212
Cyclopropane	397.8	54.9	163.	0.130
Ethane	305.4	48.8	148.3	0.099
Ethylene	282.4	50.4	130.4	0.089
n-Heptane	540.3	27.4	432.	0.349
n-Hexane	507.5	30.1	370.	0.299
Isobutane	408.2	36.5	263.	0.183
Isobutylene	417.9	40.0	239.	0.194
Isopentane	460.4	33.9	306.	0.227
Methane	190.4	46.0	99.2	0.011
Naphthalene	748.4	40.5	413.	0.302
n-Octane	568.8	24.9	492.	0.398
n-Pentane	469.7	33.7	304.	0.251
Propadiene	393.	54.7	162.	0.313
Propane	369.8	42.5	203.	0.153
Propylene	364.9	46.	181.	0.144
Toluene	591.8	41.0	316.	0.263
m-Xylene	617.1	35.4	376.	0.325
o-Xylene	630.3	37.3	369.	0.310
p-Xylene	616.2	35.1	379.	0.320
Miscellaneous inorganic compounds				
Ammonia	405.5	113.5	72.5	0.250
Carbon dioxide	304.1	73.8	93.9	0.239
Carbon disulfide	552.	79.0	160.	0.109
Carbon monoxide	132.9	35.0	93.2	0.066
Carbon tetrachloride	556.4	45.6	275.9	0.193
Carbon tetrafluoride	227.6	37.4	139.6	0.177

	T_c (K)	P_c (bar)	v_c (cm^3 mol^{-1})	ω
Chloroform	536.4	53.7	238.9	0.218
Hydrazine	653.	147.	96.1	0.316
Hydrogen chloride	324.7	83.1	80.9	0.133
Hydrogen fluoride	461.	64.8	69.2	0.329
Hydrogen sulfide	373.2	89.4	98.6	0.081
Nitric oxide	180.	64.8	57.7	0.588
Nitrous oxide	309.6	72.4	97.4	0.165
Sulfur dioxide	430.8	78.8	122.2	0.256
Sulfur trioxide	491.0	82.1	127.3	0.481
Water	647.3	221.2	57.1	0.344
Miscellaneous organic compounds				
Acetaldehyde	461.	55.7	154.	0.303
Acetic acid	592.7	57.9	171.	0.447
Acetone	508.1	47.0	209.	0.304
Acetonitrile	545.5	48.3	173.	0.278
Aniline	699.	53.1	274.	0.384
n-Butanol	563.1	44.2	275.	0.593
Chlorobenzene	632.4	45.2	308.	0.249
Dichlorodifluoromethane (Freon 12)	385.0	41.4	216.7	0.204
Diethyl ether	466.7	36.4	280.	0.281
Dimethyl ether	400.0	52.4	178.	0.200
Ethanol	513.9	61.4	167.1	0.644
Ethylene oxide	469.	71.9	140.	0.202
Isobutanol	547.8	43.0	273.	0.592
Isopropyl alcohol	508.3	47.6	220.	0.665
Methanol	512.6	80.9	118.0	0.556
Methyl chloride	416.3	67.0	138.9	0.153
Methyl ethyl ketone	536.8	42.1	267.	0.320
Phenol	694.2	61.3	229.	0.438
1-Propanol	536.8	51.7	219.	0.623
Pyridine	620.0	56.3	254.	0.243
Trichlorotrifluoroethane (Freon 113)	487.3	34.1	325.5	0.256
Trichlorofluoromethane (Freon 11)	471.2	44.1	247.8	0.189
Trimethylamine	433.3	40.9	254.	0.205

Index

A

D

E

F

G

H

I

Q

R

S

T

U

V

W